IFAC WORKSHOP ON MANUFACTURING SYSTEMS: MODELLING, MANAGEMENT AND CONTROL

Sponsored by
International Federation of Automatic Control (IFAC)
Technical Committee on Manufacturing, Modelling, Management and Control (MIM)

Co-sponsored by
IFAC Technical Committees on
- Robotics (MIR)
- Architecture for Enterprise Integration (MIA)
- Advanced Manufacturing Technology (MIT)
- Business and Management Techniques (SMB)

International Federation for Information Processing (IFIP)
International Federation of Operational Research Societies (IFORS)
Vienna University of Technology
Creditanstalt Bankverein (Austria) (CA)

Organized by
Institute for Handling Devices and Robotics, Vienna University of Technology

International Programme Committee (IPC)
A. Villa (I) (Chairman)
S. Bansal (SGP)
A.W. Chan (CND)
M. Deistler (A)
J.T. Deng (PRC)
G. Feichtinger (A)
F. Gomide (BRA)
P. Kopacek (A)
G.L. Kovacs (H)
L. Nemes (AUS)
J.L. Nevins (USA)
F. Nicolo (I)
G. Olsson (S)
J. Ranta (FIN)
U. Rembold (GER)
J.E. Rooda (NL)
F.B. Vernadat (F)
T. Watanabe (J)
M.B. Zaremba (CND)

National Organizing Committee (NOC)
P. Kopacek (Chairman)
I. Nemetz
W. Schachner

Preface:

The IFAC TC on Manufacturing, Modelling, Management and Control (MIM) was founded on the IFAC World Congress Sydney 1993. According to the IFAC regulations each TC has to sponsor one event in a triennium at least. This workshop is the first scientific event of this TC.

The goals of the workshop are concerning the development, the comparison and the classification of formal models in the field of Computer Integrated Manufacturing Systems in a descriptive as well as prescriptive way. The Computer Integrated Manufacturing Systems are able to integrate the optimization methods, simulation models, procedures and knowledge-based tools.

The target for the workshop activities will be related to the specification of requirements for new models which are used in simulating and designing manufacturing management and control strategies, including discrete-event and continuous representations.

Technical areas of interest include:

- at the system level: tools for plant layout design, process planning, production planning and control;
- at the component level: models for functional description of flexible manufacturing and assembly systems, oriented to production activity control, process supervision and maintenance.

The organizers received more than 130 abstracts; 107 were conditionally accepted for presentation. 82 full papers were included in the preprints. Finally 75 papers were presented at the workshop and they are collected in this volume.

Many thanks to the IPC members for supporting us and we hope that this workshop might stimulate scientific research in this new field of automation.

Vienna, Torino, February 1997

P.Kopacek
NOC - Chairman

A. Villa
IPC - Chairman

MANUFACTURING SYSTEMS: MODELLING, MANAGEMENT AND CONTROL

(MIM'97)

A Proceedings volume from the IFAC Workshop, Vienna, Austria, 3 - 5 February 1997

Edited by

P. KOPACEK
Institute for Handling Devices and Robotics, Vienna University of Technology, Vienna, Austria

INTERI

PERGAMON
An Imprint of Elsevier Science

UK	Elsevier Science Ltd, The Boulevard, Langford Lane, Kidlington, Oxford, OX5 1GB, UK
USA	Elsevier Science Inc., 660 White Plains Road, Tarrytown, New York 10591-5153, USA
JAPAN	Elsevier Science Japan, Tsunashima Building Annex, 3-20-12 Yushima, Bunkyo-ku, Tokyo 113, Japan

First edition 1997

Library of Congress Cataloging in Publication Data

A catalogue record for this book is available from the Library of Congress

British Library Cataloguing in Publication Data

A catalogue record for this book is available from the British Library

ISBN 0-08-042616 6

This volume was reproduced by means of the photo-offset process using the manuscripts supplied by the authors of the different papers. The manuscripts have been typed using different typewriters and typefaces. The lay-out, figures and tables of some papers did not agree completely with the standard requirements: consequently the reproduction does not display complete uniformity. To ensure rapid publication this discrepancy could not be changed: nor could the English be checked completely. Therefore, the readers are asked to excuse any deficiencies of this publication which may be due to the above mentioned reasons.

The Editor

Printed in Great Britain

CONTENTS

SURVEY PAPERS

NEURAL NETS

ASSEMBLY SYSTEMS

PETRI NETS

DESIGN AND CONTROL

FUZZY AND PLC CONTROLLERS

PRODUCTION PLANNING AND MANAGEMENT

SCHEDULING

DATA MANAGEMENT IN MANUFACTURING

INVITED PAPERS: 2D SYSTEMS IN MANUFACTURING

CONTROL OF PRODUCTION SYSTEMS

PRODUCTION DESIGN AND CAD

CONTROL OF SPECIAL MANUFACTURING PROCESSES

MANUFACTURING SYSTEM DESIGN

ENTERPRISE INTEGRATION I

MODELLING OF MANUFACTURING SYSTEMS II

MACHINING PROCESSES

MODELLING OF MANUFACTURING SYSTEMS III

PERFORMANCE EVALUATION

VARIOUS ASPECTS

A MODEL FOR MANAGING INNOVATION PROGRAMS IN MANUFACTURING PLANTS

Agostino Villa and Sergio Rossetto

Dipartimento di Sistemi di Produzione ed Economia dell'Azienda
Politecnico di Torino, c.so Duca degli Abruzzi 24, 10129 Torino

Abstract: Manufacturing systems are pushed towards relevant technological innovations mainly concerning internal logistics systems. Actual evolution tries to improve the logistics system efficiency and effectiveness either by frequent modifications and re-design of system structure and management or by increasing automation. Today, the first design approach seems to be more promising because it allows the logistics system to be adaptable to large demand variations, provided that a careful management of the re-design process is applied. The paper will analyse a new methodology for an efficient, effective and economic organization of the re-design process for an internal logistics system.

Keywords: Innovation, management, manufacturing systems.

1. INDUSTRIAL LOGISTICS NETWORKS AND RELATED DESIGN PROBLEMS

Industrial logistics networks refer to different types of services in a manufacturing plant, either internal in the shop floor or external, to connect the plant with customers places. In the first case, a transportation system devoted to move and store items, tools and fixtures, to supply manufacturing and assembling cells. In the latter, a long-range transportation network aims to move raw materials towards manufacturing plants and final products towards distribution centers.

The following considerations will be centred on internal logistics systems, also denoted *Material Handling System - MHS*, when operating in a discrete manufacturing plant (producing either mechanics or electronics or electrical items).

In industrial practice four different tipical configurations of manufacturing plants and related internal logistics systems can be envisaged (*G. Chryssolouris, 1992*):

- *job shop*, i.e. plant composed by a network of manufacturing/assembling cells, in principle all connected together;
- *project shop*, also denoted *one-of-a-kind-system*;
- *cellular shop*;
- *flow line*, sometimes denoted *channel*.

In a job-shop (*French, 1982*) machining units able to perform similar operations are grouped together. Then, a part or a part batch has to be moved in the shop floor in such a way to be progressively addressed to the different sets of machining units according to the sequence of operations which must be executed on the part itself. The internal logistics system should be characterized by relevant flexibility, with a large number of admissible movements. This feature can assure all the machining units could be well balanced while the whole network could be rapidly reorganized at the occurrence of undesired events, such as local failures.

In a project shop (*Wortmann, 1990*) a part is processed for a long time by the same manufacturing unit whilts materials, tools and fixtures are addressed to units depending on their requests. Such a plant requires to planning operations and to accurately scheduling resources utilization as for example through an *MRP - Material Requirement Planning* (*Orlicky, 1975*; *Vollman e altri, 1992*). Then the internal logistics management has to perform accordingly.

A cellular plant is usually organized by direct application of Group Technology concepts (*J. Burbidge, 1989*), as the "product family" concept, being a "product family" a set of products similar one to the others, such to be processed by similar sequences of operations. In this case, it looks convenient gouping together different machining units such that the union of their processing capabilities cover all the sequences of operations required by the products included in a same family. As a consequence, organization of a local logistics system for each group of machining units, related to a product family, appears to be necessary. All local logistics systems will be related together by a higher layer MHS.

A flow line shows a structure conceptually very closed to that of a cellular plant (*Lucertini et al. 1996*), but each product family is processed by a proper line (i.e. a physical sequence of machining units, serially connected). In this situation, the layout organization of each line is based on the sequence of operations to be executed for the related product family. The internal logistics system can be one more composed by a two-layer network, one associated to each line, the other connecting inputs and outputs of the lines.

In the industrial environments, the four types of MHS's are usually mixed for generating an actual MHS. The choice of the structure, inteed, depends on the types of objects to be produced, the expected production volumes, the opportunities offered by the market of new technologies, and the economic constraints. This short list of constraints can give an idea of the complexity of the MHS design problem.

In particular, the strongest constraints arise from the organization of the manufacturing plant. In fact, both the structure and the management of an internal logistics system has to verify several requirements directly imposed by the manufacturing process to be supplied (*J.P. Tanner, 1985*):

a) assure required volumes and delivery punctuality;
b) minimize the average levels of storages contained in internal buffers, i.e. the *Work-In-Process (WIP)*;
c) assure a high turnover of materials;
d) assure a sufficient flexibility in the part supply, such that a corresponding manufacturing flexibility could be implemented;
e) assure a high utilization of internal spaces devoted to storage;
f) assure the best possible utilization of personnel at disposal.

All these requirements can greatly affect the design proposals for both the components of an internal logistics system, namely the *network structure* and the *material handling management*.

The definition of the network structure traditionally consists of the physical organization of the network of admissible movements and storages, i.e.:

I. the definition of the *graph* of connections among the machining units;
II. the estimation of the dimensions of *spaces* to be devoted to logistics opeations, either movements (e.g. roads) or storages or loading/unloading;
III. the selection of *devices* to be used for logistics operations;
IV. the selection and training of *personnel* to be employed.

For what concerns the graph of admissible movements, the above introduced classification of logistics systems suggests to consider three different types of structures (layout), according to the different production work-programmes:

a) *fixed-positions structure*, in which working operations are accomplished by a unique machining center whilst materials, tools and fixtures are addressed towards machining units upon requests of the different working operations to be performed;
b) *process-oriented structure*, where all working operations of the same type are implemented in the same cell, which includes similar machining units;
c) *product-oriented structure*, when all working operations required for completion of a product family are applied in a same cell, which now includes complementary machining units.

Technical literature contains several contributions analyzing convenience in applying either one or the other type of structure (*G. Chryssolouris, 1992*; *J.P. Tanner, 1985*; *J.A. Buzacott e J.G. Shanthikumar, 1992*; *N. Viswanadham e Y. Narahari, 1992*).

Referring to the problem of an efficient utilization of the internal space of an industrial department reserved for the shop floor, two approaches can be found (*A. Kusiak, 1990*):

a) allocating the logistics network in a pre-subdivided space, each subdivision being a part of space where a cell or a machining unit can be placed;
b) allocating the cells or machining units such as nodes of a standardized network (e.g., a line, a ring, a U-type route).

The selection of devices to be used in logistics operations aims to completely define all the components of the MHS, and must be performed after the final definition of the network of connections, the identification of all functions to be performed, and an estimation of the expected performance of each component.

The organization of the system for managing all logistics operations requires to design the management architecture devoted to drive the logistics system when executing its own operations. This implies to define a set of procedures for:

a) computing the best possible transportation programmes for supplying parts, tools, fixtures according to working programmes;
b) monitoring the current state of the logistics system, time by time, and then comparing state measures with planned state values;
c) selecting and applying on-line control actions (as re-addressing of parts) when significant shift is detected between actual and planned state values.

All above considerations suggest to introduce a clear classification of the main design objectives which an effective design of a logistics system must reach, objectives to be related with the main functions a MHS should be able to execute:

⇒ an effective logistics system must operate such as the true *connective system* in a set of technological processes, that means, to assuring feasibility of working programmes to be implemented;

⇒ an efficient logistics system must operate such as the guaranteed *supply service* in a set of working centers, that means, to assuring all internal production orders can be satisfied;

⇒ a functional logistics system must operate such as a *modularity factor* in the shop floor: the internal logistics network should be arranged in order to be, on one hand, the material handling service assuring to all centers the most balanced supply for the most efficient utilization, on the other, the re-addressing service able to isolate centers to be turned off and substituted and to move residual loads towards the most efficient remaining units.

The three complementary ways of describing main attributes of a logistics system give direct suggestions for classifying the most significant design objectives:

a) the network must be sufficient to assure connectivity required by the product mix to be processed (graph connectivity target);
b) the network must be modified at low cost either by introducing or removing links and by displacing storages (graph modularity target);
c) the network must assure a sufficient transportation capacity, mainly when demand peaks occur (capacity planning target);
d) the network must supply all production units by preventing any idle time (fault tolerability target)
e) the network must move items to keep the internal storages at the lowest level possible (WIP minimization target)
f) the network must be able to respond very quickly to delivery requests from production units (transient management target)

2. OUTLINE OF THE "BLACKBOARD-BASED" DESIGN APPROACH

A design process which could be able to reach all the objectives previously stated cannot be neither fast nor simple. It looks obvious to think said design process such as a decision-making process evolving through progressive development and refinement of an initial hypothesis of logistics network. Moreover, it looks natural to think that the design of an industrial logistics system can require to organize a development procedure not stated in a sequential form but according to a *"generate and test"* form. This organization includes synthesis phases, devoted to the proposal of new system configurations, to be alternated with analysis phases dedicated to verifying if the proposed system configuration could be accepted or not (Fig. 1).

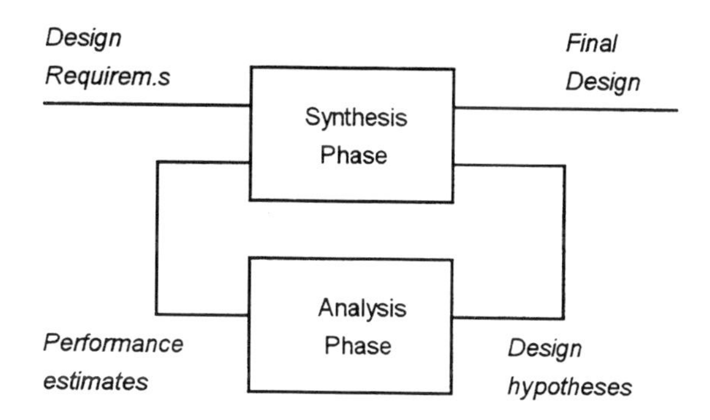

Fig. 1. Logical scheme of a "generate and test" design approach.

Fig. 1 shows that the results obtained by the analysis of a proposed hypothesis of innovated logistics network gives new suggestions for applying further synthesis procedures, with the aim of further improving the analyzed network proposal. This way of doing makes evidence of the *progressive refinement loop of design*, which specifies the concept of "generate and test" approach.

At each iteration of the design cicle, the synthesis phase results to be the creative step. It can be developed having at disposal prescriptive models of logistics systems, computation, optimization and choice tools, as well as Decision Support Systems - DSS. To this extent, several books and papers, including these types of models, tools and provedures, exist (see the extended bibliography in *Brandimarte & Villa eds., 1995*).

The analysis phase, on the contrary, respresents a step dedicated to reflect upon the utility and effectiveness of the design proposals. It has to be performed by applying descriptive models of a logistics network, able to give accurate description of the logistics system operations, analyze the efficiency of movements and the utilization of storages, verify how the MHS operation reflects on the production system efficiency.

Execution of said two phases must be driven by a specific design protocol, that means a set of rules supporting the designer in choosing the synthesis and analysis tools at each design iteration. This design protocol should be interactive, such as to allow direct intervention of the designer at each decisional step depending on the results obtained at the previous design step and on their difference with respect to the design target.

The preliminar analysis of the logistics design problem allows to recognize the set of synthesis and analysis tools (denoted in the following *modules*) which a designer should have at disposal in order to generate an effective *design support system* (i.e. a DSS containing an organized set of formal models and procedures, that means a knowledge base, as well as a corresponding set of software tools for implementing said models and procedures). Exam-

ples of these DSS can be found in the books of *Kusiak (ed), 1988* and of *Tzafestas (ed), 1993*.
At the same time, analysis modules suited to the specific class of logistics systems to be developed has to be selected. A good solution is that of implementing a general purpose simulator, to define proper scenarios representing both production demands and the production environment, and to identify specific performance measures (see suggestions proposed by *Viswanadham and Narahari, 1992*).

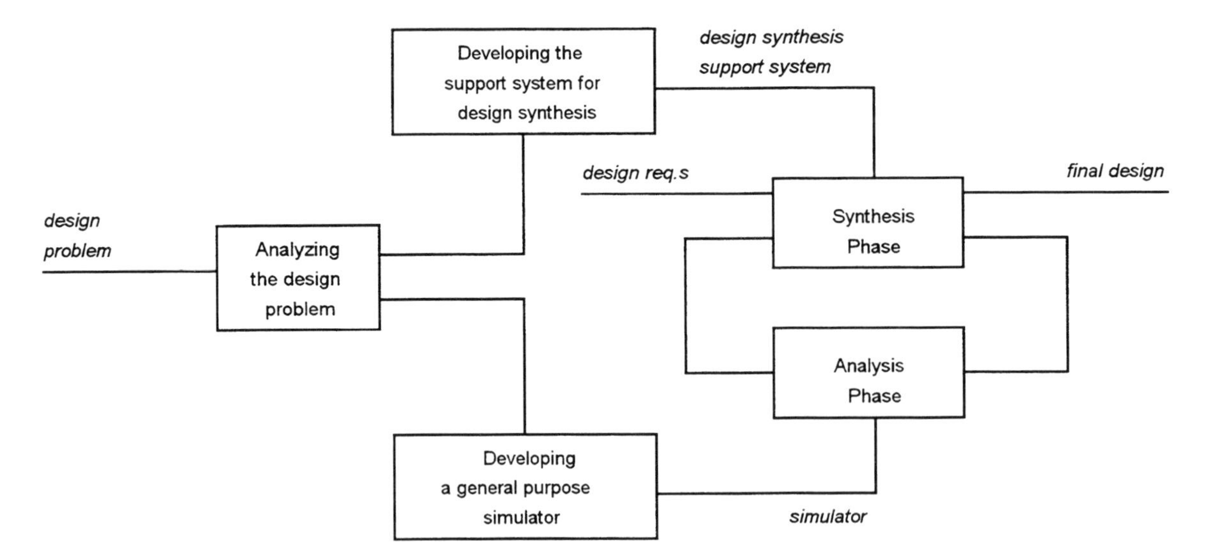

Fig. 2. Logical scheme for generating the complete design workbench..

Taking into account the above stated more accurate description of the two design phases, the "generate and test" design approach outlined in Fig. 1 can be more detailed in Fig. 2, in which the sequence of "design protocol generation" and then of "design procedures utilization" is evidenced.
In effect, Fig. 2 shows how a design process should be approached, by first selecting design tools for both the synthesis and the analysis phase, then supplying said tools to the design phase, and finally including both design phases into the refinement loop. In other words, Fig. 2 illustrates the main concept of a real design workbench, that means the complete set of tools which allows an industrial designer to approach the logistics system design problem (what is sometimes called "development system").An application of this idea is reported in the book by *Villa et al, 1996*. The design workbench illustrated in Fig. 2 has been described, till now, only in conceptual terms. Its realization obviously requires further efforts, by defining

- on one hand, a detailed model of the various steps of the design process, such to specify how the "generate and test" approch could be implemented;
- on the other hand, an accurate model of the class of logistics systems to be innovated, their components, physical connections and functional interactions among them.

The former point is of preminent interest in this contribution.

The non-procedural design approach here proposed implies to introduce a new formulation of the design process model. To this aim, let us refer to a group of designers, each one endowed with a proper technical knowledge, and all allowed to use a common blackboard on which the attributes of the system to be designed are listed. The designers can communicate together only through the blackboard. Knowledge of each designer allows him to compute values of some attributes of the system to be developed, and to write them on the blackboard, only if other attributes have already been computed and their values communicated on the same blackboard. All designers contributions have to be regulated by a coordinator.
In terms of the above introduced concepts motivating the "generate and test" approach, this illustration of a designers team can be used for detailing the design workbench of Fig. 2. Each designer corresponds to a *design module*, i.e. a specific procedure either for computing values of some attributes of the logistics system under innovation or for estimating some performance measures. Design modules can either read or write attributes values on a common blackboard (i.e., an array of attributes). The ability of each design module in computing attributes as well as the input values required are stated in an access table, each raw being related to an attribute and each column being related to a design module. A conceptual example of access table is illustrated in Fig. 3.
The blackboard concept is the key to realize how the "progressive refinement design approach" here proposed can operate. The core of the design protocol is the set of criteria and rules for the progressive application of the different design modules, provided that each time (i.e., at each *refinement step*) the most effective module has to be choosen depending on the values of system attributes till then computed.

Design modules: Blackboard	Modul 1	......	Modul i	......	Modul n
Attribute 1	r/w/p		r		w
.....					
Attribute j			r/w		
.....					
Attribute n			p		r

Fig. 3. Example of access table ("r" means input value; "w" output value).

To this aim the design protocol must be able to understand, at each refinement step, the actual state of the design process (i.e., the values of the system attributes and their residual difference with respect to target values, a priori estimated during the feasibility step) and to accordingly suggest modules which could be applied by the designer.
The "generate and test" design process, now detailed by evidencing the blackboard and its use, is illustrated in Fig. 4, in which the practical components of the design workbench are represented:
(a) the library of synthesis and analysis modules to be implemented;
(b) the blackboard, that means the array summarizing, at each design step, the proposal of new logistics systems;
(c) the control of execution of each new module to be activated;
(d) the function of the designer, i.e. the decision-maker who must choose the next module to be activated, among those proposed by the function devoted to control execution of each module after its activation.

This scheme should clearly show how the design protocol can operate, by applying procedures for controlling the execution of each activated module, and by using a set of procedures for managing information exchanges among modules, from modules to the blackboard, between the blackboard and the designer.
As it can be seen in Fig. 4, the designer has the responsibility of choosing the module to be activated by selection over a menu which, each step, is updated according to the evolution of the design state.

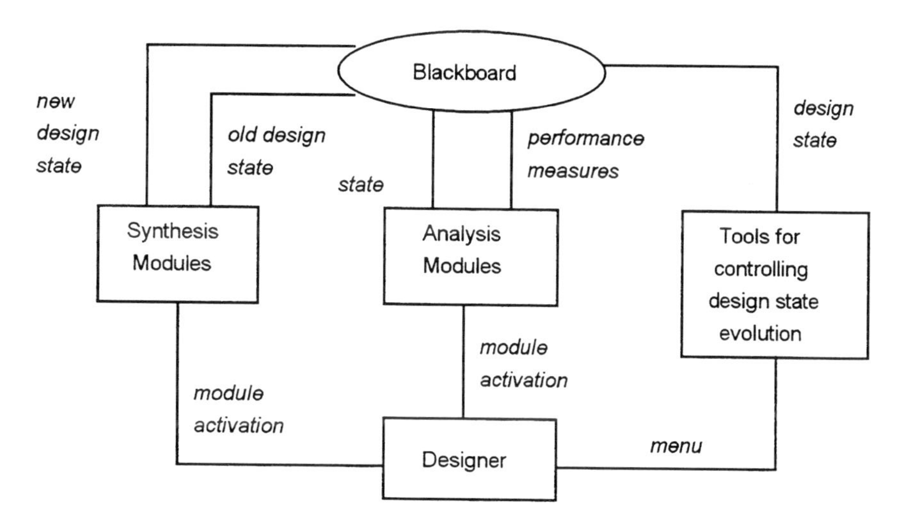

Fig. 4. Scheme of the "generate and test" design process detailing the design protocol components.

This step-by-step selection allows to take care of design constraints, such as bounds for the system attribute values.
Knowledge of which attributes have still to be computed, at each design step, allows to recognize alternative "design routes" over the access table (ecah design route being a sequence of modules which can be activated).
To each design route a proper cost could be associated, as either time or cost to complete the design. The recognition of a minimum cost design route could be a criterion for helping the designer in taking his decisions.

3. OUTLINE OF A METHODOLOGY TO APPLYING THE BLACKBOARD-BASED DESIGN APPROACH

This section is devoted to describe how the proposed design method can be applied,

- by first specifying activities required for organizing the design workbench,
- then illustrating the application of the workbench in selecting modules to be included in the access table of Fig. 3.

First organization activity is dedicated to the selection of:

a) model of the logistics system to be developed and array of attributes (blackboard) allowing a sufficient description;
b) procedures for implementing each synthesis and analysis phase, as well as the formal tools required;
c) access table for the blackboard;
d) design protocol.

Output of this first organization activity is the dedicated workbench. Inputs are the customer needs and experience gained in previous design of similar systems.

Second organization activity consist in the preliminar selection of the main characteristics of the new logistics system, namely:

a) preliminar formulation of an initial configuration of the logistics system layout by identifying and by modelling the customer needs;
b) execution of the sequence of design activities included in the *generate and test loop*:
   - ⇒ identification of modules which can be activated according to the current menu, depending on the design state described by the blackboard currently updated (this last activity has to be performed by module controlling the design state)
   - ⇒ choice of the next module to be activated, among those proposed in the menu;
   - ⇒ validation of the proposal of new system configuration, to be performed by a proper analysis tool;
   - ⇒ choice among design alternatives, each time more than one proposal seems to be acceptable (i.e., proposals are comparable); this task has to be performed by the designer.

To completely realize the applicability and novelty of the blackboard-based approach, let us simulate the activity of choosing modules (step a) in the first organization activity).

Design tools to be applied for synthesizing logistics networks differ depending on the type of application required, namely either referred to the formulation of a preliminary configuration of the network, or to the detailed design of a network under development.

In the first case, a logistics network can be outlined in two steps:

- by identifying the design requirements, submitted by the customer, and translating them into admissible operations and performance of the new system;
- by formulating an initial configuration of the new system which must be obtained by adapting previously developed layouts to the new requirements.

Identification of the design requirements implies to state the following:

- bounds for the cost for developing and implementing the network, as well for the duration of the implementaion transient;
- minimum efficiency level which the logistics network must assure to the production plant;
- target values for several effectiveness measures, such as delivery punctuality, dimensions of moved volumes, internal storage levels;
- bounds for the realiability and maintainability of the whole logistics systems and of its more critical components;
- estimates of expected production volumes and of their normal variations in time;
- characterization of the production modes, either mix-oriented or batch-oriented;
- compact description of the production cicles to be more frequently applied.

These requirements are usually defined by a strong interaction of the designer with the customer. Classification of said requirements can be obtained by existing design support procedures, such as *QFD - Quality Function Deployment* (*Akao, 1990*).

Synthesis of the preliminar configuration consist of a selection of a first-attempt system layout: attention should be paid to the design problems which could arise depending on the specific design constraints. Main steps for defining the preliminar system layout are:

a) definition of the system functional specifications: for instance, identifications of product families to be processed and handled by applying some *clustering* procedure derived from *Group Technology* (*Kusiak, 1990*);
b) *generation of a first-attempt production programme:* this second activity implies that a preliminary schedule is computed by some lot sizing procedure and then a dispatching strategy applied to verifying admissible movements of lots (*Brandimarte and Villa, 1995*);
c) *estimation of the transportation capacity on the network links:* estimates of loading conditions at machining centers should be used in order to evaluate the transportation volumes over the network connections, e.g. by applying some *capacity planning* module;
d) *estimation of the needs for storage capacity:* to this aim, preliminary evaluations can be obtained by applying discrete-event simulation modules such as the ones derived by queue theory and *Mean Value Analysis* (*Suri and Hildebrant, 1984*).

The improvement of the system configuration, to be obtained through progressive refinements in the "generate and test" loop, results from two main further design activities:

a) the proposal of improved layout configuration;
b) the synthesis of new transportation management strategies.

The definition of a new layout as well as its improvemet calls for adoption of some choices (and then development of related modules) about:

a) *verifying the convenience of "transit buffers":* in case the planned production loads for a gruop of machining centers require a buffer capacity for each center exceeding standard dimensions, inclusion of a transit buffer (i.e. a buffer to be shared by all centers in the group) can be interesting for reducing delivery times;

b) *verifying utility of local transportation systems:* sometimes the global logistics system in the shop floor could be unable to offer an effective service to some groups of centers, where movements of parts, tools and fixtures are particularly frequent; in these situations it can be convenient to arrange a local MHS, to be locally managed and to be connected to inter-cell material handling;

c) *verifying convenience of a unique transportation system for both parts and tools:* the number of all missions which the transportation system has to execute has to be estimated and compared with standard capacity offered by existing transport devices;

d) *selecting new layout configurations:* taking into account existence of local MHS's as well as proposal of using the logistics system for moving parts and tools either by same devices or separately, evaluation of the type of implementable layout and of its innovative features must be done.

Besides layout definition, the set of modules devoted to developing management strategies for the logistics network has to be organized such as a "tools library". It should give methods for optimizing utilization and efficiency of the internal logistics system at disposal of the designer, in front of each proposal of network structure. When choosing the management strategy, the following elements have to be taken into account:

a) type of production scenario, either repetitive or not;
b) types of admissible movements among various centers;
c) type of network organization, either cell based or not, either including transit buffers or not;
d) types of constraints assigned by the customer to the production programmes (as for strong due date constraints of products and raw materials).

The consequent problem to be solved typically consists of a scheduling problem. In theoretical terms, it appears to be very complex (*French, 1982; Conway, Maxwell and Miller, 1967*). Its theoretical complexity pushed reaserchers towards definition and application of heuristics methods partly based on dispatching rules partly on procedures for the search of a "satisfying" schedule (*Brandimarte e Villa eds., 1995*).

After having outlined the workbench control logics and having characterized main synthesis modules, some concepts will be addressed for what concerns analysis phase. In usual practice the design state and its adequacy is evaluated by simulation tools. Often simulation is applied during the final iteration of the "generate and test" loop. However it should be necessary to perform a sufficiently accurate analysis at each design step. This need, now recognized also by industrial designers, calls for availability of simulation tools both simplified in its structure and accurate in its result. This now appears to be the real gap between available analysis tools and designer's needs: as soon as this gap will be covered, an effective "incremental prototyping" of innovated logistics network will be possible.

4. CONCLUSIONS

The most useful conclusions which can be drawn by the above outline of a new design methodology are the following.

First, the design process has to be approached according to a structured method, clearly organized and equipped with real-utility support tools. Organization of a design method means to have at disposal a "design protocol" showing which are the design phases, which is the designer's role, which are the design support tools. Availability of a methodology appears to be more and more necessary given the increasing dimension and complexity of the industrial design problems.

Second, the design methodology has to be equipped with synthesis and analysis modules as simple as possible but such that they could be connected together into different admissible design routes. An effective innovation can only be obtained if the design can develop, propose and compare alternative configurations of the system to be innovated. This looks particularly important in case of new logistics systems, in which even small innovations in the structure and, sometimes, in the management, give rise to relevant improvement of the system efficiency and effectiveness.

REFERENCES

Akao Y., ed. (1990), "Quality Function Deployment", Productivity Press, Cambridge, MA..

Brandimarte P. (1992), Neighbourhood search-based optimization algorithms for production scheduling: a survey, in *Computer Integrated Manufacturing Systems*, Vol. 5, p. 167.

Brandimarte P., Villa A. (1995), *Advanced Models for Manufacturing Systems Management*, CRC Press, Boca Raton, CA.

Brandimarte P. and A. Villa, eds. (1995), *Optimization Models and Concepts in Production Management*, Gordon & Breach Publishers SA, Basel, Switzerland.

Burbidge J. (1989), *Production Flow Analysis for Planning Group Technology*, Clarendon Press, Oxford, UK.

Buzacott J. and Shanthikumar J.G. (1992), Models of production systems, in *Handbook of Industrial Engineering*, J. Wiley, New York.

Chryssolouris G. (1992), *Manufacturing Systems - Theory and Practice*, Springer-Verlag, New York.

Conway R.W., Maxwell W.L. and Miller L.W. (1967), *Theory of Scheduling*, Addison-Wesley, Reading, MA.

French S. (1982), *Sequencing and Scheduling: An Introduction to the Mathematics of the Job-Shop*, Ellis Horwood LTD, Chichester.

Kusiak A. (1990), *Intelligent Manufacturing Systems*, Prentice-Hall, Englewood Cliffs, NJ.

Kusiak A. ed. (1988), *Expert Systems - Strategies and Solutions in Manufacturing Design and Planning*, SME, Dearborn, MI.

Lucertini M., F. Nicolò, W. Ukovich and A. Villa (1996), Family of discrete-event models for flow line productions, in Proc. 4th Int. Conf. on Advanced Manufacturing Systems and Technology, Udine, Italy.

Orlicky J. (1975), *Material requirements planning*, McGraw-Hill, New York.

Suri e Hildebrant R.R., "Modelling flexible manufacturing systems using Mean Vallue Analysis", J. Manufact. Systems, vol. 3, 1984, p. 27.

Tanner J. (1985), *Manufacturing Engineering - An Introduction to the Basic Functions*, M. Dekker, Inc., New York.

Tzafestas S. (1993), *Expert Systems in Engineering Applications*, Springer-Verlag, Berlin.

Villa A., M. Cantamessa and S. Rossetto, Logistics newtroks: a design manual, Progetto Finalizzato Trasporti Due, CNR, 1996.

Viswanadham and Y. Narahari (1992), *Performance Modeling of Automated Manufacturing Systems*, Prentice-Hall, Englewood Cliffs, NJ.

Vollmann T.E., Berry W.L. and Whybark D.C. (1992), *Manufacturing Resource Planning and Control Systems*, Irwin, Homewood, IL.

Wortmann J.C., Towards one-of-a-kind production: the future of European industry, in Proc. Int. Conf. Advances in Production Management Systems, Espoo, Finland 20-22 August 1990.

REGULARIZATION OF A SINGULAR 2-D ROESSER MODEL BY OUTPUT-FEEDBACK

Kaczorek Tadeusz

Warsaw University of Technology
Faculty of Electrical Engineering
Institute of Control and Industrial Electronics
00-662 Warszawa, Koszykowa 75, POLAND

Abstract: The regularization method presented in (Kaczorek, 1996) is extended for output-feedbacks. Sufficient conditions are established under which the singular 2-D Roesser model can be regularized by output-feedbacks. A procedure is presented for computation of the feedback matrices and ilustrated by a numerical example.

Keywords: regularization, singular 2-D Roesser model, output feedback, sufficient condition

1. INTRODUCTION

The regularization of singular (descriptor) linear system by state and output feedbacks has been considerd in many papers (Bunse-Gestner, et.al., 1992; Bunse-Gestner, et al., 1994; Ozcandiran, et al., 1990). In (Bunse-Gestner, et al., 1994) it was shown that propotionnal and derivative output-feedback controlers can be constructed such that the closed-loop system is regular and has index at most one. The regularity guarantees the existence and uniqueness of solution to singular linear system (Campbell, 1980; Ozcandiran, et al., 1990; Yip,et al., 1981; Kaczorek, 1993). The regularization problem for singular 2-D Roesser model by state-feedbacks has been considered in (Kaczorek, 1996). The main goal of this paper is to extend the method given in (Kaczorek, 1996) for output-feedbacks.

2. PROBLEM FORMULATION

Consider the 2-D discrete Roesser model described by equations

$$Ex_{ij}^{(1)} = Ax_{ij} + Bu_{i,j}, \qquad i,j \in Z^{+} \quad (1)$$

$$y_{i,j} = Cx_{ij} + Du_{i,j} \quad (2)$$

where

$$x_{ij} := \begin{bmatrix} x_{i,j}^{h} \\ x_{i,j}^{v} \end{bmatrix}, \; x_{ij}^{(1)} := \begin{bmatrix} x_{i+1,j}^{h} \\ x_{i,j+1}^{v} \end{bmatrix}, E, A \in R^{n\times n}; B \in R^{n\times m};$$

$x_{i,j}^{h} \in R^{n_1}$ is the horizontal state vector; $x_{i,j}^{v} \in R^{n_2}$ is the vertical vector; $n_1 + n_2 = n$; $u_{i,j} \in R^{m}$ is the input vector; $C \in R^{p\times n}, D \in R^{p\times m}$; $y_{i,j} \in R^{p}$ is the output vector.

Let the output-feedback have the form:

$$u_{i,j} = Fy_{i,j} - Gx_{i,j}^{(1)} + v_{i,j} \quad (3)$$

where $F \in R^{m\times p}, G \in R^{m\times n}$; $v_{i,j} \in R^{m}$ is the new input vector.

Substituting (2) into (3) we obtain:

$$(I - FD)u_{i,j} = FCx_{i,j} - Gx_{i,j}^{(1)} + v_{i,j}$$

Assuming that the matrix $(I\text{-}FD)$ is nonsingular and defining $H:=(I\text{-}FD)^{-1}$ we get

$$u_{i,j} = HFCx_{i,j} - HGx_{i,j}^{(1)} + Hv_{i,j} \quad (4)$$

Let

$$E := \begin{bmatrix} E_{11} & E_{12} \\ E_{21} & E_{22} \end{bmatrix}, \quad A := \begin{bmatrix} A_{11} & A_{12} \\ A_{21} & A_{22} \end{bmatrix},$$

$$B:=\begin{bmatrix} B_1 \\ B_2 \end{bmatrix}, \quad G:=\begin{bmatrix} G_1 & G_2 \end{bmatrix}, \quad C:=\begin{bmatrix} C_1 & C_2 \end{bmatrix},$$

then

$$E+BHG=\begin{bmatrix} E_{11}+B_1HG_1 & E_{12}+B_1HG_2 \\ E_{21}+B_2HG_1 & E_{22}+B_2HG_2 \end{bmatrix},$$

$$A+BHFC=\begin{bmatrix} A_{11}+B_1HFC_1 & A_{12}+B_1HFC_2 \\ A_{21}+B_2HFC_1 & A_{22}+B_2HFC_2 \end{bmatrix}, \qquad (5)$$

$$B(I-FD)^{-1}=BH=\begin{bmatrix} B_1H \\ B_2H \end{bmatrix}$$

Definition 1. The system (1) is called regular if and only if

$$\det\left\{\begin{bmatrix} E_{11}z_1 & E_{12}z_2 \\ E_{21}z_1 & E_{22}z_2 \end{bmatrix} - A\right\} \neq 0 \quad \text{for some} \quad z_1, z_2 \in \mathbf{C}$$

where $\mathbf{C}$ is the field of complex numbers.
It is assumed that

1) $\det E = 0$

2) $$\det\left\{\begin{bmatrix} E_{11}z_1 & E_{12}z_2 \\ E_{21}z_1 & E_{22}z_2 \end{bmatrix} - A\right\} = 0 \quad \text{for all} \quad z_1, z_2 \in \mathbf{C} \qquad (6)$$

The problem of regularization can be formulated as follow:
Given the matrices E, A, B, C, D and a nonnegative integer r, $0 \le r \le n$,

find the matrices F, G such that the closed-loop system is regular and

$$rank(E+BHD)=r \qquad (7)$$

3. PROBLEM SOLUTION

The matrices G_1, F are choosen so that

$$E_{21}+B_2HG_1=0 \quad \text{and} \quad A_{21}+B_2HFC_1=0. \qquad (8)$$

Equations (8) have solutions if and only if

$$rank\,B_2 = rank\left[E_{21}, B_2\right] = rank\left[A_{21}, B_2\right] \qquad (9$$

$$rank\,C_1 = rank\begin{bmatrix} A_{21} \\ C_1 \end{bmatrix}$$

Let $M=\begin{bmatrix} M_1 \\ M_2 \end{bmatrix}$ be a nonsingular row-compression matrix of the matrix B_2, and $N=\begin{bmatrix} N_1 & N_2 \end{bmatrix}$ be a nonsingular column-compression of the matrix C_1 such that

$$MB_2=\begin{bmatrix} B_2' \\ 0 \end{bmatrix}, \quad C_1N=\begin{bmatrix} C_1' & 0 \end{bmatrix} \qquad (10)$$

where B_2' has full row rank and C_1' has full column rank.
Premultiplying (8) by the matrix M obtain:

$$\begin{bmatrix} M_1E_{21} \\ M_2E_{21} \end{bmatrix} = \begin{bmatrix} -B_2'HG_1 \\ 0 \end{bmatrix}, \qquad (11a)$$

$$\begin{bmatrix} M_1A_{21} \\ M_2A_{21} \end{bmatrix} = \begin{bmatrix} -B_2'HFC_1 \\ 0 \end{bmatrix} \qquad (11b)$$

Postmultiplying (11b) by the matrix N we have:

$$\begin{bmatrix} M_1A_{21}N_1 & M_1A_{21}N_2 \\ M_2A_{21}N_1 & M_2A_{21}N_2 \end{bmatrix} = \begin{bmatrix} -B_2'HFC_1' & 0 \\ 0 & 0 \end{bmatrix} \qquad (12)$$

Let

$$G_1 = H^{-1}\left(-B_2'^T(B_2'B_2'^T)^{-1}M_1E_{21} + \left(I - B_2'^T(B_2'B_2'^T)^{-1}B_2'\right)K_1\right)$$
$$F = H^{-1}Q \qquad (13)$$

where

$$Q = -B_2'^T(B_2'B_2'^T)^{-1}M_1A_{21}N_1(C_1'^TC_1')^{-1}C_1'^T$$
$$+\left(I - B_2'^T(B_2'B_2'^T)^{-1}B_2'\right)K_2 + K_3\left(I - C_1'(C_1'^TC_1')^{-1}C_1'^T\right)$$

and K_1, K_2, K_3 are arbitrary matrices.
By assumption $\det H \neq 0$ and we have

$$\det[H-QD] \neq 0 \qquad (14)$$

Substituting (13) into (5) we obtain

$$E+BHG=\begin{bmatrix} \overline{E}_{11}+\overline{B}_1K_1 & \overline{E}_{12} \\ 0 & E_{22}+B_2HG_2 \end{bmatrix},$$

$$A+BHFC=\begin{bmatrix} \widetilde{A}_{11}+\overline{B}_1K_2C_1 & \widetilde{A}_{12} \\ 0 & \widetilde{A}_{22}+B_2K_3\widetilde{C}_2 \end{bmatrix}, \qquad (15)$$

where

$$\overline{E}_{11} := E_{11} - B_1B_2'^T(B_2'B_2'^T)^{-1}M_1E_{21},$$
$$\overline{B}_1 := B_1\left(I - B_2'^T(B_2'B_2'^T)^{-1}B_2'\right),$$

$$\overline{E}_{12} = E_{12} + B_1HG_2$$
$$\widetilde{C}_2 := \left(I - C_1'(C_1'^TC_1')^{-1}C_1'^T\right)C_2 \qquad (16)$$

$$\widetilde{A}_{11} := A_{11} - B_1B_2'^T(B_2'B_2'^T)^{-1}M_1A_{21}N_1N_1'$$
$$\widetilde{A}_{12} = A_{12} + B_1HFC_2$$

$$\widetilde{A}_{22} := A_{22} - M_1'M_1A_{21}N_1(C_1'^TC_1')^{-1}C_1'^TC_2$$

where $\begin{bmatrix} M_1' & M_2' \end{bmatrix} := M^{-1}$, $\begin{bmatrix} N_1' \\ N_2' \end{bmatrix} := N^{-1}$

Lemma 1. (Miminis, 1993). If $E, A \in R^{n\times n}, B \in R^{n\times m}$ then there exist nonsingular matrices $P, Q \in R^{n\times n}$ such that:

$$PEQ=\begin{bmatrix}E_{11} & E_{12} & E_{13}\\ 0 & E_{22} & E_{23}\\ 0 & 0 & E_{33}\end{bmatrix},\quad PAQ=\begin{bmatrix}A_{11} & A_{12} & A_{13}\\ 0 & A_{22} & A_{23}\\ 0 & 0 & A_{33}\end{bmatrix},$$

$$PB=\begin{bmatrix}B_1\\ 0\\ 0\end{bmatrix},$$

where $E_{ij},A_{ij}\in R^{\bar{n}_i\times\bar{n}_j};i,j=1,2,3;B_1\in R^{\bar{n}_1\times m}$; E_{22},A_{22} are nonsingular, matrices E_{33},A_{33} are upper triangular and have all zero diagonal entries, (E_{11},A_{11},B_1) is c-controllable, i.e.

$$rank(\alpha E_1-\beta A_{11},B_1)=\bar{n}_1\forall(\alpha,\beta)\in C^2/\{(0,0)\}$$

3.1. Condition $rank(E+BHG)=r$

We apply the Lemma 1 to the matrices: $\bar{E}_{11},\tilde{A}_{11},B_1$ and $E_{22},\tilde{A}_{22},B_2$

Let P_1,Q_1,P_2,Q_2 be nonsingular matrices defined in Lemma 1, such that

$$P_1\bar{E}_{11}Q_1=\begin{bmatrix}\bar{E}_{11}^{11} & \bar{E}_{12}^{11} & \bar{E}_{13}^{11}\\ 0 & \bar{E}_{22}^{11} & \bar{E}_{23}^{11}\\ 0 & 0 & \bar{E}_{33}^{11}\end{bmatrix},$$

$$P_1\tilde{A}_{11}Q_1=\begin{bmatrix}\tilde{A}_{11}^{11} & \tilde{A}_{12}^{11} & \tilde{A}_{13}^{11}\\ 0 & \tilde{A}_{22}^{11} & \tilde{A}_{23}^{11}\\ 0 & 0 & \tilde{A}_{33}^{11}\end{bmatrix},\quad P_1B_1=\begin{bmatrix}\hat{B}_1\\ 0\\ 0\end{bmatrix},$$

$$P_2E_{22}Q_2=\begin{bmatrix}E_{11}^{22} & E_{12}^{22} & E_{13}^{22}\\ 0 & \bar{E}_{22}^{22} & E_{23}^{22}\\ 0 & 0 & E_{33}^{22}\end{bmatrix}, \tag{17}$$

$$P_2\tilde{A}_{22}Q_2=\begin{bmatrix}\tilde{A}_{11}^{22} & \tilde{A}_{12}^{22} & \tilde{A}_{13}^{22}\\ 0 & \tilde{A}_{22}^{22} & \tilde{A}_{23}^{22}\\ 0 & 0 & \tilde{A}_{33}^{22}\end{bmatrix},\quad P_2B_2=\begin{bmatrix}\hat{B}_2\\ 0\\ 0\end{bmatrix},$$

where $\bar{E}_{22}^{11},E_{22}^{22},\tilde{A}_{22}^{11},\tilde{A}_{22}^{22}$ are nonsingular, matrices $\bar{E}_{33}^{11},E_{33}^{22}$ is upper triangular and have all zero diagonal entries.

Premultiplying and postmultiplying (15) by the matrices $P=\begin{bmatrix}P_1 & 0\\ 0 & P_2\end{bmatrix}$ and $Q=\begin{bmatrix}Q_1 & 0\\ 0 & Q_2\end{bmatrix}$,

respectively we obtain

$$P(E+BHG)Q=\left[\begin{array}{ccc|ccc}\bar{E}_{11}^{11} & \bar{E}_{12}^{11} & \bar{E}_{13}^{11} & & & \\ 0 & \bar{E}_{22}^{11} & \bar{E}_{23}^{11} & & P_1\bar{E}_{12}Q_2 & \\ 0 & 0 & \bar{E}_{33}^{11} & & & \\ \hline & & & E_{11}^{22} & E_{12}^{22} & E_{13}^{22}\\ & 0 & & 0 & E_{22}^{22} & E_{23}^{22}\\ & & & 0 & 0 & E_{33}^{22}\end{array}\right]+\left[\begin{array}{c|c}\hat{B}_1'K_1Q_1 & 0\\ 0 & 0\\ 0 & 0\\ \hline 0 & \hat{B}_2HG_2\\ 0 & 0\\ 0 & 0\end{array}\right] \tag{18}$$

$$P(A+BHFC)Q=\left[\begin{array}{ccc|ccc}\tilde{A}_{11}^{11} & \tilde{A}_{12}^{11} & \tilde{A}_{13}^{11} & & & \\ 0 & \tilde{A}_{22}^{11} & \tilde{A}_{23}^{11} & & P_1\tilde{A}_{12}Q_2 & \\ 0 & 0 & \tilde{A}_{33}^{11} & & & \\ \hline & & & \tilde{A}_{11}^{22} & \tilde{A}_{12}^{22} & \tilde{A}_{13}^{22}\\ & 0 & & 0 & \tilde{A}_{22}^{22} & \tilde{A}_{23}^{22}\\ & & & 0 & 0 & \tilde{A}_{33}^{22}\end{array}\right]+\left[\begin{array}{c|c}\hat{B}_1'K_2C_1Q_1 & 0\\ 0 & 0\\ 0 & 0\\ \hline 0 & \hat{B}_2K_3\tilde{C}_2\\ 0 & 0\\ 0 & 0\end{array}\right],\quad PBH=\begin{bmatrix}\hat{B}_1H\\ 0\\ 0\\ \hat{B}_2H\\ 0\\ 0\end{bmatrix}$$

where

$$\hat{B}_1'=\hat{B}_1(I-B_2'^T(B_2'B_2'^T)^{-1}B_2'),$$
$$\hat{B}_1''=\hat{B}_1(I-B_2'^T(B_2'B_2'^T)^{-1}B_2') \tag{19}$$

Using Lemma 2 (Appendix) it is easy to show that if

$$rank\left[\bar{E}_{11}^{11},\hat{B}_1'\right]+rank\left[E_{11}^{22},\hat{B}_2\right]\geq r-rank\bar{E}_{22}^{11}-rankE_{22}^{22} \tag{20}$$

then there exist the matrices K_1,G_2 such that $rank(E+BHG)=r$.

3.2. Regularity of closed-loop system.

Theorem 2. If there exist the matrices K_2,K_3 such that

$$\det\left\{\left(\bar{E}_{11}^{11}+\hat{B}_1'K_1Q_1\begin{bmatrix}I\\ 0\\ 0\end{bmatrix}\right)z_1+\left(\tilde{A}_{11}^{11}+\hat{B}_1'K_2C_1Q_1\begin{bmatrix}I\\ 0\\ 0\end{bmatrix}\right)\right\}\neq 0$$

and for some $z_1, z_2 \subset C$ (21)

$$\det\left\{\left(\left(E_{11}^{22}+\hat{B}_2 HG_2\begin{bmatrix} I\\0\\0\end{bmatrix}\right)z_2+\left(\widetilde{A}_{11}^{22}+\hat{B}_2 K_3 \widetilde{C}_2\begin{bmatrix} I\\0\\0\end{bmatrix}\right)\right)\right\}\neq 0$$

and matrices $\widetilde{A}_{33}^{11}, \widetilde{A}_{33}^{22}$ are nonsingular then the closed-loop system is regular.

Proof. Matrix

$$\left\{P\begin{bmatrix}(E_{11}+B_1HG_1)z_1 & (E_{12}+B_1HG_2)z_2\\ (E_{21}+B_2HG_1)z_1 & (E_{22}+B_2HG_2)z_2\end{bmatrix}Q+P(A+BHFC)Q\right\}$$

is upper block triangular with nonsingular diagonal bloks for some z_1, z_2.

Therefore the condition for regularity of the pair $E+BHG, A+BHFC$ is satisfied and closed-loop system is regular. □

Therefore, the following theorem has been proved

Theorem 3. *If the conditions* (9), (14), (20), (21) *are satisfied and* $\widetilde{A}_{33}^{11}, \widetilde{A}_{33}^{22}$ *are nonsingular then the regularization problem for system* (1) *has a solution.*

If the conditions of theorem 3 are satisfied then the desired feedback matrices F, G may be computed by the use of the following

Procedure

step1. Find nonsingular matrices M and N satisfying (10).

step2. From (16) we compute the matrices $\overline{E}_{11}, \widetilde{A}_{11}, \widetilde{A}_{22}$.

step3. Compute the nonsingular matrices P_1, Q_1, P_2, Q_2 satisfying (17).

step4. Choose matrices K_1, G_2 satisfying (7).

step5. Choose the matrices K_2, K_3 satisfying (21).

step6. From (13) compute F and G_1.

step7. $G=[G_1, G_2]$.

Example Consider the model (1) with

$$E=\begin{bmatrix}E_{11} & E_{12}\\ E_{21} & E_{22}\end{bmatrix}=\left[\begin{array}{ccc|cc}0&3&2&0&1\\0&0&0&0&0\\0&1&1&0&1\\\hline 0&1&1&0&0\\0&0&0&1&-1\end{array}\right]$$

$$A=\begin{bmatrix}A_{11} & A_{12}\\ A_{21} & A_{22}\end{bmatrix}=\left[\begin{array}{ccc|cc}1&2&1&0&1\\0&0&0&0&0\\1&3&2&2&1\\\hline 1&3&-1&0&1\\0&0&0&-1&1\end{array}\right]$$

$$B=\begin{bmatrix}B_1\\B_2\end{bmatrix}=\begin{bmatrix}0&1\\0&-1\\1&0\\ \cdots&\cdots\\1&0\\0&0\end{bmatrix}\quad C=\begin{bmatrix}C_1 & C_2\end{bmatrix}=\left[\begin{array}{ccc|cc}1&0&0&0&0\\0&1&0&0&0\\0&0&1&0&0\\0&0&0&1&0\end{array}\right]$$

$$D=[0]$$

Find the feedback matrices $G, F \in R^{2\times5}$ such that the pair $E+BHG, A+BHFC$ is regular and $rank[E+BHG]=4$.

It is easy to check that $\det E=0$ and the pair (E,A) is nonregular.

Step 1. The matrices M, N satisfing (10) have the form:

$$M=\begin{bmatrix}1&0\\0&1\end{bmatrix},\ N=\begin{bmatrix}1&0&0\\0&1&0\\0&0&1\end{bmatrix}$$

Step 2. From (16) we have

$$\overline{E}_{11}=\begin{bmatrix}0&3&2\\0&-1&-1\\0&0&0\end{bmatrix},\ \widetilde{A}_{11}=\begin{bmatrix}1&2&1\\-1&-3&1\\0&0&3\end{bmatrix},\ \widetilde{A}_{22}=\begin{bmatrix}0&1\\-1&1\end{bmatrix}$$

Step 3. The nonsingular matrices P_1, Q_1, P_2, Q_2 have the form

$$P_1=\begin{bmatrix}1&0&0\\1&1&0\\0&0&1\end{bmatrix},\ Q_1=\begin{bmatrix}1&0&0\\0&1&0\\0&0&1\end{bmatrix}\ P_2=\begin{bmatrix}1&0\\0&1\end{bmatrix}\ Q_2=\begin{bmatrix}1&0\\1&1\end{bmatrix}$$

and

$$P_1\overline{E}_{11}Q_1=\begin{bmatrix}0&3&2\\0&2&1\\0&0&0\end{bmatrix},\ P_1\widetilde{A}_{11}Q_1=\begin{bmatrix}1&2&1\\0&-1&2\\0&0&3\end{bmatrix}$$

$$\hat{B}_1'=\begin{bmatrix}0&1\end{bmatrix}$$

$$P_2E_{22}Q_2=\begin{bmatrix}0&0\\0&-1\end{bmatrix},\ P_2\widetilde{A}_{22}Q_2=\begin{bmatrix}1&1\\0&1\end{bmatrix},$$

$$\hat{B}_1''=\begin{bmatrix}1&0\end{bmatrix}$$

Step 4. Taking into account that

$$rank\left[\overline{E}_{11}^{11}, \hat{B}_1'\right]+$$

$$rank\left[E_{11}^{22}, \hat{B}_2\right]=2=4-rank\overline{E}_{22}^{11}-rankE_{22}^{22}$$

we choose $K_1=\begin{bmatrix}0&0&0\\1&0&0\end{bmatrix}$, $G_2=\begin{bmatrix}1&0\\0&0\end{bmatrix}$

It is easy to check that $rank[E+BHG]=4$

Step 5. Note that (20) is satisfied for any K_2, K_3 then we choose $K_2 = K_3 = [0]$.

Step 6. From (13) we have

$$F = \begin{bmatrix} -1 & -3 & 1 & 0 & 0 \\ 0 & 0 & 0 & 0 & 0 \end{bmatrix}, \quad G_1 = \begin{bmatrix} 0 & -1 & -1 \\ 1 & 0 & 0 \end{bmatrix}.$$

Step 7. $G = [G_1, G_2] = \begin{bmatrix} 0 & -1 & -1 & 1 & 0 \\ 1 & 0 & 0 & 0 & 0 \end{bmatrix}$

REFERENCES

Bunse-Gestner A., Mehrmann V. and Nichols N.K., (1992), *Regularization of derivative and propotional state- feedback*, SIAM J. Matrix Anal. Appl., 13; pp. 46-67.

Bunse-Gestner A., Mehrmann V. and Nichols N.K., (1994), *Regularization of descriptor system by output feedback*, IEEE Transaction Automatic Control, AC-39; No. 8, pp.1743-1748.

Campbell S.L., *Singular system of differential equation*, (1980), Pitman, San Francisco.

Fornasini E. and Marchesini G., *Doubly indexed dynamical systems: state-space models and strutural propoties*, (1978), Math. Syst. Theory, vol.12, pp. 59-72.

Kaczorek T., *Two-Dimentional Linear Systems*, (1985), Springe-Verlag, Berlin-Tokyo.

Kaczorek T., *Linear Control System*, Vol. II - *Synthesis of Multivariable Systems and Multidimentional systems*, (1993), Reseach Studies Press and J. Wiley, New York.

Kaczorek T., *Regularization of a singular 2D Roesser model by state feedbacks*, (1996), Proc. of Intern. Conf. Circuits, Systems and Computers. Hellenic Naval Academy, Piroeus Greece; July 15-17.

Kurek K., *The general state-space model for a two-dimensional linear digital systems*, (1995), IEEE Trans. Autom. Contr., Vol. AC-30, No. 6, pp. 600-601.

Miminis G., *Deftation in eigenvalue assignment of descriptor systems using state feedbacks*, (1993), IEEE Trans. Autom. Contr., AC-38, No. 9,pp. 1743-1748.

Ozcandiran K. and Lewis F.L., *On the regularizabiliti of singular systems*, (1990), IEEE Trans. Autom. Contr., AC-35, pp. 1156-1160.

Roesser P.R., *A discrete state-space model for linear image procesing*, (1975), IEEE Trans. Autom. Contr., Vol. AC-20, No. 1, pp. 1-10.

Yip E.L. and Sincovec R.F., *Sovability, Controlability and observability of continuous descriptor systems*, (1981), IEEE Trans. Autom. Contr., AC-26; pp. 702-707.

APPENDIX

Lemma 2.

If $rank[E, B] = m$, $E \in R^{n \times n}; B \in R^{n \times m}$ *then for a given integer n* ($rank\, E \le n \le m$) *there exists a matrix K such that* $rank[E + BK] = n$.

Proof. From assumption $rank[E, B] = m$, it follows that from matrices E and B we can choose $m_1 + m_2 = m$ linear independent columns. Let E_{m_1}, B_{m_2} be the matrices consisting from the linear independent columns of E and B such that $E\overline{M} = [E_{m_1} \vdots E'], B\overline{N} = [B_{m_2} \vdots B']$ where $\overline{M}, \overline{N}$ are nonsingular matrices, $E' = E_{m_1}R + B_{m_2}S$, R, S are some matrices. It is easy to see that there exists nonsingular a matrix T such that for $H = \begin{bmatrix} -S \\ 0 \end{bmatrix} + \begin{bmatrix} 0 & T \\ 0 & 0 \end{bmatrix}$ we have $rank(E\overline{M} + B\overline{N}H) = n$.

Let
$$K = \overline{N}H\overline{M}^{-1} \tag{19}$$

Then
$$rank(E + BK) = rank((E\overline{M} + BNH)\overline{M}^{-1}) = n. \quad \square$$

THE ROLE OF MODELS IN FUTURE ENTERPRISES

U. Rembold, W. Reithofer, B. Janusz

*University of Karlsruhe, Institute for Real-Time Systems and Robotics (IPR)
Kaiserstr. 12, D-76128 Karlsruhe, Germany, E-Mail: rembold@ira.uka.de*

Abstract: This paper describes the general problem of manufacturing, discusses the impact and consequences of the growing internationalisation of markets and gives an overview of the trends in key technologies of manufacturing systems. Furthermore, the need for models to support the design and reorganisation of CIM systems and Virtual Factories is discussed. Finally, particular approaches to solve such problems are introduced.

Keywords: Enterprise Modelling, CIMOSA, Virtual Factory

1 OVERVIEW

For the industrialised nations, the manufacturing industries are the most important contributors to prosperity. However, it becomes increasingly difficult to meet the customers' demands and to compete on the international market. Thus, the manufacturing industries must be able to quickly react to prevailing market conditions and to maximize the utilization of resources. Conventional means of hard automation have not been able to meet these challenges any more, since they are very poor in information processing. The concept of the virtual factory will help to master the challenges in future factories.The first part of the paper deals with global trends in manufacturing and key characteristics of future enterprises and related problems, the second part explains the role of models in designing, implementing, and operating CIM systems, especially virtual factories.

2 GLOBAL TRENDS

Within recent years, the computer technology in conjunction with software technology have made available to the manufacturer tools which can greatly improve the reaction to a new market situation, to speed up design of a product, to improve process planning, to maximize resource scheduling, and to stream line the production flow through a factory. When the computer has become a major component of a manufacturing system and helps to plan and operate it, we are talking about Computer Integrated Manufacturing (CIM). The operations of most manufacturing systems follow a similar pattern. Engineering establishes the physical structure of a product and its performance parameters, process planning defines the various processing steps needed to make the product, scheduling tries to optimize the utilization of resources, manufacturing control leads the product through the plant and quality control ensures that the product adheres to the planned quality standards. All these activities or business functions are configured to an integral system for which goals are set, executed and controlled. A factory must be understood as an integral business operation consisting of many activities; it makes no sense to consider an activity by itself. In the real world, there is a strong interdependence between all manufacturing functions and information is passed back and forth between them. Every function and sub function must be operated as an integral part of the whole system. The use of the computer is associated with many problems. To build a hardware and software system for a specific factory is very expensive if done on an individual basis. In the future, there must be a strong effort to standardise hardware and software components to the extent that a factory planning and control system can be build from off-the-shelf modules and tied together with the help of configuration aids. In this article we define CIM as it is common in industry. CIM is a concept of a factory in which all processes leading to the manufacture of a product are integrated and supported by computers as far as useful. It includes computer aided design (CAD), computer aided process planning (CAPP), production planning and control (PP&C), computer aided quality control (CAQ), and computer aided manufacturing (CAM). CIM is concerned with common data models which can be used for the entire design and manufacturing cycle. Thus, CIM is centred around the decisions regarding the planning and controlling of the data flow, data processing and data dissemination in a plant. The tools are models, algorithms, Artificial Intelligence methods, software engineering aids, computers, data communication systems, interfaces to man and machines as well as interfaces between machines. Thereby, machines are either computers or manufacturing equipment.

2.1 Internationalisation: impact and consequences

When we look at the present situation where the market for products becomes increasingly internationalised, there are certain developments which are taking place and which will have an effect on individual companies and entire areas of manufacturing as such. Probably the most important development is the consolidation of the industries of many countries to free trade zones The free trade zones which are being formed or are most likely are:

- The European common market
- The North American free trade zone
- The free trade zone of the Community of Independent States (Former USSR)
- The East Asian free trade zone

When a free trade zone is realized, there suddenly will be a completely new situation for many manufacturing organisations. The first thing to be done is to internationalise if that has not already happened. But internationalisation is not something that does just occur overnight. In most cases, it is not possible to take a product and just sell it in another country. Here problems may arise that the customer of the other country does not like the product, that he needs instructions in his own language and that there is not even a sales organisation available to do the marketing. In addition there will be stiff competition from companies of the other countries who are trying to sell a similar product. Soon a company will find that the laws governing the free market are very tough and that in order to survive it becomes necessary to rationalise, consolidate, join with others or to give up certain market segments.

With this problem in mind, the manufacturers will have to measure their agility against that of competition. Agility is measured in terms of the response time to a customer's order, the time needed to develop and build a new product. Agility also is related to the reconfigurablility of the manufacturing system and the ability of the company to start a relationship with other companies. There are certainly many more measures for agility which depend on the nature and type of business a company is in.

One of the main problems of western manufacturers is the inability to react to changing market situations. This, in many cases, leads to a large fluctuation of people; often the best of them leave when a problem arises at the horizon. In contrast, the Japanese work with a well motivated permanent staff of employees who often devote their entire productive live to one organisation. Of course, fluctuation can also be of benefit to an entire economy since new ideas will be stimulating the competitive market. However, unsteady employment practices as they are prevalent in many North American organisations can also be of detriment to a company or to an entire business. For this reason, manufacturers must exercise good management practices to maintain motivated and loyal people. In this respect, some of the following points must be emphasised:

1. A company must establish long and short term goals and standards against which these goals are measured.
2. There must be a healthy employment attitude by which people are valued as the most important resources. Loyal people want to be motivated, challenged and given responsibility. A good climate produces good teamwork, sense of urgency, resourcefulness and high efficiency.
3. Another important measure is an efficient and fast internal and external communication system in an organisation. Most management structures are too deep, leading to communication with long feedback delay. A responsive and agile organisation is flat and makes extensive use of modern means of communication in a real time data processing environment.
4. Performance measures and self critique are the feedback mechanisms to take corrective actions if they are needed. For this purpose, there must be a company wide data gathering and evaluation system which is frank, open,

and generally accepted as a tool to improve efficiency and responsiveness. It is important to detect problems early to assess their seriousness and to involve all people associated with them to find solutions.

5. There must be an incentive for people to update their technical and management skills. This is particular important when new computers and communication technology is introduced.

2.2 *Trends in technical developments*

The rapid progress in technology provides new approaches but also imposes new requirements on future manufacturing systems. In this section, we will highlight the most promising technical developments in key areas. The technical trends can be divided into *general trends*, that have an impact on many areas of the business including manufacturing, and into *specific trends* that influence the development of manufacturing systems in particular.

General trends

The ongoing rapid general developments in the following information processing areas will be continuing:

- Microelectronics
- Hardware computer architectures
- Communications technology
- Software engineering
- Database technology
- Knowledge based systems and AI

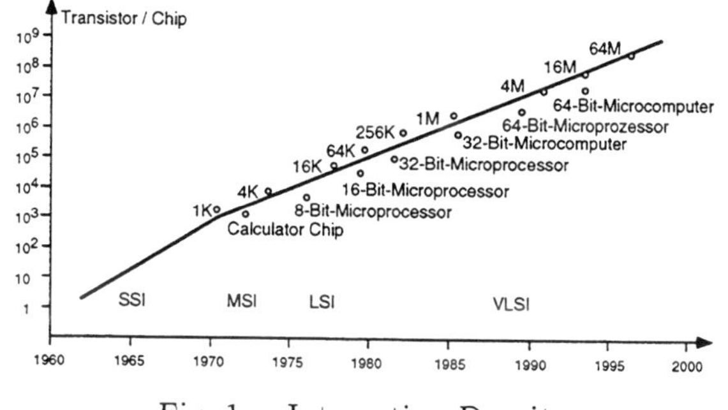

Fig. 1. Integration Density

The basic impact on future manufacturing systems will come from the developments in *microelectronics*. When we have considered as the major technological indicators over the past two decades, namely the integration density, speed, and power dissipation, we notice that the former two items have increased whereas the latter has decreased steadily. In the case of integration density, i.e. the number of transistors per chip, the increase has been exponentially, doubling every 1.5 years (Fig. 1). The same is true for the economical indicator namely the cost per bit, which has decreased exponentially (Fig. 2)

The availability of fast, highly integrated and inexpensive hardware components makes the realisation of new *hardware architectures* possible and economical. Mainly the developments in parallel architectures show promising approaches. First results can be seen in the development of multiprocessor systems, hypercube architectures, and transputers, which manage to overcome the "complexity gap", by a linear improvement of the performance with increasing number of processors.

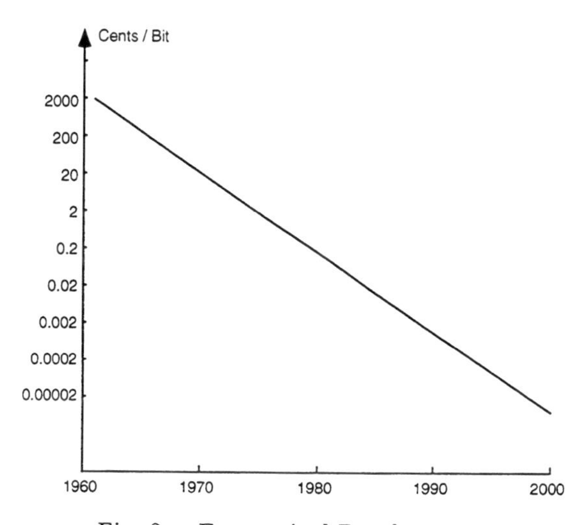

Fig. 2. Economical Development

Penetration of fault-tolerant architectures will follow that of parallelism. This increased penetration will be helped by further advances in new operating system functions in areas like fault detection, isolation and recovery and data integrity.

Another key technology which pushes the advances in manufacturing systems is *communications technoloy*. The local and wide area networks which appeared in the 70th and 80th for fast and reliable communication networks and protocols provide a sound basis for data processing. These developments were concerned with telephone circuits switching and broadcasting networks, packet data and virtual networks. The most recent of these results are MAP (Manufacturing Automation Protocol) and TOP (Technical and Office Protocol) in the U.S. and CNMA (Communications Network for Manufacturing Applications) and ISDN (Integrated Services Digital

Network) in Europe.

With the introduction of the fiber optics technology allowing broad band frequency operation of local area networks, high data transmission rates are becoming possible. The 10 Megabaud transmission rate in current local area networks, will give way to Gigabaud transmission rates, allowing e.g. the transmission of complex digitised video signals. The main emphasis in wide area networks is the global harmonisation of transmission rates and protocols and the provision of national integrated services (e.g. ISDN).

Europe		USA		Japan	
Bit Rate (MB/s)	No. VF Circuits	Bit Rate (MB/s)	No. VF Circuits	Bit Rate (MB/s)	No. VF Circuits
2.00	30	1.50	24	1.50	24
8.25	120	6.16	96	6.16	96
33.50	480	43.69	672	31.31	480
136.00	1920	267.75	4032	95.44	1440

Fig. 3. International Transmission Rates

This work has already led to a closed network of interconnected points spread throughout the world (Fig. 4), but suffers from global inconsistencies, mainly with regard to transmission rates (Fig. 3). To overcome these inconsistencies within multinational networks and to bridge the gap between local and wide area networks, various activities have been started (e.g CCITT 802, TAT-8)

The *Database technology*, as one classic research area of computer science, has already been fairly developed. Relational systems are there to stay (Lockemann et al, 1991). For new applications, database systems lose their identity; their functionality will be highly distributed, autonomous, and fault tolerant. Database and telecommunication functionality will merge. With the availability of high density and inexpensive mass media, based on laser disc and semiconductor technology, it will be possible to keep and access nearly any information at any time. Due to this on-line availability, even on the production level, database systems will have immediate contact to the physical world.

In the late 60th, it already became evident that conventional ,,kick and rush" programming led to inflexible, not maintainable and unreliable software. The term *Software Engineering* summarises the approaches to solve these problems by using engineering methods for software development.

However, it was not until the 80th, when commercial tools became available dealing with system development, quality control, documentation

Fig. 4. International Networking (Dinn, 1991)

and project management as an integrated process. Only recent developments try to extend the application area of these tools from office automation to manufacturing applications.

After a two decades period during which laboratory studies first demonstrated the technical feasibility *Knowledge-based systems* and *AI* seem to have reached the realm of possibility and applicability.

In the area of manufacturing control, expert systems support the operator in handling complex situations and the maintenance staff in diagnosis in case of failure. Knowledge based systems are main parts of the decision layer of autonomous planning and controlling, they show new solutions in robotics and automation. Neuronal-networks allow fast processing of extensive video data, Fuzzy-technology provides a basically new approach in handling of uncertain data in general. All these approaches will revolutionise those applications that deal with high complexity and uncertainty.

To sum up these discussions, it is obvious that the most important impact on future manufacturing system will result from standardisation and integration of the topics covered above. They include developments in hardware and software, including microprocessor architectures, mainframes, operating systems, data management systems, communication devices, user interfaces and applications.

2.2.1 Specific trends

Finally, important developments that influence manufacturing systems in particular are concerned with

- NC-machines
- Rapid prototyping
- Just-In-Time production

- Lean production
- Design for manufacturing and assembly
- Simultaneous engineering
- CAD-CAM integration
- Virtual factory
- Intelligent machines

New developments on the shop floor level will influence the manufacturing level as well. This in particular is true for NC machines, because these machines are main objects of computerised planning and execution. In material movement, the flow of tools and workpiece will be integrated, there will be new technological processes which will be realised with current machine concepts, also multiple processes will be integrated into one machine. The implementation of new devices and processes will be supported by efficient simulation tools that allow to forecast their effectiveness and their cost.

Shorter product life cycles and a wider differentiation of products will entail new developments in *rapid prototyping.* The keyword on this subject is "Desktop Manufacturing". It is becoming an integrated part of CAD/CAE/CAM and aims at shorter shorten life cycle, lower cost, and better quality.

New processes will allow the easy and fast production of prototypes of nearly any size and shape, without building auxiliary tools and moulds, which is time and cost consuming. These processes are based on a solid model description of the workpieces, generated by any CAD-system. The model is mapped on the prototype via "slicing". In other words the prototype is build slice by slice from cross-sections of the model (Cord, 1995).

Design for manufacturing and assembly is of particular importance to flexible factory operations. In mass production, the process is often tailored to special methods of machining or manufacturing entailing the design of special equipment. Such measures could be justified because equipment was used for many years without change. In flexible manufacturing the basic processes are standardized and configured to a specific manufacturing run. This, however, necessitates design for manufacturing so that basic processes can be used. RISC Manufacturing is also a trend in this area. In conventional assembly systems the product was designed to be assembled by workers. Man, however, has an unsurpassed dexterity, an excellent sensor system and a superb memory. For this reason he is able to assemble with few and little instructions and perform operations which machines can not do at all. If machines are used for assembly the assembly work has to be greatly simplified and adapted to the capability of the machines. In recent years, design for manufacturing and assembly has been extended to Simultaneous Engineering (otherwise known as Concurrent Engineering, or Life-cycle Engineering). Simultaneous Engineering considers the whole life-cycle: design, manufacturing, assembly, testing and maintenance of a product. For this, a close interaction between people involved in these processes is required. New methods that have been developed include team-work, checklist procedures, rule-based expert systems, and AI approaches.

Just-In-Time production (JIT) is a keyword that is based less on technical but more on organisational considerations. JIT tries to minimize the storage and investment costs by ordering material and producing goods just when they are needed. Although JIT is an organisational aspect of manufacturing it will influence the development of future manufacturing systems, because it requires flexibility and a high degree of coordination in every product line.

Lean Production is a concept to deal with the high demands of flexibility and coordination by minimising the large number of production lines and concentrating on one main line. Therefore, "Lean Production" in turn relies on JIT. Both approaches are linked together.

CAD-CAM integration is a process which has started about 20 years ago. Here an attempt is made to integrate the decision making and information processing of all business functions into one manufacturing control System (Fig 5). It is necessary to start with a product model which will contain all necessary data to generate the process plan and machine control programs. On-line scheduling will be done in real-time operations with the help of factory data acquisition. With this approach the strategy of an enterprise will be on CIM whereby operational and technical planning will be the input to factory control.

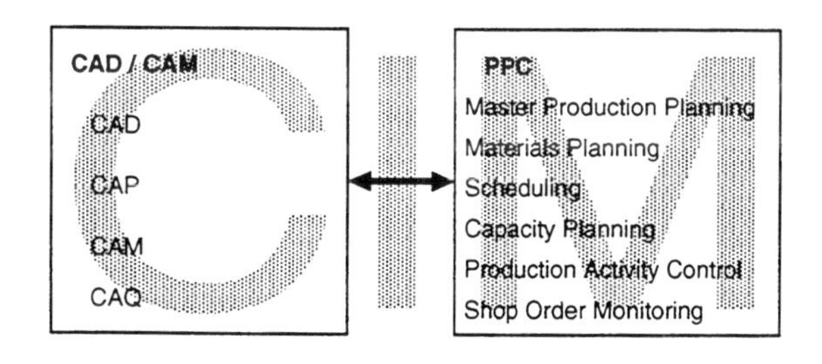

Fig. 5. The integration with CIM

A factory operation will be planned according to a virtual manufacturing concept. E.g. the CIM-OSA efforts are trying to provide standard business functions and interfaces to be able to construct for a particular product or a new manufacturing system standard functions to be integrated with the help of configuration tools.

Virtual Enterprise, also known as *Extended Enterprise* (Browne, 1994) is a new key word to master the dynamic changes of products, product components, production process and sales volume and to overcome problems resulting from manufacturing processes in different countries to reduce costs. In this concept activities of various enterprises are grouped together to form integrated operations for the functions of marketing, research, development, production, and maintenance. Such a structure can be changed easily and dynamically if there is a need for it. The same idea is already in use to group manufacturing lines together to manufacture a specific product. With the increasing capabilities resulting from the telecommunication technologies such as data highways together with virtual reality systems, the direct cooperation among teams of different companies will be possible

Intelligent manufacturing with intelligent machines. is another important challenge for the improvement of a CIM system. Modern large-scale manufacturing systems, designed by CAD/CAE techniques are often equipped with complex components like industrial robots, automatically guided vehicles, or other advanced transportation and manipulation system. Because these manufacturing systems consist of so many different components, the control of the system is very difficult. In most cases strictly hierarchical and other top-down methods are chosen to master the system control. All information is processed by a node on a high level, which has knowledge of the domain. More detailed commands are sent by this node to the components that execute the commands. Thereby, the tasks of the subcomponents are scheduled and synchronized. Manufacturing with intelligent machines will reduce the need for centralized control and synchronized scheduled tasks. To obtain the decentralized characteristic, a higer level of intelligence and autonomy is required for the controller of the machines.

CIM systems and their inherent flexibility are associated with high investment costs. The efficient utilization of such equipment can only be guaranteed if there are enough products and product variants to be manufactured. CIM, for this reason, often is unattainable for small and medium sized manufacturers. In the future, users of CIM systems may have to share facilities for similar products otherwise the necessary investment will not be affordable. It will also be necessary to use these CIM systems in a three shift operation. The return on investment (ROI) of such facilities will also be much longer than with conventional equipment.

Summarized, the key elements of the factory of the future will be:

- Existing rigid, static, centralized hierarchical organizations will be replaced by flattened, network-like organizations.
- Enterprises will be composed of widely autonomous, but cooperating work units.
- Work units distributed all over the world, possibly belonging to different enterprises, will cooperate within virtual enterprises.
- Virtual enterprises will cover the whole product cycle from the supplier to the customer.

The evolution from todays enterprises towards future enterprises is characterized by continous improvement and adaption to changing environments. This implies the continous reorganization of processes and reconfiguration of internal and external relationships. To achieve this in a systematic manner, models and methodologies are essential. In the following, we describe first a suitable model and possible applications.

3 CIMOSA

Enterprise models were developed to support the design, analysis and control of Computer Integrated Manufacturing (CIM) systems. They integrate a set of general models for representing enterprise's functions, processes, information structure, etc. The following examples of CIM systems and Virtual Factory design are based on CIMOSA. Among other enterprise models, CIMOSA offers the most comprehensive and the most formal model representation. The latter is needed for model execution. CIMOSA consists of:

- a Framework for Enterprise Modelling and
- an Integrating Infrastructure.

The *Modelling Framework* itself comprises data structures for model representation and rules which define how to instantiate those structures to model a particular enterprise. The *Integrating Infrastructure (IIS)* links the model to the real system. During enterprise operation, the IIS is needed for controlling heterogenous resources.

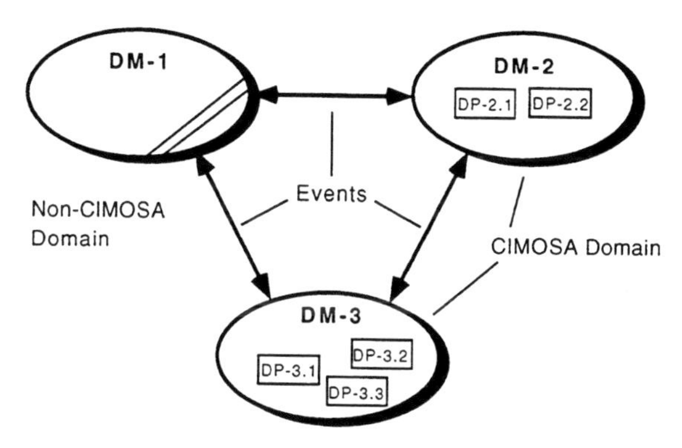

Fig. 6. Relationships of CIMOSA Domains

According to CIMOSA, an enterprise is a set of cooperating *Domains* which contain one or more *Domain Processes.* The processes communicate by message exchange (Fig. 6). *Events* are used for synchronization. The elements of Domain Processes are *Business Processes* and *Enterprise Activities.*

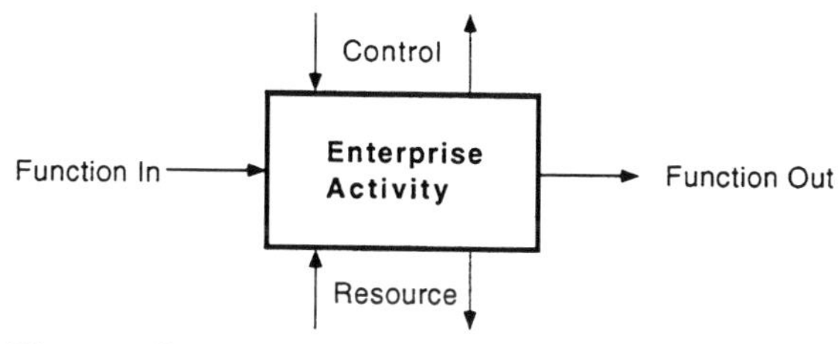

Fig. 7. Inputs and outputs of Activities of an enterprise

The function of an activity or a process is defined by specifying its inputs and outputs (Fig. 7). Function In/Output, Control Input and Resource In/Output are references to *Object Views.* Object Views are special views on *Enterprise Objects* such as products, work plans, etc.

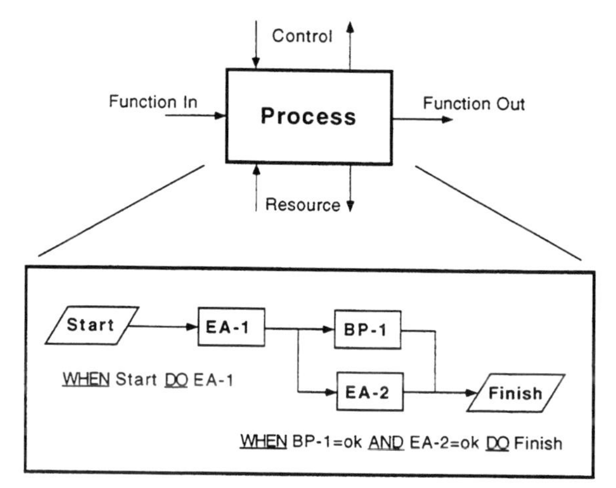

Fig. 8. Decomposition and the behaviour of a Domain Process in an enterprise

The behavior of a Domain Process is defined by *Procedural Rules.* Procedural Rules determine the execution order of the Enterprise Activities and/or Business processes (Fig. 8). A Business Process combines a section of an activity chain of another process. Its use is to allow a hierarchical structuring of the the model. The machines, computers, and men which perform the activities are modelled with the CIMOSA object *Resource.* Properties and capabilities of Resources are represented by *Capability Sets. Views* are used to emphasize particular aspects of the model: The *Function View* depicts functional and behavioural aspects of the model. The *Information View* displays the informational and physical objects as well as views on them, called the Object Views. Further views are the *Resource View* and the *Organizational View.* The most outstanding feature of CIMOSA is the integration of general models for describing function, behavior, information structure and organisation within a uniform model representation, so-called *Templates.* A Template is a table with several sections. It contains attributes and links to other objects.

Starting point of the modelling process is the *generic model* which comprises the generic building blocks for representing Domains, Domain Processes, Events, etc. The generic objects are specialized by *instantiation.* For example, an instance of the generic building block Domain Process may be used to describe a particular process in an individual enterprise. During model execution, *occurrences* of this instance are created. To reduce the modelling effort, an intermediate level among the generic and the particular model level is suggested: The *partial level* contains partially instantiated models which are applicable to a certain set of enterprises or domains within enterprises. A more detailed description of CIMOSA is given in AMICE, (1993) and AMICE, (1994).

4 SOFTWARE SYSTEMS PLANNING AND SELECTION

Today's industrial practice allows, in general, only the selection of whole planning and control systems for CIM (e.g. CAD systems and PP&C systems). These systems can be adapted in a limited range to the user requirements. Usually, the different CIM systems of a company are loosely coupled via file transfer. The software imposes a number of constraints on the organization of a company. Therefore, a company must adapt to the requirements of the software which more or less meets its needs. Tailored software is very expensive, difficult to maintain, and its implementation is very time-consuming. A solution to this dilemma is seen in the introduction of executable enterprise models into the industrial environment. The basic paradigm of this approach is the separation of functionality and behaviour. In this approach, the functionality of a CIM system has to be provided by the software, and the behaviour of the CIM system has to be defined by the enterprise model. During operation, the enterprise model decides which function has to be performed next. The 'operating system' of this CIM system calls the software packages providing the functionality. With this approach, it is expected that it will be possible to design tailored CIM-systems built of standardized software modules. A corresponding planning system consists of two "worlds". The CIMOSA models are used to define the planning problem in a way which is easily understandable by the user. The constraint network is used to solve the planning problem.

The next paragraphs will introduce the architecture of the system which is depicted in Fig. 9.

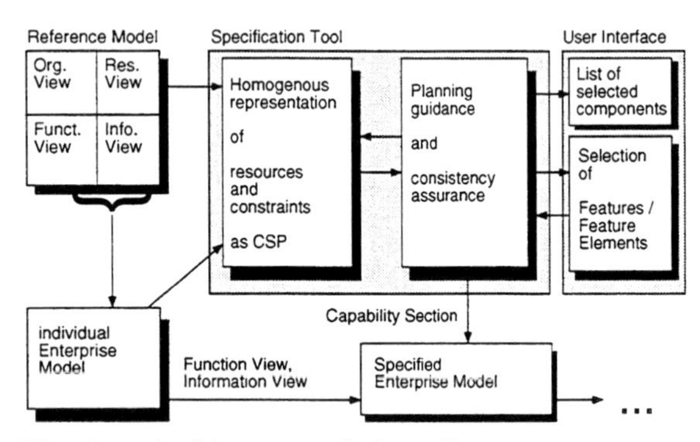

Fig. 9. Architecture of the software system planning tool

The kernel of this software system planning tool consists of the following components:

- The Reference Model plays an important role in the planning tool. It contains at least the Function View, the Information View and the Resource View. In the Function View, each Enterprise Activity contains a Capability Section. This Capability Section is instantiated while the Resources are modelled. The modeller of the Resources has to specify which Enterprise Activity can be supported by which Feature of a Resource. Therefore, the Function View of the Reference Model contains a collection of methods for each Enterprise Activity. These methods can support the execution of the Enterprise Activity. In CIMOSA terminology, this means that the methods provided by Resources are modelled as Features and Feature Elements in the Capabiltiy Section of the Enterprise Activity.

- The Individual Enterprise Model represents the enterprise for which the software system has to be planned. It is derived from the Reference Model. Therefore, the corresponding component of an Enterprise Function can be identified in the Reference Model. The constraint variables are the set of Features which are necessary to describe the requirements for Resources of the Particular Enterprise Model and a variable representing a combination of resources supporting the Enterprise Activities of the Particular Model.

- The Specification Tool contains the constraint network which represents the planning problem in an equivalent constraint network. Through the user interface, the system offers the available and still employable methods. The user enters his/her selection. If required, the Capability Section can be added to the individual enterprise model for further processing. Afterwards, it is named a Specified Enterprise Model.

Several conversion steps are required to transform the CIMOSA representation into an equivalent constraint network. This conversion process will be explained in more detail.

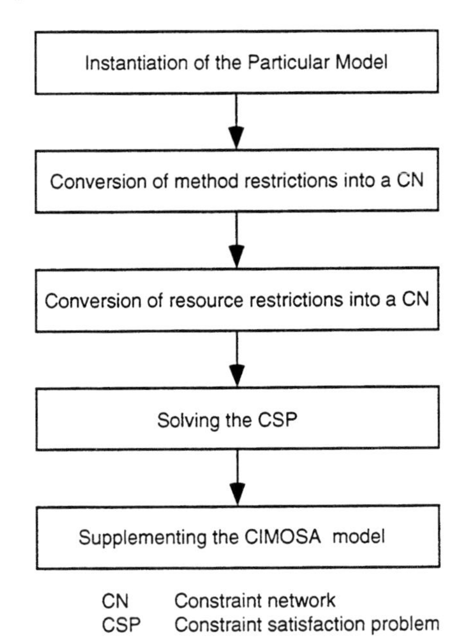

Fig. 10. The course of software systems planning and selection

First, with the instantiation of the Particular Model the process is started (Fig. 10). It is derived from the Reference Model in such a way that the origin of the Enterprise Activity can be identified in the Reference Model. As already mentioned above, the link between Enterprise Functions and Resources is provided in the Reference Model.

In the second step, the dependencies between the methods applicable in the Particular Model are converted into a constraint network. The representation of dependencies between the software methods works as follows: it assumes that a prerequisite can be a specific Function Input which must be provided by a previously executed Enterprise Activity, or it assumes that it is a specific method which must be applied by a previously executed Enterprise Activity.

Third, the allowable method combinations which are provided by Resources are also converted into a constraint network. The constraint network is now complete and a problem-solving algorithm controls the instantiation of the constraint variables. If the user selects a method which cannot be provided by a Resource, a constraint will be violated and the instantiation of the method has to be retracked.

Finally, the solution found can be added to the individual Enterprise Model.

The most difficult task is to generate the equivalent constraint network. How this can be accomplished is described in Naeger, (1994).

5 VIRTUAL ENTERPRISE DESIGN

The (re-) design of a Virtual Enterprise is a function where a consortium of partners is assembled and the common tasks are split up into subtasks and are assigned to the partners. Amongst others, the following problems have to be solved:

- Who are the right partners?
- Where are the right interfaces?

Assume a company detects a need for a new product and this product could be developed and manufactured in-house except of one particular component. Several other companies are able to fulfill this task. Solving the *partner problem* means to select one of them. Since the product should be developed concurrently, intensive data exchange is necessary. Therefore the following issue must be asked: Do the design departments of both companies use the same product model? Is it possible to transform different product representations? Is it possible to exchange the data on-line?

In some cases the partners of the intended cooperation are fixed. For example, two car manufacturers know that they want to develop and produce a special-purpose vehicle. Both have their own core competencies, but since they have similar plants, a set of tasks can be performed by both of them. Now, the problem is to determine partner to perform each task, taking into account existing constraints, such as the desired system's throughput. This problem is called the *interface problem*.

The following, systematic approach supports the solution of the problems encountered above. Assuming, an enterprise model for each partner and/or candidate exists. Components of those model which represent the relevant parts of the enterprises are seperated from the original models and linked together to form one or possibly several new models of the virtual enterprise. After designing the model, it is transformed into an executable model. This allows the evaluation of design alternatives. The procedure of creating alternative models depends on the knowledge of the designer. However, it may be supported by an algorithm which is described in the next section of this paper. Finally, since the selected model is executable, it can also be used for controlling the operation of the virtual enterprise.

The outlined approach has been implemented at the University of Karlsruhe by realizing the Virtual Factory Lab Victor. Victor is an application specific building block library for the simulation system Simple++. Its architecture is shown in Fig. 11.

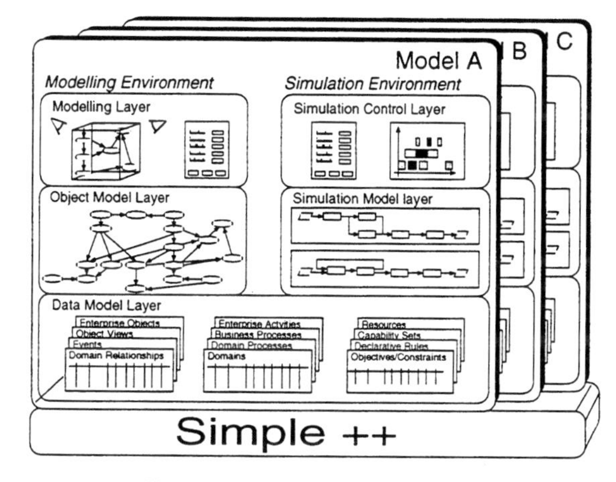

Fig. 11. Victor Architecture

By instantiating an empty enterprise model frame, it is possible to create several particular models in one application (refered to in Fig. 11 as Model A, Model B, etc.). All models have the same internal structure which is subdivided into five layers. Every layer uses objects and methods of its lower layers. Simple++ provides the mechanism for object-oriented, graphical interactive modelling and simulation.

Both, the Victor Modelling Environment and the Simulation Environment are based on the *Data Model Layer*. This layer is the lowest level of every particular model. It stores all information about the instances of the CIMOSA generic building blocks in tables.

The *Object Model Layer* transforms the data model into an object-oriented representation. It provides object specific methods for the consistent manipulation of the model, i.e. for creating, changing and removing objects.

The *Modelling Layer* is the interface between the system and the user. Its role is twofold: First, it displays the model to the user. This is done by using several views to reduce the complexity of the model. Second, it enables the user to develop the model by using design functions. The basic design functions are functions to create, define, and delete objects. Advanced functions include the conversion of objects, the aggregation and decomposition of objects as well as the export and import of Particular Model Components.

The *Simulation Model Layer* consists of an executable represention of the model. Patterns describing the processes are generated automatically. During simulation, occurrences of those processes are created and executed. The Simulation Model Layer also includes a Resource manager and an Event handler.

The *Simulation Control Layer* is the interface between the user and the Simulation Model Layer. It provides functions and methods to define the simulation parameters and to control the simulation run.

Victor is the first tool which integrates in modelling and simulation the Function, Information and Resource View. Its simulation capabilities (e.g. the possibility to take into account stochastic events, the processing of real data like workplans and orders etc.) are comparable to other existing material flow simulation systems. However, the design of the simulation models is much easier since only the *WHAT* has to be done and not the *HOW* this has to be done is modelled. More information about modelling and simulation of enterprises is given in Reithofer, (1996a) and (1996b).

6 A SOFTWARE TOOL BASED METHOD FOR DESIGNING A VIRTUAL ENTERPRISE

As mentioned in Section 5, the design of a virtual enterprise can be done intuitively or systematically using an appropriate tool. In the following, a CIMOSA based tool is presented which can be used for the design of virtual processes, i.e., processes which consist of activities performed by different partners.

It is assumed that an enterprise model for each candidate of the partnership in the virtual enterprise exists. If the partner's responsibilities for some specific activities or sub-processes are already known, this information must be given. Also, the finished product which should be produced in the new cooperation is known. Now, the problem is to choose the partners and interfaces so that the production can be performed optimaly regarding some given criteria.

From the enterprise models describing the single candidate enterprises, only the description of the Enterprise Activities, their Function and Control Inputs and their Function Outputs in terms of Object Views is considered (Fig. 7). All enterprise activities are then stored in a common pool.

Because the algorithm is used to generate models of processes which are dedicated for the manufacturing of the desired product, it works backwards from the desired end product of the virtual process to its first activity. The algorithm works as follows:

1. Define the desired output product(s) of the virtual process.
2. Look for all activities in the pool producing these products as their Function Output. They belong to the last stage of the virtual process.
3. Define the semifinished/raw products or data which are used by these activities as their Funtion or Control Inputs.
4. Look for all activities producing these products/data as their Function Output. If there are several alternative activities producing the same necessary outputs, two solutions are possible:
   (a) Once an activity is indicated as an activity which has to belong to the virtual process, then all other activities are not further considered.
   (b) All activities are considered in succeding steps. They are indicated as activities belonging to the different alternative virtual processes.

   The choosen activities are predecessors of the actually considered activities.
5. Finish, if all activities found do not have any Function and Control Input or if all activities only need inputs which are assumed to be given. Otherwise go back to Step 3.

The result of the algorithm is a precedence graph or several alternative precedence graphs which enumerate all necessary activities and the sequence of their execution needed for the production of the desired product. The implementation details of this algorithm are described in Janusz (1997).

If the procedure described above only finds one virtual process, the virtual enterprise can be described by enumerating the partners responsible for the activities belonging to the process. This can be done by considering the origin of the Enterprise Activities, i.e., the enterprise model from which they were extracted.

Otherwise, the best of the alternative virtual processes has to be chosen with regard to the given evaluation criteria. The appropriate procedure is shown in Figure 12. One of the alternative virtual process is choosen and evaluated by simulation. Because the process consists of parts of

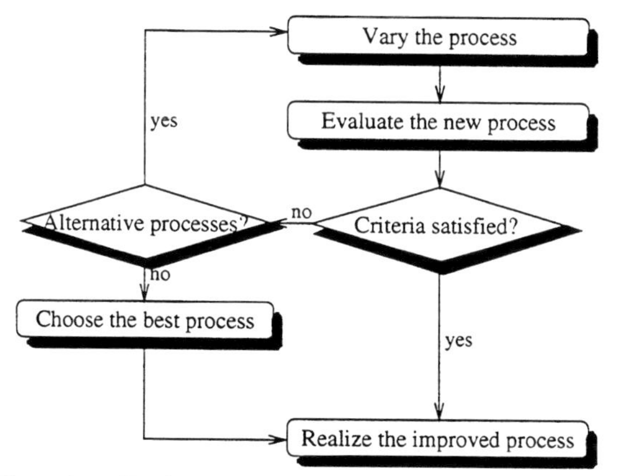

Fig. 12. Evaluation of alternative virtual processes.

the CIMOSA enterprise model and is described as a CIMOSA object itself, Simulation Model Layer and the Simulation Control Layer of the Victor architecture can be used for its simulation and evaluation. If the criteria are satisfied, the virtual process is found. If not and if not evaluated alternative process exist, another new process is evaluated by simulation.

If all alternative processes are evaluated and the criteria are not satisfied, the process which best meets the criteria can be choosen for the realization.

7 CONCLUSIONS

This paper gives an overview on the development of future manufacturing technology and systems. The computer will play an important role in the planning and controlling of the manufacturing processes. A second generation of integrated and intelligent manufacturing systems will be planned and controlled by new promising concepts in many key areas. The concept of virtual factory is a solution to the dynamically changing constraints and dependencies in today's manufacturing reality. Enterprise models will play a role of increasing importance in setting up CIM systems and virtual enterprises fast and effectively.

8 ACKNOWLEDGEMENTS

This research was performed at the Institute for Real-Time Compter Systems and Robotics (IPR), Prof.Dr.-Ing. U. Rembold and Prof.Dr.-Ing. R. Dillmann, Computer Science Department, University of Karlsruhe.

REFERENCES

Browne, J.; Sackett, P.; Wortmann, H. (1994). Industry Requirements and Associated Research Issues. *Eurepean Workshop on Integrated Systems Engineering*, Grenoble.

Cord, Th.; Reithofer, W. (1995). Micro-Stereolithography using a Liquid Chrystal Display as Exposure Mask. *Proceedings of the International Symposium on Automotive Technology and Automation, ISATA*. Stuttgart, Germany.

Dinn, N.F. (1991). Network Architecture. *Future Generation Computer Systems*. 7. North-Holland, p 79-89

ESPRIT Consortium AMICE (1993). *Open System Architecture for CIM*. Springer-Verlag.

ESPRIT Consortium AMICE (1994). *Formal Refcrence Base - ESPRIT Project 7110 AMICE CIMOSA, M-3 Deliverables - Volume 3.*

Janusz, B. (1997). A process oriented method for the reuse of CIM models. *Manufacturing Systems: Modelling, Management and Control MIM'97*, Vienna, Austria.

Lockemann, P.C., A. Kemper and G. Moerkotte (1991). Future Database Technology: Driving Forces and Directions; 7; Future Generation Computer Systems; 7; North-Holland; 1991; 41-54

Naeger, G.; Rembold, U. (1995): An integrated approach to software systems planning and selection based on CIMOSA models. Control Eng. Practice, Vol. 3, No.1, pp. 97-103

Reithofer, W. (1996a): Bottom-up modelling with CIMOSA. In *Proceedings of the Second International Conference on the Design of Information Infrastructure Systems for Manufacturing '96*, 1996.

Reithofer, W. (1996b): Virtual Enterprise Modelling and and Simulation. In *Proceedings of the IMS Next Generation Manufacturing Systems Technical Conference '96*, Magdeburg, Germany.

MANAGEMENT OF CHANGES AND DISTURBANCES IN MANUFACTURING SYSTEMS

Monostori, L.; Szelke, E.; Kádár, B.

Computer and Automation Research Institute, Hungarian Academy of Sciences
Kende u. 13-17, H-1111 Budapest, Hungary, Tel.: (36 1) 1665-644, Fax: (36 1) 1667-503

Abstract: Manufacturing systems today operate in a changing environment rife with uncertainty and disturbances. Difficulties arise from unexpected tasks/events, nonlinearities, and a multitude of interactions possible failures during attempting to control various activities in dynamic shop floors. The fundamental aim of the paper is to outline reactive and proactive approaches on the one hand and distributed control architectures on the other hand for change and disturbance management in manufacturing.

Keywords: Intelligent manufacturing systems, artificial intelligence, machine learning, scheduling algorithms, distributed control

1. INTRODUCTION

A manufacturing system is considered automated if it is to some degree self-acting, self-regulating and self-dependent. The concept and realisation of *Flexible Manufacturing Systems* (*FMSs*), and *Computer Integrated Manufacturing Systems (CIMs)* where IT plays a major role can be regarded as significant steps in this area. Hardware involved in these complexes includes (Levary, 1996):

- *Computerised numerical control* (CNC) machine tools. This technology uses software to automatically control the operations of machines.
- *Specialised robots* used in loading and unloading machine tools, inspecting workpieces, assembling products or performing manufacturing tasks.
- *A material transfer system*, such as a network of automated guided vehicles (AGVs), used in moving raw materials, workpieces, partially finished workpieces or tools, and an automated storage and retrieval system (ASRS).

Adaptivity is further emphasised in the concept and gradual implementation of *Intelligent manufacturing systems* (*IMS*s), which are expected to solve unprecedented, unforeseen situations and problems, even on the basis of incomplete and imprecise information. (Hatvany, 1983).

Growing complexity is an other characteristics of today´s manufacturing which manifests itself not only in manufacturing systems, but also in the products to be manufactured, in the processes, and in the company structures (Wiendahl, Scholtissek, 1994).

Manufacturing systems today operate in a *changing environment* rife with *uncertainty*. Difficulties arise from unexpected tasks/events, nonlinearities, and a multitude of interactions during attempting to control various activities in dynamic shop floors. The above *complexity* and *uncertainty* seriously limit the effectiveness of conventional control and scheduling approaches.

The broad goal of *manufacturing operation management*, like other resource constrained, multiagent planning/scheduling problems, is to exhibit a *co-ordinated efficient behaviour* of servicing production demands while responding to changes in shop floors in a timely and cost-effective manner (Smith, 1994). Quality of the factory scheduling process generally has a profound effect on the overall factory performance. Advance generation of the factory schedules is necessary broadly to co-ordinate the manufacturing activities in order to meet organisational objectives, and closely to anticipate potential performance obstacles (e.g., resource contention) thus minimise the disturbing effects on the overall manufacturing system operation.

In industrial practice, however, at least two factors confound the use of *predictive* (advance) *schedules* as operational guidance (Smith, 1994).

- Advance schedules result from scheduling systems running with static models that ignore important new operating constraints/objectives of live shop operation (Prosser, 1991). It is due to the lack of a close correspondence to the live status of executed processes and the data resulting from their real-time monitoring.
- They can not cope with the many environmental and executional uncertainties (Szelke and Márkus G., 1994) arising at companies today, such as unexpected production demands raised by changing market conditions, late deliveries by suppliers, failed operation/break down of machines/equipment etc., all of which work against efforts to follow predictive schedules.

The performance of manufacturing companies ultimately hinges on their ability to rapidly adapt schedules to current circumstances of production floors. This demand has brought to life the relatively new concept of *reactive scheduling* (Szelke and Kerr, 1994) which has been coupled with another pragmatic concept of proactive scheduling (Hadavi, 1994; Szelke and Márkus G., 1994; Szelke and Monostori, 1995).

The goal of *reactive operation management* in manufacturing is to achieve a *co-ordinated adaptive behaviour* during the execution of manufacturing operations, by responding to changes while servicing customer demands in a timely and cost-effective manner (Smith, 1994). For achieving this goal, it is indispensable for real-time production control and scheduling systems to be enhanced with reactive/ proactive scheduling capabilities based *on real-time shop floor monitoring.*

Management of complexity and changes/disturbances is an issue of high importance in enterprise organisation. Companies have found different ways to cope with these issues. The most important ones are (Wiendahl and Scholtissek, 1994):

- decentralisation of company functions (segmentation),
- exploitation of creativity, experience and competence of the employees,
- concentration on the core skills of the company.

In the field of manufacturing systems, decomposition of systems into smaller units (e.g. *manufacturing cells*) is a usual way to overcome this difficulty.

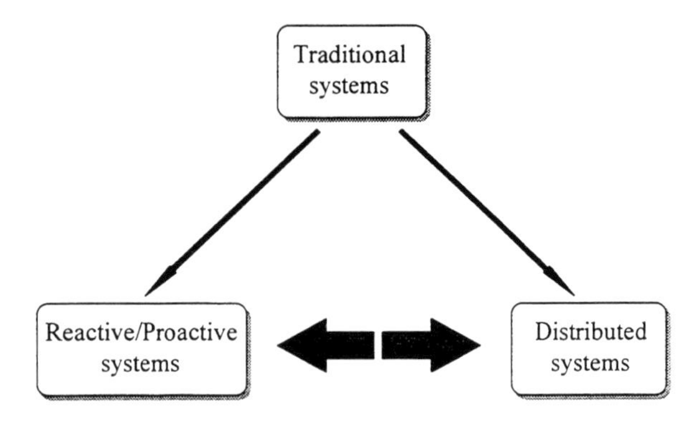

Fig. 1. Two ways for enhancing systems′ resistance against changes and disturbances

Two kinds of approaches to deal with the enumerated problems are illustrated in Fig. 1. One solution is to enhance the reactivity, proactivity of systems by sophisticated new control techniques, the other is to construct decentralised, distributed systems. (Of course these solution approaches may overlap each other.)

An other - also overlapping - way of dealing with changes and disturbances to develop *adaptive systems*, which are able to learn from past history.

As a summary, concerning *dynamic control and management of manufacturing processes*, the following new paradigms have come into the limelight:

- approaches for *reactive scheduling* of manufacturing systems (repairing/adjusting a complete but flawed schedule to keep it in-line with live system status);
- *monitoring* of production processes and their *proactive scheduling* (revision of a complete schedule which is going to be flawed at execution time, preventing an anticipated failure and minimising further performance deteriorations);
- control and management of *holonic manufacturing systems* (systems consisting of autonomous, intelligent, flexible, distributed, co-operative agents, or *holons*) (Valckenaers et al., 1994);
- application of machine learning approaches in manufacturing (Monostori et al., 1996).

The fundamental aim of this paper is to survey and compare the above approaches, to indicate the main results achieved recently and to outline further research and development issues.

2. REACTIVE AND PROACTIVE SYSTEMS

According to a recent overview (Szelke and Kerr, 1994) in the field, *reactive scheduling* is generally conceived as a real time revision or repair of a complete but execution-time flawed schedule to keep it in line with the live status of shop floor processes/events and to make it further executable near to optimum.

Reactive scheduling is generally an event-driven incremental repair/revision process carried out on a complete but flawed schedule (possibly only on its affected parts while re-using its unaffected parts as much as possible). It should meet global management objectives by globally satisfying some major shop performance criteria such as, e.g., to minimise job tardiness, while maximise capacity utilisation and throughput rate at an acceptable product quality.

In addition, it has been recognised increasingly important for reactive operation management in practice to ensure a *stable behaviour* of manufacturing system operation (keep the manufacturing processes smoothly moving).

Proactive behaviour of the scheduler enables to timely prevent some anticipated disturbances, as early as they are foreseeable from monitored/sampled performance trends of schedule execution.

Proactive scheduling is essentially a data-driven re-optimisation of the schedule in order to save/improve the schedule quality (Szelke and Márkus G., 1994), and to preserve stability/continuity in manufacturing activity, while making those schedule changes that are necessary to ensure continued feasibility and adherence to overall performance objectives. It is carried out as an incremental adjustment of the current schedule, in line with evaluation of regularly sampled statistics on the moving averages of some performance measures (like mean tardiness of jobs, mean resource utilisation, mean throughput rate etc.), based on real-time monitoring of executed processes (Szelke and Monostori, 1995).

2.1 Production monitoring

All of the above scheduling decisions should be based on *real-time monitoring* and a continual data-acquisition in the shop floor to trace state changing events (such as the arrival of a new order, start or finish-time of an operation, availability/ unavailability of a resource, etc.).

Proactive scheduling is assumed (Szelke and Márkus G., 1994;, Szelke and Monostori, 1995) to be based on the sampling of statistics (moving averages/standard deviations) at regular discrete time spans in the manufacturing shop. Statistics are continually collected on the work-load of the shop, the machine/resource utilisation, jobs tardiness, the work remained on the jobs, etc., that reflect major performance trends of real-time production.

By using the above statistics *as patterns for situation recognition*, a projection can be made on how to evolve the possible alternative courses of further execution of scheduled activities. Current assessment of performance trends can reflect the goal and guide the demanded route of a subsequent proactive adjustment of the current schedule under its execution. For instance, based on resource usage samples, resource usage distributions can be calculated and bottleneck regions identified (Sadeh, 1993).

Beside the bottleneck patterns of resources relevant to resource congestion, other directions of proactive schedule maintenance can be indicated by using the tardiness patterns of jobs, and breakdown patterns of machines etc., and used for preventing some anticipated failure states (see for more details (Drummond, 1994; Hadavi, 1994; Szelke and Monostori, 1995)).

It is indispensable for reactive/proactive schedulers to make decisions based on real-time monitoring the manufacturing environment from *both orders perspective* and *resource perspective*. Real-time monitoring can increase the transparency and reliability of decision making in real-time production control, for which broad categories in an integrated order and resource monitoring system are assigned by (Wiendahl and Scholtissek, 1994). Monitoring of these perspectives can stem from the different management objectives.

If such observations on the basic performance measures (Szelke and Márkus G., 1994) can be performed on a real-time basis, the different perspectives can provide immediate feedback to the scheduler and achieve better control by, e.g., dynamically *focusing attention* between the *resource* and the *order perspective of scheduling, as* being implemented in some systems (such as ReDs, Hadavi et al., 1992; OPIS, Ow and Smith, 1988; BOSS and MICRO-BOSS, Hynynen, 1990)).

In the following, a reactive cell management system will be illustrated based resource monitoring using a combined artificial neural network (ANN) and expert system (ES) approach (Fig. 2) (Kadar, et al., 1996).

The concept can be traced back to the *hierarchical structure* of intelligent machine tool controllers suggested by Matsushima and Sata (1980). In their scheme, the lower levels consist of adaptive controllers and process pattern recognisers, designed off-line. The higher levels are more global and provide data processing over a longer period of time. The results from the higher levels are manifested as changes in the lower level parameters.

This approach was generalised in the *concept of a hierarchical monitoring and diagnostic system for manufacturing cells* (Monostori *et al.*, 1990). Model based and pattern recognition based algorithms characterised the lower, machine tool level, which was connected to the cell level with symbolic knowledge representation and processing techniques.

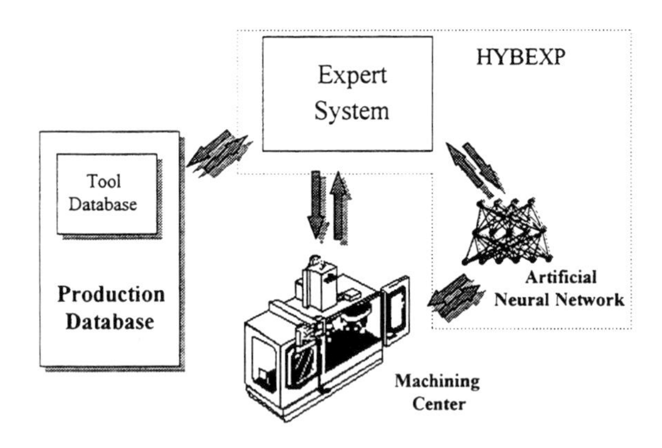

Fig. 2. Concept of the reactive, KB cell management system (Kadar et al., 1996)

As a further step, a *hierarchical structured* monitoring system, called *HYBEXP* was developed in the Computer and Automation Research Institute, Budapest (Monostori, 1996), which combines ES and ANN approaches. This system constitutes the base of the cell management system to be described here.

ES part of the cell management system (Fig. 2) receives information from the *ANN* part of the monitoring module and from a production database which includes the data about the machine, the workpieces to be manufactured, the required and available cutting tools.

Effective utilisation of manufacturing resources requires hard cutting conditions, cutting tools are driven at their technical limits, consequently tool replacements are more frequently initiated than necessary, which influences rentability adversely.

With the aid of on-line information about the tool states and the available spare tools, a more effective resource utilisation can be addressed. In the following, main parts of the developed system are described.

Reactive cell management systems are expected to cope with internal and external disturbances. The proposed treatment of external changes was described in a previous publication (Szelke and Monostori, 1995).

Tool failures (extensive wear and breakage) constitute an essential part of internal disturbances. As it was mentioned, proper tool management can significantly contribute to the effectiveness of the production.

In the reactive cell control system described here, tool data in the production database are updated by conventional analytical techniques and by results of based on sensor measurements and ANN-based estimation.

In addition of this dual tool life monitoring, fundamental feature of the solution is the hybrid AI approach, which combines symbolic and subsymbolic knowledge representation and processing techniques.

Using the information stored in the production database and supplied by the ANN monitoring system, knowledge based techniques determine the control policy to be followed in the given situation.

As an example, let us consider the situation when extensive wear is indicated by the ANN module! The expert system´s task is to take into consideration, among others, the following factors:

- estimated (remained) lifetime of the tool given the cutting tasks to be accomplished,
- available spare tools,
- the number of workpieces to be machined from the same type,
- workpieces waiting for being processed on the given machine,
- cost of cutting tools, scraps, delay in order completion, etc.

As a consequence, the cell management system can

- lower the cutting conditions,
- initiate tool replacement,
- reschedule the tasks to be fulfilled, etc.,

with appropriate messages to other parts of the system.

Significant features of this novel approach are the following:

- dual determination of available tool life (i.e. by conventional analytical techniques and by on-line monitoring),
- multisensor based monitoring of tool wear and breakage by ANN technique,
- integrated use of symbolic (ES) and subsymbolic methods,
- and as a result, reactive and adaptive behaviour.

2.2 Complexity of reactive/proactive scheduling

Over the known complexity of problems in industrial operation scheduling, *reactive/proactive scheduling* has some *additional complexity* and has been difficult to automate (Burke and Prosser, 1994; Szelke and Kerr, 1994) for a variety of reasons:

- Reactive scheduling, like *resource constrained scheduling problems* in general, is a *combinatorially complex, NP hard problem*, thus computationally unfeasible to be solved by the sole use of conventional Operations Research (OR) approaches. AI based or hybrid techniques using domain specific heuristics are necessary to guide the search and to provide satisfying good solutions timely. This demand gave a high *importance of constraint satisfaction techniques* (Minton et al., 1990; Fox et al., 1990; Fox, 1994).

- There may be *tight interactions among* the scheduling *constraints* themselves; the scheduling *objectives* (criteria) are often *ill defined, multiple and often conflicting* (e.g., to minimise WIP inventory while maximise machine/resource utilisation); thus, it is not possible to assess with any precision the impact of scheduling decisions on the global satisfaction of objectives.
- Provision of a *feasible/executable solution near real time* so to keep its validity by the time it has been computed and enacted requires a real timeness/responsiveness of the scheduler.
- For *handling uncertain/incomplete information/ knowledge*, mostly available on the controlled system status (valid goals/constraints etc.) needs special AI techniques namely pattern recognition/situation recognition techniques, fuzzy logic, neural networks etc. Combined use of these techniques is anticipated for proactive scheduling in (Szelke and Monostori, 1995).
- To *reduce problem complexity and handle uncertainty in distributed factory environment* are both indispensable for ensuring *the tractability* of reactive scheduling from an IT perspective. For coping with complexity and ensuring consistency of shop floor scheduling decisions, *distributed problem solving (DPS)* and joint *decomposition techniques* should be taken into consideration.
- Reactive schedulers *must interact and on-line communicate* with their environment.
- *Interfacing with the controlled plant, the human, and existing software* of a reactive scheduler demands to fulfil different IT concerns.
- Finally, a truly reactive scheduler *must adapt its own behaviour* (control/scheduling strategies) in response to changes in the controlled system status/valid goals. It is best facilitated by *learning from examples.* For this purpose, a case based learning (CBL) approach is used in (Sycara and Miyashita, 1994) and another CBL by (Szelke and Márkus G., 1994). Inductive learning is preferred by (Kerr and Kibira, 1994), and learning by neural networks using fuzzy logic is proposed by (Szelke and Monostori, 1995). Learning from experiments in a simulation testbed (see Fig. 3) as outlined and illustrated in (Chiuc and Yih, 1995), a repository of situation/goal dependent strategies for reactive and proactive scheduling can be established to improve the scheduler problem solving efficiency.

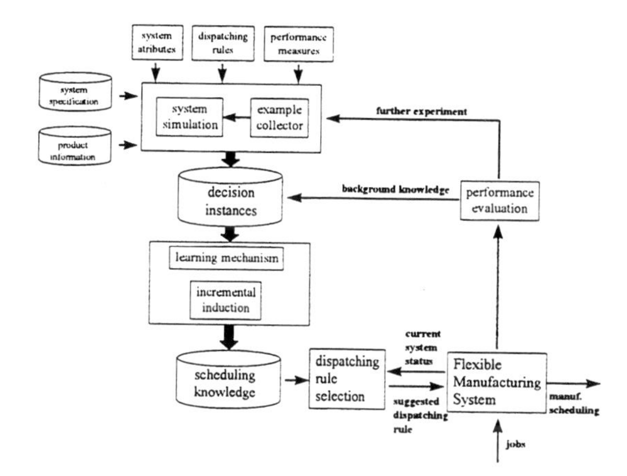

Fig. 3. Architecture of a learning-based dynamic scheduling system (Chiuc and Yih, 1995)

2.3 Modelling and solution approaches to reactive/proactive scheduling

For reactive systems like *reactive schedulers* that are embedded and operating real-time in dynamic/ stochastic production processes, the *behavioural view* is inescapable. They should use *behavioural models* (Pnueli, 1986) because their problem solving cannot be adequately supported by the sole use of conventional *relational* or *functional* models/views. Any distributed AI (DAI) based distributed scheduling systems like every concurrent system, regardless of whether it is automatically produce the final result or interactive, must be studied by behavioural means (Pnueli, 1986).

Manufacturing systems from modelling perspective are considered as dynamic discrete event systems *(DEDS)* (Ho, 1987), (Tawegoum et al., 1994). For constructing valid models of manufacturing processes by the design and simulation of the real-time discrete control/scheduling logic, the models should represent the discrete event evolution of the system as well as features of the underlying continuous processes.

With integrating on-line monitoring of the discrete event system and intelligent reactive/proactive scheduling, the monitoring method should use some heuristic or systematic *behavioural models* to characterise the valid/admissible system trajectories/ evolution graphs which are consistent with the real-time observations on the current system status, inputs/outputs (Holloway et al., 1991).

A truly reactive scheduler then revises the system inputs in response to deviations from predicted status detected by the behavioural model based monitoring system. The behavioural model can be used to experiment with selected control alternatives. The model based system determines whether there are upcoming events/observations which can be predicted from the behavioural model specifications. Any predicted deviation may trigger the operation of the reactive/proactive scheduler.

For building behavioural models, analysing the current system state and the problem situation encountered, or evaluating the performance outcome of different alternative solution scenarios to an RS problem, there are various modelling tools/techniques available and used in a variety of solution approaches to RS. A common demand against these techniques to be powerful and *capable of handling* some sophisticated *timing relationships*.

In the following, main modelling approaches to reactive/proactive scheduling are enumerated accompanied by an overview of solution approaches.

- *Graphical and simulation modelling techniques.* Graphical specification of systems is a natural way to express system functionalities
  - *Petri Nets* (PN) (Van der Aalst, 1994), Timed Petri nets (Van der Aalst, 1994)), Coloured Petri nets, (Chengen et al., 1993)
  - *CDPS nets* for modelling how the elements of a loosely coupled network can work together to solve problems like reactive scheduling that are beyond their individual capacities (Durfee et al., 1989).
  - *Contract nets* are used for multistage negotiation in multiagent-agent systems of solving distributed constraint satisfaction problems like (Conry, 1988) which extends the basic contract net protocol/negotiation paradigm (Davis et al., 1983) to allow iterative negotiation during bidding and awarding. Another extension of contract nets towards the notion of 'Consortium' (a temporary and logical grouping of agents executing the same task (e.g., a RS task as an enterprise activity) is given by (Rabelo, 1994). With a holistic view to schedule generation/execution like 'business processes'/tasks within a CIM-engaged enterprise, the latter extension (based on an integrated agent architecture called HOLOS) provides high control flexibility for negotiation, and supports to envisage virtual manufacturing.
  - *Scheduling graph (SG)* as a direct means of RS and an alternative representation structure to the known Gantt (Wu and Li, 1996). It represents the interrelations between various scheduled jobs, between operations on the same scheduled job, and between scheduled jobs and machines. The SG together with the new concepts of time effect and relationship effect of a schedule compression, as introduced by (Wu and Li, 1994), provides means of : (1) identifying those operations that require revision in a current schedule; (2) revising the identified/affected operations (via a partial change in the scheduling graph structure), and (3) updating the start/finish times of the revised operations.
  - *Simulation* (simulation languages and sytem simulators) as a major family of modelling tools, often in hybrid use with other techniques, should be taken into account for modelling manufacturing systems with RS problem situations. Some RS systems use priority dispatching rules, and based on hybrid approaches of, e.g., simulation and ANNs (Grabot et al., 1994) for selecting priority rules by a situation adequate manner to re-optimise a schedule with respect to overall system performance.
- Solution approaches to reactive/proactive scheduling viewing it as a *Constraint Satisfaction (CS) problem* (Tsang, 1993; Fox, 1994; Prosser, 1991; Minton et al., 1992), need the fast manipulation/handling of various constraints including *resource capacity constraints, temporal precedence constraints, global/local preference constraints etc.* In RS related problem solving, during the conflict resolution, a dynamic creation and propagation of constraints is accomplished (Fox, 1994). The propagation of hard constraints (e.g. completion date of customer orders) and relaxation of some soft constraints (e.g. local preferences on set-up times) can manipulate the search space. Typical example is ISIS (Fox, 1994) that uses constraints to bound, guide, and analyse the search space.
- *Artificial neural networks* (*ANNs*) can meet demands for shop floor control and scheduling due to their real-time capability and learning ability (Monostori and Barschdorff, 1992; Rabelo L.C., 1995; Arizono et al., 1992; Lo and Bavarian, 1991). (for, e.g., selecting best performing dispatching rules for the scheduling/sequencing decisions in RS systems with respect to given objectives).
- *Genetic Algorithms* based models. Genetic algorithms use randomisation techniques to find a local optimum without exhaustively searching through the state solution space. Their main advantage lies in their ability to hop randomly from schedule to schedule, allowing them to escape from local optima in which other algorithms might land. Their applied operators may take into account several real-life constraints, e.g. tool changes, material transport, etc. They are easy to implement and fast to run, and therefore can be taken into consideration as a 'reasoning' methodology for RS in live environment. One of the nice features of genetic algorithms according to (Lawton, 1992) is their ability to interface with other approaches like ANNS, simulation etc. in solving scheduling problems.
- *Stochastic and Fuzzy logic* based models. In stochastic/ dynamic manufacturing environment, the schedule execution is characterised by high degree of uncertainty in defining parameters,

e.g., the start/finish time of executed operations. The manufacturing system itself as a technical and material processing system is not faultless. There is a certain probability of faults that are foreseeable to occur during the schedule execution and may lead to disruption. To detect faults of real-time systems, it supposes a close coupling of the control and fault detection (reflected before) during schedule execution. (Holloway et al., 1995). Although probabilistic representations of scheduling uncertainty have been reported by (Berry, 1992; Sadeh and Fox, 1990), an alternative and possibly more fruitful approach lies in the use of fuzzy sets theory and fuzzy logic (Zadeh and Kacprzyk, 1992; Stoffel et al., 1993; Bugnon et al, 1995). Fuzzy rules seem appropriate to represent expert knowledge, being often confident in qualitative and imprecise terms. The main advantage of fuzzy logic based models in RS lies in the expressive language and systematic framework they provide for representation, propagation/ relaxation of imprecise constraints, and the schedule evaluation against vaguely defined goals/ multiple criteria.

3. DISTRIBUTED (AGENT BASED) SYSTEMS

Over the past years significant research efforts (Bond and Gasser, 1988; O'Hare, 1993; Uma et al., 1993; Chaib-draa et al., 1992; Durfee, 1991; Fox., 1994) have been devoted to the development and use of *Distributed Artificial Intelligence* (*DAI*) techniques. In the eighties, the Distributed AI community has mainly focused its attention on problems where agents contended for computational resources (Durfee, 1989). Less attention was paid to apply DAI to distributed factory scheduling (Sycara et al., 1991), although the importance of distributed decision making arises from the fact that factories are inherently distributed.

A distributed system is a *collection of agents* that can be viewed as an organisation (Fox, 1994). An organisation is defined as a composition of a structure and a control regime. The set of possible structures ranges from strict hierarchies to complex heterarchies. Complexity and uncertainty seem to be opposing forces in deciding how to structure an organisation for information processing. Different examples for organisational metaphor in existing systems and the scheduling problem decomposition with control mechanisms are investigated in (Gomes and Beck, 1992).

Conceptual model of RS related distributed problem solving should facilitate hybrid reasoning at the group of agents level, and deals with the interaction of agents sharing their knowledge and abilities to co-operate to solve a global RS problem split into many tasks (Sycara et al., 1991). In the model, beside the agents, there are a distinguished role of *tasks* as parts of the global RS problem; *objects* used by agents to execute tasks; the *control* that defines the co-operation between agents, the group organisation and its co-ordination problems; and the *communication* between agents depending on the selected protocol, that is the set of rules that specifies the way to synthesise messages to make them significant and correct.

The common denominator applicable to all DAI systems is the *inter agent communication protocol*. When the agent architecture maps a *'Blackboard Model'* built around a multi-agent, multi-task, real-time decision-making, reasoning capabilities are divided among several independent agents or *knowledge sources (KS)*, which co-operate by sharing results through the use of a common memory structure called the *'Blackboard' (BB)*.

The agent activities are co-ordinated by a control module *(Control KS)*, which selects the most appropriate agent to be executed given the current state of the BB. Important changes of the BB are notified to the Control KS through events. The main characteristics of BB systems: (1) independence of agents, (2) strong centralised control, (3) event-driven behaviour, all make them relatively close to classical real-time architectures.

However, the design of true real-time BB systems usually requires the introduction of concurrent agent execution, which raises a number of important consistency problems. Aspects such as data access serialisation and delays experienced by agents waiting for locked resources become fundamental in such systems (as detailed further at REAKT, Lalanda et al., 1992).

The *dominant inter agent communication protocol* here is message passing for performing negotiations between a number of agents/people involved in different complementing aspects/local goals of RS related decision making, to finally reach an agreement/a common goal. The control and co-ordination of the negotiation process often depicts the control and co-ordination of the scheduling process. The extent to which this negotiation process is comprehended, dictates the degree of *opportunism* (Ow and Smith, 1988) that one is able to design into a scheduling/RS technology to do it more flexible alike human experts' problem solving.

DAI architectures enable a scheduling system to be capable of dealing with the right information at the time and place needed. Furthermore, the information should be processed and presented as a recommendation based on incremental decisions by the agents in some framework for negotiation (like the contract net protocol of HOLOS being designed for CIM environment of enterprises, by (Rabelo, 1994)). The latter offers an infrastructure for RS-related problem solving by an agent architecture with an explicit communication mechanism.

3.1 Holonic manufacturing systems

Holonic systems, as one of the new paradigms of manufacturing automation, consist of autonomous, intelligent, flexible, distributed, co-operative agents or *holons*.

Research on *Holonic Manufacturing Systems (HMSs)* is part of the world-wide Intelligent Manufacturing Systems (IMS) project (Yoshikawa, 1992). The HMS consortium within the project formulated the following definitions of holonic concepts for the manufacturing domain (Valckenaers et al., 1994):

- *Holon*: An autonomous and co-operative building block of a manufacturing system for transforming, transporting, storing and/or validating information and physical objects. It consists of an information processing part and often a physical processing part. A holon can be a part of another holon.
- *Autonomy*: The capability of an entity to create and control the execution of its own plans and/or strategies.
- *Co-operation*: A process whereby a set of entities develops mutually acceptable plans and executes them.
- *Holarchy*: A system of holons that can co-operate to achieve a goal or objective. The holarchy defines the basic rules for co-operation of the holons and thereby limits their autonomy.
- *Holonic manufacturing system*: a holarchy that integrates the entire range of manufacturing activities from order booking through design, production, and marketing to realise the agile manufacturing enterprise.
- *Holonic attributes*: attributes of an entity that make it a holon. The minimum set is autonomy and co-operativeness.

Figure 4 depicts the generalised flow model of a holon.

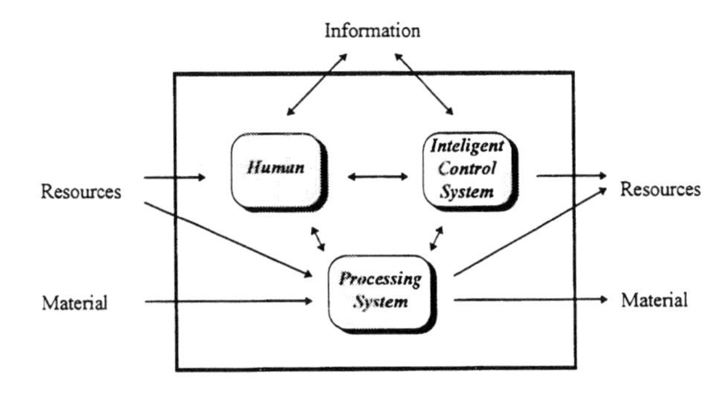

Fig. 4. Generalised flow model of a holon

A manufacturing system incorporates e.g. manufacturing holons, assembly holons, transport holons, information holons, etc. (Fig. 5).These intelligent units function nearly independently, they have their own knowledge representation, processing, decision making and communication capabilities.

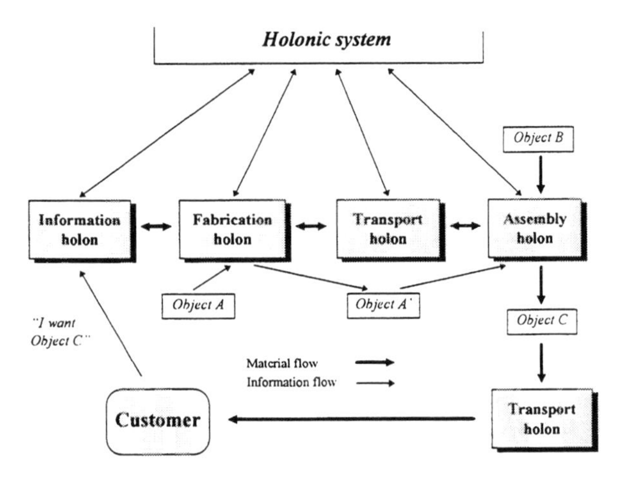

Fig. 5 Manufacturing holarchy

Valckenaers et al. (1994) compared hierarchical, heterarchical and holonic control structure for an assembly cell. They found that holonic systems delivered good performance in a wider range of situations than their more conventional counterparts.

The concept of *Random Manufacturing Systems (RMSs)* can be considered also as a holonic type approach (Iwata and Onosato, 1994). Its basic characteristics can be enumerated as follows:

- autonomous machine systems,
- dynamics machine grouping,
- tender-based task allocation,
- shop floor control by a reward and penalty system.

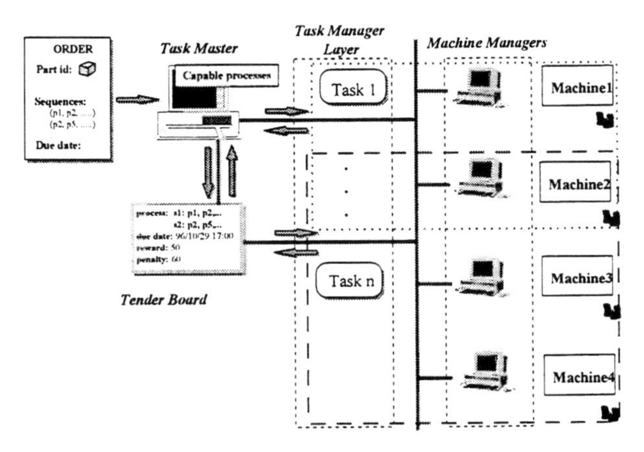

Fig. 6 Basic architecture of RMS (Iwata and Onosato, 1994)

There are four types of elements connected by an information network which are specific to RMS (Fig. 6):

- *Machine manager*: it is attached to each machine on the shop floor and incorporates the processes which the machine can deal with, the

jobs reserved for processing at the machine, assets (the index of profit which the machine has gained), behaviour pattern of the machine, capable processes of other machines, etc.
- *Task manager*: to organise a machine groups to undertake an order.
- *Task master*: it co-ordinates the activities in RMS (accepts orders from customers, calls for tenders, evaluates them, and select a task manager with the most desirable plan, etc.)
- *Tender board*: is a mean of communication among the task master and machine managers.

A *market approach* to holonic manufacturing was introduced by (Márkus A. et al. 1996a, 1996b). Main features of the proposed control mechanism are:
- support of *least commitment and parallel commitment* problem solving,
- guarantee *complete* and *feasible solutions*,
- *directability* through an incentive mechanism by the pricing scheme,
- *predictive* and *reactive* behaviour,
- openness: machines can be added or removed without any modification of the protocol.

Distributed, holonic-like systems represent viable alternatives to hierarchical and heterarchical structures and the corresponding reactive/proactive approaches. Further benchmarking activities are required - preferably using simulation tools - to run more extensive tests.

A very important feature of the holistic approach that it easily scales up to the level of the extended enterprise (Van Brussel et al., 1996)

Table 1. Solution approaches to reactive/proactive scheduling

Solution approach to RS	Accommodated model type	Learning embedded	Application system	Name
Opportunistic	Constraint Satisfaction Problem (CSP)		(Fox, 1987)	ISIS-2
			(Fargher et al., 1987)	
			(Farhoodi, 1990)	
			(Berry, 1990)	
			(Collinot & Le Pape, 1989)	SONIA
	CSP+BB model of DAI		(Smith, 1994)	OPIS
			(Beck, 1993)	TOSCA
	CSP+BB+operator model	Case-based	(Szelke & Márkus G., 1994)	SUPREACT
	CSP+human preference	Case-based	(Sycara & Miyashita, 1994)	CABINS
	CSP+ DAI framework		(Burke & Prosser, 1991, 94)	DAS
			(Hadavi et al., 1992, 1994)	ReDS
Micro-Opportunistic	CSP+BB model		(Hynynen, 1989)	BOSS
			(Sadeh, 1994)	Micro-BOSS
			(Berry, 1992)	
ANN based	ANNs+simulation	supervised	(Chryssolouris et al., 1990, 1991)	
		supervised	(Liu and Dong, 1996)	
		supervised	(Rabelo L. et al., 1990, 95)	
		supervised	(Hoong et al., 1991)	
	ANNs	supervised	(Kurbel and Ruppel, 1995)	LEISTAND
		supervised	(Willems and Brandts, 1995)	
		supervised	(Khaw et al., 1991)	
	ANNs+Semi Markov model	supervised	(Yih and Thesen, 1991)	
	Neuro-fuzzy	supervised	(Bugnon et al., 1995)	FUN
		supervised	(Grabot et al., 1994)	SIPAPLUS
		supervised	(Roy and Zhang, 1996)	
Genetic algorithms based	GA+ANNs+simulation		(Rabelo L. et al., 1995)	
			(Syswerda, 1991)	
	GA		(Fang et al., 1993)	
			(Starkwether et al., 1992)	
			(Wiendahl & Garlichs, 1994)	
Fuzzy based	Fuzzy sets	Inductive Machine Learning	(Kerr and Kibira, 1994)	
			(Dubois et al., 1993)	
			(Yuan and Wu, 1991)	
DAI	Agent architecture		(Roboam & Fox, 1992)	
			(Sycara et al., 1991)	CORTES
			(Rabelo R.., 1994)	HOLOS

4. CONCLUSIONS

With consideration to the very limited extent of this paper, solution approaches for reactive/proactive scheduling (Sect. 2) that have been currently reviewed more deeply in (Szelke and Kerr, 1994) are summarised in Table 1.

The importance of adaptive nature of reactive/ proactive systems was outlined in the introductory chapter. Concerning the different machine learning techniques and their applications in manufacturing we refer to a recently published survey (Monostori et al., 1996).

A more detailed analysis of distributed approaches to reactive/proactive scheduling is the subject of another forthcoming paper.

ACKNOWLEDGEMENTS

This work was partially supported by the *National Research Foundation, Hungary,* Grant No. T 016512 and T016475. A part of the work was covered by the *PHARE TDQM Programme of the European Union,* Grant No. H 9305-02/1071 (REMADE) and by the *National Committee for Technological Development, Hungary* which promotes Hungarian participation in the *ESPRIT 21108 Working Group on Integration In Manufacturing and Beyond* (IIMB 21108).

REFERENCES

Arizono, I., Yamamoto, A., and Ohta, H. (1992). Scheduling for Minimising Total Actual Flow Time by Neural Networks, *Int. Journal of Production Research,* **Vol.30.**, No.3, pp. 503-511.

Beck, H.A. (1993). TOSCA: A novel approach to the management of job-shop scheduling constraints. In: *Realising CIM's Ind. Potential,* (C. Kooij, P.A. Mac Conail and J. Bastos, IOS Press, (Ed.)), pp.138-149.

Berry, P.M. (1992). Scheduling: A problem of decision making under uncertainty. *Proc. 10th European Conf. on AI, ECAI'92,* Vienna, John Wiley, pp.638-642.

Bohner, P. (1995). A multi-agent approach with distributed fuzzy logic control. *Computers in Industry,* **Vol. 26**, pp. 219-227.

Bond, A.H. and Gasser, L. (Eds.), (1988). *Readings in DAI,* Morgan-Kaufmann, 1988.

Bose, P. (1992). An abstraction-based search and learning approach for effective scheduling., In: *AI Applications in Manufacturing.* (A. Famili, D. S. Nau, S. H. Kim, (Ed.)) AAAI Press / The MIT Press, pp. 187-197.

Bugnon, B., Stoffel, K. and Widmer, M. (1995). FUN: A dynamic method for scheduling problems. *European J. of Operational Research,* **Vol. 83**, pp. 271-282.

Burke, P., and Prosser, P. (1994). The Distributed Asynchronous Scheduler. In: *Intelligent Scheduling.* (M. Zweben and M.S. Fox, Morgan Kaufmann (Ed.)), San Francisco, pp.309-340.

Chaturvedi, A. R., Hutchinson, G. K., Nazareth D. L. (1992). A synergistic approach to manufacturing systems control using machine learning and simulation. *J. of Intelligent Manufacturing,* **Vol. 3**, pp. 43-57.

Chengen, W., Jianying, Z., and Zhongxin, W. (1993). An expert system for FMS control. *Intelligent Systems Engineering,* Winter 1993, pp. 223-230.

Chiuc, C., Yih, Y.(1995). A learning based methodology for dynamic scheduling in distributed manufacturing systems. *Int. J. Prod. Res.,* **Vol. 33**, No. 11, pp. 3217-3232.

Chryssolouris, G., Lee, M., Domroese, M. (1991). The use of neural networks in determining operational policies for manufacturing systems. *J. of Manufacturing Systems,* **Vol. 10**, No. 2, pp. 166-175.

Chryssolouris, G., Lee, M., Pierce, J. and Domroese, M. (1990). Use of Neural Networks for the Design of Manufacturing Systems, *Manufacturing Review,* **Vol.3**, No.3, pp.187-194.

Chu, C., Portmann, M.-C. (1993). Application of the artificial memory approach to multicriteria scheduling problems, *J. of Int. Manufact.,* **Vol. 4**, pp. 151-161.

Collinot, A., and Le Pape, C. (1989). Testing and comparing reactive scheduling strategies. *Proc. of 1989 AAAI-SIGMAN Workshop on Manufacturing Production Scheduling,* Publ., Detroit, USA.

Conry, S., Meyer, R., and Lesser, V., (1988). Multistage negotiation in distributed planning. *Readings in Distributed Artificial Intelligence,* Morgan-Kaufmann, 1988, pp. 367-384.

Davis, Randall, Smith, Reid (1983). Negotiation as a metaphor for distributed problem solving, *Artificial Intelligence,* **No. 20**, 1983, pp. 63-109.

Dubois, D., Fargier, H. And Prade, H. (1993). Handling flexibility and uncertainty in job shop scheduling. *Proc. FLAI'93 Workshop on Fuzzy Logic in AI,* Linz, pp. 13-17.

Durfee, E.H. (Ed.) (1991). Distributed Artificial Intelligence, *IEEE Trans. on Systems, Man and Cybernetics,* **Vol. 21**, No. 16, pp. 1167-1183.

Durfee, E.H. et al. (1989). Trends in co-operative distributed problem solving. *IEEE Trans. on Knowledge and Data Engineering,* **Vol. 1**, No. 1, March 1989, pp.63-83.

Fang, H-L., Ross, P. and Corne, D. (1993). A Promising Genetic Algorithm Approach to Job-Shop Scheduling, Rescheduling, and Open-Shop Scheduling Problems. *Proc. of the Fifth Int. Conf. on Gas.,* pp. 375-382.

Fargher, H.E., Elleby, P., and Addis, T.R. (1987). A reactive scheduling system. *Proc. of IEE Colloquium on Expert Systems in Production Control,* May.

Farhoodi, F. (1990). A knowledge based approach to dynamic job-shop scheduling. *Int. J. of Computer Integrated Manufacturing,* **Vol.3**, 1990, pp. 84-95.

Foo, S. Y., Takefuji, Y., Szu, H. (1994). Job-shop scheduling based on modified Tank-Hopfield linear programming networks. *Eng. Appl. of Artificial Intelligence,* **Vol. 7**, No. 3, pp. 321-327.

Fox, M.S. (1987). Constraint-Directed Search: A case study of Job-Shop scheduling. *Intelligent Scheduling,* (M. Zweben and M.S. Fox, (Ed.)), Morgan Kaufmann, San Francisco, California.

Fox, M.S. (1994). ISIS: A Retrospective. In: *Intelligent Scheduling,* (M. Zweben and M.S. Fox, (Ed.)), Morgan Kaufmann, San Francisco, California, pp. 3-28.

Fox, M.S., Sadeh, N., Baykan, C. (1990). Constrained heuristic search. Proc. Workshop of Innovative Approaches to Planning, Scheduling and Control, San Diego, California, DARPA Publ., pp. 309-315.

Grabot, B., Geneste, L. and Dupeux, A. (1994). Multi-heuristic scheduling in SIPAPLUS: three approaches to tune compromises. *J. of Intelligent Manufacturing*, **Vol.5**, pp. 303-313.

Hadavi, K. (1994). ReDS: A Real Time Production Scheduling System from Conception to Practice. In: *Intelligent Scheduling*, (M. Zweben and M.S. Fox, Morgan Kaufmann (Ed.)), San Francisco, 1994, pp.581-604.

Hadavi, K., Hsu, W-L., Chen, T., and Lee, C-N. (1992). An architecture for real-time distributed scheduling, *AI Magazine*, Fall, pp.46-56.

Hatvany, J. (1983). The efficient use of deficient information. *CIRP Annals*, **Vol. 32/1**, pp. 423-425.

Ho, Y.C. (1987). Performance Evaluation and Perturbation Analysis of Discrete Event Dynamic Systems. *IEEE Trans. on Automat. Contr.*, AC-32,7, pp. 563-572.

Holloway, L.E., Paul, C.J., Strosnider, J.K. and Krough, B.H. (1991). Integration of behavioral fault-detection models and an intelligent reactive scheduler. *Proc. of IEEE Int. Symp. on Intelligent Control*, Aug 13-15, Arlington, Virginia, USA, pp. 134-139.

Hoong, L.W., Yeo, K.T., and Sim, S.K. (1991). Application of neural networks in a job-shop environment. *Proc. of Int. Conf. on CIM, ICCIM'91, 'Manufacturing enterprises of the 21st Century'* , (World Scientific Publ., Singapore), pp. 559-562.

Hynynen, J.E. (1989). BOSS-An artificial intelligent system for distributed factory scheduling. *Proc. CAPE'89 IFIP Conference on Computer Application in Production and Engineering*, pp. 667-677.

Iwata, K., Onosato M., Koike, M. (1994). Random manufacturing system: a new concept of manufacturing systems for production to order. *CIRP Annals*, **Vol. 43/1**, pp. 379-383.

Jones, A., Rabelo, L., Yih, Y. (1995). A hybrid approach for real-time sequencing and scheduling. *Int. J. Computer Integr. Manuf.*, **Vol. 8**, No. 2, pp. 145-154.

Kádár, B., Markos, S., Monostori, L. (1996). Knowledge Based Reactive Management of Manufacturing Cells. In: *IT and Manufacturing Partnerships*, Conf. On Integration in Manuf., Galway, Ireland, pp. 197-205.

Kerr, R.M., and Kibira, D. (1994). Simulation and Machine Induction with Fuzzy set theory. *Proc. of the 2nd IFAC/IFIP/IFOR Conf. on 'Intelligent Manufacturing Systems'- IMS'94*, June 1994, Vienna, Austria, Techn. Univ. of Vienna, pp.397-404.

Khaw, J., Siong, L.B., Lim, L., Yong, U., Jui, S.K. and Fang, L.C. (1991). Shop floor scheduling using a three-dimensional neural network model. *Proc. Int. Conf. Computer Integrated Manufacturing*, World Scientific Singapore, pp. 563-566.

Kurbel, K., and Ruppel, A. (1996). Integrating intelligent job-scheduling into a real-world production scheduling system, *J. of Intelligent Manufacturing*, **Vol. 7**, pp. 373-377.

Kwok, A., Norrie, D. (1993). Intelligent agent systems for manufacturing applications. *J. of Intelligent Manufacturing*, **Vol. 4**, pp. 285-293.

Lalanda, P., Charpillet, F., and Haton, J-P. (1992). A real-time blackboard-based architecture, *Proceedings 10th European Conf. on Artificial Intelligence - ECAI'92*, Vienna, Austria, August (B. Neumann (Ed.)), Wiley & Sons, Vienna, pp. 262-266.

Lawton, G. (1992). Genetic algorithms for schedule optimization. *AI Expert*, **Vol.5**, pp. 23-27.

Levary, R. K. (1996). Computer Integrated Manufacturing: a complex information system. *Production planning & Control*, **Vol. 7**., No. 2, pp 184-189.

Liu, H., Dong, J. (1996). Dispatching rule selection using artificial neural networks for dynamic planning and scheduling. *J. of Intelligent Manufacturing*, **Vol. 7**, pp. 243-250.

Lo, Z. and Bavarian, B. (1991). Scheduling with Neural Networks for Flexible Manufacturing Systems. *Proc. of the 1991 IEEE Conference on Robotics and Automation*, Sacramento, California, pp. 818-823.

Márkus, A., Kis, T., Váncza, J., Monostori, L. (1996). A market approach to holonic manufacturing, *CIRP Annals*, **Vol. 45/1**, pp. 433-436.

Márkus, A., Váncza, J. (1996). Are manufacturing agents different? In: *Eropean Workshop on Agent-Oriented Systems in Manufacturing*, Berlin, pp. 42-59.

Matsushima, K. and T. Sata (1980). Development of intelligent machine tool, *Journal Faculty Eng., Univ. Tokyo*, **Vol. 35**, No. 3, pp. 299-314.

Matsushima, K., Sata, T. (1980). Development of intelligent machine tool, *J. Faculty Eng., Univ. Tokyo*, **Vol. 35**, No. 3, pp. 299-314.

Mezgár, I., Egresits, Cs., Monostori, L. (1995). Re-engineering of robust manufacturing system configurations. *Preprints of the Second Int. Workshop on Learning in IMSs*, Budapest, Hungary, April 20-21, pp. 260-280.

Minton, S., Johnston, M.D., Philips, A.B. and Laird, Ph. (1992). Minimising conflicts: a heuristic repair method for constraint satisfaction and scheduling problems, *Artificial Intelligence*, **Vol. 58**, 1992, pp.161-205.

Monostori, L. (1996). From pattern recognition techniques through artificial neural networks to hybrid AI solutions in manufacturing. *Proc. of the 1996 Japan-USA Symp. on Flexible Automation*, Boston, Massachusetts, USA, July 7-10, Vol. II, pp. 1453-1460.

Monostori, L., Barschdorff, D. (1992). Artificial neural networks in intelligent manufacturing. *Robotics and Computer-Integrated Manufacturing*, **Vol. 9**, No. 6, Pergamon Press, pp. 421-437.

Monostori, L., Bartal, P., Zsoldos, L. (1990). Concept of a knowledge based diagnostic system for machine tools and manufacturing cells. *Computers in Industry*, **Vol. 15**, pp. 95-102.

Monostori, L., Egresits, Cs. (1995). On hybrid learning and its application in intelligent manufacturing. *Preprints of the Second Int. Workshop on Learning in IMSs*, Budapest, Hungary, April 20-21, pp. 655-670.

Monostori, L., Márkus, A., Van Brussel, H., Westkamper, E. (1996). Machine learning approaches to manufacturing, *Annals of the CIRP*, **Vol. 45/2**, pp. 675-712.

O'Hare, (1990). Designing intelligent manufacturing systems: a distributed artificial intelligence approach. *Computers in Industry*, **No. 15**., pp. 17-25.

Ow, P.S., Smith S.F. and Howie, R. (1988). A co-operative scheduling system. *Expert Systems and Intelligent Manufacturing*, (M.D. Oliff (Ed.)), Elsevier, Amsterdam, pp. 43-55.

Pnueli, A. (1986). Specdification and development of reactive systems. In: *Information Processing* (Kugler, H.-J. (Ed.)), Elsevier, Holland, pp. 845-858.

Prosser, P. (1991). Reactive scheduling. *Proceedings of SIGMAN Workshop of AAAI-IJCAI'91*, Detroit, American Association of Artificial Intelligence, pp. 57-59.

Rabelo, L., Yih, Y. and Jones, A. (1995). Knowledge Based Reactive Scheduling using the integration of Neural Networks, Genetic Algorithms and Machine Learning, *Proc. of the IFIP Int. Working Conference on 'Managing Concurrent Manufacturing to Improve Industrial Performance'*, Sept 11-15, Seattle, USA, Washington Univ. Press, pp.444-455.

Rabelo, R. and Camarinha-Matos, L.M. (1995). A Holistic Control Architecture Infrastructure for Dynamic Scheduling, In: *Artificial Intelligence in Reactive Scheduling*, (R. Kerr and E. Szelke (Ed.)), Chapman & Hall, London, pp.78-94.

Rabelo, R. J., Camarinha-Matos, L. M. (1994). Negotiation in multi-agent based dynamic scheduling. *Robotics & Computer-Integrated Manufacturing*, **Vol. 11**, No. 4, pp. 303-309.

Roboam, M., and Fox, S.M. (1992). Enterprise Management Network Architecture - A Tool for Manufacturing Enterprise Integration. In: *AI Applications in Manufacturing*. (Famili, A.F., Nau D.S. and Kim, S.H. (Ed.)), AAAI Press, California.

Roy, U. and Zhang, X. (1996). A heuristic approach to n/m job shop scheduling: fuzzy dynamic scheduling algorithms, *Int. Journal of Production Planning & Control*, **Vol.** 7, No. 3, pp. 299-311.

Sadeh, N. (1994). Micro-Opportunistic Scheduling: The Micro-Boss Factory Scheduler. In: *Intelligent Scheduling*. (M. Zweben and M.S. Fox, Morgan Kaufmann (Ed.)), San Francisco, pp. 99-136.

Sadeh, N. and Fox M.S. (1990). Variable and value ordering heuristics for activity based job-shop. *Proc. of the Fourth Int. Conference on Expert Systems in Production and Operation Management*, Hilton Head Island, USA, pp.134-144.

Smith, S.F. (1994). OPIS: A Methodology and Architecture for Reactive Scheduling. In: *Intelligent Scheduling*, (M. Zweben and M.S. Fox, Morgan Kaufmann (Ed.)), San Francisco, 1994, pp. 29-66.

Starkwether, T., Whitney, T., and Cookson, B. (1992). A Genetic Algorithm for Scheduling with Resource Consumption. *Proc. of the Joint German/US Conference on OR in Production Planning and Control*, Springer-Verlag, Berlin, pp. 567-583.

Stoffel, K., Law, I., and Hirsbrunner, B. (1993). Fuzzy logic controlled dynamic allocation system. In: *Parallelism and Artificial Intelligence*, (B. Hirsbrunner, M.Courant and M. Aguilar (Eds.)), Series in Computer Science, University of Fribourg, Fribourg,.

Sycara, K.P., and Miyashita, K. (1994). Adaptive Schedule Repair, *Proceedings First IFIP Workshop on 'Knowledge Based Reactive Scheduling'*, (E. Szelke and R.M. Kerr (Ed.)), Elsevier (North-Holland), Amsterdam, pp.107-123.

Sycara, K.P., Roth, S., Sadeh, N., and Fox, M. (1991). Resource Allocation in Distributed Factory Scheduling, *IEEE Expert*, Feb., pp.29-40.

Syswerda, G. (1991). Schedule Optimization Using Genetic Algorithms. *Handbook of Genetic Algorithms*. (L. Davis (Ed.)), Van Nostrand Reinhold, New York.

Szelke, E., and Kerr, R.M. (1994). Knowledge Based Reactive Scheduling - State-of-the-Art. *Int. Journal of Production Planning and Control*, **Vol.5**, March-April, 1994, pp.124-145.

Szelke, E., and Márkus, G. (1994). Reactive scheduling - An intelligent supervisor function, *Proceedings First IFIP Workshop on 'Knowledge Based Reactive Scheduling'*, (E. Szelke and R.M. Kerr, (Ed.)) Elsevier (North-Holland), Amsterdam, pp.125-142.

Szelke, E., Monostori L. (1995). Reactive and proactive scheduling with learning in reactive operation management, Proc. of *IFIP Int. Working Conf. on Managing Concurrent Manufacturing to Improve Industrial Performance*, Sept. 11-15, Seattle, Washington, USA, pp. 456-483.

Tawegoum, R., Castelain, E.,and Gentina, J.C. (1994). Dynamic Operations Control in Flexible Manufacturing Systems(FMS). *Proceedings of the* IFAC/IFIP/IFORS *Conf. on 'Int.t Manufact. Systems'* IMS'94, June 1994 Vienna, Austria, pp.285-290.

Tsang, E. (1993). *Foundations of Constraints Satisfaction*. In: Series for Computation in Cognitive Science, Academic Press.

Uma, G., Prasad, B.E., and Kumari, O.N. (1993). Distributed intelligent systems: issues, perspectives and approaches. *Knowledge Based Systems*, **Vol. 6**, No. 2, June, pp. 77-86.

Valckenaers, P., F. Bonneville, H. Van Brussel and J. Wyns (1994). Results of the holonic system benchmark at KULeuven, Proc. of the *Fourth Int. Conf. on Computer Integrated Manufacturing and Automation Technology*, Oct. 10-12, Troy, New York, pp. 128-133.

Van Brussel, H., Valckenaers, P., Wyns, J., Bongaerts, L. And Detand, J. (1996). Holonic Manufacturing Systems and IiM. In: *IT and Manufacturing Partnerships*, Conf. On Integration in Manufacturing, Galway, Ireland, pp. 185-196.

Van der Aalst, W.M.P. (1994). Putting high-level Petri nets to work in industry. *Computers in Industry*, **25**, pp. 45-54.

Wiendahl, H.-P. And Scholtissek, P. (1994). Management and control of complexity in Manufacturing. *Annals of the CIRP*, **Vol 43/2**, pp. 533-540.

Wiendahl, H.-P., Garlichs, R. (1994). Decentral Production Scheduling of Assembly Systems with Genetic Algorithm, *Annals of the CIRP*, **Vol.43/1**, pp.389-395.

Willems, T.M., Brandts, L.E.M.W. (1995). Implementing heuristics as an optimization criterion in neural networks for job-shop scheduling. *J. of Intelligent Manufacturing*, **Vol. 6**, pp. 377-388.

Yih, Y., (1992). Learning real-time scheduling rules from optimal policy of semi-Markov decision processes. *Int. J. Computer Integrated Manufacturing*, **Vol. 5.**, No. 3, pp. 171-181.

Yih, Y., Thesen, A. (1991). Semi-Markow decision models for real-time scheduling. *Int. J. Production Research*, **Vol. 29**, No. 11, pp.2331-2346.

Yoshikawa, H. (1992). Intelligent Manufacturing Systems Program (IMS), *Technical Cooperation that Transcends Cultural Differences*, University of Tokyo, Japan.

Yuan, Y., and Wu, Z. (1991). Algorithm of fuzy dynamic programming in AGV scheduling. *Proc. Int. Conf. on Computer Integrated Manufacturing, ICCIM'91, 'Manufacturing Enterprises of the 21st Century'*, World Scientific Publ., Singapore, pp.405-408.

Zadeh, L.A., and Kacprzyk, J. (1992). *Fuzzy Logic for the Management of Uncertainty*, Wiley and Sons, New York.

REGULATION OF GMA WELDING PROCESS FEATURES USING NEURAL ADAPTIVE CONTROL

S.G.Tzafestas, G.G. Rigatos and E.J. Kyriannakis

Intelligent Robotics and Automation Laboratory
National Technical University of Athens
15773, Zografou Campus, Athens, Greece

Abstract : *Neural adaptive control (NAC)* is naturally a very good candidate for the regulation of the geometrical and thermal characteristics of the welding process which is subject to strong variations and disturbances. Here two different versions of neural adaptive control are tested. The applied control scheme is the Feedback Error Lerning Architecture (FELA). In its first version a *Multilayered Perceptron (MLP) Neurocontroller* is used while its second version is based on a *Radial Basis Function (RBF) Neurocontroller.* Extensive numerical results are provided along with a discussion of the relative merits and limitations of the above techniques.

Keywords : neural adaptive control, back propagation algorithm, radial basis function network

1. INTRODUCTION

Arc Welding is one of the primary applications of intelligent robotic systems (Hunt,1983 ; Tzafestas , 1991). Actually, arc welding is the third largest job class behind assembly and machining in the metal fabrication industry.Automation of the welding process opens very challenging avenues of research in automatic control, robotics, sensor systems, signal and image processing, computer vision, and artificial intelligence (Tzafestas 1995a).

This paper examines the application of neural network-based adaptive control techniques to one of the principal types of arc welding, namely the *Gas Metal Arc (GMA)* welding. GMA welding is a complex physical process that needs careful control for minimizing the defects caused by the improper regulation of parameters like arc voltage and current, or travel speed of the torch. (Doumanidis , 1989 ; Doumanidis , 1991)

- Defects related to the geometrical characteristics. This means that the weldment's width, height or depth differs from the desired one.
- Thermally induced discontinuities in the weldment's structure such as porosity, incomplete fusion, inadequate penetration, undercutting and cracking.

The task of automated welding is to produce weldments of high-quality and strength. This can be achieved through proper on-line control of the geometrical and thermal properties of the process. Among the various control techniques available, the adaptive ones seem to be the most appropriate to face the parameter variations and the disturbances of GMA welding, and satisfy the weld quality requirements (Doumanidis ,1989 ; Lightbody ,et al.,1995).

Conventional adaptive control techniques require the availability of models that describe adequately the geometry of the weld, as well as of models that represent the thermally activated properties of the material during the process. These models are usually determined by open-loop tests using step input currents.

An alternative adaptive control approach which does not use an explicit mathematical model of the welding process is the one that employs some kind of neural network. The purpose of the work described in this paper is to provide a study of the capabilities of both the MLP and RBF neural network-based adaptive controllers in GMA welding.

Before proceeding to the neural adaptive control (Sec. 4), the standard linearized models of arc welding characteristics are first summarized (Sections 2 and 3).s

2. A MODEL FOR THE GEOMETRICAL CHARACTERISTICS OF ARC WELDING

The outputs that describe the geometrical characteristics of a welding process are (Doumanidis , 1994): (i) The width W (ii) The height H (iii) The penetration depth D.

Open loop experiments have shown that the geometrical characteristics can be satisfactorily described by first order linear models :

$$K / (\tau s + 1) \quad (1)$$

Both the gain K and the time constant τ depend on the conditions of the welding and the amplitude of the welding's inputs. This means that the welding process which is highly nonlinear can only be locally approximated by its linear equivalent (Doumanidis , 1994) .
The linearized first-order model was fitted to the experimental data and expressed in the form of transfer functions :

$$\frac{W(s)}{U(s)} = \frac{K_w}{\tau_w s + 1}, \quad \frac{H(s)}{U(s)} = \frac{K_h}{\tau_h s + 1}, \quad (2)$$
$$\frac{D(s)}{U(s)} = \frac{K_d}{\tau_d s + 1}$$

where the values of the constants K and τ_i (i=w,h,d) are given in Nishar, et al., (1994) .

3. A MODEL OF ARC WELDING THERMAL CHARACTERISTICS

The following selection of welding outputs has been suggested (Doumanidis ,1989 ; Doumanidis , 1991) :
(a) The weld nugget cross section NS defined by the solidus isotherm Tm.
(b) The heat affected zone HZ defined by an enveloping isotherm Th. This may indicate the extent of weak zones, such as the recovery, recrystallization and grain growth areas.
(c) The centerline cooling rate CR defined at the critical temperature Tc. This may determine the crystallization of undesirable, kinetically favoured phases, or supply a measure of the cracking tendency of the weldment caused by thermal stresses

$$(CR = \partial T / \partial t \mid_{T=T_c}) \quad (3)$$

The definition of the welding outputs is illustrated in Fig.1

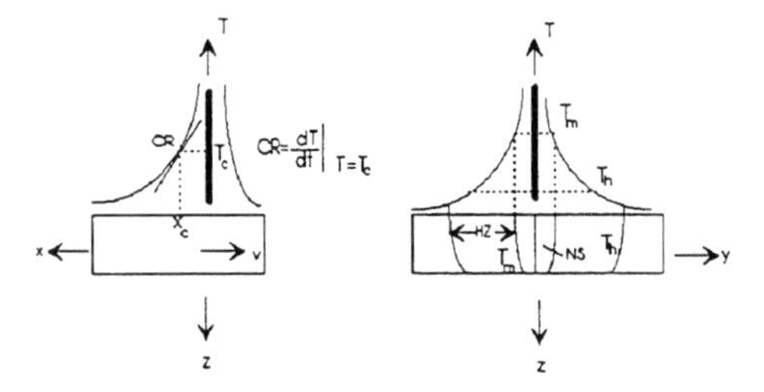

Fig.1 Definition of the welding outputs NS, HZ, CR

The response of the bead cross section area NS to step inputs (either Q_1 or V) can be approximated by an overdamped second-order behavior, in which one pole clearly dominates over the other, so that NS may be adequately modelled by a first order transfer function, with respect to either Q_1 or V:

$$\frac{NS}{Q_1}(s) = \frac{K_a}{\tau_a s + 1}, \quad \frac{NS}{V}(s) = \frac{K_a'}{\tau_a' s + 1}, \quad (4)$$
$$\frac{NS}{Q_2}(s) = 0$$

The response of the heat affected zone HZ to steps in either Q_1 or V has a nonminimum phase form and, can be closely approximated by a nonminimum phase second- order behavior.The sensitivity of HZ to the third input Q_2 is almost insignificant and can be approximated by an almost zero, first order, transfer function. Thus :

$$\frac{HZ}{Q_1}(s) = \frac{K_b(\tau_b s + 1)}{(\tau_1 s + 1)(\tau_2 s + 1)} \quad (5)$$

$$\frac{HZ}{V}(s) = \frac{K'_b(\tau_b' s + 1)}{(\tau'_1 s + 1)(\tau'_2 s + 1)} \quad (6)$$

$$\frac{HZ}{Q_2}(s) = \frac{K''_b}{\tau_2'' s + 1} \approx 0 \quad (7)$$

Finally the response of the centerline cooling rate CR to steps in either Q_1 or V may be approximately described by an over-damped second-order behavior owing to the existence of two dominating real poles. The dependence of CR on the third input Q_2 is also of second order, but with the one pole dominating over the other, so that it can be reduced to a first-order transfer function :

$$\frac{CR}{Q_1}(s) = \frac{K_c}{(\tau_a s + 1)(\tau_b s + 1)} \quad (8)$$

$$\frac{CR}{V}(s) = \frac{K'_c}{(\tau'_a s + 1)(\tau_b' s + 1)} \quad (9)$$

$$\frac{CR}{Q_2}(s) = \frac{K''_c}{\tau''_a s + 1} \quad (10)$$

The calculated values of the gains and the time constants of the linearized model are presented in Doumanidis ,(1989).
This table reveals the strong nonlinear dependence of the parameter values on the input magnitude. Thus the table provides ranges of variation for the nonstationary model parameters and indicates the need for their in-process identification.

4. APPLICATION OF NEURAL CONTROL

4.1. General Issues

Neural networks are systems that involve a large number of special type nonlinear processors called "neurons" (Haykin , 1994). A neural network-based controller (called briefly neural controller or neurocontroller) performs some kind of adaptive control, with the controller taking the form of a feedforward or a feedback neural network. This is done through learning which is actually an adaptation of the strengths of the interconnections among the neurons.
Neurocontrollers are distinguished in *supervised* and *unsupervised* neurocontrollers. The structure of a supervised neural control system involves a *teacher* (i.e. a human controller or an other automated controller), a *trainable controller* (i.e. a neural network that can be trained receiving examples of proper control signals from the teacher), and the *system (process) under control*.In unsupervised neural control no external teacher is available and the dynamics of the process under control are unknown and/or involve severe variations or uncertainties (as is the case of the welding process).
From among the various unsupervised neurocontrol architectures, we select Feedback Error Learning Architecture (FELA) for application to the welding process which is used in the control of the nervous system and has been proposed by Kawato and coworkers for voluntary movement control (e.g. robot motion control) (Kawato, et al., 1987 ; Kawato, et al. 1988).

4.2. The Feedback Error Learning Architecture

The feedback error learning architecture (FELA) can be very useful for the welding control problem (Kawato, et al. 1988 ; Tzafestas , 1995b) and has the form shown in Fig. 9.

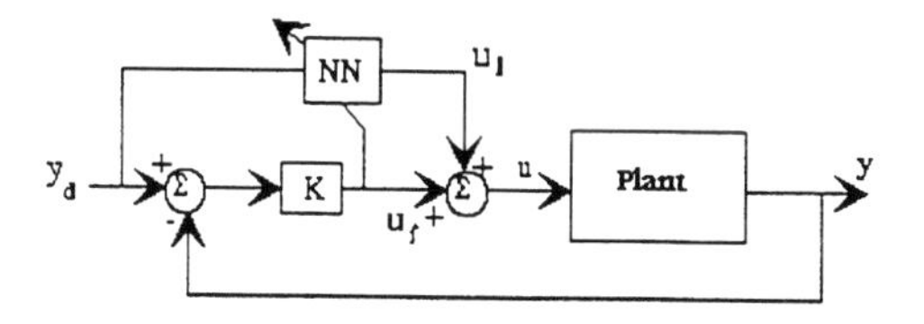

Fig.2 FELA Neural Network Controller

The total control input which is applied to the plant is $u(t) = u_f(t) + u_i(t)$. The inverse dynamics neural network (NN) receives as input the desirable output of the plant $y_d(t)$ and monitors the feedback signal $u_f(t)$ as the error signal. The error signal tends to zero while the NN's training goes on.
Compared to other schemes of neural control the FELA technique has several important advantages (Tzafestas, et al. 1995b;).

4.3. Welding control by the FELA controller

In order to train the neural controller NC we apply as inputs the desirable geometrical or thermal characteristics of the welding process. The recorded outputs of the neural controller could be either the velocity of the torch, or the thermal power or the wirefeed rate. In this way NC learns gradually the inverse dynamics of the process.
The total input that is applied to the plant is

$$u(t) = u_i(t) + u_f(t) \qquad (11a)$$

Here $u_i(t)$ is the output of the neural controller and

$$u_f(t) = K_p e(t) \qquad (11b)$$

where :
e(t) = deviation of the plant's output from the desirable value
K_p = amplification or attenuation coeeficient of the error e(t).
The neurocontroller consists of a multilayer perceptron and its training is based on the Back Propagation (BP) algorithm. The neurocontroller takes as input the desirable response of the plant and in the phase of forward computation creates the signal $u_i(t)$. In the phase of backward computation the weights of the neural controller are corrected according to the size of the error e(t) :

$$w_{ij}(k+1) = w_{ij}(k) - \eta e_j(k) \frac{\partial y_j}{\partial w_{ji}} \qquad (12)$$

4.4. The Multilayer Perceptron

A multilayer perceptron (MLP) has the following distinctive characteristics :
(i) Each neuron includes a nonlinearity at the output end. A commonly used form is the sigmoidal nonlinearity

$$y_j = \frac{1}{1+e^{-u_j}} \qquad (13)$$

where u_j is the net internal activity level of neuron j and y_j is the output of the neuron.
(ii) The network has one or more layers of hidden neurons. The universal approximation theorem (based on Kolmogorov's superposition theorem) states that a single hidden layer is sufficient for a multilayer perceptron to approximate any nonlinear function.
The architectural graph of an MLP with two hidden layers has the form shown in Fig.10 :

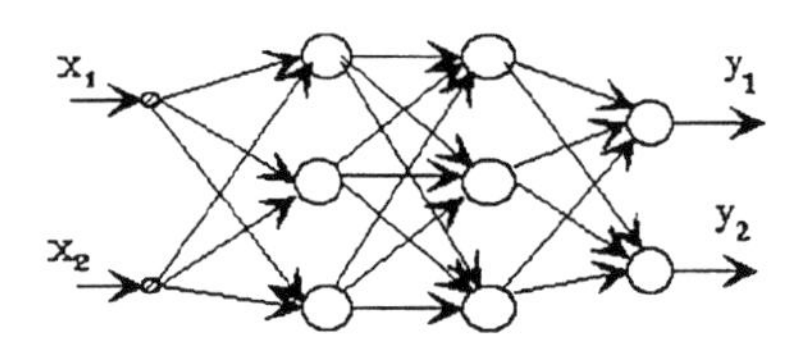

Fig.3 A multilayer perceptron structure

4.4. The Back Propagation Algorithm

The Back Propagation algorithm was first described by Werbos and was rediscovered by Rumelhart. It involves the following five steps (Haykin, 1994) :

Step 1: Initialization : Start with a reasonable network configuration and set all the synaptic weights and threshold levels of the network to small random numbers that are uniformly distributed.

Step 2 : Presentation of the training examples : Present the network with an epoch of training examples.

Step 3 : Forward computation : Let a training example be denoted by [x(n),d(n)] with the input vector x(n) applied to the input layer of sensory nodes and the desired response vector d(n) presented to the output layer of computation nodes. Compute the activation potentials and function signals of the network by proceeding forward through the network, layer by layer. The net internal activity level for neuron j and i is

$$u_j^{(l)}(n) = \sum_{i=0}^{p} w_{ji}^{(l)}(n) y_j^{(l-1)}(n) \tag{14}$$

where $y_j^{(l-1)}(n)$ is the function signal of the neuron i in the previous layer l-1 at iteration n, and $w_{ji}^{(l)}(n)$ is the synaptic weight of neuron j in layer l which is fed from neuron i in layer l-1. For i=0 we have $y_0^{(l-1)}(n) = -1$, and $w_{j0}^{(l-1)}(n) = \theta_j^{(l)}(n)$ is the threshold applied to neuron j in layer l. Using a logistic function for the sigmoidal nonlinearity, the function (output) signal of neuron j in layer l is

$$y_j^{(l)}(n) = 1 / [1 + e^{-u_j^{(l)}(n)}] \tag{15}$$

If the neuron j belongs to the first hidden layer (l=1) we set $y_j^{(l)}(n) = x_j(n)$,where $x_j(n)$ is the j-th element of the input vector **x(n)**. If neuron j belongs to the output layer (l=L) we set $y_j^{(l)}(n) = o_j(n)$. Hence we compute the error signal as

$$e_j(n) = d_j(n) - o_j(n) \tag{16}$$

where $d_j(n)$ is the j-th element of the desired response vector d(n).

Step 4: Backward computation : Compute the δ's (local gradients) of the network by proceeding backwards, layer by layer .
For neuron j in the output layer L, we have

$$\delta_j^{(L)}(n) = e_j^{(L)}(n) o_j(n)[1 - o_j(n)] \tag{17}$$

and for neuron j in the hidden layer l, we have

$$\delta_j^{(l)}(n) = y_j^{(l)}(n)[1 - y_j^{(l)}(n) \sum_k \delta_k^{(l-1)}(n) w_{kj}^{(l-1)}(n) \tag{18}$$

Thus we adjust the synaptic weights of the network nodes in layer l according to the generalized delta rule :

$$w_{ji}^{(l)}(n+1) = w_{ji}^{(l)}(n) + a[w_{ji}^{(l)}(n) - w_{ji}^{(l)}(n-1)] + \eta \delta_j^{(l)}(n) y_j^{(l-1)}(n) \tag{19}$$

where η is the learning rate parameter and α is the momentum constant.

Step 5 : Iteration : Iterate the computation by presenting new training examples to the network until the free parameters (weights) of the network stabilize their values and the average squared error is at an acceptably small value.

As it has already been mentioned, the MLP FELA technique does not wait until the network has been trained by a batch of examples, but performs the training and control simultaneously.

4.5. The Radial Basis Function Network Architecture

The RBFN is used to approximate multivariable nonlinear functions using sparse sample data. The simple RBFN consists of an input layer, a hidden layer and an output layer (Haykin , 1994). The input layer consists of a number of units clamped to the input vector. The hidden layer is composed of units which are driven by radial basis activation functions. The input units are fully connected by unit weighted links to the receptive fields in the hidden layer, and the receptive fields are fully connected by weighted links to the output units. The RBFN architecture is described in Fig. 11.

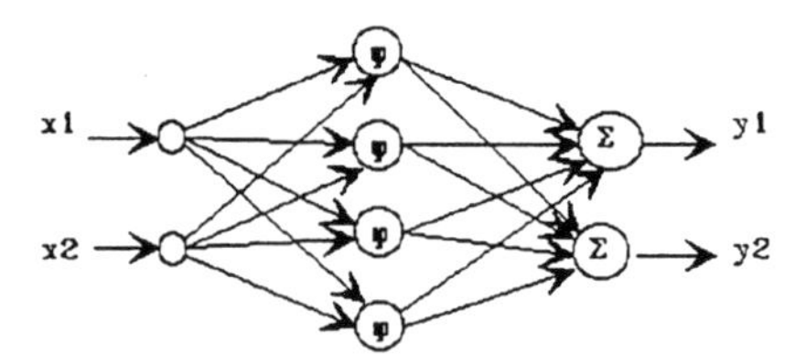

Fig. 4 RBFN Architecture

The radial basis functions used for the receptive field activation functions can take many forms. Here we use an N-dimensional Gaussian. The activation function for the receptive fields is

$$\phi_i(\mathbf{x}) = \exp(-\frac{1}{2\sigma_i}\|\mathbf{x} - \mathbf{t}_i\|^2) \qquad (20)$$

where $\mathbf{t}_i$ is the center of the i-th basis function and $\mathbf{x}$ is the input vector.

The i-th output of the network is defined by

$$y_i(x) = \sum_{j=1}^{M} w_{ji}\phi_j(x) + b_i \qquad (21)$$

where w_{ji} is the weight of the link from the j-th receptive field to the i-th output unit, b_i is a threshold term, and M is the number of hidden neurons.

4.6. RBFN Learning

There are three phases of learning in the RBFN. Learning can be performed on 1) the center locations of the receptive fields 2) the widths of the receptive fields, and 3) the layer of weights between the hidden layer and the output layer (Haykin,1994).

Training the weight terms between the receptive fields and the output units is accomplished by minimizing the sum squared error of the output layer

$$E = \frac{1}{2}\sum_{i=1}^{N}(y_{d_i} - y_i)^2 \qquad (22)$$

where N is the number of nodes at the output layer, y_{di} is the desired output of the i-th node and y_i is the real output of the i-th node.

Any optimization technique can be used to adjust the weight settings in the network (Haykin ,1991). Two common iterative optimization techniques are the Newton's method and the backpropagation algorithm. We chose to use Newton's method in our system. The weight updating equation for the weight term for the j-th receptive field to the i-th output unit is

$$w_{ji}(n+1) = w_{ji}(n) + \eta[yd_i(\mathbf{x}) - y_i(\mathbf{x})]\frac{\phi_j(\mathbf{x})\sum_{k=1}^{M}\phi_k(\mathbf{x})}{\sum_{k=1}^{M}\phi_k^2(\mathbf{x})} \qquad (23)$$

where η is the learning rate for the weight term, $y_{d_i}(\mathbf{x})$ is the target output for the unit i. The weights represent the contribution a receptive field makes to the activation level of an output unit. Thus, a large positive weight means that a receptive field plays a large role in an output unit's activation, a small weight indicates that a receptive field plays a small role in an output unit's activation, and a large negative weight indicates that a receptive field has a large inhibitory effect on an output unit's activation.

4.7. The Application of the RBFN to the Welding Control

In our application an RBFN of one input,10 hidden units with fixed centers, and one output was used. Initially the network learns the inverse dynamics of a reference plant that is assumed to have the ideal response during the welding process.

This means that the input of the network is a geometrical or thermal characteristic of the welding, and the output can be the corresponding torch velocity or the thermal power or the wirefeed rate. The RBFN neurocontroller tries to compensate on-line the disturbances of the parameters of the welding system by adapting it's weights.

The RBFN neurocontroller is used in the previously described FELA architecture. The total input that is applied to the plant is again

$$u(t) = u_i(t) + u_f(t) \qquad (24)$$

5. SIMULATION RESULTS

Here we present a set of representative results.

5.1. Application of the MLP neurocontroller

Here, reference models that are only locally linear (but globally nonlinear) are used. We examine the following input-output pairs :

a. Input Q [2.0-3.0 Kw] - Output W
b. Input Q [2.0-3.0 Kw] - Output D
c. Input V [4-6 mm/s] - Output HAZ
d. Input V [4-6 mm/s] - Output CR

The results obtained by the MLP FELA neurocontroller are shown in Figures 5 through 8.

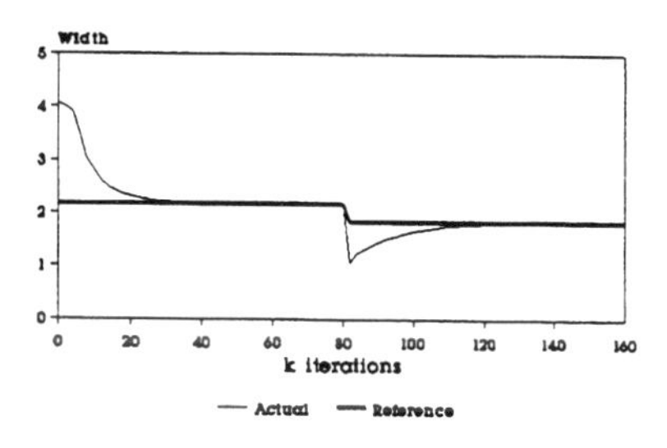

Fig. 5 : Input Q [2.0-3.0 Kw] - Output W

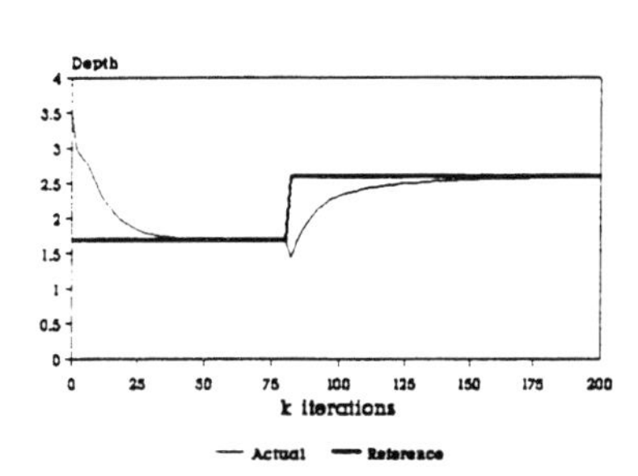

Fig. 6 : Input Q [2.0-3.0 Kw] - Output D

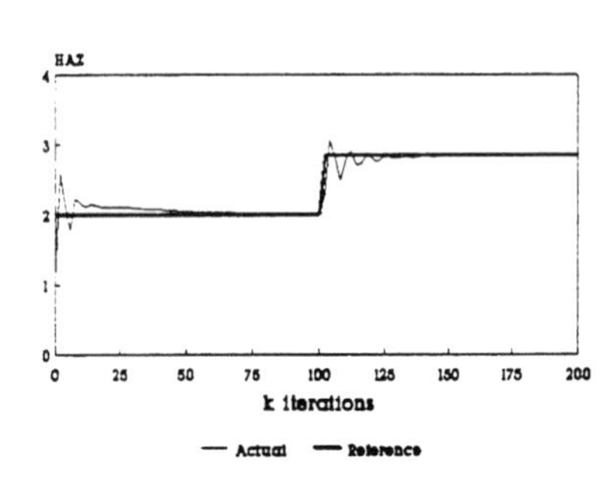

Fig. 7 : Input V [4-6 mm/s] - Output HAZ

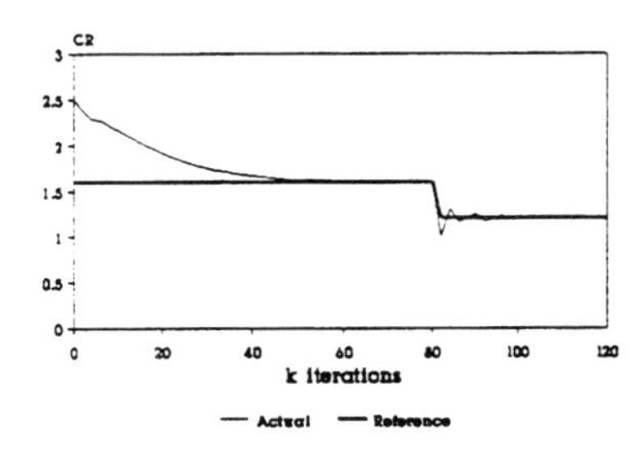

Fig. 8 : Input V [4-6 mm/s] - Output CR

5.2. Application of the RBFN neurocontroller

We examine the same input-output pairs as in the case of the MLP neurocontroller:

a. Input Q [2.0-3.0 Kw] - Output W
b. Input Q [2.0-3.0 Kw] - Output D
c. Input V [4-6 mm/s] - Output HAZ
d. Input V [4-6 mm/s] - Output CR

The results obtained by the RBFN FELA neurocontroller are shown in Figures 9 through 12.

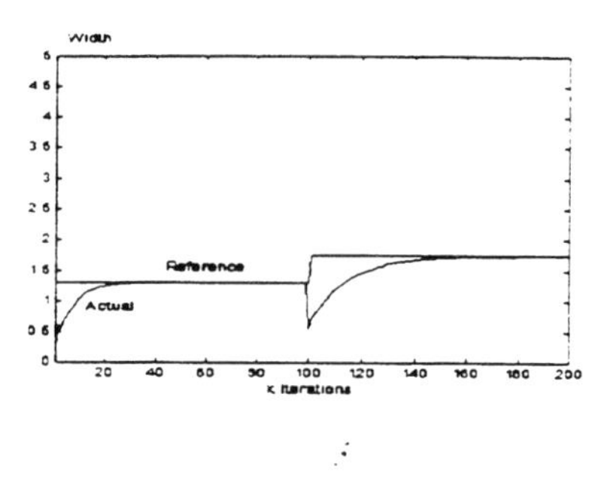

Fig. 9 : Input Q [2.0-3.0 Kw] - Output W

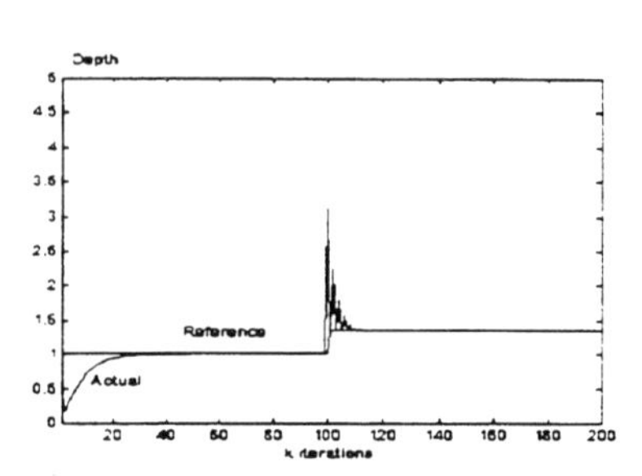

Fig. 10 : Input Q [2.0-3.0 Kw] - Output D

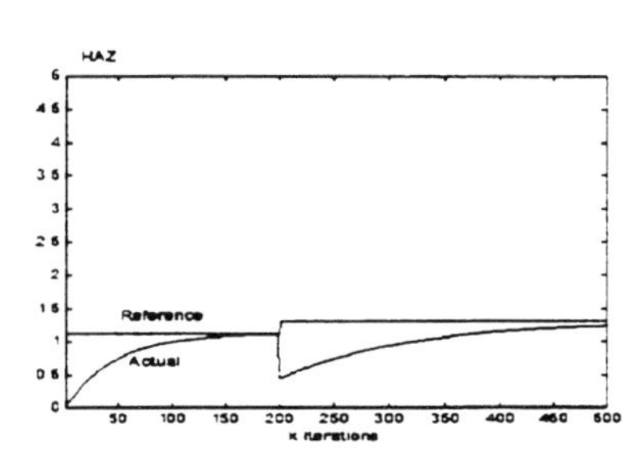

Fig. 11 : Input V [4-6 mm/s] - Output HAZ

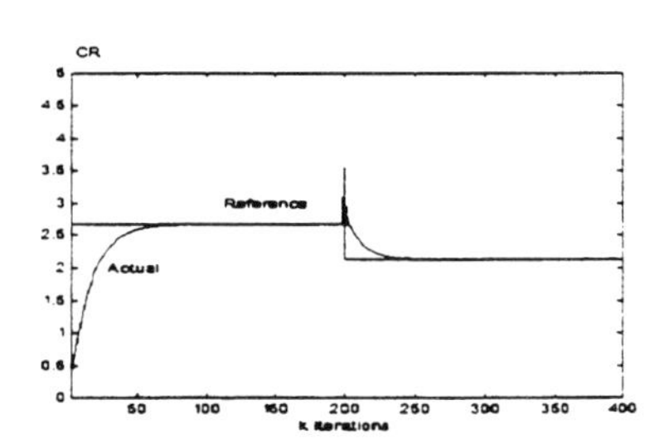

Fig. 12 : Input V [4-6 mm/s] - Output CR

6. EVALUATION OF THE IMPLEMENTED CONTROL TECHNIQUES

The desired closed-loop specification in all cases was to achieve zero steady state error while obtaining as fast transient response as possible. The present approach to the welding neural control problem is to control separately each one of the SISO systems in which the welding procedure can be analysed. This strategy was adopted for both the MLP and the RBFN neurocontrol techniques and is capable to produce a weld of high quality in some welding situations. However the complete approach to welding features regulation would be to control the process using the its MIMO model . Concequently the multivariable neural control of arc welding is a subject for further research. Our simulation results have shown the effective applicability of neural control to the welding process. This can be considered as an alternative methodology to the existing conventional adaptive control techniques.

In all simulation results for the SISO case, after 400 iterations, the steady-state of the output did not deviate from the target setpoint more than 0.001-0.01 %.

Both of the above mentioned neural control architectures have shown a remarkable ability to reject parameter disturbances of the plant.They can also handle successfully a change of the process setpoint and a simultaneous parametric perturbation of the plant.

A slight difference can be noticed between these two FELA neurocontrollers. The MLP neurocontroller treats every distinct operating state as a novel one, thus learning and control are performed simultaneously. On the other hand the RBFN neurocontroller is initially trained with some input-output pairs of the desired behavior so it exploits some of its past experiences to produce a satisfactory control signal at the beginning of the control process.

7. CONCLUSIONS

This paper has presented a unified study of the application of two neural adaptive control techniques for the regulation of the geometrical and thermal properties of the GMA welding process. After a short discussion of the basic linearized models of the geometrical and thermal characteristics, the following neural adaptive controllers were considered and applied : (i) MLP FELA neurocontroller and (ii) RBFN FELA neurocontroller.

The MLP and RBFN control methods are applied to the welding process for the first time in the present paper.

The simulation results of these controllers are encouraging and their implementation in actual robotic systems is expected to be beneficial. However, since each technique has been tested only for the SISO subsystems that represent the welding process, the selection of a neural control technique must be done with care in order to fulfill most of the requirements of a specific weld. Perhaps a hybrid control technique that appropriately combines one of the above techniques with a conventional adaptive control scheme may be proved more effective.

REFERENCES

Doumanidis , C. , Hardt ,D.E. (1989), A model for In-Process Control of Thermal Properties During Welding, *ASME J. Dynamic Syst. Meas. and Control* , **vol. 111**, pp. 40-50 .

Doumanidis ,C. , Hardt ,D.E. (1991), Multivariable Adaptive Control of Thermal Properties During Welding , *ASME J. Dynamic Syst. Meas. and Control* , **vol. 113**, pp. 82-92

Doumanidis ,C. (1994), Multiplexed and Distributed Control of Automated Welding , *IEEE Control Systems Magaz.*, pp. 13-24, Aug .

Haykin ,S. (1991), *Adaptive Filter Theory* , Prentice Hall , New Jersey

Haykin ,S. (1994), *Neural Networks : A Comprehensive Foundation* , Macmillan College Publishing Company, Toronto .

Hunt ,V.D. (1983), *Industrial Robotics Handbook*, Industrial Press Inc., New York .

Kawato , M., Furukawa ,K. and Suzuki, R. (1987), A Hierarchical Neural Network Model for Control and Learning of Voluntary Movement , *Biol. Cybern.*, **Vol. 57**, pp. 169-185.

Kawato ,M., Uno ,Y., Isobe ,M., and Suzuki , R. (1988), Hierarchical Neural Network Model for Voluntary Movement with Application to Robotics , *IEEE Control Systems Magaz.* , pp. 8-16, April .

Lightbody ,G. , Irwin ,G.W. (1995), Direct Neural Model Reference Adaptive Control , *IEE Proc.-Control Theory and Appl.*, **vol. 142**, pp. 31-43.

Nishar, D.V., Schiano J.L., Perkins, W.R., and Weber, R.A., (1994), Adaptive Control of Temperature in Arc Welding, *IEEE Control Systems*, pp.4-12, Aug.

Tzafestas , S.G. (ed.) (1991), *Intelligent Robotic Systems*, Marcel Dekker, New York .

Tzafestas , S.G. , Verbruggen , H.B. (eds.) (1995a), *Artificial Intelligence in Industrial Decision Making, Control and Automation* , Kluwer Academic Publishers , Dordrecht/Boston .

Tzafestas ,S.G. (1995b), *Neural Networks in Robot Control, In : Artificial Intelligence in Industrial Decision Making , Control and Automation* , Kluwer, Dordrecht/Boston , pp. 327-387 .

MODELING OF THE LEAVENING PROCESS IN A CONTINUOS BREAD-MAKING INDUSTRIAL PLANT USING NEURAL NETWORKS

A.Ferroni(*), M. L. Fravolini(+), A. Ficola(+), M. La Cava(+)

(*) *Colussi Perugia S.r.l. Via dell'Aeroporto, Petrignano di Assisi Perugia (Italy)*
(+) *Istituto di Elettronica - Università di Perugia, Via Duranti - Perugia (Italy)*

Abstract: In this paper we propose a model of the leavening process in an industrial plant for bread-making production. The model is implemented by means of a mix of neural networks and linear ARX systems. The model can be used by a human operator to predict the effect of manual control action; furthermore, it could be used to implement a predictive automatic control in order to achieve a certain mean quality of the products. The identification has been carried out by means of measurements on the plant and on experiments in laboratory; in operating mode the model needs of some measurements which can be easily acquired from the plant. The good fitting of the results shows that the proposed architecture is well suited for the considered plant.

Keywords: Plants, Models, Neural Networks, System Identification.

1. INTRODUCTION

The industrial production of yeast leavened food is one of the most widespread industrial process all over the world. This process is carried out by a series of well defined working phases. After mixing the recipe ingredients with a mixing machinery, the product is kneaded, partitioned and shaped in loaves, which are subsequently inserted inside a proofing chamber. Once the desired volume is achieved by the dough, the proofing is stopped introducing the products in the oven for the baking; the last phases are conditioning and, for some productions, cutting and toasting.

In a continuos production plant all these processes are linked together to generate a complex processing line. An efficient conduction of such a plant should be able to guarantee both high quality products and few production discards. In order to achieve these results, a careful control of the product quality in each working phase must be effected. It is well known that the most critical phase is represented by the proofing process inside the leavening chamber. [1]During this phase the CO_2 delivered inside the dough causes the product to rise and a porous structure is created. A bad leavening phase can generate a lack of quality of the final product. For each kind of recipe many factors influence the proofing process: temperature and humidity, proofing time, quality and quantity of yeast, etc.. For a large class of industrial productions proofing takes place inside a large number of closed metallic moulds, carried inside the proofing chamber by a conveyor belt, where both temperature and humidity are controlled.

A rich literature exists about proofing process and leavening chambers, testifying a great research interest in order to achieve good proofing results. Pinter (1988), Maklyukov and Puchkova (1983) and Yurchak and Kishen'ko (1984) have studied extensively the relations between the porosity of the finished bread and the values of temperature, humidity and proofing time. They derived a set of regression equations which can determine optimum set-points. Dixon and Kell (1989) and Kuznetsov and Vasin (1989) have investigated how the temperature and the humidity inside the leavening chamber influence the rate of production of CO_2

(*)To whom the correspondence should be addressed

and the rheological characteristics of the dough. Chernikh and Puchkova (1983) proposed to measure the porosity by means of the cooling rate of a proofing dough subjected to a pulse of electrical heating.
All these models have been developed on the base of measurements carried out on laboratory chambers, in which environmental variables are easily controllable and measurable with a high precision degree; moreover, not appreciable external disturbances exist.
From an industrial point of view, it is not easy to regulate the temperature and humidity of each zone of the leavening chamber to a prescribed set point value, because of environmental disturbances; the most important ones are seasonal and daily temperature variations inside the factory and the initial mean temperature of the metallic moulds inside which proofing takes place. These disturbances, even if of small amplitude, can generate unacceptable variations in the final volume, affecting the overall quality of the product. To reduce the effects of the disturbances a control scheme must be introduced; for this reason it is necessary to have an efficient model which relates the temperature and humidity fields inside the chamber (input variables) to the rising of the loaves inside the moulds (output variable).
Since in real plants it is very difficult to get direct measures from the moving moulds, it is virtually impossible to apply an on-line feedback control based on measures on the proofing loaves; moreover feedback control could be not effective because the dynamics involved in the thermal exchange between air, moulds and loaves is very slow compared with the usual proofing times.
Currently, for the lack of information on the state of the rising loaves, the control of this process is manually performed by a skilled operator, who, in order to reduce the effects of the disturbances, uses his experience and intuition varying the amount of yeast in the recipe and the permanence time of the moulds inside the leavening room. The temperature of the chamber is usually set to a fixed value, while the temperature field into the loaves cannot be directly controlled. The effects of any control action can be evaluated only at the end of the proofing process, when the moulds come outside the chamber. Such control strategy involves long time delays (some hours); therefore, the classic try and error correction scheme could be inefficient for continuos high production industrial plants. For these reasons a model which could predict the effect of the corrections on the final product would be very useful for the operator, carrying out a sort of predictive control.
In the present work we develop a model which describes the rising process of the loaves inside the metallic moulds on the base of the temperatures field of the leavening chamber. The relations between these variables involve biochemical, mechanical and thermodynamic processes, which are non linear phenomena hardly to be modeled by mathematical differential equations; therefore the model we propose has been developed using black-box structures (Sjoberg et al., 1995) implemented by means of Neural Networks and linear ARX blocks (Ljung, 1987, 1994; Chen and Billings, 1992).

2. MODEL STRUCTURE

A model useful to achieve the predictive control for a such industrial process must be able to give the correct rising of the volume of the loaves inside the moulds moving in the leavening room; moreover it should use easily measurable input variables. The output variable which best characterize the quality of the product is the evolution of the volume of the loaf (Yurchak and Kishen'ko, 1984). The aim is to predict the effects of the disturbances and of the control inputs (duration and yeast quantity) on the volume. The volume is mainly influenced by two groups of variables and disturbances: the dough composition and the environmental variables.
The first group includes the proportions of ingredients used in the recipe, the characteristics of the ingredients (for example the strength of the flour), the energy supplied during kneading, etc. During the production of a specific product the recipe and the kneading time are constant and the ingredients are strictly selected and controlled by means of laboratory analysis; for this reason we can suppose that it is sufficient to consider only one variable parameter of the first group: the quantity of yeast in the dough.
About the second group, literature reports that the environmental variables which influence the process are the relative humidity and the temperature of the dough (Quaglia, 1990); however for productions in closed moulds, the relative humidity inside them is slightly influenced by that of the chamber. The temperature field in the loaves is determined by the temperature of the moulds, which is related with the temperatures found by the moulds during their trip inside the leavening chamber.
On the industrial plant it is easier to get continuous measurements of temperatures in fixed points inside the room than to get directly the temperatures of the moving moulds; therefore the proposed architecture of the model (Fig. 1) is based on two main blocks representing respectively the model of the leavening room and the model of the proofing process inside the closed moulds.
The first block determines the time evolution of the temperature $TM(k)$ of the mould in function of the temperature field inside the leavening room and on the speed of the moving mould itself. This speed can be varied by the operator, defining the total transit time (t_pr) of the mould through the chamber;

besides the evolution of $TM(k)$ depends on the initial value TMi of the temperature of the moulds.

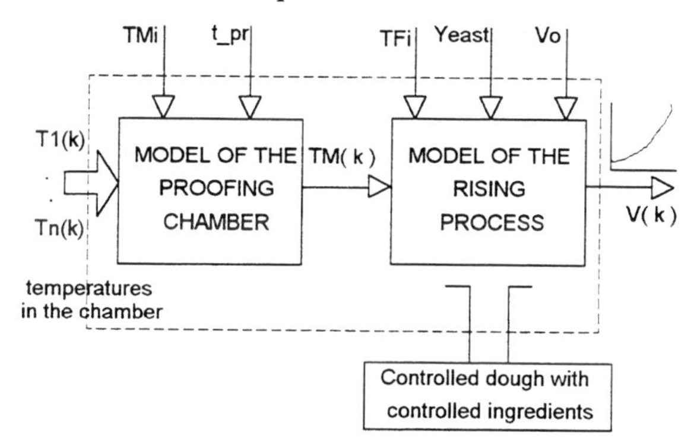

Fig. 1 - Structure of the model.

The second block provides the temporal evolution of the volume V of the loaves in function of the evolution of the mould temperature $TM(k)$, the initial quantity of the yeast (*Yeast*), the initial temperature TFi and the initial volume Vo of the loafs. Since the effective volume of a loaf cannot be easily measured, in our analysis it is replaced with the measure of the thickness of the loaf itself, being volume and thickness strictly related; therefore in the following the terms volume and thickness are considered equivalent.

In all the subsequent considerations a sampling time of 30 seconds was considered as a good choice for the dynamics involved.

3. MODEL OF THE PROOFING CHAMBER

A complete thermodynamic model for the determination of the temperature field inside the chamber must consider the effects of the conditioning system and all the involved thermal exchanges. The regulation of the temperatures of the chamber is carried out by a cooling-heating system which causes also a quasi stationary air flows in the chamber. The temperature field of the room has been characterized by means of the temperatures measured by eight fixed thermometers (see $T1,\ldots,T8$ in Fig. 2).

The relationship between the temperature field in the chamber and the temperature of a generic moving mould has been determined by means of the architecture reported in Fig. 3. The scheme is composed of a static Neural Network which computes the temperature $TA(k)$ of the air near a mould in function of its position in the chamber ($x(k)$ and $y(k)$) and of the eight temperatures $T1(k),\ldots,T8(k)$ of the fixed points. The mould temperature $TM(k)$ in function of $TA(k)$ is obtained by means of a linear dynamic model, which describes the thermodynamic behavior of the mould itself.

This scheme was based on the following considerations:

- the first process (from $T1,\ldots,T8$ to TA) is highly non linear; its dynamic is negligible, due to the fast flows of the air inside the chamber; therefore an instantaneous Neural Network represents a good choice to implement it;
- the second process is very slow compared with the first one and it must represents the thermal exchange; it can be supposed linear for the ranges of the involved temperature and therefore a linear dynamic model can be considered well suited.

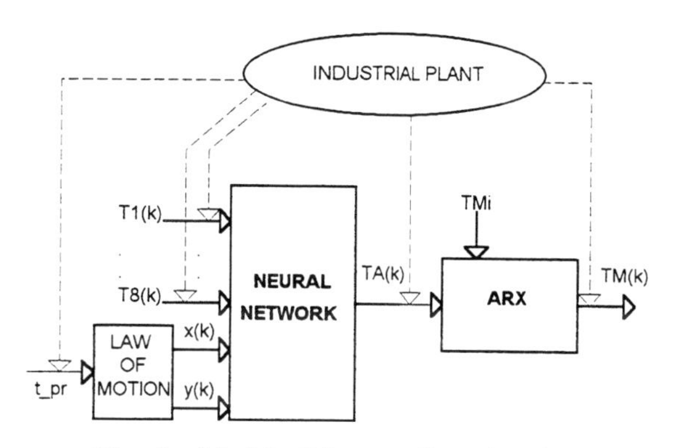

Fig. 3 - Model of the proofing chamber.

The parameters of the blocks were identified by means of measurements carried out directly on the industrial plant. Data loggers were fixed to some moulds in order to get the temperatures $TA(k)$ of the air around the mould and the temperature $TM(k)$ of the mould itself. The data loggers were synchronized with the acquisitions of the fixed

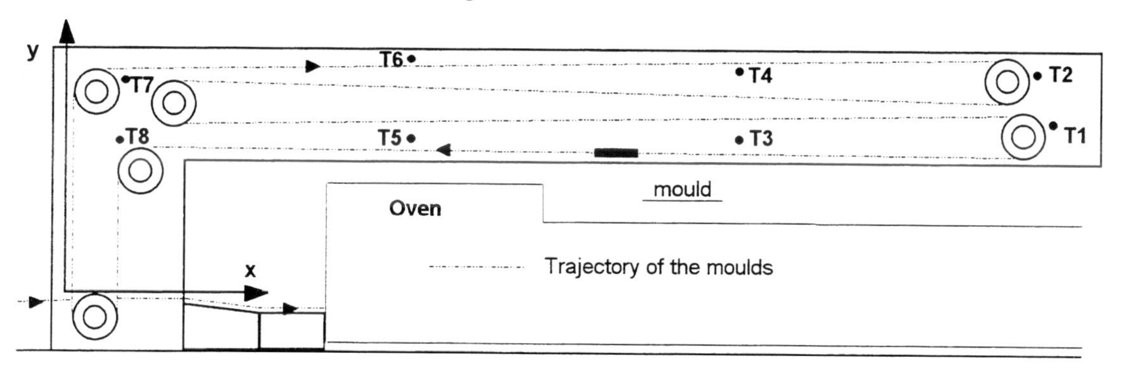

Fig. 2 - Leavening chamber with the positions of the thermometers.

thermometers *T1,...,T8* and a rich data set was acquired.

The Neural Network was trained with an input data set composed of the eight temperature profiles of the fixed thermometers and the position of the mould (Fig. 4); the target set is the air temperatures *TA* provided by the data loggers. We used a Feed Forward NN with ten inputs, two hidden layers with ten sigmoidal neurons and an output layer with one linear neuron. A batch Levenberg-Marquardt training algorithm was used;

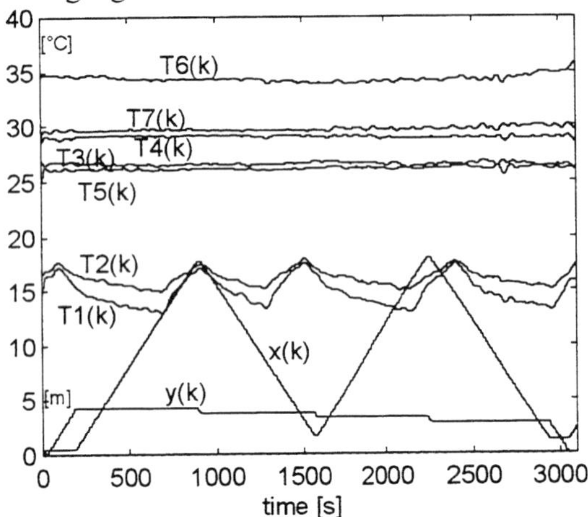

Fig. 4 - Example of input data for the NN block.

this block can be represented by means of the following expressions:

$$[x(k), y(k)] = g(t\_pr, k) \quad (1)$$

$$TA(k) = f(T_1(k), \cdots, T_8(k), x(k), y(k)) \quad (2)$$

The thermodynamic model of the mould has been identified using a linear ARX model. The regressor vector is composed of the air and mould temperatures *TA* and *TM*, measured by the data loggers (Fig. 5). The structure of the ARX model is the following:

$$TM(k) + a_1 TM(k-1) = b_0 TA(k) + b_1 TA(k-1) + b_2 TA(k-2) + \\ + b_3 TA(k-3) + b_4 TA(k-4) + b_5 TA(k-5) + e(k) \quad (3)$$

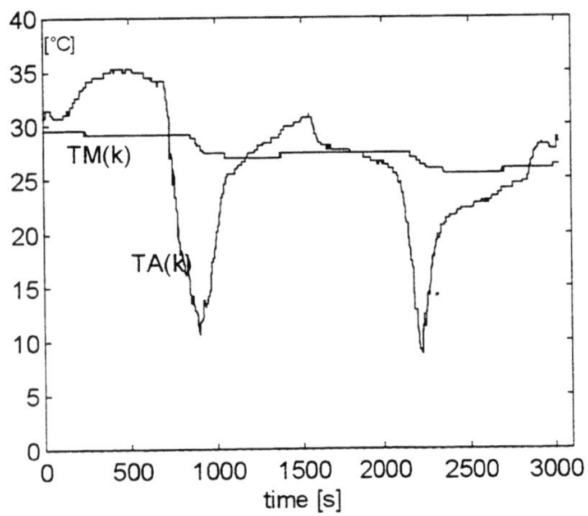

Fig. 5 - Input and output of the ARX block.

4. MODEL OF THE RISING PROCESS

The second model is composed of two blocks (Fig. 6). The first one evaluates the temperature *TF(k)* in the center of the leavening loaf basing on the temperature of the mould *TM(k)*; the second one computes the volume rising of the dough *V(k)* in function of both temperatures.

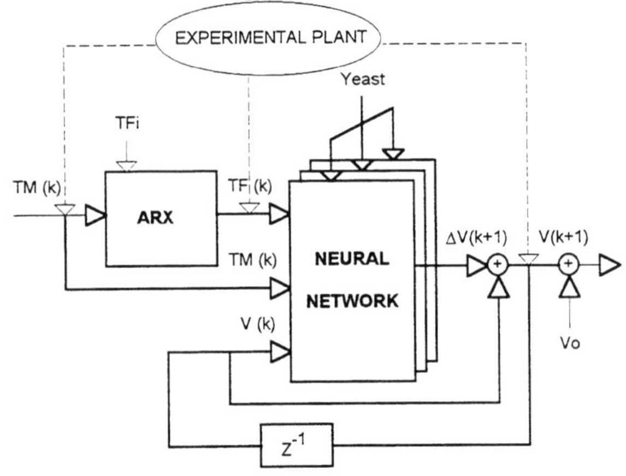

Fig. 6 - Structure of the model of the rising process.

This scheme was based on the following considerations:

- the first block models the thermodynamics of the loaf; in particular it should relate the temperature of the metallic mould *TM(k)* with the field of temperature inside the loaf; due to the small dimension of the loaf and the slow dynamic of the input, we reduce the complexity of the model using only two temperatures: *TM(k)*, which represents also the surface temperature of the loaf, and *TF(k)*, which is the temperature in the middle of the dough; being this process quasi-linear and dynamic we decided to model it as an ARX block;
- the second block, which computes the rising of the loaf in function of the temperature field, is time dependent and non linear; therefore it is necessary to characterize it by means of a non linear dynamic model obtained using a bank of tree-layer Neural Networks with external feedback. Each network is used for a certain quantity of *Yeast* and computes the instantaneous rising speed as a function of the temperature field in the loaf and of the actual volume:

$$\dot{V} = \Im(TM, TF, V, Yeast) \quad (4)$$

The integration of the rising speed produces the volume evolution of the loaf that is used as a feedback of the Network itself.

To get the data for the identification of such blocks it was necessary to reproduce the proofing process in a laboratory plant. Inside a small climatic controlled chamber it was possible to use low friction sensors able to get measurements of the volume rising (Fig. 7).

The results of the experiments for two different temperatures laws are displayed in Fig. 8, where the input temperatures and the resulting volumes are reported.

The ARX model is

$$\begin{aligned} TF(k) + a_1 TF(k-1) + a_2 TF(k-2) = \\ = b_1 TM(k-1) + b_2 TM(k-2) + e(k) \end{aligned} \tag{5}$$

The neural network was trained with an input data set composed of the measures of *TM(k)*, *TF(k)* and the actual volume *V(k)*; the target was the variation $\dot{V}(k)$ of the volume; therefore the network is trained to work as an one-step predictor:

$$\begin{aligned} \hat{V}(k+1) = V(k) + \tau \cdot f(V(k), TM(k), TF(k)) \\ \hat{\dot{V}}(k) = f(V(k), TM(k), TF(k)) = \left[\hat{V}(k+1) - V(k)\right]/\tau \end{aligned} \tag{6}$$

A skilful training can achieve a small enough prediction error: $|\hat{V}(k) - V(k)| < \varepsilon; \quad \varepsilon > 0$.

Unfortunately, during the current operation the actual volume *V(k)* is not available; therefore we use as input the estimation $\hat{V}(k)$, which is feedback to the algorithm; so, the system works as a parallel model:

$$\begin{aligned} \hat{V}(k+1) = \hat{V}(k) + \tau \cdot f(\hat{V}(k), TM(k), TF(k)) \\ \hat{\dot{V}}(k) = f(\hat{V}(k), TM(k), TF(k)) = \left[\hat{V}(k+1) - \hat{V}(k)\right]/\tau \end{aligned} \tag{7}$$

It is well known that this scheme does not guarantee that the estimation error is bounded by an arbitrary small positive number (Sjoberg et al. 1995; Narendra and Levin, 1996; Piuri and Alippi, 1996); however, in our case, due to the slow dynamics and to the limited time horizon during which the process must be observed, we could expect a sufficiently correct estimation.

We used a Feed Forward NN with three inputs, one hidden layer with ten sigmoidal neurons and an output layer with a linear neuron. A batch Levenberg-Marquardt training algorithm was used.

Some test where carried out to validate this assumption, using some validation data sets. Some results are reported in the following figures (Fig. 9), where the estimation $\hat{V}$ of the volume is compared with the actual measured volume *V*.

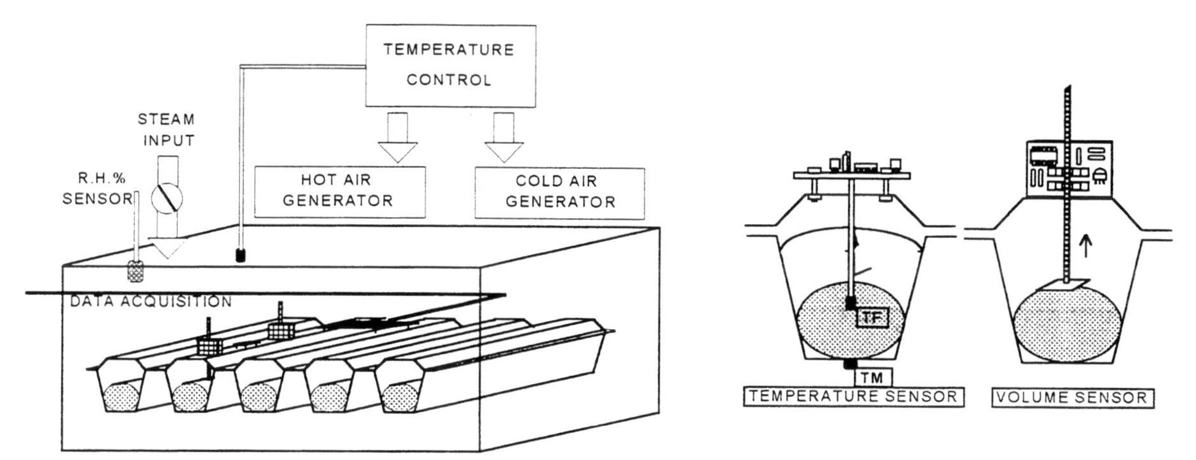

Fig. 7 - Laboratory plant and sensors.

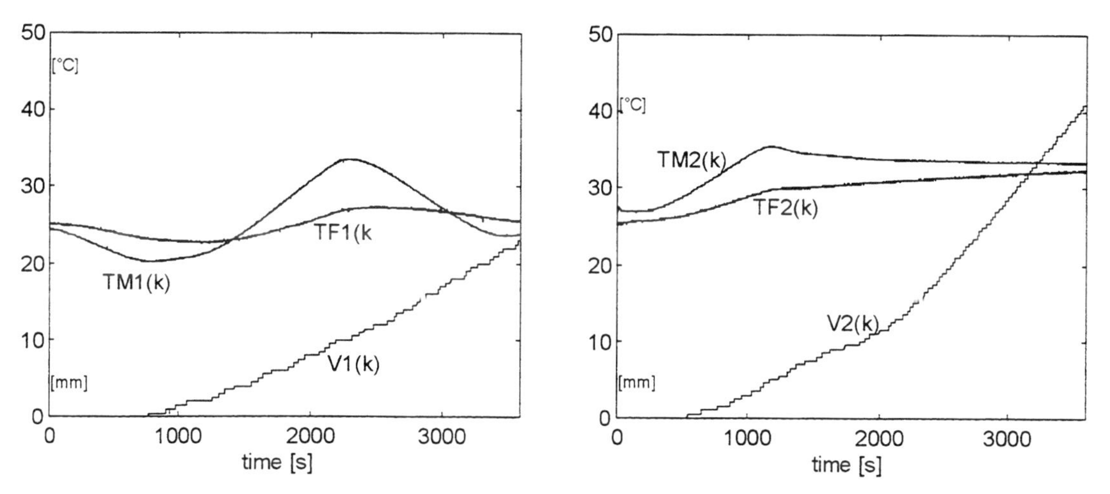

Fig. 8 - Volume rising for two different temperature laws.

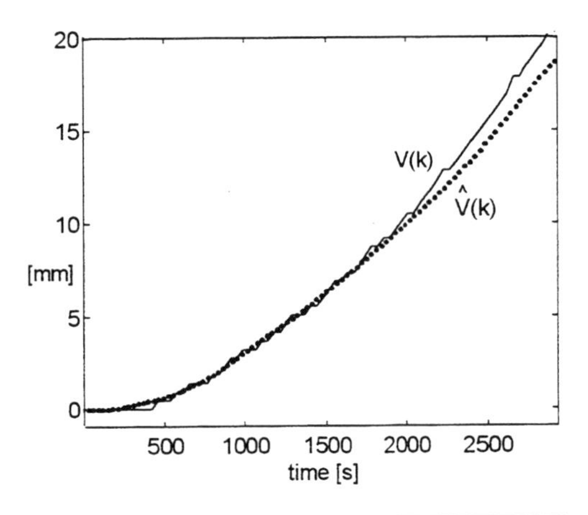

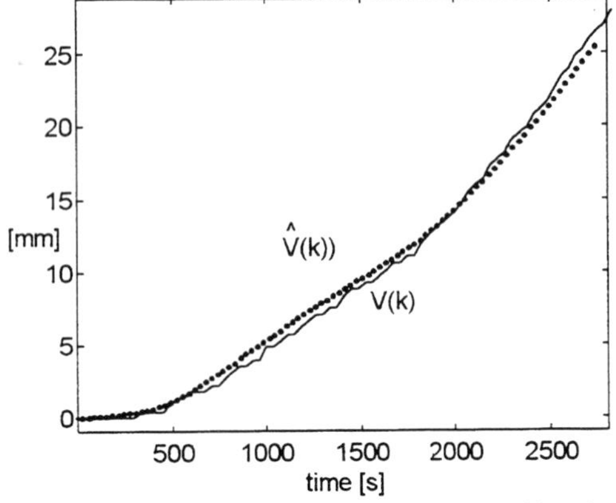

Fig. 9 - Comparison between the predicted and measured volume.

5. CONCLUSIONS

In this paper we have presented a model of the proofing process in an industrial plant. The model needs of some measurements which can be easily acquired from the plant in order to evaluate the rising evolution of the dough.

The model can be used by a human operator to predict the effect of manual control actions ; furthermore, it can be employed to implement a predictive automatic control in order to achieve a certain mean quality of the products.

The model has been implemented by means of a mix of neural networks and linear ARX systems.

The identification has been carried out by means of measurements on the plant and of experiments in laboratory. The good fitting of the results show that the proposed architecture is well suited for the considered plant.

Further developments are concerned with the improvement of the model, in order to predict not only the final volume of the dough, but also the final quality of the loaves.

ACKNOWLEDGEMENTS

This work was supported by the *Colussi Perugia S.r.l.* The authors wish to thank Dr.Ing. Valerio Capodaglio and Dr. Federico Ghirelli for the useful suggestions and discussions.

REFERENCES

Chen S. and Billings S. (1992), Neural Networks for nonlinear dynamic system modelling and identification, *Int. Jour. Control,* **56**, 319-346.

Chernikh V. and Puchkova L. (1983), Automatic control of leavening during proofing, *Baecker und Konditioner*, **31**,(**7**), 215-217.

Dixon N. and Kell D. (1989), The control and measurement of CO2 during fermentations, *Journal of microbiological methods,* **10**(**3**), 155-176.

Kuznetsov Y. And Vasin M. (1989), Measuring the temperature and relative humidity regimes in the proofing chamber, *Klebopekarnaya i Konditerskaya promishlennost'*, **6**, 29-30.

Ljung L. (1987). *System Identification : Theory for the user*, Prentice Hall, Englewood Cliffs, NJ.

Ljung L. and Glad T. (1994), *Modelling of dynamic systems*, Prentice Hall, Englewood Cliffs, NJ.

Maklyukov V. and Puchkova L. (1983), Effects of proofing conditions on the quality of finished bread. *Klebopekarnaya i Konditerskaya promishlennost'*, **3**, 28-30.

Narendra K. S. and Levin A. U. (1996), Control of nonlinear dynamical systems using neural networks - part II : obsevability, identification and control, *IEEE Trans. Neural Networks*, **7**(**1**).

Pinter J. (1988), In line control in the baking industry based on instrumental measurements. *Elelmezesi Ipar*, **42** (**8**), 294-299.

Piuri V. and Alippi C. (1996), Experimental neural networks for prediction and identification, *IEEE Trans Instrumentation and Measurement*, **45**(**2**).

Quaglia (1990), *Scienza e tecnologia della panificazione*, Chiriotti Editore, Pinerolo.

Sjoberg J., Zhang Q., Ljung L. et alii (1995), Nonlinear black-box modeling in system identification: a unified overview, *Automatica*, **31**(**12**).

Yurchak V. and Kishen'ko V. (1984), Assessment of the effectiveness of the control of dough production processes, *Izvestiya Veysshikh Uchebnykh Zavedenii Pishchevaya Tekhnologiya*, **4**, 60-62.

MARKOV PROCESSES APPLICATION FOR ESTIMATION OF PARAMETERS OF ASSEMBLY LINES

Jan Kałuski

*The Silesian Technical University,
Department of Automatic Control,
44-100 Gliwice, Poland*

Abstract: In the paper consider an industrial discrete process which includes a system of two assembly lines with a semi-products store placed between the lines. The number of semi-products in the store during each cycle-time depends on the random parameters of input and output lines. Only the randomness of the input lines is considered. The output of the system is deterministic and its intensity is constant. Under the conditions mentioned above a mathematical model of discrete process in the form of a Markov chain in discrete time with a countable number of states is presented.

Keywords: Markov models, approximate diffusion, assembly line, Fokker - Planck equation.

1. INTRODUCTION

Industrial discrete processes are distingushed by the fact that, as complexes of operations in space and in time on condition of indivisibility of processing single operation, by their nature they are described by mathematical models in the form of logically determined mathematical relationships between operations parameters and parameters of work stations (machining stations, assembly lines, transport systems or all of them simultaneously) in discrete moments of time. As a result, any discrete processis are specified with accuracy to a single operation of various types and time accuracy to determined time interval called a cycle.

Another characteristic feature of an industrial discrete process is mass character of homogenous products being the output of particular technological processes.

It is a large number of ready homogenous products or semi-products that in some cases allows us to describe discrete industrial processes in the form of discrete or continuous mathematical models. One of the possible approaches is employing discrete Markov processes , aiming at probabilistic estimation of some parameters of manufacturing processes.

Taking the above into account, in section 2 of this work a discrete Markov chain with countable number of states is presented as a model of number of semi-products in the store between two assembly line syste.....ms. In section 3 diffusion approximation of number of semi-products in the store is given. Secton 4 contains some final conclusions.

2. DISKRETE MARKOV CHAIN AS A MODEL OF A NUMBER OF DETAILS IN THE STORE

Let us consider a system of two assembly lines with a store for semi-products placed between them, as shown in Fig.1.

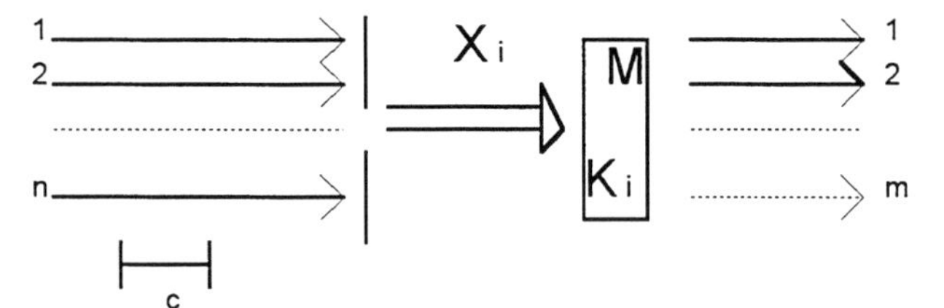

Fig.1. System of assembly lines with a store

In this system n- identical assembly lines provides synchronously a certain number X of semi-products with the time interval c equal to a line cycle. They are consequently stocked in the store **M** of unlimited capacity. Thus it can be said that in the moment $t=ci\cdot$, $i \geq 0$, X_i well assembled products are ready. With the same time cycle c, $m<n$ semi-products are being taken from the store **M** for further assembling on m lines X_i is a random variable which depends on defectiveness of n - assembly lines and their reliability. It is easy to notice that X_i is a discrete random variable whose value is from $[0, n]$ range, i.e.

$$P\{X_i = k\} = p_k, \quad k = 0,1,2,\dots,n \tag{1}$$

It is obvious that after the time interval $t=i\cdot c$, $i\geq 0$, depending on lines system parameters n, m, c and X_i, the number K_i of semi-products remaining in the store is also a random variable and can take any value $K_i = 0,1,2,\dots$. From Fig.1. it is clear that for K_i the following recurrent relationship is satisfied:

$$K_{i+1} = \max\{K_i + X_{i+1} - m, 0\} \tag{2}$$

Assuming that process of defects arising and reliability of one line is independent of the faults on other lines and reliability, it can be proved that the sequence $\{X_i\}$, $i\geq 0$ is a sequence of independent random variables. This in turn implies that the sequence $\{K_i\}$, $i\geq 0$ is a discrete Markov chain with a countable number of states.

On the basis of a relationship (2) and the Bernoully distributions elementary consideration it can be proved that the elements of the one-step probability matrix of transitions for such a chain for determined m, n $(m<n)$ and r, $s=0,1,2,\dots$ given (3).

$$p_{r,s}^{m,n} = \begin{cases} p^{s-r+m} \, q^{n-s+r-m} \binom{n}{s-r+m}; & r-m \leq s \leq r-m+n \\ 0 & \text{otherwise} \end{cases} \tag{3}$$

In (Kimmel, 1982), analyzing a similar assembly lines system and assuming Bernoulli distribution of $\{P_k\}$ with parameters np and $np(1-p)$, where p is probability of correct assembly on every of n lines, it is shown that there exists stationary distribution function of a chain $\{K_i\}$, $i\geq 0$ in two cases: 1) when $m=1$, $p=1/n$ and 2) when $m=1$, $p=2/3$.

Hitherto the author has not found any works which prove that those particular cases can be generalized for any m and n and $p<m/n$, where X_i is of Bernoulli distribution or other distribution function.

In the following sections diffusion approximation of the aforementioned problem will be considered, i.e. there will be shown an approximation using continuous diffusion-type Markov processes, and subsequently a method of estimating number K_i of semi-products in the store for any m, $n>0$ $(m<n)$ by means of these processes.

3. DIFFUSION APPROXIMATION OF NUMBER OF SEMI-PRODUCTS IN THE STORE

Now let us see the diffusion process with which the number K_i of semi-products in the store can be approximated.

Let intensity of increase and decrease of the stock in the store (the average number of increase and decrease of the stock in the time unit in relation to the stosk unit) depend on general number $K(t)$ of semi-products in the store (this number determines if there are enough semi-products in the store for supporting m output lines), equal to $\lambda(k)$ and $\mu(k)$ respectively. It can be assumed that $K(t)\cong k$ for small time intervals. Hence λ and μ are approximately constant, too, and are independent of time. In our case $\mu(k)$ is also independent of k, i.e. $\mu(k)=m$, because it is constant decrease of semi-products stock in the store in every cycle.To make approximation close to reality, it should be assumed that number of stock increase in a short time interval, in one cycle, is subject to Poisson distribution with parameters $k\lambda(k)$ (in Poisson process -λt).

Let us calculate the expected value $E_k[K(t)-k]$ and variance $V_k[K(t)-k]$ of random variable $(K(t)-k)$. Regarding that $K(t)$ has Poisson distribution with expected value $k\lambda(k)$ and equal variance and that $k\cong mt$ for $t=ic$, $i\geq 0$, we obtain:

$$E_k[K(t)-k] \cong [k\lambda(k) - m]\cdot t, \tag{4}$$

and

$$V_k[K(t)-k] \cong V_k[K(t)] = k\lambda(k)\cdot t. \tag{5}$$

Thus we obtain that the transfer coefficient of the generated diffusion process is equal to

$$b(k)=k\lambda(k) - m \tag{6}$$

and the diffusion coefficient for this process is as follows:

$$a(k)=k\lambda(k) \tag{7}$$

Corresponding differential opereator L of the considered diffusional process is as follows (Schuss, 1980):

$$Lf(k) = \frac{1}{2}k\lambda(k)\frac{\partial^2 f(k)}{\partial k^2} + [k\lambda(k) - m]\frac{\partial f(k)}{\partial k}. \tag{8}$$

Now let us introduce function $f(k) = P_m(t,k,s,r) = P_m(K_t = k / K_s = r)$, expressing probability of continuous time Markov process transition from the state $K_s=r$ in the moment s to the state $K_t=k$ in the moment t, $t>s$, when m is definite.

It can be shown that this probability can be found as a solution to the Cauchy boundary problem

$$\frac{\partial P_m(t,k)}{\partial t} = LP_m(t,k) \tag{9}$$

for initial conditions $P_m(0,k)=1$ for $k \leq 0$ and $k<m$.
Thus the diffusion equation which relates number K of semi-products in the store for the assembly lines system takes the following form:

$$\frac{\partial P_m(t,k)}{\partial t} = \frac{1}{2}k\lambda(k)\frac{\partial^2 P_m(t,k)}{\partial k^2} + [k\lambda(k) - m]\frac{\partial P_m(t,k)}{\partial k}. \tag{10}$$

Further on we will consider three particular cases:
1) $k\lambda(k)=m$, $m=1$; 2) $k\lambda(k)=m$, $m>1$; 3) $k\lambda(k)\neq m$, $m>1$.
In the first case equation (10) is reduced directly to the heat equation

$$\frac{\partial P(t,k)}{\partial t} = \frac{1}{2}\frac{\partial^2 P(t,k)}{\partial k^2}, \quad t>0, \quad k<+\infty \tag{11}$$

and $P(0,k)=\varphi(k)$.
The solution to the equation (11) is, of course, as follows:

$$P(t,k) = \frac{1}{\sqrt{2\pi t}}\int_{-\infty}^{\infty} \exp(-(k-r)^2/2t)\varphi(r)dr, \tag{12}$$

where φ is tentative function (Schuss, 1980).

Let us consider the diffusion process on the semi-axis $k \geq 0$ with reflection in zero.

If the transition probability $P_1(k,t)$ satisfied the equation (11) for $m=1$, then its distribution function, (denoted by $g(k,t)$), should satisfy similar heat equation in the form of:

$$\frac{\partial g(k,t)}{\partial t} = \frac{1}{2}\frac{\partial^2 g(k,t)}{\partial k^2}, \quad k\rangle 0, \tag{13}$$

where

$$P(k,s,B,t) = \int_B g(k,s,r,t)dr, \quad r \in B \tag{14}$$

for boundary-initial values $\frac{\partial g(k,t)}{\partial k}/_{k=0} = 0$, $g(k,s,r,t) \to \delta(k-r)$, $when\ t \downarrow 0$, $k,r < +\infty$.
The solution to (13) is a distribution function (see e.g. Dinkin and Yushkevich (1970))

$$g(k,s,r,t) = \frac{1}{\sqrt{2\pi t}}\left[\exp(-{(k-r)^2}/{2t}) + \exp(-{(k+r)^2}/{2t})\right]$$

(15)

This is the distribution function of the process $\{K_t\}$, $t \geq 0$, transition from the state $K_s=r$ in the moment s to the state $K_t=k$ in the moment t, with the shield reflecting in zero.

Now let us consider the case when $k\lambda(k)=m$, $m>1$.
Then, the equation (13), according to the equation (10) takes the following form:

$$\frac{\partial g_m(t,k)}{\partial t} = \frac{1}{2}m\frac{\partial^2 g_m(t,k)}{\partial k^2}. \tag{16}$$

It is known (see e.g. Budak and Samarski (1965)), that the former boundary problem with the same boundary - initial value conditions as before in its general form is as follows:

$$\frac{\partial g}{\partial t} = a^2\frac{\partial^2 g}{\partial k^2}. \tag{17}$$

The solution to (17) is the following distribution function:

$$g(x,t,r) = \frac{1}{2a\sqrt{\pi t}}\left[\exp(-\frac{(x-r)^2}{4a^2t} + \exp(-\frac{(x+r)^2}{4a^2t})\right]. \quad (18)$$

Hence, the solution to the eqation (16) is distribution function

$$g_m(k,t,r) = \frac{1}{\sqrt{2\pi tm}}\left[\exp(-\frac{(k-r)^2}{2mt}) + \exp(-\frac{(k+r)^2}{2mt}\right], \quad k\langle 0, \quad r\langle +\infty$$
(19)

for $a^2 = \frac{1}{2}m$.

Finally, without any detailed explanations, let us point out, that if $k\lambda(k)\neq m$, then from the equation (10) the following can be written for distribution $g_{m,\lambda}(k,t)$:

$$g_{m,\lambda}(k,t) = \frac{1}{\sqrt{2\pi k\lambda(k)}}\left\{\begin{matrix}\exp\left[-\frac{(k-k\lambda(k)+m)^2}{2k\lambda(k)t}\right]+ \\ \exp\left[-\frac{(k+k\lambda(k)-m)^2}{2k\lambda(k)t}\right]\end{matrix}\right\}. \quad (20)$$

Hence, the probability of $(k_1 \le K_i(t) < k_2), \quad k_1, k_2 > 0$ can be calculated from the relationship presented below:

$$P_{m,\lambda}(k_1 \le K_i(t)\langle k_2) = \int_{k_1}^{k_2} g_{m,\lambda}(k,t)dk. \quad (21)$$

Having introduced the Laplace function in the form:

$$\Phi(u) = \frac{2}{\sqrt{2\pi}}\int_0^u e^{-\frac{y^2}{2}}dy \quad (22)$$

we obtain computational formula for $P_{m,\lambda}(\cdot)$

$$P_{m,\lambda}(k_1 \le K_i(t)\langle k_2) = \left[\Phi_{12}(u_{12}) - \Phi_{11}(u_{11})\right] + \left[\Phi_{22}(u_{22}) - \Phi_{21}(u_{21})\right],$$
(23)

where

$$u_{11} = \frac{k_1 - k\lambda(k) + m}{\sqrt{k\lambda(k)t}} \quad (24)$$

$$u_{12} = \frac{k_2 - k\lambda(k) + m}{\sqrt{k\lambda(k)t}} \quad 25)$$

$$u_{21} = \frac{k_1 + k\lambda(k) - m}{\sqrt{k\lambda(k)t}} \quad (26)$$

$$u_{22} = \frac{k_2 + k\lambda(k) - m}{\sqrt{k\lambda(k)t}} \quad (27)$$

It would be also interesting to calculate probabilites $P_{m,\lambda}(K_t \le m)$ lub $P_{m,\lambda}(K_t \rangle m)$. On the basis of the above formulas, relationships for these probabilities can be easily derived. Moreover it should be noticed that putting $f=i\cdot c$, $i\geq 0$ in appropriare formulas we obtain the final formulas for calculation of various probabilities of number of semi-prod. in the store.

4. FINAL CONCLUSIONS

Diffusion approximation in presented above comes from Fokker and Planck. They obtained the equation for distribution of diffusion process transitoin, which describes the development of the article system in the phase space with random particle impacts. Precise mathematical formulation on the basis of a diffusion approximation problem was done by A.V.Kolmogorov. He proposed so called retrospective equations and prospective for transition distribution. Since then, the diffusion approximation has been applied in many areas, even seemingly very remote from diffusion processes; in particular, these processes have existed as boundary processes for descrete models describing various biological phenomena such as change of species population or a gene concentration in a population.

Later works which were presented in the field of diffusion approximation applications concern the modeling of changes is assessing the accuracy of instrument (Kaluski, 1988) and computer systems operations (Czachorski, 1987). This paper is an approach to such approximation in the domain of discrete industrial processes.

REFERENCES

Budak, B. M., Samarski A.A. and Tikhonov A. N.(1965). *Problems in mathematical physics.* PWN, Warszawa (in Polish)

Czachórski, T.(1987). Elementary returns processes in improving difusion approximation as modelling of computer systems operations. *ZN Politecniki Slaskiej, seria: Informatyka,* **z.9,** Gliwice (in Polish).

Dinkin, F.B. and Yushkevich A.A.(1970). *Thesis and problems of Markov processes.* PWN, Warszawa 1970 (in Polish).

Kaluski, J.(1988). Zuverlaessigkeit der Messgeraete. . *Schriftenriehe, WDK* **15**: *Zuverlaessigkeit technischer Systeme. ETH*, Zurich, pp.149-155

Kimmel, M.(1982). Applying the theory of Markov processes to the investigation of boundary discrete processes. *ZN Politechniki Slaskiej, seria: Automatyka,* z.**63,** Gliwice (in Polish).

Schuss, Z.(1980). *Theory and applications of stochastic differential equations.*John Wiley & Sons, Inc.

TRANSFORMING APNC INTO IEC 1131-3 INSTRUCTION LISTS USING THE TPL METHODOLOGY

A.H. Jones, D. Karimzadgan

Intelligent Machinery Division
Research Institute for Design, Manufacture and Marketing
University of Salford
SALFORD M5 4WT, UK

fax: 44 161 745 59 99
e-mail: d.karimzadgan@aeromech.salford.ac.uk

Abstract: To date there are no general techniques available to convert ordinary and/or timed control Petri nets into IEC1131-3 Instruction Lists. In this paper, the Token Passing Logic technique is proposed to solve the problem of converting Automation Petri Nets for Control into Instruction Lists. The generality of the technique means that the method can be applied to any APNC and implemented on any IEC1131-3 compliant PLC. The resulting methodology provides a straight forward and powerful method by which to implement IEC1131-3 IL code. The technique is described by considering standard primitive structures for six different validation conditions inherent in APNCs and shows how to convert these into Instruction Lists.

Keywords: Manufacturing systems, Petri-nets, Programmable logic controllers.

1. INTRODUCTION

Today's automation requirements are in general met by the Programmable Logic Controller (PLC). Indeed, the PLC has played a major role in the standardisation of the factory hardware, and the majority of PLC's are now hardware compatible. However, the same can not be said for the software. Each PLC manufacturer provides different development software tools which are only compatible with their equipment. As automated manufacturing system become more complex the need for effective design tools which can be used on any PLC platform becomes even more important. This shortfall in software compatibility has been recognised, and an international standard IEC1131-3 has been defined to provide a set of international standard PLC programming languages.

IEC1131-3 describes the following programming languages:

Graphical Languages:
- Sequential Function Chart(SFC)
- Ladder Diagram(LD)
- Functional Block Diagram(FBD)

Textual Languages:
- Instruction List(IL)
- Structured Text(ST)

The syntax of these programming languages has been defined in detail in the IEC1131-3 standard, to ensure that users find the same syntax in all programming packages.
The majority of PLC's can be programmed in simple low level language called Instruction Lists (ILs). The very simplicity of this language which makes the PLC easy to program is also their greatest downfall. This is because when developing complex control

systems involving parallel tasks which interact periodically these programs offer little in the way of structural constructs to deal with problem

For simple problems it is easy to design programs using heuristic methods. However, as problems get more complex it becomes very difficult to handle the problem effectively. This difficulty is compounded when multi-product systems are considered. In fact, it is almost impossible to write down IL programs for multi-product systems using heuristic approaches.

The most successful solutions to the design of sequential control system have involved the use of Petri nets (David and Alla, 1992) for the conceptual design. Because of the success of Petri net designs there have been some attempts to produce methods to convert Petri nets into a graphical language, (Jafari and Boucher, 1994; Satoh, 1992; Rattigan, 1992). However, until the advent of TPL (Jones, et al., 1996a; Jones, et al., 1996b; Jones and Uzam 1996) none of these methods have produced a technique that is easy to apply and general, in the sense that it can deal with Automation Petri Net for Control (APNC) incorporated timers, and coloured tokens. The TPL methodology uses an alternative interpretation of the original Petri net analysis, so as to enable the Petri net analysis method to be used to produce sequential logic.

2. TOKEN PASSING LOGIC (TPL) METHOD

The prime feature of the Token Passing Logic (TPL) technique is that it facilitates the conversion of any Automation Petri Nets for Control into sequential logic. This is achieved by adopting the Petri net concept of using tokens as the main mechanism for controlling the flow of the sequential logic. Hence, each place within the Petri net corresponds to a place within the TPL program. Each action at a Petri net place corresponds to an action at a logic place, and each transition within the Petri net involving a movement of tokens corresponds to a simulated movement of tokens between logic places. This simulated movement of tokens is achieved by deploying separate memory bits or memory bytes, for one bounded places or k bounded places, at each logic place within the TPL program and, setting and resetting or incrementing and decrementing these memories to simulate token flow. Thus, each logic place within the TPL program has an associated memory, and the current value at the logic place represents the number of tokens that would be at the corresponding place within the Petri net. Furthermore, in consonance with the Petri net approach, if the memory associated with the place is non-zero then any actions associated with that place are activated. Finally, to complete the Petri net synergy, if the memory associated with a place is non-zero and a Petri net like transition associated with that place becomes active, then the memory at the place is reset or decremented by m (the input arc weight), and the subsequent place linked by the transition is set or incremented by n (the output arc weight). Moreover, the PLC *on delay timer* can be readily used to model P timed or T timed Petri nets.

In theory, the methodology can cope with any number of tokens and provide a visual description of the sequential logic program which has all the advantages of a full Petri net analysis. It is believed that this new technique provides a tool which is a simple, but sophisticated way of developing complex sequential logic programs. The technique for the design of IEC 1131-3 IL programs using the TPL method is illustrated by considering the following Petri net structures for different elementary validation conditions (α, β, γ, δ, ε, ς) which has been defined to encompass elementary leading edge, level and trailing edge validation conditions and ($\alpha+\varsigma$) as a combination of elementary validation conditions:

1. Standard Transition
   - with validation condition α
   - with validation condition β
   - with validation condition γ
   - with validation condition δ
   - with validation condition ε
   - with validation condition ς
   - with validation condition ($\alpha+\varsigma$)
2. Inhibited Transition
3. Enabled Transition
4. Action at a Place
5. Timer at a Place
6. Timed Action at a Place
7. Conflict

In this paper 2, 3, 4, 5, 6 and 7 are only considered for ($\alpha+\varsigma$) validation condition.

2.1 *Standard Transition*

A *Standard Transition* is shown in Figure 1. In Petri net theory, a transition can only be fired if the number of tokens at the input place is greater or equal to 'm' and the firing condition associated with the transition occurs (David and Alla, 1992). In the TPL method, each transition withdraws 'm' token from the current logic place and adds 'n' token to the next logic place.

This is achieved by using a memory word at each place to represent the tokens. When a transition is fired, to simulate the passing of a token the input memory value is decremented by 'm' and the output counter is incremented by 'n'. The IEC1131-3 IL program for different validation condition of the *Standard Transition* is given accordingly.

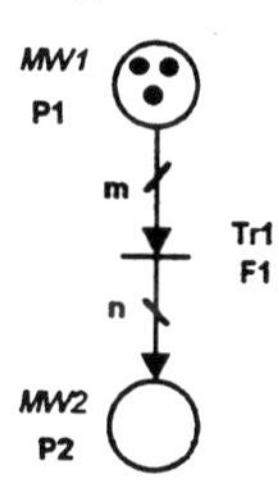

Fig. 1 .Standard Transition.

Validation condition α: Leading edge of enabling condition and simultaneous leading edge of firing condition. The IEC1131-3 IL for the transition is as follows:

```
Tr1:    LD      F1
        EU
        S       M3.1,1
        LDW>=   MW1,m
        EU
        S       M3.2,1
        LD      M3.1
        A       M3.2
        -I      m,MW1
        +I      n,MW2
        LD      M3.1
        R       M3.1,1
        LD      M3.2
        R       M3.2,1
        MEND
```

Validation condition β: Leading edge of enabling condition and simultaneous trailing edge of firing condition. The IEC1131-3 IL for the transition is as follows:

```
Tr1:    LD      F1
        ED
        S       M3.1,1
        LDW>=   MW1,m
        EU
        S       M3.2,1
        LD      M3.1
        A       M3.2
        -I      m,MW1
        +I      n,MW2
        LD      M3.1
        R       M3.1,1
        LD      M3.2
        R       M3.2,1
        MEND
```

Validation condition γ: Level of enabling condition and simultaneous leading edge of firing condition. The IEC1131-3 IL for the transition is as follows:

```
Tr1:    LD
        EU
        S       M3.2,1
        LDW>=   MW1,m
        AN      M3.1
        S       M3.3,1
        LD      M3.2
        R       M3.2,1
        LDW>=   MW1,m
        S       M3.1,1
        LD      M3.2
        A       M3.3
        -I      m,MW1
        +I      n,MW2
        LD      M3.2
        R       M3.2,1
        LD      M3.3
        R       M3.3,1
        MEND
```

Validation condition δ: Level of enabling condition and simultaneous trailing edge of firing condition. The IEC1131-3 IL for the transition is as follows:

```
Tr1:    LD      F1
        ED
        S       M3.2,1
        LDW>=   MW1,m
        AN      M3.1
        S       M3.3,1
        LD      M3.2
        R       M3.2,1
        LDW>=   MW1,m
        S       M3.1,1
        LD      M3.2
        A       M3.3
        -I      m,MW1
        +I      n,MW2
        LD      M3.2
        R       M3.2,1
        LD      M3.3
        R       M3.3,1
        MEND
```

Validation condition ε: Leading edge of enabling condition and level of firing condition. The IEC1131-3 IL for the transition is as follows:

```
Tr1:    LD      F1
        AN      M3.1
        S       M3.2,1
        LD      M3.1
        R       M3.1
        LD      F1
        S       M3.1,1
        LDW>=   MW1,m
        EU
        S       M3.3,1
        LD      M3.2
        A       M3.3
        -I      m,MW1
        +I      n,MW2
        LD      M3.2
        R       M3.2,1
        LD      M3.3
        R       M3.3,1
        MEND
```

Validation condition ς: Level of enabling condition and level of firing condition. The IEC1131-3 IL for the transition is as follows:

```
Tr1:    LD      F1
        AN      M3.1
        S       M3.2,1
        LD      M3.1
        R       M3.1
        LD      F1
        S       M3.1,1
        LDW>=   MW1,m
        AN      M3.3
        S       M3.4,1
        LD      M3.3
        R       M3.3
        LDW>=   MW1,m
        S       M3.3,1
        LD      M3.2
        A       M3.4
        -I      m,MW1
        +I      n,MW2
        LD      M3.2
        R       M3.2,1
        LD      M3.4
        R       M3.4,1
        MEND
```

In the Automation Petri Nets for Control these six validation conditions can be ORed to make new validation condition. Among the 57 possible combination, the (α+ς) is the one which was addressed by early works on TPL (Jones et al., 1996). This validation condition (α+ς) is leading edge or level enabling condition and leading edge or level firing condition. The IEC 1131-3 IL for the transition is as follows:

```
Tr1:    LD      F1
        AW>=    MW1,m
        -I      m,MW1
        +I      n,MW2
        MEND
```

If the APNC can be designed such that validation condition for any transition is (α+ς), the converted IEC1131-3 IL code will be compact and extra clear.

In the case of both leading edge and trailing edge of either enabling or firing condition, the order of the IEC1131-3 IL must follow the flow of token through the PN.

2.2 *Inhibited Transition*

An *Inhibited Transition* is shown in Figure 2(a) and consists of a *Standard Transition* and an *inhibitor arc*. An inhibitor arc has one end marked by a small circle as shown in the figure. The transition Tr1 has two input places P1 and P2, where P2 has an inhibitor arc. The transition Tr1 will be fired if P1 has at least 'm' token and P2 has no token (David and Alla, 1992). When it is fired 'n' token passes to the output place P3. There could be more than one output place in which case a single token would be passed to every output place. The transition Tr1 is inhibited from firing if there is a token in place P2. In the TPL method, the transition Tr1 will be enabled if the memory word of MW1 is greater than 'm' and the memory value of M2 is zero. When it is fired it will decrement the memory word MW1 associated with place P1, thus withdrawing 'm' token from this place, and increment the memory word MW3 associated with place P3, thus adding 'n' token to that place. If the memory associated with place P2 is set it will inhibit the transition from firing. The IEC1131-3 IL for the *Inhibited Transition* described is given in Figure 2(b).

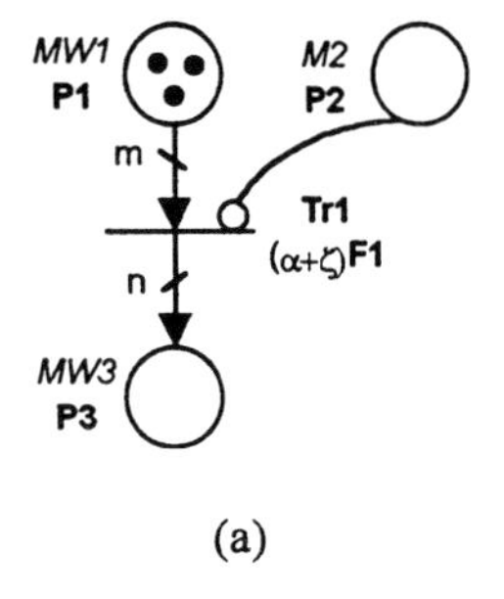

(a)

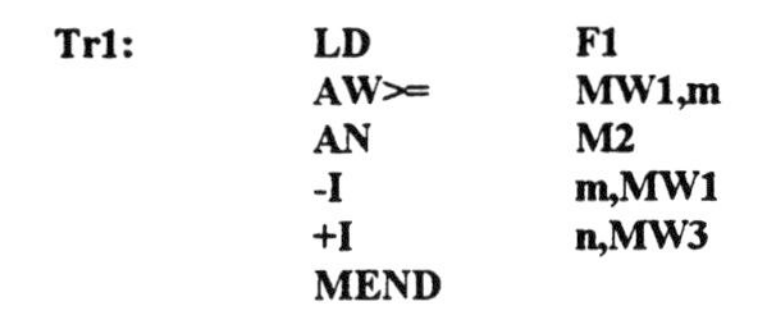

```
Tr1:    LD      F1
        AW>=    MW1,m
        AN      M2
        -I      m,MW1
        +I      n,MW3
        MEND
```

(b)

Fig. 2. (a) Inhibited Transition (b) IEC1131-3 Instruction List for Inhibited Transition.

2.3 *Enabled Transition*

In some Petri nets the firing of a transition is conditioned by the markings at a place, without modifying the markings at the place. This has been referred to as 'reading' (David and Alla, 1992). In this case, when the transition Tr1 is fired the token is removed from the place P2 and immediately replaced at the same place. If this Petri net is converted into IEC1131-3 IL via the TPL methodology, superfluous code is generated. In order to avoid this problem and to simplify the original Petri net a new *Enabling arc* has been proposed by Jones et al.(1996a) as shown in Figure 3(a). The *Enabling arc* can be used to enable a transition without modifying its markings. In the TPL method, the transition Tr1 will be fired if the memory word MW1 is greater than 'm' and memory M2 is set. When it is fired it will decrement the MW1 associated with place P1, thus withdrawing 'm' token from this place, and increment the MW3 associated with place P3, thus adding 'n' token to that place. After the transition Tr1 is fired the memory M2 will remain the same. If the memory value of the M2, associated with place P2, is zero it will not enable the transition to fire. The IEC1131-3 IL for the *Enabled Transition* is shown in Figure. 3(b).

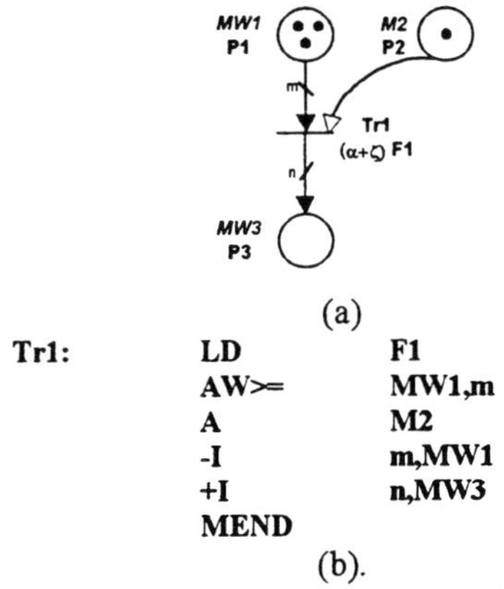

(a)

```
Tr1:    LD      F1
        AW>=    MW1,m
        A       M2
        -I      m,MW1
        +I      n,MW3
        MEND
```

(b).

Fig. 3. (a) Enabled Transition. (b) IEC1131-3 Instruction List for Enabling Transition

2.4 *Action at a Place*

In TPL the PLC outputs are termed *Actions*. These Actions are controlled by the counters at associated Petri net places. Figure 4(a) shows a Petri net place for which an action is assigned. An Action at a given place within a Petri net occurs only if the number of token at the place is non-zero. In TPL, if the value of a memory at the place is greater than zero then any actions associated with the place are enabled. The action at a place can accommodate both impulse and level action. An *Impulse Action* is obtained if the condition on Tr2 is 1 and a *Level Action* is obtained if Tr2 is conditioned with an event. The IEC1131-3 IL program for *Action at a Place* is given in Figure 4(b).

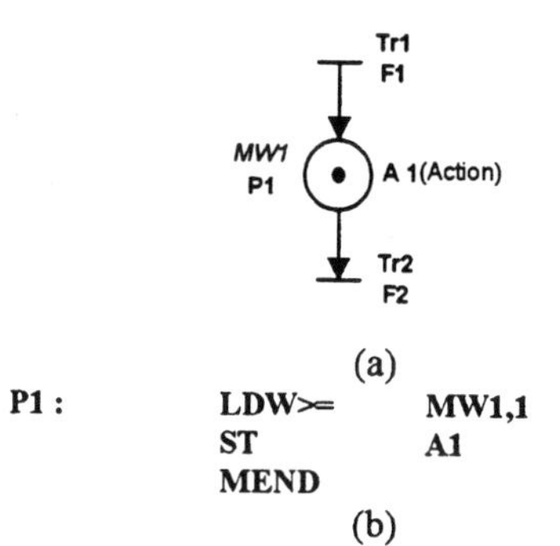

(a)

```
P1 :    LDW>=    MW1,1
        ST       A1
        MEND
```

(b)

Fig. 4. (a) Action at a Place. (b) IEC1131-3 Instruction List for Action at a Place.

If an action can be activated from more than one place, the places which activate the action have to be ORed together.

2.5 *Timed Place*

Although there are two types of timed Petri nets, namely P-timed Petri nets and T-timed Petri nets, in this paper only P-timed Petri nets are described. From the definition of P-timed Petri nets; when a token is deposited in a place, the token must remain in the place for at least a time *td*. During this time the token is said to be *unavailable*. When the time *td* has elapsed, the token becomes *available* to any associated firable transition (David and Alla, 1992).

(a)

```
Tr1 :   LD      F1
        S       M1,1
Tr2 :   LD      T1
        A       M1
        R       M1,1
        S       M2,1C
P1 :    LD      M1
        TON     T1,td
        MEND
```

(b)

Fig. 5. (a) Timed Place. (b) IEC1131-3 Instruction List for Timed Place.

In Figure 5(a), P1 is a timed place. When transition Tr1 is fired, a token is deposited in place P1 and remains unavailable for a time *td*. When this token becomes available, transition Tr2 is enabled. When it is fired (not necessarily immediately), the token will be deposited in P2. In the TPL method, when Tr1 is fired, the memory M1 will be set. If the memory M1 is greater than zero, then an on-delay timer T1 will be started for a time *td*. When the time *td* has elapsed, transition Tr2 becomes enabled, and if it is fired, it will reset M1 and set M2. The IEC1131-3 IL is given in Figure 5(b).

2.6 *Timed Action*

In some control applications it may be necessary to activate an actuator for a specific time. The solution to this problem is to introduce two places P1 and P2. P1 is a level action at place and P2 is a P-timed place. These two places are then linked in series as shown in Figure 6(a). The solution ensures the level action is started immediately the transition Tr1 fires, and not when the token in the P-Timed place P1 becomes available. The action is terminated when the token of this P-Timed place becomes available, by assigning this condition to Tr2. Figure 6(b) shows the IEC1131-3 IL program for Timed Action at a Place.

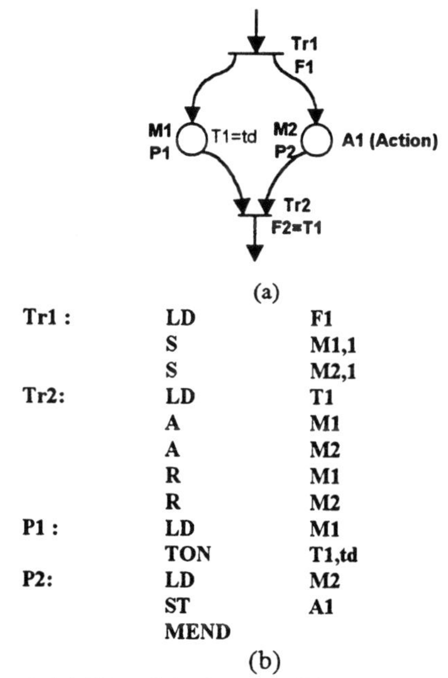

(a)

```
Tr1 :   LD      F1
        S       M1,1
        S       M2,1
Tr2:    LD      T1
        A       M1
        A       M2
        R       M1
        R       M2
P1 :    LD      M1
        TON     T1,td
P2:     LD      M2
        ST      A1
        MEND
```

(b)

Fig. 6. (a) Timed Action at a Place IEC1131-3. (b)Instruction List for Timed Action at a Place.

2.7 *Conflict*

A *conflict* corresponds to the existence of an input place which has at least two output transitions. According to the definition of Petri net, only one output place can receive a token in the case of conflict. One simple way to resolve the conflict is to assign a priority to each of the transitions, i.e. a technique is used to resolve the conflict by choosing which transition is to be allowed to fire (Desrochers and Al-Jaar, 1995). This choice is often based on a priority scheme. The Petri net in Figure 7(a) is called a *free-choice* net. A token in place P1 enables transitions Tr1 and Tr2. The conflict in Figure 7(a) can be resolved by assigning a priority between this two transitions. In the TPLL method, conflict resolution is achieved by firstly deciding the order of priorities for the conflicting transitions. Once this has been done, each IL of transition is then written in the same order. Because of the nature of IL this process will automatically resolve any conflict such that the chosen priorities are met. Therefore, as can be seen from Figure 8(a) transition Tr1 has a priority over transition Tr2. If the IL program was written the other way around, transition Tr2 would have the priority over Tr1. The Petri net in Figure 8(b) has two additional places P4 and P5 which are used to have a conflict free Petri net. In this case one token has to be put in either place P4 or place P5 indicating the priority of the conflict resolution. In Fig. 8(b) priority has been given to transition Tr1 by putting a token in place P4. This technique can also be used if the PLC only updates counters at the end of the scan-time. The corresponding IEC1131-3 IL program for Fig. 7(b) is given in Fig. 8(b).

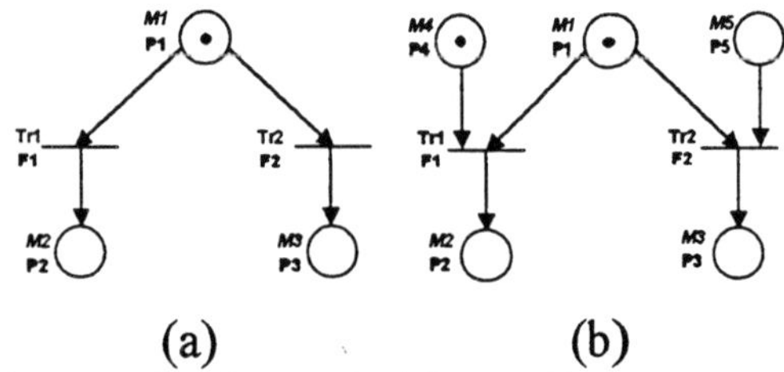

Figure 7. Conflict or decision-making structure.

Tr1	LD	M1
:	A	F1
:	R	M1,1
:	S	M2,1
Tr2	LD	M1
:	A	F2
:	R	M1,1
	S	M3,1
:	MEND	

Figure 8(a) IEC1131-3 IL for conflict resolution.

Tr1	LD	M1
:	A	M4
:	A	F1
:	R	M1,1
:	R	M4,1
:	S	M2,1
Tr2	LD	M1
:	A	M5
:	A	F2
:	R	M1,1
:	R	M5,1
:	S	M3,1

Figure 8(b).IEC1131-3 IL for conflict resolution.

3. CONCLUSION

To date there are no general techniques available to convert ordinary and/or timed control Petri nets into IEC1131-3 Instruction Lists. In this paper, the Token Passing Logic technique is proposed to solve the problem of converting Automation Petri Nets for Control into Instruction Lists. The generality of the technique means that the method can be applied to any APNC and implemented on any IEC1131-3 compliant PLC. The resulting methodology provides a straight forward and powerful method by which to implement IEC1131-3 IL code. The technique is described by considering standard primitive structures for six different validation conditions inherent in APNCs and shows how to convert these into Instruction Lists.

REFERENCES

David, R and Alla, H (1992). Petri Nets and Grafcet, Tools For Modelling Discrete Event Systems. *Englewood Cliffs,NJ:Prentice Hall Inc.*

Desrochers, AA and Al-Jaar, RY (1995). Application of Petri Nets in Manufacturing systems-Modelling,Control,and Performance Analysis. *IEEE press, Piscaway, NJ.*

Jafari, MA and Boucher, TO (1994). A Rule-Based System For Generating Ladder Logic Control Program From a High Level System Mode. In: *Journal of Intelligent Manufacturing*, **vol 5**, pp. 103-120.

Jones, AH; Uzam, M; Khan, AH; Karimzadgan, D; Kenway, S (1996a). A General Methodology for Converting Petri Nets Into Ladder Logic: The TPLL Methodology.In: *Proceedings of the CIMAT'96*, France, May 29-31,pp. 357-362.

Jones, AH; Uzam, M; Karimzadgan, D (1996b). Design of Knowledge Based Sequencial Control Systems. In: *First International Symposium on Computing for Industry - ISSCI'96*, Montpellier, France, May 27-30.

Jones, AH and Uzam, M (1996). Towards a Unified Methodology for Converting Coloured Petri Net Controllers into Ladder Logic Using TPLL: Part II- An Application. In: *WODES"96*, Edinburgh, UK, August 19-21.

Rattigan, S (1992). Using Petri Nets to Develop Programs for PLC Systems. In: *Lecture Notes in Computer Science 616: Application and Theory of Petri Nets*, pp. 368-372.

Satoh, T; Oshima, H; Nose, K; Kumagai, S (1992). Automatic Generation System of Ladder List Program By Petri Net. In: *IEEE International Workshop on Emerging Technologies on Factory Automation - Technology For The Intelligent Factory - Proceedings*, pp. 128-133.

STRUCTURAL MODELLING WITH PETRI NETS

Ilona Bluemke

Institute of Computer Science
Warsaw University of Technology
Nowowiejska 15/19, 00-665 Warsaw, Poland

Abstract: An example of structural modelling with Petri nets with refinements is discussed. A simple assembly line consisting of two machines is considered. The assembly line model is made more precise by refining each transition representing the machine, by the Petri net defining its operation. The Petri net model of the assembly line is used as a building block while modelling an entire factory unit. The descriptions of all Petri nets used are given in the new high-level language.

Keywords: Petri-nets, modelling, hierarchies, structural, system models

1. INTRODUCTION

For many years Petri nets have been widely used as a formal modelling tool. Petri nets are employed to model systems in which events can occur concurrently like in multiprocessing systems, manufacturing systems, distributed databases. The use of Petri nets is not limited only to modelling systems but also to their analysis and specification. Although it is a very convenient and elegant tool, problems may arise while modelling complex systems such as factory, production systems or assembly lines. Petri net models of complex systems tend to be too complicated to be handled efficiently. In order to bypass these obstacles the structuring methods have been proposed in the literature (Huber *et al.*, 1990), (Vogler, 1992). Unfortunately these methods are not commonly used in the modelling. The structuring methods allow the modular approach to the modelling. A modular approach allows the modeler to consider different parts of the system independently and also to reuse the same module in several different systems. A Petri net consists of nodes of two distinct types: places, denoted by circles, and transitions, denoted as bars. A set of directed arcs connects places with transitions and transitions with place. The connections are performed in such a way that for each transition there exists at least one place being connected to it. In marked Petri nets the initial marking function is defined to assign a nonnegative number of tokens to each place. A transition is enabled provided that its every input place has at least one token. The enabled transition can fire. When a transition fires a token is removed from each of its input places and is added to each of its output places. This process produces the new marking of the net, the new set of enabled transitions and so on. Places and transitions can be the substitution nodes i.e. can be replaced by a net. This substitution process is called place or transition refinement. A transition or a place is then deleted from the host Petri net and replaced by some net called the refinement net. The refined net (the net after the refinement process) constitutes the more precise system model.

In general, a Petri net can be defined by a graphic editor or described in a dedicated description language. The language is often a sequence of integers describing the connections between places and transition, the incidence matrix. Such a form is very inconvenient for the modeler and many errors may occur in the description of a net.

In this paper a new high-level language for the description of Petri nets with refinements is proposed and used. The new language consists of declarations of Petri nets, places, transitions, arcs and markings. Nets, places and transitions can be referred to by names, allowing the modeler to use names corresponding to the elements in the modelled system. The description of the net may be followed by the refine section, declaring substitution nodes, refinement nets and the way in which the refinement net is connected. This language can be used to the description of various Petri net types eg. timed (Zuberek,1986), and/or colored (Jensen, 1990). The language is formally described in (Bluemke, 1996).
The structural approach to the modelling with Petri nets is shown on a simple example. First the Petri net model of an assembly line is presented. The assembly line consists of two machines and three intermediate buffers. Next, the Petri net modelling the operations of each machine is introduced. The assembly line model is made more precise by refining each transition, modelling the machine, by the Petri net defining the operations of this machine. The Petri net model of the assembly line is next used as a building block for the modelling of a factory unit. The advantages of using the proposed high-level language are demonstrated using the discussed example.

2.REFINEMENTS IN PETRI NETS

Structuring methods provide an abstract mechanism to simplify large models by identifying submodels. The complexity of the model is reduced by dividing it into the submodels. The structuring methods support the top-down or bottom-up approach.
First approaches to the structuring of Petri nets appeared in the literature in the late '80, early '90. For colored Petri nets five structuring constructs were proposed (Huber *et al.*, 1990) : transition substitution, place substitution, transition invocation, place fusion and transition fusion. The transition and places substitutions are similar to the transition and place refinements discussed in this paper. The difference is in the way in which the substitution net is connected into the host one. Fusion of places and transitions are used for reducing several models into a single one. They support the bottom-up modelling technique. Substitution or fusion are performed statically. Transition invocation is executed dynamically. The concept is similar to the concept of calling a procedure in programming languages.

Many structuring methods are formally described and its properties are discussed in (Vogler, 1992). Introduction to the formalization and unification of the refinement process can be found in (Zuberek and Bluemke 1996)

2.1 Transition refinement

Refinement in Petri net means replacing an element of the host net by another net, called the refinement net. Various techniques for inserting the refinement net have been proposed. In the literature, mostly the refinement of a transition has been studied. The proposed refinements differ by the definition of the refinement net and by the way in which the substitution is performed. Some refinements preserve the net properties such as liveness and boundedness while others, like transition invocation do not. In the following the transition refinement implemented in the high-level language for Petri net description, and used in structural modelling in section 3, is described. Other transition refinements can be found in (Vogler,1990, Bluemke 1996).

A refinement net ***R*** that is going to substitute a transition ***t*** has some initial transitions representing the begin of ***t*** and some final transitions representing the end of ***t***. The refined net (refined model) is constructed by removing a transition ***t*** from the host net (rough model) and by inserting the refinement net. The input arcs of the transition ***t*** are directed to the initial transitions in the refinement net ***R***. The output arcs of the transition ***t*** are directed to the final transitions. It means that every input place of ***t*** is connected with an initial transition in ***R*** with the same weight as in the host net *N*. Similarly, final transitions and output places are connected. Fig. 1 illustrates the refinement process. The information about which place is connected to which transition is given in the high level language describing the net (refine instruction).

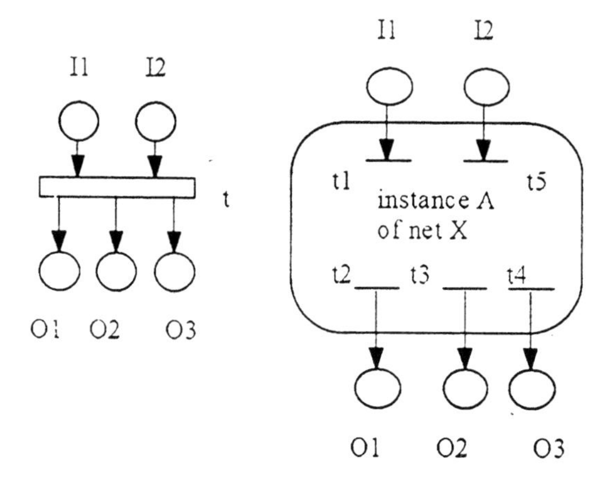

Fig. 1 Transition refinement
refine t by X as A[I1:t1,I2:t5;t2:O1,t3:O2,t4:O3];

2.2 Place refinement

Description of several place refinements can be found in (Vogler,1990). The place refinement used in this paper introduces a different way in which the refinement net is connected to the host net (different interface).

A refinement net ***R*** that will substitute a place ***p*** has some initial places and some final places. The refined net (refined model) is constructed by removing the ***p*** place from the host net (rough model) and inserting into its place the refinement net. The input arcs of the place ***p*** are directed to the initial places in the refinement net ***R***. The output arcs of the place ***p*** are directed to final places. It means that every input transition of *p* is connected with an initial place in ***R***. Similarly final places and output transitions are connected. Fig. 2 illustrates the place refinement. The connections between places in the refinement net and transitions in the host net are shown in the refine instruction.

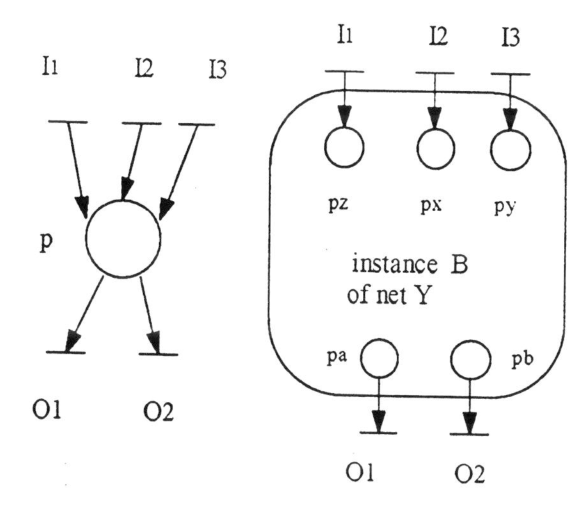

Fig. 2 Place refinement
refine p by Y as B[I1:pz,I2:px,I3:py;pa:O1,pb:O2];

3. STRUCTURAL MODELLING

The idea of structural modelling with Petri nets with refinements will be explained on a simple example. Let us consider a simple assembly line consisting of two machines ***M1*** and ***M2***. Machine ***M1*** gets parts from the input buffer, performs some actions, puts the manufactured elements into the output buffer. This buffer is also the input buffer for the machine ***M2***. Machine ***M2*** having completed its operation puts the elements into the output buffer, from which they are removed. This trivial assembly line is modelled by a Petri net shown in Fig. 3. Transitions named **M1** and **M2** model the machines. Transition **request** is added to show that the assembly begins when requested i.e when a token is in the place **start** Places named **bufA**, **bufB** and **bufC** are modelling the input buffer of the machine ***M1***, the intermediate buffer and the output buffer of the machine ***M2*** respectively. Transition **remove** stands for the action of removing ready elements and puts a token in the place **stop** when the assembly is completed.

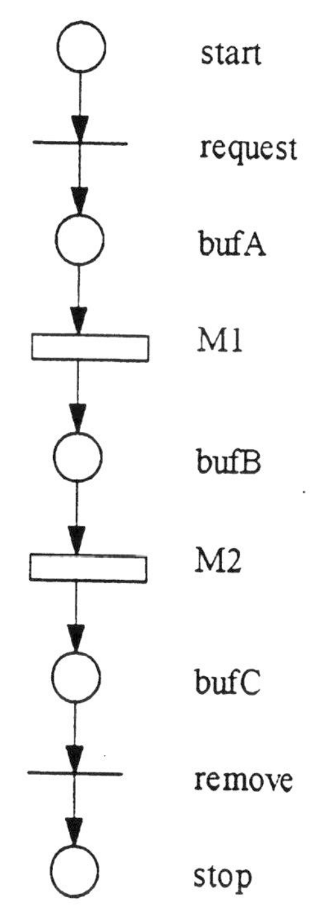

Fig. 3 Petri net modelling assembly line

This Petri net modelling the very simple assembly line using the high-level language (Bluemke, 1996) is as follows:

pnnet Assembly;
places start, bufA, bufB, bufC, stop;
transitions request, M1, M2, remove;
arcs (start; request), (request; bufA), (bufA; M1), (M1; bufB), (bufB; M2), (M2; bufC),
(bufC; remove), (remove; stop);
end;

The keyword **pnnet** begins while the keyword **end** ends the description of a Petri net called "Assembly". Keywords **places** and **transitions** start the declarations of places and transitions. In this simple Petri net declaration contain only names. For colored Petri nets (Jensen, 1990) the declaration of a place may contain also the place attribute, color. In colored Petri nets the transition declaration may also show the transition guard function. In timed Petri nets (Zuberek, 1986) the transition attribute is the firing time. In the specification of free-choice Petri nets yet

another transition attribute may show the probability of transition firing.
The keyword **arcs** introduces the definitions of connections between places and transitions eg. arc outgoing from transition M1 to place bufB is denoted as '(M1;bufB)'. Each arc may have attributes like: arc weight, or/and arc function for colored Petri nets (Jensen, 1990).

The simple model of the assembly line can now be refined. Let us assume that both machines perform two operations. Machine ***M1*** executes operations called ***Op1*** and ***Op2***. Machine ***M2*** as the first operation also executes operation ***Op1*** and as the second one ***Op3***. The machines can be modelled by the same net shown in Fig. 4. The difference in the executed by machines operations is achieved by initial marking. For the machine ***M1*** the token in the place called **enabOp2** is needed. To enable the operation ***Op3*** performed by machine ***M2*** a token must be placed in **enabOp3** and the place **enabOp2** must be empty.

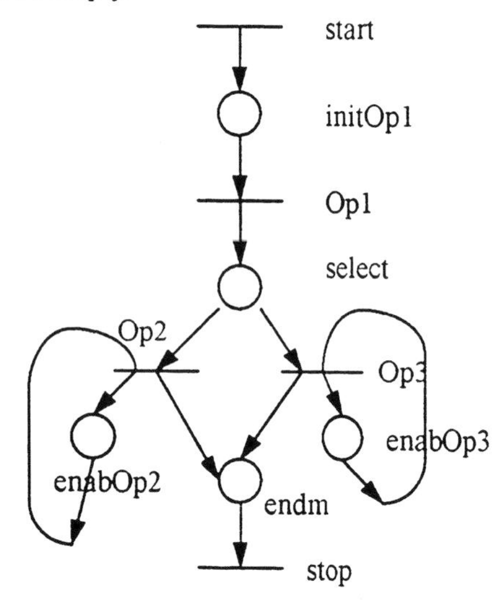

Fig. 4 Petri net modelling the machine

The description of Petri net modelling machine is following:
pnnet Machine;
places initOp1, enabOp2, enabOp3, endm, select;
transitions start, Op1, Op2, Op3, stop;
arcs (start; initOp1), (Op1; select), (initOp1; Op1), (select;Op2,Op3), (enabOp2;Op2), (enabOp3; Op3), (Op2; enabOp2, endm), (Op3; enabOp3, endm), (endm; stop);
end;

In order to get Petri net model of the assembly line more precise we can refine the transitions representing machines **M1** and **M2** by a subnet and define how this refinement net should be connected into the host net. In the high-level language we need to write:
refine M1 **by** Machine **as** RM1 [(bufA; RM1.start), (RM1.stop; stop)];
refine M2 **by** Machine **as** RM2 [(bufA; RM2.start), (RM2.stop; stop)];
marking RM1.enabOp2:1, RM2.enabOp3:1 ...;

The keyword **refine** is followed by a name of transition/place to be refined by the net. The name of this net is placed after the keyword **by**. The keyword **as** is followed by the name of the refinement process, eg. RM1. This name is used for prefixing the names of transition and places in the refinement net. Inside brackets [] the connecting arcs are defined. These arcs describe how the refinement net is connected with the host net, eg. place bufA (from the net Assembly shown in Fig.3) is connected to the transition start (Fig.4) prefixed by the name RM1 of the refinement process to have different names. From transition RM1.stop there is an arc to the place stop in the host net. The number of arcs defining the connections of the refinement net must be the same and of the same type as is in the host net. The appropriate checking is performed by the compiler of the language. The net after refinements is shown in Fig.5.
The keyword **marking** begins the definition of marked places, containing tokens. The number of tokens is written after the colon eg. in place RM2.enabOp3 there is one token.

The Petri net model of the assembly line can be used as a building block while modelling a factory unit. Let us assume that the factory unit has two assembly lines working concurrently. The Petri net model of this simple factory unit is shown in Fig. 6.

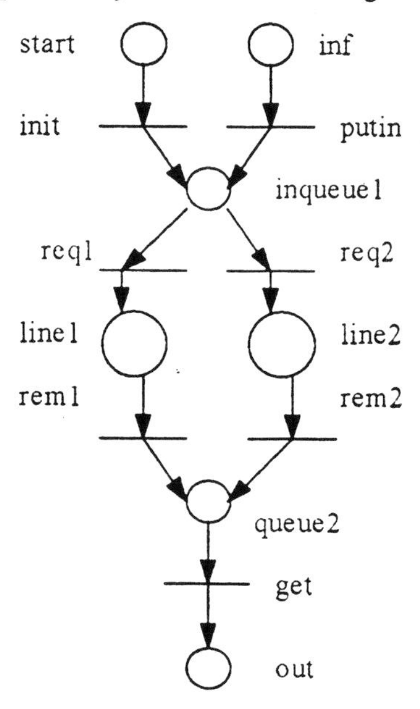

Fig. 6 Petri net modelling factory unit

The two assembly lines can work concurrently if there are elements in the input queue, modelled by the tokens in the place **inqueue1**. Transitions **req1** and **req2** represent the request of assembly lines. The assembly lines are represented by places **line1** and **line2**. Transitions **rem1** and **rem2** are modelling removing products from assembly lines into the output buffer represented by the place **queue2**. The description of this net is following:
pnnet Factory;
transitions init, putin, req1, req2, rem1, rem2, get;
places start, inf, inqueue1, line1, line2, queue2, out;
arcs (start;init), (init; inqueue1), (inf; putin), (putin; inqueue1), (inqueue1; req1,req2); (req1; line1), (line1; rem1), (req2; line2), (line2,rem2), (rem1, rem2; queue2), (queue2; get), (get; out);
end;

Depending on the level of abstraction we can have more adequate model of the factory unit. The places **line1** and **line2** representing assembly lines can be refined by net Assembly.The refined model is shown in Fig.7. The refinements are following:
refine **line1** by Assembly as AL1 [(req1; AL1.start), (AL1.stop; rem1)];
refine line2 by Assembly as AL2 [(req1; AL2.start), (AL2.stop; rem1)];

Adding time as an attribute to transitions (timed Petri nets) the duration of activities can be taken into account and performance analysis of the system can be performed.

4.CONCLUDING REMARKS

The transition and place refinements can be used very efficiently in structural models of complex systems such as manufacturing systems, assembly lines. The structural approach allows to consider the parts of the system independently, and to reuse some parts in different models.
A high level language is advantageous and necessary for the specification of complex Petri nets. In this language libraries of the descritions of elements often used in modelling can be kept.

REFERENCES

Bluemke, I. (1996). *Refinements in Petri Nets*, Institute of Computer Science Research Report no 38/96

Huber, P. Jensen, K. Shapiro, R.M. (1990), Hierarchies in colored Petri nets. In: *Advances in Petri Nets 1990* (Lecture Notes in Computer Science **483**), Springer Verlag 1991, 281-305

Jensen, K. (1990), Colored Petri nets: a high-level language for system design and analysis. In: *Advances in Petri Nets 1990* (Lecture Notes in Computer Science **483**), Springer Verlag 1991, 342-416

Vogler, W. (1992). *Modular construction and partial order semantics of Petri nets*, (Lecture Notes in Computer Science **625**), Springer Verlag 1992

Zuberek, W.M. Bluemke, I. (1996). Hierarchies of Place/Transitions Refinements in Petri Nets. *Proceedings of ETFA'96*, Hawaii, 355-360.

Zuberek, W.M. (1986). M-timed Petri nets, priorities, preemptions and performance evaluation of systems. In: Lecture Notes in Computer Science **222**, Springer Verlag 1986, 478-498

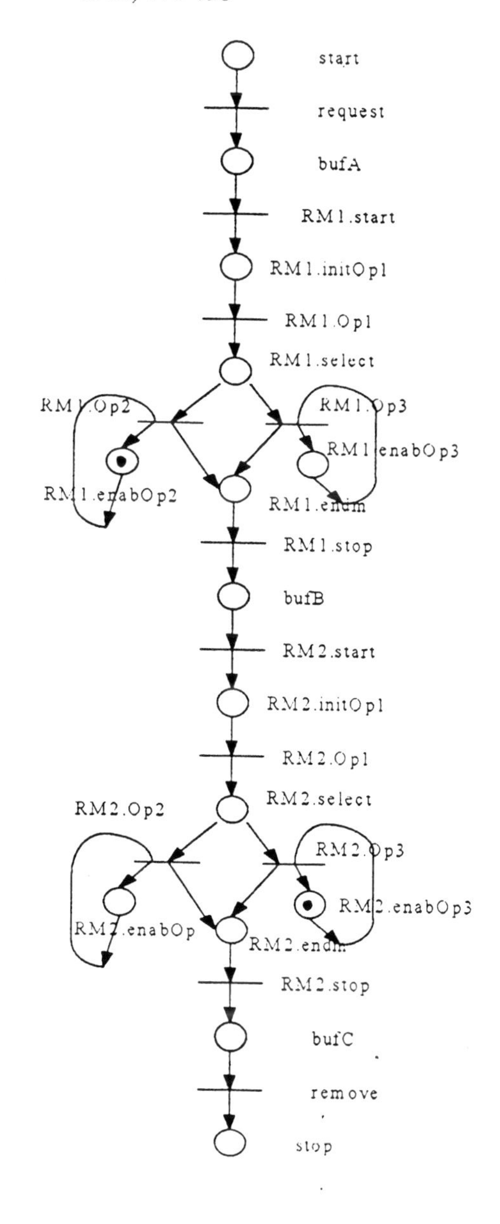

Fig. 5 Petri net modelling refined assembly line

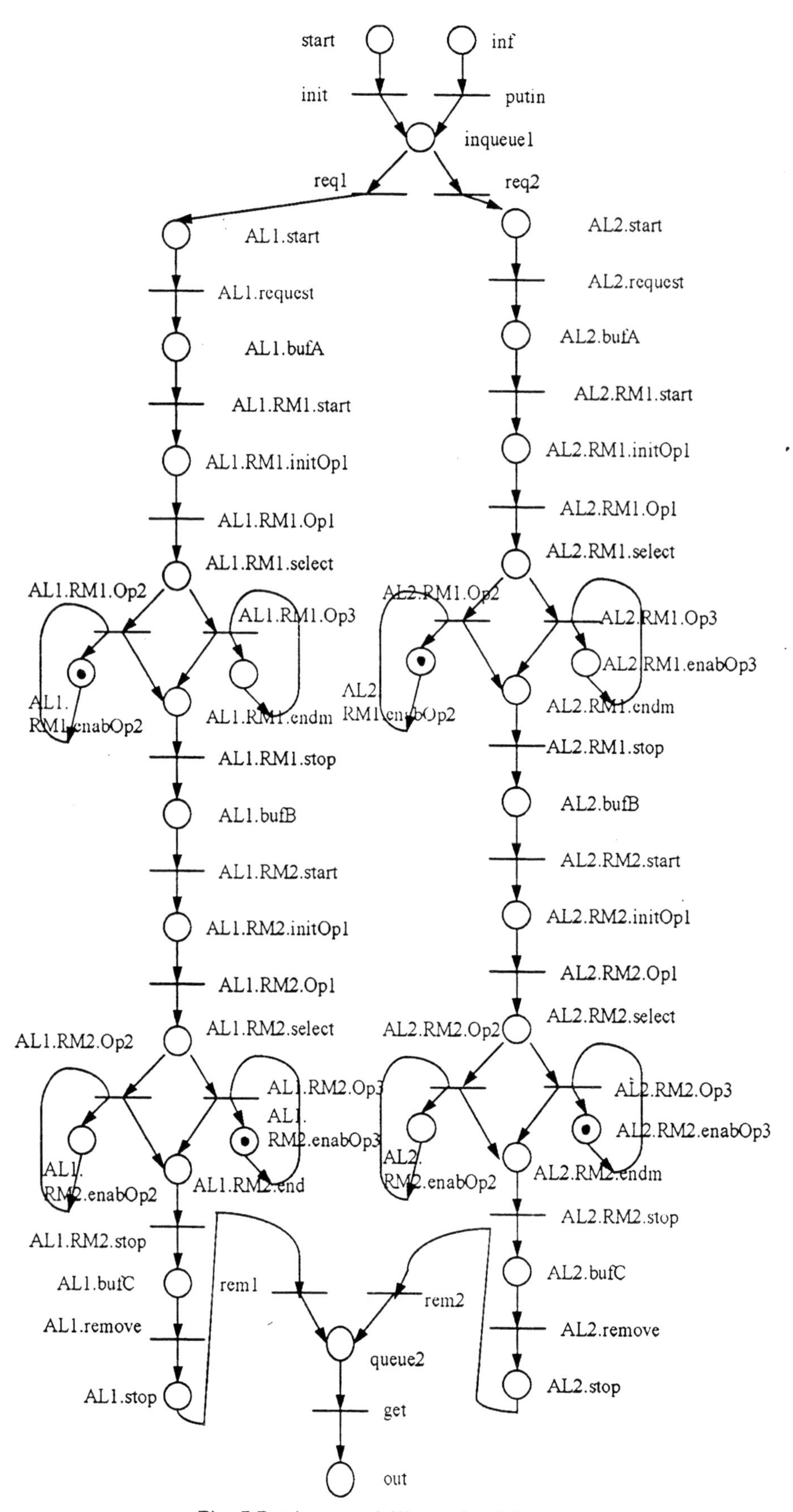

Fig. 7 Petri net modelling refined factory unit

USING P-TIME PETRI NETS FOR ROBUST CONTROL OF MANUFACTURING SYSTEMS

Pascal Aygalinc, Soizick Calvez
Wael Khansa, Simon Collart-Dutilleul

LAMII-CESALP, ESIA, Université de Savoie
41, avenue de la Plaine, BP 806. F 74016 Annecy

Abstract : Constraints such as zero stock and/or just in time require a robust control in order to maintain specified properties (sequence, throughput rate,...) although disturbances appear in the manufacturing system. In this paper, a method to determine a robust schedule of a manufacturing cell with staying time constraints, in regard of disturbances on the input rate of the parts is proposed. This method uses P-time Petri net as model and the periodic control functioning mode. The schedule for a definite throughput rate is obtained by solving a linear program which constraints are all stemming from the model.

Keywords : Time Petri-Net, Control, Robustness, Manufacturing systems, Performance Evaluation

1. INTRODUCTION

P-time Petri Nets (p-TPN on short) (Khansa, *et al.*, 1996a, b) are convenient tools to model manufacturing systems whose operations times are included between a minimum and a maximum value. For example, this is the case for electroplating lines or chemical industry. So, the typical feature of these nets is that a control of the firing dates of the transitions is required in order to ensure that the tokens remain in the places a duration included in the specified intervals. Consequently, the model gives the set of constraints on the control of the firing dates in order to ensure that the potential constraints are enforced.

In this paper, this tool is used to determine a robust control of a manufacturing cell (a single hoist electroplating line) when the resources allocation and choices have been resolved. In the second paragraph, you are reminded the definitions of P-time Petri net, and the linear program formulation in order to obtain the performances (minimum and maximum average production rates). In the third one, the manufacturing cell is briefly described and the P-time Petri Net model for a particular control is given. In the fourth paragraph, the external robustness of the control is illustrated by means of periodic functioning mode of P-time Petri net. Using this property in the last paragraph, we express the linear program in order to make a robust schedule in regard of disturbances on the input rate of the parts in the case of a definite throughput rate.

2. P-time PETRI NET

2.1 Formal Definition

The formal definition of a p-TPN (Khansa, *et al.*, 1996a,1996b) is given by a pair < N ; I > where :

- N is a marked PN (Brams, 1983)
- I : $P \rightarrow (Q^+ \cup 0) \times (Q^+ \cup \infty)$
  $p_i \rightarrow I_i = [a_i , b_i]$ with $0 \leq a_i \leq b_i$

I_i defines the static interval of residence time (duration) of a token in place p_i (Q^+ is the set of positive rational numbers). A token in place p_i participates in the enabledness of output transitions of this place if it has stayed at least duration a_i in this place and b_i at the most. Consequently, the token must leaves p_i at the latest when its residence duration becomes b_i. After this duration b_i, the token will be "dead" and will not participate any more in the enabledness of the transitions.

2.2 Structural Analysis and Performances Evaluation

The aim of the structural analysis is to determine the set of constraints ensuring the liveness of tokens (Khansa, *et al.*,1996b ; Calvez, *et al.*,1997). As this kind of PNs requires a control of the firing dates of the transitions (the earliest functioning mode is not always feasible), the structural analysis is closely linked with the performances evaluation.

The structural analysis provides exactly the performances evaluation if the P-time Petri net model is a p-time State Machine or a P-time Strongly Connected Event Graph (P-time SCEG on short) iff its underlying graph is live and bounded. In order to compute the minimum and maximum average throughput rate of P-time SCEG, two linear programs can be expressed (Calvez, *et al.* ,1997). Their common set of constraints is stemming from the model :

a) potential constraints : the staying time in the place p_i (denoted by q_i) must be included between a_i and b_i ($a_i \le q_i \le b_i$).
b) constraints on average cycle times : the average cycle times of the elementary circuits must be equal to themselves.

The criterion to minimise (or to maximise) is the average cycle time of one of the elementary circuits.

Thus, using the equivalence between P-semi flows and elementary circuits (Murata, 1989), the linear program for the lower bound of the average cycle time can be stated as follows :

$$\mu_{min} = \min\left(\frac{\mathbf{X}_j^T.\mathbf{Q}}{\mathbf{X}_j^T.\mathbf{M}_0}\right) \quad \text{with } j \in [1, m-n+1] \qquad (1)$$

subject to :

$$\begin{cases} \mathbf{A}.[\mathbf{1}] \le \mathbf{Q} \le \mathbf{B}.[\mathbf{1}] \\ \dfrac{\mathbf{X}_i^T.\mathbf{Q}}{\mathbf{X}_i^T.\mathbf{M}_0} = \dfrac{\mathbf{X}_j^T.\mathbf{Q}}{\mathbf{X}_j^T.\mathbf{M}_0} \quad \forall\, i = 1,\ m-n+1,\ i \ne j \end{cases}$$

where :
m : number of places of the P-time SCEG
n : number of transitions
m-n+1 : number of P-semiflows or number of elementary circuits
$\mathbf{Q}$: column vector [m x 1] of average staying times in the places
$\mathbf{X}_i$: P-semiflows [m x 1] associated with the i-th elementary circuit ($\mathbf{X}_i^T\mathbf{W} = 0$ and $\mathbf{X}_i \ne 0$ where $\mathbf{W}$ denotes the incidence matrix)
$\mathbf{M}_0$: column vector [m x 1] associated with the initial marking
$\mathbf{A}$ ($\mathbf{B}$) : diagonal matrix [m x m] of the minimum (maximum) staying times a_i (b_i)
[**1**] : column vector [m x 1] where each component is a unit one

The linear program for the upper bound uses as criterion :

$$\mu_{max} = \max\left(\frac{\mathbf{X}_j^T.\mathbf{Q}}{\mathbf{X}_j^T.\mathbf{M}_0}\right) \quad \text{with } j \in [1, m-n+1] \qquad (2)$$

subject to the same set of constraints.

Notice that : the liveness of the underlying SEGC guarantees that $\mathbf{X}_i^T.\ \mathbf{M}_0 > 0$, $\forall i \in [1, m-n+1]$. If the set of admissible solutions is empty, the P-time SCEG is a dead-token net, else it is live.

2.3 Periodic Functioning Mode in P-time SCEG

The behaviour of this mode (Ramchandani, 1974 ; Laftit, 1991) is fully determined by :

$$\forall k \ge 1 \quad S_i(k) = S_i(1) + (k-1).\pi \qquad (3)$$

where $S_i(k)$ is the k-th firing date of the transition t_i and π the functioning period.

That means that the dates of the first firing of the transitions and the period of the functioning are sufficient to describe entirely the periodic functioning mode. Notice that, in such a functioning mode, the places must be First In First Out (FIFO).

To build a periodic schedule for a P-time SCEG, you have only to determine the first firing date of each transition. In other words, it suffices to determine the typical staying duration of each place in order to obtain the definite period.

Theorem 2.1 : the average staying durations values q_i, solutions of the following system :

$$\begin{cases} \mathbf{A}.[\mathbf{1}] \le \mathbf{Q} \le \mathbf{B}.[\mathbf{1}] \\ \dfrac{\mathbf{X}_i^T.\mathbf{Q}}{\mathbf{X}_i^T.\mathbf{M}_0} = \dfrac{\mathbf{X}_j^T.\mathbf{Q}}{\mathbf{X}_j^T.\mathbf{M}_0} \quad \forall\, i = 1,\ m-n+1,\ i \ne j \end{cases} \qquad (4)$$

define a periodic schedule (Calvez, *et. al*, 1997).

Corollary 2.1 : the periodic control functioning mode of a P-time SCEG has the same performances as a k-periodic one. Consequently, the period of that mode must verify : $\pi \in [\mu_{min}, \mu_{max}]$.

Therefore, , to obtain a periodic schedule with a throughput rate equal to $\pi = \pi_{obj}$ with $\pi_{obj} \in [\mu_{min}, \mu_{max}]$, it suffices to solve the following system :

$$\begin{cases} \mathbf{A}.[\mathbf{1}] \le \mathbf{Q} \le \mathbf{B}.[\mathbf{1}] \\ \dfrac{\mathbf{X}_i^T.\mathbf{Q}}{\mathbf{X}_i^T.\mathbf{M}_0} = \pi_{obj} \quad \forall\, i = 1,\ m-n+1 \end{cases} \qquad (5)$$

Nevertheless, this system can admit more than one solution. We have generally less equations (m-n+1) than firing dates (n) to determine. In that sense, several approaches can be employed :

i) One is to give typical values to some of the staying durations and/or to increase the inequalities system dimension with knowledge of the manufacturing system that the model does not take into account (for example, rate between treatment time values).

ii) An other is to establish a gradation of the set of admissible solutions. The control problem becomes an optimization one. It can be expressed as follows :

$$\max \text{ (or min) } [\textit{ a criterion}] \qquad (6)$$

subject to :

$$\begin{cases} \mathbf{A}.[\mathbf{1}] \leq \mathbf{Q} \leq \mathbf{B}.[\mathbf{1}] \\ \dfrac{\mathbf{X}_i^T.\mathbf{Q}}{\mathbf{X}_i^T.\mathbf{M}_0} = \pi_{obj} \quad \forall i = 1, m-n+1 \end{cases}$$

The criterion that we propose to use is the robustness of the control in regard of the disturbances on the input-rate of the parts (external robustness). Our development will be illustrated on a manufacturing cell with staying time constraints, described in the following paragraph.

3. MANUFACTURING CELL

An electroplating line consists of a sequence of chemical tanks arranged typically in a row. A loading port and an unloading one are physically located at the extremities (see figure 1). Each tank contains the chemicals required for a particular treatment. Therefore, a chemical treatment consists of the successive immersions of a product, in the order specified by a given program. The movements of products are performed by a handling hoist. This hoist can disengage and travel to some other tank to remove an other product.

In order to ensure the quality of the reactions, and furthermore of the whole treatment, the travels of the engaged hoist have to be as short as possible (no wait) and minimum and maximum immersion times are given. As this line is a manufacturing system with staying time constraints, its Petri-net model must be a P time one.

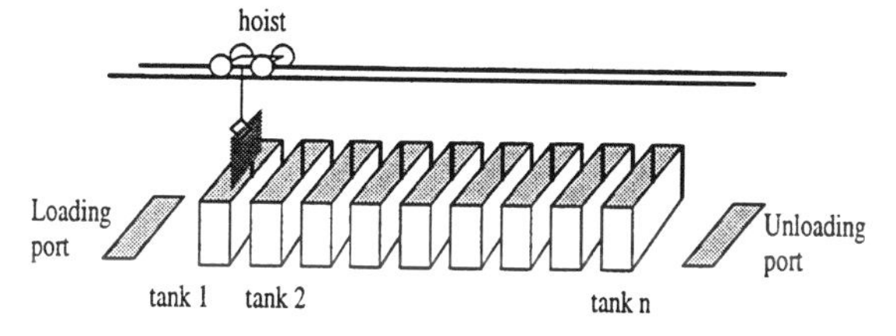

Fig. 1 : a single-hoist electroplating line

3.1 Building a model

Many approaches (Manier and Baptiste, 1994; Lei and Wang, 1989; Philips and Hunger, 1976; Shapiro and Nuttle, 1988) exist to determine a single-cyclic schedule of electroplating lines. It is known as "The cyclic Hoist Scheduling Problem". These approaches give the sequence of the movements of the hoist and their starting times. To build the P-time PN (Calvez *et al.*,1997) of the control (see figure 2), the only needed knowledge is the sequence of the movements of the hoist.

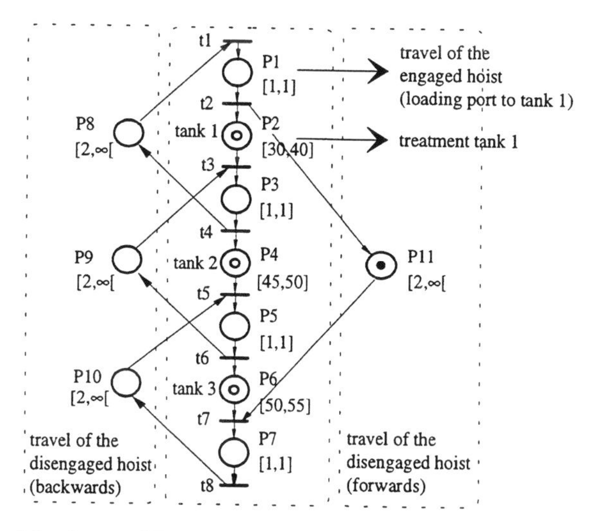

Fig. 2 : a p-TPN model associated with a particular control of the manufacturing cell.

As all choices have been decided, the structure of the Petri net model is a P-time SCEG. The places in the linear part of the graph represent alternatively :

- a travel of the hoist when carrying a part. The static interval of these places is $[a_i, a_i]$ where a_i is the time value of [loading + engaged traveling + unloading] : this ensures that the travels of the engaged hoist are as short as possible (no possibility of waiting).

- a treatment in a tank. The static interval $[a_i, b_i]$ is specified by a_i the minimum value of the duration of the treatment, and b_i the maximum one in order to respect the chemical specifications.

The right and the left hand side places of the graph model the movements of the disengaged hoist. The static interval of these places is $[a_i,\infty[$ where a_i is the smallest time value of the disengaged travel. The upper bound (∞) of the interval indicates that the hoist can wait as long as needed.

Whatever the control is, notice that :

- its underlying graph is live, bounded (in order to respect the unit capacity of each tank and the uniqueness of the hoist).

- the model describes only the steady state of the control : the marking represents one of the marking of the steady state.

Using the linear programs (1) and (2), it is possible to achieve performances evaluation of the control in steady state $[\mu_{min}, \mu_{max}]$.

4. ROBUSTNESS OF THE CONTROL

When the periodic functioning mode (K-periodicity=1) is used, the minimum and the maximum throughput rate $[\mu_{min}, \mu_{max}]$ allows to determine the earliest firing date of $S_i(k+1)$ and the latest one. These dates define the active robustness (to preserve the specified properties the schedule requires to be partly or totally modified (Collart-Dutilleul, *et al.*, 1994).

As : $S_i(k+1) = S_i(k) + \pi$ and $\pi : \mu_{min} \leq \pi \leq \mu_{max}$

then : $S_i(k) + \mu_{min} \leq S_i(k+1) \leq S_i(k) + \mu_{max}$ (7)

In the particular case of our application (the duration of the treatments in tanks and the duration of the engaged travels are fixed), the proof that the control can be maintained even if disturbances appear in the firing interval $[\mu_{PRmin}-\pi, \mu_{PRmax}-\pi]$ is given in (Collart-Dutilleul, *et al.*, 1994). These two bounds $[\mu_{PRmin}, \mu_{PRmax}]$ correspond to the passive robustness. The proof uses a P-timed SCEG model and the periodic functioning mode.

We propose to generalize this property to any control of a manufacturing system modeled by a P-time SCEG when the periodic functioning mode is used.

In order to obtain a definite throughput rate $\pi=\pi_{obj}$, it suffices, as seen above, to set the staying duration of each place. But, if two transitions are linked by several elementary directed paths, it is sufficient to set the staying durations of places along one path only, and partly along the other paths. For instance in our application, whatever the control is, the staying durations of places associated with a treatment in a tank or with an engaged travel must be set to the typical value in order to obtain $\pi=\pi_{obj}$, while the staying durations of the places associated with the disengaged travel are not fixed. Consequently, the structural analysis of the model with partly set staying durations, gives two bounds $[\mu_{PRmin}, \mu_{PRmax}]$ which correspond to the passive robustness.

Theorem 4.1 : Let <N ; I> a live and bounded P-time SCEG where for each pair of transitions (t_i, t_j), it exists at least one elementary directed path linking t_i to t_j or t_j to t_i which its places have a static interval $[a_i, b_i]$ with $a_i=b_i$. The structural analysis of this graph gives two bounds $[\mu_{PRmin}, \mu_{PRmax}]$ where the control can be maintained if the firing date of each transition t_i verifies :

$$S_i(k) + \mu_{PRmin} \leq S_i(k+1) \leq S_i(k) + \mu_{PRmax} \quad (8)$$

Proof : The liveness of tokens is guaranteed if the period π or the cycle time of each elementary circuit is included between $[\mu_{PRmin}, \mu_{PRmax}]$. As the graph is strongly connected, each transition t_i belongs to one circuit at least. Thus, the earliest firing date of $S_i(k+1)$ and the latest one is given by (8). The delay (the advance) of the firing date of $S_i(k+1)$ in regard of its typical firing date, denoted by ΔS_i is :

$$S_i(k+1)=S_i(k)+\pi+\Delta S_i \text{ with } \mu_{PRmin}-\pi \leq \Delta S_i \leq \mu_{PRmax}-\pi \quad (9)$$

But each elementary directed path from t_i to t_j imposes a constraint on the firings of t_j in relation to the firings of t_i and consequently, constraints on the staying duration of tokens for each path. Without loss of generality, let us consider the case of two transitions (t_i, t_j) linked by only two elementary directed paths (if (t_i, t_j) are linked by more than two elementary directed paths, it suffices to consider them by pairs).

For each place p_i where t_i is the input transition of p_i and t_{i+1} the output one, the staying duration of p_i (q_i) imposes to the firing dates of t_i and t_{i+1} that :

$$q_i = [S_{i+1}(k) + m_i\pi] - S_i(k) \quad (10)$$

This expression can easily been generalized for two transitions (t_i,t_j) linked by an elementary directed path l_{ij} in order to determine the staying duration of tokens in this path. Thus, let t_i and t_j be two transitions, l_{ij}^f and l_{ij}^k are elementary directed paths from t_i to t_j, for each pair of paths l_{ij}^f and l_{ij}^k, it follows :

$$\begin{cases} S_j(k)+m_{l_{ij}^f}.\pi - S_i(k) = q_{l_{ij}^f} \\ S_j(k)+m_{l_{ij}^k}.\pi - S_i(k) = q_{l_{ij}^k} \end{cases} \quad (11)$$

where $m_{l_{ij}^f}$ $(m_{l_{ij}^k})$ is the number of tokens on the path l_{ij}^f (l_{ij}^k), $q_{l_{ij}^f}$ $(q_{l_{ij}^k})$ is the summation of the staying durations of the places along the path l_{ij}^f (l_{ij}^k).

Hence :

$$q_{l_{ij}^k} = q_{l_{ij}^f} - (m_{l_{ij}^f} - m_{l_{ij}^k}).\pi \quad (12)$$

Assume that the elementary directed path along which the staying durations of the places are set is l_{ij}^f. Their static intervals are equal to $[q_i^{typ}, q_i^{typ}]$ where q_i^{typ} is the typical value in order to obtain $\pi=\pi_{obj}$ and $q_{l_{ij}^f}$ is equal to a constant value denoted by $q_{l_{ij}^f}^{typ}$. The equation (12) becomes :

$$q_{l_{ij}^k}^{typ} = q_{l_{ij}^f}^{typ} - (m_{l_{ij}^f} - m_{l_{ij}^k}).\pi \quad (13)$$

Let us consider $m_{l_{ij}^f} = m_{l_{ij}^k} = m_{l_{ij}}$. As the functioning mode is a periodic one and as the places are FIFO in such a mode, no further constraints on ΔS_i are imposed by these paths. A delay (or an advance) of $S_i(k)$ introduces only a shift of $S_j(k+m_{1_{ij}})$.

Notice that the same result is obtained when only one elementary directed path links t_i to t_j because the staying durations of its places are set.

When $m_{1^f_{ij}} > m_{1^k_{ij}}$, these relations can be written as :

$$q_{1^k_{ij}} = q^{typ}_{1^k_{ij}} + \Delta q_{1^k_{ij}} \quad (14)$$

$$\text{with} \quad \begin{cases} (m_{1^f_{ij}} - m_{1^k_{ij}}).(\pi - \mu_{PR\,max}) \le \Delta q_{1^k_{ij}} \\ \Delta q_{1^k_{ij}} \le (m_{1^f_{ij}} - m_{1^k_{ij}}).(\pi - \mu_{PR\,min}) \end{cases}$$

As $(m_{1^f_{ij}} - m_{1^k_{ij}}).(\mu_{PR\,max} - \pi) \ge \Delta S_i$, the delay (when $\Delta S_i > 0$) of the firing of t_i can be recovered along the elementary directed path 1^f_{ij}. Using the same reasoning, we deduce that it is possible to fire the transition t_i before its typical firing date.

When $m_{1^f_{ij}} < m_{1^k_{ij}}$, it follows :

$$q_{1^k_{ij}} = q^{typ}_{1^k_{ij}} + \Delta q_{1^k_{ij}} \quad (15)$$

$$\text{with} \quad \begin{cases} (m_{1^f_{ij}} - m_{1^k_{ij}}).(\pi - \mu_{PR\,min}) \le \Delta q_{1^k_{ij}} \\ \Delta q_{1^k_{ij}} \le (m_{1^f_{ij}} - m_{1^k_{ij}}).(\pi - \mu_{PR\,max}) \end{cases}$$

Using the dual reasoning of the previous case (when the transition t_i is fired later than its typical firing date, the duration of the elementary directed path 1^f_{ij} must be increased), we deduce that it is always possible to adjust $q_{1^k_{ij}}$ in regard of ΔS_i.

As ΔS_i fulfill the condition (9), this concludes the proof.

Thus, the control can be maintained even if disturbances appear in the firing interval $[\mu_{PRmin}-\pi, \mu_{PRmax}-\pi]$. In that sense, it is very interesting to have a control and a schedule which have a passive robustness $[\mu_{PRmin}, \mu_{PRmax}]$ as large as possible. The next paragraph gives the synthesis method in order to obtain this property.

5. SYNTHESIS OF THE ROBUST CONTROL

So, to obtain this robust schedule in regard of the disturbances on the input-rate of the parts, some constraints are added to the general expression of the optimization problem (6). As the criterion (the passive robustness) is a linear one, it becomes a linear program, and it can be written as :

$$\max\left[\frac{X_j^T.(Q_r + Q_{dt_{max}})}{X_j^T.M_0} - \frac{X_j^T.(Q_r + Q_{dt_{min}})}{X_j^T.M_0}\right] \quad (16)$$

$$\text{with } j \in [1, m-n+1]$$

subject to :

a) potential constraints :

$$A.[1] \le Q_r + Q_{dt} \le B.[1]$$
$$A.[1] \le Q_r + Q_{dt\,min} \le B.[1]$$
$$A.[1] \le Q_r + Q_{dt\,max} \le B.[1]$$

b) the typical staying times must reach the definite throughput ($\pi = \pi_{obj}$) :

$$\forall i = 1, m-n+1 \quad \frac{X_i^T.(Q_r + Q_{dt})}{X_i^T.M_0} = \pi_{obj}$$

c) equality of cycle time of each elementary circuit in order to evaluate the passive robustness :

$$\forall i = 1, m-n+1, i \neq j$$

$$\frac{X_i^T.(Q_r + Q_{dt_{min}})}{X_i^T.M_0} = \frac{X_j^T.(Q_r + Q_{dt_{min}})}{X_j^T.M_0}$$

$$\frac{X_i^T.(Q_r + Q_{dt_{max}})}{X_i^T.M_0} = \frac{X_j^T.(Q_r + Q_{dt_{max}})}{X_j^T.M_0}$$

where :

Q_r : column vector [m x 1] of typical staying times in the places where the staying durations are set in order to obtain $\pi = \pi_{obj}$. The other components are equal to 0 (for our application, these places are associated with a treatment in a tank or an engaged travel).

Q_{dt} : column vector [m x 1] of typical staying times in the remaining places in order to obtain $\pi = \pi_{obj}$. The other components are equal to 0 (for our application, these remaining places are associated with a disengaged travel).

$Q_{dt\,min}$ ($Q_{dt\,max}$) : column vector [m x 1] of current staying times in the remaining places in order to obtain $\pi = \mu_{PRmin}$ ($\pi = \mu_{PRmax}$). The other components are equal to 0.

Notice that : the constraints of this linear program are included in the constraints of the linear program associated with the structural analysis (1), (2). Consequently, this linear program is always feasible if the control is an admissible one and if π_{obj} is an admissible performance (π_{obj} belongs to the active robustness : $\pi_{obj} \in [\mu_{min}, \mu_{max}]$).

In order to set precisely the typical firing dates inside the passive robustness, this constraint must be added to the linear program (16) :

$$\alpha\left(\frac{X_i^T.(Q_r + Q_{dt_{max}})}{X_i^T.M_0}\right) + (1-\alpha)\left(\frac{X_i^T.(Q_r + Q_{dt_{min}})}{X_i^T.M_0}\right) = \pi_{obj}$$

$$\text{with } 0 \le \alpha \le 1 \quad (17)$$

The value of the parameter α allows to specify where the typical firing date of the transition associated with the introduction of a new part in the line is set inside the passive robustness. For example, if you take $\alpha=1/2$, you set this typical firing date in the middle of the passive robustness : Thus, the maximum delay of a part can be totally recovered at the next introduction of a part. Nevertheless, the value of the criterion established by the linear program does not correspond in general to the passive robustness. It gives only twice as big as the delay that can be recovered at the next part. For example, if $\pi_{obj} = \mu_{min}$ and $\alpha=1/2$, the value of the criterion, is equal to 0. The interpretation of this result is that it is impossible to recover a delay with this setting (the throughput rate is the highest one for this control). But this control can be maintained if the delay is less or equal to $(\mu_{PRmax}-\mu_{min})$.

7. APPLICATION

The main results of this work are illustrated here on a single electroplating line composed of one hoist and three tanks. The p-TPN of this line, including its control, is given by the figure 2. According to the duration values specified on this model (for example, the immersion time (in unit of time) of tank 2 is included between [45;50]). The performances evaluation of this control, established by the linear programs (1) (2), is : $[\mu_{min}, \mu_{max}] = [54, 72.5]$

Using the linear program (16) for the definite throughput rate $\pi = \pi_{obj} = 67$, the following results are obtained : the typical immersion values are : 40 units of time for the first tank, 50 for the second and the third one. The passive robustness is : $[\mu_{PRmin}, \mu_{PRmax}]=[54, 70]$. With this setting, the admissible delay is less than or equal to 3.

Adding the constraint (17) with $\alpha=1/2$, the typical firing dates are set in the middle of the passive robustness. The typical immersion values become : 40 units of time for the first tank, 50 for the second one, and 55 for the third one. The maximum recoverable delay is : 5.5. The passive robustness is : $[\mu_{PRmin}, \mu_{PRmax}] = [59, 72.5]$. The figure 3 shows how it is possible to recover totally an admissible delay in order to preserve the throughput rate $\pi_{obj} = 67$ (the maximum delay on the k-th introduction of a part in the line is totally recovered on the (k+1)-th one).

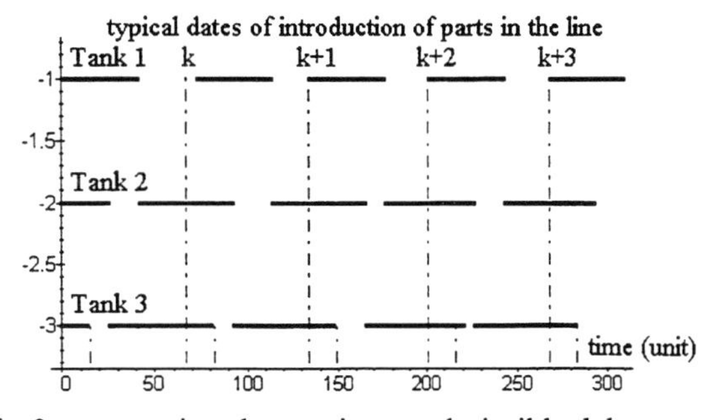

Fig 3 : recovering the maximum admissible delay

8. CONCLUSION

In this paper, we have proposed a method to design a robust schedule in regard of the disturbances on the input-rate of the parts for a definite throughput rate. It can be employed after you have resolved choices and resources allocation problems in order to obtain a P-time SEGC model of your manufacturing cell. A criterion to evaluate external passive robustness has been defined. The robust periodic schedule is obtained by solving a linear program using this criterion. Our next step is to establish a dynamic method using the active robustness : this interval is generally larger than the one of the passive robustness.

This work has received the support of the Rhône - Alpes Region project CORINE.

REFERENCES

Brams, G.W. (1983). *Réseaux de Petri: Théorie et Pratique*, Tome 1 and 2, Masson.

Calvez, S. P. Aygalinc, W. Khansa (1997) P-time Petri Nets for manufacturing systems with staying time constraints, IFAC CIS 97, Belfort France May 20- 22 1997.

Collart-Dutilleul, S. J.P. Denat, W. Khansa (1994) Commande robuste d'un atelier à flot sans stock et sans attente, *APII*, **Vol 28 n°6**, pp. 625-644.

Khansa, W. P. Aygalinc, J.P. Denat. (1996) Structural analysis of p Time Petri Nets, CESA '96 IMACS Multiconférence, Lille France July 9-12 1996, pp. 127-136

Khansa, W. J.P. Denat, S. Collart-Dutilleul (1996). P-Time Petri Nets for Manufacturing Systems, International Workshop on Discret Event SystemsWodes'96. Edinburgh UK, August 19-21 1996, pp. 94-102.

Manier, M.A., P. Baptiste (1994). Etat de l'art : ordonnancement de robots de manutention en galvanoplastie. *APII*, **Vol 28 n°1.**

Laftit, S. (1991). *Graphes d'Evénements Déterministes et Stochastiques : Application aux Systèmes de Production*, Thèse de Docteur en Mathémathiques, Université Paris- Dauphine.

Lei, L. T.J. Wang (1989). On the Optimal Cyclic Schedules of Single Hoist Electroplating Processes. *GSM working Paper, Graduate School of Management, Newark, New Jersey*

Murata, T. (1989). Petri Nets : Properties, Analysis and Applications. *Proceedings of the IEEE*, **Vol 77 n°4,** pp. 541-580.

Philips, L.W., P.S. Hunger. (1976). Mathematical Programming Solution of a Hoist Scheduling Program *AIIE Transactions* , pp. 219-225.

Ramchandani, C. (1974). *Analysis of Asynchronous Concurrent Systems Using Petri Nets*, Technical Rapport **n°120**, Laboratory for Computer Science, MIT, Cambridge, MA.

Shapiro, G.W. L.W. Nuttle (1988). Hoist Scheduling For a PCB Electroplating Facility. *IIE Transactions*, **Vol 20 n° 2**, pp. 157-167.

A FRAMEWORK FOR SIMULATION MODEL MANAGEMENT

Brian W Hollocks

The Business School
Bournemouth University
Fern Barrow, Poole, Dorset BH12 5BB, UK

Abstract: The paper notes that manufacturing industry as the most common field of application for discrete-event simulation, with impact on a wide range of business decisions, yet weaknesses have been found in simulation practice in the field. In the context of increasing use by non-specialists under time-pressure, the paper argues that more emphasis in simulation software development needs to be given to model *management*, rather than to further refinements in model *creation*, and describes a pilot framework for guiding users through the overall simulation process.

Keywords: Simulation, Methodology, Manufacturing Systems, Management, Decision Support Systems

1. INTRODUCTION

Manufacturing industry is under considerable competitive pressure which is continuing to mount. This in turn places increasing pressure on the design and control of manufacturing systems. The emergence of new technologies and operating practices is further increasing the pace of change. According to Peters (1988) and others, change is now the norm, and management, engineers and planners must be concerned with responding speedily and appropriately to that change, pursuing effectiveness and profitability in the midst of uncertainty.

Against this background, decision support systems have much to offer in raising the leverage of industrial decision makers on the issues which they face. A particularly valuable decision support technique in such circumstances is simulation. This involves the creation and use of a computer program which represents the relevant features of some part of the real world such that experiments on that "model" are a predictor of what will happen in reality. The ability to try ideas out in advance of commitment to a course of action is very attractive in an environment of change and uncertainty.

2. APPLICATION AREAS AND BENEFITS

A survey of the use of simulation in UK manufacturing industry (Hollocks 1992) identified the most common application areas to include the following: plant layout and utilisation (77%), material control rules (66%), plant scheduling (60%), capital equipment analysis (52%), inventory control (49%). A total of 14 areas were identified, with from two to six quoted by each user. The application areas are not peculiar to particular industrial sectors or to large companies, but spread widely, including SMEs (Small Manufacturing Enterprises).

Hollocks (1995a) identifies simulation applications with a five-category taxonomy, namely Facilities, Productivity, Resourcing, Training, and Operations. It is evident within that structure that users perceive significant quantitative and qualitative benefits to their companies from utilising simulation and it is also clear that simulation impacts upon a wide range of mission-critical business decisions with substantial financial implications. Benefits identified by users (Hollocks, ibid.) included: risk reduction (80%), operating cost reduction (72%), lead time reduction (72%), faster plant changes (52%), and capital cost reduction (48%). The UK Manufacturing Survey (Hollocks, 1992) identified only 3% of organisations as believing that they had obtained no benefit from simulation. The overall picture, in summary, is of a technique of wide applicability, offering a substantial range of benefits which are delivered with considerable client satisfaction.

3. USER PRACTICE

The direct users of simulation (that is, those who construct and run models) have in the past come from specialist functions such as Operational Research (OR) or Industrial Engineering (IE). However, case studies discussed by Hollocks (1995a) within his taxonomy reveal that, although contemporary use still included a number of such specialists, simulation was already in the hands of many problem-holders, for example engineers and planners, and that was a clear trend. Economic pressures on industry are forcing the reduction in size, if not indeed the closure, of specialist groups, placing responsibility for simulation increasingly with non simulation-specialists (although users who are specialists in *other* fields). An examination of trade literature for simulation software confirms this picture, for example: "without requiring a computer expert" (GENETIK, Insight Logistics), "designed for the end-user" (HOCUS, P-E International), "even for inexperienced users" (ProModel, Production Modeling Corporation), "for use by engineers and other staff directly involved" (WITNESS, Lanner Group). There appears potential for much greater use of simulation and the principal thrusts of the recommendations of the UK Manufacturing Survey were toward increasing that penetration, growth which is likely to be amongst non-specialists. However, two leading obstacles to the use of simulation, specifically identified in that Survey (whose respondents included non-specialists), were lack of training (51%) and lack of appropriate skills (43%).

Simulation, as a technique, has been available for well over 35 years (for example Brigham, 1955 and Tocher & Owen, 1960). Over that time there have been major advances in the power of the software used, for example in the facilities for colour graphics animation (for example Fiddy et al, 1981) and for interactive modelling (for example Clark, 1988). The UK Manufacturing Survey indicated that around 75% of users already employing commercial software, and that proportion will now be much higher given there are fewer in-house teams to create or support local software packages. However, despite its packaging in commercial software, simulation remains a non-trivial tool with statistical foundations. Factors within an industrial system are subject to stochastic behaviour and this has important implications: firstly, that the behaviour is adequately represented in a model so as to provide satisfactory representation (and therefore projections), and, further, that the experimental approach adopted, which is at the heart of simulation, is also adequate. Such statistical issues are well documented in the literature (for example Tocher, 1963, Law & Kelton, 1982 and Kleinjen, 1987).

Hollocks (1995b) reports surveys of actual experimentation practice amongst simulation users in a range of manufacturing and other industries in Europe and the USA. Many of the users were in specialist departments where simulation was a substantial part of their role as consultants to the actual decision makers or problem holders. The survey results should therefore show a "best-case" picture of practice in the field, as the non-specialist, more occasional, users may be expected to be less rigorous. The results, however, demonstrated weaknesses in simulation practice, illustrated here by the responses regarding Run Selection, Warm-up, and Run Length.

In *Run Selection* the general pattern was of simulation being used to evaluate known options (or confirming pre-conceived ideas) rather than exploring the problem space and evolving new options, or of runs being selected subjectively in pursuit of some best solution. Almost all respondents, at some time, simply adopted fixed options for their runs, affecting around a third of the disclosed run selection decisions. A majority of users, associated with a further third of the run decisions, determined the run selection on the basis of personal judgement. However, there were indications of method in their judgement, for example the investigation of bottlenecks, but the selection was basically subjective. Less than 20% of models employed a formal approach to run selection, either by statistical experimental design or with some formally defined search procedure. There was no mention of variance reduction in run decisions.

In *Warm-up* around a third of models, from almost half the respondents, used a fixed warm-up period. The basis was commonly subjective and, for many models, unchanging. However, for just over half of the models, there *was* a deliberate seeking of stability of some measure. This practice was followed by 75% of users at least some of the time (and by half of those every time). Unfortunately, the means to achieve this within the simulation software at their disposal was limited, typically by visual inspection of a time series graph, a statistic, or a dynamic system element in the display (such as a queue length).

In *Run Length* the most significant factor in determining run length (covering c. 40% of models and 70% of users), was the real time available, irrespective of the accuracy secured. Otherwise the picture was similar to Warm-up, with a search for perceived stability within the limited software functionality available.

4. NEED FOR FURTHER USER SUPPORT

It appears from the user survey described above, that simulation methodology in practice is, at best, variable, with subjective judgement common. This must, in turn, raise questions concerning the soundness of some simulation results. There is a prima facie case that simulation reliability has been heavily dependant on the empirical contribution of specialists in judgement and interpretation. With non-specialist users, there will be higher risk. A common problem expressed by users was time pressure. With little software support to experimentation, it is user-time intensive and the same business circumstances which encourage the use of simulation, place pressures on users to deliver results quickly. Economic trends are unlikely to allow that environment to change. With limited or no tools to increase the leverage of users on the wider process, it cannot be surprising if short cuts are taken and compromises on experimentation made. Such a picture indicates that priority in simulation software development now needs to be given to what can be termed model *management*, that is cohesive support to the wider simulation process, including experimentation.

There is some evidence from the survey reported above, in the not infrequent employment of existing graphics/statistics features to address stability, that users would employ experimentation support were it available and easy to use - and providing that timetables can be met. However, inspection of the historic development path of mainstream simulation software reveals an emphasis (albeit not wholly exclusive) on model building as distinct from model use. A 1990 review of software developer/vendor plans orchestrated by Law (1990), and covering seven leading players in the simulation market, revealed no material shift of emphasis to support of the wider simulation process. On the contrary, priority remained with ease of use and graphics, followed by what might be termed "accessibility", namely portability and integration.

Some existing software has offered limited support for some time, for example SIMAN (Simulation Modeling Corporation) has facilities to assist set up a series of simulation runs and compare outcomes. Latterly there have been further signs of response to the need, for example the OptQuest function of Micro Saint 2.0 (Micro Analysis and Design) which aims to find the best parameter combination against a defined objective function. However, against this, past research in experimentation support, for example the inclusion of Expert Systems functionality (such as in Mellichamp & Park, 1989), has shown considerable promise but has still not been taken up in mainstream software.

5. A FRAMEWORK FOR PROCESS SUPPORT

The overall simulation process can be represented as a formal methodology. The classical simulation methodology (for example Banks & Carson, 1984) reflects the philosophy of early batch computing. Hollocks (1991) proposed a wider Interactive Methodology to reflect advances in hardware and software technology and new approaches in Information Technology, for example prototyping (as in Grant, 1986) which is highly relevant to contemporary simulators, such as WITNESS (Lanner Group), ProModel (Production Modeling corporation) or SimFactory (CACI). Within this Interactive Methodology (and within the profile of a given simulation package) it is important that the software facilities encourage good practice and guide the user, particularly the non-specialist.

Each study should commence with a clear **Problem Definition**, establishing the objectives unambiguously in the context of the decision(s) to be made, be they one-off or routine. **Pre-Analysis** can then be used to confirm the need for simulation and the real problem. (At this point the resources to carry through the simulation study can be planned). **Measure(s) of Performance** should then be agreed in order to judge the behaviour of the model, and thus the system, against the operational goals. A **Start-Point** for modelling is then selected and the model "**Grown**" commencing with that area. As an area is **Proved** (ie verified and provisionally validated), contiguous areas are added one by one, each extension being proved in turn. When the model has been grown to what, from the goals and from proving appears a suitable point, full **Validation** can be carried out. Where validity is not confirmed, the growth phase is resumed to increase scope or detail. **Data Collection and Analysis** should be commenced as soon as the need is identified.

Having established a valid model with valid data, experimentation can commence focusing on the measures of performance established. This involves appropriate **Run Selection**, with each **Run** involving an appropriate **Warm-up** or Run-in period followed by a carefully chosen **Run-length**. The **Analysis** of results from the experiment may generate further requirements to investigate new ideas or alternatives. Runs have limited meaning in isolation and inter-run **Comparison** is necessary. **Communication** of the outcome should flow on naturally, but still involve formal reporting, followed by **Implementation** of the decisions made. All the assumptions and data used in a model, together with information regarding model construction, must be **Documented** together

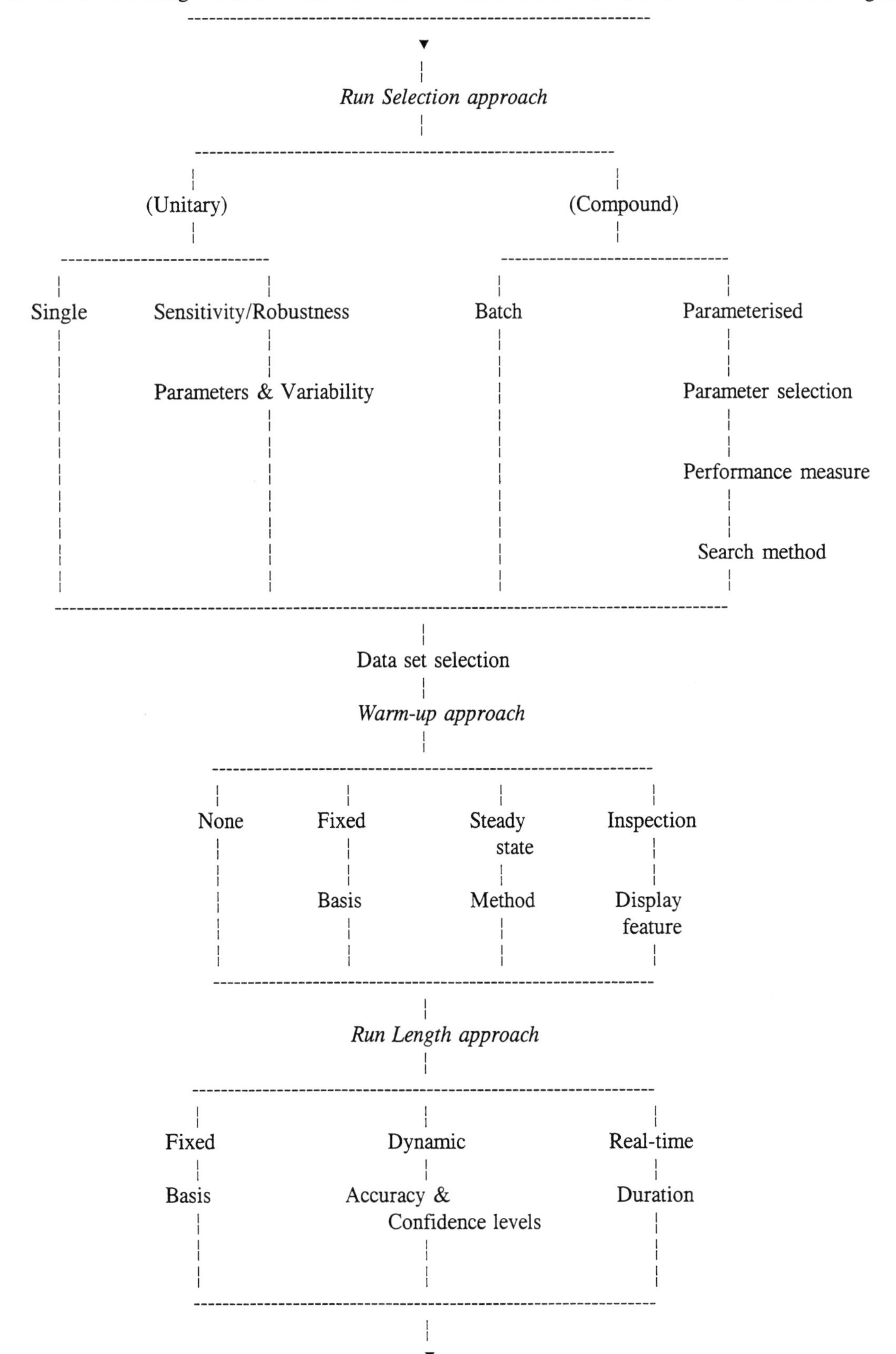

Fig. 1. Outline structure of section of framework for simulation process support.

with validation records and full results. A final, later, phase is an **Audit** of effectiveness, in particular comparison of predicted results, data and assumptions against those actually occurring when the change addressed is operational. The main goal is to identify where new facilities do not conform to specification rather than to check the accuracy of the model.

To test the basic hypothesis that a framework for sopport to the wider simulation process could be provided and is likely to be employed by users, a simple prompting structure of menus was created (independently of simulation software), employing components already proven in other contexts. This was "bench tested" on four intensive industrial users of simulation and their response was positive. Space here does not permit a full description, but the overall structure of the sections covering Run Selection to Run Length is shown in Figure 1.

This part of the process would be preceded by guidance in Problem Definition, for example with Soft Systems tools, and Pre-Analysis, with links to spreadsheets or Rough-cut Modelling facilities (for example Suri & Diehl, 1987). The framework at Run Selection offers users not only simple single or batched runs but also sensitivity or robustness analysis around a single basic data set plus a choice of parameter-driven search options. The latter could include full option evaluation, fractionalfactorial experiment (for example Hunter & Naylor, 1970), AI-based search methods (such as the breadth-first or depth-first techniques discussed by Rowe, 1988), Hill-climbing (for example Rosenbrock, 1965), Meta-models (for example Friedman, 1996), or other tools.

At Warm-up, the framework offers users choice of existing fixed and visual-inspection methods plus the option of dynamic monitoring of a performance measure based on established empirical algorithms (for example Conway, 1963 and Gafarian et al, 1977). At Run Length, the user is offered Fixed, Real-time, and Dynamic options, the latter being based on user-specified accuracy and confidence levels to be applied to a the performance measure supplied earlier in the dialogue, eg 95% confidence that the true value of the performance measure is within +/- 5%.

Prior to initiating the run(s), there is a check back to ensure that the user does not wish to change some factor. The run specification framework is then followed by the selection of Run Monitoring, for example: None, Animation, Full-trace, Cusum, and Time-series. The Experiment is then processed, with interaction on monitoring where necessary (and as permitted by the software), followed by guidance of the user through interrogation and analysis of the results, for example: variable values, queue lengths, run comparisons, significance testing. Analysis facilities can exploit products of research in data presentation (for example Ehrenberg, 1975, Tufte, 1983 & 1990, and Eick & Fyock, 1996).

At each step, the user selects from a menu the approach to be employed. Subsequent prompts are then dependent on the selection made until the framework steps converge to the next stage. For example, at Run Selection the user is offered: Single-shot ... Robustness ... Batch ... Parameterised. If, for example, Robustness were chosen, the user would be prompted for data source and then parameter variability range. The variability range reflects the user's subjective assessment of the reliability of the parameter value supplied in the base data. The support sequence then moves to Warm-up and thence to Run Length.

6. CONCLUSION

Simulation is increasingly being employed by end-user groups directly, rather than through specialists. However, there are clear indications of weaknesses in practice in the field. Simulation software development has characteristically focused on model building, leaving experimentation as user-time intensive. There is therefore an urgent need for support to users in the wider simulation process.

There is evidence that users would employ experimentation support were it available in an easy to use form. A framework has been formulated which prompts users through the key components of model management, including experimentation, and user reaction has been favourable. The next stage is to create a full prototype, integrated with a software package.

Manufacturing industry is the most common field of application of discrete-event simulation involving a wide range of design and control issues in manufacturing systems with significant quantitative and qualitative benefits to the companies concerned. Growth in the use of tools which can leverage manufacturing effectiveness is highly desirable.

REFERENCES

Banks, J. and J.S. Carson (1984). *Discrete-Event System Simulation.* Prentice Hall, New Jersey.

Brigham, G. (1955). On a Congestion Problem in an Aircraft Factory. *Journal of the Operations Research Society of America* **3**, 412-428.

Clark, M.F. (1988). WITNESS - Unlocking the Power of Visual Interactive Simulation. Paper to *International Conference on Visual Interactive Modelling*, Warwick, UK. European Federation of Operational Research Societies.

Conway, R. (1963). Some Tactical Problems in Digital Simulation. *Management Science* **10**, 47-61.

Ehrenberg, A.S.C. (1975). *Data Reduction: Analysing and Inter-relating Statistical Data*. Wiley, Chichester.

Eick, S.G. and D.E. Fyock (1996). Visualizing Corporate Data. *AT&T Technical Journal*, **75**, 74-86

Fiddy, E., J.G. Bright and R.D. Hurrion (1981). SEE WHY - Interactive Simulation on the Screen. *Proceedings of the Institute of Mechanical Engineers* **C293/81**, 167-172.

Friedman, L.W. (1996). *The Simulation Metamodel*. Kluwer Academic Publications, Massachusetts.

Gafarian, A.V., C.J. Ancker and T. Morisakn (1977). *The Problem of the Initial Transient with Respect to Mean Value in Digital Computer Simulation and the Evaluation of Some Proposed Solutions*. Technical Report No. 77-1, University of Southern California.

Grant, T.J. (1986). Lessons for OR from AI: A Scheduling Case Study. *Journal of the Operational Research Society* **37**, 41-57.

Hollocks, B.W. (1991). The Impact of Interactive Modelling on Simulation Methodology. In: *Proceedings of the European Simulation Multi-Conference, Copenhagen*. SCS, Ghent.

Hollocks, B.W. (1992). A Well Kept Secret? Simulation in Manufacturing Industry Reviewed. *OR Insight* **5**, 12-17.

Hollocks, B.W. (1995a). The Impact of Simulation in Manufacturing Decision Making. *Control Engineering Practice* **3**, 105-112.

Hollocks, B.W. (1995b). Experimentation Practice and the Implications for Simulation Software. In: *Proceedings of EUROSIM'95* (F. Breitenecker and I. Husinsky, (Ed.)), 153-158. North Holland/Elsevier, Amsterdam.

Hunter, J.S. and T.H. Naylor (1970). Experimental Designs for Computer Simulation Experiments. *Management Sciences* **16**, 422-434.

Kleinjen, J.P.C. (1987). *Statistical Tools for Simulation Practitioners*. Marcel Dekker, New York.

Law, A.M. and W.D. Kelton (1982). *Simulation Modelling and Analysis*. McGraw-Hill, New York.

Law, A.M. (1990). Simulation Software for Manufacturing Applications: The Next Few Years. *Industrial Engineering*, June/July.

Mellichamp, J.M. and Y.H. Park (1989). A Statistical Expert System for Simulation Analysis. *Simulation* **52**, 134-139.

Peters, T. (1988). *Thriving On Chaos*. Macmillan, London.

Rosenbrock, H.H. (1965). An Alternative Method for Finding the Minimum of a Function of Several Variables without Calculating Deviations. *Computer Journal* 7, 155-162.

Rowe, N.C. (1988). *Artificial Intelligence Through Prolog*. Prentice-Hall, New Jersey.

Suri, R. and G.W. Diehl (1987). Rough Cut Modelling: An Alternative to Simulation. *CIM Review*, **2**, 2.

Tocher, K.D. (1963). *The Art of Simulation*. English Universities Press, London.

Tocher, K.D. and D.G. Owen (1960). The Automatic Programming of Simulations. In: *Proceedings of the Second Conference of the International Federation of O.R. Societies*, Aix-En-Province (J. Banbury and J. Maitland (Ed.)), 50-60. EUP, London.

Tufte, Edward R. (1983). *The Visual Display of Quantitative Information*. Graphics Press, Connecticut.

Tufte, Edward R. (1990). *Envisioning Information*. Graphics Press, Connecticut

All Trade Marks are acknowledged.

A COMPREHENSIVE APPROACH FOR PLANNING AND CONTROL IN MANUFACTURING SYSTEMS

B. Janusz, W. Reithofer, G. Grütter

University of Karlsruhe, Institute for Real-Time Systems and Robotics (IPR)
Kaiserstr. 12, D-76128 Karlsruhe, Germany, E-Mail: bjanusz@ira.uka.de, rhofer@ira.uka.de

Abstract: This paper describes results of the ESPRIT project HIMAC (Hierarchical Management and Control in Manufacturing Systems). Within this project, several algorithms were developed and implemented: The planning algorithm concerns the minimization of the number of tools in a manufacturing system. The second algorithm is used for the definition of part families and resource groups. Regarding scheduling, an existing algorithm has been adapted and further improved to be applicable to the HIMAC specific representation. The implementation of the algorithms is based on the HIMAC theory, hence they use a common representation of a manufacturing system.

Keywords: Algebraic approach, hierarchies, modelling, optimization, scheduling algorithms

1 INTRODUCTION

The aim of the workpackage "Hierarchical theory of production management and control" in the HIMAC (Hierarchical Management and Control in Manufacturing Systems) project is the development of a unifying theory and methods for designing, realizing and tuning various management and control sub-systems in charge of governing production at different levels. Central steps are the development of an aggregation theory and methods for production planning and scheduling, and the development of control modules.

This paper describes three algorithms for production planning and scheduling. The algorithms are based on a common, algebraic description of a manufacturing system:

- The planning algorithm concerns the minimization of storage and tool utilization for a given production system regarding a specific production plan.
- The second algorithm deals with the definition of part families and groups. The basis for such a grouping is a structural hierarchy which was defined in HIMAC (1995). Its goal is to decrease the complexity of the subsequent scheduling.
- Regarding scheduling, an algorithm and its adaption to the HIMAC theory is presented.

This paper is is structured as follows: First, the aspects of the HIMAC theory are summarized. Then the developed applications are presented: the planning algorithm in section 3, the hierarchy forming algorithm in section 4, and the scheduling algorithm in section 5.

2 MANUFACTURING ALGEBRA

The manufacturing algebra is intended to describe products and related manufacturing operations in a formal way. The algebra should be used when the main concern is *what* and *how* to produce.

The evolution of a manufacturing system is described by the factory state equation. This equation allows to express mathematically, *where* and *when* the manufacturing operations defined by the algebra are performed in order to produce the products at the right time and in the right quantities. For this purpose, the elements of a manufacturing system, i.e., the stores and the machines, must be described.

In this paper, the framework of the theory is presented. More details can be found in Canuto *et al.* (1993) and Donati *et al.* (1995).

2.1 The Algebra

Objects: The set O of objects consists of all types of movable objects in a factory, i.e., raw material, semifinished and finished products, fixtures, tools etc. The number of different object types is n_O. Hence, a n_O-dimensional vector can be defined over the set O, which denotes the quantities of all possible object types.

Manufacturing Operations: Given a set of manufacturing objects O, a manufacturing operation a is defined by an ordered pair $a = (v, y)$ of two quantity vectors where v describes the objects used, y the objects produced by operation a. Each operation has also associated values like working time, cost or probability of failing, and a list of machines which can perform this operation.

Based on all feasible operations, the balance matrix of the manufacturing system can be defined. For each operation $a = (v, y)$ a column with $y - v$ is introduced. The operations and hence the balance matrix can be used in a store independent representation or in a space-representation which considers object types as well as stores, not only object types.

2.2 The State Equation

Store State: The state of a store is characterized by the n_O-dimensional quantity vector, each quantity denoting how many objects of the corresponding type are stored in the store.

Manufacturing System State: The state of the whole manufacturing system is given by a vector of dimension $n_O \times k$ (k denotes the number of stores) which is yielded by concatenating the state vectors of the storing units.

The Factory State Equation: Looking at the manufacturing system state vector at two discrete points of time t and $t+1$ one can see that all manufacturing and transport operations performed within the time interval $[t, t+1]$ transform the state vector X_t into the state vector X_{t+1} according to the following equation:

$$X_{t+1} = X_t + B \cdot u_t. \qquad (1)$$

The control vector u_t selects the manufacturing operations executed at the time t from matrix B.

The algebraic description is suitable for performing computations, but is not suitable for storing data about a manufacturing system. Therefore, an additional formal description of a manufacturing system is derived from the Manufacturing Algebra. The following syntax is given in Backus-Naur Form as it is introduced in Gries (1981).

```
<mfgSys>    ::= <row>|<row><mfgSys>
<row>       ::= <op>|<store>
<op>        ::= <opSpec><inOb><outOb><defOb>
                <execInf>;
<opSpec>    ::= (<machName>,<opName>,<opID>)
<machName>  ::= <string>
<opName>    ::= <string>
<opID>      ::= <string>
<inOb>      ::= [<obList>]
<outOb>     ::= [<obList>]
<defObs>    ::= |[<obList>]
<obList>    ::= <obDescr>|<obList>
<obDescr>   ::= <obID>,<obNb>,<storeName>)
<obID>      ::= <string>
<obNb>      ::= <integer>
<storeName>::= <name>
<execInf>   ::= (<execTime>,<faultProb>,<cost>)
<execTime>  ::= <integer>
<faultProb>::= <float>
<cost>      ::= <float>
<store>     ::= s <capSpec>|S <capSpec>
<capSpec>   ::= (<storeID>,<cap>);
<cap>       ::= -1|<integer>
<string>    ::= <character>|<string><character>
<integer>   ::= <digit>|<integer><digit>
<float>     ::= <digit>|<integer>.<integer>
<digit>     ::= 1|2|3|4|5|6|7|8|9|0.
```

A description of a manufacturing system according to this grammar is the common input of the three algorithms presented in the following. Such a description can be easily transformed into the vector notation introduced in Section 2.1.

3 PLANNING

Two planning areas can be distinguished:

1. The plant, layout or equipment planning usually performed before the operating of a manufacturing system. These tasks can be performed independently of current orders.

2. The production planning, i.e., the assignment of machines to operations and the determination of processing sequences which both depend on the current orders.

The equipment planning is subject of this section. Using the planning algorithm, the "optimal" number of circulating tools for a manufacturing system can be calculated, whereby an optimal system maximizes the capacity utilization with a minimal number of circulating objects.

In the following, the layout of the manufacturing system and the material flow are known. Beside this, a fixed control strategy (push strategy) and a known strategy for solving conflicts are given.

Optimization problems can be expressed as a function which has to be minimized by taking into account a given set of constraints. The constraints for the considered problems are given in the following: The number of circulating objects in a cycle must be higher then the quotient of the time a circulating object needs to run through the cycle and the time t_{max}, which is the processing time of the critical machine. Thus, it must hold:

$$\frac{x_l}{\sum_j t_j} \geq \frac{1}{t_{max}} \Longleftrightarrow x_l \geq \frac{t(l)}{t_{max}}, \quad (2)$$

where x_l is the number of tools in a cycle l, t_j the processing time of machine j in cycle l, and $t(l)$ the sum of the processing times in the cycles.

In order to solve this optimization problem, several sets must be defined:

1. the set of all circulating object types in the manufacturing system, (the number of objects of these types has to be minimized)
2. for each cycle, the set of all circulating object types in this cycle (the number of these objects is limited by the processing times of the involved machines and t_{max}).

The first step of the following provide the input of the algorithm. In the last step, the optimization problem is solved with convetional methods. In steps 2-4, Manufacturing Algebra is used in order to gain the required information for the last step:
1. Establish the set of operations

Describe the m operations in the system using the space-representation of the balance matrix B. Define the processing time t_j of each operation j.

2. Define the set of cycles

For this purpose, an auxilary matrix is derived from matrix B describing which stores belong to which cycle. In a next step, it is extended to a matrix C describing the cycles using the space representation:

$$c_{ij} = \begin{cases} 1, & \text{the store in which object } i \\ & \text{is stored belong to cycle } j \\ 0, & \text{otherwise.} \end{cases} \quad (3)$$

The detailed algorithm can be found in Janusz and Reithofer (1996).

3. Define the set of circulating objects in cycles

In order to define the set of possible circulating object types, the balance matrix B is multiplied with the $(n_O \times (n_O * k))$-dimensional matrix I:

$$I = \begin{pmatrix} I_1 & I_2 & \cdots & I_k \end{pmatrix} \quad (4)$$

where k is the number of stores. The blocks I_b $(1 \leq b \leq k)$ are $n_O \times n_O$ identity matrices. The result of the multiplication $I * B$ is a $n_O \times m$ matrix B^o, which is the store independent balance matrix.

Only an object type q for which $b^o_{qj} = 0$ for all operations j of a cycle can be a circulating object type. This can be expressed by an i-dimensional vector $R^{v,o}$:

$$r^{v,o}_q = \begin{cases} 1, & b^o_{qj} = 0, 1 \leq j \leq n \\ 0, & \text{otherwise.} \end{cases} \quad (5)$$

By multiplying I^T and $R^{v,o}$, the space representation R^v of $R^{v,o}$ is obtained. It is extended to a matrix R by comparing R^v with each coulumn of matrix C: $R_l = R^v \cap C_l$. The operation $\cap$ is defined as follows:

$$\begin{aligned} V &= X \cap Y \\ v_a &= \begin{cases} 1, & x_a = 1 \wedge y_a = 1, a \leq \dim(V) \\ 0, & \text{otherwise.} \end{cases} \end{aligned} \quad (6)$$

R is a $(n_O * k) \times c$-dimensional matrix which describes which objects can be reused in the individual cycles.

4. Define the set of circulating objects by taking into account the operations

In order to count the circulating objects in the system, the state vector X must be multiplied with the $(n_O * k)-$dimensional vector N^v, whereby N^v is subject to the constraint:

$$R_l \cap_N N^v = R_l (1 \leq l \leq c). \quad (7)$$

where $\cap_N$ means:

$$\begin{aligned} V &= X \cap_N Y \\ v_a &= \begin{cases} 0, & x_a = 0 \vee y_a = 0, a \leq \dim(V) \\ y_a, & \text{otherwise.} \end{cases} \end{aligned} \quad (8)$$

N^v must also fulfil a second constraint: X_0 is the $i*k$ dimensional vector which describes the allocation of all objects in the system at time 0. In terms of Manufacturing Algebra, X_0 is the system state at time 0. X_t is the system state at time t if at time d $(0 \leq d < t)$, operations u_d were executed:

$$X_t = X_0 + B * (\sum_{d=0}^{t-1} u_d). \quad (9)$$

For all circulating objects, the following equations must hold independently of u_d, because the number of circulating objects must not change:

$$N^{vT} * X_0 = N^{vT} * X_0 + N^{vT} * B \sum_{d=0}^{t-1} u_d$$
$$\leftrightarrow 0 = N^{vT} * B. \qquad (10)$$

5. Solve the optimization problem

A vector X must be determined, such that:

- $(\sum_{l=1}^{c} N_l)^T * X$ is minimal, and
- $\forall l : 1 \leq l \leq c$: $N_l^T * X \geq \frac{t(l)}{t_{max}}$.

Special cases, e.g., cycles with common machines or loops in superordinate cycles, are treated in Janusz and Reithofer (1996).

4 STRUCTURAL HIERARCHY

The design of the structural hierachy depends on the bills of materials (BOMs) of the products. An appropriate algorithm based on the generic BOMs was presented in HIMAC (1996). It is mainly based on a pseudo-distance matrix which compares two objects with each other. The distance between two objects increases with increasing number of different machines used by these objects. According to this definition, object groups are defined which consist of objects with the pseudo-distance 0. Then, the resources used by the objects can be grouped.

Concerning the Manufacturing Algebra for the implementation of this algorithm, an extension is necessary. A new matrix, the $n_O \times r$ dimensional resource matrix R (n_O is the total number of operations, r the number of resources), is introduced, which describes which operations can be executed on a resource:

$$r_{i,j} = \begin{cases} 1, & \text{if resource } j \text{ is used for operation } i \\ 0, & \text{otherwise.} \end{cases} \qquad (11)$$

Hence, two inputs are assumed to be given in the algebraic description, the balance matrix B, and the resource matrix R. Using these inputs, the algorithm is executed in following way:

1. Definition of the input and output matrices:

Define the input and output matrices B_I and B_O with $B_I = \frac{|B-|B||}{2}$ and $B_O = \frac{B+|B|}{2}$.

2. Definition of the product precedence matrix:

Using the input matrix B_I and the output matrix B_O, the sequence of objects which must be produced in order to produce a finished product is defined: $PP = (B_O * B_I^T)$. PP is the $n_O \times n_O$ dimensional product precedence matrix with:

$$pp_{ij} = \begin{cases} 1, & \text{object } i \text{ is predecessor of object } j \\ 0, & \text{otherwise.} \end{cases} \qquad (12)$$

3. Computation of the usage matrix:

Compute the usage matrix U (dimension $n_O \times r$) by multiplying the output matrix B_O with R: $U = B_O * R$. The usage matrix indicates which objects use which resources:

$$u_{ij} = \begin{cases} 1, & \text{object } i \text{ uses resource j} \\ 0, & \text{otherwise.} \end{cases} \qquad (13)$$

4. Definition of the pseudo-distance matrix:

Based on the usage matrix, the pseudo-distance matrix D (dimension $n_O \times n_O$) can be defined: $d_{ij} = \sum_{k=1}^{e} |u_{ik} - u_{jk}|$.

5. Determination of object groups:

The objects must be grouped into sets, whereby all objects in a set use exactly the same resources. For this purpose, the rows of the pseudo-distance matrix D are compared with each other. The $g \times n_O$ dimensional matrix H is defined (g is the initialy unknown number of groups), whereby:

$$h_{i,j} = \begin{cases} 1, & d_{ij} = 0 \wedge i \leq j, \\ 0, & \text{otherwise.} \end{cases} \qquad (14)$$

All objects i with $h_{ij} = 1$ belong to group j.

6. Determination of resource groups:

The multiplication of the matrices H and U, results in the $r \times g$ dimensional matrix $S = H * U$ indicating which resources are used in which group:

$$s_{i,j} \begin{cases} > 0, & \text{group } i \text{ uses resource } j \\ = 0, & \text{otherwise.} \end{cases} \qquad (15)$$

7. Determination of group sequences:

The aim is to define the sequence (hierarchy) of the resource groups. The sequence of products is already given in the precedence matrix PP. By multiplying matrices H and PP, and the precedence relationship between objects and object gruops is obtained. The resulting matrix $HP = (H * PP) * H^T$ is defined as follows:

$$hp_{ij} \begin{cases} > 0, & \text{object group } j \text{ is predecessor of object group } i \\ = 0, & \text{otherwise.} \end{cases} \qquad (16)$$

The precedence matrix for resources is obtained by multiplying S^T with HP and the result with SP: $SP = (S^T * HP) * S$.

For SP (dimension $r \times r$) hold:

$$sp_{ij} \begin{cases} > 0, & \text{resource group } j \text{ is predecessor of group } i \\ = 0, & \text{otherwise.} \end{cases} \qquad (17)$$

8. Elimination of cycles:

It is possible that some cycles occur in matrix SP. They can be recognized by performing the following algorithm:

1. Compute the input and output degree of each node in SP.
2. Set all entries corresponding to nodes with no inputs or no outputs to zero.
3. Repeat step 1 and step 2 until all nodes have the input and output degree not zero. They belong to cycles.

Now, all resource groups belonging to a cycle are united in one group.

The results of the algorithm are the object families (matrix H), their precedence relationships (matrix HP), and the corresponding resource groups with their precedence relationships (matrices S and SP after the elimination of cycles).

5 SCHEDULING

Finding an optimal job shop schedule is a NP-hard problem. Several algorithms exist to calculate such an optimal schedule. However, they are not applicable for complex systems, Due to this, a heuristics which allows us to calculate a suboptimal schedule with linear time complexity is used.

The HIMAC scheduler presented in the following, uses heuristics for the computation of a schedule and also optimizes a given manufacturing system by optimizing the number of tools and the stock level. This is achieved by generating a subsequent order of schedules.

The structure of the tool follows its three basic functions: the syntactic analysis, the semantic analysis, and the simulative analysis.

The *syntactical analysis* captures all explicit information on the manufacturing system such as machines, stores, objects, operations, etc.

From this, the *semantic analysis* derives the implicit information such as products, raw materials, tools, jobs, number of tools, cycle times of jobs and tools, etc. The procedure to detect the objects is simple: *Raw material* does never appear as output object of an operation. Analogously, *products* never appear as input objects. *Tools* are detected since they circle around in a constant quantity.

A job is a sequence of operations which lead from a set of raw materials to a product. Jobs of the given manufacturing system are detected backwards. For each manufacturable product, the final operation is determined. Afterwards, all operations which produce the semi finished products needed by the final operation are determined and so on.

The task of the *simulative analysis* is to calculate the machine utilization, the store utilization, and the system throughput. To fulfill this task, the schedule has to be calculated, from which the performance measures can be derivated.

5.1 The scheduler

The *scheduler* is based on the static single-pass algorithm (Baker, 1974). The basic difference between Bakers algorithm and the one used here concerns the determination of the subsequent operation. Baker uses a predecessor-relationship, i.e. stores and objects of the manufacturing system are not modelled. However, in the following, this is necessary because the flow of tools and the store content should be detected. Therefore the executability of the operations to be scheduled is not represented with a predecessor relationship but by checking the availability of the resources needed such as the input objects, the operation time on the selected machine and the available store capacity of the output stores.

5.2 Heuristics

As already mentioned, the scheduler is based on heuristics. A two-step heuristics which distinguishes between a main and a tie-break strategy is used. If after the first step more than one operation have the same priority, one operation is selected by applying the tie-break strategy. The following strategies are available as main and tie break strategy:

FCFS	First Come First Served
SPT	Shortest Processing Time
LPT	Longest Processing Time
LWKR	Least Work Remaining
MWKR	Most Work Remaining
MOPNR	Most Operations Remaining
RANDOM	Randomly selected Operation

5.3 *Scheduling Algorithm*

The modified scheduling algoritm is as follows:
`Notations:`

OP_{ex}	List of schedulable operations
OP_{unex}	List of unschedulable operations
σ_j	The earliest time at which operation $j \in OP_{ex}$ could be started
ϕ_j	The earliest time at which $j \in OP_{ex}$ could be completed

`Algorithmn Scheduler`

Inputs:	J_{list}, List of jobs
	M_{list}, List of machines
	L_{list}, List of stores
Outputs:	S_{list}, Schedule
	$Stock(L_i)$, Content of store i

{Initialization OP_{ex} and OP_{unex}}
$OP_{ex} = \emptyset$
$OP_{unex} = \emptyset$
for $i = 1$ to $\#J$
  for $j = 1$ to $\#OP(J_i)n$
    if ($OP(J_i, j)$ is executable) then
      $OP_{ex} = OP_{ex} \cup OP(J_i, j)$
    else
      $OP_{unex} = OP_{unex} \cup OP(J_i, j)$
    fi
  next j
next i
{Scheduling the operations}
while ($OP_{ex} \neq \emptyset$)
  $\phi^* = \min_{j \in OP_{ex}} \{\phi_j\}$
  $m* =$ Machine on which ϕ^* will be exuted
  $OP = \{OP_i \in OP_{ex} | \sigma_i <= \phi^*\}$
  $OP_s \in OP$ = Operation with highest priority according to heuristics
  $S = S \cup (OP_s, \sigma_s, \phi_s)$
  $OP_{ex} = OP_{ex} \backslash OP_s$
  remove all objects $EOb(OP_s, i)_{1 \leq i \leq \#EOb(OP_s)}$ from
    input strore at time σ_s
  set Maschine m^* in interval $[\sigma_s, \phi_s]$ working
  store all objects $AOb(OP_s, i)_{1 \leq i \leq \#AOb(OP_s)}$ in
    output stores at time ϕ_s
  for all $OP_i \in OP_{ex}$
    if OP_i is not executable then
      $OP_{ex} = OP_{ex} \backslash OP_i$
      $OP_{unex} = OP_{unex} \cup OP_i$
    fi
  next OP_i
  for all $OP_j \in OP_{unex}$
    if OP_j is ex then
      $OP_{unex} = OP_{unex} \backslash OP_j$
      $OP_{ex} = OP_{ex} \backslash OP_j$
    fi
  next OP_j
wend

6 CONCLUSION

In this report, three algorithms developed in the HIMAC project are described: an equipment planning algorithm, an algorithm for the design of a structural hierarchy in a manufacturing system, and a scheduling algorithm.

Some applications were described in this paper, which made it possible to use a single algebraic desription of a manufacturing system for solving different tasks. This fact is due to the universality of the algebraic description, which can be - with only few extensions - used for different purposes. All three algorithms were validated by some smaller ans some industry close examples (Janusz and Reithofer, 1996b).

7 ACKNOWLEDGEMENTS

This research was performed at the Institute for Real-Time Compter Systems and Robotics (IPR), Prof.Dr.-Ing. U. Rembold and Prof.Dr.-Ing. R. Dillmann, Computer Science Department, University of Karlsruhe. We would like to thank Prof. Rembold for his support.

REFERENCES

R. Baker (1974). *Introduction to Sequencing and Scheduling.* Wilwy, New York, 1974.

E. Canuto, F. Donati, and M. Vallauri (1993). Factory modelling and production control. In: *International Journal of Modelling and Simulation,* 13:162–166.

F. Donati, E. Canuto, and M. Vallauri (1995). Manufacturing algebra and factory dynamics - synthesis of the research progress. *Report EICAS # 6, HIMAC ESPRIT Project 8141.*

D. Gries (1981). *Science of Programming.* Springer-Verlag.

HIMAC Esprit Basic Research Project 8141 (1995). *Deliverable D3.1: Theory of hierarchical production management and control.* Technical report, INRIA Lorraine.

B. Janusz and W. Reithofer (1996). *Planning and scheduling in HIMAC.* Technical Report 3.3, University of Karlsruhe.

B.Janusz and W.Reithofer (1996b). *Planning and scheduling in HIMAC: Interfaces to the test cases.* Technical Report 3.4, University of Karlsruhe, 1996.

CUSTOMIZED FUZZY LOGIC CONTROLLER GENERATOR

Nelson Acosta[1], Jean-Pierre Deschamps, Gustavo Sutter[2]

ISISTAN - Departamento de Computación y Sistemas
Facultad de Ciencias Exactas - UNCPBA
San Martin 57 - 7000 Tandil (Buenos Aires)
Email: nacosta@tandil.edu.ar

Abstract: Fuzzy controllers can be implemented by standard hardware, dedicated microcontroller or application specific circuits. The basic fuzzy controller architecture can use one or more arithmetic and logical units (ALU). ALUs execute the following operations: fuzzyfication, defuzzyfication, lattice and arithmetic functions. The generator includes several tools that allows to translate the initial problem specification to a specific circuit implementation. From the rule based specification the generator produces a first computing scheme. By applying various transformations (lattice and arithmetic operation minimization, optimal register assignation, ...) the system produces the microprogram that drives the controller. For this purpose, the VHDL language is used to simulate the hardware controller, and the ES2 design kit is used to design the circuits to be integrated.

Keywords: computer aided control system design, fuzzy control, integrated circuit

1. INTRODUCTION

Fuzzy controllers can be implemented by standard hardware. To process a high number of rules in a short time a dedicated microcontroller can be used (see Togai and Watanabe, 1986; Watanabe, et al., 1990; Ungering, et al., 1993; Sasaki, et al., 1993). Another possibility is to use specific circuits (ASIC, FPGA, PLD; see Costa, et al., 1995; Mendel, 1995; Dechamps and Martinez Torre, 1995; Hung, 1995): this approach is considered in this paper. The basic fuzzy architecture can use one or more arithmetic and logical units (ALU). ALUs execute the following operations: fuzzyfication, defuzzyfication, lattice and arithmetic functions.

The section 2 formalizes the fuzzy control algorithm concept, and it deals with the computing scheme translation problem by using unary and binary operations. The unary operations are the fuzzyfication and defuzzyfication functions; they can be implemented by means of memory blocks; the binary operations are the lattice (maximum, minimum) and the arithmetic functions. These functions are made by one or more arithmetic and logical units. In section 3, the synthesis tools developed to reach this purpose are shown.

---

[1] Becario Perfeccionamiento Secretaría de Ciencia y Técnica, UNCPBA, Tandil.
[2] Becario Iniciación, Comisión de Inv. Científicas de la Prov. Bs. As., Tandil.

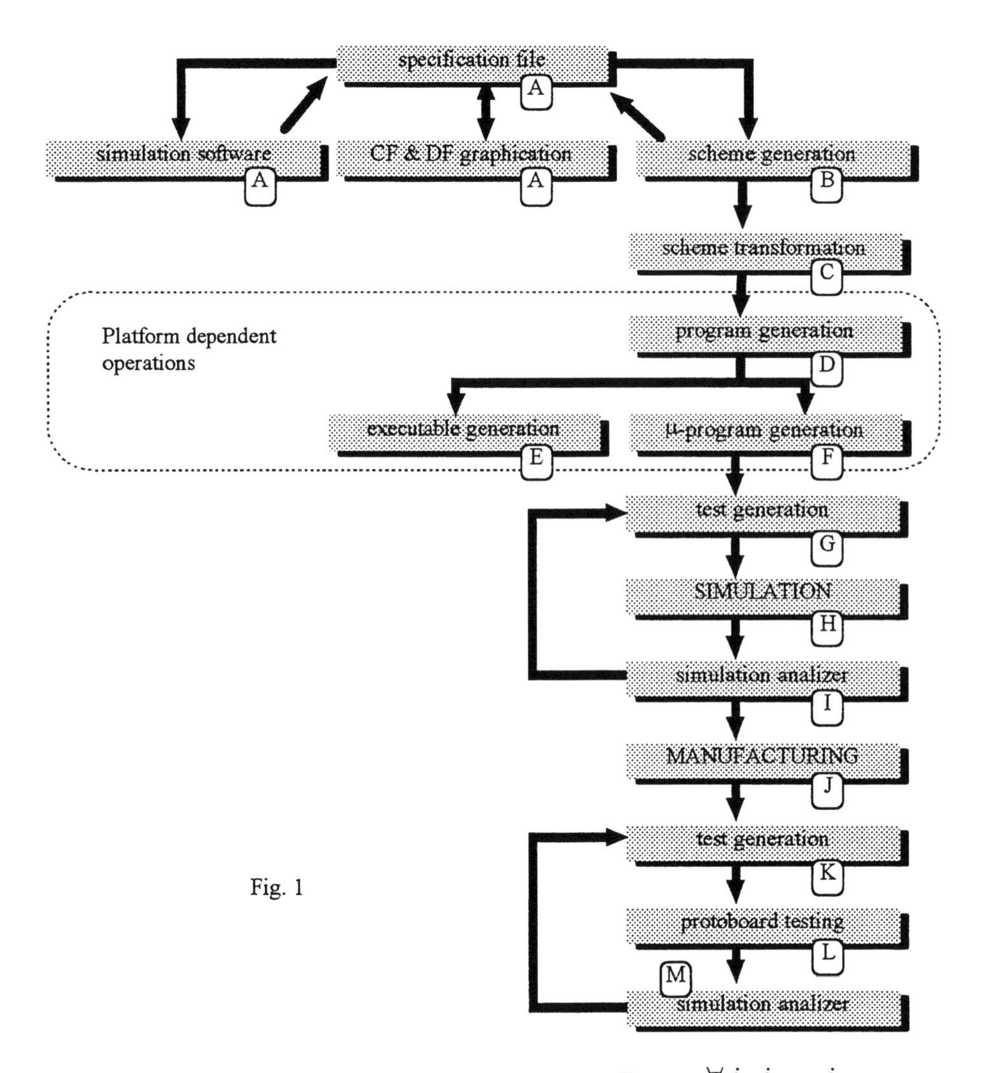

Fig. 1

2. FUZZY LOGIC ALGORITHM

In this section an algorithm to compute the m functions $f_0, f_1, \ldots, f_{m-1}$ of n variables $x_0, x_1, \ldots, x_{n-1}$ is depicted. The value of x_i belongs to a set S_i, and the value of f_k to the set of the real numbers R. To each variable x_i correspond p_i membership functions: $A_{ij}: S_i \rightarrow [0,1]$, $0 \leq j \leq p_i-1$; to each function f_k correspond q_k decodification functions: $B_{kl}: [0,1] \rightarrow R$, $0 \leq 1 \leq q_k-1$; the computation of f_k is based on sentences (inference rules) such as: "IF x_0 is $A_{0\,j0}$ AND x_1 is $A_{1\,j1}$ AND ... AND x_{n-1} is $A_{n-1\,jn-1}$ THEN f_k is B_{kl}". A set of coefficients r is defined: $r(j_0, j_1, \ldots, j_{n-1}, k, 1) = 1$ if the previously mentioned rule applies; else $r(j_0, j_1, \ldots, j_{n-1}, k, 1) = 0$.

The computing algorithm of the f_k functions is:

$\forall i: 0 \leq i \leq n-1$, $\forall j: 0 \leq j \leq p_i-1$,

compute

$y_{ij} = A_{ij}(x_i)$;

$\forall k: 0 \leq k \leq m-1$, $\forall l: 0 \leq l \leq q_k-1$,

compute

$w_{kl} = \vee r\,(j_0, j_1, \ldots, j_{n-1}, k, 1) \,.\, y_{0\,j0} \,.\, y_{1\,j1} \,.\, \ldots \,.$

$\ldots\, y_{n-1\,jn-1}$, $\forall\, j_0, j_1, \ldots, j_{n-1}$

(where the dot symbol "." means the minimum operator and "$\vee$" the maximum);

$v_{kl} = B_{kl}(w_{kl})$; $\forall k: 0 \leq k \leq m-1$,

compute

$N_k = v_{k\,0} + v_{k\,1} + \ldots + v_{k\,qk-1}$,

$D_k = w_{k\,0} + w_{k\,1} + \ldots + w_{k\,qk-1}$,

$f_k = N_k / D_k$

3. PROGRAM GENERATION SYSTEM DESCRIPTION

The developed system, Fig. 1, tries to fill in the gap between the problem specification (rule based) and its implementation by a specific circuit (microprogram controlled). A specification language using rules and fuzzy sets was defined to describe the problem. From this specification file, the system produces a first computing scheme. By applying several translations (lattice and arithmetic operation minimization, register assignation, ...) the system produces a microprogram that drives the controller hardware.

The development cycle to design a fuzzy controller is depicted in Fig. 1. It shows the following steps:

a) Text file controller definition, by using rules, and codification and decodification functions. Two tools are used to verify the specification, to get a membership and decodification function graph, and a source code generator to simulate the controller.
b) Computing scheme generation obtained from the rules that define the controller behavior.
c) Computing scheme transformation, by using lattice optimization.
d) Program generation, by using an optimal register assignation.
e) Executable generation to simulate or execute on 80x86 platforms. This executable has an optimal lattice operations computing with a minimal amount of memory. It is possible to configure the code generator to obtain a maximum execution speed software.
f) μ-program generation to the selected hardware platform (ALUs number, word length, ...).
g) Testing example generation to simulate the hardware platform.
h) Hardware platform simulation using the μ-program in some hardware simulator (e.g. ViewLogic).
i) Simulation analysis to match the controller input/output behavior. The executables generated can be used to compare the outputs.
j) Hardware platform manufacturing (FPGA, ASIC, ...).
k) Testing example generation to verify the hardware platform chip.
l) Chip testing in a workbench PC connected, where the software simulated process activates the controller and gets the outputs back.
m) Simulation analysis to match the controller chip input/output behavior. The executables generated can be used to compare its outputs with the chip outputs.

The program generation system is described by using a short example. The goal is to control a valve (*f0*). The system input variables are: level (*x0*), speed (*x1*) and weight (*x2*). Six rules are used to control *f0*.

3.1. Specification Language

This *input language* allows to define the problem specification (Fig. 1, step A). A language file must include the following information: a) hardware feature definitions (word length, amount of input and output signals). b) signals and fuzzy sets (alias and table name) definitions. c) membership and decodification functions associated with its inputs and outputs. d) set of rules.

The language syntax is:

```
invar( "IVN",IVI,MinX,MinY,MaxX,MaxY )
f_p( "CFN", CFI, "IVN", Polygon )
outvar("OVN",OVI,MinX,MinY,MaxX,MaxY)
f_d( "DFN", DFI, "OVN", Polygon )
si([is( "IVN","CFN"),...], is("OVN","DFN" ) )
```

Where:

- ***CFI*:** codification function number.
- ***CFN*:** codification function name.
- ***DFI*:** decodification function number.
- ***DFN*:** decodification function name.
- **f_d( ):** decodification function.
- **f_p( ):** codification function.
- **invar( ):** input variable definition.
- **is( ):** condition definition.
- ***IVI*:** input variable number.
- ***IVN*:** input variable name.
- ***MinX, MinY, MaxX, MaxY*:** define the function area, where (*MinX,MinY*) is the left-lower point and, (*MaxX,MaxY*) is the right-upper point.
- **outvar( ):** output variable definition.
- ***OVI*:** output variable number.
- ***OVN*:** output variable name.
- ***Polygon*:** define the *n* function points for membership or decodification functions. It is a (*X, Y*) point sequence , that describes the function values. These points make a poligon that defines the function. There are two implicit points defined in the variable, such as the minimum and maximum points over the X axis.
- **si( ):** rule definition.

The definition file example is:

```
invar("level",0,0,0,255,255 )
invar("speed",0,0,0,255,255 )
invar("weight",0,0,0,255,255 )
f_p("high",0, "level",0,255,127,0)
f_p("low",1, "level",128,0,255,255)
f_p("slow",2, "speed",0,0,64,255,127,0)
f_p("fast",3, "speed",64,0,128,255,192,0)
f_p("light",4, "weight",0,0,64,255,127,0)
f_p("hard",5, "weight",64,0,128,255,192,0)
outvar("valve",0,0,0,255,255)
f_d( "open", 0, "valve", 64,255)
f_d( "close", 1, "valve",0,0,64,0,255,255)
si([is("level", "high"),is("speed", "fast")],
        is("valve", "close" ) )
si([is("level", "high"),is("weight", "hard")],
        is("valve", "close" ) )
si([is("level", "low"),is("speed", "slow")],
        is("valve", "open" ) )
si([is("level", "low"),is("weight", "light")],
        is("valve", "open" ) )
```

si([is("speed", "fast"),is("weigth", "hard")],
is("valve", "close"))
si([is("speed", "slow"),is("weigth", "light")],
is("valve", "open"))

3.2. Translation to a computing scheme

Each operation result is assigned to an internal variable (Fig. 1, step B). The *scheme generator* produces the following computing scheme:

```
y[0][0] = A0(x0);
y[0][1] = A1(x0);
y[1][2] = A2(x1);
y[1][3] = A3(x1);
y[2][4] = A4(x2);
y[2][5] = A5(x2);
w[0][0] = y[0][1].y[1][2] |
        | y[0][1].y[2][4] | y[1][2].y[2][4];
w[0][1] = y[0][0].y[1][3] |
        | y[0][0].y[2][5] | y[1][3].y[2][5];
v[0][0] = B0( w[0][0] );
v[0][1] = B1( w[0][1] );
N[0] = v[0][0]+v[0][1];
D[0] = w[0][0]+w[0][1];
F[0] = N[0] / D[0];
```

A *membership and decodification functions editor* produces the contents of each of the memory blocks, both in Jedec format (PROM) and in ES2 memory compiler format.

3.3. Scheme Transformation

The computing circuit includes the tables (A_i,B_j) and a two-input arithmetic and logic unit. Then, it is convenient to reduce the lattice and arithmetic expressions to two-operand expressions (Fig. 1, step C). For this reduction a program based on an heuristic algorithm is used (Deschamps and Acosta, 1996). An inference rule analysis is performed to generate a computing scheme based on two- operand instructions (max, min, add, ...).

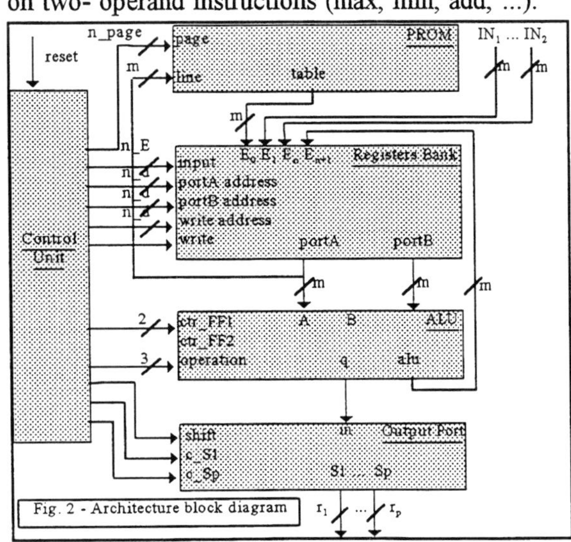

Fig. 2 - Architecture block diagram

The system reorders the computing scheme to let each expression being computed only once.
The following computing scheme is the output of the *scheme transformer* system:

```
y[0][0] = A0(x0);
y[0][1] = A1(x0);
y[1][2] = A2(x1);
y[1][3] = A3(x1);
y[2][4] = A4(x2);
y[2][5] = A5(x2);
w[0][0] = ec5 . ec6;
w[0][1] = ec3 . ec4;
ec3 = ec7 | y[0][0];
ec4 = y[1][3] | y[2][5];
ec5 = ec9 | y[0][1];
ec6 = y[1][2] | y[2][4];
ec7 = y[1][3] . y[2][5];
ec9 = y[1][2] . y[2][4];
v[0][0] = B0( w[0][0] );
v[0][1] = B1( w[0][1] );
N[0] = v[0][0]+v[0][1];
D[0] = w[0][0]+w[0][1];
F[0] = N[0] / D[0];
```

3.4. Computing Program

The data path to execute the calculus is made of the following blocks: (1) a (P)ROM to store the membership tables and the decodification functions, (2) a register bank, (3) an arithmetic and logic unit, (4) output ports.

This data path is controlled by a microprogram stored in another (P)ROM. The circuit is depicted in Fig. 2. To each block corresponds a parametrized VHDL description. For integrating a particular controller, it is necessary to define the parameters (*n,p, m, n_page, n_E* and *n_d*) and to generate the (P)ROM data files (all of them are parameters of the VHDL description).

The translation of the last scheme into a microprogram is done by the following three phases (Fig. 1, step D):
a) optimal register assignation to execute the scheme with a minimal amount of registers.
b) a program based on instructions of table 1.
c) table identification assignation.

The *program generator* produces a program that can be seen in table 2.

Table 1. Instruction set.

INPUT i, j	R(j) <= INi
OUTPUT i	OUTi <= shift register Q (division output)
TABLE i, j, k	R(i) <= row number R(k) of the table number j
ZERO i	R(i) <= 0
MIN i, j, k	R(i) <= R(j) . R(k)
MAX i, j, k	R(i) <= R(j) v R(k)
ADD i1, ..., ip; j1, ..., jp; k1, ..., kp	R(i1)&....&R(ip) <=R(j1)&....&R(jp) + R(k1)&....&R(kp) (two p words operands)
DIV i1,..., ip; j1,..., jp; k	Q(k-1:0) <= R(i1)&....&R(ip) / R(j1)&....&R(jp) (division of p words operands, with k bits precision; the quotient is stored in a shift register in the output block).

Table 2. The program produced by the *program generator*.

0: INPUT 0,0	13: MAX 0,5,4
1: INPUT 1,1	14: MAX 4,2,6
2: INPUT 2,2	15: MIN 2,0,3
3: TABLE 3,2,1	16: MIN 0,4,1
4: TABLE 4,3,1	17: TABLE 4,6,0
5: TABLE 5,4,2	18: TABLE 5,7,0
6: TABLE 6,5,2	19: TABLE 1,8,2
7: MAX 1,3,5	20: TABLE 3,9,2
8: MIN 2,3,5	21: ZERO 7
9: MAX 3,4,6	22: ADD 5,4,3;7,3,1;7;5,4
10: MIN 5,4,6	23: ADD 2,0;7,2;7,0
11: TABLE 4,0,0	24: DIV 5,4,3;2,0,7
12: TABLE 6,1,0	25: OUTPUT 0

The table identification assignation for the membership and decodification functions is depicted in table 3.

Table 3. Table identification assignation

0: A0	5: A5
1: A1	6: $B0_{(7:0)}$
2: A2	7: $B0_{(15:8)}$
3: A3	8: $B1_{(7:0)}$
4: A4	9: $B1_{(15:8)}$

Let's point out that the decodification tables readings are done in single precision, implementing double precision through double memory access.

3.5. Microprogram Generation

To generate the microprogram it is necessary to split the multiple precision operations and the divisions in elementary operations (add and subtract of a single word) executed in one clock cycle (Fig. 1, steps E and F).
For example, the block ADD 2,0;7,2;7,0 will produce the following 3 microinstructions:
a) x <= 0, c <= 0, S/R <= 0;
b) R(0) <= ADD/SUBTRACT(R(2), R(0)),
x <= xsigu, c <= csigu, S/R <= S/R;
c) R(2) <= ADD/SUBTRACT(R(7), R(7));
(where *x*, *c* and *S/R* are internal flip-flops of the ALU while *xsigu* and *csigu* are outputs of the ALU).
The produced microprogram (to be stored in a memory) is obtained both in Jedec format (PROM) and in ES2 memory compiler format. A source code generator can be used to get a executable controller version, and it can be set to yield memory and/or speed optimization.

3.6. Testing data generator and analyzer

During the simulation and testing development steps it can be necessary to use a kit of tools (Fig. 1, steps G, I, K, L and M) to generate testing data inputs and to analyze the simulation outputs.
Test data input sets can be generated by using a small specification language. This tool generates a simulator command file. The simulator executes the command file and generate a textual trace file to be analyzed. The analyzer tool is an information system with SQL queries to analyze the simulation signals.

4. OTHER EXAMPLES

The well known Kosko (1992) controller for parking a truck in a load zone has been entirely developed with the *generator*. The initial specification is based on 12 membership functions, 7 decodification functions and 35 rules. The *scheme generator* produces a 28 instructions computing scheme. The *scheme transformer* generates a 69 instructions computing scheme (one- and two- operand operations). The*program generator* produces a 58 instructions program. Finally, the *microprogram generator* produces a 135 microinstructions microprogram.

Some experiences were done by using classic examples from the literature or environment manufacturers to develop controllers (table 4). The following table shows main features of 10 examples (Acosta, et al., 1997) treated with the development cycle.

Table 4. Controller developed examples

N°	# Rules	# Func.	# Var.	Time (sec.)	Cycles per sec.
1:	20	14	4	8	62500
2:	16	20	6	11	45454
3:	35	19	3	13	38461
4:	18	19	6	7	71428
5:	30	17	3	10	50000
6:	64	22	3	20	25000
7:	22	18	3	9	55555
8:	24	17	3	8	62500
9:	236	31	5	98	5102
10	9	11	3	4	125000

Where, time refers to 500000 cycles execution. The respective controllers apply to: 1) assembly optic fiber, 2) car driving in a track with obstacles, 3) parking a truck in a loading zone, 4) autofocus of a photo camera, 5) double inverted pendulum, 6) servomotor, 7) temperature to glass manufacturing with error conditions, 8) glass manufacturing temperature, 9) plasma reactor temperature, 10) washing machine.

The outputs were obtained from the execution of 500000 cycles, on a PC-Pentium 100Mhz, within DOS available memory. Depending on experience, memory requirements and execution time consumptions for simulation software (Fig. 1, step A) are about 50% higher than for executable generated at step E (Fig. 1).

5. SUMMARY AND FUTURE WORKS

The development of a fuzzy controller by means of a specific circuit (ASIC, FPGA) is performed according to the following steps:

- Definition of rules and membership functions.
- Description of the controller with the *input language*.
- Generation of the Jedec or ES2 ROM files for the membership and decodification functions.
- Translation to a computer scheme (*scheme generator*).
- Scheme transformation (*scheme transformer*).
- Generation of the computing program (*program generator*).
- Generation of the microprogram (*microprogram generator*).
- Circuit synthesis. Thanks to the selected structure (Fig. 2), this problem is reduced to: computing the parameter values (n, p, m, n_page, n_E and n_d) and instantiating several parametrized VHDL entities.

As a future work, the synthesis tools will be adapted in order to support multi-resource architectures (multi-ALU pipelined data path).

REFERENCES

Acosta N., Deschamps JP, Sutter G: "Herramientas para la materialización de controladores difusos µ-programados", Iberchip, DF Mexico, '97.

Costa A., De Gloria A., P.Faraboschi, A.Pagni and G.Rizzotto, "Hardware Solution for Fuzzy Control", Proceedings of the IEEE, vol. 83, n° 3, March 1995, pp. 422 - 434.

Deschamps J.-P. y Acosta N., "Algoritmo de optimización de funciones reticulares aplicado al diseño de controladores difusos", 25as JAIIO, Buenos Aires, sep.'96.

Deschamps J.-P. y Martínez Torre J.I., "Metodología para la materialización de algoritmos borrosos: de una especificación de alto nivel a un ASIC", ESTYLF'95, Murcia, sep.'95, pp. 251-256.

Hung L., "Dedicated digital fuzzy hardware", IEEE Micro, v15, n° 4, Aug 95, pp.31-39.

Kosko B., "Neural network and fuzzy systems", Prentice Hall, 1992.

Mendel M., "Fuzzy logic systems for engineering: a tutorial", Proc. of the IEEE, vol. 83,n° 3, March 1995, pp. 345-377.

Sasaki Mamuro, Fumio Ueno & Takahiro Inoue, "7.5 MFLIPS fuzzy microp. using SIMD and logic-in-memory structure", Proc. IEEE Int.Conf. Fuzzy Systems, pp.527-534, '93.

Togai Masaki and Watanabe Hiroyuki, "Expert system on a chip: an engine for real-time approximate reasoning", IEEE EXPERT, vol.1, n°3, pp.55-62, Aug'86.

Ungering A.P., Thuener Karsten and Goser Karl, "Architecture of a PDM VLSI fuzzy logic controller with pipelining and optimized chip area", Proc. IEEE Int. Conf. on Fuzzy Systems, pp. 447-452, 1993.

Watanabe Hiroyuki, Dettlof D. and Kathy Yount, "A VLSI Fuzzy Logic Controller with Reconfigurable, Cascadable Architecture", IEEE Journal on Solid-State Circuits, vol. 25, n° 2, April 1990.

FUZZY–ADAPTATION OF PI–CONTROLLERS FOR PNEUMATIC SERVO–DRIVES

Michael Berger, Thomas Bernd and Holger Hebisch

*Department of Measurement and Control, Faculty of Mechanical Engineering,
University of Duisburg, D–47048 Duisburg, Germany,
Phone: +49(203)379–3423, Fax: +49(203)379–3027,
(e–mail: berger@mobi.msrt.uni–duisburg.de, bernd@uni–duisburg.de,
hebisch@uni–duisburg.de)*

Abstract: In this paper, a self–tuning procedure for the automatic adaptation of the parameters for the PI controller of a pneumatic system is proposed. The automatic adaptation is documented on the test bed of a pneumatic servo-drive system. The self–tuning procedure is used for the commencement of operations of the process with the PI controller, and for the adaptation of the parameters to changes to the working point, as part of the self–tuning procedure of the closed loop.

Keywords: Fuzzy logic, adaptive control, fuzzy hybrid systems, knowledge acquisition, process control, pneumatic systems

1. INTRODUCTION

Automatic commencement of a process operation helps to shorten the start–up time and also lowers the initial costs. The automatic commencement of operation must be

- simple to handle,
- a simple and intelligible concept,
  and must have
- few or no operating parameters, and
- no or little demand for the dynamics of the process.

The self–tuning procedure is intended for the commencement of process operation with the PI controller and for adaptation of the parameters to changes to the working point, as part of the self–tuning procedure of the closed loop. The step response of the process and three marks are evaluated. The parameters of the PI controller are adapted independently of their initial parameters. The self–tuning procedure of Berger (1996) for a hydraulic servo–drive system has been applied for the pneumatic servo-drive system. An adaptation of the parameters of the PI controller is independent of their initial parameters, and after a few iterations, good control performance results.

Pneumatic actuators are mainly applied in fast working handling systems (Hebisch, 1996). For a long time, the nonlinear characteristics of these drives have prevented control engineers from making use of their common methods of linear control to operate them. Especially methods, that require a linear model will fail, because the nonlinear characteristics of the system demands for a nonlinear model e. g. a fuzzy model. This includes the servosystem, which shall be under study here (Fig. 1). Though several models have been identified

Fig. 1. The pneumatic servo–drive system

for that drive (Hebisch, 1995, Kroll et al., 1995, Bernd et al., 1996), it shall be controlled without the need of any model.

2. DESCRIPTION OF THE TEST BED

The major components of the pneumatic servosystem are depicted in figure 2. Its modular structure is based on the integration of several peripherals in the casing of the actual drive, a rodless pneumatic linear cylinder with a stroke x of 800 mm. These integrated components are first of all the two three-way electro-pneumatic servovalves. They are connected via voltage/current converters to a PC/AT control unit. An additional three-way pneumatic switching valve is included to activate the cylinder carriage's locking brake. Moreover, several feedback elements are installed in the housing of the cylinder. In detail, there are two pressure sensors with the task of measuring the pressure in the two chambers of the cylinder, which are separated by the piston. Furthermore, an incremental position sensor is fitted into the cylinder's shell and into the carriage, which is linked to the piston by a tie.

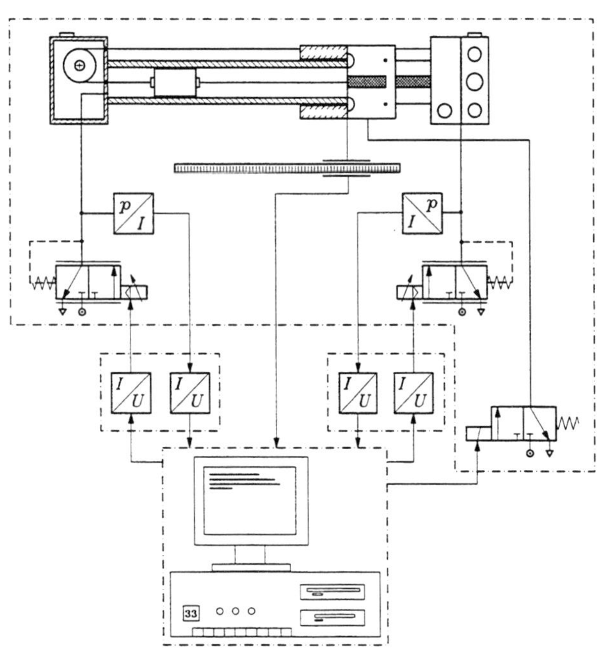

Fig. 2. Diagram of the system

3. SELF–TUNING PROCEDURE

The only requirement for the self–tuning procedure is a stable process. There is no need for a mathematical model or other a–priori information about the process. The goal is independent of the initial parameters of the PI controller; the parameters depend on the step response to adapting until a good control performance results. The parameters of the PI controller depend on the step response computed by means of fuzzy logic. The marks are (Pfeiffer, 1994, Pfeiffer and Isermann, 1994)

- *Overshoot* o_v:
  This is the biggest deviation of the step response from the steady state y_∞ after the step response reaches the tolerance band $y_\infty \pm \varepsilon$ for the first time. Overshoot is divided by the height of the step demand to obtain a relative quantity.

- *Settling ratio* $\gamma = t_r/t_s$:
  This describes the ratio between the rise time t_r and the settling time t_s. At $t = t_r$ the step response reaches the tolerance band for the first time, and for $t > t_s$ the deviations from the steady state stay inside the tolerance band $\pm\varepsilon$ (Fig. 3).

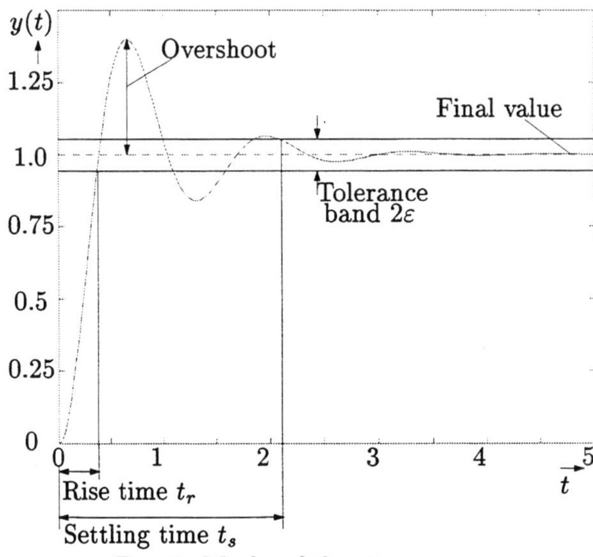

Fig. 3. Marks of the step response

The crisp variables of the two marks, overshoot o_v and settling ratio γ, are computed for all rates of the step command fuzzification (about singletons). Because of this, the correction of the parameters for the next step command is computed. The outputs of the fuzzy algorithms are the factors kr_{Tuning} and cir_{Tuning} for adaptation of the parameters for the i–th iteration step

$$k_c(i) = k_c(i-1)\; kr_{Tuning} \tag{1}$$

and

$$c_I(i) = c_I(i-1)\; cir_{Tuning} \tag{2}$$

where $c_I = \frac{q_0 + q_1}{k_c}$, $k_c = p_0$. c_I is the relative integral action and k_c is the proportional gain of the z–transfer–function

$$G_R(z) = k_c \left(1 + c_I \frac{z^{-1}}{1 - z^{-1}}\right) \tag{3}$$

with the z–transfer of the control algorithms

$$u(k) = u(k-1) + q_0\, e(k) + q_1\, e(k-1) \tag{4}$$

where $q_0 = k_c$ and $q_1 = -k_c(1 - T/t_i)$ (Isermann, 1977).

The marks are limited to the space $\mathbb{D}$. The fuzzification using singletons leads to the fuzzy marks O_v and Γ. The reference fuzzy sets (Zadeh, 1965, 1973) are described by piecewise linear functions.

Reference fuzzy sets as triangular^R:

$$\mu_{\mathbb{A}}(x) = \begin{cases} 0 & \text{for } x \notin]m, c_2[\\ 1 & \text{for } x = m \\ \dfrac{c_2 - x}{c_2 - m} & \text{for } x \in]m, c_2] \end{cases} \tag{5}$$

where $\mathbb{A} = A^{\Gamma}_{1(\Gamma)}$ and $x = \Gamma$.

Reference fuzzy sets as trapezoid^L:

$$\mu_{\mathbb{A}}(x) = \begin{cases} 0 & \text{for } x < c_1 \\ \dfrac{x - c_1}{m - c_1} & \text{for } x \in [c_1, m[\\ 1 & \text{for } x \geq m \end{cases} \tag{6}$$

where $\mathbb{A} \in \left\{A^{\Gamma}_{4(\Gamma)}, A^{O_v}_{4(O_v)}\right\}$ and $x = (O_v, \Gamma)$.

Reference fuzzy sets as trapezoid^R:

$$\mu_{\mathbb{A}}(x) = \begin{cases} 0 & \text{for } x > c_2 \\ \dfrac{c_2 - x}{c_2 - m} & \text{for } x \in [c_2, m[\\ 1 & \text{for } x \leq m \end{cases} \tag{7}$$

where $\mathbb{A} = A^{O_v}_{1(O_v)}$ and $x = O_v$.

Reference fuzzy sets as trapezoid:

$$\mu_{\mathbb{A}}(x) = \begin{cases} 0 & \text{for } x \notin]c_1, c_2[\\ \dfrac{x - c_1}{m_1 - c_1} & \text{for } x \in [c_1, m_1[\\ 1 & \text{for } x \in [m_1, m_2] \\ \dfrac{c_2 - x}{c_2 - m_2} & \text{for } x \in]m_2, c_2] \end{cases} \tag{8}$$

where $\mathbb{A} \in \left\{A^{O_v}_{2(O_v)}, A^{O_v}_{3(O_v)}, A^{\Gamma}_{2(\Gamma)}, A^{\Gamma}_{3(\Gamma)}\right\}$ and $x = (O_v, \Gamma)$.

According to (Pfeiffer, 1994), the length of the step command must be big enough that the process reaches a stationary value for computing the settling ratio γ. Pfeiffer makes no statements about the largeness of the tolerance band $\pm\varepsilon$. The settling ratio cannot be computed if the step response is under the crisp tolerance band. To avoid this the tolerance band is $\pm\varepsilon \neq$ *const.* (Berger, 1996). In Figs. 4 and 5 the reference fuzzy sets for the overshoot O_v, $A^{O_v}_{i(O_v)}$ and the settling ratio Γ, $A^{\Gamma}_{i(\Gamma)}$ are depicted.

The modal value m (Bhme, 1993, Pedrycz, 1993) is different from that of Pfeiffer (1994) and the bounds of the sphere of influence c (Kruse, *et al.*, 1994) of the reference fuzzy sets (for the input and output variables) are changed. In addition, the reference fuzzy set $A^{\Gamma}_{1(\Gamma)}$ was developed (Fig. 5). With these modifications an overshoot of about 1% - 2% and an adaptation independent of the initial parameters of the PI controller result.

By this process, through an enlargement of the integral action time the overshoot can be compensated for. The proportional gain in this situation should be slightly changed. This is realized by means of the additional reference fuzzy set $A^{\Gamma}_{1(\Gamma)}$ (Fig. 5).

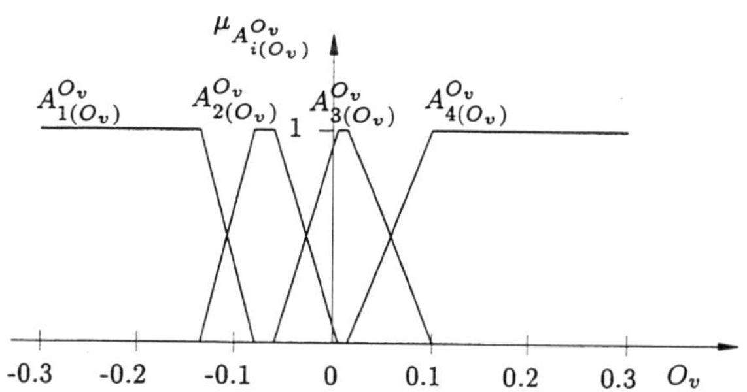

Fig. 4. Modified partition of the reference fuzzy sets $A^{O_v}_{i(O_v)}$ with $i(O_v) = 1, \ldots, 4$ about the space $\mathbb{D}$

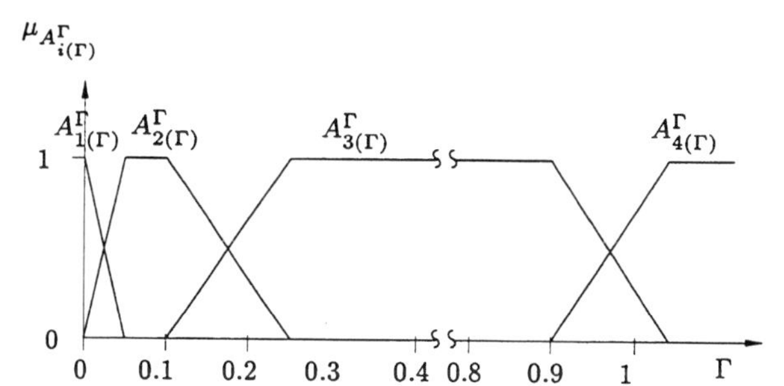

Fig. 5. Modified partition of the reference fuzzy sets $A^{\Gamma}_{i(\Gamma)}$ with $i(\Gamma) = 1, \ldots, 4$ about the space $\mathbb{D}$

Table 1 Parameters of the reference fuzzy sets

Reference fuzzy sets	Parameter
$A^{O_v}_{1(O_v)}$	$m = -0.13; c_2 = -0.08$
$A^{O_v}_{2(O_v)}$	$c_1 = -0.13; m_1 = -0.08$ $m_2 = -0.06; c_2 = 0.01$
$A^{O_v}_{3(O_v)}$	$c_1 = -0.06; m_1 = 0.01; m_2 = 0.02$ $c_2 = 0.1$
$A^{O_v}_{4(O_v)}$	$c_1 = 0.01; m = 0.1$
$A^{\Gamma}_{1(\Gamma)}$	$m = 0; c_2 = 0.05$
$A^{\Gamma}_{2(\Gamma)}$	$c_1 = 0; m_1 = 0.05; m_2 = 0.1$ $c_2 = 0.25$
$A^{\Gamma}_{3(\Gamma)}$	$c_1 = 0.1; m_1 = 0.25; m_2 = 0.9$ $c_2 = 1.1$
$A^{\Gamma}_{4(\Gamma)}$	$c_1 = 0.9; m = 1.1$

The conclusions were modified and adapted during the process. Table 1 shows the modal values m and the bounds of the spheres of influence c of the reference fuzzy sets $A^{O_v}_{i(O_v)}$ and $A^{\Gamma}_{i(\Gamma)}$.

The basis of the fuzzy algorithms consist of 16 relation rules, R_z:

$$\text{IF} \quad (O_v \text{ IS } A^{O_v}_{i(O_v)}) \text{ AND } (\Gamma \text{ IS } A^{\Gamma}_{i(\Gamma)}) \text{ THEN} \quad (Kr_{Tuning} \text{ IS } Z^{Kr_{Tuning}}_{i(Kr_{Tuning})}) , \tag{9}$$

and

$$\text{IF} \quad (O_v \text{ IS } A^{O_v}_{i(O_v)}) \text{ AND } (\Gamma \text{ IS } A^{\Gamma}_{i(\Gamma)}) \text{ THEN} \quad (Cir_{Tuning} \text{ IS } Z^{Cir_{Tuning}}_{i(Cir_{Tuning})}) \tag{10}$$

where $i(O_v) = 1,\ldots,4$, $i(\Gamma) = 1,\ldots,4$, $i(Kr_{Tuning}) = 1,\ldots,13$ and $i(Cir_{Tuning}) = 1,\ldots,13$. Figs. 6 and 7 illustrate the rule–matrix of the relation rules. The modal values (crisp values) of the singletons are entered in the rule–matrix.

		Γ			
		$A^{\Gamma}_{1(\Gamma)}$	$A^{\Gamma}_{2(\Gamma)}$	$A^{\Gamma}_{3(\Gamma)}$	$A^{\Gamma}_{4(\Gamma)}$
O_v	$A^{O_v}_{1(O_v)}$	2.0	2.1	2.2	2.5
	$A^{O_v}_{2(O_v)}$	1.1	1.2	1.5	1.8
	$A^{O_v}_{3(O_v)}$	1.0	1.0	1.0	1.0
	$A^{O_v}_{4(O_v)}$	0.4	0.9	0.65	0.7

Fig. 6. Modified rule–matrix of the relation rules for the factor kr_{Tuning}

		Γ			
		$A^{\Gamma}_{1(\Gamma)}$	$A^{\Gamma}_{2(\Gamma)}$	$A^{\Gamma}_{3(\Gamma)}$	$A^{\Gamma}_{4(\Gamma)}$
O_v	$A^{O_v}_{1(O_v)}$	2.5	1.3	1.3	1.2
	$A^{O_v}_{2(O_v)}$	2.2	1.2	1.1	1.0
	$A^{O_v}_{3(O_v)}$	2.0	0.8	1.0	0.9
	$A^{O_v}_{4(O_v)}$	2.3	0.4	0.75	0.7

Fig. 7. Modified rule–matrix of the relation rules for the factor cir_{Tuning}

The degree of fulfillment α_r is computed by means of the algebraic product (t–norm) (Mizumoto, 1989)

$$\top_{ap,z}(O_v,\Gamma) = \mu_{A^{O_v}_{i(O_v)}} \; \mu_{A^{\Gamma}_{i(\Gamma)}} \tag{11}$$

for each premise z, $z = 1,\ldots,16$, because the algebraic product is interactive, strict (Bandemer and Nther, 1992), fit the logical valuation (Berger 1997a), and a smoother defuzzified output is obtained. The compositions are computed about the sum–prod–operator. For the defuzzification, height defuzzification is used (Driankov, *et al.*, 1993):

$$kr_{Tuning} = \frac{\sum_{z=1}^{16} \top_{ap,z}(O_v,\Gamma)\, m_z}{\sum_{z=1}^{16} \top_{ap,z}(O_v,\Gamma)} \tag{12}$$

and

$$cir_{Tuning} = \frac{\sum_{z=1}^{16} \top_{ap,z}(O_v,\Gamma)\, m_z}{\sum_{z=1}^{16} \top_{ap,z}(O_v,\Gamma)} . \tag{13}$$

By means of the convex (Definition 1.), normalized ($\max_{x\in\mathbb{D}} \mu(x) = 1$) and orthogonal (Definition 2.), here

$$\sum_{i(O_v)=1}^{4} \mu_{A^{O_v}_{i(O_v)}} = 1.0 \tag{14}$$

and

$$\sum_{i(\Gamma)=1}^{4} \mu_{A^{\Gamma}_{i(\Gamma)}} = 1.0 \tag{15}$$

reference fuzzy sets $A^{O_v}_{i(O_v)}$ and $A^{\Gamma}_{i(\Gamma)}$ and the use of the algebraic product we obtain

$$\sum_{z=1}^{16} \top_{ap,z}(O_v,\Gamma) = 1.0 , \tag{16}$$

and get a simplified description of the output values:

$$kr_{Tuning} = \sum_{z=1}^{16} \top_{ap,z}(O_v,\Gamma)\, m_z \tag{17}$$

and

$$cir_{Tuning} = \sum_{z=1}^{16} \top_{ap,z}(O_v,\Gamma)\, m_z . \tag{18}$$

Definition 1. (Lowen 1980)

A fuzzy subset $\mu : \mathbb{D} \to \mathbb{D}^{+}_{N}$ is an convex fuzzy set if for all $x, y \in \mathbb{D}$ and $a \in \mathbb{D}^{+}_{N}$

$$\mu(a\,x + (1-a)\,y) \geq \min(\mu(x), \mu(y)) .$$

Definition 2. (Berger 1997b)

A set $\mathbb{A} = \{A^{(n)}_{i(n)}\}_{i=1,\ldots,K}$ of fuzzy sets $A^{(n)}_{i(n)} = \{(x^{(n)}, \mu^{(n)}_{A^{(n)}_{i(n)}}(x^{(n)})) \mid x^{(n)} \in \mathbb{D}_N\}$ will be called orthogonal about $\mathbb{D}_N$, if

$$\sum_{i=1}^{K} \mu^{(n)}_{A^{(n)}_{i(n)}}(x^{(n)}) = 1 \qquad \forall \; x^{(n)} \in \mathbb{D}_N ,$$

where $x^{(n)}$ is the n–th input variable of the fuzzy system.

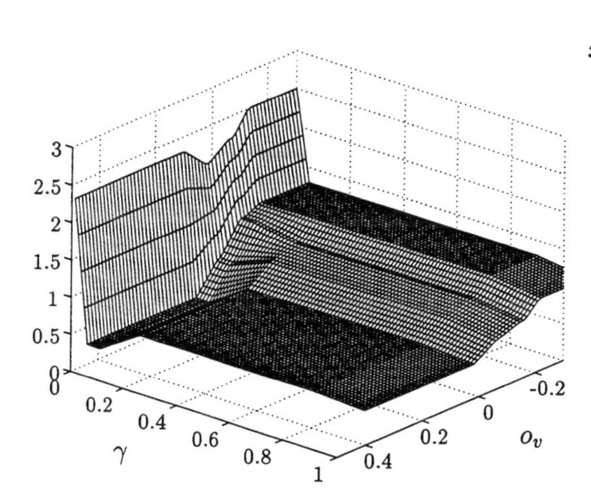

Fig. 8. Control surface of the factor cir_{Tuning}

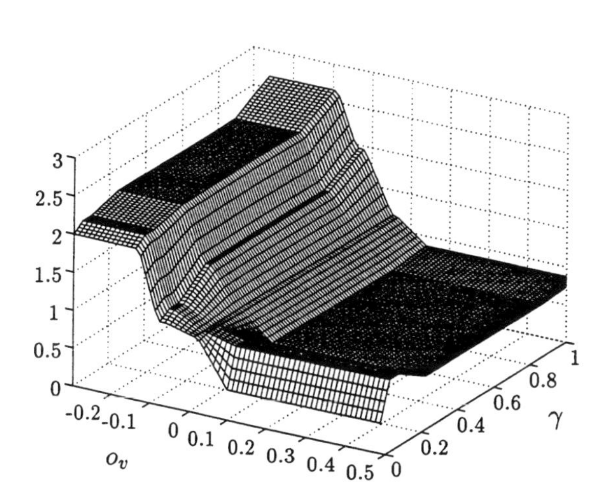

Fig. 9. Control surface of the factor kr_{Tuning}

The control surface is demonstrated in Figs. 8 and 9. In Fig. 10, the progress of the adaptation of the parameters of the PI controller is demonstrated through the measured values of the pneumatic servo-drive system. With the parameters of the PI controller from $k_c = 0.05$ and $t_i = 10000$ s the parameters in 10 iteration steps were adapted so that the step response to the step command $x_d = 0.2$ m was reached without overshoot and with little ramp time. In Fig. 11 the parameters of the PI controller k_c and t_i over the iteration steps are indicated. Figs. 12 and 13 show the progress of the adaptation of the PI controller with initial parameters $k_c = 8$ and $t_i = 2000$ s.

4. CONCLUSIONS

The ability to commence process operation automatically helps to shorten the start–up time and further lowers the initial cost. In this paper, a self–tuning procedure for automatic adaptation of the parameters for a PI controller for a pneumatic

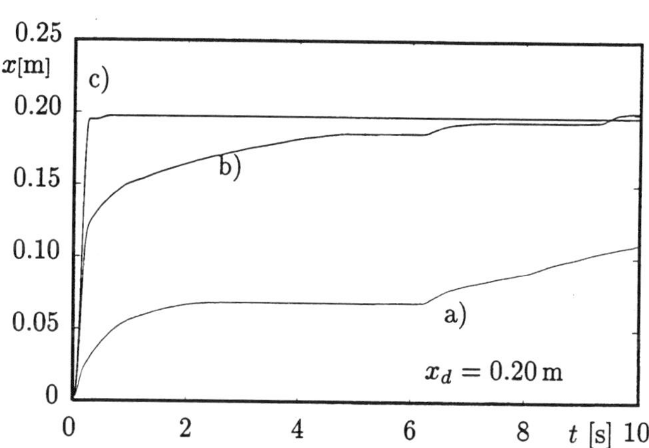

Fig. 10. Progress of the adaptation of the parameters of the PI controller demonstrated through the measured values of the pneumatic servo–drive system after a) 2, b) 4 and c) 10 adaptions

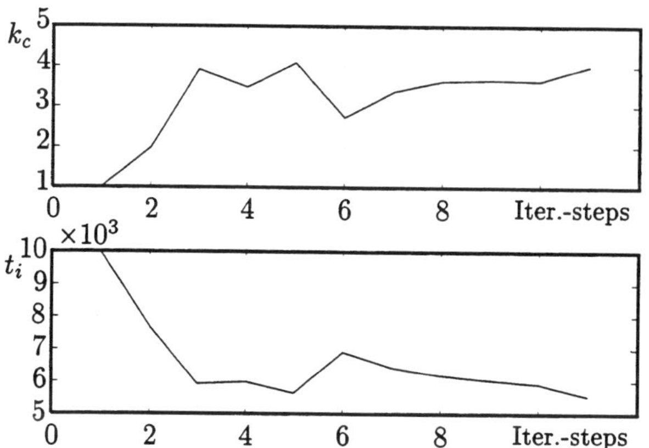

Fig. 11. Rate of the parameters of the PI controller over the iteration steps with initial parameters of $k_c = 0$ and $t_i = 10000$ s

servo-drive system is proposed. The automatic adaptation is documented on the test bed of the pneumatic servo-drive system. The self–tuning procedure is shown for the commencement of process operations with the PI controller, and for the adaptation of the parameters by changes to the working point, as a self–tuning procedure of the closed–loop. The adaptation shows very good results for the process of the pneumatic servo-drive system.

Acknowledgement: The work was supported by the Deutsche Forschungsgemeinschaft (DFG) under grant Schw 120/53-3.

REFERENCES

Bandemer, H. and W. Näther (1992). *Fuzzy Data Analysis.* Kluwer Academic, Boston.

Berger, M. (1996). Self–tuning of a PI controller using fuzzy logic, for a construction unit testing apparatus. *IFAC J. of Control Eng. Practice*, **6**, 785–790.

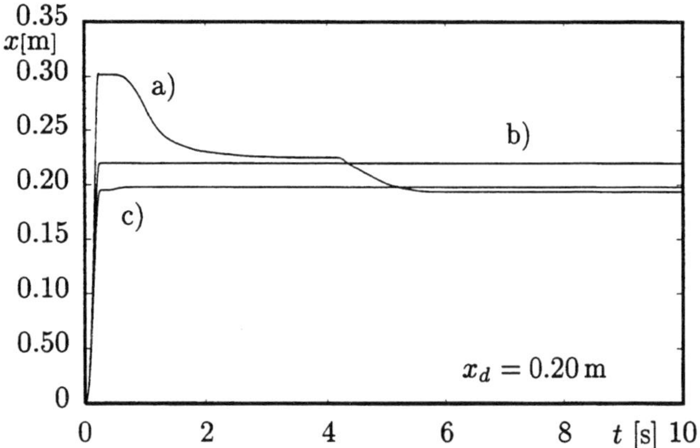

Fig. 12. Progress of the adaptation of the parameters of the PI controller demonstrated through the measured values of the pneumatic servo–drive system after a) 2, b) 4 and c) 10 adaptions

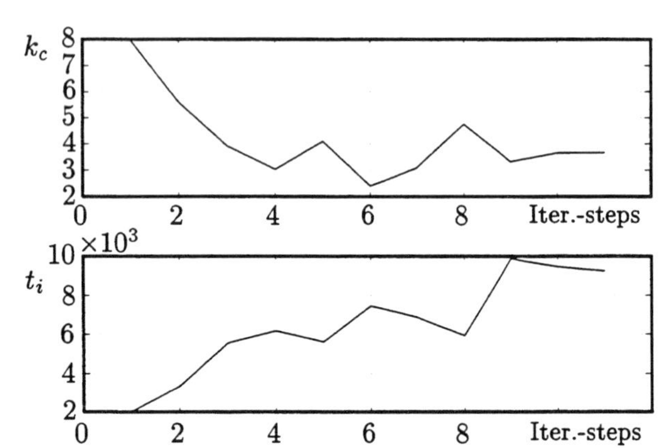

Fig. 13. Rate of the parameters of the PI controller over the iteration steps with initial parameters of $k_c = 8$ and $t_i = 2000$ s

Berger, M. (1997a). T–norm and averaging operators for fuzzy systems: A new valuation about partial differential equations with general variables. *J. of Intelligent Automation and Soft Comp.*.

Berger, M. (1997b). A new parametric family of fuzzy connectives and their application to fuzzy control. *Fuzzy Sets and Systems.*

Bernd, T., A. Kroll and H. Schwarz (1997). *LS-optimal fuzzy-modelling and its application to pneumatic drives.* ECC97 (accepted paper). Brussels, Belgium.

Böhme, G. (1993). *Fuzzy-Logik: Einfhrung in die algebraischen und logischen Grundlagen.* Springer, Berlin.

Driankov, D., H. Hellendoorn and M. Reinfrank (1993). *An Introduction to Fuzzy-Control.* Springer, Berlin.

Hebisch, H. (1995). *Modelling of a pneumatic cylinder drive*, Dep. of Measurement and Control, University of Duisburg. Research report 7/95 (in German).

Hebisch, H. (1996). Sliding–mode control of a rodless pneumatic servo–drive. *CESA '96 IMACS Control, Optimization and Supervision, Lille, France,* 446–449.

Isermann, R. (1977). *Digitale Regelsysteme.* Springer, Berlin.

Kroll, A., H. Reuter and H. Hebisch (1995). *Experimental fuzzy modelling of a pneumatic drive.* Dep. of Measurement and Control, University of Duisburg. Research report 13/95 (in German).

Kruse, R., J. Gebhardt and F. Klawonn (1994). *Foundations of Fuzzy Systems.* John Wiley & Sons, New York.

Lowen, R. (1980). Convex Fuzzy Sets. *Fuzzy Sets and Systems*, **3**, 291–310.

Mizumoto, M. (1989). Pictorial Representation of Fuzzy Connectives, Part I: Case of t–Norms, t–Conorms and Averaging Operators. *Fuzzy Sets and Systems*, **31**, 217–242.

Pedrycz, W. (1993). *Fuzzy Control and Fuzzy Systems.* New York: John Wiley & Sons.

Pfeiffer, B.–M. (1994). Selbsteinstellende klassische Regler mit Fuzzy–Logik. *Automatisierungstechnik at*, **2**, 69–73.

Pfeiffer, B.–M. and R. Isermann (1994). Criteria for successful applications of fuzzy control. *Engineering Applications of Artificial Intelligence*, **3**, 245–253.

Zadeh, L.A. (1965). Fuzzy Sets. *Information and Control*, **8**, 338–353.

Zadeh, L.A. (1973). Outline of a new approach to the analysis of complex systems and decision processes. *IEEE Transactions on Systems, Man, and Cybernetics SMC*, **3**, 28–44.

DESIGN OF DISCRETE EVENT CONTROL SYSTEM FOR IEC1131-3 COMPATIBLE PLCS

D. Karimzadgan, A.H. Jones

Intelligent Machinery Division
Research Institute for Design, Manufacture and Marketing
University of Salford
SALFORD M5 4WT, UK

fax: 44 161 745 59 99

e-mail: d.karimzadgan@aeromech.salford.ac.uk

Abstract: To date there are no general techniques available to convert ordinaryd and/or timed control Petri nets into IEC1131-3 Instruction Lists (IL). In this paper, the Token Passing Logic technique is used, to solve the problem of converting an Automation Petri Net for Control (APNC) of a manufacturing system into IEC1131-3 IL. The generality of the technique means that the method can be applied to any APNC and implemented on any IEC1131-3 compliant PLC. The technique involves, converting the APNC into IL using the TPL methodology. The method provides a straight forward but powerful technique by which to implement IL code.

Keywords: Manufacturing systems, Petri-nets, Programmable logic controllers.

1. INTRODUCTION

IEC1131-3 is claimed to be the first standard to cover PLC programming at international level. This standard describes the Graphical as well as Textual programming languages. Although the Instruction List (IL) is among the textual language covered by IEC1131-3 there are no formal design method for the production of IL. A solution to the design problem, is to deploy Petri net techniques to design Petri net control systems, and use the TPL (Jones, et al. 1996a; Jones, et al. 1996b; Jones, et al. 1996c; Uzam and Jones, 1996a; Uzam and Jones, 1996b; Jones and Uzam, 1996) methodology to convert from PNC into IL. The TPL method is derived from the original work of Petri and embraces the concept of token passing in the design of Petri Net controller.

The purpose of this paper is to show how the TPL method can be applied to design IEC1131-3 IL for a manufacturing system. To do this:

- an Automation Petri Net for Control (APNC) of a Manufacturing System, which shows the working sequence of the control system, is described.
- the TPL method is applied to this APNC by assigning memory bits (for one-bounded places) and memory words (for k bounded places), actions and timers to the appropriate places.
- by using a straight forward mapping from TPL to IEC1131-3 IL, the final IL program is obtained.

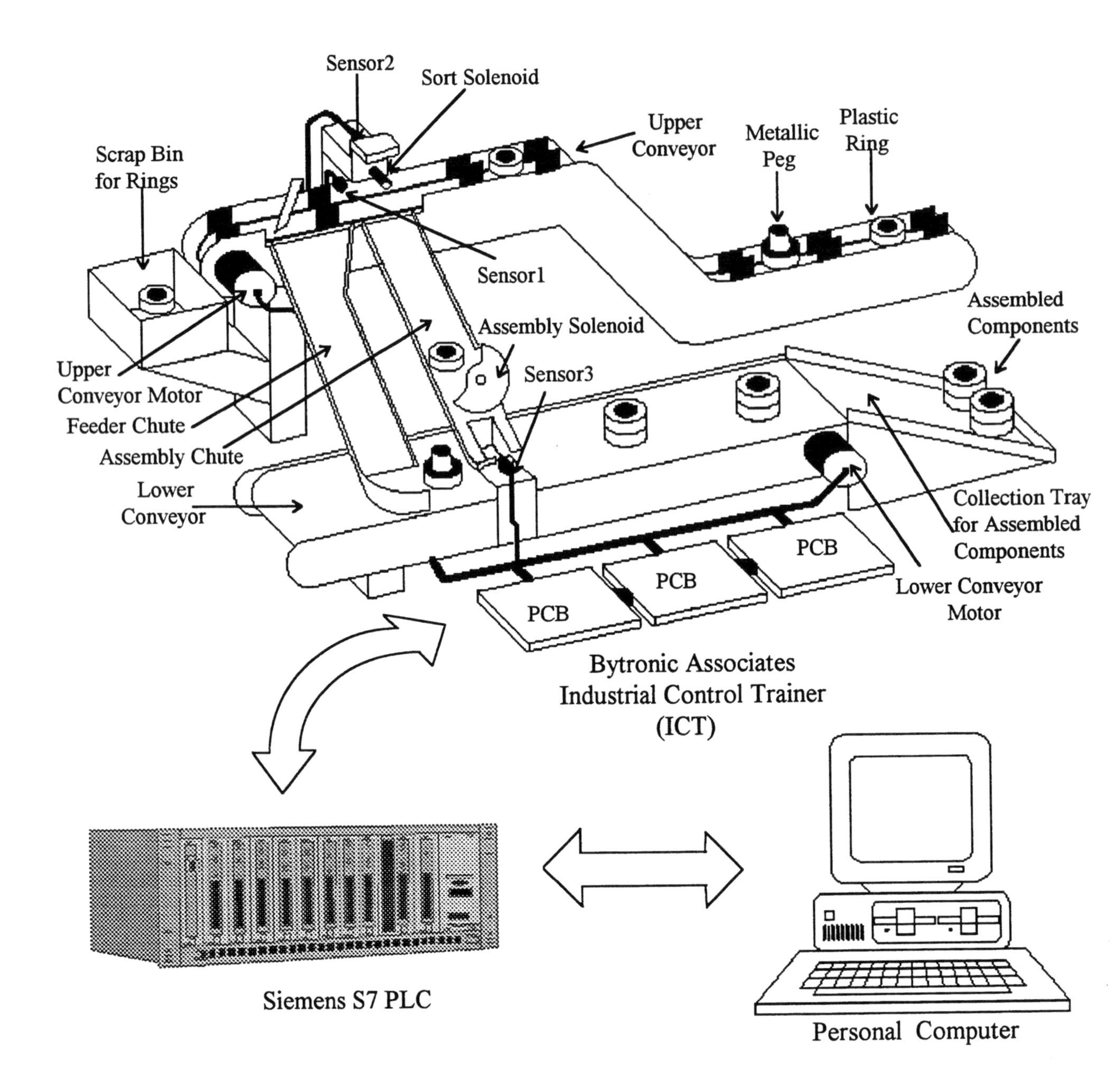

Fig. 1. Manufacturing system.

Table 1: PLC's input assignment description.

PLC Inputs	Sensor No.	Definition
I0.0	Sensor1	Detects a ring or a peg at sort area
I0.1	Sensor2	Detects a peg at sort area
I0.2	Sensor3	Detects a ring in assembly area
I1.0	-	Stops the system
I1.1	-	Starts the system

Table 2: PLC's output assignment description.

PLC Outputs	Actuator No.	Definition
Q2.0	Actuator1	Upper conveyor motor
Q2.1	Actuator2	Lower conveyor motor
Q2.2	Actuator3	Sort Solenoid
Q2.3	Actuator4	Assembly Solenoid

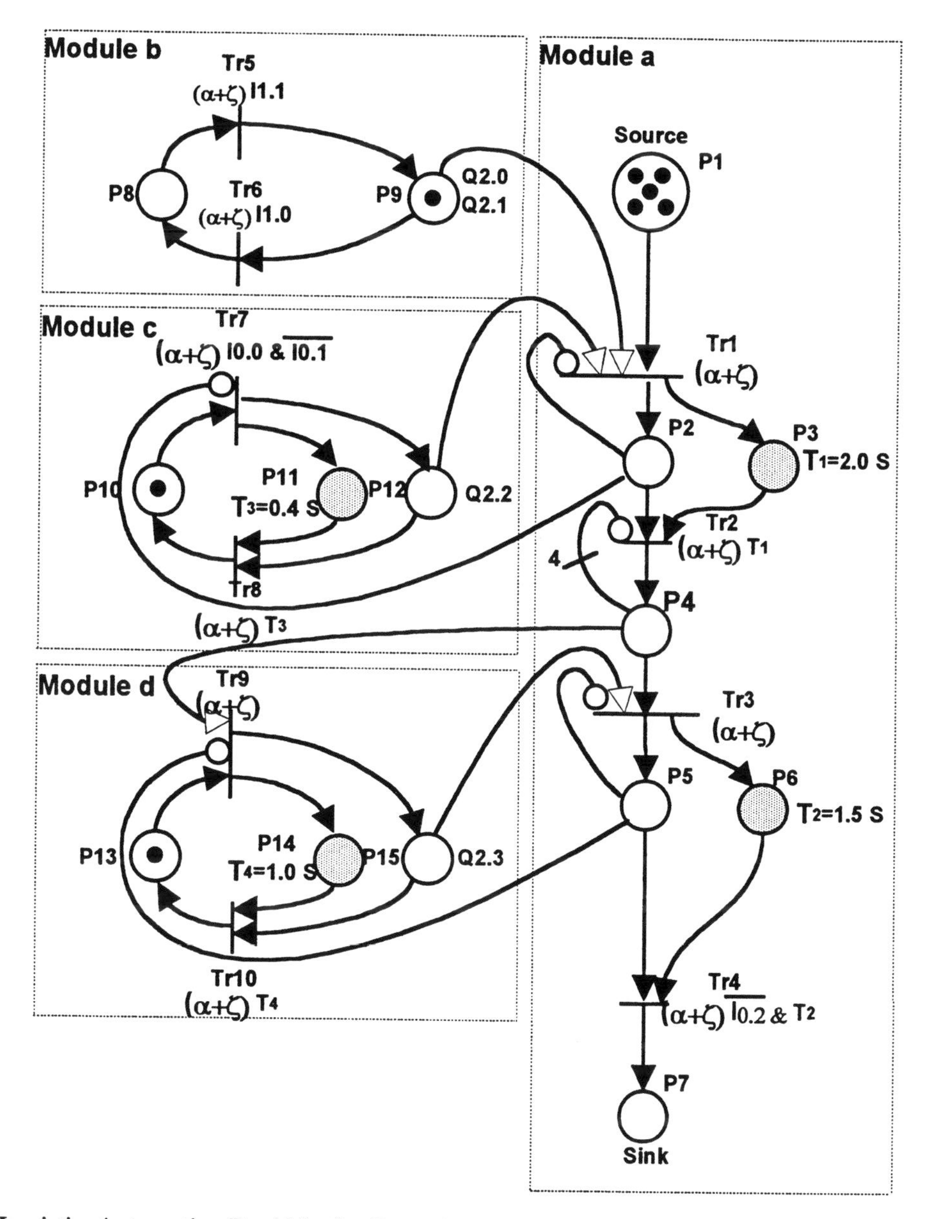

Fig. 2. Heuristic Automation Petri Net for Control (APNC) of manufacturing system.

Table 3 Place description

Place No.	Manufacturing System Interpretation
P1	Source for Ring and Peg components
P2	Fifth place in assembly chute
P3	Timed place (2 sec for ring to take the first place in the chute).
P4	First four place in assembly chute
P5	Assembly area
P6	Timed place (1.5 sec to clear the assembly area before arrival of the next ring).
P7	Complete assembly component
P8	Conveyor motor 1 and 2 off
P9	Conveyor motor 1 and 2 on
P10	Sort solenoid off
P11	Timed place (0.4 sec for proper pushing action).
P12	Sort solenoid on
P13	Rotary solenoid off
P14	Timed place (1 sec for proper routing action).
P15	Rotary solenoid on

Table 4 Transition description

Transition No.	Condition Occurrence
Tr 1	Ring component is detected
Tr 4	No ring in assembly area.
Tr 7	Ring is detected

2. MANUFACTURING SYSTEM

The Manufacturing System, shown in Figure 1, represents a component sorting and assembly processes that can be controlled by virtually any PLC.

The upper conveyor and the lower conveyor are driven by the upper conveyor motor (actuator 1) and the lower conveyor motor (actuator 2) respectively. A random selection of metallic pegs and plastic rings are placed on the upper conveyor. The rings and pegs need to be identified and separated. This is done by two sensors, a proximity sensor (sensor 1) and an infra-red reflective sensor (sensor 2). By using these two sensors a distinction can be made between the peg and the ring. By means of the sort solenoid (actuator 3), plastic rings can be ejected down the assembly chute which can hold a maximum of five plastic rings. Metallic pegs, meanwhile, continue on the upper conveyor and are deflected down the feeder chute. The feeder chute automatically feeds pegs onto the lower conveyor. An infra-red emitter/detector (sensor 3) is used to determine whether or not the assembly area is empty. If it is, the assembly solenoid (actuator 4) is used to dispense a ring from the assembly chute into the assembly area. The assembly area is positioned just above the belt conveyor and, when a metallic peg passes, the peg engages with a hole in the ring and the two components are assembled. The lower conveyor is used to carry completed components into the collection tray.

For the purpose of clearly describing the operation of both the manufacturing system and Automation Petri Net for Control (APNC) the system has been split into four separate areas, namely;

- Conveyor operation.
- Sort area.
- Assembly chute.
- Assembly area.

A Siemens PLC S7-200 (S7 200, 1995) is used to program the PLC. PLC inputs and outputs are given in Table 1 and in Table 2 respectively.

3. PETRI NET CONTROLLER

A heuristically designed Automation Petri Net for Control (APNC) of manufacturing system is presented in Figure 2. The APNC contains four separate modules connected to each others through indirect arcs (i.e. enabling and inhibitor arcs).
The function of each module is as follows:

Module 'a' controls the flow of components through sort area, chute area and assembly area.
Module 'b' controls the two conveyor motors.
Module 'c' controls the sort solenoid.
Module 'd' controls the rotary solenoid

The APNC of the manufacturing system, is designed such that the validation condition for all transitions are ot the type ($\alpha+\varsigma$) (Jones, et al., 1997) or in other words the transition is fired on leading edge or level of enabling condition and leading edge or level of firing condition.

4. TOKEN PASSING LOGIC (TPL) METHOD

The TPL method is applied to this APNC by assigning memory bits (for one-bounded places) and memory words (for k bounded places), actions and timers to the appropriate places.

In order to use TPL method, a memory bit or word is deployed at each IL place to represent the tokens at that place. The TPL method is then applied to the APNC shown in Figure. 2. Detailed information about APNC can be obtained from Table 3 and Table 4.

5. IEC1131-3 INSTRUCTION LIST PROGRAM

The IEC1131-3 IL program is obtained for the Automation Petri Net for Control (APNC) by using the TPL method. The resulting IL program for a Siemens S7 PLC is shown in Table 6. The IEC1131-3 IL symbols (IEC1131-3, 1993) used in Siemens S7 PLC are defined in Table 5.

The IL program has been structured in such a way that at the label *Init.* to initialise the system, the labels *Tr1* to *Tr9* represent the transitions Tr1 to Tr9 respectively and finally, the labels *P3, P6, P11, P14, 98, P12* and *P15* represent the timer and the action places. By adopting this concept further clarity can be added to the system documentation.

Table 5 IEC1131-3 IL used symbols and definitions

IEC1131-3 IL Used Symbols	Definition
LD	Set current result equal to operand
LDN	Set current result to NOT operand
ST	Store current result to operand
S	Set Boolean operand to 1
R	Reset Boolean operand to 0
A	Boolean AND
AN	Boolean ANDN

NOT	Invert power flow
MOVW	Copy input word to output word
LDW>=	Load word integer greater than or equal
LDW<=	Load word integer less than or equal
LDW=	Load word integer equal to
-I	Subtract word integer
+I	Add word integer
TON	On delay timer

Table 6 IEC1131-3 IL program for the APNC.

```
Init
        LDN     M0.1
        MOVW    5,MW1
        MOVW    0,MW4
        MOVW    0,MW7
        S       M9,1
        S       M10,1
        S       M13,1
        R       M0.1,1
Tr 1
        LDW>=   MW1,1
        A       M9
        A       M12
        AN      M2
        -I      1,MW1
        S       M2,1
        S       M3,1
Tr 2
        LDW>=   MW4,4
        NOT
        A       M2
        A       M3
        A       T1
        R       M2,1
        R       M3,1
        +I      1,MW4
Tr3
        LDW>=   MW4,1
        A       M15
        AN      M5
        -I      1,MW4
        S       M5,1
        S       M6,1
Tr4
        LD      M5
        A       M6
        A       T2
        AN      I0.2
        R       M5,1
        R       M6,1
        +I      1,MW7
Tr5
        LD      M8
        A       I1.1
        R       M8
        S       M9
Tr6
        LD      M9
        A       I1.0
        R       M9
        S       M8
Tr7
        LD      M10
        AN      M2
        A       I0.0
        AN      I0.1
        R       M10
        S       M11
        S       M12
Tr8
        LD      M11
        A       M12
        A       T3
        R       M11
        R       M12
        S       M10
Tr9
        LD      M13
        A       M4
        AN      M5
        R       M13
        S       M14
        S       M15
Tr10
        LD      M14
        A       M15
        A       T4
        R       M14
        R       M15
        S       M13
P3
        LD      M3
        TON     T1,2000
P6
        LD      M6
        TON     T2,1500
P11
        LD      M11
        TON     T3,400
P14
        LD      M14
        TON     T4,1000
P9
        LD      M9
        ST      Q2.0
        ST      Q2.1
P12
        LD      M12
        ST      Q2.2
P15
        LD      M15
        ST      Q2.3
        MEND
```

6. CONCLUSION

To date there are no general techniques available to convert ordinary and/or timed control Petri nets into IEC1131-3 Instruction Lists (IL). In this paper, the Token Passing Logic technique is used, to solve the problem of converting a Petri net Controller for a manufacturing system into IEC1131-3 IL. The generality of the technique means that the method can be applied to any Automation Petri Net for Control and implemented on any IEC1131-3 compliant PLCs. The technique involved, converting the Automation Petri Net for Control, into IL using the TPL methodology. The method has provided a straight forward but powerful technique by which to implement IL code.

REFERENCES

IEC1131-3 :1993 = BS EN 61131-3 : 1993 (Programming Languages)

Jones, AH; Karimzadgan, D; Kenway SB (1997). A Formalisation of Automation Petri Nets for Control and Matrix Logic Controllers. To be published in: *Proceedings of the Control of Industrial Systems (CIS'97)*, Belfort (FRANCE) May 20-22

Jones, AH; Uzam, M; Khan, AH; Karimzadgan, D; Kenway SB (1996a). A General Methodology for Converting Petri Nets Into Ladder Logic : The TPLL Methodology. In: *Proceedings of the CIMAT'96* ,France, May 29-31, pp.357-362.

Jones, AH; Uzam, M; Karimzadgan, D (1996b). Design of Knowledge Based Sequential Control Systems. In *First International Symposium on Computing for Industry - ISSCI'96*, Montpellier, France, May 27-30.

Jones, AH; Uzam, M (1996). Towards a Unified Methodology for Converting Coloured Petri Net Controllers into Ladder Logic Using TPLL: Part II- An Application. In *WODES"96*, Edinburgh, UK, August 19-21.

S-7 200 Programmable Controller Manual, Siemens AG 1995.

Uzam, M; Jones, AH (1996a). Design of a Discrete Event Control System for a Manufacturing System Using Token Passing Ladder Logic. In *CESA'96 IMACS Multi Conference, Symp. on Discrete Events and Manufacturing Systems*, Lille, France, July 9-12, pp.513-518.

Uzam, M; Jones, AH (1996b). Towards a Unified Methodology for Converting Coloured Petri Net Controllers into Ladder Logic Using TPLL: Part I- Methodology. In *WODES'96*, Edinburgh, UK, August 19-21.

DESIGN RECOVERY FOR PLC-CONTROLLED MANUFACTURING SYSTEMS

Jacqueline Jarvis[a] and Dennis Jarvis[b]

[a] *University of South Australia, School of Computer and Information Science, The Levels, South Australia, 5095, Australia*
[b] *CSIRO Division of Manufacturing Technology, P.O. Box 4, Woodville South Australia, 5011, Australia*

Abstract: Design recovery has a potentially important role to play in the maintenance and redesign of complex PLC controlled manufacturing systems. We believe that if the finite state behaviour of such systems can be efficiently constructed, we get a representation which is closer to the designer's original intent. Thus the understanding of an existing system which is an essential component of maintenance and redesign is facilitated. The opportunity also exists to use the finite state representation as the basis for formal analysis of system behaviour and also to replace the existing control system with one based directly on the finite state representation. In this paper we describe an approach which has enabled the finite state behaviour to be extracted from a behavioural model of a 700 i/o point, 3 station assembly line.

Keywords: programmable logic controllers, maintenance, distributed control

1. INTRODUCTION

While PLCs have played a major role in the implementation of cost-effective automation, the style of programming that is generally adopted is not conducive to the development of modular control systems. Complex manufacturing systems (such as assembly lines) normally have a well defined hierarchical structure. However, it is difficult to construct PLC based control systems that reflect that structure. If one could do this, we believe that the verification and maintenance tasks for such systems would be significantly reduced. This problem is recognised by industry, and one solution that is emerging is to use separate PLCs for separate functions. However, this solution is only being used for new systems and the problem of maintaining the integrity of the total manufacturing system still remains.

The concept of a holonic manufacturing system has recently emerged as a realistic candidate for the creation of the flexible and modular systems that the manufacturing sector will increasingly require (McFarlane et al, 1995; Deen, 1994). Holonic manufacturing systems will be constructed from entities known as holons, which will exhibit the dual characteristics of autonomous behaviour and the ability to function cooperatively. As such, they are similar to the multi-agent systems of distributed AI (Jennings, 1995). While holonic architectures offer the potential for systems which exhibit improved flexibility, reconfigurability and fault tolerance, most manufacturers are unlikely to embrace this new technology if its adoption requires them to discard their existing manufacturing and control systems. Manufacturers have an enormous investment in existing control technology, and what is required is a methodology which enables existing systems to be progressively evolved into holonic systems. A major motivation for the work described in this paper was to demonstrate that the first step in this process, namely the creation of skeletal holonic subsystems which co-exist with conventional control systems is indeed feasible and that software can be developed which will enable the holonic subsystems to be automatically generated from a model of the existing manufacturing system. In our initial formulation, the control function for each holon is achieved by the holon maintaining knowledge of the states that it can adopt (e.g. clamp open, clamp closed) and the preconditions for its state transitions to occur. Preconditions are specified in terms of holon states. Notification of state transitions is achieved by a broadcast mechanism - when a holon changes state, all holons are notified. The key issue in the conversion of the PLC-based model to a holonic model thus becomes the extraction of preconditions for state transitions from the PLC-based model.

Our work can also be viewed as design recovery[1]. Given a model of a PLC controlled manufacturing system which captures the behaviour of both the control system and the manufacturing system, we generate the detailed finite state behaviours of the elements of the manufacturing system. Note that the resulting model represents a higher level description of the behaviour of the system and better reflects the control system designer's original intent. Being at a higher level of abstraction than the underlying PLC code, the finite state model is easier to understand and it is also easier to check the model for completeness and consistency. The opportunity also exists to use the finite state model as the basis for a formal analysis of system behaviour using a process algebra such as Circal (McCaskill and Milne, 1992).

We have observed with interest recent developments which use finite state representations as the basis for control system design (Brandin and Charbonnier, 1994; Otto and Rath; 1996; Brandl et al, 1996). Brandin and Charbonnier specify the behaviour of the manufacturing system in terms of finite automata; this specification is then implemented on a particular PLC using Grafcet. Otto and Rath describe a graphical programming environment known as HiGraph which was developed by Siemens and is available for their S7 range of PLCs. It directly supports the specification of system behaviour using finite automata (these are referred to as state diagrams) and PLC code is automatically generated from the graphical specification. The graphical representation can also be used to monitor the system during operation, thus facilitating testing and fault finding. Brandl et al have recognised that control system development involves more than PLC programming - mechanical design and electrical design is involved as well, and these activities require design descriptions, albeit at differing levels of detail. They advocate the use of state graphs as the underlying description form and have developed a CASE tool called ASPECT which supports this cross-functional design process. These developments are of interest because they support our belief that finite state automata are a natural way to specify the behaviour of manufacturing systems. Furthermore, that specification appears in a form which is understandable by cross-functional design teams and maintenance personnel.

1. Design recovery originated in the software engineering community from the need to understand and modify poorly documented legacy systems. Design recovery attempts to extract a higher level description of a software system from its source code. The belief is that the new representation will better capture the designer's original intent, thereby facilitating understanding and maintenance of the system. In manufacturing, legacy systems are common and the need for design recovery is exacerbated (particularly in the domain of complex PLC controlled manufacturing systems) because of lax documentation practices and the low level of the languages typically used to implement these systems.

2. PREVIOUS WORK

Falcione and Krogh (1993) describe an algorithm for converting ladder logic programs for PLCs into Sequential Function Chart (SFC) programs. Since an SFC better represents the sequential flow of control logic, it can be viewed as a representation which is closer to the control system designer's intent and its extraction from a PLC program can therefore be viewed as design recovery. However, the scope of their work is limited to the control system - an SFC program interfaces to the manufacturing system in the same way that PLC program does - through physical input and output lines. Not being familiar with their application domain (batch chemical processes), we are uncertain whether extension of the scope of their work to include the state behaviour of the entities being controlled would yield a better understanding of design intent. However, in our domain of interest (complex discrete manufacturing systems such as assembly lines and transfer lines) it does appear to be the case. The reason for this is that an understanding of state behaviour is implicit in the design of such systems - for example, if we were to consider an automotive assembly line then roof clamps must be open before an overhead frame is raised, side clamps must be open before side frames are advanced and so on. A PLC program that embodies such intent will typically control an entity (such as a clamp) by turning an output line on or off. Unfortunately, except in entities which exhibit a very simple state behaviour (such as lamps) this information is insufficient to determine what the resultant state of the entity will be - what is needed is knowledge of the current state. Certainly if that information was incorporated in the PLC program, we would be able to extract state behaviour from a consideration of PLC logic alone. However in our domain, to incorporate such information would greatly increase the size and complexity of the PLC programs and is consequently not standard practice.

The starting point for our work is a behavioural model of the total manufacturing system.The modelling methodology is described elsewhere (Cirocco et al, 1995; Jarvis and Jarvis, 1996). This model captures both the control logic and the finite state behaviour of the entities being controlled in a single model of the complete manufacturing system. We are unaware of any other attempts to construct such models, although as noted in (Jarvis and Jarvis, 1996) there has been some interest in expressing PLC ladder logic using representations other than PLC opcodes. It should be noted that a feature of our modelling approach is that the control logic in our models is generated automatically from PLC program listing files. This is particularly important given the frequency of PLC program modifications in our domain of interest.

3. MODELLING PLC CONTROLLED MANUFACTURING SYSTEMS

3.1 Requirements

A PLC will typically execute the following algorithm:

```
for (ever)
{ Read all PLC inputs;
  Evaluate the PLC program;
  Set all PLC outputs;
}
```

One loop is known as a *scan*. Our objective is not to emulate the detailed operation of the PLC so that accurate scan times can be determined, but rather to simulate its input / output behaviour. That is, having read all the PLC inputs, we need to determine what PLC outputs should be set by evaluating a representation of the PLC program.

A PLC controls a manufacturing system which we view as consisting of 2 types of entities:

a. Agents

An agent is a collection of electromechanical devices that can exist in one of several states. State selection is controlled by one or more PLC outputs. One or more events in the manufacturing process are associated with each agent state.

b. Sensors

Sensors enable us to determine whether a particular manufacturing event has occurred. Examples include limit switches, proximity switches and palm buttons.These entities typically interact in the following manner::

```
Set a PLC output;
The agent changes state and associated
manufacturing events occur;
if (this event/state has a sensor)
{ The new state is detected;
  The associated PLC input is set;
}
```

Any attempt to model the behaviour of a PLC controlled manufacturing system must therefore not only model the behaviour of the PLC and its control program, but also model the state behaviour of the entities that the PLC controls.

3.2 The Target System

In (Jarvis and Jarvis, 1996), we described an approach which enabled us to develop a behavioural model of a PLC controlled assembly line. The assembly line consisted of 3 assembly stations (Stations 10, 20 & 30) linked by a transfer line. Two types of products (referred to as Style A and Style B) are assembled on the line. The operations performed at each station are an alternating sequence of automatic and manual steps. Automatic operation at a station is initiated by an operator (or operators) depressing one or more sets of palm buttons. Those buttons remain depressed for the duration of that step. The activities performed during automatic operation typically involve the opening and closing of clamps. On completion of an automatic step (indicated by a lamp being illuminated on the station control panel), the operator removes his or her hands from the palm buttons and initiates a sequence of manual activities. This may involve welding, the loading of panels, or perhaps removal of the assembly from the station using a manually operated hoist.

When the assembly operations have been completed at a station, the operator needs to wait until assembly has been completed in the other stations. At that point, the transfer line is activated (by the operators at each station simultaneously depressing their palm buttons) and the cycle begins again with a new product. Note that the number of stages in a cycle depends on the style of product that we are building.

4. DESIGN RECOVERY

The model developed in (Cirocco et al, 1995) reflected the structure of the existing PLC controlled manufacturing system. As such, the following entities were explicitly modelled:

- the PLC (code, program execution cycle)
- PLC inputs
- PLC outputs
- sensing devices (limit switches etc.)
- actuated devices (lamps, clamps, motors etc.)
- manufacturing events

In the finite state representation that we want to achieve, the only entities that will appear are agent states and transitions. Furthermore, the preconditions for state transitions to occur will be expressed in terms of agent states. The key issue in the extraction of the finite state representation from the PLC-based model thus becomes the extraction of preconditions for state transitions. The approach that we adopted involved the following steps:

- transformation of the PLC-based model into an equivalent agent state model.

- generation of a complete simulation trace for the agent state model
- extraction of the preconditions for every state transition from the simulation trace

We shall now look in more detail at each of these steps.

4.1 *Creation of the Agent State Model*

The first step in creating the agent state model was to expand each rung of PLC ladder logic (which we had represented in the form of an AND / OR graph) into a corresponding sequence of AND paths. An example of this transformation can be seen in Figure 1.

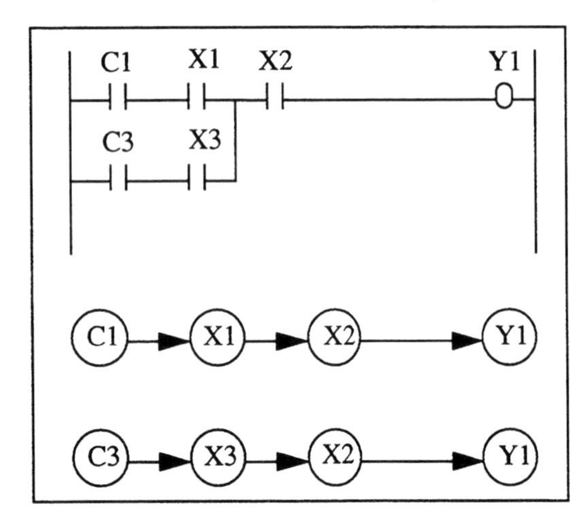

Fig. 1. A PLC program fragment and its corresponding AND path representation

The second step involved the explicit incorporation of agent states into our PLC ladder logic representation. In the actual manufacturing system, PLC inputs correspond directly to agent states. For example, clamp closure may be detected by one or more limit switches which are directly connected to PLC inputs. Software was written to replace all occurrences of PLC inputs in the AND path representation of the PLC logic with their associated agent states. On the other hand, PLC outputs are associated with transitions between states, not the states themselves. The association between agent states and PLC outputs is therefore more complex, as in general it requires knowledge of the current state of the agent as well as the value of the PLC output. Therefore we do not replace PLC outputs when they appear as rung outputs. However, it is common practice to use PLC outputs as rung inputs, (as in the case of latches and interlocks). In the case of a latch, the latched path corresponds to a state preserving transition In such situations, the PLC output which is used as a rung input can be unambiguously associated with an agent state and substitution made. Interlocks are used to ensure mutual exclusion of agent states. In an agent state representation, mutual exclusion is implicit, so we simply remove references to interlocks. We also encountered situations where the PLC output corresponded to a transition between two states, but that transition took a significant amount of time and needed to be referred to as an input in PLC program. This behaviour was observed in the raising and lowering of the overhead frame and the advancement and retraction of the shuttle mechanism. If we model those situations with three states - an initial state, a final state and an in motion state, then the PLC outputs can be unambiguously associated with an agent state.

The resulting agent state model has all PLC inputs and all PLC outputs which appeared as rung inputs replaced by agent states. PLC outputs which appeared as rung outputs are retained. The control loop for the agent state model now becomes

```
for (ever)
{ Read all agent states;
  Evaluate the PLC program;
  Set all PLC outputs;
}
```

Note that modelling of the physical entities is now simplified; we no longer have to associate PLC inputs with agent states.

4.2 *Extraction of Preconditions*

The objective of this phase is to generate for each agent in the system the guards for each state transition associated with that agent. These guards are to be specified in terms of agent states. The procedure is as follows:

```
for (each PLC output)
{ identify the corresponding agent
  state transition;
  for (each time the transition was
  activated)
  { determine the agent states
    necessary for the activation;
    record the preconditions;
  }
}
```

Note that the preconditions (or guards) can be simplified considerably if we introduce additional agents to capture key event occurrence (such as stage completion) and commonly occurring preconditions (such as automatic operation and the type of product being assembled). These "meta-agents" are associated with PLC memory locations and their guards are determined in the same manner as described above.

Generation of the guards was implemented as a three stage process. In stage 1, the agent state model was executed and for every rung output the value for each of its AND paths was recorded. This information was captured for every scan. Note that rung outputs include both PLC outputs and PLC memory locations. In stage 2, dependency records for each change in value of rung outputs associated with either an agent or a meta-agent were generated. The following algorithm was used as the basis to generate the dependency records:

```
for (each rung output to be analysed)
{ outputNode = rung output;
  outputNode = rung output;
  for (j = 0; j <= numberOfScans; j++)
  { outputNodeValue = value(
    outputNode);
    if (any AND path associated with
        outputNode changed its value
        during scan j)
    { print details of the transition;
      examine(outputNode,
      outputNodeValue);
    }
  }
}
```

Note that the process of examining nodes is recursive:

```
examine(outputNode,outputNodeValue)
{ for (each AND path associated with
       outputNode)
  { pathValue = value(AND path);
    for (each node in the AND path)
    { nodeValue = value(node);
      if (relevant(outputNodeValue,
                   pathValue,nodeValue))
        if (node is a terminal node)
          print terminal node details;
        else
        { print node details;
          examine(node,nodeValue);
        }
    }
  }
}
```

`relevant()` determines whether or not we need to examine a particular node. If `outputNodeValue` and `pathValue` are 1, then `nodeValue` must be 1 and the node is relevant. If `outputNodeValue` and `pathValue` are 0, the node is relevant only if `nodeValue` is 0.

The above algorithm generates dependency records for both agents and meta-agents. When generating dependency records for agents, the rung outputs to be analysed will be PLC outputs and terminal nodes will consist of agent states and meta-agent states. When generating dependency records for meta-agents, the rung outputs to be analysed will be PLC memory locations and terminal nodes will be agent states. Note that the presence of mutually dependent nodes and latches required modifications to the above algorithm.

Each dependency record that we generate is a precondition or guard for a particular agent or meta-agent. All that needs to be done in stage 3 is to associate the dependency records with the appropriate agent or meta-agent and to convert the dependency record layout into a more appropriate form (i.e. one line per record). We also add the starting state for the guard to the guard. Typical output is shown in Figure 2.

```
SAV31a: C544b C258b C249b Hoist1b
        SAV29a SAV31b

SAV31a: C545b C258b C249b Hoist1a
        SAV31b

SAV31b: C543b C258b C249b Hoist1a
        SAV31a

SAV31b: C545b C258b C249b Hoist1b
        SAV31a
```

Fig. 2. Guards for the agent `SAV31` when product B is present. States are designated by suffixes; in this case all agents and meta-agents have 2 states, a and b. `SAVxx` are clamps, `Hoist1` is the hoist motor, `C54x` are stage ready indicators, `C249` indicates automatic operation and `C258` indicates that product B is present.

The output in Figure 2 is a much more concise, understandable and complete representation of the behaviour of a clamp than the corresponding ladder logic which consisted of 11 rungs distributed through a 63 page listing file. In particular, duplicated preconditions and redundant behaviours (e.g. manual operations and latches which are used in conjunction with interlocks to ensure mutual exclusion) are removed.

The guard generation process was verified by constructing a distributed model of the assembly line and comparing its behaviour with that of the original model. In the distributed model, the only entities that appeared were agents and meta-agents. The control function was achieved by each entity maintaining knowledge of the states that it can adopt and the preconditions for its state transitions to occur (which were generated using the process described above). In the current model, notification of state transitions is achieved by a broadcast mechanism - when an entity changes state, all entities are notified. The model was constructed using the IDPS distributed programming environment[1].

5. CONCLUSION

The purpose of this exercise was to demonstrate that it is possible to recover a higher level representation of the behaviour of a PLC controlled manufacturing system, given that we have a closed loop model of the system. We believe that feasibility has been clearly shown through our test case, which was a working assembly line of significant complexity. The representation that we have extracted uses finite state automata - we have also shown how these entities can form the basis of a distributed control system.

Further work now needs to be initiated to determine the limitations of our approach and to formalise the process of transforming our original model to an agent state model. We also have the opportunity to formally verify the behaviour of the control system through the use of a process algebra. Other approaches to the generation of guards also need to be considered. We had access to a suitable symbolic representation of the control system, so a dependency-directed approach was attractive. However, if this is not the case for a particular application, then other approaches will need to be used. Recording system states when interesting events occur and then eliminating agent states which do not impact on the event in question is an approach that we are currently exploring.

REFERENCES

Brandin, B. and Charbonnier, F. (1994), The Supervisory Control of the Automated Manufacturing System of the AIP, In *Proc. of the 4th. International Conference on Computer Integrated Manufacturing and Automation Technology*, Troy New York, pp. 319-324.

Brandl, T., Lutz, R. and Reichenbacher,J. (1996), ASPECT - a CASE Tool for Control Functions Originating from Mechanical Layout In Storr, A. and Jarvis, D.H. (Eds) *Software Engineering for Manufacturing Systems: Methods and CASE Tools*, Chapman & Hall,1996.

Cirocco, L.R., Jarvis, D.H., Jarvis, J.H. and Ryan A. (1995), Simulation of a PLC Controlled Assembly Station. Technical Report MTA 333, CSIRO Division of Manufacturing Technology, Adelaide, South Australia.

Deen, S.M. (1993), Cooperation Issues in Holonic Manufacturing Systems. In Yoshikawa, H. and Goosenaerts, J. (Eds.) *Information Infrastructure Systems for Manufacturing (B-14)*, Elsevier Science, pp. 401-412.

Falcione, A. and Krogh, B. (1993), Design Recovery for Relay Ladder Logic. *IEEE Control Systems*, **13**, pp. 90-98.

Jarvis, J.H. and Jarvis, D.H. (1996), Life Cycle Support for PLC Controlled Manufacturing Systems. In Storr, A. and Jarvis, D.H. (Eds) *Software Engineering for Manufacturing Systems: Methods and CASE Tools*, Chapman & Hall, 1996.

Jennings, N.R. (1995), Controlling Cooperative Problem Solving in Industrial Multi-Agent Systems Using Joint Intentions. *Artificial Intelligence*, **75**, pp. 195-240.

McFarlane, D., Marett, B., Elsley, G. & Jarvis, D. (1995), Application of Holonic Methodologies to Problem Diagnosis in a Steel Rod Mill. In *Proc. of the 25th. IEEE Conf. on Systems, Man and Cybernetics*, Vancouver, pp. 940-945.

McCaskill, G.A. and Milne, G.J. (1992), Hardware Description and Verification Using the Circal System, Research Report HDV-24-92, University of Strathclyde, Department of Computer Science, Glasgow, Scotland.

Otto, H. and Rath, G. (1996), State Diagrams: A New Programming Method for Programmable Logic Controllers. In Storr, A. and Jarvis, D.H. (Eds) *Software Engineering for Manufacturing Systems: Methods and CASE Tools*, Chapman & Hall, 1996.

Seki, T., Hasegawa, T., Okataku, Y. and Tamura, S. (1991), An Operating System for the Intellectual Distributed Processing System - An Object Oriented Approach Based on Broadcast Communication, *Journal of Information Processing*, **14**, pp. 405-413.

1. IDPS is a programming environment which supports the development of prototype IDPS-OS applications. IDPS operates in a UNIX / TCP/IP environment. IDPS-OS (Seki et al, 1991) is a distributed, fault tolerant operating system which was implemented using reliable broadcasting between objects.

MULTIMEDIA REQUIREMENTS FOR MANAGEMENT AND CONTROL OF MANUFACTURING SYSTEMS

Géza Haidegger, George L. Kovács

CIM Research Lab., Computer & Automation Institute, Budapest, Hungary
Kende u. 13, H-1111 Budapest, Hungary, Phone: (36-1) 1811-143, Fax: (36-1) 1667-503
e-mail:haidegger@sztaki.hu

Abstract: There is a clear trend in computer communications to apply fast and high bandwidth networking methods and elements. Their spread in the office environment is very promising and straightforward, but their direct application in manufacturing environment needs some prior investigation. How will it effect the management and control tasks in manufacturing systems? Besides other effects, it will enable the wide-spread use of multimedia features, and the authors focus on that issue. After giving a bunch of application oriented discussion topics, an experimental pilot system is detailed, that should be oriented in the future to make some application experiments with manufacturing system.

Keywords: multimedia, manufacturing systems, LAN, ISDN, control, management

1. INTRODUCTION

This paper discusses some aspects of industrial networking features, discusses the advent of ATM technology and the wide-spread use of ISDN at the shop-floor level, and investigates the chances for using efficiently the higher bandwidth's possibilities in manufacturing.

The future management and control systems will definitely require the use of multimedia techniques. Local multimedia functions are already appearing in systems under implementation, and future systems will need to establish even more multimedia features integrated with remote resources.

The paper discusses the potential power, benefits associated with real-time sound and video features implemented in shop-floor systems. It has already been investigated, that video signal transmission can significantly help human participants in a hybrid automated system, but some results in picture recognition suggest, that highly automated computer control systems could also make use of remotely stored video data.

Before applying new technologies in a production/manufacturing environment, it is needed to test it with trials, and/or with pilot, experimental systems. The authors have been participating in a Copernicus EC project developing a test demonstration environment for distributed multimedia (ISLAND Project). Presently the aim is to apply the underlying technology in an experimental CIM environment.

2. MULTIMEDIA REQUIREMENTS

There are 3 basic requirement categories for multimedia (MM) technology applications. We describe MM as an underlying technology that equally allows and facilitates the use of data, voice and pictures (still images, sequences of images and real-time video images up to full-screen full-motion video on demand) among various platforms and sites almost regardless of physical distance.

With that definition, the basic requirement categories, such as networking, contents generation and MM utilization will be discussed in the following chapters.

3. NETWORKING

It has already been proven and accepted, that in manufacturing systems, the large number of computer-based equipment and the large number of persons taking part in the manufacturing process are located in a distributed, rather large space, and only *Open Systems* based networking technology can cope with the data communication tasks among the participating nodes.

Point-to-point communication is overtaken by bus (or ring) topology based technologies, and LANs (local area networks) are accepted as fundamental building blocks for factory automation projects. The basic issue on LANs now refer to the available bandwidth, that they can offer. Ethernet based LANs can deliver information packets with 10 Mbits/sec. In a well designed environment, Ethernet should not be loaded over 60% for efficient use. There is a new version of the Ethernet technology, the Fast-Ethernet, that allows the increase of bandwidth up to 100 Mbits/sec.

Presently there is another type of available technology to offer higher bandwidth, that is the ISDN. Most of the European countries are ready to offer services of Euro-ISDN, that use the twisted pair cables of the telephone lines, and the exchange centres can run connections for 2 to 30 times 64 Kbit/sec data rates (Basic rate or Primary rate ISDN).

The technology advancement introduced the ATM, for offering high bandwidth transfer mechanism, and it will supply almost unlimited bandwidth, according to individual user demand. Presently this technology is not available on a large scale, though each country in our neighborhood had already initiated some sort of pilot. Due to the very high initial costs, end-users are not yet prepared to apply ATM. The authors believe, that on long term, however, ATM will be dominating in the telecommunication area.

For Multimedia applications, one of the most concerned issue is the available bandwidth. To transmit a still picture, the pixels of the full picture must be transmitted, thus the size of the picture (64x64, 640x480, 800x600, 1024x768, 1280x1024) and the number of the bytes for the pixel color (4 bits, 8 bits, 15-16 bits or 24 bits) are determining factors.

To transmit a video signal with voice, it consumes around 10-12 Mbauds. Clearly, it is too much for the presently widely applied media technologies, (e.g. Ethernet). When video signals can be compressed, the required bandwidth can significantly be reduced. If small picture size is enough, and the refresh rate can be less than needed for „continuous" moving effect, again, the transmitting demands are less burning.

The networking requirements of office or shop-floor environment are similar, but not totally equivalent. Let us point out just two of the most important features, worth for considerations: the real-time issue and the noise tolerance. The MAP technology has proven, that real-time features with properly designed loads could be utilized with Ethernet based cabling technology for applications in large-scale production systems, like automotive factories all across the globe, and that noise tolerance can also be dealt with by applying industry-hardened Ethernet cables (having an additional shielding coating) for example.

4. MM CONTENTS

To apply MM features, the systems must provide them. Now, we discuss primarily the images, the video and the sound media types, as additional items above the already utilized data (numeric) type of information. Similarly to the data type, the pictures, the video and the sound could either come from a remote data (MM-data)-base, or from a source node either pre-stored on generated on-spot in real-time. The functions of a previously stored MM information is dealt with a new technology termed as „Video-on-demand" and with traditional Remote Data-base technology. With the previously stored video, the content is already compressed and packed, for efficient transmission and efficient storage requirements.

When picture, video and sound is to be generated on-line and real-time for MM transmission, the basic requirement is on the processing power needed to code and compress the signals. Presently this is the most demanding functionality for such services.

5. USING AND BENEFITING OF MM ACCESS

Replaying or rebuilding the video pictures on-line from a compressed format requires a pentium chip, in contrast to the compression protocols, which need about 5 times higher computing power. The source of MM information can be local, or LAN-accessible or remote. We consider MM source to be local, when it can be accessed by the computer from its own hard-disc drive or CD drive. It is LAN accessible, when a Local Area Network can deliver the information with at least Ethernet bandwidth.

(e.g. shared CD drives, server PC). When the required MM information is remote, it can be accessed by linked networks, or gateways. The open networking has the advantage of reaching any destination on the globe, but the distance correlates with longer access times and delivery times. Also the available bandwidth is likely to be narrower for continuous MM information transfer.
It must clearly be considered, that distributed MM is not to be handled just as the data networking. Its use and miss-use can be very costly and dangerous for the complete IT environment. Applying the benefits can bring great results (see prototype applications), but when not used correctly, it can overload most computer nodes and networks by the high traffic. It is also important to design the application environment such a way, that the human operators working with those systems should not get bored or blind.

6. MM APPLICATIONS IN MANUFACTURING

It is commonly estimated, that MM will spread into every aspect of our life, let it be the child's education, the grown-up person's working scenario or his or her entertainment. As television and computers became part of our everyday life, so will the MM integrate them with tele-communication (e.g. telephone and computer net or „internet"). In a factory environment let us now focus on the design and operation engineers' specific roles, and not forget that each individual will need to use and apply MM for functions like tele-education, video-conferencing, tele-shopping, etc.

The design engineer will need to use his MM terminal to compare his animated, simulated manufacturing process data with those of other, stored and retrieved pictures, he will need to talk to his colleagues and customers while exchanging computer data and images, or while analyzing video recordings at the same time at different locations, or when opening a window to monitor the presently running factory processes, while the other window represents the animated, simulated manufacturing process. The means to have a direct comparison between the virtual reality (computer generated environment of the manufacturing process) and the real-life (video-camera transmitted) picture appearing in an adjacent display window allows to test and verify the computer models and enhance the design and operation process.

On the shop-floor, the local operators (we strongly believe, that even with highly integrated and automated manufacturing systems, skilled technicians or engineers will be needed to run the factory operations) will have access to MM type of information to talk (phone) with each other, while having pictures shared or running shared computer application programs. Display windows can insert real-time video images (e.g. camera signals from various shop-floor locations, and also from in-process technology view-points. (When using cooling liquid at machining centres or lathes, the closed doors do not allow the operators to have a good, close-up view on the chip-removal process itself.) These pictures can be stored locally or at a video-server, can be transmitted to the design-engineer's workstation, etc.

The operation management staff can have access to any information happening at the plant. Let it be numeric data on the machines, status signals, or sound or video images on the production. These again can be stored on special technologies, they can be compared to previously run technologies. Picture processing computers can analyze and evaluate them. In the case of production problems, the engineers can apply tele-presence or tele-operation functions. With robots and visual sensor equipped manipulators hazardous operations can easily be performed. Such functionalities are already used in underwater or space construction projects, but they will be widely used in almost all factories. Catastrophic events will be handled by extensive use of tele-presence, tele-operation and international real-time video-conferencing. This will highly increase safety.

7. PROTOTYPE APPLICATION OF THE MM TECHNOLOGY

The EU COPERNICUS project ISLAND (Island, 1995) aimed to provide a prototype demonstration for the integrated services implemented in a local area network. ISDN network, as a widely utilized and fast spreading communication means may and should be enhanced to allow LAN functions and services. In other words the main objective of the prototype demonstration is to transport multimedia information in a LAN environment over a single physical bearer. This means data and isochronous information (for instance video, voice) transfer over a standard twisted pair cable system. The international standardization activities for such solutions have resulted in a document referenced as the OSI IEEE 802.9.

The demonstration had to be done in an efficient and affordable way. It is shown that a technical solution is feasible without demanding a great investment in a LAN infrastructure, and without losing compatibility with existing LAN hard- and software. Thus, we tried to match descriptions and procedures as laid down in the approved draft 'IEEE 802.9 Standard, Integrated Services LAN Interface at the MAC and PHY Layers'. However, due to limitations in the project resources we could not develop a full implementation of the IEEE 802.9 standard. By using well-known modules and defining sensible interfaces we tried to be prepared for technical innovations in the future, and make the demonstration set-up flexible enough to enhance it towards a more complete implementation of the IEEE 802.9.

7.1 Data communication

The demonstration system is able to transport isochronous information and data simultaneously. In accordance with the IEEE 802.9 the data channel is called P-channel, and the isochronous channel is referenced as C-channel. In the 802.9 a C-channel can be any combinations of B-channels (nx64 Kbits). The ISLAND demonstrator can support a dynamic bandwidth allocation only in a limited scale. A D-channel is being emulated through the P-channel.

The data on the P-channel can be of any sort, all standard applications must be able to work with this channel. The isochronous channel will be for video and voice information.

7.2 TOPOLOGY

The demonstration setup is able to support at least four workstations. These workstations are connected to a central Access Unit (AU). The AU acts as a bridge or a hub for the data information. It also connects the video channels of the workstations and (for now) it is able to set up the connection and to switch on and off the required C-channel bandwidth.

7.3 HARDWARE DESCRIPTION

The ISLAND project demonstrates the development results of a new telecommunication capability, i.e. the use of Primary-rate ISDN for the transmission of both isochronous and data packets to enable distributed multimedia applications.
The hardware architecture for the workstation nodes consists of the following main functional blocks (independent blocks and/or boards):

- Basic IBM PC module
- Network interface module
- Video interface module
- CODEC (COmpressing DECompressing)
- Switch module

A detailed description of these modules can be found in Haidegger,1995.

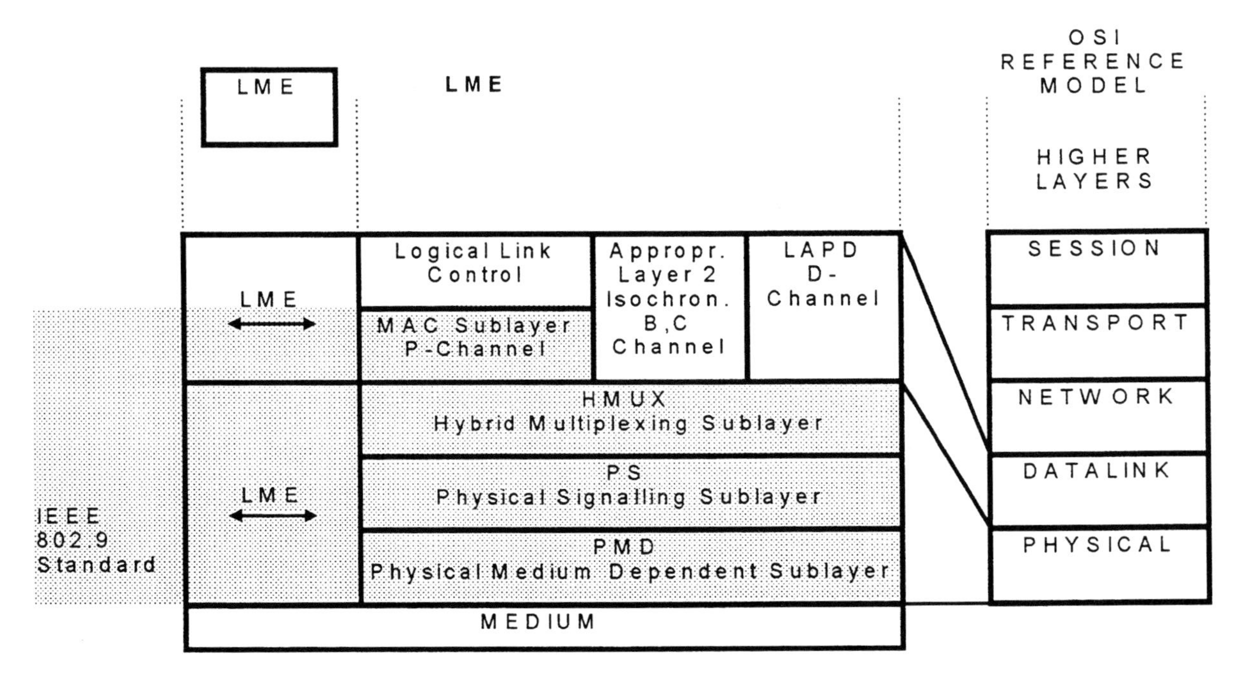

Fig. 1. Relationship of ISLAN Interface

7.4 NETWORK DESCRIPTION

The environment of the demonstration is a local area network, which follows the IEEE 802.9 Integrated Services LAN (ISLAN) recommendation as closely as possible. In both directions there is synchronous data transmission, in which the data is sent in frames. The frames are built up according to the time division multiplexing principle (TDM frame). Independent 64 Kb/s data channels share the bits of the frames.

The Payload information field contains the information octets of any additional C-channel, and the P-channel. In the packet channel the data are transmitted in ISLAN MAC Frames. The relationship of ISLAN (Integrated Services Local Area Network) Interface Model to the OSI Reference Model is illustrated in Fig. 1.

8. THE HARDWARE ELEMENTS OF THE WORKSTATIONS

The ISLAND workstation is given in Fig. 2.

8.1 The I-Adapter Card

This enhanced card was redesigned for the ISLAND project to accommodate the basic needs of the 802.9 requirements. The I-Adapter card consists of a basic board and a piggy-back board - called CEPT module - that allows transmission at 2 Mbps.

Through the MVIP BUS INTERFACE 8 serial synchronous channels (ST-BUS) each with 2Mbps bandwidth and the synchronous clock signals are connected.

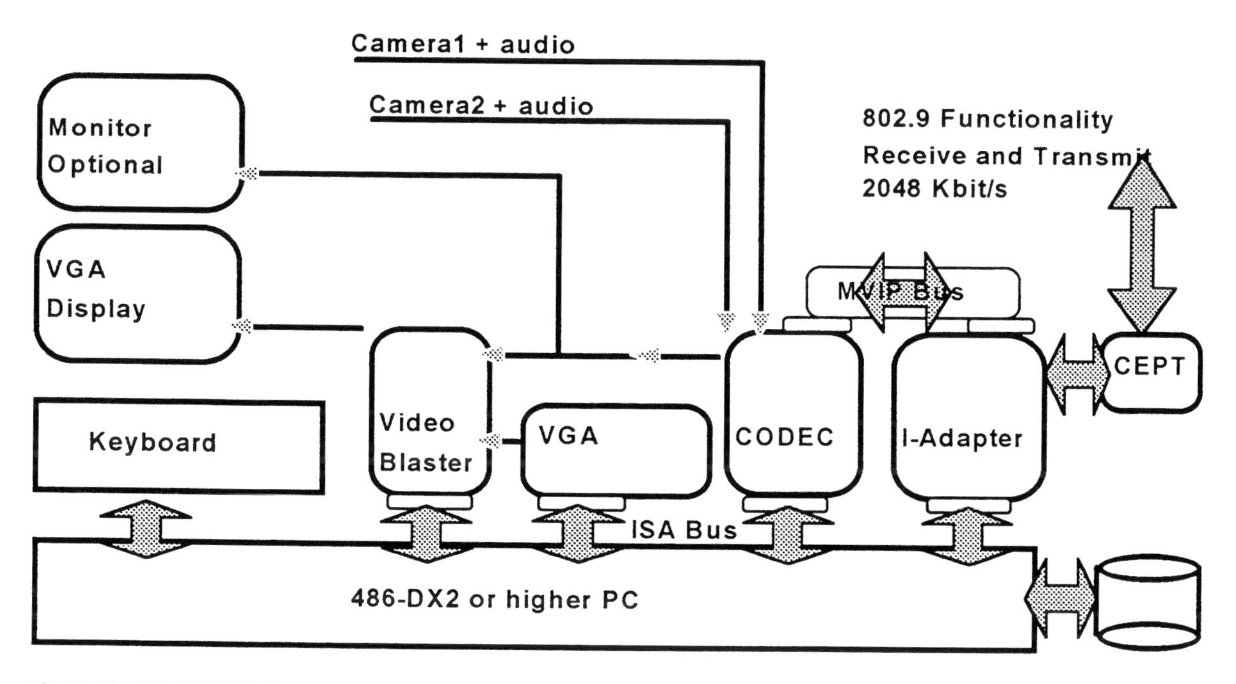

Fig. 2. The ISLAND Workstation

8.2 The CODEC card

The video image compression-decompression is made by the Bitfield H.261 Video Compression Board. This board has two software selectable video inputs and one output. Both PAL and NTSC standards either in composite or Y/C form are supported. Inside the board the video is digitized and compressed according to the CCITT recommendation H.261.

8.3 The hardware elements of the Access Unit

The SWITCH within the ISLAND project is built up of a PC module with 4 network interface modules integrated into it. Neither CODEC nor video cards are present within the switch.

8.3.1 The software structure of the Access Unit

The Access Unit has two main tasks: the video data switching and the packetised data distribution.

The packetised data distribution is made by the HUB function. The HUB resides over the packet driver interfaces of the I-Adapter cards. It gets every data packet, analyses the IP addresses and forwards the packets to the appropriate destination through the proper packet driver.

A TCP/IP protocol stack connects the system to the HUB. The switching module, which controls the video data connections, lays over it.

9. TYPICAL APPLICATIONS

The availability of real-time video, sound communication together with computer data transfer gives new challenges to automation engineers. In manufacturing or assembly automation (see Haidegger, Kuba, 1995, Erdélyi, 1994 and Greguss, 1995) with the intensive use of robotic systems, *the integration of vision* has already started to get high acceptance in the last decade. With the advent of the technical feasibility to implement integrated services on a single network, the importance of remote vision and tele-control has grown significantly. We envisage the short term future also pointed out in robotic trends (see Rudas, 1995, and Rudas, 1995a and Kovács, 1995 and 1996) that intelligent, smart sensor techniques together with vision features will change the applicabilities of the robotic systems. Any application, where data, video and voice in real-time can enhance the computer data communication means, will deeply benefit by the developed and demonstrated technology.

The ISLAND project partners have developed an application software package to demonstrate the applicabilities of the implemented 802.9 features.

Teleconferencing and tele-working are further progressive application areas. The next generation of office environment will allow workers to perform their job functions at least partially from a remote site, with the use of tele-presence facilities, like the demonstrated ISLAND environment.

10. CONCLUSIONS

The idea to apply MM services at the factory environment is reasonable, and within a short time, end-users will ask for it. Not only the office engineers with high-performance workstation computers, but also the shop-floor operators and plant managers will use the technology just as openly, as they are to use normal computers nowadays.

An IBM PC-based possible solution to fulfill such requirements is demonstrated within the ISLAND project. Robotic applications are in need of such features, that allow human intervention to robotic operations, control and supervision, but from a safe distance, or an indefinitely remote site. By linking several of these intelligent vision systems, a range of new applications areas emerge. Our task is to get prepared to take full advantage of these new possibilities. Our present work is focusing on finding real industrial environment to initiate test implementations.

11. ACKNOWLEDGMENTS

The team of the ISLAND project has supplied the fundamental results to reach the demonstration phase. Project members from Belgium, Poland, France and Holland had contributed to the success (Island, 1995 and Haidegger, 1995).

REFERENCES

ISLAND (1995): Technical Descript., European Commission, COPERNICUS Project #9499.

Haidegger, G. al.(1995): *Distributed Multimedia for Advanced Applications.* AUTOMATION Conference, MATE Budapest. Proceedings. pp. 263-272. Sept 5-7. 1995.

Rudas, I.J. et al. (1995): *Hungarian R&D Efforts in Robotics: Trends and International Impact,* 4th International Workshop on Robotics in Alpe-Adria Region, 6-8 July, 1995. Pörtschach, Austria, pp.205-208.

Haidegger, G. Kuba, R. (1995) highly automated flexible assembly system. MATE Automation '95 Proc. pp. 347-352.

Erdély, F. (1995): *Hierarchy of Manufacturing Systems Control,* AUTOMATION '95 Budapest, pp. 383-391.

Somló, J. Podurajev (1994), *Optimal Cruising Trajectory Planning for Robots,* Mechatronics, 1994.Vol.4. No.5. pp.517-538

Somló, J. Buzidi, A. (1996): *Reverse Scheduling,* RAAD96 Conf., 1996 Budapest, pp. 417-422.

Rudas, I. J. (1995): *Introduction to Robot Control.* 11th ISPE/IFAC Intl Conf. on CAD/CAM, Robotics and FOF, Pereira, Colombia, August, 1995.pp.191-223.

Kwon, W.H.(1993): *Implementation of a Parallel Algorithm for Event Driven Programmable Controller,* 1993 Pergamon Press, Control Eng. Practice Vol.1. No.4. pp.663-670

Greguss, P., Alpek, F. (1995): *Further development in a PAL-optic based vision system for robocars.* DAAAM Symp. on Int. Manuf. Systems, Krakow, 1995. pp.119-120.

Haidegger, G., Kopácsi, S. (1996): *Telepresence with remote vision.* RAAD 1996. Budapest pp.: #110. pp. 557-562.

Kovács, G. L (1996): *Changing Paradigms in Manufacturing Automation,* Proc. IEEE Robotics and Automation, Minneapolis, 1996 Vol. 4. pp. 3343-3348.

Kovács, G. L., Nacsa, J. (1995): *A PC-based OSI System to Assist Robotized Welding.* In: Advances in Design and Manuf. Vol.6. IIM Conference ESPRIT Austria, 1995. Ed. Barisani, et al. Pp.: 159-165.

PRODUCTION PLANNING SUPPORT SYSTEM OF REPAIR PARTS FOR HOME ELECTRIC APPLIANCES

Tetsuji Oishi, Satoshi Hori, Masao Tsuda

Manufacturing Engineering Center, Mitsubishi Electric Corporation
E-mail: oishi@int.mdl.melco.co.jp

Abstract: After-sales service practice is important at the home electrical appliance market in Japan. We developed a service parts supply system which forecasts the demand of service parts and can control service parts inventory better than conventional manual control. We are deploying this system to Mitsubishi's home electrical appliance factories. The system has been achieving the stable supply and inventory reduction of service parts. It also contributes to improving business efficiency.

Keywords: Inventory control, Production control, Computer integrated manufacturing, Prediction, Forecasts

1. INTRODUCTION

The home electrical appliance market in Japan has been matured, saturated in volume and open to foreign manufactures. This harsh competition is leading manufactures to differentiate themselves by the quality of their service and support to customers, in addition to lower price and high-tech products. MITSUBISHI ELECTRIC CORPORATION considers after-sales service practice is one of central corporate functions that will determine its future. The point of the reinforcement of after-sales service is a system that can perform fine service by a low cost and the system is composed of the following three functions.

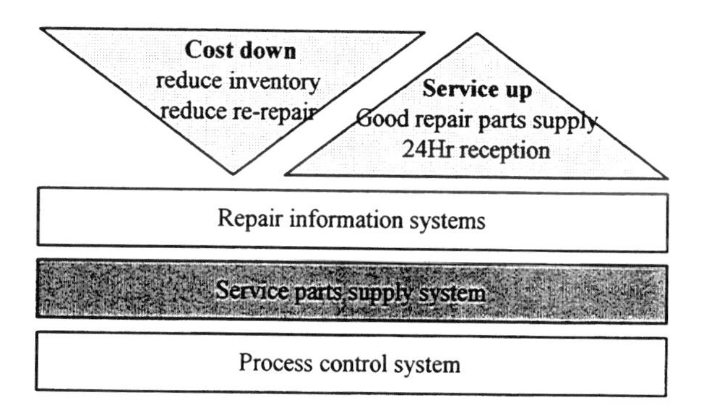

Fig. 1 Concept of after-sales service.

(1) Repair information systems
Service technicians are required to possess a lot of knowledge and information to repair thousands of types of home electrical appliances in the field. The necessary information includes parts configuration of a product under repair, past experience of a failure and a repair etc. Repair information systems can provide young technicians with experienced service technicians knowledge and experience and help them to perform better and quicker repairs.
(2) Service parts supply system
Service parts are the parts stored for repair. It is important to supply repair parts to users who need and exactly when they are demanded.
(3) Process control system
Process control system manages of repair requests from customers. The system dispatches service technicians and stores the repair requests in a database, with which a service status query can be answered anytime.

We have developed a demand forecast algorithm and an inventory control method that solves the service parts inventory problems mentioned above. This paper explains the algorithm and describes the overview of the production planning support system that employs the newly developed method and are able to achieve smooth supply and reduce the inventory of service parts.

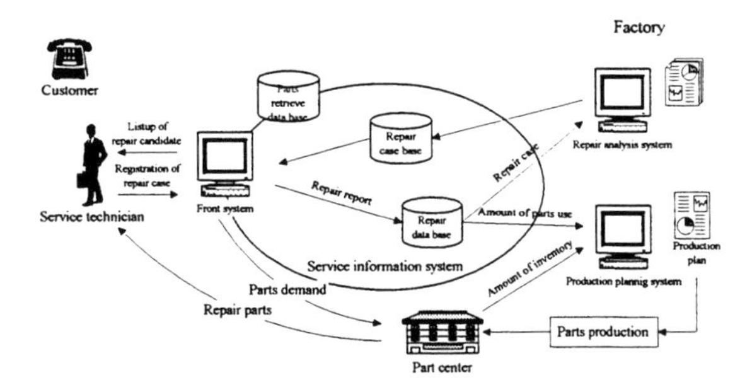

Fig. 2 System configuration.

2. PROBLEM IN SERVICE PARTS SUPPLY

One of the problems in the service parts supply is the inventory control. The inventory control of service parts has been becoming very difficult due to the following reasons:

(1) Number of service parts has increased very big as a large number of home electrical appliance models are produced to meet various market needs.

(2) Demand of service parts largely varies along with product life-cycle and season.

(3) Inventory volume should consider parts delivery period.

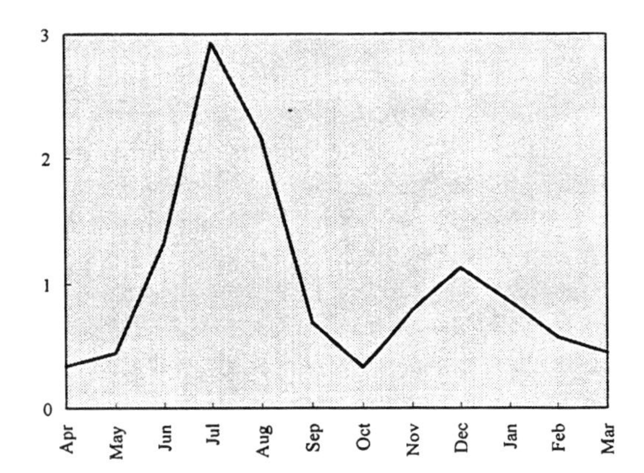

Fig. 3 Monthly Demand of parts.

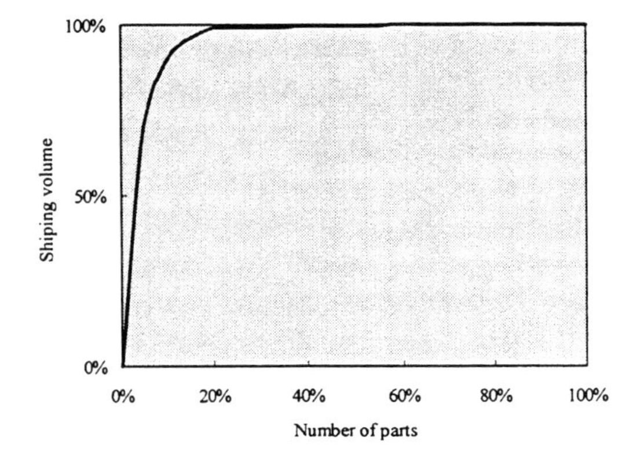

Fig. 4 Parate chart of service parts.

The demand forecast of each parts might be possible if we knew the market circulation quantity of the product from purchase to disposal and the failure rate which contains secular change of each parts. However, an accurate forecast based on this modeling is extremely difficult because some service parts are shared for several product models and parts specifications are often changed to increase product quality.

3. METHOD OF CONTROLLING SERVICE PARTS INVENTORY

To solve these problems, we developed two new control methods.

3.1 Real-demand synchronized production, shipment, inventory control

The control of the service parts inventory was previously operated as the multistep replenishment type and logistics positions, e.g., service stations, the parts center, and the factory had respectively large inventories and the supply action was fulfilled independently by each inventory replenishment order. This conventional system had a difficulty to follow a big change of parts demand and the total inventory increased. Therefore, we have developed the real-demand synchronized production, shipment, inventory control system. This system is able to conform to the seasonal change of the parts demand, as shown in figure 3, by developing the infrastructure of that information uniform management with which we can monitor service parts flow in the market.

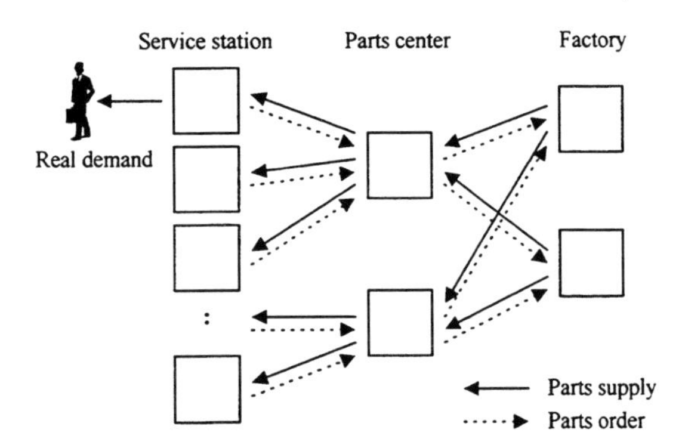

Fig. 5 parts supply system.

3.2 Classification inventory control method

The inventory control method proposed here is a classification control method, which clusters service parts according to their annual shipping volume and their characteristics, e.g., price, lead-time. We control the inventory of large shipping-volume parts based on their demand forecasting. The demand forecast subsystem computes monthly parts demand by considering each part's value and coefficients of seasonal variation. In the case of small shipping-volume parts, their inventory control is automatically operated by the reorder-point system.

Class	Annual shiping Volume	Number of parts	System function item
A	over 100 (80%)	5.5%	Demand forcast Field inventory plenning Production planning
B	50~99 (10%)	4.5%	Demand forcast Production planning
C	0~49 (10%)	90%	Automatic ordering

Fig. 6 Classification inventory control.

4. METHOD OF DEMAND FORECAST

Fortuin, Yamashina et al. proposed demand forecasting methods of service parts, which consider a failure rate and product life, and showed some evaluation with past results. We adopt the technique of a standard time series forecast to predict service parts demand. As for the service parts, the monthly shipping patterns depends on the characteristics. However, the error grows if the coefficients of seasonal variation are estimated for each individual parts. Therefore we classified service parts into groups so that the parts in a group have similar characteristics, e.g. price, lead-time, seasonal change of demand. Then the coefficients of seasonal variation are computed for each group and more accurate demand forecast is calculated by putting each tendency value on these coefficients of seasonal variation.

5. EVALUATION BY APPLICATION TO AIR CONDITIONER

We developed a system which employs the methods described in section 3 and 4. The demand forecast function runs on PC and Windows. The whole system was installed at air-conditioner factory of Mitsubishi Electric Corp.

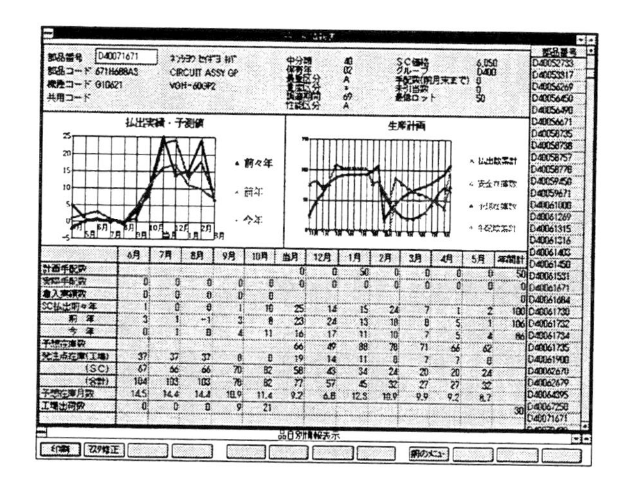

Fig. 7 Demand forecast system screen.

Air conditioners are mainly used in summer hence the service parts demand has large seasonal change. We classified the service parts of the air conditioners into seven groups by the pattern of the season change. For instance, there are different patterns of printed circuit boards for cool/heat type and one for cool only. The system contributed to the inventory reduction and the steady supply.

6. CONCLUSIONS

This paper proposed inventory control methods and demand forecasting method for service parts. These methods are proven to work effectively in a real-world problem, i.e., air-conditioner service parts supply. We will reinforce after-sales service and make the service business practice more competitive because the importance of the after-sales service will be increasing in the future.

REFERENCES

Sugimatsu, H., S.Hori, et al. (1994). A CBR Application: Service Productivity Improvement by Sharing Experience. *Proceeding of Emerging Technologies and Factory Automation*, 132-141.

Fortuin, L. (1984). Expected demand for service parts when failure times have a Weibull distribution. *European Journal of Operational Research* **17**, 266-270.

Yamashina, H., M.Sharaf, et al. (1987). A new forecasting method of service parts demands. *Journal of the Japan Society for Precision Engineering* **53-6**, 983-989

DEVELOPMENT OF A PROCESS- MANAGEMENT AND INFORMATION SYSTEM (PMIS)

Michael Reimann* Andrea Arenz* U. Harten**
Karsten Lemmer* Eckehard Schnieder*

** Institut für Regelungs- und Automatisierungstechnik,*
Technische Universität Braunschweig,
Langer Kamp 8, D-38106 Braunschweig, Germany
*** Nordkristall GmbH,*
Verbindungschaussee, D-18273 Güstrow, Germany

Abstract: In the present paper a process- management and information system for a sugar producing plant is shown. First the final aims are described. In a first step a data base to support the maintenance management is presented. The next chapter deals with PMIS itself. The system is based on a hypertext net. Following the requirement specifications for the system are derived. Now the main installation components and their interaction are described. The paper concludes with a summary and with a short outlook.

Keywords: computer aided manufacturing, computer networks, data handling systems, databases, information systems, local area network, maintenance engineering, management systems

1. INTRODUCTION

At the Institute of Control and Automation Engineering at the Technical University of Braunschweig a concept for the design of a Process-Management and Information System (PMIS) for the sugar plant 'Nordkristall GmbH' in Güstrow is developed.

The goals are

- to increase the cost transparency within the enterprise,
- to raise the productivity and
- to improve the economy balance.

These evolutions are especially important as the competition will intensify due to the ending guarantees of the european quota system. To meet presentday's requirements a modern and uptodate management is necessary.

In order to reach the desired goals mentioned above a Process- Management and Information System is developed within this project. The aim is to combine the different existing information and data flows in a suitable way, so that finally each piece of information from any possible location in the plant or even in the enterprise is visible and modifiable on request.

1.1 *Data type structures*

The information itself has not only a numeral format but also complex data types like structered text documents, images and mutimedia objects. In general three important data types are distinguished:

- **data:** numbers, registrations, text entries like name, address or description of less than a page (process data, maintenance data)
- **business documents:** notes, correspondence, drafts, offers, requests, e-mail (if interesting for the plant itself)

- **strategical documents:** business transactions, solutions of customer problems, process documentation, plant security, patent specifications, administrative regulations, international quality standards

The first data type is predestinated for the use in a data base system. The different entries have mostly a fixed length and a fixed format. The data set structures are not regarded to be changed much. Furthermore the data relations among one another are predefined. Therefore the given structures can be taken into consideration and programmed.

Regarding more complex data types like documents including formatted text and images (e.g. business reports) these facts are rather different. In this case the structures and relations are directly depending on the content of the documents or single extracted keywords. These context sensitive coherences render the data base design more difficult especially since the structures are subject to a permanent change due to additional documents. Even an object orientated design would not simplify the problem since not only the documents themselves but also text blocks or words would be independant objects.

One main aspect is the presentation of different types of information within one uniform (graphical) user interface. Keeping this in ones mind the acceptance will surely increase.

2. THE DATA BASE SYSTEM

In the following subsections the different developing steps are outlined. At first the design of a data base system to manage maintenance data was derived. Following the implementation of the information system and its interface with the data base was investigated.

2.1 *Object identification and organisation*

The data base system was designed and programmed for recording, processing and archiving occuring data. Primarily the maintenance data (repairs, costs, operating times, installation places, ...) was entered and linked together. Therefore an object orientated concept was chosen in order to model the plant down to the machine level. The characteristics and identification methods are based on (*DIN 28004*, n.d.). The whole identification structure is shown in figure 1. An object is classified using different criterias (like plant, section, installation area, process section, group, machine, part). Following a consecutive number is chosen. Redundant objects (if occuring) are distinguished by an additional number.

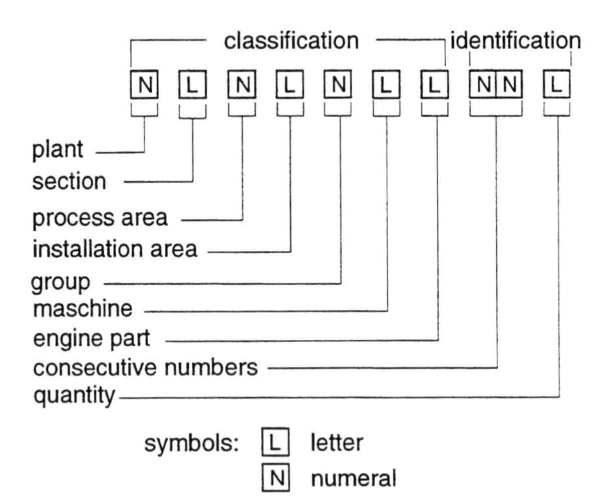

Fig. 1. The identification system

It was kept in mind to display the code of a single object not only in a sequence of numerals but instead in a combination of numerals and letters. Advantages of this method are that the code is easier legible and recognizable. Furthermore the user acceptance will rise since the new system resembles the old incoherent system in many ways. By means of the identification key an experienced user can easily detect both the type and the exact position of the machine (part).

2.2 *Results*

Different search methods in the data base allow a fast and efficient tracing of the entries. Using suitable data filtering techniques it is for example possible to assign increased costs to a certain machine part (low quality product) or to a certain location (construction errors, faulty machine dimensioning, ...). Storing both the occuring running times and the carried out repairs and routine repair work the whole machine histories are known. This is the first step towards an object orientated maintenance. Upkeep gets more planable. It may e.g. be more efficient not just to control and replace an object in the case of failure but when a defect due to high working time and ageing is likely. Necessary repairs are carried out when the machine group belonging to the substituted part is not in action. The results are directly reduced maintenance costs and indirectly reduced following costs by interrupted processes.

2.3 *Implementation*

During the testing period the data base software Access 2.0 (Microsoft) was used in order to avoid prime and developing costs. Therefore the proposed object orientated structure was transfered to the relational data base. In addition to the maintenance data more general data can be assigned to the objects as well (like mounting steps, repair instructions or particularities). Furthermore pull down menus are used to cross refer from one plant object to an address entry. A comfortable access to important data from component suppliers, manufacturers and partners is realized.

3. THE PROCESS- MANAGEMENT AND INFORMATION SYSTEM (PMIS)

The information system has the task to provide the connection of different plant parts and information sources. The final aim of PMIS is the design of a homogeneous supervision and information providing system. The chosen solution results a company intranet with a hypertext system as essential part. Several advantages are given: The linking even of different types of information is easy and comfortable. The system is able to fit its size automatically depending on the demanded complexity and information capacity. Moreover the user interface has global uniform textual and graphical layout. The possibility of additional personal features is given as well. The system is (nearly) completely location and platform independent.

3.1 *Definition of a hypertext net*

The information system is designed as a hypertext network. Hypertext is defined by its structure, by performed operation, by the representation media and by interaction (Hoffmann, 1995). Its structure is a tree consisting of knots and branches. The knots include the information. By referencing from one knot to other knots a netlike structure is obtained. In figure 2 the hypertext net structure is visualized. Proceeding from an initial starting point (on the highest level) it is possible to descend in a hierarchical manner down to the bottom layer. The information is shown as a knot in the net. Lines between the single dots represent links to other documents. Clearly visible is that caused by the cross references (dashed lines) the tree structure is expanded to a net structure. The result is a very complex information structure.

The main operations in a hypernet deal with the programming, modification and navigation in the hypertext net. For a presentation media the computer presents itself automatically. One distinguishes between static media (like text and images) and dynamic media (such as video, audio). Based on a graphical user interface interaction can be performed. The knots represent in this case the information input-output media.

Of course there are both advantages and disadvantages with a hypertext net (Conklin, 1987; Latz,

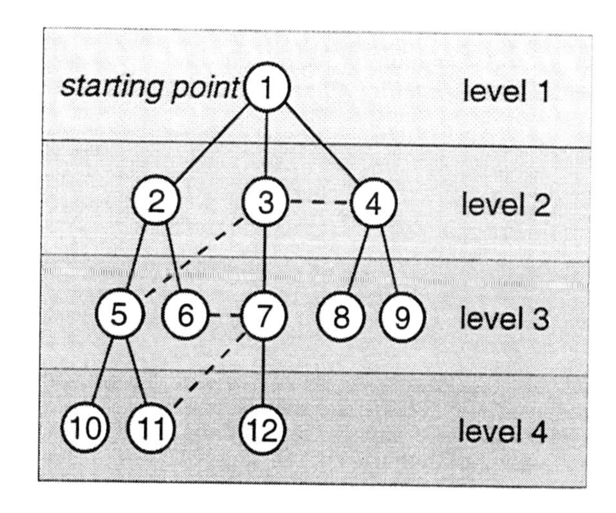

Fig. 2. The hypertext net structure

1992). Positive effects are the interactive use of the documents. It is not only possible to read and search but also to collect and arrange documents. With references within the documents any kind of information may be linked together. Furthermore many different formats can be implemented. The author on the one hand has the freedom not to program only sequentiell documents. Graphical connections are very simple. Furthermore the structure is easily extendable. The reader on the other hand has the freedom to choose his own document order. He is able to generate his own remarks and references. Of course the complexity of hypernets involves some negative aspects as well. The user has to select his own order, which may get the more difficult and tiresome the lower he gets in the structure. The author has the duty to program closed loop structures. This is especially complicated as there are most of the time more than one entry to a certain document. It is advisible to use thoughtful names to increase the recognition factor. Recommendable one chooses the directory structure similar to the hypertext net so that an easy overview and an easy data maintenance is guaranteed. Probably the biggest problem is a loss of orientation, commonly known as 'lost in hyperspace'. The user does not know, where he exactly is, how he got to this point or how he could arrive at a certain information.

3.2 *Implementation specs*

The hypertext system was developed for the use in the sugar manufacturing plant 'Zuckerfabrik Nordkristall GmbH Güstrow'. PMIS was developed in several steps (see 3). Having performed the requirement/performance specs and the functional/data flow analysis the following main demands for PMIS were extracted:

(1) **computer hardware:** Two different network structures exist, one is a novel net (with IBM-compatible PCs, the other is a Unix net

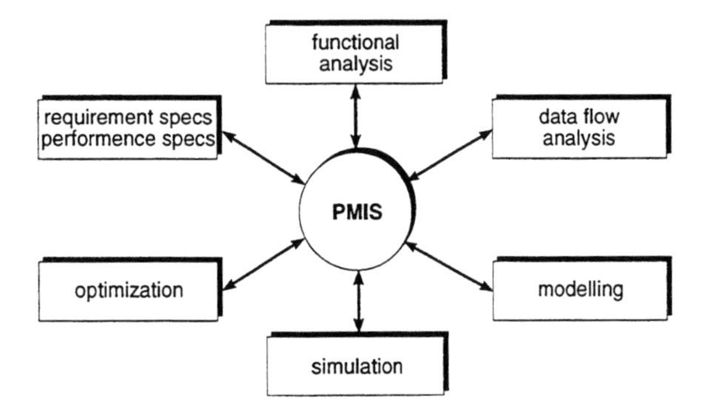

Fig. 3. PMIS developing steps

with one HP server and some PCs). It should be possible to run the hypertext system on both nets. Furthermore several users should be able to use it at the same time.

(2) **software tying:** Very important points are the gateways to other tools. It is e.g. necessary to interface both the process controlling system Contronic P together with the data base system CIM/21 and the Access data bases. Furthermore the programs Microsoft Excel and the industrial management tool SAP should be included, too.

(3) **navigation tools:** Different search mechanisms should be provided for the user such as contents directory, index, notes, history list, bookmarks, hotlists, go back/forward, guided tours, hierarchical layers. A higher user acceptance will be aided with the possibility of individual configurations.

(4) **programming:** Programming additional or given system functions and including external programs the hypertext has to be adjusted to the given environment.

(5) **developing system for authors:** The author needs an efficient system in order to develop a hypertext net (like a programming language, ...).

(6) **structure and maintenance tools:** There is the need to develop sensible complex net structures. Furthermore the possibility to analyse and correct the network structure can be very helpful.

(7) **user interface:** A graphical user interface has the advantages of direct input control (e.g. mouse click) and is easily understandable and changeable.

(8) **presentation:** Besides including images other mutimedia presentations should be offered as well.

(9) **information quantity:** The information system is designed for a longer time period and therefore grows permanently. A hypertext system should not force internal limitations.

(10) **security concepts:** Data has to be protected against illegal use from any side. Especially in multi user net structures this topic is very important, suitable passwords may be necessary.

Different hypertext systems, like WorldWideWeb (WWW), Guide3.1/Idex or ToolBook3.0/Multimedia ToolBook, were compared using the demands described above. Finally the WWW came out to be best suited for the purpose. It was able to fulfil all criterions and impresses through its modular concept. It is a service offer in the internet and composed out of different software components. Contrarily the other products are often complete software packages.

3.3 *The main installation components of PMIS*

As mentioned above two separate network (one Unix net and one Novel net) exist. Information can be exchanged via an serial interface or via floppy. The (smaller) network consisting of a Unix-Workstation and several PCs was chosen to install PMIS. For the planned fusion of both nets the expansion of the intranet could be done with little effort.

The intranet consists of four main components:

- the intranet server
- the browser
- the data base server
- cgi scripts

The intranet server is the essential part of the process, management and information system. It not only combines the different information sources but also organizes the existing data flows. Among many possible servers the Cern Server (public domain) was chosen. Selection criterions were e.g. the desired platform (Unix or Windows) and the different server functionalities. Furthermore using the Cern server a full compatibility is given between the sugar plant network and the institute network, which simplifies the programming, the implementation and the servicing. Some of the servers functionalities are e.g. to call external programs via an integrated interface (like external viewers). By consequence many software components (data bases, ...) and data formats (clickable images, ...) can be included. Furthermore password and security aspects are managed by server configuration files.

The browser is the graphical interface software with which the user of PMIS directly interacts. The browser both loads documents, which in general have HTML format and displays them textually or graphically. The abbreviation HTML means hypertext markup language and is one commonly used programming language for hypertext nets. In a document links to other documents can be programmed using text or

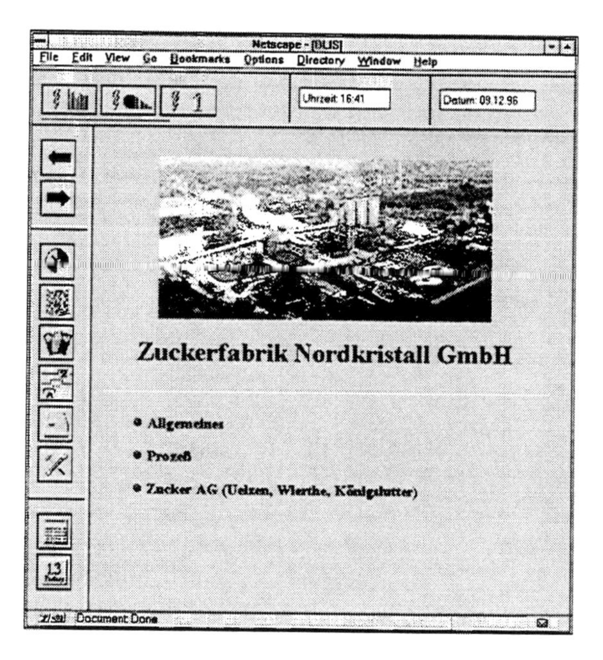

Fig. 4. The PMIS user interface

images. The links are displayed in a special text colour or with an image box. They are clickable (by mouse click). The referenced document is directly loaded in the browser. Among a variety of browser software the tool netscape (Netscape Navigator 3.0) was chosen. It can be used both on Unix and on Windows platforms provided that there exist both versions of Netscape. The performance and the design on both platforms is nearly identical, so that the demand of a uniform graphical user interface is fulfilled. Figure 4 shows exemplarily the homepage of the intranet. Base adjustments of the browser like window size, letter type, size and colour, homepage (document which is loaded when starting the srver), memory, ... are adjustible to personal needs and preferences.

The data base server is responsible for transfering request parameters and for generating SQL commands (SQL: structured query language). This SQL statement is transmitted to the data base.
A very important selection criteria for the data base server was the use of Access as data base system. The chosen server Axorion (from Internet Database Consultants) is a software product, which is directly programmed under Access. Axorion itself is a part of the data base application and uses given developing and programming facilities. To start the server the program Access has to run, too. A disadvantage is that the server gets activ and stops running jobs when a request occurs. Therefore a 'normal' PC use is not advisible. The hardware demands are quite high as well. For a sensible response time a PC 486 should be used.

CGI is the programmable interface of the intranet server. It is an abbreviation of common gateway interface. The CGI script is a file

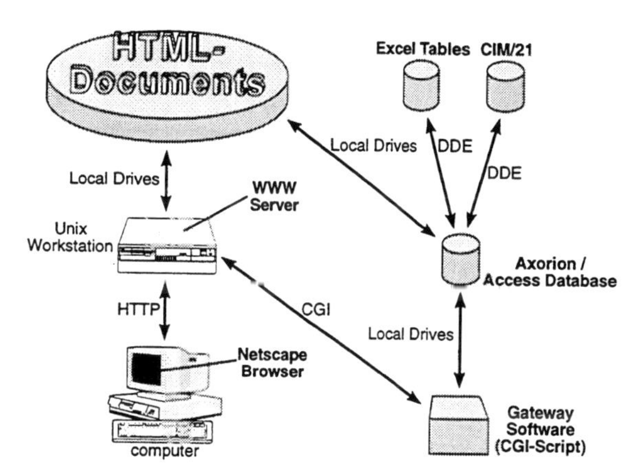

Fig. 5. The PMIS net structure

which can be started under Unix. Generally it is called by the intranet server. When finished the program returns the resulting file (e.g. HTML file) to the server. For PMIS shell scripts were implemented. These files are (batch) files containing a sequence of Unix commands. On call the commands are executed sequentially.

3.4 *The interaction of the software components*

The following section describes the interaction of the different software tools, how they communicate and how data is exchanged. In figure 5 the basic intranet structure is shown. A request or command is started by the user of the process-management and information system. Hence the starting point is the user computer where netscape is installed. The shown surface allows many possibilities to navigate within the information system. Via a CGI script the database Access can be questioned. Programs like Excel or CIM/21 are interfaced by DDE. Furthermore it is possible just to ask for HTML documents. Obviously all actions require the intranet WWW server. Request are generally created using forms in HTML documents.

To load a new document e.g. the following action takes place. The user clicks on a link in a document. The browser sends this information via HTTP (hypertext transfer protocol) to the intranet server. The server evaluates the information and loads the requested HTML file via NFS (network file system) protocol from a local drive. Now the file is transmitted to the original sender and automatically loaded in the browser. If the file happens to be some other but not HTML format the file is displayed in a separate window using external viewers.

3.5 *The gateway to the data base server*

A little more complicated is the entry in a data base or the request. Generally the user has to fill out a form in the information system, which includes entries to change fields of the corresponding data base. The input in the different form fields are transmitted to the intranet server as parameter values (via HTTP). The server evaluates the file and calls a program, a CGI script, with the given parameters (instead of asking for a HTML file). Which program to call is specified by the browser. The server now waits for the program to return a (HTML) file in order to send it back to the browser. For the data base server the CGI script has the duty to save a file 'A' in a special directory under a special name and to wait for a returning file 'B'. The data base server works in the polling mode, that means Axorion tests in cycles for the existence of the file 'A'. In the case of existence the file is evaluated and the data base is modified or searched for an answer. The result is saved in file 'B', the CGI program then reads it and gives it to the server. Finally the answer is shown on the graphical user display.

Advantages of the chosen data base server are a high flexibility to interface many different data bases supporting SQL and ODBC. It can be programmed and formated in many different ways, too. By definition other interfaces can be included. To dynamically generate HTML documents (on the fly) the capability to explicitly read and write many file formats is used. A programmable DDE (dynamic data exchange) interface realises the data exchange with programs like Excel or CIM/21 (for archiving process data).

4. CONCLUSIONS

The paper describes the several processing steps to design and implement a process- management and information system for the plant 'Nordkristall GmbH' in Güstrow. The final aim is a homogeneous supervision and information providing system. Any kind of information should be accessible from any location.

In a first step a data base model was programmed which serves for storing machine data. Classification and identification methods were based on (*DIN 28004, n.d.)*. The concept of an object orientated structure was transfered to a relational data base and implemented in Access 2.0. The data base serves for supporting the maintenance management, so that action on the one hand are more planable and on the other hand can be performed working condition dependent.

The company intranet was developed using a hypertext system as essential part. This solution offers an easy and comfortable linking and referencing of many different types of information. Furthermore it is able to grow with the complexity and information capacity of the plant. Different hypertext systems were compared using the derived requirement specifications. The WorldWideWeb came out to be best suited for this given purpose. Following the main installation components are described like intranet server, browser, data base server and scripts. Especially interesting are the interactions of the different components and the possibilities of interfacing different tools. The gateway to the Access data base was implemented using an additional data base server.

With PMIS information can be displayed, stored, linked, evaluated and distributed easily. It is possible to get a quick and efficient overview over the current activities. All in all the information system serves to store and reference the knowledge of the staff. It is possible to spread information over all parts of the factory. By implementing a comprehensive knowledge base the information system acts as a kind of expert system.

There exist many possibilities to extend and refine the process- management and information system. E.g. the main process data should be included and displayed in the information system. Interfacing the industrial management tool SAP is important, too. The staff would be responsible to extend the process knowledge basis. Finally the whole group of factories can be joined in a combined intranet.

REFERENCES

Conklin, J. (1987). Hypertext: An introduction and survey. *IEEE Computer.*

DIN 28004 (n.d.).

Hoffmann, M. (1995). *Problemlösung Hypertext - Grundlagen, Entwicklung, Anwendung.* Hanser-Verlag.

Latz, H.W. (1992). Entwurf eines Modells der Verarbeitung von SGML-Dokumenten in versionsorientierten Hypertext-Systemen - Das HyperSGML-Konzept. PhD thesis. TU Berlin, FB Informatik.

TEMPORAL MANAGEMENT IN COMMUNICATION MODELS: ATM CASE STUDY

P. Lorenz

IUT/GTR 34, rue du Grillenbreit 68008 Colmar - FRANCE
Email: lorenz@colmar.uha.fr

Abstract: The evolution of the various automated systems, the needs of tools and methods to satisfy reliability, safety and response time constraints leads to introduce new communication models. This paper presents different communication models (producer(s)/consumer(s) and (multi)client(s)/(multi)server(s) models) and describes how the time is handled by these communication models. An analyze about communication models used by ATM network is done.

Keywords: Distributed models, Real-time computer systems, Models, Communication systems, Model management, Networks.

1. INTRODUCTION

A great number of manufacturing companies problems may be solved by integrating the computer in process control, product design, etc ... This computer integration is called "Computer Integration Manufacturing" (or CIM). A CIM architecture allows to introduce the notion of functional, material and operational architecture (Didic, *et al.*, 1993), (Lorenz, *et al.*, 1993), (Kotsiopoulos, 1993).

In the operational architecture, some communication models enabling communication between tasks in distributed applications will be studied.

This paper is structured as follows: first, some definitions about communication architectures are introduced. In the second section, different communication models are presented and in the last section, communication models used in ATM networks are described.

2. COMMUNICATION ARCHITECTURE

A functional architecture is composed by different functions, relationships between these functions, constraints, properties and expected characteristics of the solution.

The material architecture is composed by the list of the different equipments (processors, networks, ...) used by the functions, characteristics resulting of material choices, networks used, etc ....

The operational architecture comes from the mapping of functional architecture on the material architecture. The expected characteristics and properties are verified thanks to the choices done in the material architecture.

In the functional architecture, the temporal requirements can be translated into temporal constraints (Lien and Yang, 1992), (Shaw, 1992), (Northcutt and Kuernet, 1993), (Lorenz, *et al.*, 1994).
In the operational architecture, the functional temporal requirements can be expressed in terms of communication models, services and protocols

which can be translated, in the operational architecture, into temporal mechanisms to verify functional temporal constraints (Godefroid and Holzmann, 1993) (Raju, *et al.*, 1992).

The following scheme summarizes the different temporal concepts introduced in the functional and operational architectures:

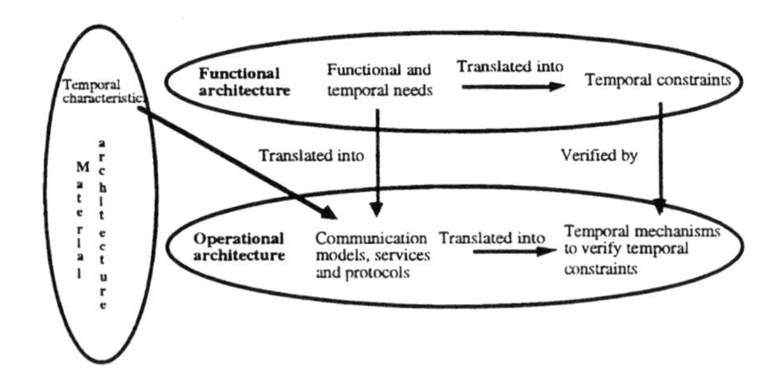

Fig. 1. Temporal aspects in functional and operational architectures

In the next section, communication models (as producer(s)/consumer(s) and (multi)client(s)/(multi)server(s) models) will be studied because time management is different according to the selected communication model (ISO TR12178, 1993).

3. COMMUNICATION MODELS

The need of data transmission to one or more destinations introduce the concept of multipeer/multicast data transmission.
A multipeer/multicast group is a group of entities which are senders or receivers of data transmissions with other members of the group.

Distributed applications require point-to-multipoints (multicast, 1 to N) as well as multipoints-to-multipoints (multipeer, M to N) communications. If multipoints communication is not supported by the network or by the protocol, then multiple point-to-point connections can be used but the QoS (Quality of Service) will not be the same (Heinrichs, *et al.*, 1993).

Multipeer or multicast communication can be used with or without acknowledgement then,

- for multicast communication, the communication is defined from one to N entities. The transmission is done between one publisher to N subscribers. If a publisher knows the subscribers, and/or the subscribers know the publisher, then confirmation mechanisms can be introduced.

- for multipeer communication, the communication is defined between M publishers which produce data to N subscribers. If a relation between M to N entities exists, then it is possible to assume relations between one to N entities (with M=1). But M times a relation between one to N entities is different from a relation between M to N entities, because the first case does not assume properties as grouped acknowledgement which can be necessary for the latter case.

The term of broadcast communication is used if all entities interconnected to a communication system receive the same produced information. The multipeer/multicast term is used if only a part of all entities (or a group) receives the same produced information. Broadcast communication is always used without acknowledgement, because the receivers number are unknown.

Two mechanisms exist for the communication between publisher(s) and subscribers :
- the publisher(s) know the address of the different subscribers: (multi)client(s)/(multi)server(s) models,
- the publisher(s) do not know the address of the different subscribers, it sends the information on the medium and subscribers know what type of information they wish: producer(s)/consumer(s) models.

3.1 Producer/consumer model

In the producer/consumer model, the producer produces an information and sends this later to the consumers which can consume or not the information. Producer and consumer entities are located at the same level. This model supposes a network configuration because the producer knows often the needs of the consumer.

3.2 Producer/consumers model

In the producer/consumers model, an information can be consumed in the same time, and at a given time, by several consumers. The exchange of the information is not a point-to-point communication but a multipoints (or multicast) communication.

It is possible to introduce a distribution function allowing to a producer to know when it must produce a given information to the consumers. The distributor/producer/consumers model may be used for this communication.

This model can be represented as follows:

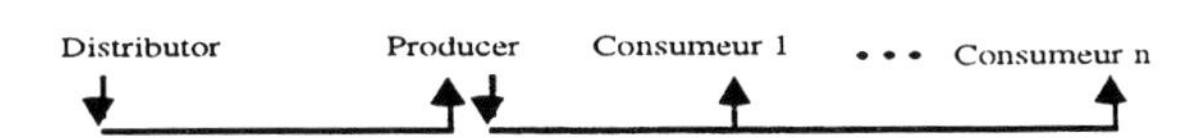

Fig. 2. Distributor/Producer/Consumers model

The distributor allows to schedule data communication between a producer and the consumers.

3.3 Producers/consumers model

When a lot of rules link together different producers, it is necessary to introduce a new communication model: the producers/consumers model.
The producers produce information (for example describing the state, at a given date, of a manufacturing system) to consumers which receive, at the same time, the same information.

As previously, it is possible to introduce a distributor function allowing to producers to know when they must produce their information. The distributor sends production orders to the different producers. Then, at the same time, all producers do a copy of their local data to their buffers and then, they can send sequentially their information to the consumers.

The distributor/producers/consumers model can be represented as follows:

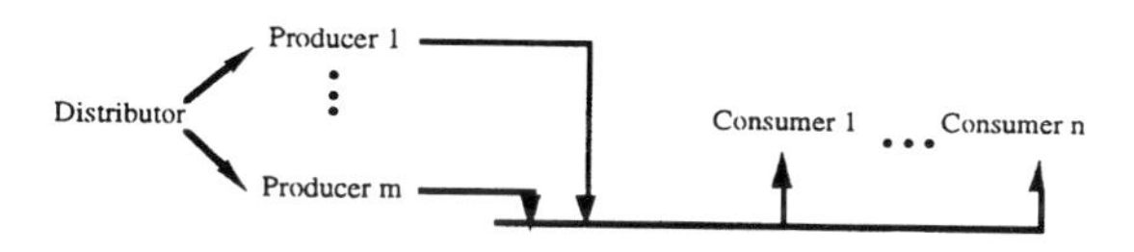

Fig. 3. Distributor/Producers/Consumers model

The producers/consumer model can be directly deduced from the producers/consumers model with the number of consumer equal to one.

3.4 Client/server model

The client/server model is a point-to-point model. Two Application Process are concerned: the client and the server (ISO 9506, 1989). The client sends a request to the server. According to its resources, the server performs the request and sends the response to the client. The processing time of the server is unknown because it depend of the server load. The transaction duration is unforeseeable except by making hypothesis on the server and on the network availabilities.

The client/server model can be represented as follows:

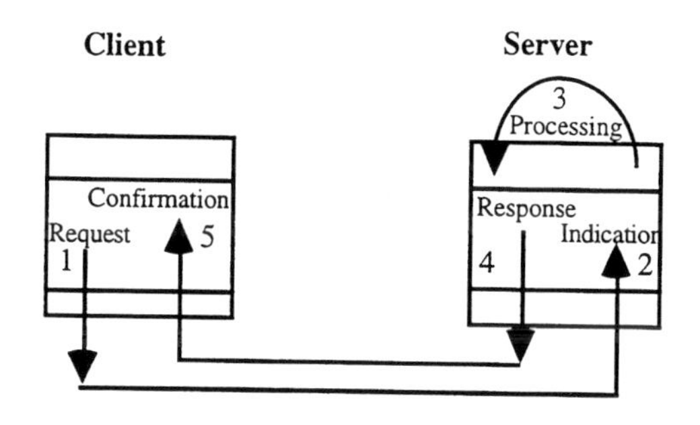

Fig. 4. Client/server model

If at a given time, several clients want to read the same value from a server, then each client must establish an association with the server. Due to the sequential management of the requests by the server, it is not possible to know if the responses received by the clients are identical or not, because each request is performed by the server at different dates.

When a client sends a request to a server, the client is considered as the producer and the server as the consumer. Likewise, when a server sends the response to the client, the server can be considered as the producer and the client as the consumer. The distributor function can be considered as a particular client which must synchronize the different servers (i.e. the producers).

A representation of these concepts can be represented as follows:

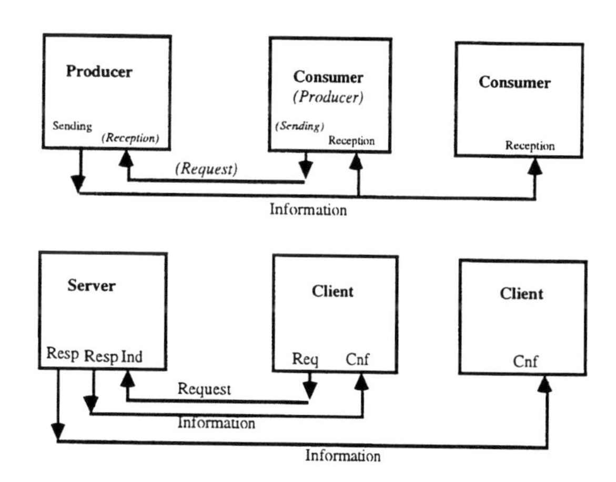

Fig. 5. Representation of producer/consumer and client/server models

The major difference between producer/consumer and client/server models is that in the producer/consumer model, a given data can be consumed at the same time by different consumers. Thus, in the client/server model, it is not possible to have a temporal statuses allowing to know if a given value has been received at the same time by each different clients.

3.5 Client/multiservers model

In the client/multiservers communication model, different types of communication are possible:

1°) Several requests from a same client are sent to several servers unknown by the client,
2°) Several requests from a same client are sent to several servers known by the client,
3°) A request from a client is sent to several servers unknown by the client,
4°) A request from a client is sent to several servers known by the client.

Only the case 1, with the stronger constraints, will be studied because the other ones can be deduced from this later.
In the client/multiservers model, a client 1 sends requests to a known server 1 which becomes a client 2 for other servers 2.n. This case can be represented as follows:

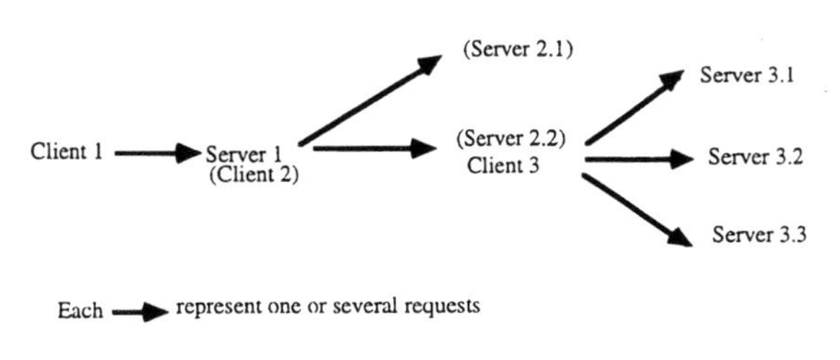

Fig. 6. Case where the server 2.n and 3.n are unknown from the client 1

Then, when a server is unknown from a client, it is necessary to add at least a step for the exchange of information between a client and a server.

An example of the different communication steps between a client and two servers, in the client/multiservers model, can be represented as follows:

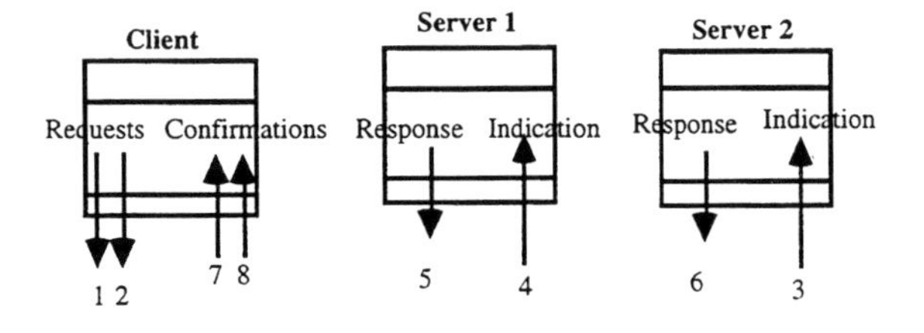

Fig. 7. Exchange of information between a client and two servers

In our example, the client sends two requests (number 1 and 2) to two different servers. The responses order from the servers are unknown because each server has a different load.

If the communication must be performed with respect of timing constraints, it is necessary to verify that each server has responded in the fixed delays.

3.6 Multiclients/multiservers model

When a lot of rules links together the different clients, it is necessary to introduce a new communication model, as the multiclients/multiservers model allowing to different clients to execute an atomic operation.

In the multiclients/multiservers communication model, different types of communication are possible:

1°) Different clients send different requests to several servers unknown by the clients,
2°) Different clients send different requests to several servers known by the clients,
3°) Different clients send identical requests to several servers unknown by the clients,
4°) Different clients send identical requests to several servers known by the clients.

To illustrate this mode, an example of communication between two clients and two servers is presented:

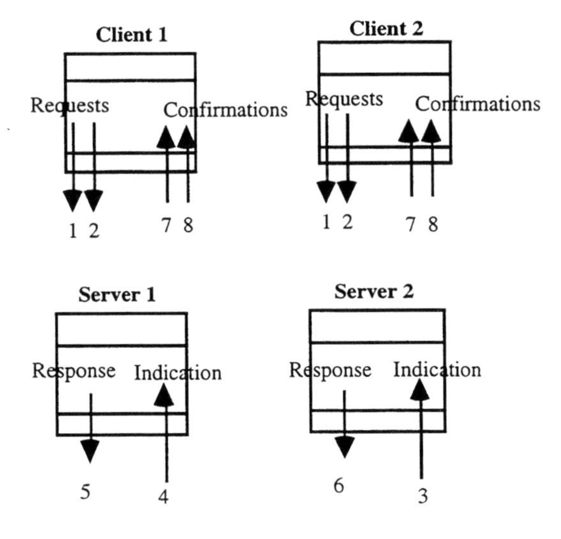

Fig. 8. Exchange of information between two clients and two servers

At the same time, two clients send two identical requests (number 1 and 2) to two servers. Server 1 and 2 responses are sent sequentially to the two clients.
It is possible to add mechanisms allowing to know if the responses received by the clients are identical or not.

The multiclients/server model can be directly deduced from the multiclients/multiservers model with the number of server equal to one.

4. COMMUNICATION MODELS IN ATM

ATM (Asynchronous Transfer Mode) is a high-speed (155 Mb/s) connection-oriented network allowing to transmit video, voice and data on the same medium. An ATM network is built by a set of ATM switches interconnected by links or interfaces (Carle, 1994), (Pokman, et al., 1995), (Nyong, 1994), (Luckenbach, 1994), (Conway, 1994).

An ATM station is composed by the four following layers:
- Higher level protocols (from OSI layer 3)
- ATM Adaptation Layer (OSI level 2.2),
- ATM layer (OSI level 2.1),
- Physical layer (OSI level 1).

There are two fundamental types of ATM connections:
- point-to-point connections, which connect two unidirectional or bidirectional end-system,
- point-to-multipoints connections, which connect a single source end-system to multiple destination end-systems.

ATM signalling is a protocol for establishing connections across the network. Signalling requests are used by nodes to establish or break down connections or relay transmission characteristics. At this time, only one formal standard exists for ATM signalling between end-systems and switches: User Network Interface (UNI, 1994). ATM specification allows management of point-to-multipoints connections. For multipoints-to-multipoints connections, each node in a group that wishes to communicate must establish a point-to-multipoints connection with all other nodes of the group. N point-to-multipoints are required for a group with N stations. If one ATM station joins or leaves a group, the multicast tree must be modified.

MARS (Multicast Address Resolution Server) allows multicast and broadcast services for IP (Internet Protocol) over ATM. Then, when an end-system wants transmit data to a particular multicast group, it opens a connection with the MARS.

A comparative representation of the multicast server, point-to-point and point-to-multipoints connections can be represented as follows:

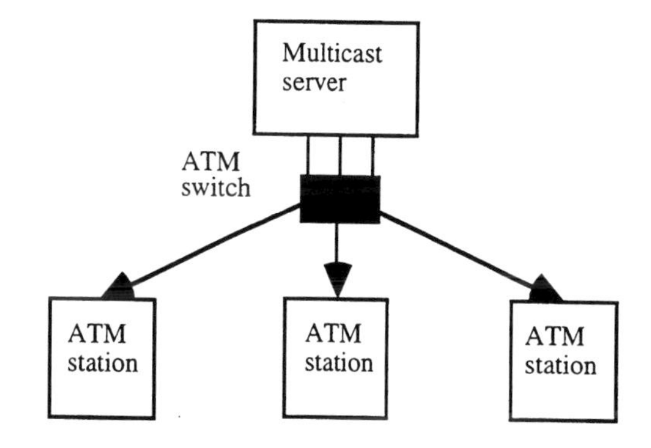

Fig. 9. ATM point-to-point connection

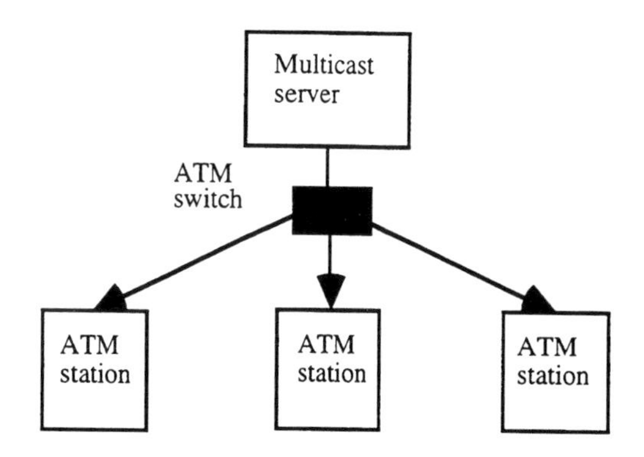

Fig. 10. ATM point-to-multipoints connection

Point-to-multipoints connections, from the multicast server to every member of the group, are used to transmit messages to every members of a group.

ATM point-to-point connection can be directly deduced from the server/multiclients model. ATM point-to-multipoints connection can be directly deduced from the producer/consumers model.

ATM Adaptation Layer (AAL) is a protocol that encapsulate data to be carried in ATM cells and determining how data should be sent and segmented into cells for transmission by ATM layer. Upon receipt of ATM cells from the ATM layer, the destination AAL reassembles information into proper order and form before passing it to higher network layer. In AAL5, the most used ATM adaptation layer used to transmit data across ATM network, point-to-multipoints connections are unidirectional, because there is no provision within the cell format for the interleaving of cells from different packets on a single connection. All packets send to a particular destination across a connection must be received in sequence with no interleaving between cells of different packets on the same connection. AAL5 does not provide a field for demultiplexing cells.

To solving this problem in ATM, different methods can be used:

- the use of a multicast server which sends information across a point-to-multipoints connection and receives packets across another point-to-point connections from the destination stations,
- all nodes establish a point-to-multipoints connection with each other node in the group and then, all nodes can transmit and receive information to all other nodes,
- a unique identifier, the VCI (Virtual Channel Identifier), is allocated to each connection with others stations and then, interleaving packets can be identified by an unique value.

5. CONCLUSION

Different communication models have been presented, showing their capabilities to manage temporal constraints.

The choice of a communication model is done according to the temporal constraints in the functional architecture, the used material and the needed QoS.

The introduced models can be applied to different networks as the high-speed ATM network which can be used for control and management of manufacturing systems.

REFERENCES

Carle, G. (1994). "Reliable group communication in ATM network", 12th annual conference on European fibre optic communications and networks - ATM and networks, 21-24 June 1994, Heidelberg, Germany, pp 30-34.

Conway, A.E. (1994). "Forward delay protection of time-critical traffic in ATM networks with FEC", IEEE Globecom, 28 Nov-2 Dec 1994, San Francisco, pp 1200-1206.

Didic, M., F. Neuscheler, L. Bogdanowicz (1993). "McCIM: execution of CIMOSA Models", Proceedings of the Ninth CIM-Europe, Amsterdam, 12-14 May 1993.

Godefroid, P., G.J. Holzmann (1993). "On the verification of temporal properties", 13th IFIP Symposium on Protocol Specification, Testing and Verification, Liège, Belgium, 25-28 May 1993, pp. B3.1-B3.16.

Heinrichs, B., K. Jakobs, A. Carone (1993). "High performance transfer services to support multimedia group communication", Computer Communication, Vol 16, number 9, September 1993, pp 539-547.

ISO 9506 (1989). "Manufacturing Message Specification (MMS), Part 1: Service Definition, Part 2: Protocol Specification", 1989.

ISO TR12178 (1993). "User requirements for systems supporting time-critical communications", 1993.

Kotsiopoulos, I.L. (1993). "Theatrical aspects of CIM-OSA modelling", Proceedings of the Ninth CIM-Europe, Amsterdam, 12-14 May 1993.

Lien, C.C., C.C. Yang (1992). "Reduction of useless services with timing constraints", Proceedings of the 3rd IEEE workshop on future trends of distributed computing system, Taipei, 14-16 April 1992, pp. 226-231.

Lorenz, P., Z. Mammeri, J.P. Thomesse (1993). "Time management in communication for real time applications", Proceedings of the Ninth CIM-Europe Annual Conference, Rai Amsterdam, The Netherlands, 12-14 May 1993, pp. 161-169.

Lorenz, P., Z. Mammeri, J.P. Thomesse (1994). "A State-Machine for Temporal Qualification of Time-Critical Communication", 26th IEEE Southeastern Symposium on System Theory, Ohio, USA, 20-22 March 1994, pp. 654-658.

Luckenbach, T. (1994). "Performance experiments within local ATM networks", 12th annual conference on European fibre optic communications and networks - ATM and networks, 21-24 June 1994, Heidelberg, Germany, pp 150-154.

Northcutt, J.D., E.M. Kuerner (1993). "System support for time-critical applications", Computer communications, Vol 16, n°10, October 1993, pp. 619-636.

Nyong, D. (1994). "Timer related performance issues of system functions within ATM sub-networks", IEE Colloquium on 'Time critical data communications', 9 Dec 1994, pp 2/1-4.

Pokam, M.R., J.F. Guillaud, G. Michel (1995). "Integrate multimedia in manufacturing network using ATM", 20th Conference on local computer networks, 16-19 October 1995, Minneapolis, USA, pp 307-316.

Raju, C.V., R. Rajkumar, F. Jahanian (1992). "Monitoring timing constraints in distributed real-time systems", Proceeding of the real-time system symposium, Phoenix, 2-4 December, 1992.

Shaw A.C. (1992). "Communicating real-time state machine", IEEE Transactions on software engineering, Vol 18, n°9, September 1992, pp. 805-816.

UNI (1994). "ATM User-Network Interface Specification", Version. 3.1, ATM Forum, September 1994.

AN ALGORITHM FOR SCHEDULING TASKS ON MOVING EXECUTORS IN COMPLEX OPERATION SYSTEM

Jerzy Józefczyk

Institute of Control and Systems Engineering, Technical University of Wroclaw, Janiszewski St. 11/17, 50–370 Wroclaw, Poland

Abstract: The problem of scheduling tasks on moving executors in complex operation system with application to discrete manufacturing system is considered. The minimization of makespan for unrelated executors and nonpreemptive, independent tasks is investigated in detail. It is assumed that tasks are performed at the stationary workstations by moving executors. This leads to the new optimization problem, which is solved using the algorithm based on the branch and bound approach.

Keywords: scheduling algorithms, operations research, optimization problems, moving objects.

1. INTRODUCTION

The paper concerns a class of complex systems called complex operation systems. Operations treated as elements of such systems are connected by time restrictions. Scheduling tasks is an example of decision problems which are formulated and considered for complex operation systems. The applications of such problems in discrete manufacturing systems, on the one hand, and in designing computer systems, on the other, cause the necessity to investigate the new and more complicated problems for complex operation systems in general and for scheduling tasks in particular. These applications require new assumptions for decision problems in comparison with the assumptions for the classical ones. In the area of manufacturing systems, the participation of more than one executor performing the manufacturing operation may be treated , for example, as such a new assumption, see (Błażewicz, *et al.*, 1994). The modifications refer also to performance indices the new forms of which often lead to multi–objective problems, as well as to solution methods and to solution algorithms, see (Nelson, *et al.*, 1986; Willon, 1989). The methods based on artificial intelligence techniques have also been intensively studied over the last years, see (Bubnicki, 1992; Prosser and Buchanan, 1994). The new scientific trends which have been outlined are true also for problems of scheduling tasks. This fact is more and more important when manufacturing systems are taken into account. Only such applications are considered in the paper. Then, the exemplary problems of scheduling tasks are the following: the determination of routes for plants to be produced, sequencing problems for executors, different problems in which additional resources, i.e. tools, pallets, fixtures, transport facilities are taken into account.

The movement of different elements of manufacturing systems or strictly speaking the necessity of taking it into account in different problems of scheduling tasks is a very important aspect of the problem being introduced. The movement in discrete manufacturing systems may refer to different objects and may be understood and considered in different ways. One can speak about the movement of plants to be produced, and it is obvious. But one can also consider the movement of executors to perform the manufacturing operations on the plants located at the stationary workstations. Moreover, the term "movement" is perceived in two ways. In the first sense the movement of plants or executors is treated as their displacement to perform the consecutive operations. In the case of the movement of plants the displacement includes the determination of the elements of manufacturing system (usually the executors) among which the plant should be trans-

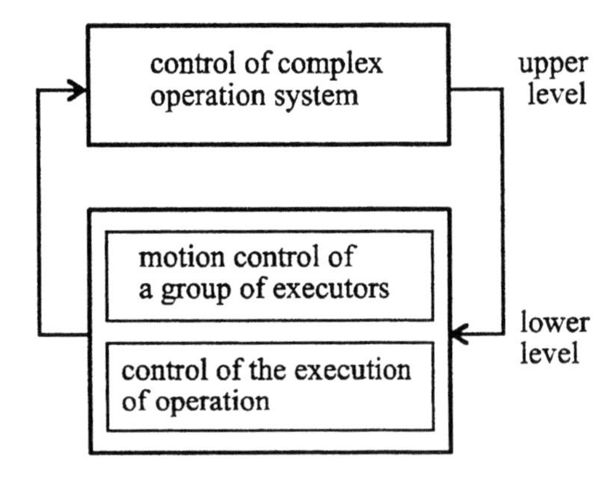

Fig.1. Block scheme of the two–level control system.

ported, the number of transport facilities which ought to do it and the moment when the displacement should be performed. In the case of the movement of executors the displacement comprises the determination of the routes for executors, among workstations with plants loaded for performing the manufacturing operations. The second sense of the term "movement" is quite different. It is the driving between two given positions: the initial and the final ones. The decision problem is then the movement (motion) control problem which consists in controlling the driving mechanism of the moving executor, or the transport facility in the case of the movement of the plant. This problem is connected with the traffic control problem. The solution of the traffic control problem enables us to obtain the movement without collisions. In discrete manufacturing systems and in flexible manufacturing systems in particular, both decision problems connected with different senses of the term "movement" are interconnected and should be considered together. For the movement of executors, they create the two–level control system (fig.1). At the upper level the selected control problem for complex operation system including the displacement of executors is solved. The result in the form of routes for executors are the data for the control problems located at the lower level. At the lower level the motion control problem for a group of executors comprising the traffic control subproblem and the control of the execution of manufacturing operations should be solved. The results from the lower level in the form of the execution times and the driving times are the data for the upper level decision problem. The two–level control system is the basis for the considerations presented in the paper. However, they are limited to the problem from the upper level. The lower level problems are not taken into account. It is assumed that the results obtained there are given. Moreover, as mentioned in the previous part of the introduction, the movement of the executor performing manufacturing operations is investigated and as the decision problem for complex operation system, the problem of scheduling tasks is assumed. As the problem of scheduling tasks a simple assignment problem is considered. In the assignment problem under consideration the unrelated executors should perform the set of nonpreemptive, independent tasks (operations) to minimize the makespan. The ready times for all tasks are assumed to be the same.

To determine more precisely the scope of the paper the main notions will now be explained. The main notion is the *task* which can be understood like in the scheduling theory but having its own meaning by reason of the fact that the movement of executors is taken into account. The *executor* is the subject performing the tasks. Usually it is a technological device. The executor performs the task at a place called the *workstation*. The workstation is located at the plane or in space. Among all workstations the depot is distinguished. It is the workstation where each executor starts and finishes its work. At the depot any task is performed. The main idea of the problem presented in the paper is the following. To perform the task the executor should drive–up to the workstation. Therefore, each task consists of two parts: the driving–up to the workstation and the performing of the job at the workstation. The generalization of the term "task" leads to the generalization of the execution times which are the main data for every scheduling problem. Then the execution time is the sum of the driving–up time and the job performing time. Such generalization defines the new decision (scheduling) problem in which not only the subset of tasks for each executor should be derived but also the routes for the executors should be determined. The necessity of determining the route results from the fact that the order of performing the tasks by the executor has the influence on the execution time for the whole set of tasks. It is obvious that the problem under consideration, i.e. problem of scheduling tasks with moving executors has its application to the class of flexible manufacturing systems. The examples are given in Józefczyk (1996).

In chapter 2, the problem of scheduling tasks is stated as the optimization problem which is NP-hard from the point of view of computational complexity. Therefore, in chapter 3, the solution method using the time decomposition is discussed. Then, in chapter 4, the solution algorithm based on the branch and bound approach is presented. In conclusions the main directions for further investigations are outlined.

2. PROBLEM FORMULATION

All assumptions formulated for the classical tasks scheduling problem are valid. For example, at the same moment one executor can perform only one task and a single task is performed by one executor only. Moreover, since the movements of executors are taken into account it is assumed that at each workstation only one job can be performed and the route for each executor starts and finishes at the depot. In the notation the set of tasks and the set of workstations are not distinguished. Both sets are denoted by $\boldsymbol{H} = \{1,2,\ldots,H\}$, where H is the number of tasks to be performed and the number of workstations. Only the depot is distinguished and defined as $h = H+1$. Then $\overline{\boldsymbol{H}} = \boldsymbol{H} \cup \{H+1\}$ is the set of workstations with the depot. Analogically, $\boldsymbol{R}$ is the set of executors and R – the number of executors which may be used to perform the tasks. The execution times τ_h are the main description of the set $\boldsymbol{H}$. It is well known that

$$\tau_h = \left[\tau_{1,h}, \tau_{2,h}, \ldots, \tau_{r,h}, \ldots, \tau_{R,h}\right]^{\mathrm{T}}, \quad h = 1,2,\ldots,H, \tag{1}$$

where $\tau_{r,h}$ is the execution time for the task h performed by the executor r, and r, h are the current indices of the executor and the task, respectively. The time $\tau_{r,h}$ is the sum of two elements: $\bar{\tau}_{r,h}$ – the time of performing the job h by the executor r, $\hat{\tau}_{r,g,h}$ – the driving-up time of the executor r from the workstation g to the workstation h. Thus,

$$\tau_{r,h} = \bar{\tau}_{r,h} + \hat{\tau}_{r,g,h}, \quad r = 1,2,\ldots,R, \quad h = 1,2,\ldots,H, \quad g = 1,2,\ldots,H+1 \tag{2}$$

It is also assumed that $\hat{\tau}_{r,h,h} = +\infty$, $h = 1,2,\ldots,H+1$ and $\hat{\tau}_{r,g,H+1}$ denotes the time of driving-up of the executor r to the depot. Let us introduce the matrix of decision variables which determine the connection between the set of tasks and the set of executors in the form

$$\gamma = \left[\gamma_{r,g,h}\right]_{\substack{r=1,2,\ldots,R \\ g,h=1,2,\ldots,H+1}}, \tag{3}$$

where

$$\gamma_{r,g,h} \begin{cases} 1, & \text{if executor } r \text{ performs the task } h \\ & \text{after driving-up from the workstation } g \\ 0, & \text{otherwise} \end{cases}$$

The following constraints allow us to derive the admissible matrix γ, i.e. the solution of the problem of scheduling tasks:

$$\gamma_{r,h,h} = 0, \quad r = 1,2,\ldots,R, \quad g,h = 1,2,\ldots,H+1, \tag{4}$$

$$\sum_{r=1}^{R} \sum_{g=1}^{H+1} \gamma_{r,g,h} = 1, \quad h = 1,2,\ldots,H, \tag{5}$$

$$\sum_{g=1}^{H+1} \gamma_{r,g,p} = \sum_{h=1}^{H+1} \gamma_{r,p,h}, \quad r = 1,2,\ldots,R, \quad p = 1,2,\ldots,H+1, \tag{6}$$

$$\sum_{h=1}^{H} \gamma_{r,H+1,h} = 1, \quad r = 1,2,\ldots,R, \tag{7}$$

$$\gamma \in \boldsymbol{S}, \tag{8}$$

where

$$\boldsymbol{S} = \{\gamma : \sum_{g \in \boldsymbol{H}_S} \sum_{h \in \boldsymbol{H}_S} \gamma_{r,g,h} \le H_S - 1,$$

$\boldsymbol{H}_S$ – any non empty subset of H, $r = 1,2,\ldots,R\}$.

Among other things, the constraints assure that the route for each executor has the form of Hamilton cycle with the beginning and the end at the depot. The purpose of the problem is to minimize the makespan which can be expressed in the form

$$Q(\gamma) = \max_{r=1,2,\ldots,R} \left[\sum_{h=1}^{H+1} \sum_{g=1}^{H+1} \gamma_{r,g,h} (\bar{\tau}_{r,h} + \hat{\tau}_{r,g,h}) \right], \tag{9}$$

where $\bar{\tau}_{r,H+1} = 0$, $r = 1,2,\ldots,R$ because at the depot any job is executed. So, one can now present the formulation of the problem of scheduling tasks with moving executors called *initial problem* (IP).

Given:

1. The set of tasks $\boldsymbol{H}$ performed at the workstations denoted also by $\boldsymbol{H}$ which together with the depot create the set $\overline{\boldsymbol{H}} = \{1,2,\ldots,H,H+1\}$.
2. The set of executors $\boldsymbol{R} = \{1,2,\ldots,R\}$.
3. The times of job performing $\bar{\tau}_{r,h}$, $r = 1,2,\ldots,R$, $h = 1,2,\ldots,H$ and $\bar{\tau}_{r,H+1} = 0$, $r = 1,2,\ldots,R$
4. The driving-up times $\hat{\tau}_{r,g,h}$, $r = 1,2,\ldots,R$, $g,h = 1,2,\ldots,H+1$ and $\hat{\tau}_{r,h,h} = +\infty$, $r = 1,2,\ldots,R$, $h = 1,2,\ldots,H+1$.

Determine:

The matrix γ admissible in the sense of (4)–(8) to minimize (9).

3. TIME DECOMPOSITION OF THE PROBLEM

To solve the problem formulated in chapter 3, the time decomposition method has been applied. In this method the time horizon needed for executing all tasks from the set $\boldsymbol{H}$ and the driving-ups to the depot is divided into shorter time intervals. The decisions are made at the beginning of the time intervals. The time intervals are constituted by *events* which are determined by the end of the execution of the tasks. The length of the time intervals is variable. The consecutive events (the time intervals) and the number of all events (all time intervals) are denoted by n and N, respectively. In the time decomposition method the complex operation system under consideration which is the decision making object, is treated as the event system. The description of such a system in the state space is given. The matrix

$$\mathbf{z}(n) = \left[z_{r,h}(n)\right]_{\substack{r=1,2,\dots,R \\ h=1,2,\dots,H+2}} \quad (10)$$

is the state for the interval $n = 1,2,\dots,N$. The elements of $\mathbf{z}(n)$ have different interpretation. The sets of values for elements of $\mathbf{z}(n)$ are as follows

$$Z_{r,h} \triangleq \left[0, \tilde{\tau}_{r,h}\right] \cup \{-1,-2,\dots,-(H+R)\}, \quad r = 1,2,\dots,R,\ h = 1,2,\dots,H+1, \quad (11)$$

where

$$\tilde{\tau}_{r,h} = \bar{\tau}_{r,h} + \max_{\substack{g=1,2,\dots,H+1 \\ g\neq h}} \left\{\hat{\tau}_{r,g,h}\right\},$$

$$Z_{H+2} \triangleq \{1,2,\dots,H+1\},\ r = 1,2,\dots,R. \quad (12)$$

Then $z_{r,h}(n) \geq 0, r = 1,2,\dots,R, h = 1,2,\dots,H+1$ denotes the time remaining in n to end the performance of the task h by the executor r. This element is the index informing about the order of performing the tasks by the executor r when $z_{r,h}(n) < 0$. The value of $z_{r,H+2}(n) \in Z_{H+2}$ denotes the number of the workstation from which the executor r has been started to perform the current task. The initial and the final states have the respective forms

$$\mathbf{z}(0) = \left[z_{r,h}(0)\right]_{\substack{r=1,2,\dots,R \\ h=1,2,\dots,H+2}}, \quad (13)$$

where

$$z_{r,h}(0) = \begin{cases} \tilde{\tau}_{r,h}, & \text{for } r = 1,2,\dots,R,\ h = 1,2,\dots,H+1 \\ H+1, & \text{for } r = 1,2,\dots,R,\ h = H+2 \end{cases} \quad (14)$$

and

$$\mathbf{z}(N) = \left[z_{r,h}(N)\right]_{\substack{r=1,2,\dots,R \\ h=1,2,\dots,H+2}} \quad (15)$$

with conditions

$$\sum_{r=1}^{R} \mathbf{1}(-z_{r,h}(N)) = 1,\ h = 1,2,\dots,H, \quad (16)$$

where $\mathbf{1}(\cdot)$ is zero-one function ($\mathbf{1}(x) = 1$ for $x \geq 0$ and $\mathbf{1}(x) = 0$ for $x < 0$),

$$z_{r,H+1}(N) \leq 0,\ r = 1,2,\dots,R, \quad (17)$$

$$z_{r,H+2}(N) = H+1,\ r = 1,2,\dots,R. \quad (18)$$

As already mentioned, the decision about mapping the tasks to executors and the current form of the routes should be made for every n. Therefore, the decision matrix

$$\mathbf{v}(n) = \left[v_{r,h}(n)\right]_{\substack{r=1,2,\dots,R \\ h=1,2,\dots,H+1}} \quad (19)$$

is introduced. Its elements are zero-one, i.e.

$$v_{r,h}(n) \in \{0,1\} = \begin{cases} 1, & \text{if executor } r \text{ performs task } h \text{ in interval } n \\ 0, & \text{otherwise} \end{cases} \quad (20)$$

The matrix $v(n)$ is constrained as follows:

$$\sum_{r=1}^{R} v_{r,h}(n) \leq 1,\ h = 1,2,\dots,H+1, \quad (21)$$

$$\sum_{h=1}^{H+1} v_{r,h}(n) \leq 1,\ r = 1,2,\dots,R, \quad (22)$$

$$\left[0 < z_{r,h}(n) < \bar{\tau}_{r,h} + \hat{\tau}_{r,z_{r,H+2}(n),h}\right] \Rightarrow \left[v_{r,h}(n) = 1\right], \quad r = 1,2,\dots,R,\ h = 1,2,\dots,H+1. \quad (23)$$

The implication (23) ensures the indivisibility of tasks. The state equations allow us to calculate new values of the state in consecutive intervals. For the state variables from the first H+1 columns of $\mathbf{z}(n)$

$$\begin{aligned} z_{r,h}(n+1) &= \left[1 - \delta(z_{r,h}(n) - \tilde{\tau}_{r,h})\right] \\ &\times [z_{r,h}(n) - v_{r,h}(n)\Delta(n) - 1(v_{r,h}(n)\Delta(n) \\ &- z_{r,h}(n))] + \left[1 - \delta(h - z_{r,H+2}(n))\right] \\ &\times \delta(z_{r,h}(n) - \tilde{\tau}_{r,h}) \times [\left[1 - v_{r,h}(n)\right]\tilde{\tau}_{r,h} \\ &+ v_{r,h}(n)\left[\bar{\tau}_{r,h} + \hat{\tau}_{r,z_{r,H+2}(n),h} - \Delta(n)\right]], \\ n &= 0,1,\dots,N-1, \end{aligned} \quad (24)$$

where $\delta(\cdot)$ is zero-one function ($\delta(x) = 1$ for $x = 0$ and $\delta(x) = 0$ for $x \neq 0$) and

$$\Delta(n) \triangleq \min_{\substack{s=1,2,\dots,R \\ p=1,2,\dots,H+1}} \left[z_{s,p}(n): v_{s,p}(n) = 1\right] \quad (25)$$

is the length of the n-th time interval. For the state

variables from the last column of $z(n)$

$$z_{r,H+2}(n+1) = z_{r,H+2}(n)$$
$$\delta\left(\sum_{h=1}^{H+1} h \cdot v_{r,h}(n)\delta(z_{r,h}(n) - \Delta(n))\right)$$
$$+\sum_{h=1}^{H+1} h \cdot v_{r,h}(n)\delta(z_{r,h}(n) - \Delta(n)), \qquad (26)$$
$$n = 0,1,2,\ldots, N-1.$$

The performance index is the sum of the lengths of all intervals, i.e.

$$Q_N(\bar{v}(N-1)) = \sum_{n=0}^{N-1} \varphi_0(z(n), v(n)). \qquad (27)$$

To avoid the case when at the same interval more than one task is finished additional conditions have been imposed on the intermediate states

$$\sum_{r=1}^{R}\sum_{h=1}^{H+1}\left[1(z_{r,h}(n)) - 1(z_{r,h}(n+1))\right] = 1. \qquad (28)$$

Finally, the problem after the time decomposition can be formulated as follows.

Given:

1. The sets $\overline{\boldsymbol{H}}$ and $\boldsymbol{R}$ as well as the times $\bar{\tau}_{r,h}$ and $\hat{\tau}_{r,g,h}$ like for IP in chapter 2.
2. Model (24), (26) and the initial state $z(0)$.
3. Constraints on the state trajectory (16)–(18), (28).

Determine:

The sequence of decision matrices $\bar{v}(N-1) = (v(0), v(1), \ldots, v(n), \ldots, v(N-1))$ admissible in the sense of (21)–(23) to minimize (27).

After the analysis of this problem and the problem IP their equivalence can be stated. The argumentation is as follows. The performance indices for both problems are the time of performing all tasks. Thus, it is enough to argue that constraints (4)–(8) are fulfilled for the second problem too. The necessity of performing all the tasks is given in (5) and (16). The constraints (6)–(8) ensure the form of the routes as the Hamilton cycles. For the second decision problem the continuity of routes results from the way of their consecutive determination in the time intervals. The terms (14) and (22) demand the depot to be the beginning and the end of all routes. The last condition imposed on the solution for IP assures the lack of sub-cycles. Let us assume that the executor can return to the depot many times. Then the route for the executor couldn't be optimum because at the depot any task is executed and such return causes the waste of time.

4. SOLUTION ALGORITHM

To solve the problem after the time decomposition the *branch and bound* approach has been applied. The approach has geometric interpretation. The process of routes generating is presented as the moving along the tree. Each node of this graph represents the subset of solutions possible to obtain in the part of tree beginning at this node. Each final node corresponds to a single solution. The layers of the tree accord with the time intervals. The root lies at the layer n=0 and the final nodes – at the layer n=N. The movement is allowed only between two neighboring layers and denotes the change of the state of the object, i.e. the matrix $z(n)$. The solution algorithm consists of two main parts: generation of successors and determination of lower bounds.

Generation of successors. Let us denote by $l(n)$, $L(n)$, and $M(l(n))$ the current node, the number of nodes, and the number of successors of $l(n)$ in the n-th layer, n=0,1, ..., N, respectively. The value of $M(l(n))$ is the function of $R'(z(n))$ and $\overline{H}'(z(n))$, i.e. of the number of free executors and the number of tasks whose performance has not started till n. Then

$$M(l(n)) = \prod_{i=0}^{R'(z(n))-1}\left[\overline{H}'(z(n)) - i\right]. \qquad (29)$$

The values of $R'(z(n))$ and $\overline{H}'(z(n))$ are the number of elements of the sets $\boldsymbol{R}'(z(n))$ and $\overline{\boldsymbol{H}}'(z(n))$, respectively. The time of reaching the successor of $l(n)$ can be calculated from the recursive term

$$t(m(l(n))) = t(l(n+1)) = t(l(n)) + \Delta(n). \qquad (30)$$

Determination of lower bound. The following lower bound of the performance index (27) is used

$$Q_{N,LB}(l(n)) = t(l(n)) + \frac{1}{R}$$
$$\sum_{h \in \overline{H}'(z(n))} \min_{r=1,2,\ldots,R}\left[\bar{\tau}_{r,h} + \min_{\substack{g \in \overline{H}'(z(n)) \\ g \neq h}} (\hat{\tau}_{r,g,h})\right]. \qquad (31)$$

It is assumed that each task is performed in shortest possible time and the continuity of routes is neglected. The lower bound is used at each node. This is possible when the first admissible solution is known. To derive such a solution an assumption is made that at each layer the successor is chosen ensuring the minimum value of $t(m(l(n)))$, i.e.

$$\begin{aligned} &\min_{m(l(n))\in M(l(n))}\left[t(m(l(n)))\right] \\ &= \min_{m(l(n))\in M(l(n))}\left[t(l(n))+\Delta(n)\right] \qquad (32) \\ &= t(l(n)) + \min_{m(l(n))\in M(l(n))}\left[\Delta(n)\right]. \end{aligned}$$

The solution algorithm can be presented in the following six steps.

1. Find the first admissible solution and denote the path to it as $l(n)=1$, $n=1,2,\ldots,N$. Then calculate $M(l(n))$, $n=0,1,\ldots,N-1$ and maintain: the value of the performance index as $\tilde{Q}_N$, the final state as $\tilde{z}(N)$, and all states $z(n)$, $n=0,1,\ldots,N-1$, which correspond to the solution. Next set $n=N-1$.
2. Verify all solutions being the successors of $l(n-1)$ and choose the best one on the basis of their comparison. Modify $\tilde{Q}_N$ and $\tilde{z}(N)$.
3. Set $l(n)=l(n)+1$. If $l(n)\le M(l(n-1))$ go to step 4, else go to the step 6.
4. Determine $\overline{H}'(z(n)), R'(z(n))$ and calculate $M(l(n))$, $Q_{N,LB}(l(n))$. If $Q_{N,LB}(l(n))\ge\tilde{Q}_N$ go to step 3, else go to the step 5.
5. If $n=N-1$ go to step 2, else set $l(n+1)=l(n+1)+1$, calculate $z(n+1)$ for $v_{r(l(n+1)),h(l(n+1))}=1$, set $n=n+1$ and go to 2.
6. Set $n=n-1$. If $n>0$ go to step 3, else stop the algorithm with the optimal value of the performance index $\tilde{Q}_N$ and the optimal routes for executors comprised in $\tilde{z}(N)$.

Example: Given $H=5$, $R=2$ and the initial state determined on the basis of $\bar{\tau}_{r,h}$ and $\hat{\tau}_{r,g,h}$ in the form

$$z(0)=\begin{bmatrix}126 & 127 & 122 & 118 & 117 & 65 & 6\\ 126 & 111 & 118 & 119 & 122 & 63 & 6\end{bmatrix}.$$

As the result of application of the presented solution algorithm in the form of the computer program one obtains $\tilde{Q}_N=379$ and

$$\tilde{z}(7)=\begin{bmatrix}-6 & 127 & 122 & -4 & 117 & -3 & 6\\ 126 & -2 & -5 & 119 & -6 & -1 & 6\end{bmatrix}.$$

From the matrix $\tilde{z}(7)$ the following routes for both executors can be acquired

$r=1$ – route: $6\to1\to4\to6$,

$r=2$ – route: $6\to5\to3\to2\to6$.

5. CONCLUSIONS

In the further investigations of the presented problem two directions can be distinguished. In the first one the improvement of the described solution algorithm will be sought, first of all in the area of new, more effective lower bounds. The problem under consideration after the time decomposition is NP-hard one. Therefore, it is necessary to evaluate the solution algorithm by means of computer simulation experiments. The performance of such experiments is the second direction for the researches to be planned.

ACKNOWLEDGEMENT

The paper was supported by the State Committee for Scientific Research under Grant no 8 T11A 046 08.

REFERENCES

Błażewicz, J., M.Drozdowski, J.Węglarz, (1994). Scheduling multiprocessor tasks – a survey. *Microcomputer Appl.*, **13,** 89–97.

Bubnicki, Z. (1992). Algorithms of the knowledge based decision making for complex operation systems. *Proc. of XI IASTED Int. Conf: Modelling, Identification and Control*, pp.164–167. Acta Press, Zurich.

Józefczyk, J. (1996). *Tasks scheduling on moving executors in complex operation systems.* Wroclaw Technical University Press, Wroclaw (in Polish).

Nelson, R.T., R.F.Sarin, R.L.Daniels, (1986). Scheduling with multiple performance measures: the one machine case. *Manag. Sc.* **32**, 464–479.

Prosser, P., I.Buchanan, (1994). Intelligent Scheduling: Past, present and future. *Intelligent Systems Eng.* **3**, 67–78.

Willon, J.,M. (1989). Alternative formulations of a flow shop scheduling problem. *J. of the Oper. Res. Soc.*, **40**, 395–399.

EXPERT SYSTEM FOR FMS SCHEDULING BASED ON THE REVERSE SCHEDULING METHOD

Somlo, J. and Elbuzidi, A.S.

Department of Manufacture Engineering
Faculty of Mechanical Engineering
Technical University of Budapest

Abstract: In flexible manufacturing systems the job assignments problem can be formulated as job-shop type scheduling problems. Nowadays, it has an effective way for the solution. This is the use of expert systems. When developing an expert system, one is faced with the problem of the generation of proper knowledge base. One way of providing knowledge bases is the reverse scheduling proposed by the authors.

For describing the use of the reverse scheduling approach in the present paper the following problems are covered:
a.) Formulation of the problem of the job assignment. b.) The ideal schedule and fictitious process plans generation. c.) Priority rule based forward scheduling and goodness estimation. d.) FMS scheduling expert system structure and knowledge base generation. e.) Input statistics investigation and expert system use.

Keywords: Reverse Scheduling, FMS, Heuristic Methods, Priority Rules, Forward Scheduling, Goodness of Schedules, Expert Systems, Knowledge Base, Input Statistics.

1. INTRODUCTION

Flexible Manufacturing Systems (FMS) are such kind of automated systems which produce many kinds of manufacturing products in small lots. FMS is becoming increasingly popular mode of automating small and medium size batch type production.
Job-shop type operations are mainly performed by such systems, that is the flow of work materials is not fixed while that of mass production is. For producing many kinds of products in FMS efficiently not only good hardware is needed, like numerically controlled machines, robots, material manipulation systems, etc., but good software is also important to manage FMS effectively. Then, appropriate scheduling is needed to realise FMS actions.
To solve the scheduling problem in FMS is complicated task. It plays important role as a software for realising efficient production line. The most difficult phase of job-shop scheduling is the formulation of the scheduling criteria, suitable for different production environments, because:

a) Scheduling objectives may change depending upon the decision makers even for similar production situations.
b) Scheduling objectives are multiple and some of them are conflicting.
c) Scheduling objective measures of importance in multiple objectives could change in time affected by the production environment.

The exact algorithms for solving job-shop scheduling problem can be based on Branch and Bound methods. However, these methods require significant computing times, and they are not able to solve realistic size problems.

For this reason, heuristic procedures are developed. These procedures can be classified in to three main classes:

List scheduling algorithms:
They are simple to implement. Operations are assigned to machines according to some priority rule.
One-machine scheduling:
In this class the multi-machine problem is solved by iteratively solving one-machine scheduling problems. The shifting bottleneck procedure (Adams et al 1988) is a good example of this class.

Local search algorithms:
Simulated annealing and taboo searches are among the most popular methods in this class. They are very simple to implement, but require often a huge amount of computation in order to yield a good solution.
Multiple objective job scheduling is too complex to solve as a combinatorial optimisation problem.
In such situations expert systems approaches are used in scheduling (ISIS, Fox and Smith, (1984), Kim and Fichter (1986), Grant (1986), Kanet and Adelsberger (1987)).
The above methods and works concerned the problem of scheduling in general. The scheduling problems were investigated also with special emphasis on FMS aspects.
The review of only a few works in this direction will be given bellow.

In Hitomi's work (1979) there was an earlier attempt made to formulate and solve flexible manufacturing systems scheduling including optimisation problems for cutting conditions, too.

In Rembold, Nngaji, Storr (1993) general issues for computer integrated manufacturing including scheduling were given. Many other works were discussing these and the connected topics.

In Somlo, Nagy (1976) the secondary optimisation problem was formulated which connects the manufacturing data optimisation and FMS scheduling.

System level optimisation problem were formulated and solved in Somlo (1986), and Horvath, Somlo (1981).

Kim, Fichter, Funk (1988) proposed and built an expert system for FMS scheduling. The essence of this approach was to use different priority rules and to estimate the goodness of obtained schedules by the human operator. After a time, the expert system, learning from the human decisions , takes over the task and proposes the appropriate priority rules.

T.Watanabe and others in a series of works, analysed the FMS scheduling problem (Watanabe and Sakamoto, (1984), Watanabe et al (1990). They were using computer simulation and heuristic dispatching rules for solving the job-shop scheduling problems. Following the idea of secondary optimisation, they have shown that the change of the operation times leads to much more effective utilisation of manufacturing resources.

The paper Somlo, Watanabe (1992) presented a rather general formulation for intelligent scheduling of FMS. Formula for the goodness of schedules, priority rule variation and effectiveness estimation, improvement of schedules by discrete changes of process parameters were proposed.

In the following a new approach to the FMS scheduling problem will be proposed.

2. PROBLEM FORMULATION

Let us investigate a job-shop type scheduling problem with the following conditions.

* It is supposed that the manufacturing sequences planning, the operation planning and operation element planning have been performed. This provides an engineering data base containing for all of the parts to be produced the processing sequences and the manufacturing capacity requirements.

** It is supposed that in a higher than the scheduling level of the production planning system decision had been taken about the technological variants proposed by the process planning system. So, in this level, for all of the batches, the number of parts, routs, availability and due date are known.

*** No batch dividing, no overlapping production is considered. It is supposed that the transport and manipulation operations should not be specially scheduled because these are built in into the capacity requirements. (When this last requirement is not valid extensions of the applied method or simulation approaches should be used.)

According to the above the mathematical model of the scheduling problem can be formulated as follows:
Let:
T- is the scheduling period, I- is the number of batches, i-index of batches, (i = 1,2,...,I),
e_i - identification tag, n_i - number of parts in a batch, τ_{ai} - earliest availability time, τ_{di} - due date.

Δt_{ik}^{h} -are the capacity requirements of the ith batch on the hth machine group (i=1,2,...,I; k = 1, 2,... K_i- the serial number of operation of the given part h=1,2,...H the index of machine groups). To every ik, pair belongs some h value. It is the most important constraint that when processing the order of sequences (the routs) should strictly be kept.
H- is the overall number of machine groups,
t_h (h=1, 2,.. H) are the processing capacities of the machine groups during the T period.
The task is to assign for every ik pair (i = 1,2,..I, k=1,2,.. K_i) a section included in t_h satisfying some goodness criterion.
To this model, for the development of the present version of the reverse scheduling algorithm, the following restriction has been added: a batch processed on one machine group does not return to the same machine group.

3. REVERSE SCHEDULING (RS)

3.1. Ideal Schedule

The idea of reverse scheduling is very simple. Using some stochastic law a schedule is generated for the available manufacturing capacity of the machine groups, which fully utilises these capacities. The job assignment obtained in this way is named: ideal schedule.
From the ideal schedule, it is possible to reconstruct fictitious manufacturing capacity requirements for fictitious batches. These capacity requirements will have the properties given by the stochastic law applied for ideal schedule generation. Details about the ideal schedule generation and the computer programs for investigation of the properties of the method are given in Somlo, Elbuzidi (1996).

3.2. Forward Scheduling

A second step in the use of the method is to solve the scheduling problem for the fictitious batches using heuristic methods based on priority indices. A number of different priority indices can be used (for example: FIFO, SLACK, SPT, LPT, MAXNOP, MAXOTR, EDD, MINIT, etc.)

3.3. Goodness Estimation

The next step is the estimation of the goodness of the job assignment obtained by use of different priority indices. A number of well known goodness criterions are used, e.g. MAXCT, AVCT, LMAX, AVL, NTJ, TMAX, AVTR, EMAX, AVER, MUR. A complex criterion (ELR) proposed by us was estimated, too.
The abbreviations used are:
Priority Rules:
FIFO = higher priority for the first arrived job,
SLACK = minimum slack job assigned first,
SPT = shortest processing time job first,
LPT = longest processing time job first,
MAXNOP = job with higher number of operation remaining assigned first,
MAXOTR = job with maximum operation times remaining first,
EDD = earliest due date job assigned first,
MINIT = job with minimum idle time first

Goodness Criterions:
MAXCT = maximum completion time.
AVCT = average completion time.
NTJ = number of tardy jobs, LMAX = maximum lateness, AVL = average lateness,
EMAX = maximum earliness, AVER = average earliness, TMAX = maximum tardiness.
AVTR = average tardiness, MUR = machine utilisation ratio
ELR = Combined criterion, represents a weighted sum of number of single goodness criterions
(e.g. ELR = w1*MAXCT + w2*NTJ + w3*TMAX).
Other combinations can be formed according to the scheduler objectives.
At the evaluation of the results the multicriterion (PARETO) optimisation ideas and heuristics can be used with good results.
In this way for given statistic laws of manufacturing capacity requirements the best, or the several best scheduling algorithms can be selected.
Together with the proposed algorithms the goodness criterions values can be indicated, too.
On Fig. 1a and b the results for an example are given for demonstration.

The results of scheduling 20 jobs and 6 machines simulated 100 times.

Priority\Goodness Rule \Criterion	MAXCT	ACT	LMAX	AVL	NTJ	TMAX	AVTR	EMAX	AVER	MUR
Ideal Sch.	818	734	-70	-97	0	0	0	116	97	1.00
SPT	1156	750	370	-81	8	371	65	544	147	0.77
MAXNOP	875	783	189	-49	5	189	23	206	72	0.97
SLACK	938	728	25	-104	3	29	4	370	108	0.91
LPT	1106	739	325	-93	7	325	53	505	146	0.81
MAXOTR	882	794	209	-37	6	209	28	213	65	0.97
EDD	1219	694	290	-138	6	290	41	466	179	0.73
MIN_IT	996	678	232	-154	5	232	26	518	180	0.89

Fig. 1.a Goodness criterions values for ideal schedule and number of priority rules.

Priority\Criterion Rule \Ratio	MAXCTratio (RGC-ISGC/ISGC)	NTJratio (NTJ/MAXNJ)	MURratio (TPTJIS/TPTJR)	TMAXratio (TMAX*MAXNM/MAXNJ)	ELR (Combined Criterion)
SPT	0.41	0.21	0.40	0.37	-0.04
MAXNOP	0.07	0.25	0.29	0.19	0.16
SLACK	0.15	0.15	0.91	0.03	0.23
LPT	0.35	0.35	0.81	0.33	0.01
MAXOTR	0.08	0.30	0.97	0.21	0.15
EDD	0.29	0.30	0.73	0.29	-0.01
MIN_IT	0.23	0.25	0.89	0.23	0.11

Fig. 1.b Relative ratios of the priority rules to show their goodness.

4. EXPERT SYSTEM FOR FMS SCHEDULING

When using the RS method for the solution of practical tasks the following procedure can be applied. First the statistical properties of the manufacturing capacity requirements should be estimated. Then, from a data base obtained using the RS method the suitable priority rule (rules) are obtained and applying that scheduling is performed. The following idea (presentiment) is formulated by us: it is supposed that if at a given input data statistics and at the applied scheduling priority rule the RS method performance measure is favourable then applying the same rule for real FMS scheduling task with the same input statistics the result will be also favourable.

4.1. Knowledge Base:

Accordingly, a part of the knowledge base proposed for FMS scheduling can be constructed in the following way: For different statistical properties of the input statistics the RS method is used. Determining the ideal schedule, the fictitious process plans and performing forward scheduling the goodness belonging to different priority rules are estimated. Then, for example, at normal distribution, in the plane one axis of which is the mean value the

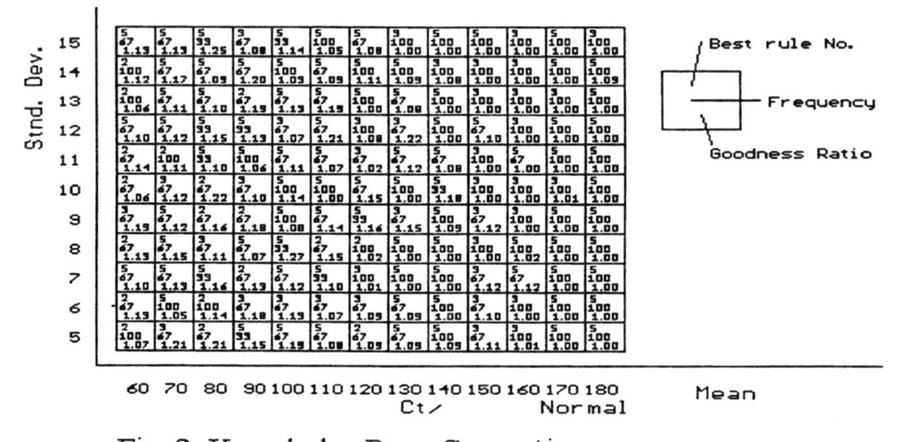

Fig. 2. Knowledge Base Generation.

other is the standard deviation, the best priority rule (and goodness belonging to that) is indicated. In fact the above procedures are made for discrete points of the plane, but the result are supposed to be valid for small domains.
An example is shown on Fig. 2. It is easy to understand from the figure the basic properties of this knowledge base generation.

4.2. Expert System

A general configuration of the proposed intelligent scheduling system is given on fig. 3.

The upper part inside the dotted box represents the knowledge acquisition system, which in turn has two approaches to get the required scheduling knowledge (the best scheduling rule). The first approach is the RS method, where ideal schedules, fictitious process plans and forward scheduling are generated for different input data sets. Estimations of the schedules goodness are performed also. In RS the scheduling data are statistically obtained. The schedules generation and its goodness estimation are performed with help of computer program coded in Pascal.
The second approach for intelligent scheduling is done by generating the forward schedules directly using priority rules and given scheduling data (processing times and job routs).
Both approaches are merging to the same knowledge base which going to be utilised by the main expert system. This scheduling knowledge base contains mainly the best scheduling rule for different input data either its statistically generated as in RS or as a result of some practical cases. This knowledge base is stored in the memory of computer and it can be updated with new knowledge or consulted for use through a data base written using Delphi programming environment.
Users can get the proper scheduling knowledge for specific problem through inference engine, scheduling algorithms and supported by explanation system.
At the application of the proposed expert system it is supposed, not only the use of the knowledge base obtained by RS, but after selecting the proper priority rule, other expert features may be used to improve the schedules efficiency.

5. CONCLUSIONS

The reverse scheduling method seems a very effective mean to solve FMS scheduling problem. It was developed to use input statistics related to the overall system, but, it can be developed into the direction of using input statistics for machine groups. At the solution of practical problems it turned out that there exist several good solutions. But, there are catastrophically bad solutions, too. It is important to avoid those in all of the cases.

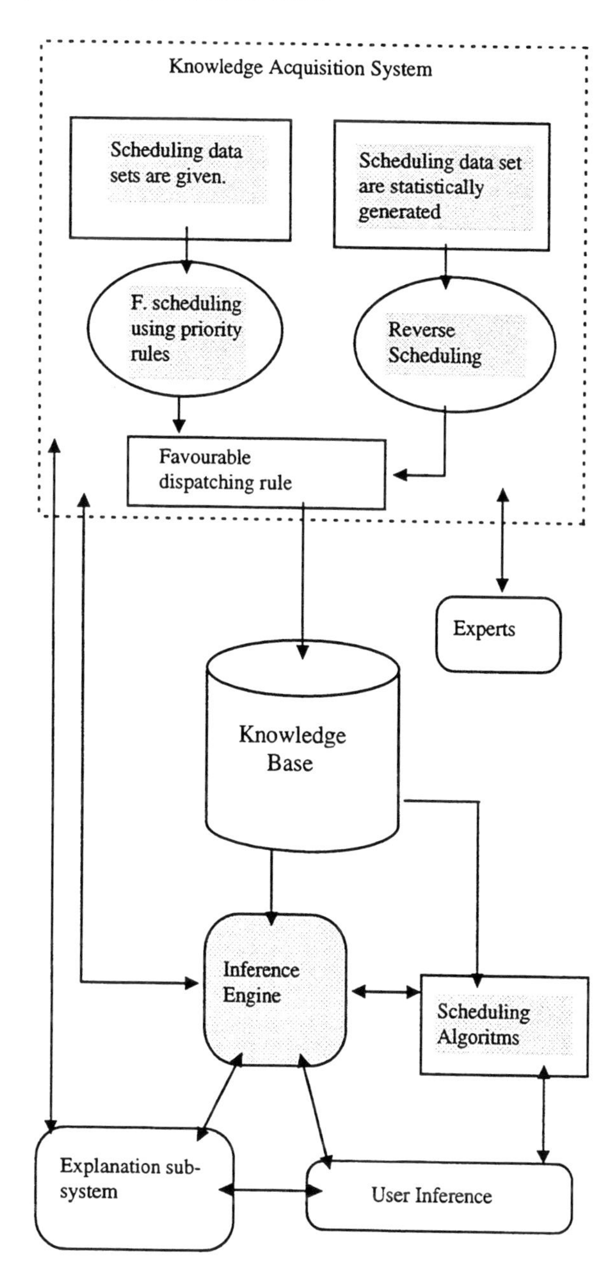

Fig. 3 Knowledge Base Generation for Expert Systems Using Reverse Scheduling method.

ACKNOWLEDGEMENT

The research work the result of which are presented in this paper was supported by the Research Found of the Hungarian Academy of Science (OTKA-T 014327).

REFERENCES

ElBuzidi, A. (1995). Application of expert systems in manufacturing scheduling. Second International Workshop on Learning in Intelligent Manufacturing Systems LIMS '95'. Budapest, Hungary.

Elbuzidi, A and F. Petrucci.(1996). Reverse scheduling and its use for intelligent FMS Job assignments. *Periodica polytechnica* Faculty of Mechanical Engineering, Technical University of Budapest (To be published).

Fox, M.S. and Smith. (1984). A knowledge based system for factory scheduling *Expert systems Journal*. **vol 1**, No. 1. July .pp 25-49.

Grant, T.J. (1986) Lessons for O.R. from A.I. A case study, J. Opl Res. Soc. **37**, 41-57.

Hitomi, K. (1979) *Manufacturing Systems Engineering*. Taylor, Francis. London.

Horvath, M. and Somlo, J. (1981) The hierarchical system, optimisation and adoptive control of machine tools. IFAC 8-th World congress Kyoto.

Kanet, J.J and Adelsberger, H. (1987) Expert Systems in production Scheduling. Eur.J. Opl Res. **29**, 52-59.

Kim , E.F.Fichter and K.H.Funk.(1988). Building an expert system for FMS scheduling. USA-Japan Symposium on Flexible Automation. San Francisco.

Rembold, U, B.O. Nnaji and A. Storr. (1993).*Computer integrated manufacturing and engineering*. Addison publishing Company.

Somlo, J. and Elbuzidi, A. (1996). Reverse scheduling, a new approach to the solutions FMS scheduling problems. RAAD'96. (The Fifth International Workshop on ROBOTICS IN ALPE-ADRIA-DANUBE REGION (Budapest, Hungary).

Somlo, J. (1986). Some problems of optimisation in computer integrated manufacturing Systems. IFAC Symposium on Large scale systems. Zurich.

Somlo, J. and M. Girnt. (1989). Optimisation aspects of experimental FMS at the Technical University of Budapest. IFAC-International workshop on decisional structures in automated manufacturing. Genova, Italy.

Somlo, J. and Nagy, J. (1976) A new approach to cutting data optimisation problem. PROLAMAT'76 symp. Stirling.

Somlo, J. and Watanabe, T.(1992). Intelligent scheduling of FMS. the optimisation approach, USA -Japan Symposium. on Flexible Automation San Francisco.

Watanabe, T and Kido, Fujii. (1990). The automatic improvement of Job-shop schedules and new dispatching rules. Japan-USA Symposium on Flexible Automation. Kyoto ,Japan.

Watanabe, T and Sakamoto. (1984). On line scheduling for adaptive control machine tool in FMS. IFAC. 9th Triennial world congress. Budapest, Hungary.

A HIERARCHICAL APPROACH TO AN FMS SCHEDULING PROBLEM

Marino Nicolich, Piero Persi, Raffaele Pesenti, Walter Ukovich

University of Trieste, Italy

Abstract: The problem of improving the saturation of an FMS cell is considered. Limited tool buffer capacity turns out to be a relevant constraint. A hierarchic approach is formulated, contemplating at the higher level the determination of "clusters", i.e., sets of parts that can be concurrently processed. At lower levels, clusters are sequenced, linked, and scheduled. While for the solution of the latter subproblems methods taken from the literature are used, two original mixed integer programming problems are formulated to determine clusters. The proposed methods are then discussed on the basis of computational experience carried out on real instances.

Keywords: Production Systems, Flexible Manufacturing Systems, Mathematical Programming.

1. INTRODUCTION

The present paper discusses some production programming and scheduling problems studied for a flexible cell which operates in the Diesel Engine Division of Fincantieri, Trieste (Italy), the major Italian state company in the shipbuilding sector.

The Diesel Engine Division produces and assembles diesel and gas engines for ship propulsion and electricity generation. Its normal production is about 50 engines per year. Typical costs for a single product range from about 3 to 6 million dollars for fast and slow engines, respectively. Orders arrive about one year before due date; then supplying row parts requires six months, production and assembly three months each.

The FMS cell under study is composed of four machines: three machining centers, one of which is a vertical lathe, and a washer. Each machine has an input/output buffer for parts. Machining Centers (MCs) also have a tool buffer with 144 positions each. Tools are loaded and unloaded automatically. Pallets carrying parts or tools are moved automatically too.

The problem considered in this paper stems from the difficulty that production managers experienced in balancing workload on the MCs of the system, with the consequence of raising production costs in the FMS cell. On these bases, it was required to investigate the reasons of such a drawback, and to devise proposals to improve plant productivity.

A first goal of the study was identified in proposing a methodology to find the most sensible ways to cluster parts which can be concurrently processed and can better saturate the machining resources. The operational constraints are the technology of the parts and the capability of the MCs in terms of the number of tools required by each part and the capacity of the local tool buffer. Operations Research techniques provided sound means to explain the present situation and to reveal convenient ways to improve the performance of the system.

The paper describes a hierarchic approach to the problem, which considers at a first level the part clustering problem from the monthly production plan. Two original models are formulated for this problem: the first one achieves the minimum production time, and the other one yields the minimum production time with the best workload balance on the MCs. At lower levels, cluster sequencing, cluster linking, and part scheduling within each cluster are problems which can be easily approached using methods taken from the literature.

2. THE SATURATION LEVEL OF THE FMS CELL

The FMS cell is a crucial resource for the whole production system, as it can process parts as much as three times faster than ordinary job shops, due to its ability to comply with changes both in production mix and in lot size. However, the cost per hour of its machines is sensibly higher (about 1.5 times) than the cost of a machine of an ordinary job shop. As a consequence, the FMS cell requires to be saturated as much as possible. The cell started production in 1991. It operates on three shifts, eigth hours each, from 6 a.m. of Monday to 2 p.m. of Saturday; the night shift is not supervised by operators, thus giving about a 50% yield.

The nominal saturation level of the cell (defined as the ratio between the produced parts, measured in processing hours, and the available hours by the three operating machines) has a value of 70%, as indicated by the supplier of the cell, versus a recorded value of about 50%. To identify the reasons of this difference one should consider:

(1) machine stops, due to various casualties;
(2) defective parts;
(3) limited efficiency of the night shift;
(4) late supplies and uncertainties in processing times, either of the upstream production and of the cell production;
(5) frequent tool changes in the machine tool buffers;
(6) unbalanced working phases among the machines of the cell.

The first two sources of inefficiency have been succesfully contrasted with a quality project (CEDAC — Cause Effect Diagram with Additional Cards). As for the third issue, increasing the efficiency of the night shift would require too large costs, both in investments and in operating expenses. From the point of view of decisions concerning planning, scheduling and dispatching, the most sensible issues to consider in order to improve the saturation level of the cell are therefore the last three.

3. IMPROVING THE SATURATION LEVEL

In this section the original approach devised to improve the saturation level of the FMS cell is discussed.

3.1 *Problem formulation*

The problem of improving the saturation level of the FMS cell has the following elements:

Input data: the monthly production plan, which specifies parts to be produced, assigns operations to machines and determines operation times;

Output decisions:
- parts to be produced concurrently;
- operation scheduling;

Objective: maximize the saturation level of the cell;

Constraints: complying with the operational rules of the cell, in particular with its limited resources, such as buffer capacities.

Concerning constraints, it turned out that a critical resource is the tool buffer capacity: since each part normally requires several dozens of tools on each machine, the possibility of concurrently working different part types is severely affected, thus reducing in practice the flexibility of the cell. This is the central issue of the problem considered.

3.2 *A hierarchical approach*

A hierarchical approach has been devised for the just formulated problem. It decomposes the whole problem in four simpler subproblems, that, when solved in turn, give the solution of the whole problem. The subproblems are arranged in a hierarchic order (see for instance Bitran and Tirupati, 1993), according to their detail level, from the most to the least aggregate. The solution of each level subproblem provides the data for the subproblem of the following level. The subproblems are:

(1) **clustering:** in this phase the set of all parts required by the production plan is partitioned into subsets ("clusters") of parts that may be processed concurrently;
(2) **cluster sequencing:** in this phase the most appropriate sequence of the clusters determined in the previous level is sought;
(3) **cluster linking:** in this phase the transition from each cluster to the following one is determined, according to the sequence decided in the previous level;
(4) **scheduling parts within each cluster:** in this phase each part is scheduled within the cluster to which it has been assigned in the first level.

The merit of the proposed hierarchical approach is to allow to tackle with a complex problem through the sequence of simpler subproblems; of course, this entails some degree of suboptimality, as the solution provided at higher levels affects the performance obtainable at lower levels, and, as a consequence, for the whole problem. However, this is the price to be paid in order to deal with the whole problem, which is so complex to be unaffordable by any global approach.

The subproblems 1 and 4 have been considered in detail, and will be discussed here. The other two subproblems seemed to have less relevance, as far as their impact on the global solution is concerned.

4. CLUSTERING MODELS

In this section the two models are presented, which have been formulated for the clustering problem. As it has been pointed out, concurrent processing of different part types is limited by the capacity of the tool buffers of the operating machines. Two different models have been devised for this problem. They basically have the same decision variables and constraints. They only differ for the objective function: for a given production plan, the former minimizes the global lead time, while the latter seeks to produce a good machine balance.

The problem data for either model are as follows:

- the set P of part types to be produced in the considered horizon;
- the set M of machines;
- the set U of available tools;
- the set $C = \{1, 2, \ldots, |C|\}$ of required clusters;
- the aggregate production plan, in terms of the number $N(p)$ of parts that must be produced for each part type $p \in P$;
- the processing times, in terms of the time $T(p, m)$ required to process a part of type $p \in P$ on machine $m \in M$;
- the tools needed for each part and each machine, in terms of a matrix $a(p, u, m)$, defined as

$$a(p, u, m) = \begin{cases} 1 \text{ if tool } u \in U \text{ is required to} \\ \quad \text{process a part of type} \\ \quad p \in P \text{ on machine } m \in M \\ 0 \text{ otherwise;} \end{cases}$$

- the number of positions $k(m)$ of the tool buffer size for each machine $m \in M$;
- the tool size, in terms of the buffer positions $b(u)$ taken by each tool $u \in U$.

4.1 *Minimizing the global lead time*

This model has the following (nonnegative integer) decision variables:

- the number $x(p, c)$ of parts of type $p \in P$ assigned to the cluster $c \in C$;
- the tool loading matrix $z(c, u, m)$, defined as

$$z(c, u, m) = \begin{cases} 1 \text{ if tool } u \in U \text{ is loaded} \\ \quad \text{in the buffer of machine} \\ \quad m \in M \text{ for cluster } c \in C \\ 0 \text{ otherwise} \end{cases}$$

Furthermore, the following auxiliary (nonnegative continuous) variables are defined:

- the load (in terms of processing time) $\tau(m, c)$ of each cluster $c \in C$ on each machine $m \in M$;
- the load $\tau(c)$ of the most loaded machine for each cluster $c \in C$.

Then the problem of determining the $|C|$ clusters minimizing the global lead time is formulated as a mixed integer programing problem:

$$\min \sum_{c \in C} \tau(c) \qquad (1a)$$

$$\sum_{c \in C} x(p, c) \geq N(p) \quad \forall p \in P \qquad (1b)$$

$$\sum_{p \in P} a(p, u, m) x(p, c) \leq K z(c, u, m) \quad \forall u \in U, m \in M, c \in C \qquad (1c)$$

$$\sum_{u \in U} z(c, u, m) b(u) \leq k(m) \quad \forall m \in M, c \in C \qquad (1d)$$

$$\tau(m, c) = \sum_{p \in P} x(p, c) T(p, m) \quad \forall m \in M, c \in C \qquad (1e)$$

$$\tau(c) \geq \tau(m, c) \quad \forall m \in M \qquad (1f)$$

In practice, problem (1) takes a given number of clusters and assigns each part of the production plan to a cluster; furthermore, the tools to be loaded on each machine buffer are determined, complying with their capacities. Among the several possible assignments, the one minimizing the global lead time is sought. The lead time is given by processing times only, thus disregarding transportation, fixing and washing times, as they are at least of one order of magnitude smaller. Thus the lead time of a cluster is the total processing time of the most loaded machine. Note that at this aggregation level idle times due to the processing sequence are not considered.

In particular, (1e) expresses the total processing time for each cluster and each machine as the sum of the processing times of all parts assigned to that cluster on that machine. Then (1f) takes the lead time of the most loaded machine for each cluster. The global lead time is then minimized by (1a). Furthermore, (1b) guarantees that at least as much parts are produced as required by the production plan. Also, (1c) guarantees that all necessary tools are loaded on each machine (K is a suitably big number), and (1d) expresses the tool buffer capacity.

The model (1) allows to find the minimum number of clusters by solving it for increasing values of $|C|$: the first feasible solution gives the minimum number of clusters. Increasing the number of clusters over the minimum is not profitable, since:

- idle times are expected to grow:
- transients increase;
- the model size increases, thus requiring larger data processing capacities.

Although model (1) correctly formulates the problem of minimizing the global leadtime, it is rather heavy for the computational resources it requires. In particular, too many integer variables $z(c,u,m)$ may result, even for practical problems of limited size. To overcome this drawback, another version of problem (1) has been devised, by introducing new decision variables $y(p,c)$ expressing the type–cluster assignment:

$$y(p,c) = \begin{cases} 1 \text{ if part of type } p \\ \quad \text{is assigned to cluster } c \\ 0 \text{ otherwise} \end{cases}$$

Then a new constraint is introduced:

$$x(p,c) \leq N(p)y(p,c) \qquad \forall p \in P, c \in C, \quad (2)$$

stating that if $x(p,c) > 0$, then part type p must be assigned to cluster c. Furthermore, the constraint (1c) is modified as

$$a(p,u,m)y(p,c) \leq z(c,u,m)$$
$$\forall p \in P, m \in M, u \in U, c \in C. \quad (3)$$

Note that variables z need not to be integer anymore, since the l.h.s. of (3) can assume only 0 or 1 values. Furthermore, there is no need for the big K constant.

Thus the new mixed integer programming problem results:

$$\min \sum_{c \in C} \tau(c) \qquad (4a)$$

$$\sum_{c \in C} x(p,c) \geq N(p) \qquad \forall p \in P \quad (4b)$$

$$x(p,c) \leq N(p)y(p,c)$$
$$\forall p \in P, c \in C \qquad (4c)$$

$$a(p,u,m)y(p,c) \leq z(c,u,m)$$
$$\forall p \in P, m \in M, u \in U, c \in C \qquad (4d)$$

$$\sum_{u \in U} z(c,u,m)b(u) \leq k(m)$$
$$\forall m \in M, c \in C \qquad (4e)$$

$$\tau(m,c) = \sum_{p \in P} x(p,c)T(p,m)$$
$$\forall m \in M, c \in C \qquad (4f)$$

$$\tau(c) \geq \tau(m,c)$$
$$\forall m \in M \qquad (4g)$$

Since the number of integer variables in (4) is now much smaller than in (1), the former model turns out to be much more convenient that the former as far as computation times are concerned.

4.2 *Balancing workloads*

The model (4) can be easily modified in order to balance machine workloads. It suffices to introduce another constraint

$$\tau(m,c) \geq \alpha\tau(c) \qquad \forall m \in M, c \in C,$$

expressing that, for each cluster, the workload for each machine cannot be larger than the load of the most loaded machine, multiplied by the "balance factor" α. Considering α as a new variable to be maximized, a two–objective nonlinear programming problem results. For ease of reference it is reported below.

$$\max \alpha \qquad (5a)$$

$$\min \sum_{c \in C} \tau(c) \qquad (5b)$$

$$\sum_{c \in C} x(p,c) \geq N(p) \qquad \forall p \in P \quad (5c)$$

$$x(p,c) \leq N(p)y(p,c)$$
$$\forall p \in P, c \in C \qquad (5d)$$

$$a(p,u,m)y(p,c) \leq z(c,u,m)$$
$$\forall p \in P, m \in M, u \in U, c \in C \qquad (5e)$$

$$\sum_{u \in U} z(c,u,m)b(u) \leq k(m)$$
$$\forall m \in M, c \in C \qquad (5f)$$

$$\tau(m,c) = \sum_{p \in P} x(p,c)T(p,m)$$
$$\forall m \in M, c \in C \qquad (5g)$$

$$\tau(c) \geq \tau(m,c) \quad \forall m \in M \quad (5h)$$

$$\tau(m,c) \geq \alpha\tau(c)$$
$$\forall m \in M, c \in C. \qquad (5i)$$

This problem may be conveniently tackled ranging the objectives in a lexicographic way, with the balance factor at the higher level; then α can be seen as a parameter, and several problems (4) with the additional constraint (4.2) can be solved for increasing values of α, until no feasible solution is found. Then the last found solution minimizes the global leadtime subject to the maximum balance factor.

4.3 *Computational experience*

Problems (4) and (5) have been solved using Cplex (cf. Cplex) on a DEC Alpha 3000 workstation, for several instances taken from practical cases. Of course, problem (5) produces higher balance factors, at the expense of the total leadtime.

5. CLUSTER SEQUENCING AND LINKING

The subproblem of the second level in the hierarchy of Section 3.2 consists in determining the most appropriate sequence in which clusters, determined at the first level, have to be processed. In order to determine this sequence, some factors are considered, that were neglected at the upper, more aggregated, level. In particular, tool loading times appear to be the appropriate performance index for cluster sequencing.

More specifically, for any pair of clusters, the number of non common tools (i.e., tools that are used only by one of them) is a sensible way to assess the burden of tool changing when they are processed one after the other. Minimizing the total number of tool changes leads to a Travelling Salesman Problem (see for instance Lawler et al., 1985), that can be conveniently solved by approximate methods (see for instance Jünger, Reinelt, and Rinaldi, 1995). However, due to the fact that the number of clusters is in practice always very low for our application, no particular importance has been attributed to this problem.

Once clusters have been sequenced, they could be "linked", in the sense of processing in a concurrent way the ending transient of the first one with the beginning transient of the following one. However, also in this case this subproblem appeared to be quite irrelevant, due both to the fact that transients are relatively short, and clusters are quite different, thus leaving few margins to the possibility of concurrent operations.

Nevertheless, it is worth to notice that this is a consequence of having produced, at the upper level, few clusters of quite large size, which is exactly what the iterative (on $|C|$) procedure introduced in Section 4.1 is able to do well.

6. DETAILED SCHEDULING WITHIN CLUSTERS

In this section the problem is faced of finding the most appropriate scheduling of the parts assigned to each cluster, as they have been determined at the upper level.

For each part type, the production plan specifies the operations that have to be performed on each machine, and their sequence. Due to te reduced number of parts involved, it is natural to model the sequencing problem as a job shop problem (see for instance Blazevicz et al., 1993). Note that also in this case, being at a lower level of our hierarchic approach, more detailed elements are considered than in upper levels: in particular, the processing sequence on machines, that was not contemplated to determine clusters, is now taken into account.

For this subproblem, local dispatching rules have been considered (see for instace Panwalkar and Iskander, 1977). In particular, six of them turned out to be quite interesting:

(1) Erliest Due Date (EDD);
(2) Minimum Lateness (ML);
(3) Random;
(4) Starvation Avoidance (SA);
(5) First In First Out (FIFO); Shortest Remaining Processing Time & Starvation Avoidance (SRPT&SA).

The performance of the schedule has been evaluated according to different criteria:

- the completion time t_e;
- a global saturation index σ, defined as
$$\sigma = \frac{\text{total workload}}{t_e - t_i \times |M|},$$
where the total workload is the sum of the processing times of all operations on all machines, and t_i is the time instant at which processing starts;
- a measure of the work–in–process (WIP), evaluated as
$$\text{WIP} = \frac{\sum \text{flow times}}{\sum \text{processing times}}$$
where sums are taken over all operations.

It turned out that SA maximizes the saturation index, while SRPT minimizes WIP. As a natural consequence, SRPT&SA provides the most equilibrate results. However, it must be pointed out that the performances showed by all these dispatching rules are quite similar (including Random). This suggests that, after all, scheduling has not a strong effect on the overall performance of the obtainable solutions.

7. CONCLUSIONS

Motivated by the problem of improving the saturation level of a flexible cell of the Diesel Engine Division of Fincantieri, Trieste (Italy), a general hierarchic approach has been proposed to the production planning and scheduling problem when tools are a scarce resource.

The subproblem corresponding to the upper level of the proposed approach has been modelled as a mixed integer programming problem, and solved using a general purpose mathematical programming package. The other subproblems have also been modelled and standard solution methods have been proposed for them.

Several computational experiments have been performed on real instances, showing the effectiveness of the proposed approach.

The calculations performed with the adopted models have produced clear evidence of the reasons of the low saturation level of the plant. As a consequence, the results suggest some measures that are oriented in the following directions:

- the review of the working cycles of the parts and new parts to be processed in the plant;
- the standardization of the tools which should minimize the tool number and the availability of the local tool buffers;
- a part clustering algorithm devoted to improve the performance and time calculation in planning the monthly production.

Each of these activities should improve the plant balance and saturation in order to increase the productivity and to reduce the leadtime in the FMS cell.

REFERENCES

Bitran, G.R. and D. Tirupati (1993). Hierarchical Production Planning. In: *Handbooks in Operations Research and Management Science, Vol. 4: Logistics of Production and Inventory* (S.C. Graves, A.H.G. Rinnooy Kan and P.H. Zipkin (Eds.)), 523–568. North–Holland, Amsterdam NL.

Błazevicz, J., K. Ecker, G. Schmidt and J. Węglarz (1993). *Scheduling in Computer and Manufacturing Systems.* Springer–Verlag, Berlin D.

CPLEX Optimization, Inc. (1995). *Using the CPLEX Callable Library — Version 4.0.* CPLEX Optimization, Inc., Incline Village NV.

Jünger, M., G. Reinelt and G. Rinaldi (1995). The Traveling Salesman Problem. In: *Handbooks in Operations Research and Management Science, Vol. 7: Network Models* (M.O. Ball, T.L. Magnanti, C.L. Monma and G.L. Nemhauser (Eds.)), 225–330. North–Holland, Amsterdam NL.

Lawler, E.L., J.K. Lenstra, A.H.G. Rinnooy Kan and D.B. Shmoys (1985). *The Traveling Salesman Problem.* Wiley, Chichester UK.

Panwalkar, A. and W. Iskander (1977). A survey of scheduling rules. *Operations Research* **25**, 45–46.

ADVANCED TECHNIQUE FOR MANUFACTURING SYSTEM MANAGEMENT

Andrea D'Angelo, Massimo Gastaldi and Nathan Levialdi

Università degli Studi di Roma "Tor Vergata"
Dipartimento di Informatica, Sistemi e Produzione
Via della Ricerca Scientifica
00133 Roma Italia

Abstract: Artificial Neural Networks are able to analyse the data distribution obtained by experimental simulations. In this paper Neural Network is utilised as black box tool to provide a low cost control procedure able to examine system events related to a Printed Circuit Board Assembly (PCBA) plant. The obtained results are compared with those arising from a traditional statistical method.

Keywords: Production Systems, Performance functions, Simulation, Neural Networks, Statistical Inference.

1. INTRODUCTION

Our study focuses on the typical Job Shop plant configuration for semiautomatic manufacturing of parts produced in limited unitary quantities. The treatment consists of the quantitative evaluation of the technological performance of a manufacturing system with particular reference to the Printed Circuit Board Assembly (PCBA) sector (Crama et al., 1990; McGinnes et al., 1992; D'Angelo et al., 1996a and 1996b). We present a method for monitoring two types of plant performance: Lead Time and Work In Process. This method is based on the capability of Neural Networks to map an m-dimensional input vector to an n-dimensional output vector (Hecht-Nielsen 1987, Caudill 1987), and is applied to the reality of a leading industry in the production of PCBA and electronic modules for telecommunications. System events are investigated through a numerical simulation model. In fact, the complexity due both to the large number of variables and to the non linearity of the correlation involves the inconvenience of an analytical approach. Generally, the data so obtained, are processed through statistical methods in order to highlight the correlation among such variables and system performances.
The inference can be alternatively proposed by a Neural Network approach in order to compare the results obtained through the two methods. The simulation experiments provide data for the Network training which learns the implicit correlations between input and output factors. In the presented analysis, a backpropagation algorithm (Werbos 1988, Rumelhart et al. 1986) is utilised. So, after the learning phase, the network has reconstructed the mapping function as a black box. To check the effectiveness of this function, a second set of data is submitted for a test phase in which the network response is compared with the one of a statistical model.
This paper is organised as follows. Section 2 describes the reference system for the manufacturing of printed circuit boards. Section 3 covers the implementation of the model built for the mathematical simulation. In section 4 an advanced backpropagation algorithm is applied to train a neural network on the data obtained by simulation experiments. The results arising from the test phase are compared with the outcomes of statistical regressions, in section 5. Finally, section 6 concludes the study.

2. PHYSICAL SYSTEM DESCRIPTION

This study is dealt with in the context of a PCBA system. The process of assembling electronic components is by its nature a flexible one, as the manufacturing process involves a great number (some one thousand) of items (part types), being produced in limited unitary quantities. It follows that the potential influence of events which are both endogenous and exogenous to the system is of a clearly stochastic nature. For example, in the short term there may be variations in the size, quantity and frequency of the lots to the system, either separately or jointly, in relation to the exogenous fluctuations in demand. This condition translates into an extreme segmentation of the manufacturing process, which, in our case, is composed of nine types of processing, performed at an equal number of Work Centres (WC).

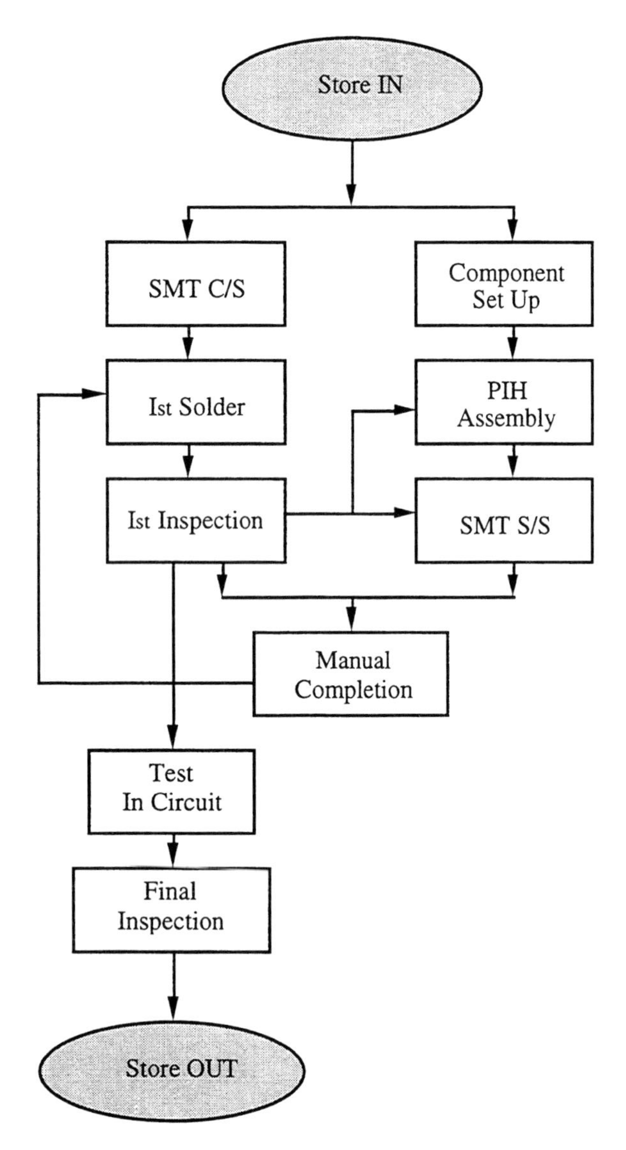

Fig. 1. System description

The complexity caused by the large number of part types makes it impossible to dynamically follow the system events due to the lot relating to the single part type. Hence, we need to use a clustering criteria. That is to say, the part types are grouped together on the basis of the component assembly sequence. This may be of the standard Pin In Hole (PIH) type or of the Surface Mounted Technology (SMT Component Side C/S or Solder Side S/S) type. In this way we can identify seven assembly sequences (families) for the part types and it is with reference to these that we conduct a study of system performances.

Let us consider the manufacturing process depicted in Figure 1, where the boxes represent Work Centres and the arrows represent the possible processing paths (sequencing) of the products being manufactured. The starting point of the process is an entrance gate ("IN" store, for the storing of the boards to be assembled), while the lots finally exit the system through an exit gate ("OUT" store, for storing of finished products). The set of part types undergoing the same sequencing represents a family of products. The layout in Figure 2, highlights:

- the processing path for each family (the number in the box is Work Centre defining)
- the number of machines present at each node (the left hand-side number of each box)
- the average processing time in minutes at each Work Centre for each family (the right hand-side number of each box).

Within each family the part types undergo, at each Work Centre, operations which require different amounts of time for each part type. That is to say, the times spent at each Work Centre are normally distributed around an average value which varies from family to family. The paths identified represent an outline sequencing as, actually, not all part types belonging to the family undergo all the operations envisaged by the sequencing itself. The overall output of the system is divided up among the various families in accordance with a set mix of specific percentage values. Specifically the seven different families are processed with part mix ratios 39:27:24:5:2:2:1.

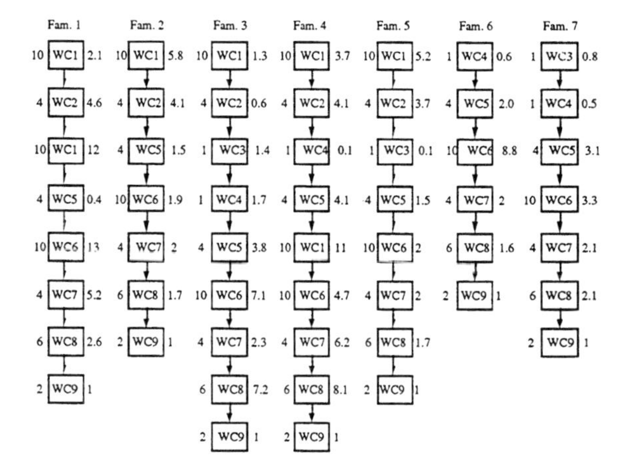

Fig. 2. System technological sequencing

The existence of a non-constant distribution of the average hourly productivity values among the various Work Centres leads to a non-homogeneous exploitation of system's resources. Indeed, the global productive capacity depends on the minima values of this distribution which characterises the Work Centres defined as the "bottlenecks". Thus, these Work Centres guide the dynamics of the production flow by determining a partial underutilization of other resources. On the other hand, the values of this distribution depend on the variables which define the average hourly productivity. For the reference manufacturing system, these variables can have values in a very wide domain. The result is that there can be a fluctuation of the bottleneck within the system. Moreover, such a fluctuation also depends on other variables which have a significant impact on the dynamics of the system, in particular, the lots size of the individual families.

3. THE SIMULATION MODEL

In order to quantify the criticality dynamics it is necessary to observe the behaviour of the system in terms of parameters variation. So, first of all we have to select the significant variables impacts on system performances. The mathematical simulation of events which can significantly disturb system dynamics enables us to quantify in precise terms the variation in the following performance indicators:

- average Work In Process
- average Lead Time.

The implementation of the reference system in a mathematical simulation model has been conducted through WITNESS (1994), an appropriate software for simulation of industrial processes. The presented analysis is referred to the system steady state. Each simulation is run for 960 hours; the first 100 hours of each run is truncated to eliminate initialisation bias. We consider constant interarrival times of lots to the system (for each product family), according to the original part mix ratio. Besides we assume:

- Machine breakdowns (exponentially distributed)
- Total system load (50.000 pieces in 960 hours)
- Zero Transportation time of parts between WCs
- Set Up times for part type changing (20% of mean processing time)
- WCs' Input Buffers capacity (200).

Pilot runs highlight the presence of bottlenecks in WC1, WC6 and WC8 and hence individuate three variables, the number of machines in each work centre, which exert a systematic influence on the performance indicators; a forth one is identified in lot size. Once these variables have been identified, a discrete interval of variation is defined for each of them, as shown in Table 1.

Table 1. Significant variables levels

		Variables			
		X_1	X_2	X_3	X_4
	-1	15	6	6	3
Levels	**0**	25	8	8	4
	1	35	10	10	5

From the combination of the possible levels for each variable we can define a series of $3^4=81$ system configurations. For each configuration a simulation experiment is run to calculate the value of the two performance indicators.

4. NEURAL NETWORK ANALYSIS

Neural networks are a form of computing inspired by the structure and learning ability of the human brain. They consists of networks of a number of simple, highly-interconnected processing elements, which process information in a vaguely analogous way to biological neurones. Neural networks work quite differently then conventional computing and consist of many processing units (nodes) that are interconnected with the capability of parallel processing and self-learning. For project purposes, a backpropagation model (Rumehart et al. 1986), is here selected to rebuild the implicit relation between the output factors Y_1, Y_2 (lead time and work in process) and input variables (X_1, X_2, X_3, X_4). In particular we choice a four perception network consisting in an input layer, two hidden layers and an output layer. The input layer is formed by four nodes, one for each input variable whereas the output layer consists of two nodes used to represent lead time and work in process levels. Notice that in this paper two neural networks are considered; the first one, described in Figure 3, contains six nodes on the first hidden layer and four on the second hidden layer whereas the second network has respectively four and two nodes (see Figure 4).

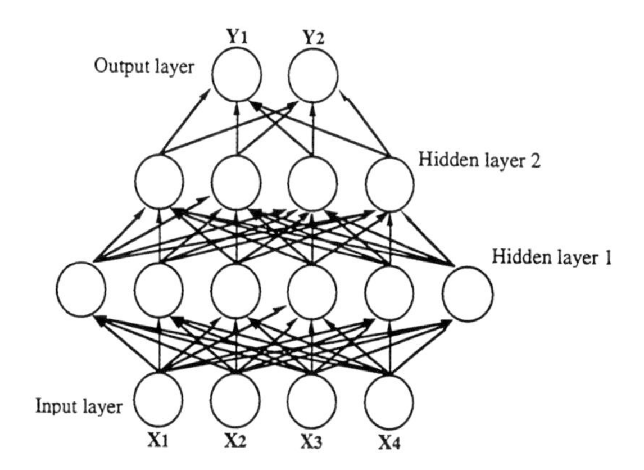

Fig. 3. The "6-4" neural network

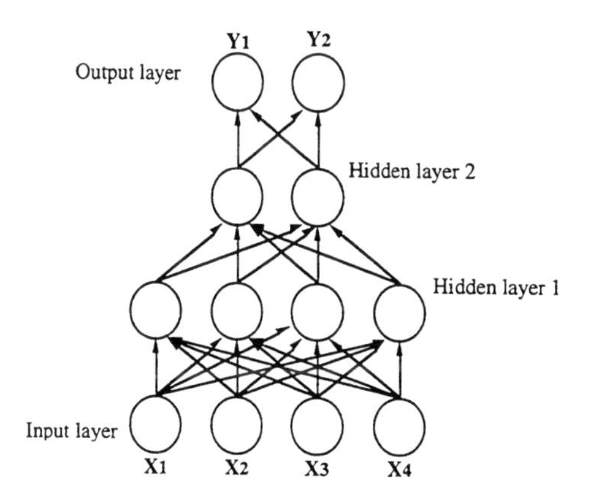

Fig. 4. The "4-2" neural network

The training phase consists on 64 data series (10 relative to replications on central point and 54 = 2/3 of total system configurations). During the learning phase, the feed-forward output state calculation was combined with the backward error propagation and weight adjustment calculations. In the test phase the remaining 27 data series are used. The aim of this choice is to verify the learning effectiveness of the network about the relation between the input and output factors. At this purpose pilot experiments show that the evaluation is better performed by "6-4" network than by the "4-2" one.

5. RESULTS ANALYSIS

In this section results of the performed analysis are discussed. In particular data obtained by the test phase of neural networks are compared with those obtained by the application of statistical model. In Tables 2-3 the outcome of the statistical regression on 64 data series (the same used in the neural network training) is shown.

Table 2. Lead time second order regression model

Lead Time	Number of Observations = 64			
Variable	Coeff.	Std error	T	α
X_1	-190.262	545.315	-0.349	0.729
X_2	14979.994	3315.243	4.519	0.000
X_3	-3956.096	3202.926	-1.235	0.223
X_4	-14519.812	6630.485	-2.190	0.033
X_1*X_1	9.639	7.155	1.347	0.184
X_2*X_2	-376.792	203.694	-1.850	0.070
X_3*X_3	880.994	178.423	4.938	0.000
X_4*X_4	2769.110	814.774	3.399	0.001
X_1*X_2	-21.325	27.504	-0.775	0.442
X_1*X_3	-15.493	26.903	-0.576	0.567
X_1*X_4	3.594	55.008	0.065	0.948
X_2*X_3	-859.340	131.900	-6.515	0.000
X_2*X_4	-43.376	264.661	-0.164	0.870
X_3*X_4	-1243.846	263.801	-4.715	0.000
	F= 62.836		R^2= 0.932	

Table 3. Work in process first order regression model

Work in process	Number of Observations = 64			
Variable	Coeff.	Std error	T	α
X_1	35.849	23.962	1.496	0.140
X_2	896.997	100.166	8.955	0.000
X_3	-71.052	109.505	-0.649	0.519
X_4	-899.494	200.332	-4.490	0.000
	F= 126.442		R^2= 0.889	

Notice that, for goodness of fit point of view, based on the 27 data series used for neural network testing, a first order model has been selected to represent work in process function whereas a second order model has been used for lead time. In Figure 5 and 6 the 27 test values for the observed performance functions are presented with respect to the simulation data (values are normalised by absolute maximum value). Even if lines seem to be not so close, they comply with an evident periodicity. Hence, neural network approach, as well as statistical model, understands the relations between factors in relative terms but not in absolute ones. A quantitative measure VC (Variation Coefficient), in order to compare distance between lines, can be defined as:

$$VC_{nn} = \frac{\sqrt{\frac{\sum_{i=1}^{27}\left(Y_i^{sim}-Y_i^{nn}\right)^2}{27}}}{Y_m}$$

$$VC_{Stat} = \frac{\sqrt{\frac{\sum_{i=1}^{27}\left(Y_i^{sim}-Y_i^{Stat}\right)^2}{27}}}{Y_m}$$

where

$$Y_m = \frac{\sum_{i=1}^{27} Y_i^{sim}}{27}$$

and Y_i^{sim} represents the i-th value of the simulated performance functions (lead time or work in process). Y_i^{nn} and Y_i^{Stat} are the i-th test values relative to neural network and statistical model respectively. In aggregated terms the variation coefficients are shown in Table 4.

Table 4. VC results comparison

	lead time	work in process
Neural Network	0.658	0.363
Statist. model	0.505	0.353

Such table shows that the two approaches are practically equivalent for work in process level prediction whereas the performance of statistical inference is higher when lead time factor is considered.

6. CONCLUSIONS

This paper is focused on monitoring two types of a job shop plant performance: lead time and work in process. In order to capture the relation between input and output factor, two alternative approaches have been applied. The first one is based on the application of a classical statistical regression model whereas the second one is built on the advanced neural network technique. The former provides the existence of an explicit polynomial relation between variables and the latter, by means of a backpropagation algorithm, maps as a black box the m-dimensional input vector to the n-dimensional one. The data utilised on both inferences are obtained by a simulation model reproducing dynamics of the system under study. The obtained results show that neural network is a low-cost procedure to understand and forecast system events with an acceptable performance level. Such results seem to be related to the particular system complexity.

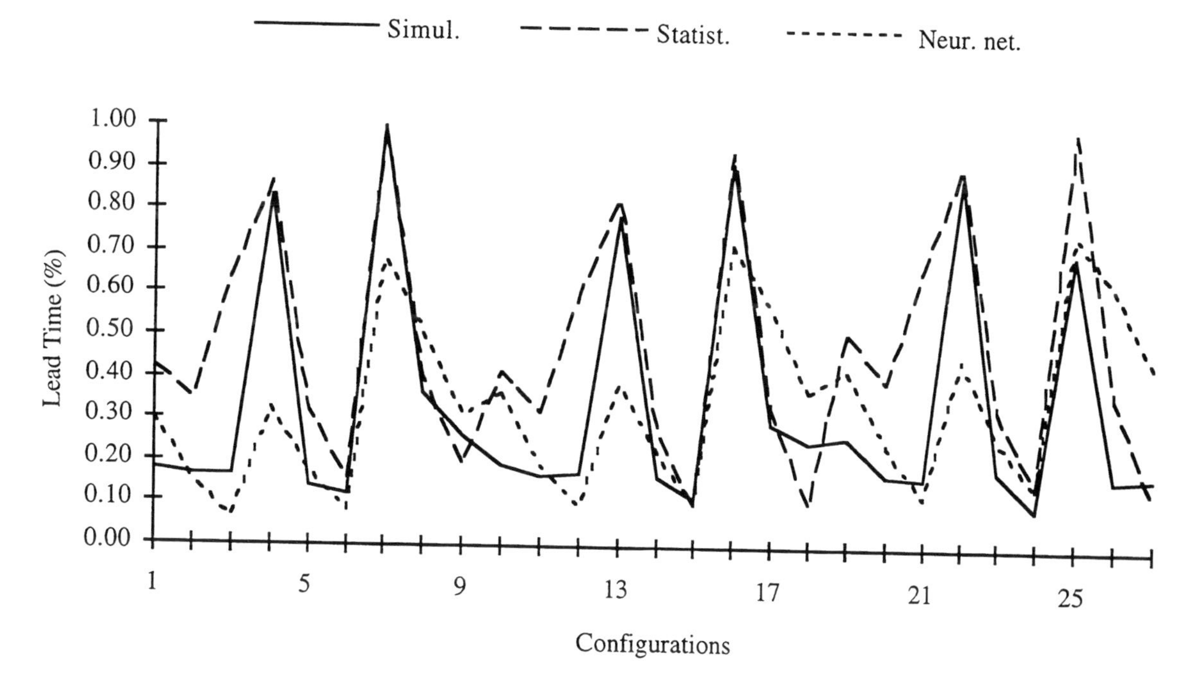

Fig. 5. Lead time results analysis

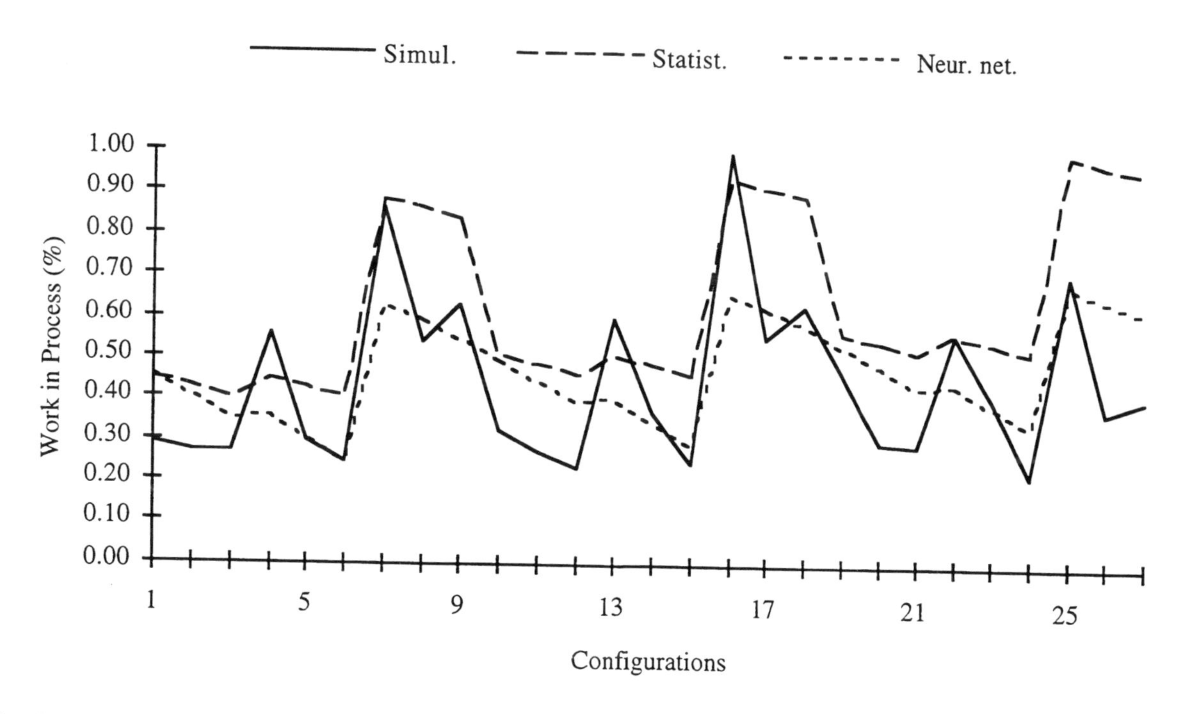

Fig. 6. Work in Process results analysis

Since the number of total possible configurations exponentially grows with the number of considered input variables, the authors are currently working on the performance evaluation of neural network approach for systems with an higher input vector size. Preliminary analysis have shown that neural networks results seem to be more significant when the training data set is reduced with respect to the entire phenomenon especially when observed functions cannot be a-priori represented by polynomial models.

REFERENCES

Caudill, M., (1987). Neural Networks primer Parts 1-8. *AI Expert*, **2**, 12.

Crama, Y., A.W.J. Kolen, A.G. Oelermans and F.C.R. Spieksma, (1990). Throughput rate optimization in the automated assembly of printed circuit board. *Annals of Operations Research*, **26**, 455-480.

D'Angelo A., M. Gastaldi and N. Levialdi, (1996a). Dynamic analysis of the performance of a Flexible Manufacturing Systems: a real case application. *Computer Integrated Manufacturing Systems*, **9**, 2, 101-110.

D'Angelo A., M. Gastaldi and N. Levialdi, (1996b). Performance analysis of a Flexible Manufacturing Systems: a statistical approach. In press: *International Journal of Production Economics*.

Hecht-Nielsen R., (1987). Kolmogorov's mapping neural network existence theorem. *Proceedings of the First IEEE International Conference on Neural Networks*, (M. Caudill and C. Butler Eds), IEEE Service Center, Piscataway, New Jersey, 11-14.

McGinnis, L.F., J.C. Ammons, M. Carlyle, L. Cranmer, G.W. Depuy, K.P. Ellis, C.A. Tovey and H. Xu, (1992). Automated process planning for printed circuit card assembly. *IEEE Transaction on Industrial engineering R&D*, **24**, 4, 18-29.

Rumelhart, D.E., G.E. Hinton and R.J. Williams, (1986). Learning internal representation by error propagation. *Parallel Distributed Processing*, **1**, (David E. Rumelhart and James L. McClelland Eds.), MIT Press, Cambridge, Massachusetts, 318-362.

Werbos, P.J., (1988). Generalisation of back propagation with application to a recurrent gas market method. *Neural Networks*, **1**, 4, 339-356.

KNOWLEDGE-BASED TOOL FOR INSTRUMENTATION'S MAINTENANCE ON A DATA ACQUISITION

Widrianto Sih Pinastiko

The Agency for the Assessment and Application of Technology
UPT. LAGG BPPT, PUSPIPTEK
Serpong - Tangerang 15310, Indonesia

Abstract: The knowledge-based expert system is a computer program designed to act as an expert to solve a problem in particular domain. Building a knowledge-based tool helps to bring the expert knowledge in the domain to the user and thus enhances the user's skills in problem solving. This paper discussed about a design knowledge-based tool for maintenance of instrumentation controlled by a Data Acquisition (DA) sub-system's computer. Using the tool makes a good and reliable instrumentation that is one of the requirements to get an optimal result of the DA Sub-system.

Keywords: Knowledge-based systems, Expert systems, Data acquisition.

1. INTRODUCTION

The Data Acquisition Sub-System usually is a part of the main system likes manufacturing system, machining controlling system, electronic monitoring system, data processing system and others. (Fig.1)
The Data Acquisition Sub-system (DAS) consists of several basic elements, those are sensors (transducers), electronics instrumentation (measurement devices), interfaces and computer. The good DA subsystem will support the quality to the main system. A good condition of the elements are needed to get a good result quality of the data acquisition subsystem. So, it needs a good management and maintenance of electronic instrumentation. This maintenance will be more and more complicated following the increasing of the number and kind of instruments.

The qualified maintenance's people or technicians are needed to get a good condition of DA subsystem specially on instrumentation. So it has high dependency to the skilled or qualified technicians. To support this condition needs a knowledge-based tool for supporting to the instrumentation's maintenance.

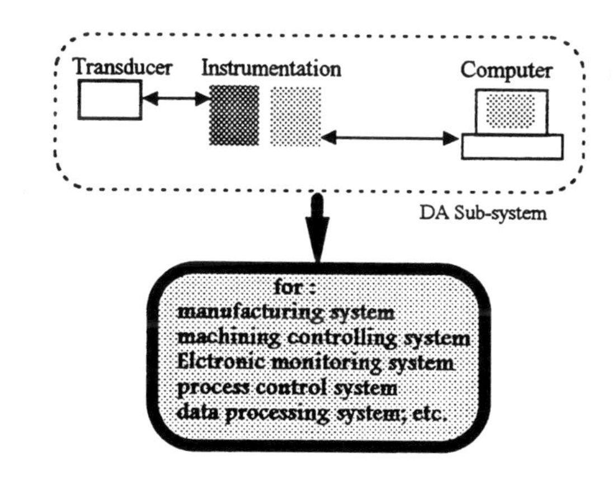

Fig. 1. Usage of a Data Acquisition Sub-system

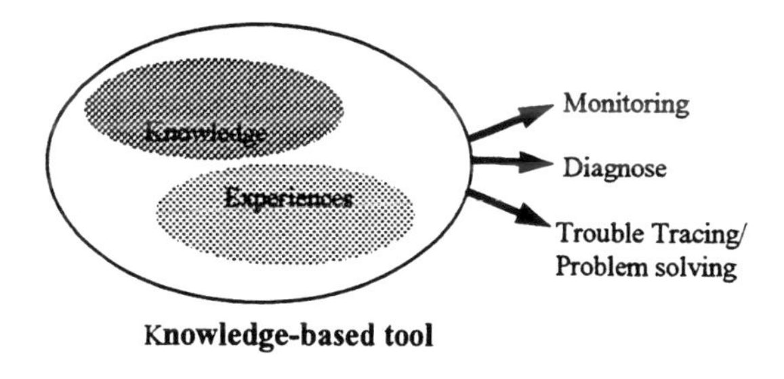

Fig. 2. Function of Knowledge-Based Tool

The designed knowledge-based tool is a smart software that runs on a DAS's computer. The system consists of three parts, those are condition of measurement device's monitoring, automatic trouble tracing, and the diagnoses and advise toward the trouble or dysfunction of the measurement devices (Fig.2). Two important aspects to provide the maintenance systems of instrument are hardware aspect and software aspect. For the first aspect, computer and measurement devices should be well connected and communicated, and for the second aspect, computer has to be able to control the measurement devices and get some data of measurement device's condition completely. In this case, condition of devices or instruments could be good or not good (have troubles) condition. The condition of the instruments acts as the *fact* that will be processed using Expert System algorithm.

Using expert system algorithm, the knowledge about electronic-mechanic, instrumentation, control, computer was integrated with engineer's experiences, which be used to process the *fact* for monitoring and trouble tracing the instruments. So, the system design has an analytic and inference ability toward the condition and trouble. A simulation was done on :

- 40 units of the Conditioning Unit instruments (CU),
- 7 units of Low Level Interface (LLI) and
- 12 units of High Level Interface (HLI).

Testing and validation were carried out to increase the ability and performance the designed knowledge-based tool.

2. EXPERT SYSTEM

Expert system is a branch of AI (Artificial Intelligence) that makes extensive use of specialized knowledge to solve problems at the level of a human **expert**. An expert is a person who has **expertise** in certain area. That is, the expert has knowledge or special skills that are not known or available to most people. An expert can solve problems that most people cannot solve at all or solve them much more efficiently (but not as cheaply). The expert system technology may include special expert system languages, programs, and hardware designed to aid in the development and execution of expert systems.

The knowledge in expert systems may be either expertise, or knowledge that is generally available from books, magazines, and knowledgeable persons. The term expert system, knowledge-based system or knowledge-based expert systems are often used synonymous.

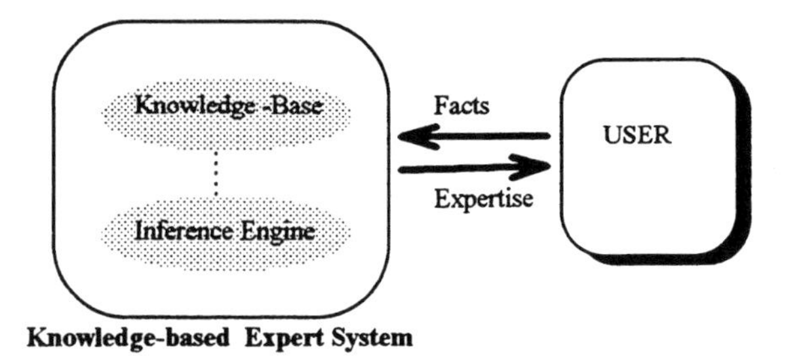

Fig. 3. Basic concept of an Expert System Functions

Fig. 3 shows the basic concept of a knowledge-based expert system. The user supplies *facts* or other information to the expert system and receives expert advice or *expertise* in response. The knowledge-based contains the knowledge with which the inference engine draws conclusions. These conclusions are the expert system's responses to the user's queries for expertise.

Useful knowledge-based systems also have been designed to act as an intelligent assistant to a human expert. These intelligent assistants are designed with expert systems technology because of the development advantages. As more knowledge is added to the intelligent assistant, it acts more like an expert.

Developing an intelligent assistant may be useful milestone in producing a complete expert system. In addition, it may free up more of the expert's time by speeding up the solution of problems.

2.1 Elements of an expert systems

The elements of a typical expert system are shown in Fig.4. In a rule-based system, the knowledge base contains the domain knowledge needed to solve problems coded in the form of rules. While rules are a popular paradigm for representing knowledge, other types of expert systems use different representations.

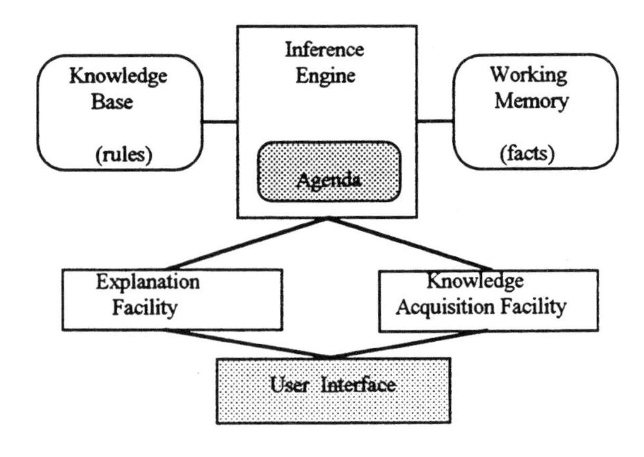

Fig 4. Structure a rule-based Expert System

An Expert system consists of the following elements :

- **User Interface**; the mechanism by which the user and the expert system communicate.
- **Explanation facility**; explain the reasoning of the system to a user.
- **Working memory**; a global data base of acts used by the rules.
- **Inference engine**; makes inference by deciding which rules are satisfied by facts, prioritizes the satisfied rules, and executes the rule with the highest priority.
- **Agenda**; a prioritized list of rules created by the inference engine, whose patterns are satisfied by facts in working memory.
- **Knowledge acquisition facility**; an automatic way for the user to enter knowledge in the systems rather than by having the knowledge engineer explicitly code the knowledge.

The user of the expert system, interacts with it through the user interface and the expert's knowledge is collected and compiled through knowledge acquisition facility. The knowledge acquired from the expert is store in the knowledge base using various formal knowledge representation techniques.
A rule base representation of knowledge consists of a set of rules, which may consist of both procedural or control knowledge and heuristic knowledge. The knowledge-base in this study was performed from the experiences of the engineers and the knowledge of the CU, instrumentation, control and computer. The knowledge-base (*rules*) processes and analyzes the *fact* . The *fact* is an information or condition about object and *rule* is procedure to process the fact. In this paper, the *fact* is condition of CUs, LLIs and HILs. that collected or measured by the software.

2.2 The advantage of knowledge-based tool

The tool has a number attractive features :

- *Increase Availability*, expertise is available on any suitable computer hardware. In a very real sense, an expert system is the mass production of expertise.
- *Reduced cost*, the cost of providing expertise per user is greatly lowered.
- *Reduced danger*, expert systems can be used in environments that might be hazardous for a human.
- *Multiple expertise*, the knowledge of multiple experts can be made available to work simultaneously and continuously on a problem at any time of day or night.
- *Intelligent tutor*, the expert system may act as an intelligent tutor by letting the student run sample programs and explaining the system's reasoning.
- *Intelligent Database*, expert systems can be used to access a database in an intelligent manner.
- *Explanation*, the tool can explicitly explain in detail the reasoning that led to a conclusion. A human may be too tired, unwilling or unable to do this all the time
- *Fast response*, fast or real-tile response may be necessary for some applications. depending on the software and hardware used, the tool may respond faster and be more available than a human expert.

3. DATA ACQUISITION SUBSYSTEM

DA subsystem is a system to measure, collect and process the data to special purpose. Fig.5 shows a data acquisition sub-system where the designed knowledge-based tool is running.

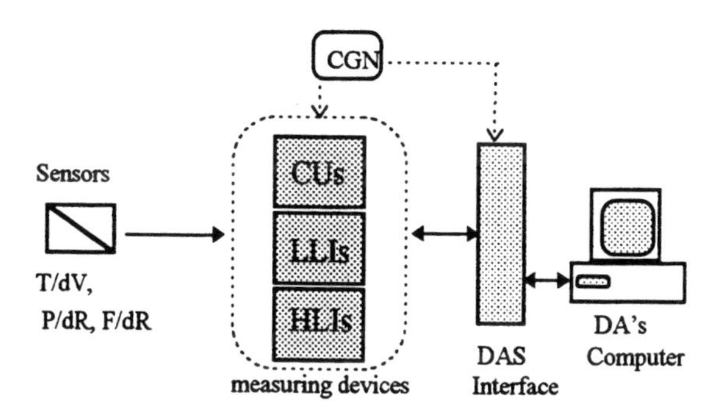

Fig. 5. Data Acquisition Sub-system

There are two main components in the data acquisition system, the first is hardware (computer, interfaces, measurement devices, transducers etc.) and the second is the software of data acquisition subsystem. The exists data acquisition sub-system's software can control the instruments and collect the data into the computer. This software also can do calibration check to the connected instruments, measuring and collecting the data by remote. Data are sensed by sensors or transducers from physical unit (Temperature, Pressure, Force) to electric signal (Volt, Resistance). This analog data is conditioned and digitized by the all measuring devices (CU, LLI and HLI). The digital data are accepted and processed by a computer trough Digital Acquisition Sub-system Interface (DAS Interface). About 40 units of CU, 7 units of LLI and 12 units of HLI were connected to the DAS-Interface using digital bus, allowing computer controlled operation and data handling. DAS Interface forms connection between a data acquisition computer system and all measuring devices such as CUs. A command word from the computer is sent to the measuring devices trough the DAS interface and in the other direction data from the devices are transferred to the computer.

A Calibration GeNerator (CGN) is used to calibrate the measuring devices and other instruments. The CGN is used as a reference generator for calibration of the conditioning units. The CGN can generate dc-signals from 0 up to +20 Volt. The voltage is selectable either by the computer through the DAS interface or by means of the front panel switches. The CGN can calibrate the connected measuring devices (CUs, LLIs, HLIs) under manual control or under remote control. The CGN can calibrate up to 5 other units of measuring devices at the same time.

A CU is one integrated package that performs all the function required for conditioning and digitizing the signal of one analog sensor or transducer. The CU is composed of a number of functional modules. These have been defined in such a way that each one of them can be deleted, replaced or modified without influencing the operation of the other modules. Each module is mounted on a separate PCB. The CU's modules are;

a) Instrumentation Amplifier board
b) Balance board
c) Excitation board
d) Filter board
e) Overload detection & output buffer board
f) Dual slope ADC board, Binary to BCD board
g) Digital interface board and
h) Two Power supply boards.

Each module can be inspected by data acquisition software on computer to know the module condition. The dimension of the conditioning unit is 40cmx10.5cmx20.5cm, and to be placed at the racks. CU is one of the important instrument in this data acquisition sub-system. Likes the CU, the LLI and HLI are also composed of a number of functional modules. These have been defined in such a way that each one of them can be deleted, replaced or modified without influencing the operation of the other modules. Each module is mounted on a separate PCB.

The measuring devices (CUs, LLIs and HLIs) should be in a good and reliable condition to get a good quality result of the data acquisition sub-system. To keep the measuring devices in a good condition needs also good maintenance system. This paper discussed about a designed knowledge-based tool for monitoring and trouble tracing of the measuring devices. This tool is very useful on the measuring device's maintenance system.

4. KNOWLEDGE-BASED TOOL ALGORITHM

4.1 Development of an expert system

The general stages in the development of an expert system are illustrated in Fig. 6.

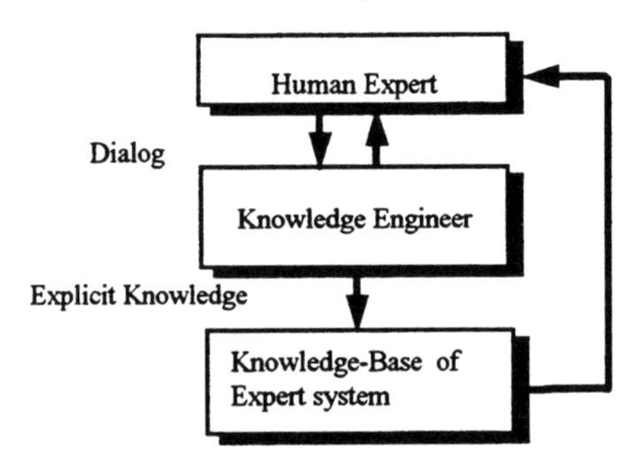

Fig. 6. Development of an expert system

The knowledge engineer first establishes a dialog with the human expert in order to elicit the expert's knowledge. This stage is analogue to a system designer in conventional programming discussing the system requirements with a client for whom the program will be constructed. The knowledge engineer than codes the knowledge explicitly in the knowledge-base. The expert then evaluates the expert system and gives a critique to the knowledge engineer. This process iterates until the system's performance is judge to be satisfactory by the expert.

4.2 Algorithm of program system

The knowledge-based tool does monitoring, inspecting condition of the measuring devices (CU, LLI and HIL) by remote and automatically. Fig. 7, shows the algorithm of program system. The knowledge of the measuring devices and experiences of engineer were integrated in knowledge base to process and analyze condition of the measuring devices (the *fact*).

Software of this data acquisition sub-system has several module programs that can be called to get the condition of the measuring devices. There are *dastest, cgntest, ccu ,contest, llitest and hiltest.* The *ccu* and *contest* are module programs to calibrate the CU with the CGN, and each module board on the CU will be calibrated/checked one by one. The result is condition of the CU, there are *a good condition* (no trouble on all the module boards) or *not good condition (has a trouble). llitest* and *hlitest* also will calibrate the LLIs and HILs. If the CU, LLI or HLI has a trouble, then the tool goes to the main steps (intelligence trouble tracing and diagnose/ suggestion) to solve the problem. After repairing the hardware problems, the tool will check the repaired problems from the beginning step until solved the problem.

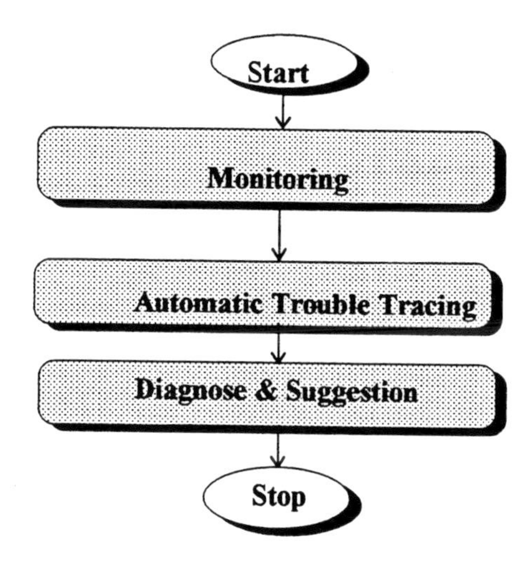

Fig. 7. System algorithm

4.2 Description of algorithm:

Monitoring (Calibration check/ Health check). Main program call module programs to calibrate all of the connected CU, LLI and HLI. And the Result is condition (*health*) of the measuring devices.

Automatic trouble tracing. Main program call *contest, llitest & hlitest,* these module programs calibrate all of module boards one by one to get the detail condition of CU, LLI and HIL. The detail condition acts as the *fact,* will be process using *rules* on the knowledge-based system. Then, call the module program *tracing* to analyze the detail condition and the trouble(s).

Diagnose & Suggestion. Call module program *diag* and *sugg.* These programs will define location of the troubles. Trouble means some components broken or the module board does not work accurately. And the result is location of the troubles and advises toward the trouble.

Repairing. Repairing the defined troubles by technician or operator at the electronic workshop. Finally, checking or testing the trouble again after repairing using the knowledge-based tool.

5. CONCLUSION

- The Knowledge-Based Tool needs *learning process* to make the system smarter. Many testing and validating the knowledge-based tool by the expert should increase the ability and performance.
- Anybody can operate this tool easily to get solution toward the problem. If the operator or engineer is not present, the smart tool can take over the job.
- The tool acts as a teacher if it is used by a junior engineer, otherwise acts as a student if it is used by a senior engineer.
- Using the knowledge-based tool can reduce the time for solving the troubles and increase the quality of result.
- A good maintenance of the data acquisition sub-system is very important to support the quality and productivity of the main system.

REFERENCES

C S Krishnamoorthy, S Rajeev (1991). *Computer Aided Design, Software and Analytical Tools,* Toppan Company(s) Pte Ltd, Delhi.

Giarratano, Riley (1989). *Expert Systems, Principles and Programming,* PWS-KENT Publishing Company, Boston.

N.N. (1986). *Calibration Generator Instruction Manual,* NLR, Amsterdam.

N.N. (1986). *Conditioning Unit Instruction Manual,* NLR, Amsterdam.

N.N. (1986). *High Level Interface Instruction Manual,* NLR, Amsterdam.

N.N. (1986). *Low Level Interface Instruction Manual,* NLR, Amsterdam.

Robert J. Schalkoff (1990). *Artificial Intelligence: An Engineering Approach,* McGraw-Hill Computer Science, Singapore.

Romli (1995). *LAN and Performance Optimization,* Master Thesis , University of Tasmania, Launceston.

Taub H., D. (1977). Schilling, *Digital Integrated Electronics,* McGraw-Hill, Singapore.

THE POTENTIAL OF TOOL INTEGRATION FOR PLANT FLOOR SOLUTIONS

Roberto Mosca, Pietro Giribone, Agostino G. Bruzzone

Department of Production Engineering
University of Genoa
via Opera Pia 15, 16145 Genoa, Italy
bruzzone@linux.it agostino@itim.unige.it

Abstract: A system can be integrated into an industrial situation not by developing new and highly-advanced tools but by using previously tested tools and machinery interlinked to generate new potential. This paper proposes the first phase of an automation project aimed at emphasizing an innovative method of integrating tools rather than developing completely new integrated systems designed to operate only in small specific applications.

Keywords: CIM, Educational Aids, Integration, Shop-Floor Oriented Systems, Simulation

1. INTRODUCTION

This paper presents the development of an automation center based on the integration of existing industrial packages for shop floor control and planning.
In fact, the SAMS (Sanna Agile Manufacturing System) is a new manufacturing laboratory. The project being carried out utilizes products from several companies involved in the automation sector, such as IBM™, DEA™, Elsag-Bailey™, Kuka Robots™, GE Fanuc™ and Digitron™. The final objective of the SAMS project is to develop a small, but advanced FMS. The authors are presenting just the first Stage of the Project in which they were involved in developing the first integration between a computer LAN and the following devices (Bina, *et al.*, 1996):

- Kuka Robot, 6 degrees of Freedom, 60 kg
- Data Terminal Collector with Laser Pencil to scan bar codes

The LAN is a PC network, based on OS/2, that uses DAE™ as the protocol to communicate with the Shop Floor. The objective is to integrate the Robot and Data Terminal with the CAPP (Computer Aided Production Planning).

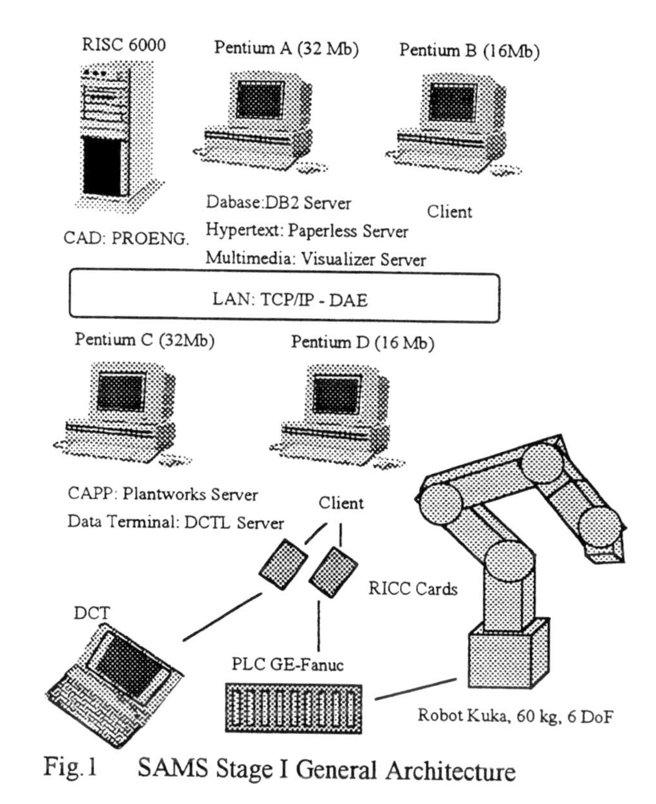

Fig.1 SAMS Stage I General Architecture

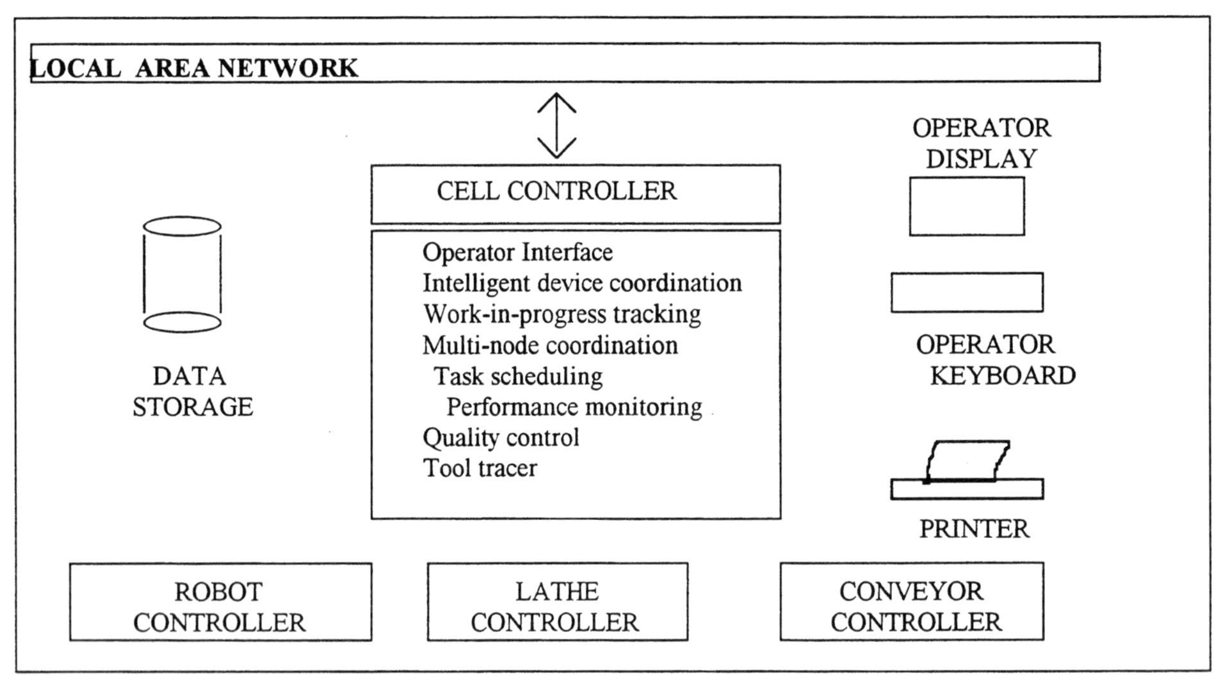

Fig.2 Tasks performed by application at the work cell level

2. GENERAL ARCHITECTURE

To achieve the SAMS objective, Plantworks™ (Software to program chains of operations in DAE™) has been integrated with Paperless™ (software to combine text, CAD files, images) to build an on-line quality manual to support all operations. Plantworks chains also link DCT-Language to automate the operations with bar-codes and to direct-link the Object-Oriented Database (DB2/2™) with the statistical analysis (Visualizer™ software). This integration is just the first step to complete the SAMS project, but currently represents one of the most advanced projects in Italy to integrate IBM™ automation software with shop floor devices. Presently, by reading a bar-code, it is possible to update the operator log file, to plan robot operations and programs and, in case of breakdown or shutdown, to utilize a complete, automated, on-line hypertext manual. The statistics of all the operations carried out are graphically presented at the end of the operator's shift. The system developed and integrated is based on the architecture presented in Fig. 1. The task performed during this project focused on the construction of Command Chains and C++ Code to link DB2/2™ (the Dbase) with DCTL™ (Data Collection Terminal Language) and Paperless™ (the package to present both CAD, text, images, etc.). In fact, Plantworks applications are event-driven programs that control manufacturing operations in response to plant events, such as signals from plant equipment, other computers, or plant operators (Gruhler & Rostan, 1996). Thus, Pantworks can run applications distributed across multiple nodes in a plant network and can integrate a host computer with plant computers that support programmable controllers, printers, robots, terminals, and other plant devices.

Fig. 2 illustrates the tasks that can be performed by an application at the work cell level. The arrows indicate the direction that data can flow between the application and the devices that the application controls. Plantworks also provides graphic displays that enable plant operators to monitor manufacturing applications very efficiently. Symbolic color representations of manufacturing activities are automatically updated to provide dynamic views of plant operations. The Plantworks software is based on a DAE™ system that could guarantee a general connection to each component (computer, software, machinery, etc.). The Distributed Automation Edition (DAE™) application is an enabling system that provides access to distributed information and allows communication between plant devices and application. The devices can be connected to nodes through co-processor adapters, which provide communication between the Distributed Automation Edition application and plant equipment. The package developed can upgrade the central Dbase by the DCT to maintain a log of DCT Operator sequences and to activate a related Paperless-based operation support manual during the procedures related directly with updated CAD Files of the item under evaluation (Balcezak, 1986). In case of any Robot Failure (Alarms), the Paperless™ is activated automatically and provides general information about that problem and an interactive, window-oriented support to solve the problem. The robot itself is controlled by Plantworks chains that activate part programs on a KUKA™ Control system through the PLC. A Graphic User Interface has been developed to keep the robot under control during all phases (Bruzzone, 1996). All commands could be introduced by DCT, PC and bar-codes (through DCT). An advanced security system guarantees the safety of all operations.

Software Packages

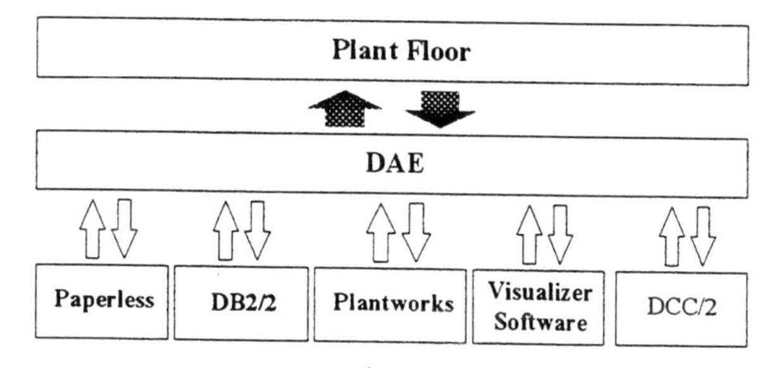

Fig 3 Integrated software in the SAMS I Project

3. THE OPERATING PROCEDURES

Phase one of SAMS, as presented in this article, is designed to integrate some of the most modern software packages that are currently available with a series of machinery arranged on the plant-shop floor (Black, 1991), as clearly illustrated in Fig. 3.
For some of these tools, this is the first time that integration has been implemented (for example the combination of Paperless™, DCC/2™ and Plantworks™). This development often required obviously the use of some C-coded parts to be implemented as links between the different systems. With this in mind, it was decided to accumulate all the information in the central database, continuously transfering what was acquired from the DCC/2™ internal database system which theoretically would be capable of performing statistical and research functions relative to the procedures managed through the DCT. This approach guarantees that the data in the central database is constantly updated and corrected with respect to the real and evolving situation in the field. In fact, with this database it was possible to establish direct connections with the various packages utilized (see Fig. 4). The entire Dbase is first integrated with Plantworks in order to enable the various operating functions. The procedures as described included the transfer of all the information sampled by the DCT and presentation of such data using tools designed to display the results: Visualizer™ with graphics integrated within the interactive Paperless™ windows.

DB2/2

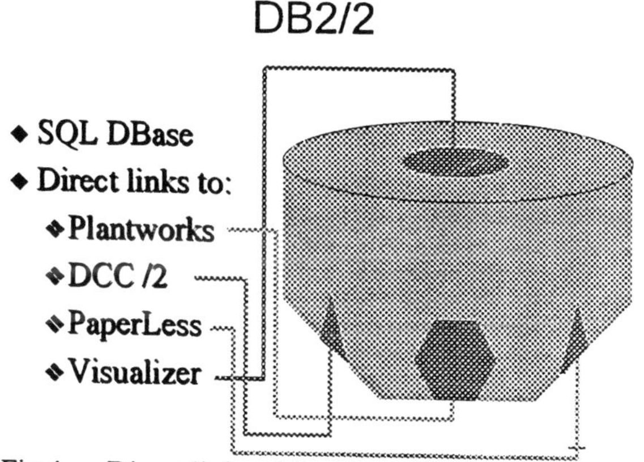

Fig.4 Direct links of the Dbase Used

The first SAMS development phase includes the construction in particular of the modules to control resources in the field, alarms and production programs. However, the final objective is aimed at constructing a complete production situation that can also independently develop the quality design and control phases. This approach is schematically represented in Fig. 5.

SAMS General Procedures

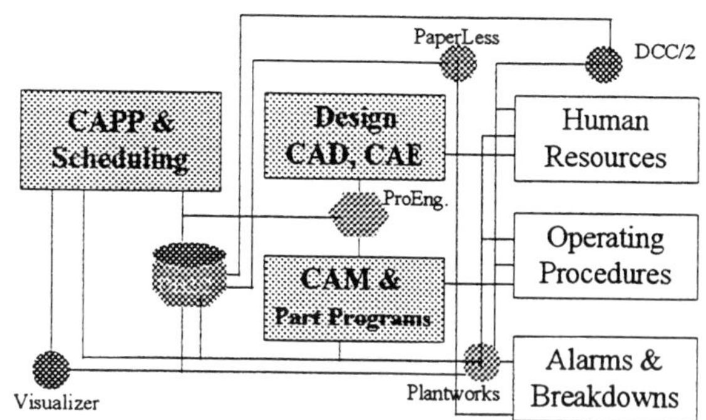

Fig.5 The Procedures Developed for SAMS together with the next Integration of the CAD/CAE & CAM modules

In fact, the design focuses in particular on the training and educational importance of allowing students and post-graduates or professional engineers to make a wide-ranging comparison of the manufacturing aspect from a modern point of view. Therefore, as previously described, this system aims at becoming a center of excellence to provide in-house training and to offer its services to the market while promoting a Concurrent Engineering approach in terms of product, process and re-engineering design of the operating procedures (Giribone & Bruzzone, 1995*a*).
To achieve these objectives, Information Technology (IT) must be used in a new manner, thus obtaining a continuous overview as the situation develops, making projections based on reliable data and acquiring decision support tools. For this reason, the experiment confirmed that the difficult task of integrating the available tools through the automation and development of a new software linking code also requires a new element to be introduced into the system architecture that is capable "ab initio" to act as a field-based decision-support tool and to acquire information directly in the field. Therefore, the authors decided from the very beginning to evaluate the development of a detailed plant simulation system.

4. THE SIMULATION

The authors' research team has been working in the simulation development sector for many years. Therefore, by utilizing the acquired know-how, it was decided to develop these modules based on an approach that is different than the one used to create

Plant-Floor Planning & Control. In fact, the simulation systems available in today's market are very inflexible languages/tools that, in particular, have not yet been integrated at the industrial application level in terms of standard DIS (Distributed Interactive Simulation). Therefore, in creating this type of system, it was preferred to develop specially-designed modules based on the proprietary libraries of the research group and to use multi-purpose languages (C/C++). In fact, this choice is also supported, on the basis of the same observations, by the current trend to use simulation in large manufacturing plant automation projects (i.e. new Volkswagen production lines for the Polo car model). In this environment, the simulation is understood not only as a tool to reproduce the behavior of the plant and to evaluate, for example, the consistency of the logic programmed on the PLCs or the machinery alarm and/or operating procedures, but the simulation system must also be capable of performing a predictive campaign on the global behavior of the production system.
To achieve real innovation in this field it is important that the simulation model be developed based on modern modeling designs (OODA Object-Oriented Design & Analysis) and in a distributed fashion. According to this approach, each entity can be reproduced by reconstructing a virtual object that reproduces its behavior. By respecting appropriate integration standards it becomes possible to link the various objects and to reproduce even very complex structures without having to change the logic of the single components (Selic, *et al.*, 1994). Thus, in this way it is possible to integrate control system, robot dynamic and cell logic simulators to verify their general coherence and to evaluate their performance by changing the production mix and/or type of product while it is being worked. It is evident that continuing the development based on these guidelines requires a rigorous design of the conceptual model of the simulator understood as the composition of different objects. In addition, the development time increases significantly. On the other hand, the modularity guaranteed by this approach allows the simulator to continue to develop over successive generations, but without continuing to re-validate and re-verify the single modules that have already been developed. In this paper the authors summarize some experiments performed with this system starting from the problems arising from the SAMS project and in particular with reference to the evaluation of a new control system for the assembly robot and for determining the procedures for using the AGV system.

4.1 CONTROL SYSTEM REDESIGN

The control system of the KUKA™ robots available in the project, as presented in this paper, is by now relatively obsolete. Therefore, it was decided to consider the possibility of replacing it with some new, specially developed systems based on new models. In particular, this research aimed at evaluating how a new integrated system can support control system development.
Starting from a model of robot dynamics it was possible to identify some terms resulting from the shift between the real position of the arms and the angle measured by the control system. These terms can be evaluated according to the following expressions:

$$\alpha_i = s_i \gamma_i + \mu_i \sin(2\gamma_i + \varphi_i) \qquad (1)$$

- α_i angular position of the i-th shafts of the harmonic drive
- γ_i i-th angular position of the links
- μ_i amplitude of the i-th disturbance
- φ_i phase of the i-th disturbance

These terms tend to introduce oscillations in precision positioning that affect the quality of the robot assembly operations. To resolve this problem it was decided to evaluate the introduction of a new fuzzy control system (Leporati & Bruzzone, 1995).
Said control was developed only in terms of a simulation module to be connected with the robot model and thus it was possible to obtain an immediate preliminary validation of the increase in achievable performances and in particular to support the FAM design (Fuzzy Allocation Matrix) and membership grades needed to optimize the system.
The analysis of the performances obtained in the specific case is based on an optimization of the trajectories of the robot-control system of the state space (Accacia, *et al.*, 1994).

4.2 AGV ENGINEERING PROCEDURES

Internal transport systems are notoriously difficult to evaluate in production systems. Each time an attempt is made to automate these tools, powerful modeling tools must be developed that can help designers correctly define the specifications of this component, its limits and especially the most economical operating logic. Two AGV that can operate on the shop-plant floor are activated by SAMS. While waiting to re-activate the single components already available and involved with requests relative to other industrial operating situations, it was decided to develop a model capable of reproducing the behavior of the internal transport systems.
This model also respected the approach described above and thus was integrated into the global simulator. The model is obviously built according to an object-oriented hierarchical base and defines the vehicles and their logic as well as the entire layout of the handling network based on dynamically connected links and joints (Giribone & Bruzzone, 1995b).

The graphic interface of this module is reported in Fig. 6. It is used to verify, validate and test the logic implemented as well as to provide the specifications with diagrams and demonstrations for the different possible situations. Once said logic is implemented, evaluations can be performed using an experimental analysis of the general performances of the transport system, the mean picking time from the warehouse, the use coefficients, etc. (Bruzzone, 1994).

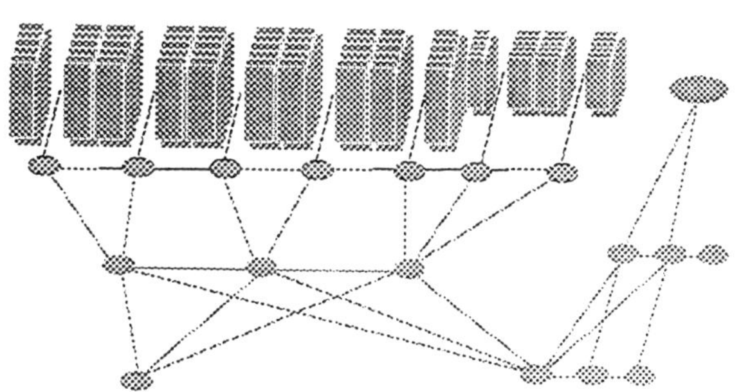

Fig.6 Development starting from the simulation models developed for SAMS: automated warehouse management

5. CONCLUSIONS

In this work, the authors are carrying out a traditional integration project, but the innovative aspect is related to the development of new inter-links between professional packages. In effect, this research highlights the real potential of this kind of tool when full integration is possible. Normally, in industrial environments, integrated functions are not available for users due to the costs and development times, but university research can be a very useful test area to develop new links to improve the shop-floor efficiency.

Thus, the project may become an excellent opportunity to evaluate the new potential of existing automation tools as well as a source for the development of new operating procedures which enhance the quality, safety and efficiency of the production system, making it flexible and reliable. In addition, by integrating the simulation into the system, new components and/or new products can be introduced into the system in a rational manner, verifying the impact a priori in terms of global production efficiency and improvement in product quality. Proper integration with economic evaluators automatically ensures a correct valorization of the new choices, thus transforming the planning system into a very powerful support for strategic company management.

REFERENCES

Accacia G.M., A.G.Bruzzone, M.Callegari, R.C.Michelini,R.M.Molfino (1994) Phase Space Analysis as a Tool for Control System Design, in Proc. Concurrent Engineering and Electronic Design Automation, Poole, UK

Balcezak W.K. (1986) "Paperless System Tracks and Controls Assembly at tandem Computers", J.Industrial Engineering, September, pp 66-72

Bina M.A., Bortoletto V., Bruzzone A.G., Gestro Sobrio M.A., Tremori A.(1996) SAMS 1st Stage, Technical Report of ICAMES, Istanbul

Black J.T. (1991) *The design of a Factory with a Future*, Mc-Graw Hill

Bruzzone Agostino G. (1994) Potential of Reduced Experimental Designs in Object-Oriented Modeling of a Manufacturing Department for Film Production, in Proc. of European Simulation Multiconfernece, Barcelona, Spain

Bruzzone A.G. (1996), Smart Integrated Interface for Evaluating and Supporting Human Performances, ESPRIT Information Technology: Basic Research Working Group 8467 Simulation for the Future: New Concepts, Tools and Application (SiE-WG), Brussels, Belgium

Giribone P. & A.G.Bruzzone (1995) Training System through Simulation and AI Techniques of Process Plant Management, in Proc. of Flexible Automation and Intelligent Manufacturing, Stuttgart, Germany

Giribone P. & A.G.Bruzzone (1995) Object-Oriented Modeling of an AGV System for a Harbor Environment, in Proc. of Western Simulation MultiConference, Las Vegas, Nevada

Gruhler G. & Rostan M. (1996), A New Generation of Control Systems for Production Cells, in Proc. of Flexible Automation and Intelligent Manufacturing , Stuttgart, Germany

Leporati C. & Bruzzone A.G. (1995) Non-Conventional Design of a Welding Control System by use of Simulation Approach, in Proc. of Advances in Materials and Processing Technologies, Dublin, Ireland

Selic B. Gullekson G and P.T. Ward (1994) *Real-Time Object Oriented Modelling*, John Wiley & Sons, New York

A NETWORK-BASED MANUFACTURING MANAGEMENT SYSTEM

István Mezgár, George L. Kovács

CIM Research Laboratory,
Computer and Automation Research Institute,
Hungarian Academy of Sciences, Internet: mezgar@sztaki.hu

Abstract: Today the industrial production steps over the walls of the individual enterprises, so new management approaches are needed to be able to co-ordinate the work of the different manufacturing organisations. A new manufacturing network model is proposed for balancing the production while maximising the profit of a group of SMEs, based on the holonic paradigm. The model has been developed in the frame of an ESPRIT Project by four countries. The paper describes an application of the model under realisation in Hungary, in an agricultural SME network, producing seeds and oil from seeds.

Keywords: computer networks, enterprise modelling, holonic manufacturing, management systems.

1. INTRODUCTION

In order to fulfill the market demands the flexible, effective manufacturing system architectures become more and more popular around the world. Manufacturing enterprises have a geographically distributed nature, so computer networks for production management are important feature of their operation.

There are different approaches, different names that basically cover the same idea; a flexible network of co-operating autonomous manufacturing units. The holonic manufacturing systems are an approach for a theoretical framework for autonomous and decentralized manufacturing organizations (e.g. agile manufacturing enterprise - AME). There are different projects and developments on AME (AIMS/Stanford & EIT (Park *et.al.*, 1993), A-PRIMED/SNL, FFC/MIT, etc.) that use different names for the AME, e.g. "extended enterprise" or "virtual enterprise".

The structure of a virtual enterprise can be seen as a holarchy, in that it is a temporary, goal-oriented aggregation of several individual enterprises. Each virtual enterprise is created to pursue a specific business objective, and remains in life for as long as this objective can be pursued. After that, the individual nodes resume their independence from each other. The resource of a node that were previously allocated to the expired business are re-directed toward the node's individual goals, or toward other virtual enterprises it may have joined.

The original virtual enterprise (VE) model has been developed for decomposing large companies that have enough financial, human and organizational resources. The decomposition is based on functions that can result in replications across the network.

Many geographic regions can be characterised with special industrial organisational structures called SME (Small- and Medium-Sized Enterprise). These SMEs are well-known for their dynamic behaviour, but their prosperity, and sometimes their existence, is continuously exposed to risks of the limited investment capability, the difficulty of diverting skilled personnel from day-by-day activities, to undertake process re-engineering initiatives, the lack of advanced planning support tools specifically conceived for them. Because of these reasons it is even a harder problem for the SMEs to give proper answer for the market challenges. The co-ordination of the different SME's production in case of producing a product jointly generates similar problems as in the case of large firms, but the solution is different (Bonfatti, *et.al.*, 1996).

In a distributed manufacturing environment the production planning has a key role. The co-ordination of the orders, the optimal assignment of the different resources in a co-operative production of several SMEs is a very difficult task. In an SME network there is a strong need of a network model that is based on the planning strategy (Bonfatti, *et.al.*, 1994). An other important goal during the software development is the product independence, i.e. the network model has to be general enough, to be able to apply it in a wide range of different products, in different production environments.

The SME network development work is done in the frame of the "PLENT"- ESPRIT Project No. 20723 with the participation of four countries (Italy, Greece, Hungary, Spain). The main goal of the project is to support organisations that manufacture mechanical parts and products in SME-like production environments. The paper shortly describes a special application of the manufacturing network model for agriculture production (oil and seed production) to demonstrate the applicability of this management philosophy and network model in fully different fields.

2. THE PLENT PROJECT

2.1 Comparison of the holonic/virtual and the PLENT approach

In the PLENT project a more practical, more appropriate approach of the virtual enterprise has been realized for SMEs. In order to give a better understanding the PLENT approach will be shortly introduced in comparison with the classical holonic theory as it was discussed by Koestler (1989).

The holonic manufacturing system is a system of co-operating holons that are organized to achieve a production goal. This system integrates the whole range of manufacturing activities from order booking through design, production and marketing to form an autonomous and decentralized manufacturing organization.

The basic building unit of holonic systems is the holon. A holon is an autonomous and co-operative building block of a manufacturing system for transporting, transforming, storing information and material. A holon consists of a physical and an information processing part. A holon can be the part of an other holon. The holarchy is the organizational structure of holons. It defines the basic rules for cooperation of the holons and in this way limits their autonomy. The holarchy is not a fixed organizational structure, but organizes itself dynamically to meet its goals and adapts itself to changes in its environment.

According to the holonic organization paradigm the manufacturing environment is going to transform itself into a holonic system as the openness, the flexibility and the similarity in building blocks are vital advantages. The virtual enterprise is a practical instantiation of the holonic paradigm in the form of a multilayer, open, flexible (through continuous, dynamic re-configuration) network.

As the PLENT project aims to coordinate the production of SMEs the pure holonic approach had to be modified. The SMEs have limited financial and human resources, usually they have no advanced planning support tools that are essential for a fully open VE. There is also a historical background namely the traditional individualistic attitude of SMEs. This means the facing of marketing, design, engineering, technological innovation and purchasing problems alone. Handling the SME network as a limited number of independent nodes have the following advantages:

- More complex and technologically advanced products can be realised by putting together skills and design capabilities of the network nodes.
- Higher manufacturing volumes are obtained by cumulating node capacities, especially if two or more nodes concur in performing the same manufacturing phases.
- Fluctuations of market demand volumes are better borne by sharing workload peaks and shortages among the network nodes.
- The limited resources of each SME are better spent in technological innovation and process re-engineering rather than in hard competition with other SMEs.

In consideration of the requirements expressed by SMEs, it is possible to compare the PLENT idea of virtual enterprise with that currently derived from the holonic theory. This allows to determine if, and to which extent, the current virtual enterprise approach can be applied to the PLENT network (Bonfatti, *et.al.*, 1995).

Indeed, a number of properties commonly attributed to the virtual enterprise are no longer valid when transferred to the PLENT organisation model. The most significant differences between the two approaches are listed in the followings:

* Functional specialization versus flexibility.

* System duplication versus identity preservation.
* Temporary versus stable aggregation.
* Local versus global network co-ordination.

A detailed description can be found in Deliverable D2 (1996).

Having in mind the above distinctions, the PLENT network can still be considered as a special case of virtual enterprise organisation. With respect to traditional co-operation models, the PLENT network has the following distinctive features:

- The network results from the aggregation of a number of independent, small-medium enterprises.
- The network is constructed and organised to reach a well defined, long-term objective, common to all member enterprises (e.g. designing, constructing and selling a certain family of products.)
- The network nodes have basically equal rights regarding the definition of network policies and exploitation of joint business opportunities.

In conclusion, the PLENT project focuses on distributed production management in a stable network of co-operating SMEs. The initial phases of network design and creation, as well as the pre-existing relationships that made co-operation possible, are outside the project scope.

2.2 PLENT manufacturing network model

In order to achieve the required network timeliness and co-ordination neutrality, the PLENT network organisation must be supported by expressly designed planning tools. In particular, three software modules are required: a co-ordination module, a local planning module and a performance evaluation module.

* Co-ordination module
  The co-ordination module realises the network planning policy. The co-ordination module is responsible for controlling the connection outside the network, for keeping the contact with the customers, for selecting the proper nodes (production units) and for distributing production tasks to them.

* Local planning module
  The local planning module supports the network-related activities in charge of each node, aimed at managing and communicating delivery plans and delivery conditions with the co-ordination unit and the supplier and client nodes. The local planning modules realise the network controlling tasks of the node, in order to fulfil the goals of the management and communication, and deliver the status information of the node for the other nodes.

* Performance evaluation module
  The performance evaluation module is used by the co-ordinating unit to calculate node reliability starting from measures of their past behaviours, and network reaction capability by simulating possible scenarios. The performance evaluation module supports the work of the co-ordination unit in selecting the best node for fulfilling a task. This node generates different measures, statistics, estimations based on the past behaviour of the nodes.

Based on the PLENT VE approach four different national networks are under development: the Italian, the Greek, the Hungarian and the Spanish SME network. The product profile of the national networks are different, demonstrating by this fact that this type of virtual manufacturing approach can be applied flexible, under different circumstances. In the following chapter the main characteristic and the organization of the Hungarian Network will be shortly presented.

3. DESIGN AND MANAGEMENT OF THE HUNGARIAN NETWORK

3.1 Description of the Hungarian Network

The main activities in the Hungarian Network (HN) are financing oil seed growing on plantations, seed-oil production and trading with the products which are the seeds and the seed-oils. The products as physical objects are seeds, seed-oils and contracts for production with farms and factories, and contracts for trading with trading houses. The market of the products is in Hungary and abroad. The main functionality is contracting partners to produce and to sell and buy. There is a management and a staff to prepare contracts using advanced forecasting and trend analysis tools. The product delivery cycle is unique: no cycle for the contract, however there is a special cycle for the seeds and oils covered by the contracts, as financing starts very early, 3-12 months before harvesting.

The co-ordination unit for the Hungarian Network is an Investment Co. Its main fields of activity are financing oil seed growing on plantations, seed-oil production and trading. The products of the Investment Co. are contracts, i.e. pieces of signed papers, for financing growing and production of seeds and oils and to sell and buy them.

There are three different profiles of individual nodes in the network :

- Big farms and oil pressing factories. Their activities include market research, approval, contract preparation, contracting, growing the seeds on the plantations and delivery of seeds or pressed oils.
- Small farms and seed producers. Their activities are basically the same that above.
- Trading houses. Their activities include the seed and oil purchasing, the selling of the purchased goods and the cashing of the money.

The functional overview of the HN can be seen on Fig. 1. Detailed description and analysis of the HN please see in Deliverable D01 (1996).

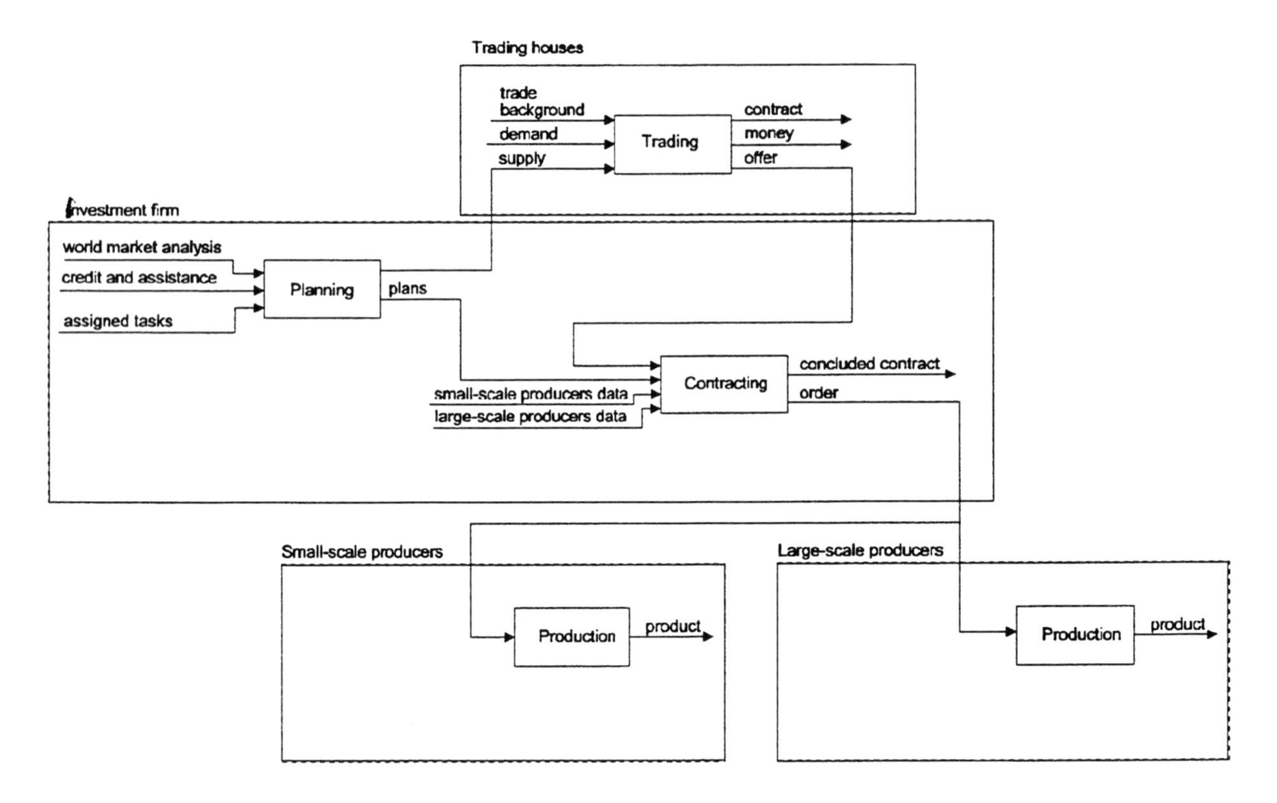

Fig. 1. Functional description of the Hungarian SME network for seed and oil production and marketing

3.2 Requirements specification and design criteria of the HN

Summarising the users requirements, it is possible to identify the different aspects of the necessities for the planning activities (in the co-ordination unit as well as in the nodes), the communication and information exchange needs, supervision and control mechanisms, and software requirements.

The critical factor in the Hungarian Network is the timing. There are big shifts among the different production phases and big differences in the duration of the phases. There is a magnitude difference between production phases duration (contract preparation - 6 days, seed growing - 180 days) and between the two sequencing phases there is a 180-360 days gap, shift. This time characteristic results in special needs in the network management. The administration process is fully separated from the flow of materials in time, that means the contract preparation takes place 1/2-1 year before the seed growing (SG) or/and the oil pressing (OP). The resource allocation, scheduling of activities (phases) are based on the actual information (at the date of contracting) that may be different from capacities during the realisation (at the date of SG, OP). Originating from this characteristics it is important to have a three-level planning possibility for the co-ordination unit:

- long term planning (breadth-first)
- medium-term planning
- operational planning in case of disturbances, re-planning.

Basic mechanism. The backward sequential planning method is the way how the HN works. The definition of the latest start-time is appropriate because of the existing uncertainties in the network. An additional problem in HN that the co-ordination unit will be integrated with a „manufacturing" node, so there will be some slight differences between the functions, operation of the HN co-ordination unit and the original PLENT co-ordination unit. Basically the HN is a simple network without complex product routes.

Order selection. In the HN the breadth-first planning method is in use in the long-term planning. First the capacities of all seed growing nodes (farmers) are planned then the capacities of all oil pressure factories. The contract preparation and trading phases are not really significant from this aspect. Because of the large time shift between contracting SG and OP during the realisation some problems may arise. In such cases an operative re-planning is needed using the actual existing free capacities. Re-planning can be needed in the following cases:

- significantly less (or more) crop at an SG node,
- longer disturbance at an OP node,
- storage or transportation problems at an SG or OP node,

Priorities during the planning are important. Product quality and node (farmer) reliability are very important in capacity selection.

Phases assignment. Phase due dates have different importance with different level of uncertainty. Due date uncertainty of the phases:

* Contract Preparation -> very low uncertainty, exact dates
* Seed Growing phase -> high uncertainty, needs flexible dates,
* Oil Pressing -> medium uncertainty
* Trading -> very low uncertainty, exact dates
* Subcontracting -> very low uncertainty, exact dates

Transportation. The given transportation time for seed transportation (1 day) means that when the transportation starts within 1 day there is enough quantity of seed to be able to start oil pressing at the factory. The transportation is continuous from the starting date.

Importance of storage. Using the storage for balancing the peaks/gaps of production is vital in the HN. The storage of seed can belong both to the SG phase and to the OP phase. There is a competition among the nodes to avoid seed storage because of the high costs and the risk of rotting. On the other side OP nodes are interested in having a minimum stock of seed in order to have a safe continuous production. The final product - oil - is delivered as soon as possible without storage. To calculate with stocks on the network level by the co-ordination unit doesn't fit into the original PLENT approach. Stock management is planned to be handled by each node separately, but in the HN there is need for a hybrid solution in order to have high level information for the co-ordination unit about seed stock levels. The node fulfilment index has to be re-interpreted in case of the HN. „Daily" time-spans, exact, fixed dates are hard to interpret in some cases in the HN (because of the uncertainty).

Network Workflow. The re-planning in the Hungarian Network will be an important action. A three-level planning mechanism is needed because of the significant differences in phase duration, and the mixture of discrete and continuous manufacturing phases. In the long-term planning the high-level co-ordination of discrete (contracting, seed growing, trading, subcontracting) and continuous (oil pressing) phases takes place well before (1 year) the real production. The medium-term planning occurs before finishing seed growing phase in order to co-ordinate the smooth transient of the most important discrete and the continuous phases taking into consideration the actual possibilities, resources etc. of the network. The re-planning situations are the usual ones plus adding the situations listed in the previous chapter. Short-term planning deals with solving the operative co-ordination problems of the network.

Performance Evaluation. The evaluation of node performance has a very important role in the HN. Important evaluation aspects are the reliability of the node and the quality of the product. As there are very big quantities in a batch, seed quality is a basic evaluation factor. A very useful characteristic of the evaluation method is that the nodes in the same phase can be compared based on objective evaluation factors. This possibility has outstanding importance in the HN in the long-term planning phase as the selection of the most reliable nodes and the nodes producing the best quality product can determine the success of the network for a long time.

3.3 Expected benefits

The main advantage of using the described network is the more effective usage of each SMEs (farms, oil pressing factories) resources, faster reaction for market demands while keeping the individual characteristic of each node (farm). Through this coordination the production of a very unstable, unreliable (working with high risk) production field, the agricultural production can become more reliable, more safe both for the customer and the farmers that are members of the network. The reliability results more orders, stabilized and higher income, less costs i.e. more profit.

The expected costs of the investment (network installation and software) are low, to learn the applied information technology needs only short time, so the PLENT manufacturing network has good chances for success in agricultural field as well.

4. CONCLUSIONS

In the PLENT approach, the network is conceived as a set of operational nodes with equal rights, co-ordinated by a central unit that interacts with customers, distributes tasks to nodes on the basis of prefixed criteria and up-to-date information on node state and reliability, and informs each node on its position in the network with respect to the manufacturing process. The nodes are responsible of executing the assigned tasks within given time intervals and, to this purpose, they maintain direct relations with the respective suppliers and clients and try to solve possible network perturbations.

This new approach for SME network coordination is very promisable, because it offers the advantages of the holonic manufacturing paradigm parallel giving the possibility of keeping the traditional individualism of SMEs. The realization of the method is going on in four countries involving SMEs from three different fields of production (in machine- and textile industry, in agriculture).

The results achieved prove that it is possible to develop a network based co-ordination software for SMEs that is general enough to be applied in fully different production fields. The early experiments show that the ISDN-based Hungarian Network has a good perspective. The program development phase has already started using Visual C++ and SYBASE.

REFERENCES

Bonfatti, F., P.D. Monari and P. Paganelli (1994). Modeling manufacturing resources and activities: an ontology. In: *Proc. of ICCIM '95 Intern.Conf., Computer Integrated Manufacturing, Singapore, 1995,* Edited by J.Winsor, A.I.Sivakumar, R.Gay, p.p. 101-108, Published by World Scientific, Singapore.

Bonfatti, F. and P.D. Monari (1995). Planning small-medium enterprise networks. In: *Proc. of the IMPDE Internl. Conf., Edinburgh, 1995,* p.p. 13-22, Sponsored by European Commission DGIII, ESPRIT Consortium 9245.

Bonfatti, F., P.D. Monari and P. Paganelli (1996). Co-ordination Functions in a SME Network. In: *Proc. of BASYS '96 Intern. Conf., Balanced Automation Systems II, Lisbon, 1996*, Edited by L.M. Camarinha-Matos and H.Afsarmanesh, p.p. 383-390, Published by Chapman & Hall, London.

Koestler A. (1989) *The Ghost in The Machine*, Arkana Books.

Park, H., J.M. Tenenbaum and R. Dove (1993). Agile Infrastructure for Manufacturing Systems (AIMS), *http://www.eit.com/creations/papers/ DMC93/ DMC93.html*

PLENT ESPRIT project 20723, Deliverable D01 (1996). Requirement Specification and Analysis.

PLENT ESPRIT project 20723, Deliverable D02 (1996). Planning Policy Formalisation.

DEVELOPING AND LINKING THE TOOL DATABASE FOR TOOL MANAGEMENT IN FMS

Mahbubur Rahman[1]
Jouko Heikkala[2]

[1]*Researcher, Production Technology Laboratory, University of Oulu, PL 444, 90571 Oulu, Finland, E-mail: mrahman@me.oulu.fi*
[2]*Laboratory Manager, Production Technology Laboratory, University of Oulu, PL 444, 90571 Oulu, Finland, E-mail: jouko.heikkala@oulu.fi*

Abstract: To ensure effective tool management in a Flexible Manufacturing System (FMS) an effective tool database with interaction with other modules of the production environment is needed. With help of this data-base users (from designers to shop floor operators) can select right tools at the right time. The effective tool database system reduces the tool inventory thus saving money. This papers describes the development of a tool database system and linking with other modules of the FMS at the Production Technology Laboratory in the University of Oulu.

Keywords: Tool Database, Flexible Manufacturing Systems, Tool Measurement

1. INTRODUCTION

Tooling management has been crucial problem in the last decade because of increasing automation in production systems. Tool management is as critical as the flow of workpieces in FMS. The performance of a flexible manufacturing system depends on the availability and quality of tools needed to machine the parts according to production plan. Many investigations show that downtimes of the FMS are because of tooling. In many companies managers concentrate on materials flow and/or optimisation of cutting processes and they largely overlook and ignore the tooling resources. It has been shown by Torvinen, S. J. et al. (1989) that

- Typically 30-60% of a shop's tooling inventory is on the shop floor.
- Operators spend up to 20% of their time for searching tools.
- Typically 16% of schedules can not make because of tooling is not available.

Considering all these problems and to improve the machine utilisation rate manufacturing industries should supply tools to the production stage at the right time with right amounts. In many cases enterprises increase their tooling inventory to provide right tools at right time. In manufacturing industry up to 25% of costs can be of tooling inventory. Considering a single tool, the inventory is very small but considering total tooling the inventory is high. Unfortunately this high inventory is still neglected by many manufacturing enterprises. A study of the tooling function should be done to obtain an integrated solution to the problem from tool delivery to the plant to final write-off. Considering these, a tool management system is under development at the Production Technology Laboratory in the University of Oulu. In this paper a simple data structure for tool management system and use of this tool data with other modules of the FMS has been discussed.

2. TOOL MANAGEMENT SYSTEM

Tooling is defined as all equipment and special fixtures that the system can draw on and use during the setup and operation of a machine or assembly process (Steven and Lyman, 1993). This definition includes such as jigs, fixtures, pallets, cutting tools, dies, moulds etc. All these tools affect directly and indirectly:

- Capabilities of a given machine, materials, type of products produced
- Capacity requirements of the manufacturing system
- Inventory system in the manufacturing system.

Tool management is a process resulting from the interaction of planning, executive and controlling functions in the tool related information flow. The objective of tool management is to ensure optimal deployment of correct tools in the right place just in time (Eversheim et al 1991).

2.1 *Benefits of tool management*

Tool management has lots of benefits in the production industries. Some of those are mentioned here:

- Optimum tool inventory control:
  Tool management prevents duplicate tool inventory and keeps the inventory in optimum level.
- Reduction of machine idle time:
  Efficient tool supply is provided to machine products based on requirement planning at right time, at right machine and in right amount thus reducing machine idle time.
- Improvement of flexibility in the system:
  Tool sets are prepared based on rational approach which helps to increase flexibility.
- Improving tool flow in the manufacturing system:
  Tool flow from different tooling department is monitored and optimised.
- Optimisation of tool selection:
  Tools are selected in an optimum way from available resources.
- Evaluation of tooling performance:
  Performance of the system is evaluated based on monitoring tooling life, costing of tools etc.
- Automatic data transfer between different subsystems (including machine tools):
  Tooling data is transferred among different modules and machine tools automatically thus reducing human errors.
- Provides tool status display and tool life history reports.

2.2 *Tool flow and data flow within the system*

Tool management should be an integrated process in manufacturing system. Tooling data in FMS may be used by the following subsystems:

- The production planning subsystem
- Process control
- Part programming
- Tool pre-setting and tool maintenance
- Tool assembly
- Stock control and materials storage

A tool list can be generated for every workstation based upon the process plan of the required production. This tool list can help to find the common tools between various parts which help to decide tool down/uploading from/to tool magazines.

Based on the production plan, for a given production horizon, a required tool list can be generated based on available tool life. This helps to decide about the backup tool in the tool magazine. When the tool life of particular tool ends the back up tool can be activated automatically. Thus it reduces interruptions in the production.

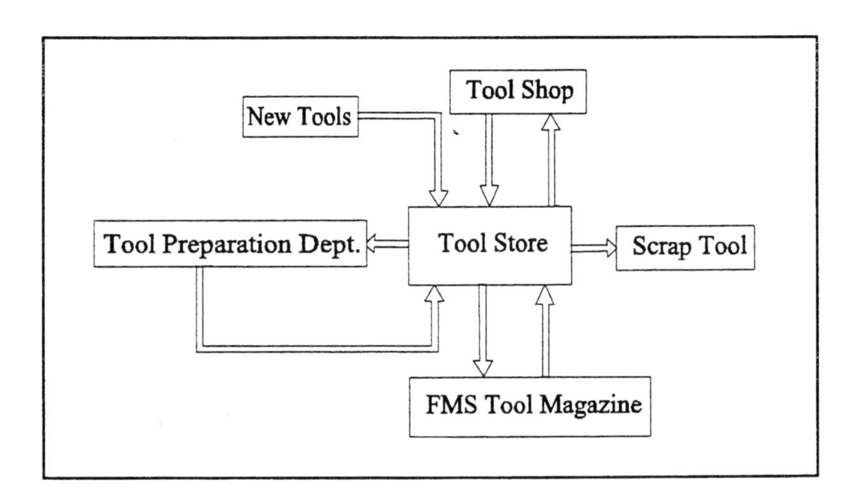

Fig. 1. Tool flow in the FMS environment.

All data regarding tools should flow automatically and reliably within all subsystems of FMS. For example the production planning and control system should be informed in the real time about the availability of tools and the contents of the tool magazine. When some tools are physically moved from one place to another the corresponding data should be upgraded. Different part programs needs different sets of tools. All data (tool life, radius, length, etc.) regarding these tool should be updated with physical tool flow.

3. TOOL DATABASE

Tool management system should have a smooth and distributed database to make the FMS work smoothly. Tool data flows within the FMS by a Local Area Network (LAN) in such a way that all subsystems can access all tooling data whenever they need. Essential data components which are included in the tool database are:

- Tool number / Identifier
- Tool location
- Definition of the machines in which it can be used
- Tool life
- Tool dimensions
- Tool holder and insert geometry
- Tool adjustability

Main components of assembled tools are toolholders and inserts. The geometric data of each toolholder and insert can be described by standard ISO codes which standardise the data base. In case of modular tool systems the different modules can also be unambiguously standardised. Another important factor is tool material data. Based on the geometric data graphics could be generated to give visual information to user.

Tool number/identifier can be based on tool's operation, group or tool type. For example MIL1234 can express certain kind of milling tool. Additional information could be expressed by integer 1, 2, 3 and 4 like standard mill with shank, end mill, plain mill etc. The advantages of this approach is that user can select or search from data-base milling tools in general or milling tools for specific purposes.

Tool life can be expressed in terms of percentage or in term of timing unit. Tool life should be updated continuously and warning should be given to the user when tool should be replaced. By two ways tool life database could be maintained. Each time the part program runs, the recalculated wear time could be accumulated to the tool life. Another dynamic way to update tool life is sensory feedback from tool/ workpiece interface. Different kinds of cutting sensors like coustic emission (Eversheim et al 1991), spindle force monitoring, current sensors or plate sensors can be employed to observe the tool wear or tool breaking. In this way tool life could be continuously updated in data-base.

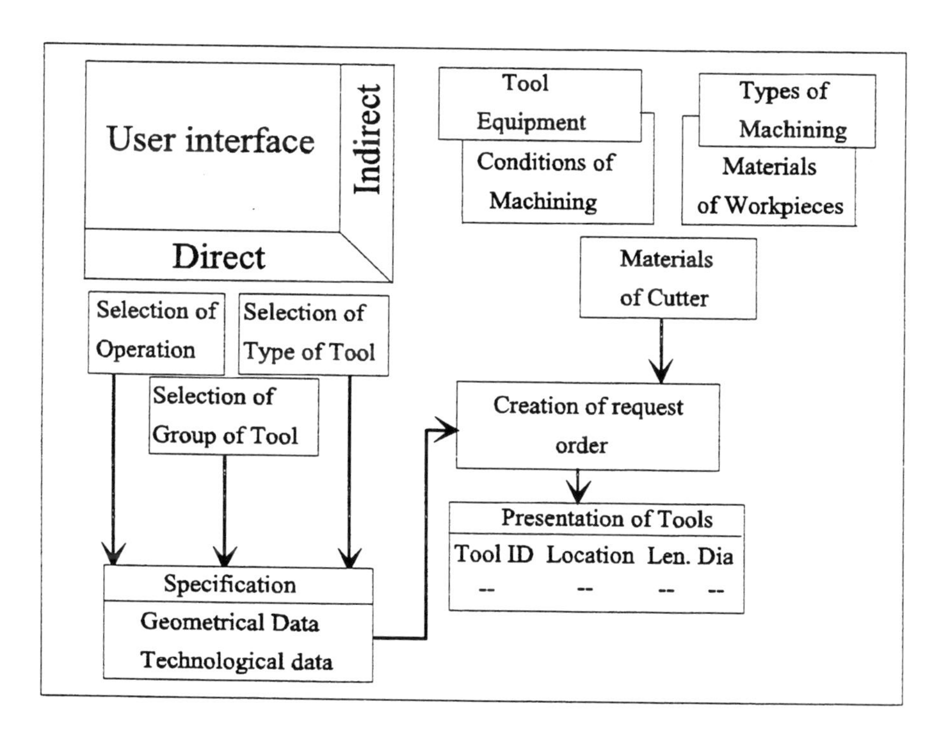

Fig. 2. Tool search (adopted from Eversheim et al 1987).

3.1 Searching tools from database

Tool data base should be distributed and accessed by all necessary subsystems. Analysis of tooling data flow within any FMS shows that part programmers needs both geometric and tool material data as well as part related data (from CAD) and manufacturing resources, or machine related data (from the FMS) to be able to select the most appropriate tools for the required operations (Ranky, 1988).

By direct search we search a particular type of tools directly where as indirect search supports the direct search for additional constraints.

3.2 *Tool management in the FMS in the Production Technology Laboratory*

As previously mentioned, to take the advantage of a tool management system and to increase the productivity of the FMS at the laboratory, a tool management system is under development. The total idea described so far is not implemented totally yet. Here in this section we focus on some of the functions already implemented:

- Tool measurement
- Tool data transfer
- Tool search
- Tool data storage

After arrival of new tools from the tool setting department, tools are measured and the tool database updated automatically. New tools can be added or scrap tools can be taken away. If the operator puts the tool directly to the machine tools, corresponding tool data (including offsets) for the machine tools is updated through local area network and work station computer. When a tool is downloaded or uploaded from/to the machine all data are updated accordingly to keep track of all tool data.

Tools can be searched based on length, diameter, tool type etc. Tool life for particular tools are displayed also to give the user idea about tool life.

Tool Measurement			
Tool Number	NC Machine (1, 2 or 3)	Nominal Length	Nominal Diameter
1. ----	-	---.---	---.---
2. ----	-	---.---	---.---

Measure	Display	Save	Cancel	Help

Fig. 3. Tool measurement interface at the Production Technology Laboratory.

4. CONCLUSIONS

To achieve full advantage of tooling resources, tool management should be integrated with other subsystems of the manufacturing system. For small to medium sized FMS's, a tool management system can easily be developed according to needs as described in this paper. For large FMS's all subsystems in the tool management should be automatic. Manual or semiautomatic solutions should be avoided for safety in operation and for reliability

Tool delivery, assembly and pre-setting all should be done by automatic means. Of course this will increase the capital investment in the system but the possibility for unmanned operation can give good return on investment. This automatic handling updates all subsystems as necessary. By this measure the productivity will increase and the number of tools and thus the costs will be reduced.

REFERENCES

Eversheim, W. et al (1991) Tool Management: The present and the Future. *Annals of the CIRP* Vol. 40/2/1991.

Eversheim, W. et al. (1987) Structure and Application of a Universal Company-Independent Data Bank for Tools. *Annals of the CIRP* Vol. 36/1/1987.

Hartly, John, (1984), *FMS at Work*, IFS (publications) Ltd, UK, pp 286.

Lenz, E. John (1989) *Flexible Manufacturing Benefits for the Low-Inventory Factory*, 1989 MARCEL DEKKER, INC pp 49-147.

Melnyk, Steve, A. CPIM, and Lyman, Steven B. (1993). Tool management and Control: Developing an Integrated Top-Down Control Process. *Annual International conference Proceedings-* American Production and Inventory Control Society, 1993.

Ranky, G. Paul, (1988) A generic tool management system architecture for Flexible Manufacturing Systems (FMS), *Robotica* (1988) v 6, pp. 221-234.

Torvinen, Seppo J. (1991). Integration of a CIM tool management system to an intelligent feature-based process planning system. *Computer in Industry* 17 (1991) 207-216

2D SYSTEMS ANALYSIS: A POLYNOMIAL APPROACH

E. Fornasini* and M.E. Valcher *

Dip. di Elettronica ed Informatica, Univ. di Padova, Padova, ITALY

Abstract: The paper stresses the relevance of polynomial matrices in three different approaches to the analysis of two-dimensional systems: input/output maps, state models and behavior descriptions. Some aspects of i/o and state models, like stability definitions and their algebraic characterizations, as well as conditions for the existence of stabilizing/dead-beat controllers are surveyed. Finally, some preliminary results about stability and the design of stabilizing controllers in the context of 2D behavior theory are presented.

Keywords: Two-dimensional systems, input/output models, state-space models, behaviour, stability, stabilizing controllers

1. INTRODUCTION

The foundations of the polynomial matrix approach to linear systems, along with its applications to industrial control design, were laid down in the early seventies by H.H.Rosenbrock (Rosenbrock, 1970). Since that time the theory has been developed by a large number of researchers, and nowadays polynomial matrix methods have gained a large acceptance within the control community. Among the others, one of the main advantages of this conceptual framework is the possibility of analysing and designing two-dimensional (2D) systems, both in their classical structure of quarter plane causal filters and in the modern behavioral setting, introduced by J.C.Willems. In fact, (Laurent) polynomial and rational matrices in two variables often constitute the only tool available for formulating and solving 2D problems, as state space methods, based on the geometric theory of finite dimensional spaces, do not extend to multidimensional systems.

The aim of this paper is to give some flavour of the kind of questions which have already been answered, and to indicate some directions for future investigations. We shall assume the reader to be acquainted with the material on 2D polynomial matrices presented in (Morf *et al.*, 1977) and to have an elementary knowledge of basic 2D system theory, as presented, for instance, in (Kaczorek, 1985).

Before proceeding, we introduce some notation. Throughout the paper we will denote by $\mathbb{R}[z_1, z_2]$, $\mathbb{R}[z_1, z_2, z_1^{-1}, z_2^{-1}]$ and $\mathbb{R}(z_1, z_2)$ the rings of polynomials and Laurent polynomials (*L-polynomials*) and the field of rational functions, respectively, in the indeterminates z_1 and z_2, with coefficients in $\mathbb{R}$. For any two-dimensional sequence $\mathbf{w} = \{\mathbf{w}(h,k)\}_{h,k\in\mathbb{Z}} \in (\mathbb{R}^q)^{\mathbb{Z}\times\mathbb{Z}}$, the *support* of $\mathbf{w}$ is the set of points $\text{supp}(\mathbf{w}) := \{(h,k) \in \mathbb{Z}\times\mathbb{Z} : \mathbf{w}(h,k) \neq 0\}$. As it is customary, any two-dimensional sequence $\mathbf{w}$ will be identified with the corresponding formal power series $\sum_{h,k\in\mathbb{Z}} \mathbf{w}(h,k)\, z_1^h z_2^k$, thus exploiting the bijective correspondence between sequences taking values in $\mathbb{R}^q$ and formal power series with coefficients in $\mathbb{R}^q$. For sake of brevity, the sequence space $(\mathbb{R}^q)^{\mathbb{Z}\times\mathbb{Z}}$ will be denoted by $\mathcal{R}^q_\infty$.

2. 2D SYSTEMS MODELLING

The simplest class of 2D systems are those in *input/output (i/o) form*, which represent the natural generalization of 1D input/output linear systems. Partitioning the system variables into "inputs" and "outputs", $\mathbf{u}$ and $\mathbf{y}$ respectively, aims to distinguish between "causes" and "effects" within the phaenomenon to describe. Usually, inputs are assumed m-dimensional and completely free, while outputs are p-dimensional and uniquely

Under the linearity, shift-invariance and finite-dimensionality assumptions, i/o maps are expressed by means of a $p \times m$ rational *transfer matrix* $W(z_1, z_2) = \sum_{(i,j)\in S} W_{i,j} z_1^i z_2^j$ as follows

$$Y(z_1, z_2) = W(z_1, z_2)U(z_1, z_2), \tag{1}$$

where $U(z_1, z_2)$ and $Y(z_1, z_2)$ are the formal power series corresponding to the input and output sequences $\mathbf{u}$ and $\mathbf{y}$. The set S is a suitable, generally infinite, subset of $\mathbb{Z}\times\mathbb{Z}$. A typical assumption on this class of models is "quarter-plane causality", which amounts to saying that S is included in the positive orthant $\{(h,k) \in \mathbb{Z}\times\mathbb{Z} : h \geq 0, k \geq 0\}$ or, equivalently, that in every left coprime matrix fraction description, $D^{-1}(z_1, z_2)N(z_1, z_2)$, of $W(z_1, z_2)$ we have $\det D(0,0) \neq 0$. Rational matrices endowed with this property are called *proper*.

An alternative approach to 2D systems is in terms of state-space description. By a *2D state-model* we mean a quarter plane causal 2D system described by the following equations (Fornasini and Marchesini, 1978):

$$\begin{aligned} \mathbf{x}(h+1, k+1) &= A_1\mathbf{x}(h, k+1) + A_2\mathbf{x}(h+1, k) \\ &+ B_1\mathbf{u}(h, k+1) + B_2\mathbf{u}(h+1, k) \\ \mathbf{y}(h,k) &= C\mathbf{x}(h,k) + D\mathbf{u}(h,k), \end{aligned}$$

where $\mathbf{u}(\cdot,\cdot)$, $\mathbf{x}(\cdot,\cdot)$ and $\mathbf{y}(\cdot,\cdot)$ are the input, state and output sequences, taking values in $\mathbb{R}^m$, $\mathbb{R}^n$ and $\mathbb{R}^p$ respectively, while A_1, A_2, B_1, B_2, C and D are real matrices of suitable dimensions. Initial conditions are assigned by specifying the state values $\mathbf{x}(h,k)$ on the *separation set* $\mathcal{S}_0 := \{(h,k) \in \mathbb{Z}\times\mathbb{Z} : h+k = 0\}$. So, when an input sequence $\mathbf{u}(\cdot,\cdot)$ has been assigned over the half-plane $\mathcal{H}_0^+ := \{(h,k) \in \mathbb{Z}\times\mathbb{Z} : h+k \geq 0\}$, the state and output sequences can be computed at every point of $\mathcal{H}_0^+$.

Upon representing input, state and output sequences by means of formal power series, model (2) is equivalently described as follows

$$X(z_1, z_2) = (I - A_1z_1 - A_2z_2)^{-1} \sum_i \mathbf{x}(i, -i) z_1^i z_2^{-i}$$

$$+(I - A_1z_1 - A_2z_2)^{-1}(B_1z_1 + B_2z_2)U(z_1, z_2)$$

$$Y(z_1, z_2) = CX(z_1, z_2) + DU(z_1, z_2),$$

and, if we assume zero initial conditions on $\mathcal{S}_0$, the associated i/o map is given by $Y(z_1, z_2) = W(z_1, z_2)U(z_1, z_2)$, where $W(z_1, z_2) := C(I - A_1z_1 - A_2z_2)^{-1}(B_1z_1 + B_2z_2) + D$ represents the (quarter plane causal) *transfer matrix* of the 2D state model.

Clearly, every 2D state model determines a unique transfer matrix, and hence a unique i/o map, while the inverse problem of "realizing" a given 2D proper rational matrix by means of a state model admits infinitely many solutions (Fornasini and Marchesini, 1976).

Quite recently, the behavior approach to the description of dynamical systems, introduced by J.C.Willems (Willems, 1988) has been extended to the multidimensional case (Rocha, 1990; Fornasini and Valcher, 1994). One of its main features is that it focuses its interest on the set of system trajectories, the *behavior*, while making no distinction between inputs and outputs when describing how the system interacts with its environment. Input/output and state space descriptions have to be deduced only later, from the mathematical equations, provided that the analysis of the behavior has enlightened some cause/effect structure.

In the 2D context a *dynamical system* Σ is defined as a triple $\Sigma = (\mathbb{Z}\times\mathbb{Z}, \mathbb{R}^q, \mathcal{B})$, with $\mathbb{Z}\times\mathbb{Z}$ as *independent variables set*, $\mathbb{R}^q$ as the set where the system trajectories take values (the *signal alphabet*) and $\mathcal{B} \subseteq \mathcal{R}_\infty^q$ as the set of admissible trajectories (the *behavior*). In order to make it possible a comparison with the previous models, we must introduce the linearity and shift-invariance hypotheses also in the behavior descriptions. A further requirement on $\mathcal{B}$ is *completeness* (Willems, 1988; Rocha, 1990), which is the possibility of checking whether a sequence $\mathbf{w}$ belongs to the behavior by simply analysing its restrictions to the finite subsets of $\mathbb{Z}\times\mathbb{Z}$. In mathematical terms, the completeness of a behavior $\mathcal{B}$ corresponds to the existence of an L-polynomial matrix H^T such that

$$\mathcal{B} = \ker H^T := \{\mathbf{w} \in \mathcal{R}_\infty^q : H^T\mathbf{w} = 0\}. \tag{2}$$

The most significant property a behavior can be endowed with is undoubtely zero-controllability (Rocha, 1990; Fornasini and Valcher, 1994) which expresses the possibility of "embedding" any portion of a behavior trajectory into a new trajectory whose support slightly exceeds that of the available portion. More precisely, we say that a behavior $\mathcal{B}$ is *zero-controllable* if there exists a positive integer δ such that, for every finite set $\mathcal{T} \subset \mathbb{Z}\times\mathbb{Z}$ and every $\mathbf{w} \in \mathcal{B}$, there is a trajectory $\tilde{\mathbf{w}} \in \mathcal{B}$, which coincides with $\mathbf{w}$ in $\mathcal{T}$ and has support included in $\mathcal{T}^\delta := \{(i,j) \in \mathbb{Z}\times\mathbb{Z} : d((i,j),\mathcal{T}) < \delta\}$, where the distance $d((i,j),\mathcal{T})$ is defined as min $\{|i-h| + |j-k|; (h,k) \in \mathcal{T}\}$. Controllability induces a very peculiar polynomial matrix description for a behavior $\mathcal{B}$. 2D controllable behaviors, indeed, are kernels of left factor prime L-polynomial matrices or, equivalently, image spaces of suitable L-polynomial operators, i.e.

$$\mathcal{B} = \mathrm{Im}G := \{\mathbf{w} \in \mathcal{R}_\infty^q : \mathbf{w} = G\mathbf{u}, \mathbf{u} \in \mathcal{R}_\infty^m\}. \tag{3}$$

for some $G \in \mathbb{R}[z_1, z_2, z_1^{-1}, z_2^{-1}]^{q\times m}$.

A property which is somehow opposite to controllability is autonomy. A behavior $\mathcal{B}$ is *autonomous* if there exists a solid cone $\mathcal{C}$ in $\mathbb{R} \times \mathbb{R}$ such that the restriction of any behavior trajectory $\mathbf{w}$ to $\mathcal{C} \cap (\mathbb{Z} \times \mathbb{Z})$ allows to uniquely retrieve the remaining portion of $\mathbf{w}$. Autonomous behaviors are kernels of full column rank L-polynomial matrices (Rocha, 1990).

Controllable and autonomous behaviors constitute the building blocks for constructing other behaviors, since every complete behavior $\mathcal{B}$ can be expressed (Fornasini *et al.*, 1993) as the sum $\mathcal{B} = \mathcal{B}_c + \mathcal{B}_a$ of its "controllable part" $\mathcal{B}_c$, which is the maximal controllable behavior included in $\mathcal{B}$, and of a suitable autonomous behavior $\mathcal{B}_a$.

3. I/O AND STATE MODELS: STRUCTURAL PROPERTIES AND CONTROL PROBLEMS

When dealing with quarter-plane causal 2D i/o models, a fundamental issue is undoubtely represented by BIBO (bounded input/bounded output) stability. An i/o model is *BIBO stable* if it produces bounded output sequences when stimulated by bounded input sequences with support in $\mathcal{H}_0^+$ or, equivalently, if the coefficients $W_{i,j}$ of its transfer matrix expansion $W(z_1, z_2) = \sum_{i,j\in\mathbb{N}} W_{i,j} z_1^i z_2^j$ constitute an ℓ_1-sequence.

It has been shown (Anderson and Jury, 1973) that a sufficient condition for BIBO stability is that $W(z_1, z_2)$ is devoid of singularities within the closed unit polydisk

$$\bar{\mathcal{P}}_1 := \{(z_1, z_2) \in \mathbb{C} \times \mathbb{C} : |z_1| \leq 1, |z_2| \leq 1\}.$$

On the other hand, when $W(z_1, z_2)$ has singularities in $\bar{\mathcal{P}}_1 \setminus \mathcal{T}_1$, where

$$\mathcal{T}_1 := \{(z_1, z_2) \in \mathbb{C} \times \mathbb{C} : |z_1| = 1, |z_2| = 1\}$$

is the distinguished boundary of $\bar{\mathcal{P}}_1$, the i/o map is not BIBO stable. When $W(z_1, z_2)$ is regular in $\bar{\mathcal{P}}_1 \setminus \mathcal{T}_1$ but exhibits (nonessential) singularities of the second kind in $\mathcal{T}_1$, a general result about BIBO stability is not available, as there are both examples of stable (Goodman, 1977) and of unstable maps with these features. In any case, as transfer matrices with second kind singularities on $\mathcal{T}_1$ cannot be realized via internally stable state models, it is often convenient to strengthen the stability definition and consider BIBO stable only rational matrices devoid of singularities in $\bar{\mathcal{P}}_1$.

Under this assumption, the feedback stabilization problem (see (Guiver and Bose, 1985) for a complete survey on this topic) has been stated in the following terms: given a "plant", described by some strictly proper rational transfer matrix $W(z_1, z_2) \in \mathbb{R}(z_1, z_2)^{p\times m}$, find a (proper) rational matrix $C(z_1, z_2) \in \mathbb{R}(z_1, z_2)^{m\times p}$ such that the resulting connected system

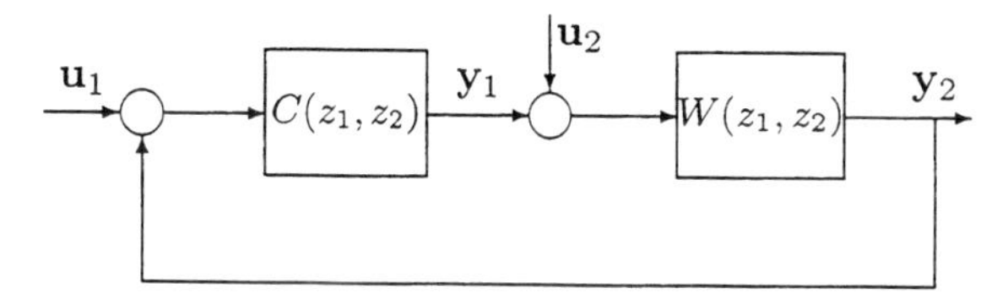

has an $(m+p) \times (m+p)$ proper stable rational transfer matrix.

If $D(z_1, z_2)^{-1} N(z_1, z_2)$ is a left coprime MFD of the plant transfer matrix $W(z_1, z_2)$, the plant is stabilizable by means of a proper rational compensator if and only if the variety of the maximal order minors of $[D(z_1, z_2) \mid N(z_1, z_2)]$ does not intersect $\bar{\mathcal{P}}_1$.

Structural properties of 2D state models constitute a much wider field of research. Actually, as stability is defined as a property of the state variables, feedback stabilization depends both on the way inputs affect the state of a system and on the possibility of reconstructing the state from the external variables. Therefore, a preliminary investigation of properties like local controllability, stabilizability, causal reconstructability and detectability is needed for dealing with the problem of designing dead-beat or stabilizing controllers.

A 2D state model is *internally stable* if the free state evolution, corresponding to every set of bounded initial conditions on $\mathcal{S}_0$, asymptotically goes to zero, i.e. $\mathbf{x}(h, k) \to 0$ as $h + k \to +\infty$. As proved in (Fornasini and Marchesini, 1978), a 2D state model is internally stable if and only if the variety of its characteristic polynomial $\det(I - A_1 z_1 - A_2 z_2)$ does not intersect $\bar{\mathcal{P}}_1$. It is clear that the internal stability of a state model ensures the BIBO stability of the associated i/o map, while the converse is not true, not even when dealing with minimal realizations.

A 2D state model is *stabilizable* if for any set of bounded initial conditions on $\mathcal{S}_0$, there exist real numbers $R > 0$ and $\rho > 1$ and an input sequence $\mathbf{u}(\cdot, \cdot)$, with support in $\mathcal{H}_0^+$, satisfying $\| \mathbf{u}(h, k) \| < \rho^{-(h+k)}$, $\forall h, k \in \mathcal{H}_0^+$, such that the corresponding state evolution satisfies

$$\| \mathbf{x}(h, k) \| < R/\rho^{h+k}, \qquad h, k \in \mathbb{N}.$$

As shown in (Bisiacco, 1985), stabilizability of 2D state models admits a polynomial matrix characterization which represents a natural extension of the well-known PBH test for 1D systems. Actually, a 2D system is stabilizable if and only if

$$\operatorname{rank}[\, I - A_1 z_1 - A_2 z_2 \mid B_1 z_1 + B_2 z_2 \,] = n, \quad (4)$$

for every $(z_1, z_2) \in \bar{\mathcal{P}}_1$. Also *detectability*, i.e. the possibility of constructing an asymptotic observer for the 2D state model, can be characterized by means of a suitable PBH matrix. In fact, a 2D system is detectable if and only if (Bisiacco, 1986)

$$\text{rank}\begin{bmatrix} I - A_1 z_1 - A_2 z_2 \\ C \end{bmatrix} = n, \tag{5}$$

for every $(z_1, z_2) \in \bar{\mathcal{P}}_1$.

The controller design problem in the context of 2D state-models is stated as an output feedback stabilization problem. More precisely, we look for a 2D state model, connected to the original plant as in the following picture

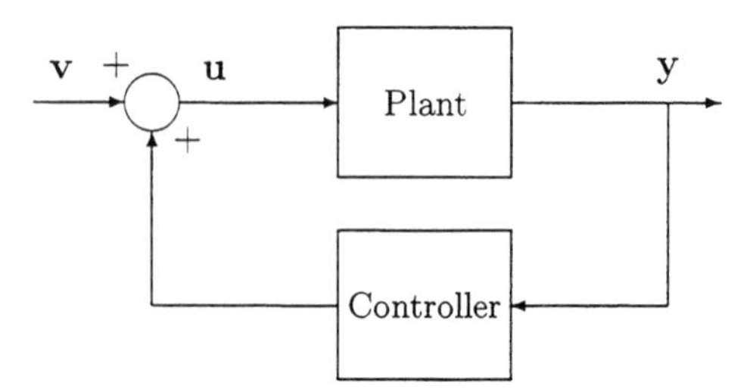

and making the overall system internally stable. A stabilizing controller exists if and only if the plant is both stabilizable and detectable, and, interestingly enough, these two conditions are the same ones guaranteeing the existence of a stabilizing regulator for a 1D system.

Local controllability and causal reconstructability appear as stronger versions of stabilizability and detectability (Bisiacco, 1985; Bisiacco, 1986). Indeed, a 2D state model is *locally controllable* if for any set of initial conditions on $\mathcal{S}_0$, there exist an input sequence $\mathbf{u}$, with support in $\mathcal{H}_0^+$, and a positive integer N, such that the state evolution satisfies $\mathbf{x}(h, k) = 0$, for $h + k > N$, while *causal reconstructability* expresses the possibility of constructing a dead-beat observer.

Also these properties admit polynomial matrix characterizations, as they correspond to the cases when matrices (4) and (5) have full rank for every $(z_1, z_2) \in \mathbb{C} \times \mathbb{C}$ (i.e. are zero prime polynomial matrices). Moreover, local controllability and causal reconstructability are necessary and sufficient for the existence of dead-beat controllers.

4. SOME RESULTS ON 2D BEHAVIORS STABILITY

The relevance of L-polynomial matrices in investigating the structural properties of 2D behaviors was pointed out by several authors (Rocha, 1990; Fornasini and Valcher, 1994). As we have seen, important features, like completeness, autonomy and controllability, have a polynomial matrix characterization, and the same holds true for observability and extendability. Instead of presenting an overview of the available results, we prefer, however, to address further aspects of 2D behavior theory which are still unexplored. So, in this section we make a first attempt to introduce the notions of stability and stabilizability for 2D behaviors, and to relate them to the algebraic features of the L-polynomial matrices involved in their description.

Introducing the stability issue in the 2D context requires some preliminary assumptions. Indeed, stability of ordinary 1D systems naturally refers to the common interpretation of the independent variable t as time coordinate, and consequently to the trajectory evolution as t goes to $+\infty$. For 2D systems, instead, there is no natural ordering in the discrete grid $\mathbb{Z} \times \mathbb{Z}$, and hence no obvious "future direction" can be singled out in the system evolution. The quarter plane causality assumption for 2D i/o maps and state models leads to a stability notion that refers to the evolution of the output and state trajectories, respectively, on the separation sets $\mathcal{S}_t := \{(h, k) \in \mathbb{Z} \times \mathbb{Z} : h + k = t\}$ as t goes to $+\infty$.

The stability definition we are going to introduce is somehow tailored for analysing the dynamics of behavior trajectories on the separation sets $\mathcal{S}_t$ for increasing values of t. Although many alternative definitions could be given, this choice has the advantage of allowing for comparisons with the previous models, considered as special cases of 2D dynamical systems. Intuitively speaking, a behavior $\mathcal{B}$ should be called stable when, for every trajectory $\mathbf{w} \in \mathcal{B}$, we have $\mathbf{w}(h, k) \to 0$ as $h + k \to +\infty$. This rules out all behaviors with a nontrivial controllable part $\mathcal{B}_c$. In fact, as $\mathcal{B}_c$ includes a finite support trajectory $\mathbf{w}_f \neq 0$, there exists a pair of positive integers (ℓ, m) such that $\sum_{i \in \mathbb{N}} z_1^{i\ell} z_2^{im} \mathbf{w}_f$ is an infinite support trajectory which does not estinguish asymptotically. Consequently, stability notion concerns only autonomous behaviors and among them, due to the assumptions on the system evolution, only those for which the cone $\mathcal{C}$, where the trajectories can be uniquely recognized, is included in $\mathcal{H}_0^- := \{(h, k) \in \mathbb{Z} \times \mathbb{Z} : h + k < 0\}$. This implies that we can always reconstruct a trajectory from its restriction to $\mathcal{H}_0^-$ and this is possible, in particular, when the behavior $\mathcal{B}$ is the kernel of a polynomial matrix $H^T \in \mathbb{R}[z_1, z_2]^{m \times q}$ with $H^T(0, 0)$ full column rank. For sake of simplicity, in this contribution we will afford only this special case.

Finally, in analogy with the definition of internal stability for state space models, it seems convenient to require that the asymptotic convergence of the local states is uniform w.r. to the separation sets. As a consequence, we will consider only behavior trajectories which are bounded on a suitable "strip" $\cup_{i=-M}^{-1} \mathcal{S}_i$.

Definition Let $\Sigma = (\mathbb{Z} \times \mathbb{Z}, \mathbb{R}^q, \mathcal{B})$ be a system, endowed with a complete autonomous behavior $\mathcal{B} = \ker H^T$, H^T an $m \times q$ polynomial matrix, with homogeneous degree M and $H^T(0,0)$ of rank q. We say that $\mathcal{B}$ is *stable* if for every sequence $\mathbf{w} \in \mathcal{B}$, which is bounded on the "strip" $\cup_{i=-M}^{-1} \mathcal{S}_i$, we have $\mathbf{w}(h,k) \to 0$ as $h + k \to +\infty$.

As in the case of state space models, stability of an autonomous behavior $\mathcal{B}$ is related to the intersections of the closed unit polydisk $\bar{\mathcal{P}}_1$ with a suitable algebraic variety.

Proposition *Let $\Sigma = (\mathbb{Z} \times \mathbb{Z}, \mathbb{R}^q, \mathcal{B})$ be a system, endowed with a complete autonomous behavior $\mathcal{B} = \ker H^T$, H^T an $m \times q$ polynomial matrix, with homogeneous degree M and $H^T(0,0)$ of rank q. If the variety $\mathcal{V}(H^T)$ of the maximal order minors of H^T does not intersect the closed unit polydisk $\bar{\mathcal{P}}_1$, then $\mathcal{B}$ is stable.*

PROOF As $\mathcal{V}(H^T) \cap \bar{\mathcal{P}}_1 = \emptyset$, there exists also $\rho > 1$ such that $\mathcal{V}(H^T) \cap \bar{\mathcal{P}}_\rho = \emptyset$, with $\bar{\mathcal{P}}_\rho := \{(z_1, z_2) \in \mathbb{C} \times \mathbb{C} : |z_1| \le \rho, |z_2| \le \rho\}$. Thus, the ideal generated by the maximal order minors $m_i(H^T)$ of H^T includes a polynomial $p_\rho = \sum_i c_i m_i(H^T)$ whose variety $\mathcal{V}(p_\rho)$ does not intersect $\bar{\mathcal{P}}_\rho$. If S_i denotes the selection matrix corresponding to $m_i(H^T)$, then $m_i(H^T)I = \text{adj}(S_i H^T) S_i H^T$, and consequently

$$L := \sum_i c_i \ \text{adj}(S_i H^T) S_i$$

is a polynomial matrix satisfying $LH^T = p_\rho I$. Consider, now, a trajectory $\mathbf{w} \in \mathcal{B}$, which is bounded in $\cup_{i=-M}^{-1} \mathcal{S}_i$, and denote by $\mathbf{w}_+$ and $\mathbf{w}_-$ its restrictions to $\mathcal{H}_0^+$ and $\mathcal{H}_0^-$, respectively. Then $H^T(\mathbf{w}_+ + \mathbf{w}_-) = 0$ implies that the sequence $\mathbf{s} := H^T \mathbf{w}_+ = -H^T \mathbf{w}_-$, and hence also $\tilde{\mathbf{s}} := L\mathbf{s}$, are bounded sequences with support included in some strip. Let $\cup_{i=0}^{N-1} \mathcal{S}_i$, $N \ge M$, be a strip including supp $(\tilde{\mathbf{s}})$ and set $S := \sup\{\| \tilde{\mathbf{s}}(i,j) \|, 0 \le i + j \le N - 1\}$. As $1/p_\rho$ admits a power series expansion $\sum_{i,j \in \mathbb{N}} q_{i,j} z_1^i z_2^j$, which is absolutely convergent in $\bar{\mathcal{P}}_\rho$, there exists a positive real K such that

$$\sum_{t \le i+j \le t+N-1} |q_{ij}| < K/\rho^t, \quad \forall t \in \mathbb{N}.$$

As a consequence, from $\tilde{\mathbf{s}} = LH^T \mathbf{w}_+ = p_\rho \mathbf{w}_+$, it follows that the sequence $\mathbf{w}_+ = \tilde{\mathbf{s}}/p_\rho$ satisfies

$$\| \mathbf{w}(h,k) \| \le \frac{SK\rho^{N-1}}{\rho^{h+k}}, \qquad h + k \ge 0,$$

which proves stability. ■

Consider, now, a complete nonautonomous behavior $\mathcal{B}$ and a representation of $\mathcal{B}$ as the sum of its controllable part and an autonomous behavior $\mathcal{B}_a$, i.e. $\mathcal{B} = \mathcal{B}_c + \mathcal{B}_a$. Then, every trajectory $\mathbf{w}$ of $\mathcal{B}$ decomposes into a trajectory $\mathbf{w}_c \in \mathcal{B}_c$ and a trajectory $\mathbf{w}_a \in \mathcal{B}_a$. Replace now $\mathbf{w}_c$ with a trajectory $\mathbf{w}'_c \in \mathcal{B}_c$ which coincides with $\mathbf{w}_c$ in $\mathcal{H}_0^-$ and is zero for all (h,k) with $h + k \ge \delta$, δ a suitable integer. If we assume, as before, that every trajectory in the autonomous part is uniquely determined by its restriction to the halfplane $\mathcal{H}_0^-$, it is immediate to realize that the possibility of asymptotically driving to zero a behavior trajectory only depends on the autonomous part $\mathcal{B}_a$. These arguments lead to the following definition of stabilizability.

Definition Let $\Sigma = (\mathbb{Z} \times \mathbb{Z}, \mathbb{R}^q, \mathcal{B})$ be a system, endowed with a complete behavior $\mathcal{B} = \ker H^T$, H^T an $m \times q$ polynomial matrix of rank q, with homogeneous degree M, and let $\mathcal{B}_c$ be the controllable part of $\mathcal{B}$. We say that $\mathcal{B}$ is *stabilizable* if there exists a decomposition $\mathcal{B} = \mathcal{B}_c + \mathcal{B}_a$, such that for every sequence $\mathbf{w} \in \mathcal{B}$, which can be expressed as $\mathbf{w} = \mathbf{w}_c + \mathbf{w}_a$, with $\mathbf{w}_c \in \mathcal{B}_c$ and $\mathbf{w}_a \in \mathcal{B}_a$ bounded on the "strip" $\cup_{i=-M}^{-1} \mathcal{S}_i$, there exists $\tilde{\mathbf{w}} \in \mathcal{B}$, which coincides with $\mathbf{w}$ on the halfplane $\mathcal{H}_0^-$ and satisfies $\tilde{\mathbf{w}}(h,k) \to 0$ as $h + k \to +\infty$.

Proposition *Let $\Sigma = (\mathbb{Z} \times \mathbb{Z}, \mathbb{R}^q, \mathcal{B})$ be a system, endowed with a complete behavior $\mathcal{B} = \ker H^T$, H^T an $m \times q$ (strictly) polynomial matrix of rank r, with homogeneous degree M, and assume that H^T factors as $H^T = F\bar{H}^T$, with F full column rank and $\bar{H}^T$ left factor prime polynomial matrices. If the variety $\mathcal{V}(H^T)$ and the variety of some maximal order minor of $\bar{H}^T$, $m_i(\bar{H}^T)$, do not intersect the closed unit polydisk $\bar{\mathcal{P}}_1$, then $\mathcal{B}$ is stabilizable.*

PROOF Let $\bar{H}_1^T$ be a maximal order submatrix of $\bar{H}^T$ such that $\mathcal{V}(\det \bar{H}_1^T) \cap \bar{\mathcal{P}}_1 = \emptyset$, and assume, for instance, that it consists of the first columns of $\bar{H}^T$. Then, it is well-known (Fornasini *et al.*, 1993) that $\mathcal{B}$ can be expressed as $\mathcal{B} = \mathcal{B}_c + \mathcal{B}_a$, with $\mathcal{B}_c = \ker \bar{H}^T$ and

$$\mathcal{B}_a = \left\{ \begin{bmatrix} \mathbf{w}_1 \\ 0 \end{bmatrix} : F\bar{H}_1^T \mathbf{w}_1 = 0 \right\}.$$

Clearly, as $\mathcal{B}_a$ is a stable autonomous behavior, $\mathcal{B}$ is stabilizable. ■

5. OPEN PROBLEMS AND CONCLUSIONS

In this paper we have pointed out the relevance of the polynomial matrix approach in the study of 2D systems. To this end we have surveyed the three main models adopted for 2D systems description, and we have shown that, as far as i/o and state models are concerned, stability notion and other features, which arise in the stabilizing controller design problem, are completely captured by the algebraic properties of suitable polynomial matrices.

In the 2D behavioral setting several issues are still unexplored: in the previous section we have introduced the notions of stability and stabilizability, and we have made a first attempt to relate them to the polynomial matrices involved in the kernel descriptions of the behaviors.

The next goal is that of designing a stabilizing controller. This problem can be afforded, as shown by J.C.Willems (Willems, 1996) in the 1D case, as an interconnection problem. More precisely, if $\Sigma = (\mathbb{Z} \times \mathbb{Z}, \mathbb{R}^q, \mathcal{B})$ is a *plant* and $\Sigma_c = (\mathbb{Z} \times \mathbb{Z}, \mathbb{R}^q, \mathcal{B}_c)$ a *controller*, we define the controlled system as the system

$$\Sigma \wedge \Sigma_c := (\mathbb{Z} \times \mathbb{Z}, \mathbb{R}^q, \mathcal{B} \cap \mathcal{B}_c), \tag{6}$$

obtained by interconnecting Σ and Σ_c, namely the system whose trajectories obey to the laws of both systems simultaneously. The controlled system is described in the following picture, and it is important to notice that plant and controller play equivalent roles in the resulting connected system.

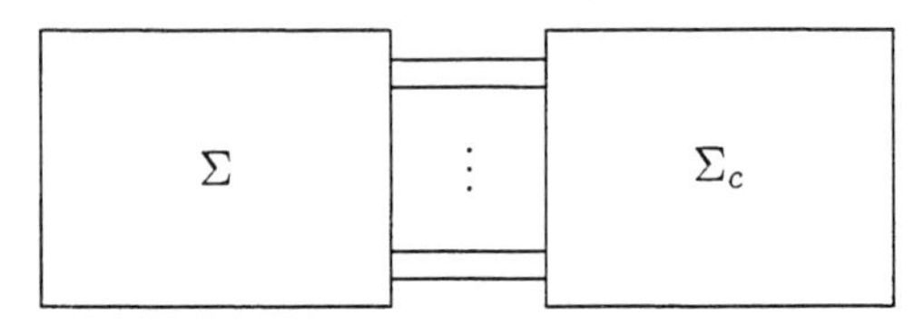

Although the above scheme may appear somehow restrictive, as, in general, plant and controller are connected only through certain terminals, all possible situations are easily reduced to the one just described, by resorting to a suitable redefinition of the system variables of plant and controller (Willems, 1996). In this setting the stabilizing control problem is naturally stated in the following terms: *given a plant* $\Sigma = (\mathbb{Z} \times \mathbb{Z}, \mathbb{R}^q, \mathcal{B})$ *is it possible to design a controller* $\Sigma_c = (\mathbb{Z} \times \mathbb{Z}, \mathbb{R}^q, \mathcal{B}_c)$ *such that the resulting interconnected system is autonomous and stable?*

For this and other problems connected with stability and stabilizability we refer the interested reader to a forthcoming paper (Fornasini and Valcher, 1997).

REFERENCES

Anderson, B.D.O. and E.I. Jury (1973). Stability test for two-dimensional recursive filters. *IEEE Trans. Audio Electroacoustics* **AU-21**, 366–372.

Bisiacco, M. (1985). State and output feedback stabilizability of 2D systems. *IEEE Trans. Circ. and Sys* **CAS 32**, 1246–1249.

Bisiacco, M. (1986). On the structure of 2D observers. *IEEE Trans. Aut. Contr.* **AC-31**, 676–680.

Fornasini, E. and G. Marchesini (1976). State space realization theory of two-dimensional filters. *IEEE Trans. Aut. Contr.* **AC-21**, 484 -92.

Fornasini, E. and G. Marchesini (1978). Doubly indexed dynamical systems. *Math.Sys. Theory* **12**, 59–72.

Fornasini, E. and M.E. Valcher (1994). Algebraic aspects of 2D convolutional codes. *IEEE Trans.Info.Th.* **IT 33**, 1210–1225.

Fornasini, E. and M.E. Valcher (1997). Stability and stabilizability of 2D behaviors. *in preparation.*

Fornasini, E., P. Rocha and S. Zampieri (1993). State realization of 2D finite dimensional autonomous systems. *SIAM J. Contr. Optimiz.* **31**, 1502–1517.

Goodman, D. (1977). Some stability properties of two-dimensional linear shift-invariant digital filters. *IEEE Trans. Circ. Sys.* **CAS-24**, 201–208.

Guiver, J.P. and N.K. Bose (1985). Causal and weakly causal 2-D filters with applications in stabilization. In: *Multidimensional Systems Theory* (N.K.Bose, Ed.). pp. 52–100. D.Reidel Publ. Co., Dordrecht (NL).

Kaczorek, T. (1985). *Two-dimensional linear systems.* Lecture Notes in Control and INformation Sciences, Springer verlag, Berlin.

Morf, M., B.C. Lévy, S.Y. Kung and T. Kailath (1977). New results in 2D systems theory, part I and II. *Proc. of IEEE* **65, no.6**, 861–872;945–961.

Rocha, P. (1990). Structure and Representation of 2-D Systems. PhD thesis. University of Groningen, The Netherlands.

Rosenbrock, H.H. (1970). *State-space and multivariable theory.* J.Wiley & Sons, New York.

Willems, J.C. (1988). Models for dynamics. *Dynamics reported* **2**, 171–269.

Willems, J.C. (1996). On interconnections, control, and feedback. *to appear in IEEE Trans. Aut. Contr.*

CONTROLLABILITY OF NONLINEAR 2-D SYSTEMS

Klamka Jerzy

Institute of Automation, Technical University, 44-100 Gliwice, Poland

Abstract: In the present paper local constrained controllability problems for nonlinear discrete 2-D system with constant coefficients are formulated and discussed. Using some mapping theorems taken from functional analysis and linear approximation methods sufficient conditions for constrained controllability are derived and proved. The present paper extends in some sense the results given in the previous papers to cover the nonlinear discrete 2-D systems with constrained controls.

Keywords: Two-dimensional systems. Controllability. Nonlinear systems. Discrete-time systems. Linearization.

1. INTRODUCTION

Controllability is one of the fundamental concept in mathematical control theory. Roughly speaking, controllability generally means, that it is possible to steer dynamical system from an arbitrary initial state to an arbitrary final state using the set of admissible controls. In the literature there are many different definitions of controllability which depend on a class of dynamical system (Kaczorek 1985; Kaczorek 1993; Kaczorek 1995; Klamka 1988b; Klamka 1991a; Klamka 1991b).

Up to the present time the problem of controllability in continuous and discrete time linear dynamical systems has been extensively investigated in many papers (see e.g. Klamka 1991b) for the extensive list of publications). However, this is not true for the nonlinear dynamical systems specially with constrained controls. Only a few papers concern constrained controllability problems for continuous or discrete nonlinear or linear dynamical systems (Klamka 1988a; Klamka 1992; Klamka 1994; Klamka 1995). In the paper (Klamka 1988b)] the relationships between local and global controllability for linear 2-D systems with control values in a given neighbourhood of zero are investigated. The paper (Klamka 1992) contains results concerning local controllability of nonlinear 2-D systems without differentiability assumptions. In the paper (Klamka 1992) global controllability of linear 2-D systems with controls taking their values in a given cone is discussed. Finally, paper (Klamka 1995) concerns local controllability of nonlinear continuous-time dynamical systems.

In the present paper local constrained controllability problems for nonlinear discrete 2-D system with constant coefficients are formulated and discussed. Using some mapping theorems taken from functional analysis Graves (1950), (Robinson 1986) and linear approximation methods (Klamka 1992; Klamka 1995) sufficient conditions for constrained controllability are derived and proved. The present paper extends in some sense the results given in papers (Klamka 1992; and Klamka 1995) to cover the nonlinear discrete 2-D systems with constrained controls.

2. PRELIMINARIES

Let us consider general nonlinear discrete 2-D system with constant coefficients described by the following difference equation

$$x(i+1,j+1) = f(x(i,j),x(i+1,j),x(i,j+1),u(i,j)) \qquad (1)$$

where: $(i,j)\in Z^+\times Z^+$, Z^+ is a set of nonnegative integer numbers,
$x(i,j)\in R^n$ is a state vector at the point (i,j),
$u(i,j)\in R^m$ is a control vector at the point (i,j),
$f: R^n \times R^n \times R^n \times R^m \to R^n$, is a given function.

Let $U\subset R^m$ be a given arbitrary set. The sequence of controls $u = \{u(i,j);\ (0,0)\le(i,j),\ u(i,j)\in U\}$ is called an admissible sequence of controls. The set of all such admissible sequences of controls forms so called admissible set of controls. In the sequel we shall also use the following notations: $\Omega^0 \subset R^m$ is a neighbourhood of zero, $U^c \subset R^m$ is a closed convex cone with vertex at zero and $U^{c0} = U^c \cap \Omega^0$.

The boundary conditions for nonlinear difference equation (1) are given by

$$x(i,0) = x_{i0} \in R^n$$

$$x(0,j) = x_{0j} \in R^n \quad \text{for } (i,j)\in Z^+\times Z^+ \qquad (2)$$

where x_{i0} and x_{0j} are known vectors.

For a given boundary conditions (2) and for an arbitrary admissible sequence of controls there exists unique solution of the nonlinear difference equation (1), which may be computed by successive iterations. Instead of the nonlinear 2-D system (1), we shall also consider associated linear discrete 2-D system with constant coefficients described by the following difference equation

$$x(i+1,j+1) = A_0x(i,j)+A_1x(i+1,j)+A_2x(i,j+1)+Bu(i,j) \qquad (3)$$

defined for $(i,j)\ge(0,0)$, where A_0, A_1, A_2, are constant $n\times n$-dimensional matrices and B is $n\times m$-dimensional constant matrix.

For linear 2-D system (2.3) we can define the so called transition matrix $A^{i,j}$ as follows (Kaczorek 1985; Kaczorek 1993)

1. $A^{0,0} = I$ (the identity $n\times n$-dimensional matrix),
2. $A^{i,j} = 0$ (the zero matrix) for $i<0$ or/and $j<0$,
3. $A^{i,j} = A_0A^{i-1,j-1} + A_1A^{i,j-1} + A_2A^{i-1,j}$ for $i,j=0,1,2,\ldots$

Using the transition matrix $A^{i,j}$ we can express the solution x(i,j) of linear 2-D system (3) in a following compact form (Kaczorek 1985; Kaczorek 1993)]

$$\begin{aligned}
x(i,j) &= A^{i-1,j-1}(A_0x(0,0)+Bu(0,0)) + \\
&+\sum_{r=1}^{r=i} A^{i-r,j-1}A_1x(r,0) + \\
&+\sum_{r=1}^{r=i} A^{i-r-1,j-1}(A_0x(r,0)+Bu(r,0) + \\
&+\sum_{s=1}^{s=j} A^{i-1,j-s}A_2x(s,0) + \\
&+\sum_{s=1}^{s=j} A^{i-1,j-s-1}(A_0x(0,s)+Bu(0,s) + \\
&+\sum_{r=1}^{r=i}\left(\sum_{s=1}^{s=j}\left(A^{i-r-1,j-s-1}B\right)\right)u(r,s)
\end{aligned} \qquad (4)$$

For zero boundary conditions (x(i,0) = x(0,j) = 0), the solution x(i,j) to (1) is given by (Kaczorek 1995)

$$x(i,j) = \sum_{r=0}^{r=j-1}\left(\sum_{s=0}^{s=i-1} A^{i-r-1,j-s-1}Bu(r,s)\right) = W_{ij}u_{ij} \qquad (5)$$

where

$$W_{ij}=[A^{i-1,j-1}B \mid A^{i-2,j-1}B \mid \ldots$$
$$\ldots \mid A^{0,j-1}B \mid A^{i-1,j-2}B \mid \ldots \mid A^{1,0}B \mid B]$$

and

$$u_{ij} = [u^T(0,0) \mid u^T(1,0) \mid \ldots \mid u^T(i-1,0) \mid u^T(0,1) \mid \ldots$$
$$\ldots \mid u^T(i-2,j-1) \mid u^T(i-1,j-1)] \in U_{ij} = \underbrace{U\times U\times\ldots\times U}_{(ij-1)-times}$$

Moreover, let the cone $V^c_{ij} \subset R^n$ denotes the image of the cone U^c_{ij} under linear mapping W_{ij}: $R^{(ij-1)m}\to R^n$ and $V^{c*}_{ij} \subset R^n$ denotes the so called polar cone defined as follows

$$V^{c*}_{ij} = \{x^*\in R^n : \langle x^*, x\rangle \le 0,\ \text{for all } x \in V^c_{ij}\} \qquad (6)$$

For linear and nonlinear discrete 2-D systems it is possible to define many different concepts of controllability, analogously as for linear 2-D systems.

In the sequel we shall concentrate on local and global U-controllability in a given rectangle $[(0,0),(p,q)] = \{(i,j) : (0,0) \le(i,j)\le(p,q)\}$.

Definition 2.1. System (1) is said to be globally U-controllable in a given rectangle [(0,0) , (p,q)] if for zero boundary conditions $x_{i0} = 0$, $i=0,1,2,\ldots,p$, $x_{0j} = 0$, $j=0,1,2,\ldots,q$ and every vector $x'\in R^n$, there exists an admissible sequence of controls $\{u(i,j)\in U;\ (0,0)\le(i,j)<(p,q)\}$, such that the corresponding solution of the equation (1) satisfies condition : $x(p,q) = x'$.

Definition 2.2. System (1) is said to be locally U-controllable in a given rectangle [(0,0) ,(p,q)] if for zero boundary conditions $x_{i0} = 0$, i=0,1,2,...,p , $x_{0j} = 0$, j=0,1,2,...,q, there exists neighbourhood of zero $D \subset R^n$, such that for every point $x' \in D$ there exists an admissible sequence of controls

$$\{u(i,j) \in U \; ; \; (0,0) \leq (i,j) < (p,q)\},$$

such that the corresponding solution of the equation (1) satisfies condition

$$x(p,q) = x'.$$

Of course the same definitions are valid for linear discrete 2-D systems (3). For linear 2-D systems various controllability conditions are well known in the literature (see e.g. Kaczorek 1985; Kaczorek 1993; or Klamka 1991b).

Now, we shall recall fundamental criteria for global U^c-controllability, global R^m-controllability and local Ω^0-controllability in a given rectangle [(0,0) , (p,q)] for linear 2-D system (3).

Theorem 2.1. (Klamka 1994). Linear system (3) is globally U^c-controllable in the rectangle [(0,0),(p,q)] if and only if the following two conditions hold

$$\text{rank } W_{pq} = n \tag{7}$$

$$V^{c*}_{pq} = \{0\} \tag{8}$$

From theorem 2.1 directly follows necessary and sufficient condition for global controllability with unconstrained controls stated in the following corollary.

Corollary 2.1. (Klamka 1991b) Linear 2-D system (3) is globally R^m-controllable in the rectangle [(0,0),(p,q)] if and only if

$$\text{rank } W_{pq} = n \tag{9}$$

It is well known (Klamka 1988b), that for sets U containing zero as an interior point, local constrained controllability is equivalent to global unconstrained controllability.

Corollary 2.2. (Klamka 1988b) Linear system (3) is locally Ω^0-controllable in the rectangle [(0,0) ,(p,q)] if and only if it is globally R^m-controllable in the rectangle [(0,0) ,(p,q)].

Corollary 2.1 directly follows from the well known fact, that the range of the linear bounded operator covers whole space if and only if this operator transforms some neighbourhood of zero onto some neighbourhood of zero in the range space (Graves 1950).

3. MAIN RESULTS

In this section we shall formulate and prove sufficient conditions of local U-controllability in a given rectangle [(0,0) , (p,q)] and different sets U for the nonlinear discrete 2-D system (1).

It is generally assumed that:

1. $f(0,0,0,0) = 0$,
2. function f(x,y,z,u) is continuously differentiable with respect to all its arguments in some neighbourhood of the point zero in the product space $R^n \times R^n \times R^n \times R^m$.

Taking into account the assumption 2, let us introduce the following notations for partial derivatives of the function f(x,y,z,u).

$$A_0 = f_x'(0,0,0,0) \;, \quad A_1 = f_y'(0,0,0,0)$$
$$A_2 = f_z'(0,0,0,0) \;, \quad B = f_u'(0,0,0,0) \tag{10}$$

where A_0, A_1, A_2, are n×n-dimensional constant matrices, and B is n×m-dimensional constant matrix.

Therefore, using the standard methods (Klamka 1994), it is possible to construct linear approximation of the nonlinear discrete 2-D system (1). This linear approximation is valid in some neighbourhood of the point zero in the product space $R^n \times R^n \times R^n \times R^m$, and is given by the linear difference equation (3) with matrices A_0, A_1, A_2, B defined above. Proofs of the main results are based on some lemmas taken from functional analysis and concerning so called nonlinear covering operators (Graves 1950; Robinson 1986). Now, for convenience we shall shortly state this results.

Lemma 3.1. (Robinsosn 1986) Let F: X→Y be a nonlinear operator from a Banach space X into a Banach space Y and suppose that F(0) = 0. Assume that Frechet derivative dF(0) maps a closed convex cone $C \subset X$ with vertex at zero onto the whole space Y. Then there exists neighbourhoods $M_0 \subset X$ about $0 \in X$ and $N_0 \subset Y$ about $0 \in Y$ such that the equation y = F(x) has for each $y \in N_0$ at least one solution $x \in M_0 \cap C$.

A direct consequence of lemma 3.1 is the following result concerning nonlinear covering operators.

Lemma 3.2. (Graves 1950) Let F: X→Y be a nonlinear operator from a Banach space X into a Banach space Y which has Frechet derivative dF(0): X→Y, whose image coincides with whole space Y. Then the image of the operator F will contain a neighbourhood of the point $F(0) \in Y$.

Now, we are in the position to formulate and prove the main result on local U-controllability in the rectangle [(0,0) , (p,q)] for the nonlinear discrete 2-D system (1).

Theorem 3.1. (Klamka 1996). Let us suppose, that $U^c \subset R^m$ is a closed convex cone with vertex at zero. Then the nonlinear discrete 2-D system (1) is locally U^{c0}-controllable in the rectangle [(0,0),(p,q)] if its linear approximation near the origin given by the difference equation (3) is globally U^c-controllable in the same rectangle [(0,0) , (p,q)].

Proof. Proof of the theorem 3.1 is based on lemma 3.1. Let our nonlinear operator F transforms the space of admissible control sequence {u(i,j) : (0,0)≤(i,j)≤(p,q)} into the space of solutions at the point (p,q) for the nonlinear 2-D system (1). More precisely, the nonlinear operator

$F:R^m \times R^m \times \ldots \times R^m \rightarrow R^n$ is defined as follows

$$F\{u(0,0), u(0,1), u(1,0), u(1,1),\ldots,u(p,q-1)\} = x(p,q) \tag{11}$$

where x(p,q) is the solution at the point (p,q) of the nonlinear system (1) corresponding to an admissible controls sequence {u(i,j) : (0,0)≤(i,j)<(p,q)}. Frechet derivative at point zero of the operator F denoted as dF(0) is a linear bounded operator defined by the following formula

$$dF(0)\{u(0,0),u(0,1),u(1,0),u(1,1),\ldots,u(p,q-1)\}=x(p,q) \tag{12}$$

where x(p,q) is the solution at the point (p,q) of the linear system (3) corresponding to an admissible controls sequence {u(i,j) : (0,0)≤(i,j)<(p,q)}.

It should be stressed, that the sequences of controls in formulas (11) and (12) contain exactly (pq-1), m-dimensional control vectors localized in the rectangle [(0,0) , (p,q)]. However, it should be pointed out, that the control vectors u(p,j), j=0,1,2,...,q and u(i,q), i=0,1,2,...,p do not affect on the solution x(p,q) (see difference equations (1) and (3)). Since f(0,0,0,0) = 0, then for zero boundary conditions nonlinear operator F transforms zero into zero, i.e. F(0) = 0. If linear system (3) is globally U^c-controllable in the rectangle [(0,0),(p,q)], then the image of Frechet derivative dF(0) covers whole space R^n. Therefore, by the result stated at the beginning of the proof, nonlinear operator F covers some neighbourhood of zero in the space R^n. Hence, by definition 2.2 nonlinear system (2.1) is locally U^c-controllable in the rectangle [(0,0),(p,q)].

Corollary 3.1. Under the assumptions stated in theorem 3.1 the nonlinear system (11) is locally U^{c0}-controllable in the rectangle [(0,0),(p,q)] if the relations (7) and (8) hold.

If the relations (7) and (8) hold, then by theorem 2.1 linear system (1) is globally U^c-controllable in the rectangle [(0,0),(p,q)]. Therefore by theorem 3.1 nonlinear system (1) is locally U^{c0}-controllable in the same rectangle [(0,0),(p,q)].

In the case when set U contains zero as an interior point we have the following sufficient condition for local constrained controllability of nonlinear 2-D systems.

Corollary 3.2. Let $0 \in int(U)$ .Then the nonlinear system (11) is locally U-controllable in the rectangle [(0,0),(p,q)] if its linear approximation near origin given by the difference equation (3) is locally U-controllable in the same rectangle [(0,0),(p,q)].

Let us consider the special case of nonlinear 2-D system, namely system described by the following nonlinear difference equation

$$x(i+1,j+1) = A_0(x(i,j))x(i,j) + A_1(x(i+1,j))x(i+1,j) + A_2(x(i,j+1))x(i,j+1) + B(x(i,j))u(i,j) \tag{13}$$

where $A_0(x(i,j))$, $A_1(x(i+1,j))$, $A_2(x(i,j+1))$, $B(x(i,j))$ are nonlinear differentiable matrices of suitable dimensions.

In this case associated linear discrete 2-D system with constant coefficients is described by the following difference equation

$$x(i+1,j+1) = A'_0x(i,j) + A'_1x(i+1,j) + A'_2x(i,j+1) + B'u(i,j) \tag{14}$$

where

$$A'_0 = A_0(0), \quad A'_1 = A_1(0), \quad A'_2 = A_2(0), \quad B' = B(0)$$

Theorem 3.2. Let us suppose, that $U^c \subset R^m$ is a closed convex cone with vertex at zero. Then the nonlinear discrete 2-D system (13) is locally U^{c0}-controllable in the rectangle [(0,0),(p,q)] if its linear approximation near the origin given by the difference equation (14) is globally U^c-controllable in the same rectangle [(0,0) , (p,q)].

Proof. Left hand side of the nonlinear equation (13) satisfies all the assumptions stated for nonlinear function f(x,y,z,w). Moreover, system (14) is a corresponding linear approximation near the origin of the system (13) Therefore, by theorem 3.1 the result follows.

4. EXAMPLE

Let us consider nonlinear 2-D system described by the following set of two difference equations

$$x_1(i+1,j+1) = x_1(i,j)+x_1^2(i,j)-x_2(i+1,j) + u(i,j) + u^2(i,j)$$

$$x_2(i+1,j+1) = x_1^2(i+1,j) - x_2(i,j+1) + x_2^2(i,j+1) + u(i,j) \quad (15)$$

defined in the rectangle $[(0,0)\ ,\ (2,2)]$. Hence, in our example n=2, m=1, p=q=2. Let us additionally assume that the cone $U^c = \{ u \in R : u \geq 0 \}$, i.e. the admissible scalar controls $u(i,j) \geq 0$ are nonnegative.

Linear approximation near the origin of the nonlinear 2-D system (15) has the form (3) with the following matrices

$$A_0 = \begin{vmatrix} 1 & 0 \\ 0 & 0 \end{vmatrix}, \quad A_1 = \begin{vmatrix} 0 & -1 \\ 0 & 0 \end{vmatrix},$$

$$A_2 = \begin{vmatrix} 0 & 0 \\ 0 & -1 \end{vmatrix}, \quad B = \begin{vmatrix} 1 \\ 1 \end{vmatrix}$$

Therefore, n×pqm = 2×4 -dimensional constant matrix $W_{pq}=W_{22}$ has the following form

$$W_{22} = | \ A^{1,1}B \ \vdots \ A^{0,1}B \ \vdots \ A^{1,0}B \ \vdots \ B \ | =$$

$$| \ (A_0 + A_1A_2 + A_2A_1)B \ \vdots \ A_1B \ \vdots \ A_2B \ \vdots \ B \ | =$$

$$\begin{vmatrix} 2 & -1 & 0 & 1 \\ 0 & 0 & -1 & 1 \end{vmatrix}$$

Hence, rank $W_{22} = 2 = n$ and linear approximation of the form (3) is globally R-controllable in the rectangle $[(0,0)\ ,\ (2,2)]$. Let us observe, that the cone $V^c_{22} \subset R^2$ has the following form

$$V_{22} = \left\{ \begin{array}{l} x = \begin{vmatrix} x_1 \\ x_2 \end{vmatrix} \in R^2 : \\ x = \begin{vmatrix} 2 \\ 0 \end{vmatrix} u(0,0) + \begin{vmatrix} -1 \\ 0 \end{vmatrix} u(1,0) + \\ + \begin{vmatrix} 0 \\ -1 \end{vmatrix} u(0,1) + \begin{vmatrix} 1 \\ 1 \end{vmatrix} u(1,1) \end{array} \right\}$$

Therefore, $V^c_{22} = R^2$ and $V^{c*}_{22} = \{0\}$. Hence, by theorem 2.1 linear approximation of the form (3) is globally U^c-controllable in the rectangle $[(0,0),(2,2)]$. Since $f(0,0,0,0) = 0$ and function f is continuously differentiable near the origin, then by theorem 3.1 the nonlinear 2-D system (15) is locally U^{c0}-controllable in the rectangle $[(0,0),(2,2)]$.

5. CONCLUSIONS

In the present paper only one simple model of nonlinear 2-D systems has been considered. The results presented can be extended in many directions. For example it is possible to formulate sufficient local controllability conditions for other models of nonlinear 2-D systems (Kaczorek 1985; Kaczorek 1993; Klamka 1991b), for nonlinear 2-D systems with variable coefficients (Klamka 1991b) and for M-D nonlinear systems. Moreover, similar controllability results can be derived for very general M-D nonlinear systems with variable coefficients defined in infinite-dimensional linear spaces (Klamka 1988a).

REFERENCES

Graves L. M., (1950). Some mapping theorems, *Duke Mathematical J.*, **17**, 111-114.

Kaczorek T., (1985). *Two-Dimensional Linear Systems*, Springer-Verlag, Berlin

Kaczorek T., (1993). *Linear Control Systems*, Research Studies Press and J. Wiley, New York

Kaczorek T., (1995). U-reachability and U-controllability of 2-D Roesser model, *Bulletin of the Polish Academy of Sciences. Technical Sciences* **43**, 31-37

Klamka J., (1988a). M-dimensional nonstationary linear discrete systems in Banach spaces, *Proc. 12 World IMACS Congress, Paris,* **4**, 31-33

Klamka J., (1988b). Constrained controllability of 2-D linear systems, *Proc. 12 World IMACS Congress, Paris,* **2**, 166-169

Klamka J., (1991a). Complete controllability of singular 2-D system, *Proc. 13 World IMACS Congress, Dublin,* 1839-1840.

Klamka J., (1991b). Controllability of Dynamical Systems, Kluwer Academic Publishers, Dordrecht, The Netherlands.

Klamka J., (1992). Controllability of nonlinear 2-D systems, *Bulletin of the Polish Academy of Sciences. Technical Sciences* **40**, 125-133

Klamka J., (1994). Constrained controllability of discrete 2-D linear systems, *Proc.IMACS Int. Symp. Signal Processing, Robotics and Neural Networks, Lille*, 166-169

Klamka J., (1995). Constrained controllability of nonlinear systems, *IMA J. Mathematical Control and Information*, **12**, 245-252

Klamka J., (1996). Controllability of 2-D nonlinear systems,*Proceedings of Second World Congress on Nonlinear Analysis, Ateny*.

Robinson S. M., (1986).Stability theory for systems of inequalities. Part II: differentiable nonlinear systems, *SIAM J. Numerical Analysis*, **13**, 1261-1275

THE RESULTANT MATRIX FOR A FAMILY OF TWO-VARIABLE POLYNOMIALS

Krzysztof Gałkowski

Institute of Robotics and Software Engineering, Technical University of Zielona Góra, Podgórna Str. 50, 65-246 Zielona Góra, Poland, email:galko@irio.pz.zgora.pl

Abstract: Given a set of polynomials in one or more indeterminate, the new resultant system in the compact matrix form is provided.

Keywords: polynomials, matrix algebra, linear analysis, dynamic systems, control system theory.

1. INTRODUCTION

The interest in a simple criterion for existence of common zeros of any set of polynomials is motivated by its numerous applications in systems and control theory. From the mathematical viewpoint the problem is well known, particularly for two polynomials (see Anderson, Scott 1977, Wimmer 1990). Consider, hence, the two polynomials

$$f_1(x) = a_{11}x^n + a_{21}x^{n-1} + \cdots + a_{n+1,1},$$

$$f_2(x) = a_{12}x^m + a_{22}x^{m-1} + \cdots + a_{m+1,2}. \qquad (1)$$

The resultant of these polynomials is defined as

$$\mathrm{Res}(f_1, f_2) \triangleq \det \begin{bmatrix} a_{11} & a_{21} & \cdots & a_{n+1,1} & & 0 \\ & a_{11} & \cdots & a_{n,1} & a_{n+1,1} & \\ & & \ddots & & & \ddots \\ 0 & & & a_{11} & a_{21} & a_{n+1,1} \\ a_{12} & a_{22} & \cdots & a_{m+1,2} & & 0 \\ & a_{12} & \cdots & a_{m,2} & a_{m+1,2} & \\ & & \ddots & & & \ddots \\ 0 & & & a_{12} & a_{22} \cdots & a_{m+1,2} \end{bmatrix} \qquad (2)$$

Now, the following well known theorem is reminded.

Theorem 1. $\mathrm{Res}(f_1,f_2) = 0$ if and only if either $a_{11} = a_{12} = 0$ or $f_1(x)$ and $f_2(x)$ have a common factor of positive degree in the variable x.

The resultant matrix is closely related to the Bezout matrix and both the matrices evolve from the Euler works in the elimination theory (see Wimmer 1990).

The following extensions are of the great practical relevance. First, the problem for the family of polynomials that consists more than two elements. Now, the satisfaction of the set of conditions, as in the theorem 1, does not guarantee the positive final result. For example, the resultant of the each pair of the following polynomials

$$f_1(x) = (x-1)(x+1),$$
$$f_2(x) = (x+3)(x+1),$$
$$f_3(x) = (x-1)(x+3)$$

is zero, but they do not have any common zero. Such a problem can be solved in the framework of the algebraic geometry but is much more complicated. The following theorem highlights the problem solution.

Theorem 2. (Hodge, Pedoe 1947, Anderson, Scott 1977). Given r polynomials $f_1(x), f_2(x), \cdots, f_r(x)$ in one indeterminate x and with indeterminate coefficients, there exists a set of polynomials $d_1, d_2, \cdots, d_N$ in the coefficients of $f_1(x), f_2(x), \cdots, f_r(x)$ and with the property that for a specialisation of the coefficients the conditions $d_1 = 0, d_2 = 0, \cdots, d_N = 0$ are necessary and sufficient to ensure that either the equations

$$f_1(x) = 0,\ f_2(x) = 0, \cdots, f_r(x) = 0 \qquad (3)$$

are soluble in some extended field of coefficients, or the leading coefficients in $f_1(x), f_2(x), \cdots, f_r(x)$ all vanish.

The polynomials $d_1, d_2, \cdots, d_N$ are called the resultant system of the set of polynomials. The procedure for constructing the d_i (i=1,2,...,r) is outlined in Hodge, Pedoe 1947.

Secondly, this problem is very important for a family of multivariable polynomials, for example, in each system theoretical question for multivariable nD systems. Here, the situation is much more complicated since, as is well known, the multivariable polynomials ring is not Euclidean which yields that two polynomials do not need to have a joint factor to be mutually (zero) prime, see for example Gałkowski 1996. Exactly, the polynomials $x-1$ and $y-1$ do not have a common factor but have the common zero $x = 1,\ y = 1$.

In this paper, the new resultant matrix for a family of polynomials is proposed and it is described how to extend the result for multivariable polynomials. The only one limitation is that the number of equations must be greater than the maximum polynomial degree in this family.

2. THE NEW RESULTANT MATRIX

Consider the set of $N \geq n+1$ polynomials of the nth degree

$$f_i(x) = \sum_{j=1}^{n+1} a_{ji} x^{n+1-j} \quad i = 1,2,\ldots,N, \qquad (4)$$

such that no polynomial is a linear combination of the remaining. Let Γ and Δ denote the sets of lexicographically ordered pairs of the elements belonging to the sets $\Theta \triangleq \{1,2,\ldots,n+1\}$ and $\Xi \triangleq \{1,2,\ldots,N\}$. Hence, let

$$\Gamma \triangleq \{\gamma_k :\ \gamma_k = \{\alpha, \beta\}, \alpha, \beta \in \Theta, \alpha \neq \beta, k = 1,2,\ldots,(1/2)n(n+1)\},$$

$$\Delta \triangleq \{\delta_l :\ \delta_l = \{\varepsilon, \mu\}, \varepsilon, \mu \in \Xi, \varepsilon \neq \mu, l = 1,2,\ldots,(1/2)(N-1)N\} \qquad (5)$$

Finally, define the following matrix

$$\mathbf{R}_{n,N} \triangleq \left[r_{kl}\right]_{\frac{1}{2}n(n+1),\frac{1}{2}(N-1)N} \qquad (6)$$

where

$$r_{kl} \triangleq \det\begin{bmatrix} a_{\varepsilon\alpha} & a_{\mu\alpha} \\ a_{\varepsilon\beta} & a_{\mu\beta} \end{bmatrix}. \qquad (7)$$

The construction of the matrix $\mathbf{R}_{n,N}$ is highlighted at the Figure 1, where the bracketed elements define the determinants (7) in $\mathbf{R}_{n,N}$.

Comment 1. The matrix $\mathbf{R}_{n,N}$, clearly, is the result of multiple performing the Gauss elimination procedure (see, for example, Wimmer, 1990) to each pair of polynomials considered and to each their element $x^k, k = 0,1,\ldots,n+1$.

It turns out that the matrix $\mathbf{R}_{n,N}$ may be used for checking the existence of joint zeros for the polynomial family $\{f_i(x)\}$, i.e. it plays a role of a resultant matrix. Before we formulate the main theorem some problems should be highlighted.
Define, first, the polynomials coefficients matrix, see (4),

$$A \triangleq \left[a_{ji}\right],\ i = 1,2,\ldots,N;\ j = 1,2,\ldots,n+1;\ N > n. \qquad (8)$$

Next, apply to the polynomials of (4) the substitution

$$x \triangleq z + \alpha. \qquad (9)$$

It is a straightforward task to show that such a transformed polynomials family can be described by the coefficients matrix B

$$B = \left[b_{ji}\right],\ i = 1,2,\ldots,N;\ j = 1,2,\ldots,n+1;\ N > n \qquad (10)$$

where, from the Newton binomial,

$$b_{1i} = a_{1i},\ b_{ji} = a_{ji} + \sum_{k=2}^{j} \binom{n-j+k}{k-1} \alpha^{k-1} a_{j-k+1,i}$$
$$i = 1,2,\ldots,N,\ j = 2,3,\ldots,n+1. \qquad (11)$$

Now, the following lemmas can be stated and proved.

Lemma 1. The rank of the matrix $\mathbf{R}_{n,N}$ is invariant under the polynomial variable change of (9).

Proof. It results immediately from the definition of the matrix $\mathbf{R}_{n,N}$ and the dependence of (11) that each column of the transformed resultant matrix $\mathbf{S}_{n,N}$ is a linear combination of columns of $\mathbf{R}_{n,N}$.

Lemma 2. If the linearly independent polynomial family of (4) possesses the common root $x=0$ the resultant matrix $\mathbf{R}_{n,N}$ satisfies

$$\text{rank}\mathbf{R}_{n,N} \le \frac{1}{2}n(n-1). \qquad (12)$$

Proof. It is easy to see that the last column of the coefficient matrix A, see (8), is zero if, and only if, $x=0$ is a common root. This yields, in the sequel, that n corresponding columns of $\mathbf{R}_{n,N}$ are zero. Hence, the dependence of (12) is true since

$$\frac{1}{2}n(n+1)-n=\frac{1}{2}n(n-1).$$

Lemma 3. Let the set of linearly independent polynomials of (4) have m common zeros, $m=0,1,\ldots,n-1$, then

$$\begin{aligned}&\frac{1}{2}(n-m)(n-m-1) < \text{rank } \mathbf{R}_{n,N}\\ &\le \frac{1}{2}(n+1-m)(n-m).\end{aligned} \qquad (13)$$

Proof Suppose, the polynomial family possesses one common root x_1. Perform, hence, the variable substitution

$$x := x - x_1.$$

Thus, by Lemmas 1 and 2, the resultant matrix satisfies the dependence of (12). In what follows, suppose that there exists another common root x_2. Hence, get rid of zero columns of both the transformed matrices A (one column) and $\mathbf{R}_{n,N}$ (n columns), and perform the new substitution

$$x := x - x_2.$$

Then, again by Lemmas 1 and 2, the resultant matrix $\mathbf{R}_{n,N}$ satisfies

$$\begin{aligned}\text{rank}\mathbf{R}_{n,N} &\le \frac{1}{2}(n+1)n-n-(n-1)\\ &=\frac{1}{2}(n-1)n-(n-1)=\frac{1}{2}(n-1)(n-2).\end{aligned}$$

Repeating such a procedure several times yields that

$$\text{rank } \mathbf{R}_{n,N} \le \frac{1}{2}(n+1-m)(n-m).$$

If, there are no more common roots then the procedure described above cannot be applied again and, hence,

$$\text{rank } \mathbf{R}_{n,N} > \frac{1}{2}(n-m)(n-m-1).$$

This completes the proof.

Theorem 3. The number of common roots for the linearly independent polynomial family of (4) is given by

$$m=\frac{2n+1+\sqrt{8r+1}}{2} \qquad (14)$$

where

$$r \mathrel{\hat{=}} \text{rank}\mathbf{R}_{n,N} \qquad (15)$$

Proof The equation of (14) is a direct consequence of the right side equation of (13). It, however, is necessary to remember that $\phi \mathrel{\hat{=}} \sqrt{8r+1}$ must be natural, when m positive and unique. Hence, it is easy to note that r could not have any positive values. For example

$$r=0 \Rightarrow \phi=\sqrt{1}=1,\quad r=1 \Rightarrow \phi=\sqrt{9}=3,$$

$$r=3 \Rightarrow \phi=\sqrt{25}=5,\quad r=6 \Rightarrow \phi=\sqrt{49}=7,$$

$$r=10 \Rightarrow \phi=\sqrt{81}=9,\quad r=15 \Rightarrow \phi=\sqrt{121}=11$$

and so on. It shows also that the rank of the matrix $\mathbf{R}_{n,N}$ can be equal only to the border values in the inequality of (13).

Now, some additional comments and examples are provided.

First, note that the common zeros are considered in the same way as the different, so that m denotes the global number of joint zeros.

Corollary 1 The resultant matrix $\mathbf{R}_{n,N}$ has a maximal row rank if and only if there is no common zeros.

If polynomials are of various degrees, then it is necessary to consider them as a family of polynomials of the same degree equal to the maximum one.

Return, now, to the assumption that defines the polynomial family of (4), i.e. that no polynomial is a linear combination of the remaining. It is easy, for example, to check that for each pair of the second order polynomials $\{f_1, f_2\}$, the set of $\{f_1, f_2, af_1 + bf_2\}, a,b \in R$ satisfies, rank $\mathbf{R}_{n,1} = 1$, i.e. the theorem 3 does not work. This ban is due to the fact that all former rank considerations refer to the column rank. The possibility that polynomials are linear combinations effects in the matrices row rank and may destroy the dependence of (14). In fact, (14) holds true if the number of linearly independent polynomials exceeds or is equal to n+1.

To avoid superfluous considerations and calculations the simple criterion for linear independence of the polynomial family is needed. It is obvious that the following statement holds true:

Corollary 2. The polynomials family of (4) does not contain any linear combinations of some its members if and only if, see (8),

$$\text{rank } A = N. \quad (16)$$

Before using the developed procedure one ought to check the condition of (16).

Now two examples that highlight these results are provided.

Example 1. Consider five polynomials of the fourth degree with three common roots

$$f_i(x) = (x^3 + ax^2 + bx + c)(x + c_i), \; i\text{=1,2,3,4,5.}$$

The matrix $\mathbf{R}_{45} = [r_{kl}]_{10\times10}$ is given as

$$r_{k1} = c_\alpha - c_\beta, \quad r_{k2} = a(c_\alpha - c_\beta),$$

$$r_{k3} = b(c_\alpha - c_\beta), \quad r_{k4} = c(c_\alpha - c_\beta),$$

$$r_{k5} = (a^2 - b)(c_\alpha - c_\beta),$$

$$r_{k6} = (ab - c)(c_\alpha - c_\beta),$$

$$r_{k7} = ac(c_\alpha - c_\beta),$$

$$r_{k8} = (b^2 - ac)(c_\alpha - c_\beta),$$

$$r_{k9} = bc(c_\alpha - c_\beta), \quad r_{k10} = c^2(c_\alpha - c_\beta)$$

where

$$k = 1 \Rightarrow \alpha = 2, \beta = 1,$$
$$k = 2 \Rightarrow \alpha = 3, \beta = 1,$$
$$k = 3 \Rightarrow \alpha = 4, \beta = 1,$$
$$k = 4 \Rightarrow \alpha = 5, \beta = 1,$$
$$k = 5 \Rightarrow \alpha = 3, \beta = 2,$$
$$k = 6 \Rightarrow \alpha = 4, \beta = 2,$$
$$k = 7 \Rightarrow \alpha = 5, \beta = 2,$$
$$k = 8 \Rightarrow \alpha = 4, \beta = 3,$$
$$k = 9 \Rightarrow \alpha = 5, \beta = 3,$$
$$k = 10 \Rightarrow \alpha = 5, \beta = 4.$$

It is easy to check that rank$\mathbf{R}_{45} = 1$ which confirms the dependence of (8) due to that n=4, m=3.

Example 2. Consider five polynomials of the fourth degree with two common roots

$$f_i(x) = (x^2 + ax + b)(x^2 + c_{1i}x + c_{2i}), \; i\text{=1,2,3,4,5.}$$

The matrix $\mathbf{R}_{45} = [r_{kl}]_{10\times10}$ is given as

$$r_{k1} = c_{1\alpha} - c_{1\beta},$$

$$r_{k2} = a(c_{1\alpha} - c_{1\beta}) + (c_{2\alpha} - c_{2\beta}),$$

$$r_{k3} = b(c_{1\alpha} - c_{1\beta}) + a(c_{2\alpha} - c_{2\beta}),$$

$$r_{k4} = b(c_{2\alpha} - c_{2\beta}),$$

$$r_{k5} = (a^2 - b)(c_{1\alpha} - c_{1\beta}) + a(c_{2\alpha} - c_{2\beta}) + c_{2\alpha}c_{1\beta} - c_{2\beta}c_{1\alpha},$$

$$r_{k6} = ab(c_{1\alpha} - c_{1\beta}) + a^2(c_{2\alpha} - c_{2\beta}) - c_{2\alpha}c_{1\beta} + c_{2\beta}c_{1\alpha},$$

$$r_{k7} = ab(c_{2\alpha} - c_{2\beta}) - c_{2\alpha}c_{1\beta} + c_{2\beta}c_{1\alpha},$$

$$r_{k8} = b^2(c_{1\alpha} - c_{1\beta}) + ab(c_{2\alpha} - c_{2\beta}) + (a^2 - b)(c_{2\alpha}c_{1\beta} - c_{2\beta}c_{1\alpha}),$$

$$r_{k9} = b^2(c_{2\alpha} - c_{2\beta}) + ab(c_{2\alpha}c_{1\beta} - c_{2\beta}c_{1\alpha}),$$

$$r_{k10} = b^2(c_{2\alpha}c_{1\beta} - c_{2\beta}c_{1\alpha})$$

where

$$k = 1 \Rightarrow \alpha = 2, \beta = 1,$$
$$k = 2 \Rightarrow \alpha = 3, \beta = 1,$$
$$k = 3 \Rightarrow \alpha = 4, \beta = 1,$$
$$k = 4 \Rightarrow \alpha = 5, \beta = 1,$$
$$k = 5 \Rightarrow \alpha = 3, \beta = 2,$$
$$k = 6 \Rightarrow \alpha = 4, \beta = 2,$$
$$k = 7 \Rightarrow \alpha = 5, \beta = 2,$$
$$k = 8 \Rightarrow \alpha = 4, \beta = 3,$$

$$k = 9 \Rightarrow \alpha = 5, \beta = 3,$$
$$k = 10 \Rightarrow \alpha = 5, \beta = 4.$$

It is easy to check that $\text{rank}\mathbf{R}_{45} = 3$ which confirms the dependence of (8) due to that n=4, m=2.

3. THE RESULTANT MATRIX FOR A TWO-VARIABLE POLYNOMIALS FAMILY

The aforementioned procedure can be applied to two-variable, and more generally, to multivariable polynomials. Then, one presents the polynomials of the family $\{f_i(x,y)\}$ as, for example, polynomials in x with polynomial coefficients in y. The resultant matrix **R** has now the polynomial entries in y. The rank of such a matrix may be calculated, for example, via the canonical Smith form for the polynomial matrices.

It must be remembered that roots for multivariable polynomials, in general, are not unique. Hence, there can exist sets of possible solutions. This problem is discussed on the following examples.

Example 4. Consider the polynomials:

$$f_1(x,y) = x^2 + (2y+2)x + y^2 + 2y,$$
$$f_2(x,y) = x^2 + (y+4)x + 2y + 4,$$
$$f_3(x,y) = 0x^2 + (y+2)x + y^2 + 4y + 4.$$

The resultant matrix is as follows

$$\mathbf{R} = \begin{bmatrix} -y+2 & -y^2+4 & -y^3+4y+8 \\ y+2 & y^2+4y+4 & y^3+6y^2+12y+8 \\ y+2 & y^2+4y+4 & y^3+6y^2+12y+8 \end{bmatrix}.$$

It is easy to check that the canonical Smith matrix for **R** is equal to the matrix

$$Q = \begin{bmatrix} 1 & 0 & 0 \\ 0 & 0 & 0 \\ 0 & 0 & 0 \end{bmatrix}.$$

Hence, the condition of (8) says that there is one joint zero for each values of the variable y. It is true since the family considered here may be rewritten as

$$f_1(x,y) = (x+y)(x+y+2),$$
$$f_2(x,y) = (x+2)(x+y+2),$$
$$f_3(x,y) = (y+2)(x+y+2).$$

It is seen that, for example, the pairs (x,y) equal to (-1,-1), (0, -2), (-2,0) etc. are joint zeros.

Example 4. Consider the polynomials:

$$f_1(x,y) = x^2 + (2y+2)x + y^2 + 2y,$$
$$f_2(x,y) = x^2 + (2y+5)x + 4y + 6,$$
$$f_3(x,y) = 0x^2 + (2y+4)x + 3y^2 + 11y + 10.$$

The resultant matrix is as follows

$$\mathbf{R} = \begin{bmatrix} 3 & -y^2+2y+6 & -2y^3-y^2+10y+12 \\ 2y+4 & 3y^2+11y+10 & 4y^3+20y^2+34y+20 \\ 2y+4 & 3y^2+11y+10 & 6y^3+29y^2+47y+26 \end{bmatrix}.$$

It is easy to check that the canonical Smith matrix for **R** is equal to the matrix

$$Q = \begin{bmatrix} 1 & 0 & 0 \\ 0 & y^3 + \frac{9}{2}y^2 + \frac{13}{2}y + 3 & 0 \\ 0 & 0 & y^3 + \frac{9}{2}y^2 + \frac{13}{2}y + 3 \end{bmatrix}$$

Hence, the condition of (8) says that there may be one joint zero when the variable y is equal to '-1', '-2' and '-3/2'. Hence, note that the family considered here may be rewritten as

$$f_1(x,y) = (x+y)(x+y+2),$$
$$f_2(x,y) = (x+2)(x+2y+3),$$
$$f_3(x,y) = (y+2)(2x+3y+5).$$

It is seen that, only the pair (-1,-1) is the common zero of the considered polynomials family since the remaining roots of $y^3 + \frac{9}{2}y^2 + \frac{13}{2}y + 3$, i.e. y=-2 and y=-3/2 are not the proper choice. This is due to that

- for y=-2 the third polynomial is zero,
- for y=-3/2 one obtains the linear combination of polynomials in x, i.e. $x^2-x-3/4$, x^2+2x, $x+1/4$.

4. CONCLUSIONS

Basing on the methods presented here, Dr D. Uciński from this Institute has constructed, in the MAPPLE, the easy algorithm which enables calculating the number of common roots for a polynomial family, for the one variable as well as two-variable polynomials case.

It is expected that both the methodology and the algorithm are possible to be generalised for the case of more than two variable polynomials which is the subject of current work and will be reported in due course.

REFERENCES

Anderson, B.D.O. and R.W. Scott (1977). Output feedback stabilization - Solution by algebraic geometry methods, *Proc. of IEEE,* **6**, 849-861.

Gałkowski, K. (1996), Multi-dimensional systems theory - A common approach to problems in Circuits, Control and Signal Processing, *Proc. of the IEE Colloquium Multi-dimensional Systems: Problems and Solutions, London 10 Jan. 1996*, 5.1-5.11.

Hodge, W.V.D. and D. Pedoe (1947). *Methods of Algebraic Geometry*, Cambridge.

Wimmer, H. K. (1990). On the history of the Bezoutians and the resultant matrix, *Linear Algebra and its Applications*, **128**, 27-34.

$$\begin{bmatrix} a_{11} & \cdots & a_{\alpha 1} & \cdots & a_{\beta 1} & \cdots & a_{n+11} \\ \vdots & & \vdots & & \vdots & & \vdots \\ a_{1\varepsilon} & \cdots & \langle a_{\alpha\varepsilon} \rangle & \cdots & \langle a_{\beta\varepsilon} \rangle & \cdots & a_{n+1\varepsilon} \\ \vdots & & \vdots & & \vdots & & \vdots \\ a_{1\mu} & \cdots & \langle a_{\alpha\mu} \rangle & \cdots & \langle a_{\beta\mu} \rangle & \cdots & a_{n+1\mu} \\ \vdots & & \vdots & & \vdots & & \vdots \\ a_{1N} & \cdots & a_{\alpha N} & \cdots & a_{\beta N} & \cdots & a_{n+1N} \end{bmatrix}$$

Fig.1.

NONLINEAR ROESSER PROBLEM WITH A TERMINAL CONDITION. MAXIMUM PRINCIPLE.

Dariusz Idczak

Department of Mathematics, University of Lodz
ul. S. Banacha 22, 90-238 Lodz, Poland

Abstract: In the paper, necessary optimality conditions for an autonomous continuous Roesser system nonlinear in a state, with a pointwise terminal condition, are derived.

Keywords: Partial differential equation, terminal condition, maximum principle.

1. INTRODUCTION

Let us consider the control problem of the type

$$\begin{aligned} \frac{\partial z^1}{\partial x} &= f^1(z^1, z^2) + B_1 u^1 \\ \frac{\partial z^2}{\partial y} &= f^2(z^1, z^2) + B_2 u^2 \end{aligned} \tag{1.1}$$

a.e. on $P = [0,1] \times [0,1] \subset R^2$,

$$z^i(\cdot, 0) \equiv z^i(0, \cdot) \equiv 0, i = 1, 2 \tag{1.2}$$

everywhere on $[0,1] \subset R$,

$$z^1(1,1) = P_1, z^2(1,1) = P_2, \tag{1.3}$$

where $z^1 \in R^{n_1}, z^2 \in R^{n_2}, u^1 \in R^{r_1}, u^2 \in R^{r_2}$ $f^1 : R^{n_1} \times R^{n_2} \to R^{n_1}, f^2 : R^{n_1} \times R^{n_2} \to R^{n_2}$, $B_1 \in R^{n_1 \times r_1}, B_2 \in R^{n_2 \times r_2}, P_1 \in R^{n_1}, P^2 \in R^{n_2}$, $n_1, n_2 \in N$.

We shall consider the above control problem in the spaces $AC = AC^{n_1} \times AC^{n_2}$ and $U = U_y^{r_1} \times U_x^{r_2}$ of trajectories and controls, respectively, where $AC^{n_i} = \{z^i : P \to R^{n_i}; there\ exists\ l^i \in L^2(P, R^{n_i})\ such\ that\ z^i(x,y) = \int_0^x \int_0^y l^i(s,t) ds dt\ for\ (x,y) \in P\}, i = 1,2,$ and $U_y^{r_1} = \{u^1 : P \to R^{r_1}; there\ exists\ \mu^1 \in L^2(P, R^{r_1})\ such\ that\ u^1(x,y) = \int_0^y \mu^1(x,t) dt\ for\ x \in [0,1]\ a.e., y \in [0,1]\}$, $U_x^{r_2} = \{u^2 : P \to R^{r_2}; there\ exists\ \mu^2 \in L^2(P, R^{r_2})\ such\ that\ u^1(x,y) = \int_0^x \mu^1(s,y) ds\ for\ y \in [0,1]\ a.e., x \in [0,1]\}$.

We shall identify functions belonging to $U_y^{r_1}(U_x^{r_2})$ and equal a.e.

The above control problem will be considered with the following performance index

$$F_0(z,u) = \int_P f^0(x, y, z^1, z^2, \frac{\partial u^1}{\partial y}, \frac{\partial u^2}{\partial x}) dx dy \tag{1.4}$$

where $f^0 : P \times R^{n_1} \times R^{n_2} \times R^{r_1} \times R^{r_2} \to R$.

Some results concerning the existence of a solution to problem (1.1-1.3) (i.e. the controllability of (1.1-1.2)) were presented in (Idczak and Walczak, 1995).

In (Idczak, 1996a), by using the maximum principle for the appropriate Fornasini-Marchesini system with terminal conditions obtained in (Idczak, 1993), the maximum principle for optimal control problem (1.1-1.4) in the linear case was derived.

In our paper, basing ourselves on the extremum principle for a smooth-convex problem (Tikhomirov, 1986), we shall prove the maximum principle for problem (1.1-1.4).

2. EXISTENCE OF A SOLUTION TO A NONAUTONOMOUS LINEAR SYSTEM

One can show (Idczak, 1990) that each function $z^i \in AC^{n_i}$ possesses a.e. on P the partial derivatives $\frac{\partial z^i}{\partial x}(x,y)$, $\frac{\partial z^i}{\partial y}(x,y)$, $\frac{\partial^2 z^i}{\partial x \partial y}(x,y)$, and

$$\frac{\partial z^i}{\partial x}(x,y) = \int_0^y l^i(x,t)dt,$$

$$\frac{\partial z^i}{\partial y}(x,y) = \int_0^x l^i(s,y)ds,$$

$$\frac{\partial^2 z^i}{\partial x \partial y}(x,y) = l^i(x,y)$$

for $(x,y) \in P a.e.$, where $z^i(x,y) = \int_0^x \int_0^y l^i(s,t)dsdt, i = 1,2.$
Similarly, each function $u^1 \in U_y^{r_1}$ possesses a.e. on P the partial derivative $\frac{\partial u^1}{\partial y}(x,y)$ such that

$$\frac{\partial u^1}{\partial y}(x,y) = \mu^1(x,y)$$

for $(x,y) \in P a.e.$, where $u^1(x,y) = \int_0^y \mu^1(x,t)dt$, and each function $u^2 \in U_x^{r_2}$ possesses a.e.on P the partial derivative $\frac{\partial u^2}{\partial x}(x,y)$ such that

$$\frac{\partial u^2}{\partial x}(x,y) = \mu^2(x,y)$$

for $(x,y) \in P a.e.$, where $u^2(x,y) = \int_0^x \mu^2(s,y)ds$.
Let us introduce in $AC^{n_i}, i = 1,2$, and $U_y^{r_1}, U_x^{r_2}$ the following scalar products

$$(z^i, w^i) = \int_P \frac{\partial^2 z^i}{\partial x \partial y}(x,y) \frac{\partial^2 w^i}{\partial x \partial y}(x,y)$$

for $z^i, w^i \in AC^{n_i}, i = 1,2,$

$$(u^1, v^1) = \int_P \frac{\partial u^1}{\partial y}(x,y) \frac{\partial v^1}{\partial y}(x,y)$$

for $u^1, v^1 \in U_y^{r_1}$,

$$(u^2, v^2) = \int_P \frac{\partial u^2}{\partial x}(x,y) \frac{\partial v^2}{\partial x}(x,y)$$

for $u^2, v^2 \in U_x^{r_2}$. The norms connected with the above scalar products are given by

$$\|z^i\|_{AC^{n_i}} = \left(\int_P \left| \frac{\partial^2 z^i}{\partial x \partial y}(x,y) \right|^2 dxdy \right)^{\frac{1}{2}}, i = 1,2,$$

$$\|u^1\|_{U_y^{r_1}} = \left(\int_P \left| \frac{\partial u^1}{\partial y}(x,y) \right|^2 dxdy \right)^{\frac{1}{2}},$$

$$\|u^2\|_{U_x^{r_2}} = \left(\int_P \left| \frac{\partial u^2}{\partial x}(x,y) \right|^2 dxdy \right)^{\frac{1}{2}}.$$

It is easy to check that the spaces $AC^{n_i}, i = 1,2$, and $U_y^{r_1}, U_x^{r_2}$ are complete. So, AC and U are Hilbert spaces with the appropriate scalar products.
Now, let us consider the system

$$\begin{aligned} \frac{\partial z^1}{\partial x} &= A_{11}(x,y)z^1 + A_{12}(x,y)z^2 + v^1 \quad (2.1) \\ \frac{\partial z^2}{\partial y} &= A_{21}(x,y)z^1 + A_{22}(x,y)z^2 + v^2 \end{aligned}$$

a.e. on P.
In the proof of the maximum principle we use the following lemma.
Lemma. If the functions $A_{11} \in U_y^{n_1 \times n_1}, A_{12} \in U_y^{n_1 \times n_2}, A_{21} \in U_x^{n_2 \times n_1}, A_{22} \in U_x^{n_2 \times n_2}$ are such that $\frac{\partial A_{11}}{\partial y}, \frac{\partial A_{12}}{\partial y}, \frac{\partial A_{21}}{\partial x}, \frac{\partial A_{22}}{\partial x}$ are essentially bounded, then system 2.1 has a unique solution in the space $AC = AC^{n_1} \times AC^{n_2}$ for any fixed $v = (v^1, v^2) \in U_y^{n_1} \times U_x^{n_2}$.
Proof. It is easy to see that it suffices to prove the existence of a unique solution to the system

$$\begin{aligned} w^1 &= A_{11}(x,y) \int_0^x w^1 + A_{12}(x,y) \int_0^y w^2 \quad (2.2) \\ &\quad + v^1(x,y) \\ w^2 &= A_{21}(x,y) \int_0^x w^1 + A_{22}(x,y) \int_0^y w^2 \\ &\quad + v^2(x,y) \end{aligned}$$

in the space $U = U_y^{n_1} \times U_x^{n_2}$.
Indeed, let $(w^1, w^2) \in U_y^{n_1} \times U_x^{n_2}$ be a solution to the above system. Then there exist functions $\nu^1 \in L^2(P, R^{n_1}), \nu^2 \in L^2(P, R^{n_2})$ such that

$$w^1(x,y) = \int_0^y \nu^1(x,t)dt$$

for $x \in [0,1] a.e., y \in [0,1]$, and

$$w^2(x,y) = \int_0^x \nu^2(s,y)ds$$

for $y \in [0,1] a.e., x \in [0,1]$.
If we define

$$z^1(x,y) = \int_0^x w^1(s,y)ds = \int_0^x \int_0^y \nu^1(s,t)dsdt,$$

$$z^2(x,y) = \int_0^y w^2(x,t)dt = \int_0^x \int_0^y \nu^2(s,t)dsdt$$

for $(x,y) \in P$, then $z^1 \in AC^{n_1}, z^2 \in AC^{n_2}$ and 2.1 holds. The uniqueness of a solution to system 2.1 in $AC^{n_1} \times AC^{n_2}$ follows immediately from the uniqueness of a solution to system 2.2 in $U_y^{n_1} \times U_x^{n_2}$.
So, let us define the operator

$$S : U_y^{n_1} \times U_x^{n_2} \to U_y^{n_1} \times U_x^{n_2},$$

$$\begin{aligned} S(w^1, w^2) &= (A_{11}(x,y) \int_0^x w^1 \\ &+ A_{12}(x,y) \int_0^y w^2 + v^1(x,y), \\ &A_{21}(x,y) \int_0^x w^1 \\ &+ A_{22}(x,y) \int_0^y w^2 + v^2(x,y)) \end{aligned}$$

and introduce in $U_y^{n_1} \times U_x^{n_2}$ the norm

$$\begin{aligned} \|w\|_k &= (\int_P e^{-k(x+y)}(|\nu^1(x,y)|^2 \\ &+ |\nu^2(x,y)|^2)dxdy)^{\frac{1}{2}} \end{aligned}$$

where $w(x,y) = (w^1(x,y), w^2(x,y))$, $w^1(x,y) = \int_0^y \nu^1(x,t)dt, w^2(x,y) = \int_0^x \nu^2(s,y)ds$, k is some positive number.
The above norm is equivalent to the norm in $U_y^{n_1} \times U_x^{n_2}$ generated by $\|\cdot\|_{U_y^{n_1}}, \|\cdot\|_{U_x^{n_2}}$.
Moreover, one can check by integrating by parts that there exists a constant k>0 such that the operator S contracts in $U_y^{n_1} \times U_x^{n_2}$ considered with the norm $\|\cdot\|_k$. Consequently, system 2.2 has a unique solution in $U_y^{n_1} \times U_x^{n_2}$ and the proof is completed.

3. MAXIMUM PRINCIPLE

Let us fix a set $M \subset R^{r_1+r_2}$ and denote $U_M = \{u = (u^1, u^2) \in U_y^{r_1} \times U_x^{r_2}; (\frac{\partial u^1}{\partial y}(x,y), \frac{\partial v^1}{\partial y}(x,y)) \in M \; for \; (x,y) \in Pa.e.\}$, $F : AC \times U_M \to R^{n_1} \times R^{n_2} \times U_y^{n_1} \times U_x^{n_2}, F((z^1,z^2),(u^1,u^2)) = (z^1(1,1) - P_1, z^2(1,1) - P_2, \frac{\partial z^1}{\partial x} - f^1(z^1,z^2) - B_1u^1, \frac{\partial z^2}{\partial y} - f^2(z^1,z^2) - B_2u^2)$.
So, we may write our optimal control problem (1.1-1.4) in the form

$$F_0(z,u) \to \min. \tag{3.1}$$

$$F(z,u) = 0$$

$$z \in AC, u \in U_M.$$

In order to obtain necessary optimality conditions for the above problem, we shall use the extremum principle for a smooth-convex problem.
We say that a pair $(z_*, u_*) \in AC \times U_M$ is a local minimum point of problem (3.1) if $F(z_*, u_*) = 0$ and there exists a neighbourhood V of z_* in AC such that, for any pair $(z,u) \in V \times U_M$ satisfying the equality $F(z,u) = 0$, we have

$$F(z_*, u_*) \leq F(z,u).$$

Theorem. Let us assume that:
- $M \subset R^{r_1+r_2}$ is bounded and convex,
- f^1, f^2 are of class C^2 on $R^{n_1+n_2}$,
- f^0 is measurable in $(x,y) \in P$, continuous and convex in $v \in R^{r_1+r_2}$ and of class C^1 in $z \in R^{n_1+n_2}$,
- $\frac{\partial f^0}{\partial z}$ is measurable in $(x,y) \in P$ and continuous in $v \in R^{r_1+r_2}$,
- there exist functions $a \in C(R_0^+, R_0^+)$, $b \in L^1(P, R_0^+)$ such that, for $(x,y) \in Pa.e.$, $z \in R^{n_1+n_2}, v \in R^{r_1+r_2}$,

$$|f^0(x,y,z,v)| \leq a(|z|)(b(x,y) + |v|^2),$$

$$\left|\frac{\partial f^0}{\partial z}(x,y,z,v)\right| \leq a(|z|)(b(x,y) + |v|^2).$$

Then there exist (not simultaneously zero) $\lambda_0 \geq 0$, $\bar{\lambda} \in R^{n_1}, \bar{\bar{\lambda}} \in R^{n_2}, (\lambda^1, \lambda^2) \in U_y^{n_1} \times U_x^{n_2}$, such that the conjugate system

$$\begin{aligned} &\frac{\partial \lambda^1}{\partial y}(x,y) \\ &- \int_x^1 \int_y^1 \frac{\partial \lambda^1}{\partial y}(s,t) \frac{\partial}{\partial y}(\frac{\partial f^1}{\partial z^1}(z_*(s,t)))dsdt \\ &- \int_x^1 \int_y^1 \frac{\partial \lambda^2}{\partial x}(s,t) \frac{\partial}{\partial x}(\frac{\partial f^2}{\partial z^1}(z_*(s,t)))dsdt \\ &- \int_x^1 \frac{\partial \lambda^1}{\partial y}(s,y) \frac{\partial f^1}{\partial z^1}(z_*(s,y))ds \\ &- \int_y^1 \frac{\partial \lambda^2}{\partial x}(x,t) \frac{\partial f^2}{\partial z^1}(z_*(x,t))dt \end{aligned}$$

$$+\int_x^1\int_y^1 \lambda_0 \frac{\partial f^0}{\partial z^1}(s,t,z_*(s,t),$$
$$\frac{\partial u_*^1}{\partial y}(s,t), \frac{\partial u_*^2}{\partial x}(s,t))dsdt+ \bar{\lambda}$$
$$= 0,$$

$$\frac{\partial \lambda^2}{\partial x}(x,y)$$
$$-\int_x^1\int_y^1 \frac{\partial \lambda^1}{\partial y}(s,t)\frac{\partial}{\partial y}(\frac{\partial f^1}{\partial z^2}(z_*(s,t)))dsdt$$
$$-\int_x^1\int_y^1 \frac{\partial \lambda^2}{\partial x}(s,t)\frac{\partial}{\partial x}(\frac{\partial f^2}{\partial z^2}(z_*(s,t)))dsdt$$
$$-\int_x^1 \frac{\partial \lambda^1}{\partial y}(s,y)\frac{\partial f^1}{\partial z^2}(z_*(s,y))ds$$
$$-\int_y^1 \frac{\partial \lambda^2}{\partial x}(x,t)\frac{\partial f^2}{\partial z^2}(z_*(x,t))dt$$
$$+\int_x^1\int_y^1 \lambda_0 \frac{\partial f^0}{\partial z^2}(s,t,z_*(s,t),$$
$$\frac{\partial u_*^1}{\partial y}(s,t), \frac{\partial u_*^2}{\partial x}(s,t))dsdt+ \bar{\bar{\lambda}}$$
$$= 0$$

for $(x,y) \in P a.e.$ is satified and the following minimum condition holds:

$$\lambda_0 f^0(x,y,z_*(x,y), \frac{\partial u_*^1}{\partial y}(x,y), \frac{\partial u_*^2}{\partial x}(x,y))$$
$$-\frac{\partial \lambda^1}{\partial y}(x,y)B_1\frac{\partial u_*^1}{\partial y}(x,y)$$
$$-\frac{\partial \lambda^2}{\partial x}(x,y)B_2\frac{\partial u_*^2}{\partial x}(x,y)$$
$$= \min_{(\mu^1,\mu^2)\in M} \{\lambda_0 f^0(x,y,z_*(x,y),\mu^1,\mu^2)$$
$$-\frac{\partial \lambda^1}{\partial y}(x,y)B_1\mu^1 - \frac{\partial \lambda^2}{\partial x}(x,y)B_2\mu^2\}$$

for $(x,y) \in P a.e.$, where $z_*(x,y) = (z_*^1(x,y), z_*^2(x,y))$.

Proof. It is easy to verify that from the assumptions it follows that

- with a fixed control $u \in U_M$, the operator F and the functional F_0 are of class C^1(in the Frechet sense) on AC with differentials

$$\partial_z F(z,u) \;:\; AC \ni h \to$$
$$(h^1(1,1), h(1,1),$$
$$\frac{\partial h^1}{\partial x}(\cdot,\cdot) - \frac{\partial f^1}{\partial z^1}(z(\cdot,\cdot)h^1(\cdot,\cdot)$$
$$-\frac{\partial f^1}{\partial z^2}(z(\cdot,\cdot))h^2(\cdot,\cdot),$$
$$\frac{\partial h^2}{\partial y}(\cdot,\cdot) - \frac{\partial f^2}{\partial z^1}(z(\cdot,\cdot)h^1(\cdot,\cdot)$$
$$-\frac{\partial f^2}{\partial z^2}(z(\cdot,\cdot))h^2(\cdot,\cdot))$$
$$\in R^{n_1} \times R^{n_2} \times U_y^{n_1} \times U_x^{n_2},$$

$$\partial_z F_0(z,u) \;:\; AC \ni h \to$$
$$\int_P \frac{\partial f^0}{\partial z^1}(x,y,z(x,y),$$
$$\frac{\partial u^1}{\partial y}(x,y), \frac{\partial u^2}{\partial x}(x,y))h^1(x,y)dxdy$$
$$+\int_P \frac{\partial f^0}{\partial z^2}(x,y,z(x,y),$$
$$\frac{\partial u^1}{\partial y}(x,y), \frac{\partial u^2}{\partial x}(x,y))h^2(x,y)dxdy$$
$$\in R,$$

respectively,

- with a fixed $z \in AC$, for any $u, v \in U_M$, $\alpha, \beta \in (0,1), \alpha + \beta = 1$,

$$F(z,\alpha u + \beta v) = \alpha F(z,u) + \beta F(z,v),$$
$$F_0(z,\alpha u + \beta v) \le \alpha F_0(z,u) + \beta F_0(z,v),$$

- the set $\operatorname{Im} \partial_z F(z,u)$ has a finite codimension in $R^{n_1} \times R^{n_2} \times U_y^{n_1} \times U_x^{n_2}$ for any $(z,u) \in AC \times U_M$.

To prove the last property, it suffices to show by using the above lemma that

$$R^{n_1} \times R^{n_2} \times U_y^{n_1} \times U_x^{n_2} = \operatorname{Im} \partial_z F(z,u) \oplus L$$

where $L = \{(z^1,z^2,0,0) \in R^{n_1} \times R^{n_2} \times U_y^{n_1} \times U_x^{n_2}; (z^1,z^2) \in R^{n_1} \times R^{n_2}\}$ and $\oplus$ denotes the direct sum.

So, the assumptions of the smooth-convex extremum principle are satisfied. Consequently, there exist $\lambda_0 \ge 0, \bar{\lambda} \in R^{n_1}, \bar{\bar{\lambda}} \in R^{n_2}, (\lambda^1,\lambda^2) \in U_y^{n_1} \times U_x^{n_2}$, such that $(\lambda_0, \bar{\lambda}, \bar{\bar{\lambda}}, \lambda^1, \lambda^2) \ne 0$ and

$$\lambda_0 \int_P \frac{\partial f^0}{\partial z^1}(x,y,z_*(x,y), \frac{\partial u_*^1}{\partial y}(x,y), \qquad (3.2)$$
$$\frac{\partial u_*^2}{\partial x}(x,y))h^1(x,y)dxdy$$
$$\lambda_0 \int_P \frac{\partial f^0}{\partial z^2}(x,y,z_*(x,y), \frac{\partial u_*^1}{\partial y}(x,y),$$
$$\frac{\partial u_*^2}{\partial x}(x,y))h^2(x,y)dxdy+ \bar{\lambda}\, h^1(1,1)$$

$$
\begin{aligned}
&+\bar{\bar{\lambda}}\, h^2(1,1) + \int_P \frac{\partial \lambda^1}{\partial y}(x,y)\frac{\partial^2 h^1}{\partial x \partial y}(x,y)dxdy \\
&-\int_P \frac{\partial \lambda^1}{\partial y}(x,y)\frac{\partial}{\partial y}(\frac{\partial f^1}{\partial z^1}(z_*(x,y)))h^1(x,y)dxdy \\
&-\int_P \frac{\partial \lambda^1}{\partial y}(x,y)\frac{\partial}{\partial y}(\frac{\partial f^1}{\partial z^2}(z_*(x,y)))h^2(x,y)dxdy \\
&-\int_P \frac{\partial \lambda^1}{\partial y}(x,y)\frac{\partial f^1}{\partial z^1}(z_*(x,y))\frac{\partial h^1}{\partial y}(x,y)dxdy \\
&-\int_P \frac{\partial \lambda^1}{\partial y}(x,y)\frac{\partial f^1}{\partial z^2}(z_*(x,y))\frac{\partial h^2}{\partial y}(x,y)dxdy \\
&+\int_P \frac{\partial \lambda^2}{\partial x}(x,y)\frac{\partial^2 h^2}{\partial x \partial y}(x,y)dxdy \\
&-\int_P \frac{\partial \lambda^2}{\partial x}(x,y)\frac{\partial}{\partial x}(\frac{\partial f^2}{\partial z^1}(z_*(x,y)))h^1(x,y)dxdy \\
&-\int_P \frac{\partial \lambda^2}{\partial x}(x,y)\frac{\partial}{\partial x}(\frac{\partial f^2}{\partial z^2}(z_*(x,y)))h^2(x,y)dxdy \\
&-\int_P \frac{\partial \lambda^2}{\partial x}(x,y)\frac{\partial f^2}{\partial z^1}(z_*(x,y))\frac{\partial h^1}{\partial x}(x,y)dxdy \\
&-\int_P \frac{\partial \lambda^2}{\partial x}(x,y)\frac{\partial f^2}{\partial z^2}(z_*(x,y))\frac{\partial h^2}{\partial x}(x,y)dxdy \\
= \;& 0
\end{aligned}
$$

for any $h = (h^1, h^2) \in AC$,

$$
\begin{aligned}
&\lambda_0 \int_P f^0(x,y,z_*(x,y),\frac{\partial u_*^1}{\partial y}(x,y), \\
&\frac{\partial u_*^2}{\partial x}(x,y))dxdy \\
&\bar{\lambda}\,(z_*^1(1,1)-P_1)+\bar{\bar{\lambda}}\,(z_*^2(1,1)-P_2) \\
&+\int_P \frac{\partial \lambda^1}{\partial y}(x,y)\frac{\partial^2 z_*^1}{\partial x \partial y}(x,y)dxdy \\
&-\int_P \frac{\partial \lambda^1}{\partial y}(x,y)\frac{\partial}{\partial y}(f^1(z_*(x,y)))dxdy \\
&-\int_P \frac{\partial \lambda^1}{\partial y}(x,y)B_1\frac{\partial u_*^1}{\partial y}(x,y)dxdy \\
&+\int_P \frac{\partial \lambda^2}{\partial x}(x,y)\frac{\partial^2 z_*^2}{\partial x \partial y}(x,y)dxdy \\
&-\int_P \frac{\partial \lambda^2}{\partial x}(x,y)\frac{\partial}{\partial y}(f^2(z_*(x,y)))dxdy \\
&-\int_P \frac{\partial \lambda^2}{\partial x}(x,y)B_2\frac{\partial u_*^2}{\partial x}(x,y)dxdy \\
= &\min_{u=(u^1,u^2)\in U_M}\{\lambda_0 \int_P f^0(x,y,z_*(x,y), \\
&\frac{\partial u^1}{\partial y}(x,y),\frac{\partial u^2}{\partial x}(x,y))dxdy \\
&\bar{\lambda}\,(z_*^1(1,1)-P_1)+\bar{\bar{\lambda}}\,(z_*^2(1,1)-P_2) \\
&+\int_P \frac{\partial \lambda^1}{\partial y}(x,y)\frac{\partial^2 z_*^1}{\partial x \partial y}(x,y)dxdy \\
&-\int_P \frac{\partial \lambda^1}{\partial y}(x,y)\frac{\partial}{\partial y}(f^1(z_*(x,y)))dxdy \\
&-\int_P \frac{\partial \lambda^1}{\partial y}(x,y)B_1\frac{\partial u^1}{\partial y}(x,y)dxdy \\
&+\int_P \frac{\partial \lambda^2}{\partial x}(x,y)\frac{\partial^2 z_*^2}{\partial x \partial y}(x,y)dxdy \\
&-\int_P \frac{\partial \lambda^2}{\partial x}(x,y)\frac{\partial}{\partial y}(f^2(z_*(x,y)))dxdy \\
&-\int_P \frac{\partial \lambda^2}{\partial x}(x,y)B_2\frac{\partial u^2}{\partial x}(x,y)dxdy\}.
\end{aligned}
$$

This equality implies that

$$
\begin{aligned}
&\lambda_0 \int_P f^0(x,y,z_*(x,y),\mu_*^1(x,y), \\
&\mu_*^2(x,y))dxdy \\
&-\int_P \frac{\partial \lambda^1}{\partial y}(x,y)B_1\mu_*^1(x,y)dxdy \\
&-\int_P \frac{\partial \lambda^2}{\partial x}(x,y)B_2\mu_*^2(x,y)dxdy \\
= &\min_{\mu=(\mu^1,\mu^2)\in L_M^2}\{\lambda_0 \int_P f^0(x,y,z_*(x,y), \\
&\mu^1(x,y),\mu^2(x,y))dxdy \\
&-\int_P \frac{\partial \lambda^1}{\partial y}(x,y)B_1\mu^1(x,y)dxdy \\
&-\int_P \frac{\partial \lambda^2}{\partial x}(x,y)B_2\mu^2(x,y)dxdy.
\end{aligned}
$$

From the above, using the theorem on the passing from an integral maximum principle to a point-wise one (Idczak, 1996b), we obtain the minimum condition from the assertion of the theorem. Substituting in 3.2 points of type $(h^1, 0) \in AC$, $(0, h^2) \in AC$ and integrating by parts, we obtain the conjugate system from the assertion of the theorem.

The proof of the theorem is completed.

Remark 1. In the linear case, i.e. when the system has the form

$$
\begin{aligned}
\frac{\partial z^1}{\partial x} &= A_{11}z^1 + A_{12}z^2 + B_1 u^1, \\
\frac{\partial z^2}{\partial y} &= A_{21}z^1 + A_{22}z^2 + B_2 u^2,
\end{aligned}
$$

if we put

$$
\tilde{\lambda}^1(x,y) = \int_x^1 \frac{\partial \lambda^1}{\partial y}(s,y)ds,
$$

$$\tilde{\lambda}^2(x,y) = \int_y^1 \frac{\partial \lambda^2}{\partial x}(x,t)dt,$$

then the assertion of the proved theorem gives the assertion of the maximum principle obtained in (Idczak, 1996a).

Remark 2. The maximum principle obtained can be generalized to the case of any finite number of terminal conditions (Idczak, 1993).

REFERENCES

Idczak, D. (1990). Optimization of nonlinear systems described by partial differential equations, Doctoral thesis, Lodz.

Idczak, D. (1993). Nonlinear Goursat-Darboux problem and its optimization, *Nonlinear Vibration Problems,* **25.**

Idczak, D. (1996a). The maximum principle for a continuous 2-D Roesser model with a terminal condition, *Proceedings of MMAR'96, Miedzyzdroje.*

Idczak, D. (1996b). Necessary optimality conditions for a nonlinear continuous n-dimensional Roesser model, *Mathematics and Computers in Simulation,* **41.**

Idczak, D. and S. Walczak (1995). On the controllability of continuous Roesser systems, *Proceedings of MMAR'95,* Miedzyzdroje.

Tikhomirov, V. M. (1986). *Fundamental Principles of the Theory of Extremal Problems,* Wiley, Chichester.

A GENERIC MODEL FOR CONTROL OF HIERARCHICAL MULTI-LEVEL PRODUCTION SYSTEMS

D. Chen, P. Farthouat, F. Pereyrol, J.P. Bourrières

LAP/GRAI, University Bordeaux I
351, Cours de la Libération, 33405 Talence cedex France
Tel: (33) 05 56 84 65 30, Fax : (33) 05 56 84 66 44
E-mail : [name]@lap.u-bordeaux.fr

Abstract: The paper presents a generic model of technical data for the control of multi-level hierarchical production systems. The objective is to generalise the concepts and data of production control usually used at the shop floor level to cover the whole hierarchy of an enterprise (from the level of machine to the level of virtual enterprise). Some new emerging concepts in enterprise organisation are reviewed. The Problem statement of and the basic concepts of the model are defined. Emphasis is made on the genericity of its application and on the aggregation of technical data from one level to another. .A simplified example will show how to get aggregated data at the cell and shop levels from the data collected at the machine (work station) level.

Keyword : Production Control, Hierarchical system, Distributed System, Aggregation of information, Graphs, Networks, Matrix formulation.

1. INTRODUCTION

It is generally agreed that the production in the future will be based on smaller units with greater flexibility, autonomy and innovative organic structures (Tharumarajah, et al, 1996). The necessity to reach a global objective of the production with a minimum cost, better quality and shorter delay obliges an enterprise not only to optimize its own production system but also to integrate constraints and even the production planning of its suppliers and customers. We are moving from enterprise to extended enterprise which in turn will move to virtual enterprise. In this perspective, each enterprise being a very specialized competence unit must cooperate with others to realize a final product. Consequently the traditional production control paradigm must adapt to a more flexible and cooperative production control approach.

The review of the available approaches in the area of enterprise organisation (hierarchical structure, bionic, fractal and holonic approaches...) shows some emerging concepts. It is well known that hierarchisation of resources is necessary to reduce a complex control problem at a manageable level.

In this paper, graph and the associated matrix representation are used to model a physical production system in terms of network of resources, with structured set of data such as routing, capacity etc. Using proposed aggregation concepts, it is possible, based on the model of the lowest level (for example, work station level), to get an aggregated model of the physical system at any level of an enterprise organisation hierarchy (cell, workshop, factory, enterprise, extended enterprise, virtual enterprise). The genericity of the model is characterized by two dimensions. The first dimension is concerned with its application domain; it can be applied to any kind of manufacturing enterprise. The second dimension is concerned with the model structure. In fact, the representation of the physical production system will have the same structure whatever the level it is concerned with.

2. REVIEW OF EMERGING APPROACHES

Today, traditional hierarchical control is being challenged by some new approaches such as bionic manufacturing, fractals company, holonic manufacturing etc.

Hierarchy is usually considered as a way to control complex system. Simon (1990) defines a hierarchic system, or a hierarchy, as a system that is composed of interrelated sub-systems, each of the latter being in turn hierarchic in structure, until some lowest level of elementary subsystems is reached. According to Valckenaers (1996), the term 'hierarchy' in the domain of Production System is usually understood in a narrow sense as to represent a relation of subordination (authority) between a "boss" (control part) and a set of subordinated subsystems (controlled parts). To meet the requirements of future production systems (virtual enterprise for example), it is necessary to introduce new concepts in the hierarchic system such as autonomy and lateral cooperation.

One of the major difficulties to introduce more autonomy in a hierarchical system is concerned with how to coordinate in order to solve potential conflicts between autonomous units. Okino (1989, 1992) draw a parallel between bionic systems and manufacturing systems. He observed that in a biological system, the function of coordination is executed by enzymes, which act as catalysts to maintain harmony among the action of cells.

To increase autonomy of elementary production units which are hierarchically linked, the Holonic approach, first proposed by Koestler (1967) and further developed in IMS program (1994, 1995) suggests to replace a hierarchy by a Holarchy. A holarchy system is composed of holons that can cooperate to achieve a goal or objective. Holons in a hierarchy can be defined by their functions or tasks. A holon can be compared with an Agent. Il consists of an information processing part and often a physical processing part. A holon can be part of another holon. Compared with traditional hierarchical systems, an holonic system is distributed and more adapted to decentralised units.

Warnecke (1993) suggests the Fractal Company as the model of future Production Systems. An enterprise is decomposed on fractals. The main characteristic of fractals is self-similarity, implying recursion, pattern-inside-of-pattern. The self-similar property is increased by self-organisation. In this approach, emphasis is given to factory fractals acting independently. This means the fractals have a goal(s) which must be consistent to the global goal(s) of the enterprise. The GRAI model for production control is coherent to this concept. Global objective is decomposed into sub-objective of each decision center and all decision centers are linked by a decision frame (Doumeingts et al, 1994) However, one difficulty to build a fractals system is to define criteria to decompose the system in fractals.

All these approaches intend to introduce more autonomy and cooperation in a hierarchical organisation. Holonic approach is based on functional decomposition. Incomplete plans are transmitted top-down giving autonomy to lower levels to refine the plans. In the Bionic approach, task specification is defined top-down while in Fractal company, objectives (goals) are transmitted top-down instead of task specification. In the proposed model, the decomposition of a production system into smaller units until the elementary resources inspires the recursive decomposition of fractal factory concept. The production planning will be based on hierarchical coordination of the holonic approach. The cooperation between units/resources and the adaptation to changing environment follows the concept of bionic approach.

3. PROBLEM DEFINITION AND BASIC CONCEPTS

The problem of production system control can be defined simultaneously in space and in time :

- in space, according to the dimension of the system to be controlled: work station, workshop, factory, industrial group, extended enterprise.
- in time, from short term to long term, where one finds successively the problems of production management, of new products design and finally of projects of investments.

The space and time horizon is longer when the hierarchic level is high. The extension of the space horizon to develop world-wide network of client/suppliers based manufacturing, remains the most remarkable evolution of this century. Current organisation would be replaced by distributed system of autonomous units which operate as a set of cooperating entities. In this context, the modern production control should take into account two types of information flows :

- vertical flow of coordination between a pilot and their resources (Enzymes function as outlined in bionic manufacturing)
- horizontal flow of cooperation between resources of the same level of organisation (communication and cooperation between holons, network of communication / cooperation and fractal navigation between fractals).

A production system is defined as being a network of resources (controlled part) coordinated by a 'manager' (control part). In this way, any production system can be viewed as a network of aggregated subsystems, and this recursion till to elementary resources of the system (for example work station).

The modelling by a hierarchical decomposition of the Production System allows to identify the organisation levels of an enterprise by defining the resources belonging to each of the levels. It also allows to define the relationship between the Control Part and Controlled Part viewed at the each of these levels. A resource is only considered as a resource by its upper level pilot. On the contrary, the same resource plays the role of a pilot to resources of its lower level.

At a given level, the set of resources are modelled in a network as well. It is considered that the network modelling allows to better analyse the interconnections between production systems in their environment, between subsystems within an enterprise and their dynamic behavior (Zolhadri, Bourrières 1995).

Each resource (elementary or aggregated) has two views : external view and internal view. The planning of activities of a system gives an external view of its load in time without consider its internal organisation to perform the jobs. The internal view of the system determines the way of execution of the jobs by a detailed dispatching of sub-activities on its resources. In this way, each of these systems-resources has its own load plan and the generalized production control can be considered as a waterfall recursion of planning and dispatching decisions (cf. figure 1).

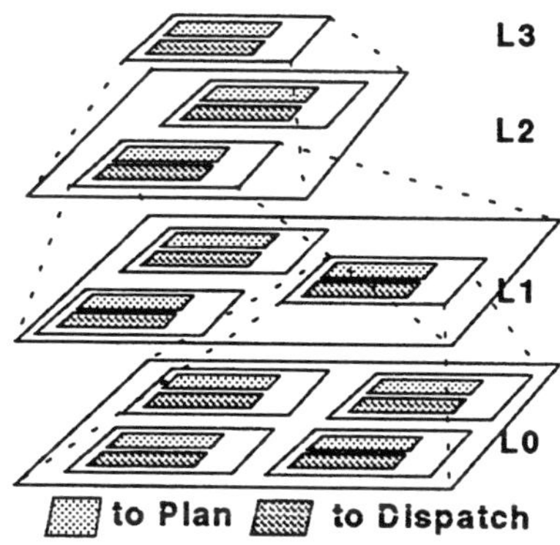

Fig. 1 : the system of control of the system of production

The generic model proposed in the paper allows to represent the set of technical data that are necessary to plan and control basic tasks of production at the lowest level (machine / workstation level). Using proposed modelling formalism, an aggregated technical data can be obtained at the upper level (cell for example) for the planning and control of macro-tasks defined at the level, and this till the level of enterprise and extended enterprise. The idea is that at the each level, a set of macro-tasks and macro-ressources can be defined, and the set of aggregated technical data can be provided by the model for the planning and control of these macro-tasks.

4. CONTEXT

4.1 *Presentation*

Our objective is to facilitate and improve the control of the system of production.

First of all the terms used are clarified. A Task (job) aims at getting a result (function) from data without prejudging the means. An Activity is the result of allocation of a task or a part of task on a resource (work station, cell, workshop).

At each level of the system of control, two different roles are finded : to plan the work and to dispatch tasks on resources (to realize the allocation of the work on the resource : to create an activity). These two roles will not use data under the same forms (cf. fig. 2).

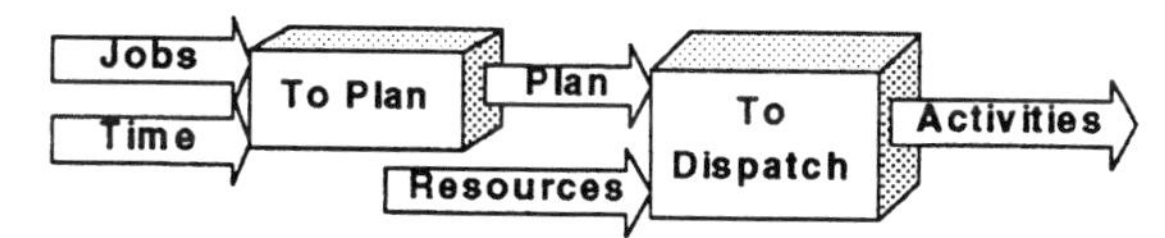

Fig. 2 : use of data

These two roles (to plan and to dispatch) situated at each level of the control in the system of production, use the tasks and resources such as they are defined on the considered level :

- at level 0 : tasks and elementary resources (work station)
- at level 1 : macro tasks (C-task) constituted of tasks of the preceding level (task), macro resources (cell) constituted of resources of the preceding level.
- at level 2 : macro tasks (W-task) constituted of tasks of the preceding level (C-task), macro resources (workshop) constituted of resources of the preceding level.

This definition of tasks and resources is generic for each level.

4.2 *Step*

The field of our approach is the production on different sites and in flexible job shop. A task (or job) can be executed by different machines that can be situated geographically on different sites and to have different run times.

Our approach is an ascending approach. The starting point is the very detailed data of the physical system (work stations) and these data are aggregated for upper levels. The system of production is perceived as a network of work stations connected by means of transport (trolley, conveyor belt..) and flows of products (routing). This notion of network is found at each considered level, inside which resources are connected physically and/or logically (product flow). Also an enterprise is seen as a set of resources (factories) forming a network in order to manufacture products.

5. PROPOSAL OF A TECHNICAL DATA AGGREGATION PRINCIPLE

5.1 *Limitations*

Only linear routings are studied (routings of manufacturing). Necessary tasks for the manufacture of a product are known and ordered.

In this paper the idea that the capacity and the load of each resource is calculated easily and always update is supposed.

5.2 Developed principles

Convention of notation :

- A matrix $M = (m_{ij})$, the transposed matrix of M is ${}^tM = (m_{ji})$,
- A matrix M characterizing the description of a level L, is noted M(L)

The Matrix of Competence (MC) and the Matrix of Temporal Competence(MTC) for level L

One knows the competence of each resource of a cell. The Matrix of Competence is a Boolean matrix, a 1 indicates that the resource can make the task, 0 indicates that it has not the necessary competence. It has the next form : Resources X Tasks

The temporal competence matrix indicates the time used by the resource to realize the task. By convention, when the resource cannot realize the task, one considers the duration is infinite.

The Matrix of Transport (MT) and the Matrix of Temporal Transport(MTT) for the level L

The Matrix of transport allows to indicate the possibility or the impossibility of interconnection between the different knots of the system. One considers that a relationship between the knot and itself is always possible.

"1" represents possible material links between resources. The matrix represents the possible transportation of articles in the workshop either by automatic means (trolley, conveyor belt..) or by manual means (men). The matrix is oriented and has the following form : Resources X Resources.

In the same way , we know the Matrix of temporal transport : time used to go from a work station to another. This approach is reducing because the transport is supposed to used the same time for each kind of product transported. Furthermore the transport means cannot carry several types of products at the same time.

Matrix of Logical Transport for the level L

The indication of the average transportation time of a task to another is necessary. So, the Matrix of Logical Transport is established.

For so doing, the number of possible different paths between tasks thanks to the Matrix Number of Paths (MNP) has to be defined.

$$MNP(L) = {}^tMC(L) * MT(L) * MC(L) \qquad (1)$$

Then, the total time of transportation between tasks is established thanks to the Matrix Times Sum (MTS) : it is the sum of times for each possible path:

$$MTS(L) = {}^tMC(L) * MTT(L) * MC(L) \qquad (2)$$

For the calculation, in the matrix of temporal transportation infinity is replaced by 0.

To get the Matrix of Logical Transport, a choice must be made. To have the average time between two tasks, the sum of the different times found is divided by the number of possible paths.

To obtain the matrix of logical transport (MLT) it is necessary to apply the following algorithm :

Algorithm 1

---

n = total number tasks
MTS = (mts(L)ij) $1 \leq i \leq n$; $1 \leq j \leq n$
MNP = (mnp(L)ij) $1 \leq i \leq n$; $1 \leq j \leq n$
MLT = (mlt(L)ij) $1 \leq i \leq n$; $1 \leq j \leq n$
For i vary 1 to n
    For j vary 1 to n
        If mnp(L)ij = 0 **then** mlt(L)ij = ∞
        Else mlt(L)ij = mts(L)ij /mnp(L)ij
        End If
    End For
End For

The aggregation of the matrix of competence for the level L+1

The matrix of competence is aggregated by keeping only competence of the resource of level L+1 without paying attention to the resource of level L. Indeed, if at least a resource of the level L is to realize the task then the macro resource of level L+1 is as competent. The matrix of intermediate competence (MIC) is obtained by applying the following algorithm :

Algorithm 2

- nm the number of matrix MC(L) of the level L (nbre of macro resources)
- the set of matrix MC(L) of the level L $= \{MC(L)^r = (mc(L)^r_{ij}), 1 \leq r \leq nm\}$
- nc number columns of MC(L) (tasks);
- nl number lines of MC(L) (resources)
- ST a set of matrix $= \{ T^u = (tu_{1j}) \; 1 \leq u \leq nm \}$

For r vary 1 to nm
    For j vary 1 to nc (for each column of $MC(L)^r$)
    $t^r_{1j} = \max(m^r_{ij}) \; 1 \leq i \leq nl$
    End For
End For

$$\mathbf{MIC(L+1)} = {}^t\left(T^1 \quad T^2 \quad \ldots \quad T^{nm}\right)$$

But the workshop level also wishes to have the vision of macro tasks. To get the values of the column of the aggregated task an "AND" is made between the columns composing the aggregated task of the matrix MIC(L+1). The result is the matrix MC(L+1)

The aggregation of the matrix of temporal competence is obtained in the same way as previously. In this case, the aggregated value of each element of the result matrix is in fact the average of times of each work station. This choice is totally arbitrary and can be modified without challenging the principle of the aggregation.

In this result the time of transportation between tasks of level L composing the task of level L+1 is not taken into account. The utilization of each matrix of logical transport of level L, allows to establish this time of transportation. So the time of transport is added to the run time.

The aggregation of matrix of temporal transport for the level L+1

The matrix of temporal transport of each resource of level L is used. They represent connections between resources of level L. They are going to allow us to estimate times between resources of level L+1.

Algorithm 3

---

p number of matrix MTT of level L.
nl number of line of the MTT level L
nc number of column of the MTT level L
For j vary 1 to p
  For r vary 1 to p
    tmp = Element Number $mtt(L)_{kl} \neq \infty$

$$mtt(L+1)_{rj} = \sum_{k=1}^{k=nc} \sum_{L=1}^{L=nl} \frac{(mtt(L)_{kl} \text{ if } mtt(L)_{kl} \neq \infty)}{tmp}$$

  End For
End For

5.3 Explanation of the mechanism thanks to an example

An example has been realized, using as a start point the technical data of the cell level. Only a part of the example is explained. (cf. fig. 3). The objective is to show clearly the passage from the cell level to the workshop level.

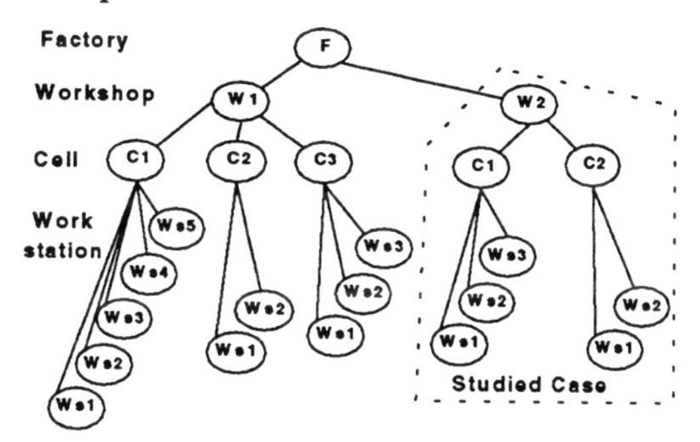

Fig. 3: Structure of the example.

Convention of notation :

- UxAyCzPv is the reference of the position of the work station v, the cell z, the workshop y, the Factory x.
- UxAyCz is the reference of the cell z, the workshop y, the Factory x.

Data of departure

The totality of achievable tasks by workstation is obtained thanks to the totality of process of production of all types of product that circulate in the physical system. A process is a logical statement of transformations to realize.

In the example, there is three products whose process is :

- Product A : WT1=CT1-CT2 ; Product B : WT2=CT3-CT4-CT1 ; Product C : WT3=CT2-CT4.
- CT1 = T2-T3 ; CT2 = T2-T6-T5 ; CT3 = T2-T5 ; CT4 = T4.

WTn : represents the task n achievable by the workshop.
CTn : represents the task n achievable by the cell.
Tn : represents the task n achievable by the position of work.

The analysis of process gives us the number of the different tasks to realize, here they are 6. The part of the example corresponding to the workshop 2 is studied (cf. fig. 3).

To the level cell

The example for the cell 2 of workshop number 2 (F1W2C2) is now explained.

MC(F1W2C2) Matrix of Competence

	T1					T6
F1W2C2WS1	0	1	0	1	0	0
F1W2C2WS2	0	1	1	1	0	0

MTC(F1W2C2) Matrix of Temporal Competence

	T1					T6
F1W2C2WS1	∞	2	∞	5	∞	∞
F1W2C2WS2	∞	7	9	4	∞	∞

MT(F1W2C2) Matrix of Transport

F1W2C2	WS1	WS2
F1W2C2WS1	1	1
F1W2C2WS2	1	1

MTT(F1W2C2) Matrix of Temporal Transport

F1W2C2	WS1	WS2
F1W2C2WS1	0	4
F1W2C2WS2	5	0

The Matrix of Logical Transport for F1W2C2 is be established thanks to the equations 1 and 2 and the algorithm 1 : MLT (F1W2C2)

	T1	T2	T3	T4	T5	T6
T1	∞	∞	∞	∞	∞	∞
T2	∞	2,25	2,5	2,25	∞	∞
T3	∞	2	0	2	∞	∞
T4	∞	2,25	2,5	2,25	∞	∞
T5	∞	∞	∞	∞	∞	∞
T6	∞	∞	∞	∞	∞	∞

To the level workshop

The information established for cells are aggregated so as to have the same type of information to manage to the level workshop.

The level workshop works on the macro tasks that are an elementary task succession. The achievable task by the cell is typical CT, so for our example :

CT1 = T2-T3 ; CT2 = T2-T6-T5 ; CT3 = T2-T5 ; CT4 = T4.

Aggregation of matrix of competence

The two following matrix are used.

$$MC(F1W2C2) ; MC(F1W2C1) = \begin{pmatrix} 0&1&0&0&1&1 \\ 1&1&0&0&0&0 \\ 0&0&1&0&1&0 \end{pmatrix}$$

Thanks to the algorithm 2, The matrix of intermediate competence of the workshop 2 is obtained.

$$MIC\ (F1W2) = \begin{pmatrix} 1&1&1&0&1&1 \\ 0&1&1&1&0&0 \end{pmatrix}$$

⇒ The matrix of competence of the workshop **2**

$$MC(F1W2) = \begin{pmatrix} 1&1&1&0 \\ 1&0&0&1 \end{pmatrix}$$

The aggregation of matrix of temporal competence is obtained in the same way that previously.

In this result, the time of transportation between elementary tasks composing a C-task is not taking into account. The utilization of the matrix of logical transport of each cell allows to establish this time of transportation, so the matrix of Temporal Competence of the workshop W2 is : MTC (F1W2)

$$\begin{matrix} CT1 & CT2 & CT3 & CT4 \end{matrix}$$
$$\begin{pmatrix} 11{,}5 & 18{,}5 & 11{,}5 & \infty \\ 16 & \infty & \infty & 4{,}5 \end{pmatrix}$$

Aggregation of matrix of transport

$$MTT(F1W2) = \begin{pmatrix} MTT(F1W2C1) & MTTC1C2 \\ MTTC2C1 & MTT(F1W2C2) \end{pmatrix}$$

But, connections between cells are lacking. The observation of the workshop gives us different possible links turned into two matrix. These Matrix are named Matrix of Temporal Transport from Cell n to Cell n+1 :

$$\mathbf{MTTC1C2} = \begin{pmatrix} \infty & 8 \\ \infty & 6 \\ \infty & \infty \end{pmatrix} ; \mathbf{MTTC2C1} = \begin{pmatrix} 3 & \infty & 5 \\ 7 & \infty & 3 \end{pmatrix}$$

then the Matrix of Temporal Transport W2 becomes

$$MTT(F1W2) = \begin{pmatrix} \begin{pmatrix} 0&4&6 \\ 4&0&3 \\ 6&2&0 \end{pmatrix} & \begin{pmatrix} \infty & 8 \\ \infty & 6 \\ \infty & \infty \end{pmatrix} \\ \begin{pmatrix} 3&\infty&5 \\ 7&\infty&3 \end{pmatrix} & \begin{pmatrix} 0&4 \\ 5&0 \end{pmatrix} \end{pmatrix} = \begin{pmatrix} 2{,}7 & 7 \\ 4{,}5 & 2{,}5 \end{pmatrix}$$

The element $mtt(F1W2)_{22}$ is obtained using the algorithm 3.

The aims is to establish the matrix of logical transport for the workshop W2. For so doing the equations 1 and 2 and the algorithm 1 described in the § 5.2 are used. The result is : **MLT (F1W2)**

$$\begin{matrix} & CT1 \ldots\ldots\ldots\ldots\ldots\ldots CT4 \\ \begin{matrix} CT1 \\ CT2 \\ CT3 \\ CT4 \end{matrix} & \begin{pmatrix} 4{,}2 & 3{,}6 & 3{,}6 & 4{,}8 \\ 4{,}9 & 2{,}7 & 2{,}7 & 7 \\ 4{,}9 & 2{,}7 & 2{,}7 & 7 \\ 3{,}5 & 4{,}5 & 4{,}5 & 2{,}5 \end{pmatrix} \end{matrix}$$

6. CONCLUSION

This paper proposed a generic model of technical data for production system control. Focus is to show how to define macro-tasks, macro-resources and to obtain the set of aggregated technical data at the upper level from the technical data collected at the level of work station. This is a first step to elaborate a complete model of production system control. Future work would be concerned with the extention of the model to a assembly workshop. In the next step, the model would include the modelling of the dynamic aspect of production control taking into account the concept of capacity, load and production follow-up.

REFERENCES

Doumeingts G., Féniè P., Chen D., (1994) GIM : GRAI Integrated Methodology : méthode pour concevoir et spécifier les systèmes productiques - Proc. of ILCE'94 , Montpellier, France.

Koestler A. (1967). The ghost in the machine, Arcana Books, London.

Mathews J.A. (1995) Holonic fundations of intelligent manufacturing systems, 5th IFAC Symposium on Automated System Based on Human Skill, Joint design of Technology and Organisation, Berlin, Germany, 25-28 sept. 1995.

Okino, N. (1989). Bionical manufacturing systems - modelon based approach, In Proceedings of the CAM.I 18th Annual Intenational Conference, New Orleans, october.

Okino, N. (1992). A prototyping of bionic manufacturing system, In: *Flexible Manufacturing Systems Past-Present-Future* (J. Peklenik (Ed.)) CIRP, pp 73-95.

Seidel, D. and Mey, M. (1994) IMS -Holonic manufacturing systems : glossary of terms. In: *IMS - Holonic Manufacturing Systems strategies* (Seidel, D. and Mey, M. (Ed.)). vol. 1, March, IFW, University of Hannover, Germany.

Simon, H.A. (1990). *The Science of Artificial.* MIT Press, Cambridge (Mass.).

Tharumarajah, A., A.J. WELLS and L. NEMES (1996) Comparison of the bionic, fractal and holonic manufacturing concepts. *International Journal of Computer Integrated Manufacturing,* **Volume 9, n° 3,** pp 217-226.

Valckenaers, P., Bongaerts, L., Wyns, J. (1996). Planning Systems in the nex century (II), Preprints, ASI'96, the annual conference of ICIMS-NOE, Toulouse, France, june 2-6.

Warnecke, H.J. (1993). The Factal Company, Springer-Verlag.

Zolhadri, M., Bourrières, J.P. (1995) Modélisation dynamique de l'entreprise en vue de l'évaluation du délai de production, *Congrès international de génie Industriel de Montréal : La productivité d'un monde sans frontière,* Montréal, 18-20 octobre (in French)

ADAPTIVE SFC BASED ALGORITHM ON FLEXIBLE PRODUCTION SYSTEMS

Ferreiro Garcia, R., Pardo Martínez, X.C., Vidal Paz, J. & Coego Botana, J.

Dept. Electrónica e Sistemas. Universidade da Coruña
Facultade de Informática. Campus de Elviña, s/n. 15071. A Coruña. Spain.
E-mail: {ferreiro, pardo}@des.fi.udc.es

Abstract: This paper presents a methodology for flexible process sequencing using SFC (sequential function charts) based approach. In this approach SFC is used as a unified framework for representing both the process sequence and transitions adaptation on a flexible application to a common industrial process.

Keywords: Adaptive Control, Adaptive Systems, Sequential Function Chart, Flexible Manufacturing Systems, Process Control, Sequential Control,

1. INTRODUCTION

It is known that there are two major knowledge representation schemes in process planing systems: a rule-based approach and a frame-based approach. In particular, a rule-based approach has been extensively used for representing process planning knowledge. There exists a strong correspondence between a rule-based and a SFC approaches. A frame based approach can be described on a SFC basis by defining its properties by means of the mathematical description of the function chart.

The mathematical model of the function chart is derived from that of Petri nets (IEC 848, 1988). A function chart is a directed graph defined as a quadruplet

$$[X, T, L, X0]$$

where:

X = (x1,..xm) is a finite, non empty, set of steps;

T = (t1,...tn) is a finite, non empty, set of transitions; X and T represent the nodes of the graph;

L = (l1,..lp) is a finite, non empty, set of directed links, linking either a step to a transition or a transition to a step.

X0 which belongs to X is the set of initial steps.

These steps are activated at the beginning of the process and determine the initial situation. Moreover, the graph is interpreted, meaning that:

with the steps, commands or actions are associated;

with each transition, a logic transition condition is associated.

Steps are represented by labelled squares, transitions by dashes. In addition to the static representation, the graph also has a dynamic aspect defined by evolution rules. Alternatively every quadruplet is a subsequence of a complete sequential task description defined by means of macrosteps. The function chart defined by means of macrosteps is a triplet such that:

$$[M, T, L]$$

where:

M = (M1,..Mm) is a finite, non empty (X>1), set of macrosteps; for X=1, M=X is just a step

T = (t1,...tn) is a finite, non empty, set of transitions between macrosteps;

L = (l1,...lp) is a finite, non empty, set of directed

links, linking either a macrostep to a transition or a transition to a macrostep. Every macrostep consist in a set of above quadruplets.

1.1. SFC description based SCADA

The abilities of SFC to manage control complex and supervision tasks comprises as most important:

1) Sequence level supervision.

2) Modelling and simulation of discrete event systems.

3) Monitoring and diagnosis of sequential processes (batch processes).

4) Operating procedures representation.

5) Large databases partitioning.

6) General procedures in control and representation of sequential reasoning.

In a general context, the global task may be described under SFC as illustrated at figure 1. In this figure it is shown how a SFC can describe a parallel computing process in which flexibility is implemented, assumed the criteria to drive the evolution of rules by updating conditions for transition or transition rules in a manner comparable to an adaptive control system.

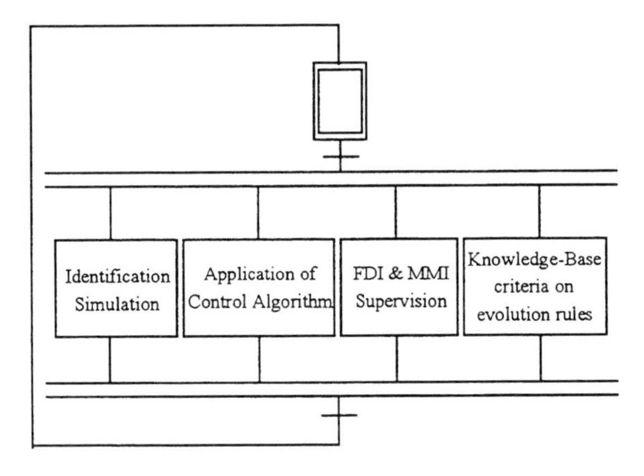

Fig. 1. SFC description of a SCADA task.

In a Petri net description, an incidence matrix is needed to describe the set of transitions and actuation. In a SFC description such incidence matrix can be splitted into two matrixes (Ferreiro, 1995; Ferreiro and Novo, 1995; Kyung-Huy and Moo-Young, 1995):

Transitions enabling matrix.

Actuating matrix.

The enabling matrix contain in every column only the necessary active steps to enable a transition. The actuating matrix contain in every column the necessary steps to be activated (set) and deactivated (reset) by every enabled transition.

Figure 2(a), 2(b) and 2(c) illustrates respectively the SFC, the transition enabling matrix and the actuating matrix of a simple control process.

The matrix of figure 2(b), (the enabling transition matrix), can be interpreted as a rule-base, which states the following rules:

IF X1 = 1
THEN T1 *is enabled* (ET1=1)
ELSE *disabled*

IF X2 = 1
THEN T2 *is enabled* (ET2=1)
ELSE *disabled*

IF X3 = 1
THEN T3 *is enabled* (ET3=1)
ELSE *disabled*

Such rules can be properly managed in closed loop mode by soft or hard depending on the architecture. The matrix of figure 2(c) (the actuating matrix), is interpreted as a set of commands to activate and deactivate the proper steps.

IF ET1 = 1 **AND** T1 = 1
THEN (X2 = 1, X1 = 0)

IF ET2 = 1 **AND** T2 = 1
THEN (X3 = 1, X2 = 0)

IF ET3 = 1 **AND** T3 = 1
THEN (X1 = 1, X3 = 0)

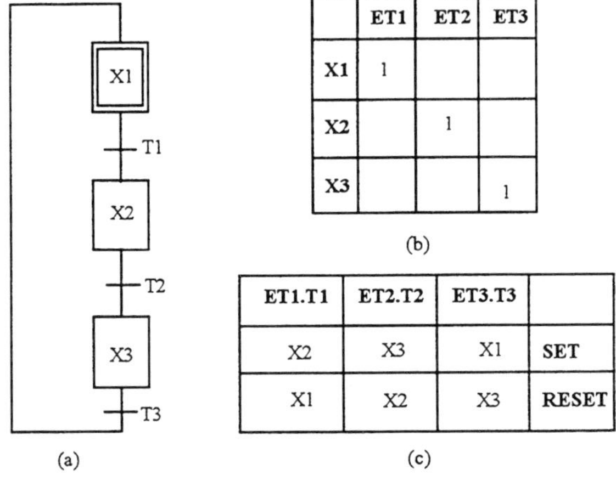

	ET1	ET2	ET3
X1	1		
X2		1	
X3			1

ET1.T1	ET2.T2	ET3.T3	
X2	X3	X1	SET
X1	X2	X3	RESET

Fig. 2. (a) SFC description; (b) transitions enabling matrix; (c) actuating matrix.

It follows that the priority of processing the second matrix depends on the status of the former; so that, to implement the algorithm, the second matrix is a consequence of the closed loop searching for enabling the conditions for transition, that is, second matrix is a set of subroutines processed from the first matrix. In figure 3 it is shown the proposed algorithm, in which structured knowledge is being processed.

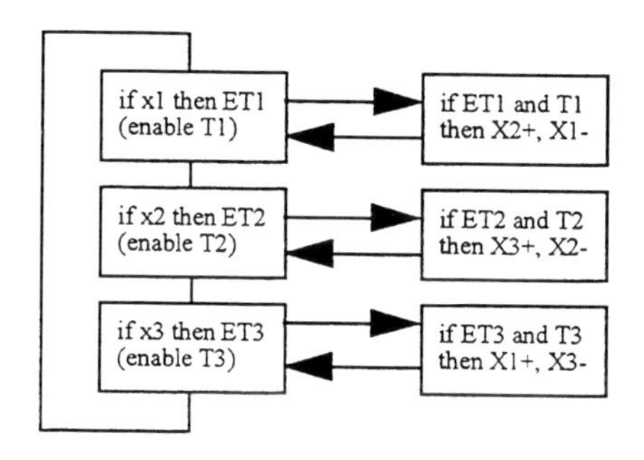

Fig. 3. SFC algorithm

2. SFC BASED FLEXIBLE ALGORITHM

An SFC flexible algorithm is based in its capacity to update the conditions for transition as a function of the plant status which is defined by external dynamic data and the proper internal data status (Ferreiro and Novo, 1995). It can be compared with an adaptive control process in which the conditions for transition are dynamically changed to satisfy performance criteria.

Lets consider the same SFC described in figure 2, with a basic difference which is the capacity of transitions to be updated as function of external data (data gathered from external events directly related with the plant), and internal data (data generated by the system control evolution). Figure 4(a), 4(b) and 4(c) illustrates respectively the SFC, the transition enabling matrix and the actuating matrix of the same simple control. The matrix of figure 3(b), (the enabling transition matrix), can be interpreted as a rule-base, which states the following rules:

IF X1 = 1
THEN T1 *is enabled* (ET1 = 1)
ELSE *disabled*

IF X2 = 1
THEN T2 *is enabled* (ET2 = 1)
ELSE *disabled*

IF X3 = 1
THEN T3 *is enabled* (ET3 = 1)
ELSE *disabled*

Such rules can be properly managed in closed loop mode by soft or hard depending on the architecture.

The matrix of figure 4(c) (the actuating matrix), is interpreted as a set of commands to activate and deactivate the proper steps as well as to update the rules defining the conditions for transition that had been immediately used.

IF ET1 = 1 **AND** T1 = 1
THEN (X2 = 1,X1 = 0), *update* T1

IF ET2 = 1 **AND** T2 = 1
THEN (X3 = 1,X2 = 0), *update* T2

IF ET3 = 1 **AND** T3 = 1
THEN (X1 = 1,X3 = 0), *update* T3

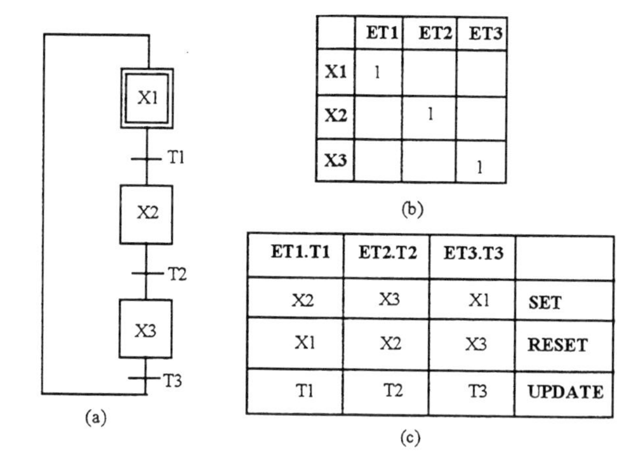

Fig. 4. (a) SFC description; (b) transitions enabling matrix; (c) actuating matrix.

It follows that the priority of processing the second matrix depends on the status of the former; so that, to implement the algorithm, the second matrix is a consequence of the closed loop searching for enabling the conditions for transition, that is, second matrix is a set of subroutines processed from the first matrix. In figure 5 it is shown the proposed algorithm, in which structured knowledge is being processed.

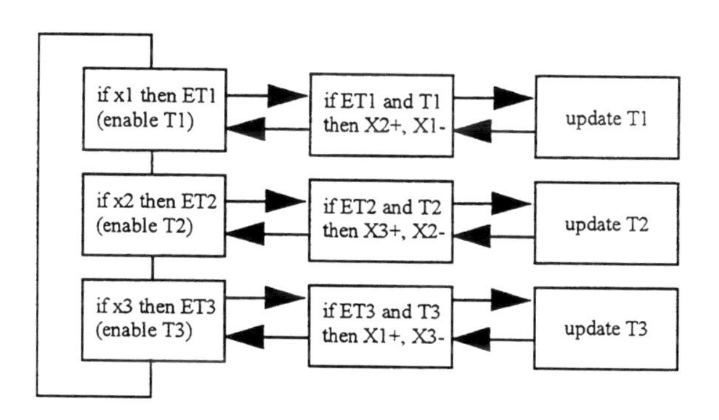

Fig. 5. SFC flexible algorithm

3. APPLICATION TO AN INDUSTRIAL CONTROL TASK

Figure 6 illustrates the structure and SFC associated to the control task of a cylinder actuator which must satisfy requirements such us adaptation on the basis of flexibility.

This application describes an actuator cylinder which can operate under the conditions for transition. The cylinder is at the cycle beginning waiting for an external command in stage E1.

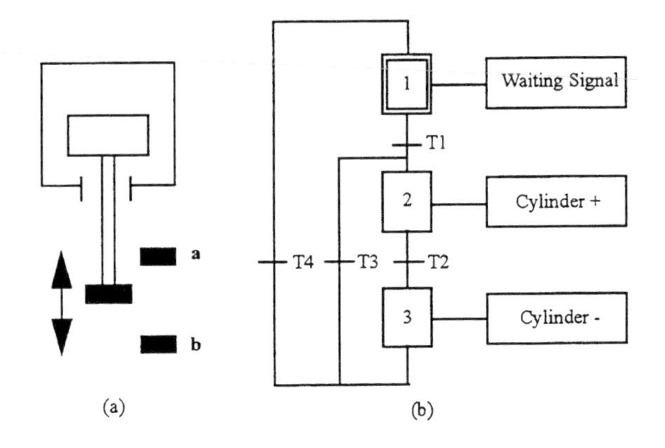

Fig. 6. Flexible application

As described in the SFC associated to that working cycle, when transition T1 or T3 is true, the cylinder moves from stage X2 to stage X3 while goes from stage X3 to X2 or X1 when transition T3 or T4 are true, achieving the initial stage X1. The associated task to achieve the objectives of global process control is defines as:

1) The cylinder is activated by an external command (c) which will be activated by T1 = a.c, where a and b are limit switches.

2) The cylinder moves towards the limit switch b and wait the time ti which is a function of the production demand. That is ti = k0+k1.S, where S is the stock of product manufactured, and k0, k1 are constant parameters to be selected properly. In this case the transition T2 is defined by T2 = b.ti

3) The cylinder returns to the original position completing a operating cycle. Such activity enables the transitions T3 and T4 where T3 = a, T4 = a.N where N is the number of cycles demanded by the production demand scheduler.

The conditions for transitions of a flexible production cycle is synthesized as,

$$T1 = a.c$$

$$T2 = b.ti \text{ with } ti = k0 + k1.S$$

$$T3 = a$$

$$T4 = a.N$$

The transitions enabling matrix and actuating matrix are shown in figures 7 (a) and 7(b). Figure 8 shows the flexible algorithm of this control task.

	ET1	ET2	ET3	HT4
X1	1			
X2		1		
X3			1	1

(a)

ET1.T1	ET2.T2	ET3.T3	HT4.T4	
X2	X3	X1	X1	SET
X1	X2	X3	X3	RESET
T1	T2	T3	T4	UPDATE

(b)

Fig. 7. (a) transitions enabling matrix; (b) actuating matrix.

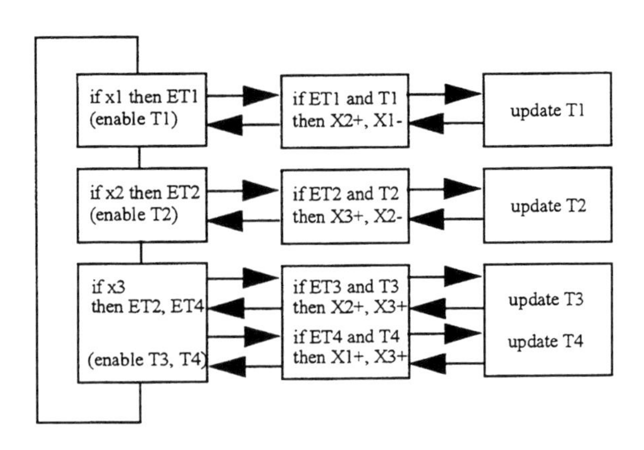

Fig. 8. SFC flexible algorithm

4. CONCLUSIONS

In this work it is shown the structured procedure to synthesis a flexible manufacturing system. From the application example, it is concluded that SFC is well suited to develop hybrid (analog and discrete event control systems) adaptive production projects with high level programming languages, and multimedia open architectures capable for massive parallel computations in order to solve parallel tasks.

REFERENCES

IEC publication 848, (first ed, 1988). *Preparation of function charts for control systems.* Atar S.A., Geneva Switzerland.

Ferreiro Garcia R, F. Novo, (1995). *Bases para el Diseño de Software Orientado al CIM.* 5as. Jornadas Nacionais de Projecto, Planeamento e Produçao asistidas por Computador. 30 maio-1 Junho de 1995. Guimaraes. Portugal.

Ferreiro Garcia R, 1995. *Nociones sobre PLC's aplicados al control industrial.* Cap.I. pp 60-100 Ed. Universidade da Coruña.

Kyung-Huy Lee and Moo-Young Jung.(1995). *Flexible process sequencing using Petri Net Theory.* Computers Ind. Engng. Vol. 28, No.2, pp. 279-290.

HETERARCHICAL CONTROL SYSTEMS FOR PRODUCTION CELLS — A CASE STUDY

J.M. van de Mortel-Fronczak and J.E. Rooda

Eindhoven University of Technology, Department of Mechanical Engineering, P.O. Box 513, 5600 MB Eindhoven, The Netherlands, E-mail: {vdmortel,rooda}@wtb.tue.nl

Abstract: Most control systems of flexible production cells have a hierarchical structure. They become very complicated and difficult to maintain and modify when the underlying production cells grow in size and complexity. Moreover, they are characterized by a relatively high sensitivity to failures. As opposed to that, heterarchical control systems are flexible, modular, easy to modify, and — to some extent — fault-tolerant. In this paper, a heterarchical control system of a flexible production cell is formally specified in the CSP-based language χ. This language is well suited for the description of autonomous components cooperating with each other by exchanging information.

Keywords: Modelling, control systems, parallel programs.

1. INTRODUCTION

Most control systems of flexible production cells have a hierarchical structure and are based on the principle of centralized control. In such a structure, different control layers form a pyramid. The lowest layer performs the commands of the layer above it by controlling the production devices. Every layer but the highest one performs the commands of the layer above it by controlling the layer below it. Furthermore, it receives status information from the layer below it and passes (a part of) it to the layer above it. The highest component in the control hierarchy takes care of the realization of system goals.

Hierarchical control systems are rigid and highly sensitive to failures. Moreover, they tend to become very complicated and difficult to maintain and modify when the underlying production cells grow in size and complexity. A change of the configuration of production facilities often requires extensive (and expensive) changes in the associated control system. This explains the growing interest for heterarchical control systems (Lin and Solberg, 1994; Veeramani *et al.*, 1993). Heterarchical control systems are based on the principle of distributed control and consist of autonomous components cooperating with each other by exchanging information. By means of cooperation, system goals are realized. Heterarchical systems are flexible and modular. They are easy to modify and to adapt in the case of re-configuration or increase in size of production facilities. To some extent, they are also fault-tolerant. However, one should be aware of a possibly high communication overhead inherent to their distributed nature.

The purpose of this paper is to present a formal specification of a heterarchical control system for a flexible production cell. The flexible production cell consists of five machining stations, one quality control station, one input-output station, and an automatic transport system. For the specification, the CSP-based (Hoare, 1985) language χ (van de Mortel-Fronczak and Rooda, 1995; van de Mortel-Fronczak and Rooda, 1996) is used. This language is well suited for the description of autonomous components cooperating with each other by exchanging information. Exchange of information is modelled by synchronous message passing. Component descriptions are concise and comprehensible, and suited for re-use. Specifications defined in χ can be validated by simulation (Naumoski and

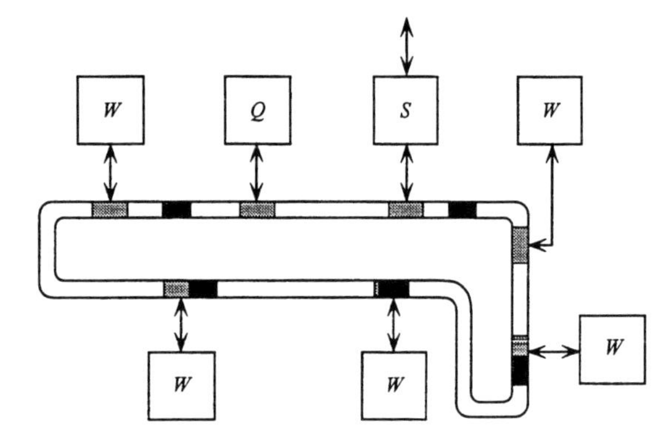

Fig. 1. Flexible production cell

Alberts, 1995). Simulation can also be used to assess communication overhead of the specified control system. Additionally, it is possible to formally verify χ specifications (Kleijn, 1996).

The paper is structured as follows. In Section 2, a heterarchical control system structure of a flexible production cell is discussed. The control components are formally described in Section 3. In Section 4, concluding remarks are presented.

2. CONTROL SYSTEM STRUCTURE

In this section, a formal specification is presented of a heterarchical control system of the flexible manufacturing cell schematically drawn in Figure 1. This cell consists of seven workstations (five machining workstations W, one quality-control station Q, and one storage station S) and an automatic transport system. The transport system consists of the conveyor and seven conveyor stations. The conveyor holds a limited number of pallets and moves at a constant speed. Each conveyor station is associated with one workstation and holds up pallets intended for it. Each workstation is equipped with a robot that moves product carriers between the pallet and the workstation. Every pallet carries a magnetic label needed for identification. The production is organized by autonomous job agents. In (Coenen, 1995), a detailed and slightly different model is presented. In this paper, a model is presented on a higher abstraction level.

For the purpose of this paper, it does not matter what kind of workstations belong to the manufacturing system. A relevant assumption is that they work independently. Moreover, for simplicity it is assumed here that storage S can hold an infinite number of parts and products. In practice, storage S also functions as the input/output station. The manufacturing cell can be used to process and/or assemble a family of products (for instance, as described in (van Brussel *et al.*, 1993) or in (Esh, 1994)).

In the model formally specified in the next section:

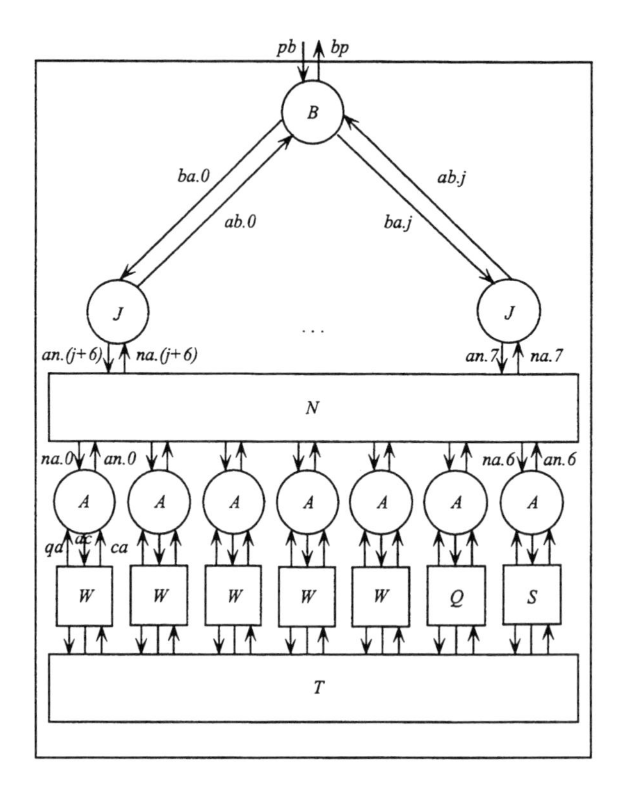

Fig. 2. Model of the cell and its control

- product carriers are universal, that is, each part or product can be placed on each product carrier,
- assembling is not treated,
- failures are not taken into account,
- by orders, released production orders are meant,
- each order applies to one product.

The structure of the model, as drawn in Figure 2, is based on the structure of the flexible production cell. In the model, the following components are distinguished: order buffer B, job agents J, communication network N, workstation agents A, transport system T, and workstations W, Q, and S. Except for the storage, the workstations can be modelled in the same way on this abstraction level. In general, several storage components are allowed.

The components B, J, N, and A belong to the control system of the flexible production cell. The job agents take care of the realization of orders that arrive and leave the cell through buffer B. Every order specifies a product to be produced by defining the raw material type (part) and the sequence of operations leading from this raw material to the product in question. The first operation of each order is a retrieve operation for a storage component.

To realize an order, every job agent repeats the following sequence of steps for every operation defined in the order. It inquires, through the communication network, which workstation is willing to perform the operation. If a few workstations respond to the inquiry, one of them is chosen accord-

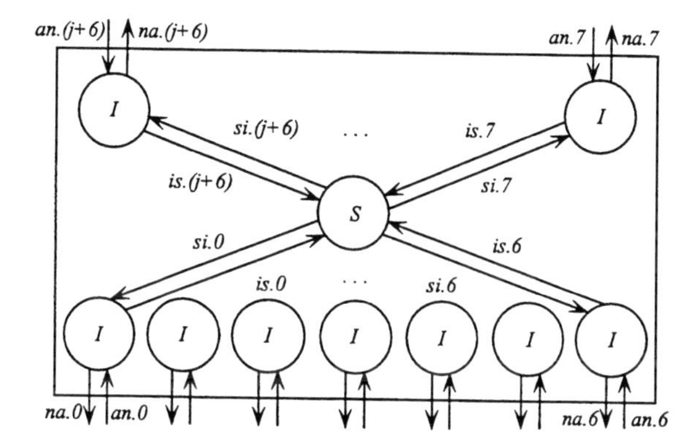

Fig. 3. Subsystem N

ing to a certain criterion. (In the model presented in the next section, the operation is assigned to the first workstation that responds to this inquiry.) The workstation chosen has to wait for the arrival of the pallet carrying the part that has to be transformed. The job agent informs the storage (if the operation is the first one of the sequence) or the workstation that performed the previous operation about the destination of the associated part, so that they can take care of passing the part (with the proper destination address) to the transport system. The transport system then delivers the part to its destination. When the workstation has completed its task, it informs the job agent about this fact and the steps described above can be repeated. The last operation specified in an order is the operation of putting the finished product in the storage.

Subsystem N consists of a number of interface components I collecting messages for and from the components connected to the communication network and switch S that takes care of passing the messages from a source to a destination. The structure of subsystem N is graphically presented in Figure 3.

The components in the model cooperate with each other by exchanging information via communication channels (arrows in the figures) according to a communication protocol.

In the sequel, only components B, J, N, and A associated with the cell control system are specified in detail. For models of workstations and of the transport system the reader is referred to (Coenen, 1995).

In the specification language χ, the following notations are used for dealing with data and defining data types:

- T^* denotes the collection of lists with elements of type T (for instance, nat* — lists of natural numbers),
- $[1, 2, 3]$ denotes the list of numbers $1, 2$, and 3,
- [] denotes the empty list,
- $xs +\!\!+ ys$ denotes concatenation of lists xs and ys,
- $\mathrm{hd}(xs)$ denotes the first element of xs (the head),
- $\mathrm{tl}(xs)$ denotes the list that results from removing the first element of xs (the tail),
- $T_1 \times T_2 \times \ldots \times T_n$ denotes the collection of n-tuples where the ith element is of type T_i (Cartesian product; for instance, nat $\times$ real — pairs of natural and real numbers),
- the elements of n-tuple p are referred to as $p.i$, for $0 \leq i < n$,
- $\langle 1, \mathbf{c}, 7, 30, 5 \rangle$ denotes a tuple of five elements: $1, \mathbf{c}, 7, 30$, and 5.

To model material and information flows, the following data types are used :

- ptype $= \{1, 2, \ldots, n\}$ — numbers corresponding to n part and product types,
- op $= \{1, 2, \ldots, m\}$ — numbers corresponding to m operations that can be performed in the production cell,
- task $=$ ptype $\times$ op — combinations of parts and operations to be applied to them,
- order $=$ ptype $\times$ ptype $\times$ task* — orders with the following contents: a product number, a part number, and a sequence of tasks that transforms the part into the specified product,
- wid $= \{0, 1, \ldots, 6\}$ — workstation identification numbers (7 workstations),
- jid $= \{7, 8, \ldots, j + 6\}$ — job agent identification numbers (j job agents),
- id $=$ wid $\cup$ jid — identification numbers of both the workstations and the job agents,
- info $=$ id $\times$ id* $\times$ char $\times$ nat $\times$ task $\times$ wid — network messages with the following contents: source, message destination list, kind of message, message number, task number, destination for a (semi-finished) product (only for **d**-messages as explained in the next section),
- cinfo $=$ id $\times$ char $\times$ nat $\times$ task $\times$ wid — information for workstation controllers: as described above except for the message destination list.

In this paper, only information flow is treated.

3. DESCRIPTION OF CONTROL COMPONENTS

The components in the model are called processes in the specification language χ. In the heading of each process the communication channels and, possibly, parameters are defined. For instance:

- *pb* : ? order in process B defines input channel *pb* over which orders are received,
- *bp* : ! ptype in process B defines output channel *bp* over which information about finished

products is sent,

- is : ? (info^{j+7}) in process S defines a bundle of $j+7$ input channels $is.0$ through $is.(j+6)$ over which network messages are received,
- si : ! (info^{j+7}) in process S defines a bundle of $j+7$ output channels $si.0$ through $si.(j+6)$ over which network messages are sent,
- id : jid in process J defines a parameter id of type jid.

The body of each process consists of variable declarations followed by executable statements modelling a components behaviour. The executable part begins with some initial communication actions and/or variable initialization. Furthermore, it usually contains an infinite repetition in which after a (conditional) communication action some other statements are performed. The enabled communication actions are executed in the first-in-first-out fashion.

Order buffer B receives orders ($pb\,?\,o$) and stores them in the list os ($os := os +\!\!+ [o]$). Orders are distributed among the free job agents ($ba.i\,!\,\mathrm{hd}(os)$) and subsequently removed from the list os ($os := \mathrm{tl}(os)$). When an order is completed the associated job agent sends the information about the finished product type back to the order buffer ($ab.i\,?\,p$). The information about finished products is stored in the list ps ($ps := ps +\!\!+ [p]$). On request, this information can be passed through ($bp\,!\,\mathrm{hd}(ps)$) and subsequently removed from the list ps ($ps := \mathrm{tl}(ps)$).

```
proc B( pb : ?order, bp : !ptype, ba : !(order^j)
      , ab : ?(ptype^j)
      ) =
|[ o : order, os : order*, p : ptype, ps : ptype*
 | os := []; ps := []
 ; *[                     pb ? o        ⟶ os := os ++ [o]
    [] i : len(os) > 0; ba.i ! hd(os)   ⟶ os := tl(os)
    [] i :                ab.i ? p      ⟶ ps := ps ++ [p]
    [] len(ps) > 0   ; bp ! hd(ps)      ⟶ ps := tl(ps)
    ]
]|
```

Job agent J takes care of the completion of each order received from the order buffer ($ba?o$). To this end, for each task in the list it inquires the workstation agents about the possibilities to perform the task ($an\,!\,\langle id, ds, \mathbf{q}, nr, \mathrm{hd}(ts), 0\rangle$). The first workstation agent that responds to this inquiry (which message p satisfies: $p.2 = \mathbf{r} \wedge p.3 = nr$) is charged with the task ($an\,!\,\langle id, [nw], \mathbf{c}, 0, \mathrm{hd}(ts), 0\rangle$). The strategy of assigning tasks to responding workstations differs from the one described in (Coenen, 1995). There, the workstation that responded in a determined time slot and that has the smallest amount of work to do is charged with the task. This results in a slightly more complicated model but does not contribute to the general idea of the model discussed in this paper. For each task of the list but the first one, the job agent also has to inform the workstation that performed the previous task about the new destination of the semi-finished product ($an\,!\,\langle id, [ow], \mathbf{d}, 0, 0, nw\rangle$). This is achieved by using variable f that changes from true to false after the first task in the list is completed. Each time a task is completed, the job agent receives message p containing this information ($p.2 = \mathbf{s}$). When all tasks in the list have been performed, the information about the finished product type is sent to the buffer B ($ab\,!\,o.0$).

```
proc J( ba : ?order, ab : !ptype, an : !info
      , na : ?info, id : jid
      ) =
|[ nr : nat, nw, ow : wid, o : order, ts : task*
 , ds : wid*, b, f : bool, p : info
 | nr := 0; nw := 0; ds := [0, 1, 2, 3, 4, 5, 6]
 ; *[ ba?o
      ⟶ ts := o.2
       ; f := true
       ; *[ len(ts) > 0
            ⟶ nr := nr + 1
             ; an ! ⟨id, ds, q, nr, hd(ts), 0⟩
             ; b := false
             ; *[ ¬b ; na ? p
                  ⟶ b := p.2 = r ∧ p.3 = nr
                ]
             ; ow := nw
             ; nw := p.0
             ; an ! ⟨id, [nw], c, 0, hd(ts), 0⟩
             ; [ f   ⟶ f := false
               [] ¬f ⟶ an ! ⟨id, [ow], d, 0, 0, nw⟩
               ]
             ; b := false
             ; *[ ¬b ; na ? p ⟶ b := p.2 = s ]
             ; ts := tl(ts)
          ]
       ; ab ! o.0
    ]
]|
```

To model the network, two types of components are defined: interface I and switch S. Interface I is just a buffer for messages sent by or destined to job agents and workstation agents. In general, it can be said that it is a message buffer for components connected to the network.

```
proc I(an, si : ?info, na, is : !info) =
|[ p : info, pi, po : info*
 | pi := []; po := []
 ; *[                  an ? p     ⟶ pi := pi ++ [p]
    [] len(pi) > 0 ; is ! hd(pi)  ⟶ pi := tl(pi)
    []                 si ? p     ⟶ po := po ++ [p]
    [] len(po) > 0 ; na ! hd(po)  ⟶ po := tl(po)
    ]
]|
```

The task of switch S is to deliver the messages sent by job and workstation agents to their destinations. A message can be received from any interface ($is.i\,?\,p$) connected to the switch. The destination list included in the message is stored in ds ($ds := p.1$). Subsequently, the message is sent to all interfaces associated with the destination list ($*[\ \mathrm{len}(ds) > 0 \longrightarrow si.\mathrm{hd}(ds)\,!\,p;\ ds := \mathrm{tl}(ds)\ ]$).

```
proc S(is : ?(info^{j+7}), si : !(info^{j+7})) =
|[ p : info, ds : wid*
 | *[ [ i : is.i?p ⟶ skip ]
    ; ds := p.1
    ; *[ len(ds) > 0 ⟶ si.hd(ds)!p; ds := tl(ds) ]
    ]
]|
```

Workstation agent A communicates with job agents. It can receive ($na\,?\,p$) one of three kinds of messages: a request ($p.2 = \mathbf{q}$), a confirmation ($p.2 = \mathbf{c}$), or a destination ($p.2 = \mathbf{d}$). It passes the message received (except for the destination list) to the workstation controller ($ac\,!\,\langle p.0, p.2, p.3, p.4, p.5\rangle$). The workstation controller uses different parts of the message depending on its kind (**q**: task number; **c**: source, message number, task number; **d**: destination for the (semi-finished) product). In the case of a request, the workstation controller gives in response the information whether or not the workstation is able to perform the specified task ($qa\,?\,b$). If true (b) the workstation agent sends a positive replay to the requesting job agent ($an\,!\,\langle id, [p.0], \mathbf{r}, p.3, 0, 0\rangle$). When the task assigned to the workstation is completed, the workstation controller informs the workstation agent about it ($ca\,?\,c$). The workstation agent passes this information to the proper job agent ($an\,!\,\langle id, [c.0], \mathrm{s}, c.1, 0, 0\rangle$).

```
proc A( na : ?info, an : !info, ac : !cinfo
      , qa : ?bool, ca : ?⟨id × nat⟩, id : wid
      ) =
|[ p : info, b : bool, c : ⟨id × nat⟩
 | *[ na?p
      ⟶ ac!⟨p.0, p.2, p.3, p.4, p.5⟩
       ; [ p.2 = q
           ⟶ qa?b
            ; [ b  ⟶ an!⟨id, [p.0], r, p.3, 0, 0⟩
              [] ¬b ⟶ skip
              ]
         [] p.2 ≠ q
           ⟶ skip
         ]
    [] ca?c
      ⟶ an!⟨id, [c.0], s, c.1, 0, 0⟩
    ]
]|
```

Components can be grouped in systems. In the heading of each system the communication channels and, possibly, parameters are defined. The body of each system consists of internal channel declarations followed by the enumeration of the components (processes or subsystems) with appropriate channel names and parameters. Internal channels connect pairs of system components and are therefore always listed in two instances. This way interaction path in the system are established. The components in a system definition are separated by $\|$ symbol which stands for parallel composition. In the case of channel bundles and various instances of one process definition an abbreviation may be used as shown below.

Formally, subsystem N graphically presented in Figure 3 is defined by:

```
syst N(an : ?(info^{j+7}), na : !(info^{j+7})) =
|[ is : (info^{j+7}), si : (info^{j+7})
 | S(is, si)
 || i : 0 ≤ i < j + 7 : I(an.i, si.i, na.i, is.i)
]|
```

Specification of the remaining components (from Figure 2) is beyond the scope of this paper.

4. CONCLUDING REMARKS

In this paper, a specification of a heterarchical control system for a flexible production cell is presented. In particular, the part that takes care of the organization of production is specified in detail. Descriptions of the components are concise and easy to change if a different strategy of assigning tasks to workstations is to be applied. The number of job agents is a parameter of the system, so it can easily be varied. However, in a particular system instance, this number is fixed. The control system is fault-tolerant to some extent: a workstation failure does not influence the functioning of the remaining parts of the production cell. It is also modular and easy to modify if the configuration of production facilities would change.

The specification language used has a parallel nature and is therefore well suited for the description of heterarchical control systems. Modelling with χ is supported by a software system including a simulator for systems defined in this language.

REFERENCES

Coenen, F.W.J. (1995). Een heterarchische besturing voor flexibele productiesystemen. Master's thesis. Eindhoven University of Technology, Department of Mathematics and Computing Science. Eindhoven, The Netherlands. (In Dutch).

Esh (1994). *Computer Integrated Manufacturing for Industrial Training Applications.*

Hoare, C.A.R. (1985). *Communicating Sequential Processes.* Prentice–Hall International. London, UK.

Kleijn, J.J.T. (1996). Verifying a χ specification with the aid of PSF and ACP. Report WPA

420104. Eindhoven University of Technology, Department of Mechanical Engineering. Eindhoven, The Netherlands.

Lin, G.Y. and J.J. Solberg (1994). Autonomous control for open manufacturing systems. In: *Computer Control of Flexible Manufacturing Systems: Research and Development* (S.B. Joshi and J.S. Smith, Eds.). pp. 169–206. Chapman and Hall. London.

Naumoski, G. and W.T.M. Alberts (1995). The χ engine: a fast simulator for systems engineering. Final report. Eindhoven University of Technology, Stan Ackermans Institute. Eindhoven, The Netherlands.

van Brussel, H., Y. Peng and P. Valckenaers (1993). Modelling flexible manufacturing systems based on Petri nets. *Annals of the CIRP* **42**(1), 479 - 484.

van de Mortel-Fronczak, J.M. and J.E. Rooda (1995). Application of concurrent programming to specification of industrial systems. In: *Proceedings of INCOM'95.* Beijing, China. pp. 421–426.

van de Mortel-Fronczak, J.M. and J.E. Rooda (1996). On the integral modelling of control and production management systems. In: *Proceedings of APMS'96* (N. Okino, H. Tamura and S. Fujii, Eds.). Kyoto, Japan. pp. 171–176.

Veeramani, D., B. Bhargava and M.M. Barash (1993). Information system architecture for heterarchical control of large FMSs. *Computer Integrated Manufacturing Systems* **6**(2), 76–91.

A PROCESS ORIENTED METHOD FOR THE REUSE OF CIM MODELS

B. Janusz

University of Karlsruhe, Institute for Real-Time Systems and Robotics (IPR)
Kaiserstr. 12, D-76128 Karlsruhe, Germany, E-Mail: bjanusz@ira.uka.de

Abstract: Enterprise models are used to support the solution of several organizational tasks. Although most of the existing modelling concepts were developed under the consideration of the aspect of process orientation, no methods exist which would prevent the user from creating a function oriented model. Beside this, once created models often cannot be reused for new tasks because the view on the enterprise described in the existing model does not correspond to the actual requirements. In the first part of this paper, a new approach is described, which helps to overcome these problems. In the second part, an optimization approach for the modelled process is proposed.

Keywords: Enterprise modelling, optimization

1 INTRODUCTION

Enterprise models are used to obtain a detailed description of the relevant aspects in the enterprise. They provide a detailed and well structured overview of what happens in the enterprise. Having such an overview, several organizational tasks can be solved in a straightforward way. Examples for this kind of tasks are, e.g., the planning of the computer usage, the enterprise wide data integration, the redesigning of business processes or the documentation according to ISO 9000.

The consideration of some existing enterprise models, e.g., Zelm *et al.* (1995); Neuscheler (1995); Hohn (1993); Schlotz and Roeck (1995), shows that often the models are designed for a specific task and cannot be reused for solving another problem. Following results were obtained:

- The models were generated for a specific task, e.g., for the design of a logistic system or for the optimization over time in the production. The existing models are not universal enough to be reused. Therefore, for each other problem, a new model of the enterprise has to be generated.
- Up to now, the completeness of a model cannot be ensured. Therefore, also a fault free extensibility method does not exist.
- No methods for the optimization of processes are provided in modelling concepts. Regarding the existing concepts, process reengineering only consists of process modelling.
- According to AMICE (1994) and Vernadat (1995), model domains have not be confused with organisational departments. They should be logical groupings of processes crossing several departments borders. Nonetheless, the existing models do not describe global processes consisting of activities leading to a common goal. Often, a process, which is global in reality, is represented by several local sub-processes and the connections between them are not obvious immediately in the model. Also, some activities may belong to the modelled processes, which lead to other goals than the considered ones.

The reason for the last aspect is the information gathering process. It is presented in Figure 1 and Figure 2. In Figure 1, the desired information gathering procedure is presented. The modeller interviews the people working in the different departments and gathers the local information about the enterprise. Then he merges and structures all the locally gathered information and generates a global model of the enterprise.

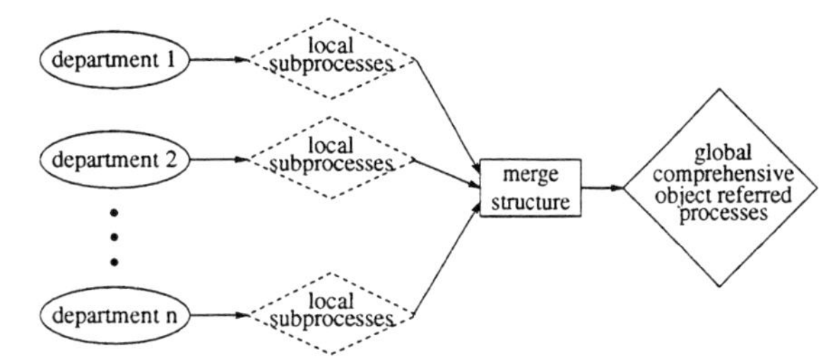

Fig. 1. Information gathering process - the requirements.

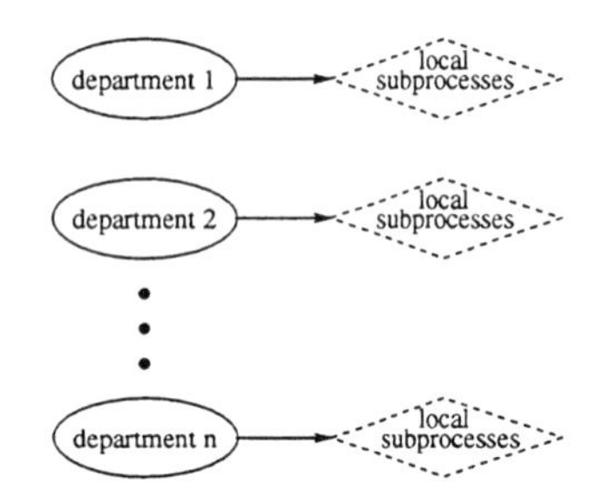

Fig. 2. Information gathering process - the fact.

But in Figure 2, the more probable information gathering process is presented. The modeller gathers the information locally and does not merge it. The amount of different information is too high, it is too complex and not clear enough to be structured by the modeller. Hence, the result is an enterprise model consisting of single, nearly independent sub-models without processes crossing the departments borders.

This is also the case, when parts of the enterprise were modelled in the past, e.g., the design department for the introduction of a CAD system, and the planning department for a PPS system. Up to now, there are no methods to connect these parts to get a global model. Considering the time the modeller needs for the creation of a new enterprise model (on average one year), it would be useful to be able to reuse once generated parts of an enterprise model.

Therefore, in the following, an approach is described which helps to overcome these problems (Section 3). It is shown, how new structured process models can be gained out of existing enterprise models and hence how to reuse once created models for new tasks.

As an example for the usage of these re-structured models, a business process redesign approach is presented in Section 4.

2 MODELLING OF ENTERPRISE ACTIVITIES

The process filtering approach described in Section 3 is realised on the basis of the modelling concept CIMOSA (AMICE, 1993), but it can be easily adapted to other modelling concepts. The most important requirement for this approach is to use a modelling concept which allows to model activities and their inputs and outputs. For instance, this can be done by the CIMOSA construct Enterprise Activity (Fig. 3). By definition, an Enterprise Activity transforms its Function Input into the Function Output using its Control Input and Resource Input (AMICE, 1994). Eventually, a Control Output and a Resource Output can be produced. Hence, an Enterprise Activity is mainly defined by its functionality, i.e., its Function, Control and Resource Inputs and Outputs. By definition, all processes in an enterprise share a single pool of Enterprise Activities.

The Enterprise Activity construct is central for the algorithm presented in the following.

3 PROCESS ALGORITHM

In the following, it is assumed that an enterprise model describing the as-is situation exists. It contains several information about the enterprise. This model is not necessarely process oriented and global, the modelled processes may be limited to departments borders. It is also possible that it was generated for the description of some other aspects and not in the context in which it should be considered actualy. Therefore, it is possible that the needed information is given but in a badly structured way and that it should be re-structured to gain a better overview.

Using the approach described in this section, a such quite arbitrarely structured model can be restructured. The result are models of processes leading to some specific objectives. The algorithm filters the needed information out of a general model of a particular enterprise.

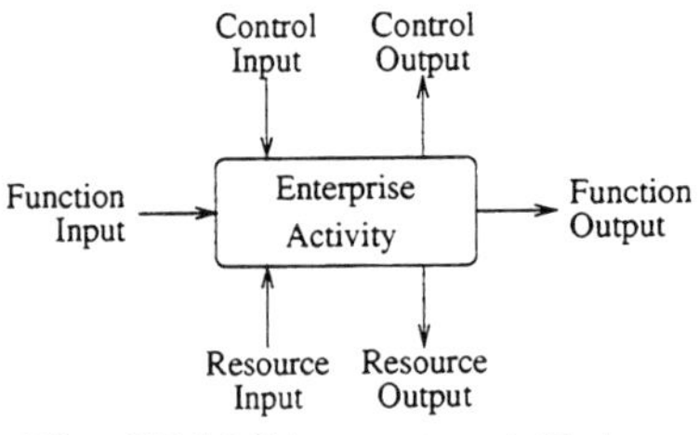

Fig. 3. The CIMOSA construct Enterprise Activity.

3.1 The algorithm

Because the algorithm is used to generate models of processes which are dedicated for the achieving of a given goal, the algorithm works backwards from the goal of a process to the first activity of the process:

1. Define the desired output objects of the process.
2. Look for all activities producing these objects as their output. They form the last stage of the process.
3. Define the objects which are used by these activities as their input.
4. Look for all activities producing these objects as their output. They are predecessors of the actually considered activities.
5. Finish, if all activities found do not have any inputs or if all activities only need inputs which are assumed to be given. Otherwise go back to step 3.

The result of the algorithm is a precedence graph which enumerates all necessary activities and the sequence of their execution needed for the achievment of a given goal. This sequence gives the predecessor relationships, i.e., it describes which activities must be completed before the execution of the current activity can be started.

3.2 The implementation of the algorithm

The algorithm presented above can be implemented in the following seven steps:

1. Transform the description of the activities into a matrix notation: For this purpose, define two $n \times m$ matrices, whereby n is the number of different input and output object types and m is the total number of different activities in the enterprise model. The matrices B_I and B_O describe, which objects are used and/or produced by which activities:

$$b_{I_{ij}} = \begin{cases} 1, & \text{object } i \text{ is used by activity } j \\ 0, & \text{otherwise,} \end{cases} \quad (1)$$

$$b_{O_{ij}} = \begin{cases} 1, & \text{object } i \text{ is produced by activity } j \\ 0, & \text{otherwise.} \end{cases} \quad (2)$$

Define also the $2 * m$ vectors In^j and Out^j with the length n:

$$In_i^j = b_{I_{ij}} \quad (3)$$

$$Out_i^j = b_{O_{ij}} \quad (4)$$

2. Compute the two auxilary matrices B'_I and B'_O:

$$b'_{I_{ij}} = \max\{0, b_{I_{ij}} - b_{O_{ij}}\} \quad (5)$$

$$b'_{O_{ij}} = \max\{0, b_{O_{ij}} - b_{I_{ij}}\} \quad (6)$$

They correspond to the matrices B_I and B_O respectively without considering the objects which belong to the input as well as to the output of an activity. These not consumed objects often occur in enterprise models. For instance, in a manufacturing process, some auxilary tools like pallets are not changed by activities. In business processes, such an object may be an information which supports the execution of an activity, for instance, it may be the identifier of an object processed by an activity. It can be assumed that no activity exists which has only the same input and output objects, because this activity would not perform any task.

3. Define the object vector *end* with:

$$end_i = \begin{cases} 1, & \text{object } i \text{ is one of the desired objects produced by the process} \\ 0, & \text{otherwise.} \end{cases} \quad (7)$$

4. Multiply matrices $B_O'^T$ and B'_I . The result is a $m \times m$ matrix PA with:

$$pa_{ij} = \begin{cases} x > 0, & \text{activity } i \text{ produces } x \text{ objects used by activity } j \\ 0, & \text{otherwise.} \end{cases} \quad (8)$$

5. Multiply *end* and B_O. The result is the activity vector *op* with:

$$op_j = \begin{cases} 1, & \text{activity } j \text{ belongs to the last stage of the process} \\ 0, & \text{otherwise.} \end{cases} \quad (9)$$

6. Perform he following modificated Breadth First Search algorithm (BFS) in the matrix PA:

1. Define the set $Stage_0$ which contains all activities whose entries are set to 1 in the vector *op*.
   Set $i = 0$.
2. Set $i = i + 1$
   Define the set $Stage_i$ which contains all predecessors of all activities in the set $Stage_{i-1}$.
3. Repeat the preceding step until no predecessors can be found.

Because matrix PA is acyclic, this modified BFS algorithm always terminates.

7. For the definition of the precedence relationships between single activities, not between sets of activities, perform the following procedure:

$i :=$ number of stages
while $(i > 0)$
  $Pre := \{Out^j$: activity j belongs to $Stage_i$ or $Stage_{i-1}\}$
  $|Pre| := k$
  $Suc := \{In^j$: activity j belongs to $Stage_{i-1}\}$
  $|Suc| := l$
  $m := 1$
  while $(m \leq l)$
   solve the inequation system:

$$\sum_{n=1}^{k} a_n * Pre_n \geq Suc_m$$

with the constraint: $a_n \in \{0, 1\}$.

Thereby, Pre_x (Suc_x) is the xth element of the set Pre (Suc) respectively. The set Pre must consists of the output objects of $Stage_i$ and of $Stage_{i-1}$ because the stages were computed without considering the not consumed but used objects. If there are several activities in $Stage_{i-1}$ which use but do not consume the same objects, they cannot be executed in parallel.

For each activity m in stage $i-1$, all activities n in stage i or $i-1$ are its predecessors, for which $a_n = 1$.

The resulting precedence graph is then transformed back into the CIMOSA notation.

3.3 Completeness of the process

The inputs and outputs of the Enterprise Activities are pointers to the three other CIMOSA modelling views, the Resource, Information and Organization View. Therefore, by guaranteeing the completeness of the Enterprise Activities in the process and by assuming that the inputs and the outputs of an Enterprise Activity are described, the completeness of a process is ensured when all Enterprise Activities belonging to it are described in the model. Using the algorithm described above, the points in the process which are incomplete are indicated to the user. This is always the case, when the process algorithm can not find an activity which produces the output needed as the input by other activities belonging to the process. Hence the user gets detailed indications to the information which is missing.

4 OPTIMIZATION APPROACH

In this section, an example for the usage of a process model obtained by executing the process algorithm presented in Section 3, is shown. The process model can be used for the process redesigning, i.e., the improvement of the process performance in the enterprise. One first redesigning step is already done by the process algorithm because it enables to recognize which activities are important to achieve a goal and which are not. But also some further improvements can be realized after a meticulous analysis of the re-structured process. A proposed procedure for the optimization of processes is presented in Figure 4. It is an extension and a precision of the approach proposed by Taudes *et al.* (1996).

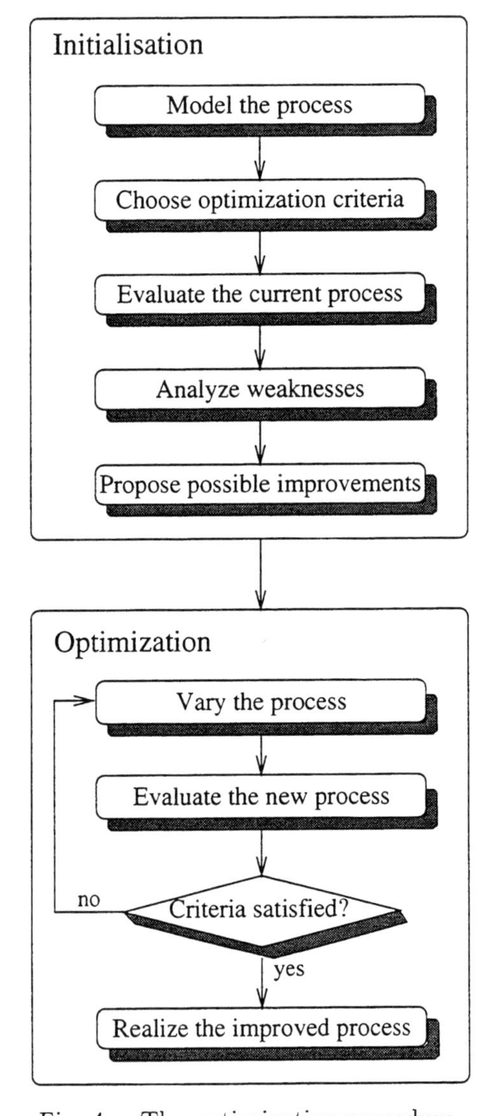

Fig. 4. The optimization procedure.

In the following, the initialisation and optimization steps are described in more details.

4.1 *Modelling of the process*

There are no modelling methods which must be followed to be able to use an enterprise model for the process optimization. Parts of the enterprise can even be modelled locally and independent on each other. If only some given name conventions are taken into account when common components are modelled in different parts of the model, the process algorithm connects all necessary model parts to one common process model. This is then used in the further steps of the optimization procedure.

4.2 *Evaluation of the process*

For the process evaluation, the evaluation criteria have to be choosen by the user. They mainly refer to the time and the cost aspects. For the time aspect, the lead time or the due date satisfaction may be the criteria. Cost criteria may refer to the stock of stores, the resource utilisation, the labour cost, etc.

Several methods for the process evaluation were developed in the past. e.g., Müller (1993) and Neuscheler (1995). Müller (1993) computes the mean values for the lead times and order processing cost, for the duration of the process, its cost and its interruption frequency.

When the process is modelled using the CIMOSA modelling concept, the method presented in Neuscheler (1995) ist the most suitable, because this simulation based method is an extension of the CIMOSA concept. It allows to evaluate processes by taking into regard user defined criteria.

4.3 *Weakness analysis*

There are several possibilities for weaknesses in business processes. Therefore, the exposing and the analysis of these weak points consists of several sub tasks, e.g., the investigation of:

- breaks in the media chains,
- changes in the automation degree in the process,
- the organizational assignments,
- the assignment of resources,
- the degree of necessity of the execution of some activities, e.g., activities which only forward some information or objects from one department to another,
- the frequency of the execution of some activities, e.g., of the the necessary transport activities,
- the spatial and temporal division of a process in eventual independent segments, and
- the abilities and the number of the people involved in the execution of the process.

Of course, this are only few examples which can be extended by some other aspects (see, e.g., Ferstl and Sinz (1996)), but they seem to be the most general and the most suitable for an algorithmic analysis based on an enterprise model.

4.4 *Variation of the process*

The optimization of processes is a creative procedure which cannot be performed automaticaly. Therefore, the optimization algorithm can discover some weak points in the process (see Section 4.3), but it cannot suggest improvements. This must be done by the user, who is supported by the results of the weakness analysis.

For each weak point in the process, the user can propose several possible alterations. The problem which has to be solved by the optimization algorithm, is to decide which of the proposed changes should be realised. This mostly depends on the criteria given in the initialisation phase. Consider for instance two mutual changes M_1 and M_2, whereby M_1 is less costly then M_2. If they result in a similar improvement of the time aspect, also if M_2 is a little better then M_1, M_1 should be prefered, when the cost should be minimized.

In the following, a simulation based method for the decision process is presented.

The dynamic, simulation based evaluation allows the investigation of different process variants and their impacts. For this purpose, from the set of all possible alterations proposed by the user, a random combination of them and their parameters is choosen and realised in the process model. Afterwards, this modified model can be transformed into the simulation model using the transformation method developed by Reithofer (1996). The simulation is then performed and the new version of the process evaluated according to the criteria used for the evaluation of the original process. If the variation does not improve the process, it is rejected. If it is an improvement, then it is used as the reference process for further variants. If the global optimization criterium is satisfied, the optimization process is finished. It is also finished, if no improvements are achieved after a given number of the investigated process variations.

This method can be quite lengthy in reality. Therefore, it is meaningful to investigate at first the impacts of the proposed alterations independently on each other by modelling one change at once and evaluating it by simulation. After approximately knowing the impacts of single changes, their common impacts can be better estimated and the direction of the process variants can be choosen less randomely than without this knowledge.

4.5 Realization of the improved process

The last step of the redesigning process is its realization in the enterprise. This is rather the task investigated in the manpower studies and therefore, it is not part of this paper.

5 CONCLUSION

In this paper, a method for the re-structuring of process models is presented. It is a new approach which ensures the ability of reusage of parts of enterprise models. By filtering task specific processes out of a general enterprise model, a process oriented model is achieved which only contains the information needed for the solution of a given task. An advantage of the approach is, that the user does not need to have a global overview on the whole enterprise in order to generate the enterprise model. It is enough to model each department or even smaller parts of an enterprise separatelly, therefore, also former, incomplete models can be reused for different new tasks, if the interfaces between the departments are clearely defined. The explicit connections between the activities in the departments are generated by the process algorithm and thus the process oriented modelling is achieved. The process can also be optimized using the optimization approach presented in Section 4.

6 ACKNOWLEDGEMENTS

This research was performed at the Institute for Real-Time Compter Systems and Robotics (IPR), Prof.Dr.-Ing. U. Rembold and Prof.Dr.-Ing. R. Dillmann, Computer Science Department, University of Karlsruhe. We would like to thank Prof. Rembold for his support.

REFERENCES

ESPRIT Consortium AMICE (1993). *Open System Architecture for CIM.* Springer-Verlag.

ESPRIT Consortium AMICE (1994). *Formal Reference Base - ESPRIT Project 7110 AMICE-CIMOSA, M-3 Deliverables - Volume 3.*

O. K. Ferstl and E. J. Sinz (1996). Geschäftsprozessmodellierung im Rahmen des Semantischen Objektmodells. *G. Vossen and J. Becker, Geschäftsprozessmodellierung und Workflow-Management,* pages 47–61.

Th. Hohn (1993) Vergleich von Unternehmensmodellen auf der Basis offener Systemarchitekturen mit einem objektorientierten Modellierungsansatz am Beispiel eines Sondermaschinenherstellers. Master's thesis, wbk, Universität Karlsruhe.

S. Müller (1993). *Entwicklung einer Methode zur prozessorientierten Reorganisation der technischen Auftragsabwicklung komplexer Produkte.* Verlag Shaker.

F. Neuscheler (1995) *Ein integrierter Ansatz zur Analyse und Bewertung von Geschäftsprozessen.* Forschungzentrum Karlsruhe, Wissenschaftliche Berichte.

W. Reithofer (1996) Bottom-up modelling with CIMOSA. In *Proceedings of the Second International Conference on the Design of Information Infrastructure Systems for Manufacturing '96,* 1996.

C. Schlotz and M. Roeck (1995). Reorganisation of a production department according to the CIMOSA concepts. *Computers in Industry,* 27:179–189.

A. Taudes, P. Cilek, and M. Natter (1996). Ein Ansatz zur Optimierung von Geschäftsprozessen. *G. Vossen and J. Becker, Geschäftsprozessmodellierung und Workflow-Management,* pages 177–189.

F. B. Vernadat (1995). CIM business process modeling and analysis. In *The 11th ISPE /IEE/IFAC International Conference on CAD/CAM Robotics and Factories of the Future'95,* pages 29–36, Columbia.

M. Zelm, F. B. Vernadat, and K. Kosanke (1995). The CIMOSA business modelling process. *Computers in Industry,* 27:123–142.

COMPUTER PERIPHERAL DEVICE FOR MODELLING 3D OBJECTS IN A CAD ENVIRONMENT

Nelson Acosta[1] & Géry Bioul

ISISTAN - Departamento de Computación y Sistemas
Facultad de Ciencias Exactas - UNCPBA
San Martin 57 - 7000 Tandil - Provincia de Buenos Aires - Argentina
Email: nacosta@tandil.edu.ar

Abstract: The complexity of 3D-object design on a CAD workstation system is mainly due to the definition of the coordinates that delimitate the object-space. The quantity of points dramatically grows with the required detail level of the object. The coordinates set may be defined and adjusted point by point, to obtain a good quality design. This task is time consuming, thus expensive. The 3D graphic definition is a hard job in which one of the most complex problems is the difficulty to represent the 3D workspace using only 2D devices (such as mouse, 2D screen, 2D scanner). The presently available specification devices are either not easy to use or very expensive. The users generally work with conventional 2D devices, but it is necessary to emulate a 3D space. This abstraction overloads the application field of each function which accordingly trends to become ambiguous. The proposal of this paper is to reduce the definition time by proposing a device design that would be able to work as a 3D digital pointer, the **BEE**. The 3D pointer (BEE) is designed to provide, in a practical and easy to handle way, 3D coordinates to the workstation which will generate the computerized graphic definition. The BEE allows the automatic capture of each object coordinate. Once the relevant points have been captured, the corresponding coordinates can be used to derive primitive objects (using lines, curves, arcs, planes, surfaces). Those primitive objects can then be used to form a complex object. The system uses the BEE as a kind of 3D mouse with the capacity to point at any point of the object space.

Keywords: CAE, controlled devices, data acquisition, measuring points, input elements

1.- INTRODUCTION

The decreasing prices together with the increasing processing capacity of modern computer systems, make CAD/CAM systems every day more attractive for industrial applications. A great amount of research and development is currently dedicated to cope with the continuously growing demand from industry. A particular aspect of this demand comes from the need of easy to handle peripheral devices for capturing physical data of objects to be sketched and saved in a computer system for further processing (Blanchard, et al., 1990; Hays, 1993; Krueger and Froehilich, 1994; Marcus, 1993).

The complexity of 3D-object design on a CAD workstation is mainly due to the definition of the coordinates that delimitate the object-space. The quantity of points dramatically grows with the desired detail level of the object. The coordinate set must be defined and adjusted point by point, to

[1] Becario Perfeccionamiento, Secretaría de Ciencia y Técnica de la UNCPBA, Tandil, Bs. As., Argentina.

obtain a good quality design. This task is time consuming, thus expensive.

3D graphic definition is a hard job in which, one of the most complex problems is the difficulty to represent the 3D work space using only 2D devices (such as mouse, 2D screen, 2D scanner). The presently available specification devices are either not easy to use or very expensive. For price reasons, the user generally works with conventional 2D devices, but it is necessary to emulate a 3D space. This abstraction overloads the application field of each function which accordingly trends to become ambiguous (Marcus and Van Dam, 1991; Mandelkern, 1993; Nielsen, 1993). The basic idea is to reduce the definition time by using a 3D device that work as a 3D digital pointer. From now on, this device is referred to, as **BEE**. The BEE allows the automatic capture of each object coordinates. Once the relevant points have been captured, the corresponding coordinates can be used to derive primitive objects (using lines, curves, arcs, planes, surfaces). Those primitive objects can then be used to form a complex object. The system uses the BEE as a kind of 3D mouse with the capacity to point at any point of the object space.

The definition of a complex 3D object, in a CAD workstation environment, consists of locating all the points necessary to delimit, as uniquely as possible, the space occupied by the object. Generally, for irregular objects, the number of points to be located can increase dramatically with the desired detail level of the representation.

2.- BEE SYSTEM ANALYSIS

The key idea of this work is to explore the capabilities of exoskeletons to work as data capturing instruments. A prototype has been designed with 5 links/4 degree of freedom (Fig. 1). One of the most simple applications to be mentioned is the successive establishing of some specific details of a three-dimensional object in real space. As far as a good precision can be achieved, such tool will provide the following advantages: increased speed in data retrieving, providing error-free dimensioning, less operator skill with respect to that required by manual instruments, and all computation efforts transferred to the computer CAD system. The application proposed will use a graphic 3D editor defined as a subset of a CAD package. The display representation, on a SVGA monitor, will be wire-frame.

The equipment is adapted for measuring 4 angles in parallel, without requesting computer assistance. Whenever needed the data will be addressed by the computer to the registers of the peripheral device. The basic pointer structure is an articulated arm similar to an exoskeleton. This device is designed to capture universal *XYZ* coordinates within the workspace range. Those coordinates are computed, with reference to joint 1 (base to link 1) location, from the links angular deviations. The operator directly handles the system, moving it manually up to the point to be captured in order to ascertain and store coordinates.

The workspace required by the application should be within the range of the BEE. Every capturing operation involves: end-effector space location adjustment, then, computer evaluation of *XYZ* coordinates through angular deviation transformations (Denavit and Hartenberg) with eventual adjustments for base translations (Critchlow, 1985; Samson, et al., 1991; Sandler, 1991). Calculation can be carried out by processing units directly attached to the device or by a remote computer connected to the BEE. This latter option has been selected in this work because, on one hand, the BEE design can be significantly simplified while, on the other hand, standard computer software is perfectly suited to handle the task.

2.1.- MECHANICAL STRUCTURE

The basic mechanical structure is made of five links with the following features (Fig. 1):

- link 0: the steady base,
- link 1: vertical rod, with rotational degree of freedom around its main axis Z (angle θ)
- links 2 to 4: articulated with respect to each other with a single angular degree of freedom (angles α, β, γ).

On every joint is mounted an *encoder* whose function is measuring and coding the related angle. In the proposed implementation, the encoder is based on mechanical-optoelectronic principles (Johnson, 1988; Lee, 1988).

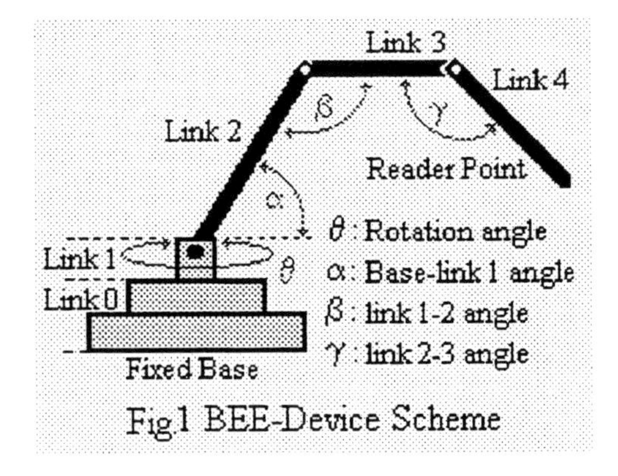

Fig.1 BEE-Device Scheme

2.2.- ENCODERS

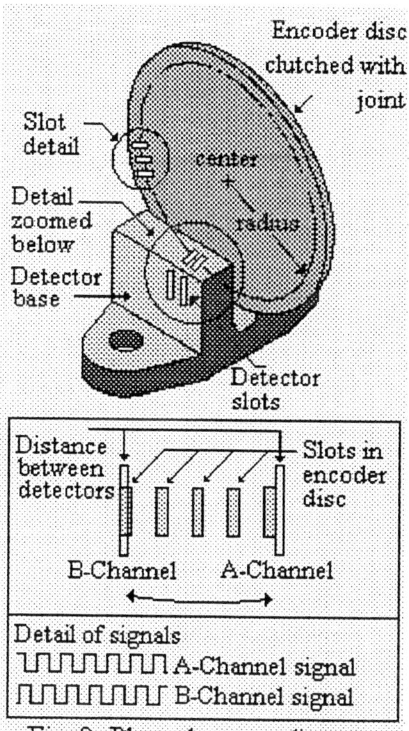

Fig. 2 - Photodetector diagram

An encoder is a transducer, whose output are pulses or binary numbers representing an attached shaft position or movement. Typical optical encoders use a transparent disc marked with concentric opaque stripes or an nontransparent disc with radial slots. This latter one be referred to as slotted-wheel. A photo emitter is placed on one side of the disc and a photo receiver on the other side. Whenever the disc rotates, the light ray is periodically interrupted. The photo receiver thus generates a pulse wave whose frequency is proportional to the rotation speed. Typically, two emitter-receiver sets are operated to generate waves with 90° phase deviation between each other. The phase deviation sign denotes the rotation direction of the disc. Accumulation (with sign) of pulse countings allows computation of a new position with respect to a known initial one (Haznedar, 1991; Sandler, 1991; Koivo, 1989).

The encoder used on each joint of the BEE device is not commercially available. It was designed at the *Instituto de Sistemas de Tandil (ISISTAN)*[1] . The design was deeply inspired by the mouse tracking system technology. The encoder has the following characteristics.

- A gear multiplication mechanism at the end of which, a slotted wheel is attached. This wheel has a diameter of 15 mm (0.59 inches) with 48 radial slots on a single track. It is made of plastic and rotates on a metallic axle.
- An optoelectronic mechanism made of 2 sets of infrared emitter-diode with phototransistor sensor. These sensors receive the light whenever they are in front of a slot. They are located at a distance computed to generate two waves, A and B, with exactly 90° phase deviation (Fig. 2).
- An electronic phase detector suited to determine the rotation direction according to the time sequence of wave A and B.
- An electronic unit to add/subtract the digitalized optical signals generated by the slotted-wheel movement. The prototype system architecture works with a word length of 16 bits. This affects the precision and/or the working mode (incremental/absolute).
- Computer interface with the following functions: click reading, reset signal, encoder (joint) selection.

[1] ISISTAN, Instituto de Sistemas de Tandil, Departamento de Computación y Sistemas, Facultad de Ciencias Exactas, Universidad Nacional del Centro de la Provincia de Buenos Aires: 57, San Martín, 7000 Tandil, Provincia de Buenos Aires, Argentina.

2.3.- BEE INTERNAL CIRCUIT

The block diagram circuit is presented in Fig. 3. The functional parts are described as follows:

a) Photodetector signal conversion to comply with the selected technology (TTL). One of the outcomes of this conversion is good discrimination between the 0 and 1 levels. Emitter currents (i_A and i_B) settings according to the sensitivity of the photodetectors and to the transparency of the slotted-wheel material.

b) Phase detector for rotation direction determination.

c) Multiplexing stage for pulse selection and counting sign (forward/reverse) transfer to the pulse counter.

d) Pulse counter, updating the coded angular position of the joint. Actually, the unit of angle variation is the slotted-wheel step period.

e) Address bus, selecting the control card device for the joint whose angle variation has to be read.

f) Data bus from the counter register to the computer. 8-bit data bus has been selected in order to reduce wiring complexity.

g) Control bus for counter enable, reset, high/low 8-bit sub-register selection, data output enable.

h) Address recognizer selecting the control card by connecting to control and data buses.

The device architecture has been thought to discharge the computer from most minor tasks, for example, excluding interruption requests whenever

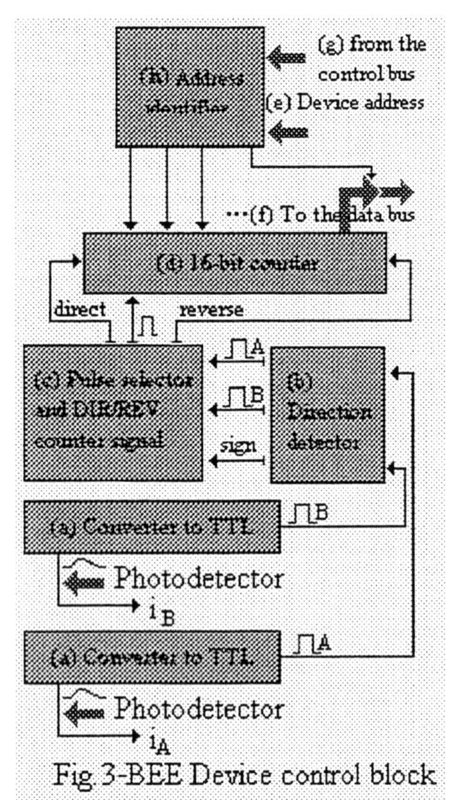

Fig. 3-BEE Device control block

a slot count has to be registered. Whenever the computer software requires a reading, it can access the control card through the address bus.

3.- SOFTWARE

3.1.- LOW-LEVEL SOFTWARE

The device driver is built using pooling techniques to transfer data from the BEE device to the computer. The pooling technique is implemented by sequences of IN and OUT operations. The driver provides preprocessed data to the high level.

When starting, the driver first task is PPI (Programmable Peripheral Interface) setting to cope with connection wiring assumptions. PPI is the classic 8255A integrated circuit. Any BEE device information to or from I/O ports passes through the 8255A. Angle information reading procedure is periodic. The real time clock (8254) of the computer is used to generate programmable interruption request pulses.

3.2.- HIGH-LEVEL SOFTWARE

The 3D editing system is a reduced CAD package tailored to handle the BEE pointer captured coordinates. Its purpose is to depict complex 3D objects, to be eventually further combined into higher level ones. As commented before, joint position data conversions to Cartesian coordinates are assigned to this level.

One of the key features of the software is its ability to move objects on the display, according to user-defined viewpoint modifications. This eases the maneuver of the BEE pointer around the object to be "sketched".

The selected display representation is the polygon wired-frame scheme. This representation is well suited for hidden-line vision generation.

The main features of the 3D editor are:

a) *elemental drawing primitives*: point, line, circle, and planar sections. Points may be combined to produce any other primitive.
b) *complex objects building:* using primitives or complex objects.
c) *embedded virtual BEE:* the exoskeleton device is represented in the virtual scene at the same object scale.
d) *objects adjustment:* adjusting an object to another one with parameters such as: element angle, common edge, relative coordinate, ...
e) *object modification on screen:* allows any modification process using classic editing procedures. Procedures selection may be operated either from a keyboard/mouse device or using the 3D pointer, directly on the real part or in combination with a pointing frame.
f) *saving and retrieving graphical objects*: custom file formats have been designed to minimize conversion time between disk and main memory.
g) *exporting to standard CAD-systems:* using DXF (Data Exchange File) from AutoCad.
h) *adjustable viewpoint:* as described above.
i) *setting or checking geometrical data:* by selecting then entering dimensions or inquiry respectively. Selection can be made either on the image or on the real object. Object dimensions can be set from data either pre-defined or acquired from an high-precision measuring instrument.

Cartesian coordinate computation involves a great quantity of trigonometric calculations. For this reason, computation time on a common PC platform is prohibitive in view of the "real time" processing needs (typically 10 readings/sec.). On the hand, the limited mechanical precision of the BEE device prototype does not provide a high number of significant figures for trigonometric function values. To accelerate the arithmetic computation, an algorithm, based on straight table look-up, with linear interpolation, has been implemented by the construction of a RAM resident trigonometric table with 0.01° resolution. The memory space requirements is easily determined as 54000 bytes, with 6 bytes floating point representation. The

RAM table was set up by storing values from Pascal trigonometric calculations.

Fig. 4- Benchmark of the Trigonometric Functions

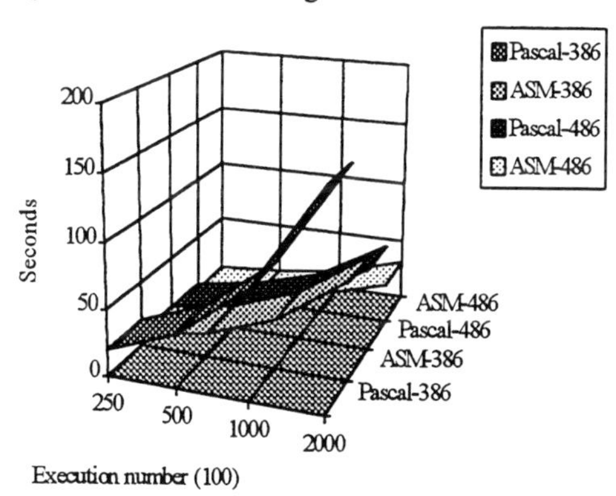

3.3.- THE CAD-SUBSET GRAPHIC EDITOR

The implemented system does not share all the features of a standard commercial CAD package. Only some basic functions have been developed to illustrate the main applications of the BEE device. The graphic editor has some features that provides the CAD operator with some additional high level functions, enhancing the 3D job performances. Those features are described as follows.

The CAD subset contains some measurement functions, thanks to which the designer can set or inquire (in the virtual or real space) about the object dimensions (Venolia and Williams, 1990; Wolf, 1992; Rubine, 1992).

As mentioned before the operator can easily select or modify the point (viewpoint) from which he wants to view the object in the virtual space. From this point he can zoom in, zoom out, rotate around pre-selected axis, translate. Those functions are useful for editing or object recognition purpose.

4.- SYNTHESIS

4.1.- CODE

The driver (≅1000 lines) and the trigonometric functions (≅1500 lines) are implemented in Assembler, while the CAD editing system (≅12000 lines) is written in Pascal.

4.2.- RESOLUTION AND PRECISION

The main quality limitations of the BEE system can be grouped into the following categories:

a) *Mechanical restrictions* dependent on the quality of the individual components of the device. Obviously, this is mainly dependent on financial considerations. Prototypes have been designed to demonstrate feasibility without aiming at physical performances.
b) *Geometrical restrictions*; clearly the respective dimensions of the links will affect the range and the resolution. Moreover the mechanical characteristics (elasticity, dilation, ...) of the link material are key factors, not considered in this work.
c) *Electro-mechanical restrictions*; the quality of the joint encoder will affect the resolution (depending on slots/revolution number) and precision (depending on reliability of light emitter/detector system, and on how well-made the slots and wheel are).
d) *3D editor limitations*. The high level software has been designed as a minimal support environment. Further enhancements should be motivated by specific application needs.
e) *Platform characteristics and performances.*

4.3.- POINTING ERROR ANALYSIS

Given the elementary error factors on joint angles($\Delta\alpha$, $\Delta\beta$, $\Delta\gamma$ and $\Delta\theta$), the errors ΔX, ΔY and ΔZ can be readily formulated applying error calculus theory. Actually, if the system would be able to measure angles in a static position, errors on these angles could be fairly assumed. The (differential) technique adopted to derive angle values makes the error assumption quite difficult. This is mainly due to the relative way of measuring (accounting angle variations), taking into account that, for a given pointing operation, the sequence of link movements is not unique, and the error depends on this sequence. The BEE operation is strictly manual, so the incertitude on movement sequences is important. The basic conclusion to draw is that angle error assumption should be derived experimentally within determined working conditions. This analysis is out of the scope of this work.

4.4.- TRIGONOMETRIC ALGORITHMS

A high-speed table look-up method has been implemented to compute sine/cosine functions. The performance of this method, including linear interpolation, has been compared to the one of algorithms implemented in the Turbo Pascal library (Fig. 4).

4.5.- SCREEN REFRESHING TIME DELAY

Depending on the platform, the number of computations required to update coordinates (capture/conversion) produces some inertial effects on the virtual object. This effect can be overlooked either by slowing down the BEE movement or increasing the platform performances.

4.6.- CIRCUIT PROTOTYPING

The circuit card was simulated in the B^2Logic® 3.0 from B-Squared Logic and in the VHDL Compiler

® 1.2a from Model Technology. The prototype circuit was designed using the SmArtWork® 1.3r4 (Artwork editor with autorouter) from Wintek.

5.- CONCLUSIONS

5.1.- INTERFACING

The essence of the user interface is a common ritual of communication and interaction between human and machine. The user interface facilitates human access to repositories of data, information, knowledge, and wisdom. Increased quality in the user interface will permit one to take advantage of the rapidly increasing power of computers (Cypher, 1991). The BEE system, with its two operation modes (space data acquisition and 3D mouse) is an innovative interface with a promising future. The BEE-CAD integrated system allows designers to achieved complex tasks such as defining a mechanical part, working with the piece, etc., with increased ease while minimizing errors. The visual effect produced by the system, representing a virtual BEE device embedded in the virtual scene, makes the designer orientation easier within the virtual workspace. Actually, the operator can watch both the design and the pointing device. Additional 3D mouse input features also provide software control abilities with virtual the aid of buttons.

5.2.- AS A MODELING TOOL

Modeling is the most expensive part of most computer graphics applications, mainly because it requires the majority of human effort. In this respect, virtual reality devices are prospective modeling tools. The BEE is believed to be one of them. The BEE system is expected to be a valuable assistant, e. g. for mechanical parts re-engineering. In this situation the availability of a sample model allows a fast capture and quick modifications of geometric specifications. The operator uses the pointer device system as an automated assistant for exploring design alternatives.

REFERENCES

Blanchard C., Burguess S., Harvil L., Lanier J., Lasko A., Oberman M and Teitel M."*Reality Built for Two: A virtual reality tool*". IEEE Computer Graphics and Applications, 12, 2, March 1990, pp. 582-583.

Critchlow Arthur J.. "*Introduction to Robotics*". Macmillan Pub. Co., New York, Collier Macmillian Pub., USA, 1985.

Cypher A.. "*Programming Repetitive Tasks BY example*". Proceedings of the ACM, CHI'91, New Orleans, LA, April 1991, pp. 33-39.

Hays Nancy. "*Graphic News, Imagination Celebration Gets a Look at High Tech*". SigGraph and SimGraph 1994, California. IEEE Computer Graphics and Applications, July 1993, page 76.

Haznedar Haldun. "*Digital Microelectronics*". Benjamin/Cummings Pub. Co., Inc., 1991.

Johnson Curtis D.. "*Process Control Instrumentation Technology*". Prentice Hall, Englewood Cliffs, NJ, 1988.

Koivo Antti J.. "*Fundamentals for Control of Robotic Manipulators*". John Wiley & Sons Inc, 1989.

Krueger Wolfgang & Froehlich Bernd. "*Visualization Blackboard: The Responsive Workbench*". IEEE Computer Graphics and Applications, May 1994, pp. 12-15.

Lee George. "*Sensor-Based Robots: Algorithms & Architectures*". Edited by C.S.George Lee, NATO Series, 1988.

Maldelkern Dave. "*Human Communications Issues in Advanced UIS*". Communication of the ACM, Apr. 1993, Vol. 36, No. 4, pp. 101-110.

Marcus Aaron & Van Dam Andries. "*User-Interface Development for the Nineties*". Computer, Sep. 1991, pp. 49-57.

Marcus Aaron. "*Human Communications Issues in Advanced UIs*". Communications of the ACM, Graphical USR Interfaces: The Next Generation. April 1993. Vol. 36, No. 4, pp. 101-110.

Nielsen Jakob. "*Noncommand User Interface*". Communication of the ACM, Apr. 1993, Vol. 36, No. 4, pp. 83-100.

Rubine D. "*Combining Gestures and direct manipulation*". Proceedings of the ACM, CHI'92, Monterrey, CA, May 3-7, 1992, pp. 659-660.

Samson Claude, Michel Le Borgne & Bernard Espiau. "*Robot Control: The Task Function Approach*". Clarendon Press, Oxford, 1991.

Sandler Ben-Zion. "*Robotics. Designing the Mechanisms for Automated Machinery*". Salomon Press, Prentice Hall Int. Inc., Englewood Cliffs, 1991.

Venolia D. and Williams L.. "*Virtual Integral Holography*". Technical report 90-10, Apple Computer Inc., Feb 1990.

Wolf C. G.. "*A Comparative Study of Gestural, Keyboard and Mouse Interfaces*". Behaviour Inf. Tech., 11, 1, January 1992, pp. 13-23.

USING THE ANALYTIC HIERARCHY PROCESS IN COMPUTER INTEGRATED MANUFACTURING

Yukiko Nishibori*, Ljubisa Vlacic and Masayuki Matsui*****

* *at present, the Japan Program Exchange Student at Griffith University; apparently with University of Electro-Communications, Tokyo, 182, Japan*
** *School of Microelectronic Engineering, Griffith University, Nathan Campus, Qld, 4111, Australia; fax: +61 7 3875 5384; e-mail: l. Vlacic@me.gu.edu.au*
*** *University of Electro-Communications, Tokyo 182, Japan*

Abstract: This research addresses the problem of the evaluation of goal achievements in an implementation of Enterprise Integration Programmes. The essence of the recommended framework is a merging of the Analytic Hierarchy Process with Nakamura's decision model which produces a set of criteria to be used to measure the economic, strategic, social, operational and organisational impacts of computer-based integration on an enterprise.

Keywords: manufacturing systems management; business process re-engineering; multiobjective decision making; computer integrated manufacturing; enterprise integration.

1. INTRODUCTION

A range of multiobjective decision modelling techniques have been proposed to address multidimensional problems. Among the more popular is the Analytic Hierarchy Process (AHP) developed by Saaty in 1986 (Saaty, 1990). The Analytic Hierarchy Process is a multiobjective decision making technique that employs pairwise comparison to rank order the alternatives of problems. It has been assumed that the objectives of a given problem can be, in the main, represented explicitly in a hierarchical structure.

Decision analysts, management scientists, operations researchers and practicing managers have shown significant interest in using the AHP. J. Shim of Mississippi State University and F. Zahedi of University of Massachusetts provide extensive bibliographies on the AHP (including books, specialised works, dissertations, journal articles and software packages). Their surveys show that little attention has been given to use of the AHP in the Computer Integrated Manufacturing/Enterprise environment (CIE/CIM) despite the facts that: a) the use of the AHP in the broad area of decision analysis has gained an increasing amount of attention among researchers at a variety of universities; and b) there have been a great number of research studies on both the theoretical development and practical applications. Flexible manufacturing and production scheduling have been only topics discussed by researchers and/or practicing engineers while dealing with the AHP. However, a recently published report by Elzinga et al. (1995) of the Business Process Management Institute, University of Florida, provides evidence on a new and increasing use of the AHP (among the other formal decision making methods) in facilitating the Critical Success Factors stage of the Business Process Management.

This paper extends previous work by Nakamura, Vlacic and Ogiwara who developed a set of critical success factors to be used to enable an enterprise to achieve its goal, and particularly shows how the AHP can be used to facilitate decision making on investment in computer-based integration of the overall enterprise as well as to facilitate the assessment of implementation results of an Enterprise Integration Programme.

A summary of decision making models within the framework of an Enterprise Integration Programme is contained in Section 2. The hierarchy of AHP and a justification on how to use AHP as a decision model solving tool for the purpose of an Enterprise Integration Programme is described in Section 3. Conclusions are drawn in Section 4.

2. AN ENTERPRISE INTEGRATION PROGRAMME

Due to the growing concern about the competitiveness on the international market, many industries are turning to the reviewing process of the quality of their products and services. That reviewing process usually begins with goal setting for the enterprise and compares the present state of the enterprise (the As-Is) with the desired future state (the To-Be) to characterise the modification path between them (the Transition), Figure 1 (Williams, 1994). In this stage, the vision, mission and goals of the enterprise are formulated and critical success factors determined. With these factors in mind, all three transition plans are to be evaluated and a specific strategy (i.e. methodology) for their implementation is to be selected. The selected strategy is then quantified, the improvements are identified and finally implemented. Following implementation of the above transition plans, which are usually split into a range of specific projects, an evaluation of achieved quality of products and services is repeatedly applied.

An implementation of the above transition plans is closely interrelated with investment decisions. To conduct repeated evaluation of the outcomes of such investment decisions is therefore of great importance for the company to succeed with its Enterprise Integration Programme.

Decision making models aimed at facilitating the processes of decision making during investment, operation and implementation stages of an Enterprise Integration Programme are presented in the paper by Vlacic (1996). Detailed presentation of an implementation decision model to be used for life cycle evaluation is explained elsewhere (Nakamura, Vlacic and Ogiwara, 1996). The reminder of this paper deals with the decision model solving methodology of Saaty pointing out how AHP can be used in a multiobjective decision making environment such as of Computer Integrated Manufacturing/Enterprise (CIM/CIE).

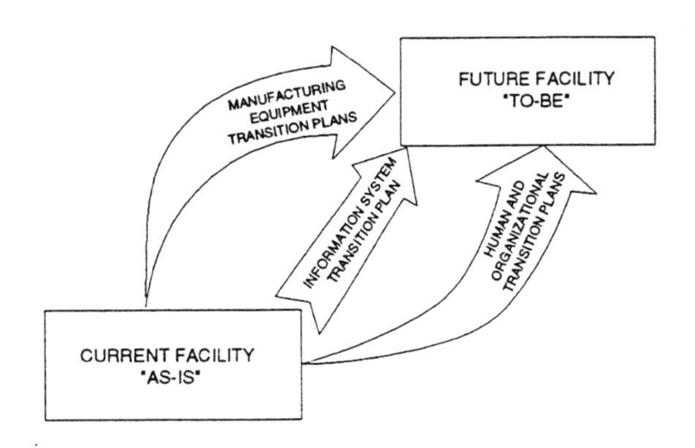

Fig. 1 Transition Plan (Source: Williams, 1994)

3. THE ANALYTIC HIERARCHY PROCESS

3.1 The Hierarchy

A hierarchy is considered as an efficient way to deal with complex systems and is efficient both structurally and functionally. This arrangement of the decision factors makes it possible for decision makers to focus on each and every part of the complex problem (Expert Choice, 1993). To solve a complex decision problem such as presented by Nakamura's decision model (Nakamura, Vlacic and Ogiwara, 1996) an Expert Choice, a decision support software the logic of which is based on the AHP, was used. An Expert Choice organises various elements of a problem, decision attributes and alternatives, into a hierarchical tree structure, Figure 2.

An Expert Choice was used to facilitate performed decision process for which a structure of a model to fit the problem as well as the judgements and the decisions were made by decision makers from a real engineering world.

3.2 The Decision Model

As part of his work on a multicriteria-based analysis of the effects of CIE/CIM operations in a range of Japanese companies from electronic, manufacturing and process industry sectors, Nakamura developed a decision model which consists of twenty criteria structured across two hierarchical levels (Nakamura, Vlacic and Ogiwara, 1996; Vlacic, 1996). The model is to be used to measure the economic, strategic, social and organisational impacts of the implementation of the Enterprise Integration Programme on the company and to assist industry decision makers while assessing the achievements of the company goals, particularly those set at a very early stage of the Enterprise Integration Programme.

Figure 3 describes Nakamura's model built in accordance with the Expert Choice rules. It consists

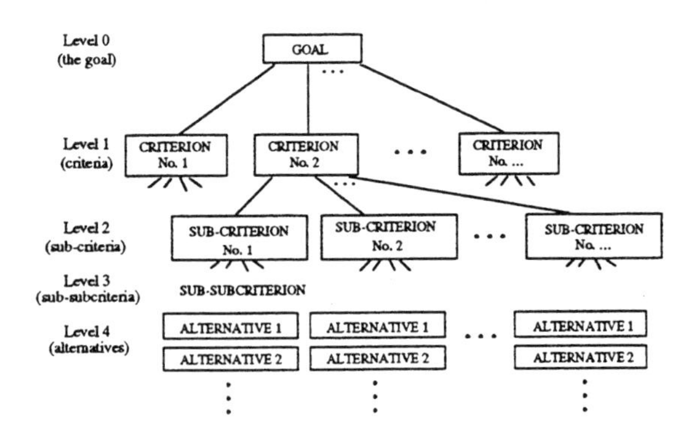

Fig. 2 A decision schema in the Analytic Hierarchy Process

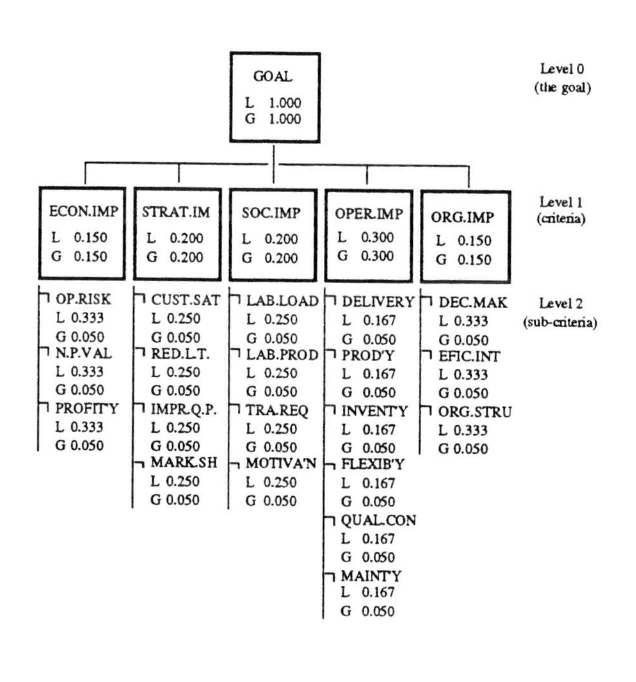

CUST.SAT — Customer Satisfaction
DEC.MAK — Decision Making
DELIVERY — Delivery Schedule Performance
ECON.IMP — Economic Impact
EFIC.INT — Efficiency of Integration
FLEXIB'Y — Flexibility
IMPR.Q.P. — Improved Quality of Products
INVENT'Y — Inventory
LAB.LOAD — Labour Loading
LAB.PROD — Labour Productivity
MAINT'Y — Maintainability
MARK.SH — Larger Portion of Market Share
MOTIVA'N — Motivation
N.P.VAL — Net Present Value
OP.RISK — Operational Risk
OPER.IMP — Operational Impact
ORG.IMP — Organisational Impact
ORG.STRU — Organisational Structure
PROD'Y — Productivity
PROFIT'Y — Profitability
QUAL.CON — Quality Control
RED.L.T. — Reduced Lead Time
SOC.IMP — Social Impact
STRAT.IM — Strategic Impact
TRA.REQ — Training Requirements
L — Local Priority
G — Global Priority

Fig. 3 Nakamura's decision model

Criteria

1. Economic Impact

OP.RISK	N.P.VAL	PROFIT'Y
-	7.5	10
-	7.5	10
-	5	7.5
-	5	7.5
-	5	5
-	5	7.5

2. Strategic Impact

CUST.SAT	RED.L.T.	IMPR.Q.P.	MARK.SH
7.5	50	10	5
7.5	66	5	5
7.5	50	0	5
7.5	25	5	7.5
5	0	5	-
5	50	5	2.5

3. Social Impact

LAB.LOAD	LAB.PROD	TRA.REQ	MOTIVA'N
10	7.5	5	7.5
5	7.5	5	5
7.5	2.5	10	7.5
5	7.5	5	5
10	7.5	-	5
7.5	7.5	7.5	5

4. Operational Impact

DELIVERY	PROD'Y	INVENT'Y	FLEXIB'Y	QUAL.CON	MAINT'Y
7.5	7.5	7.5	7.5	7.5	10
5	7.5	7.5	7.5	7.5	10
7.5	7.5	5	7.5	7.5	2.5
7.5	7.5	7.5	7.5	7.5	7.5
0	5	2.5	5	2.5	2.5
7.5	10	7.5	7.5	10	7.5

5. Organisational Impact

DEC.MAK	EFIC.INT	ORG.STRU
7.5	7.5	7.5
5	10	7.5
7.5	7.5	7.5
7.5	7.5	7.5
5	5	5
0	5	5

10 means the best of 5 criteria
7.5 means second best of 5 criteria
5 means middle of 5 criteria
2.5 means below average of 5 criteria
0 means the lowest of 5 criteria

Fig. 4 Importance of sub-criteria judged by industry decision makers

of five criteria and twenty sub-criteria. The goal node is at level 0 while criteria and subcriteria occupy levels 1 and 2 respectively. Alternatives are at level 3 of the Expert Choice model which is built for the purpose of this analysis.

Once the Expert Choice model is built, the judgement process can be undertaken. This is where a personal decision maker's judgement comes in. The judgement values are presented in Figure 4, describing a level of importance defined by industry decision makers for all the sub-criteria considered. The main criteria are then compared with respect to the goal and a relative value of one criterion versus another is then calculated by the Expert Choice software. To this end it should be pointed out that the priorities can be derived through the usual pairwise comparison process. If it is applied then the criterion having a smaller number of sub-criteria will get bigger priority than those criteria with larger number of dependent sub-criteria. A nice feature of

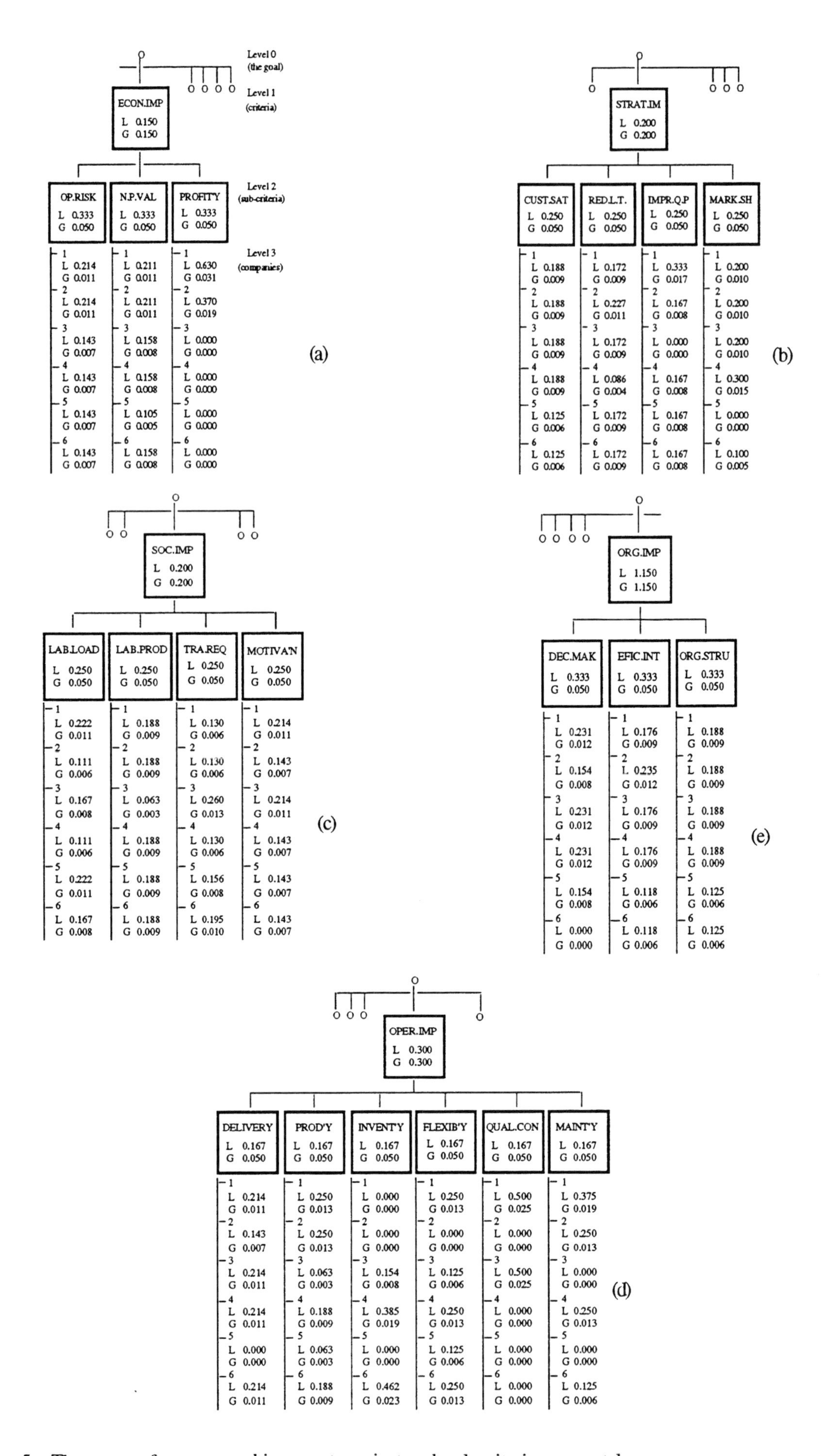

Fig. 5 The extent of company achievements against each sub-criterion separately.

the Expert Choice is that a structural adjustment can be implied so that the sub-criterion does not get penalised if it belongs to a criterion consisting of a larger group of sub-criteria. The following is how structural adjustment works: the goal priority of 1.00 is distributed among the criteria; the priority of each criterion is distributed to its sub-criteria; the priority of each sub-criterion is distributed to the sub-criteria under it (i.e. sub-subcriteria), and so on down to the alternatives (Expert Choice, 1993). Figure 3 shows, in fact, Nakamura's decision model after structurally adjusting where a letter L denotes the local priority while the letter G denotes a global priority.

Once all priorities are defined and calculated, the alternatives can be entered throughout and thereafter compared against all sub-criteria. The achievement results measured against each sub-criterion separately for all sub-criteria across six companies who participated in this analysis are presented in Figure 5.

Namely, six Japanese companies, being different in a strategy chosen to apply an Enterprise Integration Programme, are considered as alternatives within the framework of this analysis and are represented by numbers 1 to 6 in Figure 5. They are assessed and their success is measured against the same set of performance measures, decision making criteria. The resulting values (the goal achievements), judged by industry decision makers are presented at level 3 of the hierarchies in Figure 5. A detailed explanation on differences among companies can be found in the work of Nakamura (Nakamura, 1995).

Since all judgements have been made and all calculations performed, the results of the analysis can be produced. This is accomplished by synthesising the data in order to arrive at the overall achievements for the alternatives. Expert Choice offers two methods of synthesising the data, the *distributive* and *ideal* synthesis modes. In the distributive synthesis mode, Expert Choice normalises the weight of the alternatives under each criterion. That is the same normalisation method used for the rankings of all the criteria and sub-criteria.

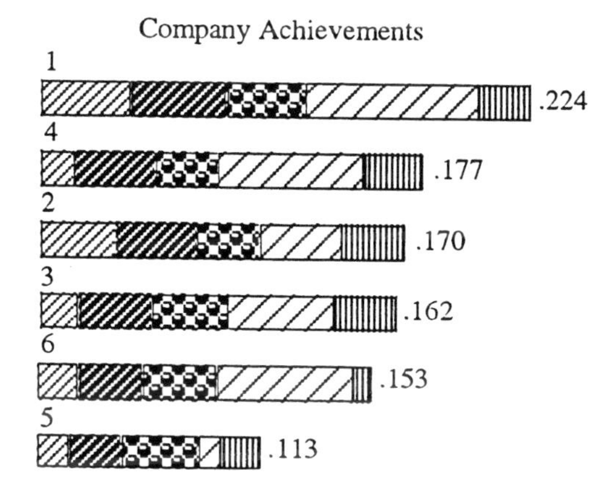

Fig. 6 Extent of achievements in applying an Enterprise Integration Programme across the six companies (1 to 6)

Figure 6 shows the results produced by the distributed synthesising mode of Expert Choice taking into account the judgement values from Figure 5. Figure 6 shows the company achievements in applying an Enterprise Integration Programme and reads that, for example, company 4 is the second top company in putting the Enterprise Integration Programme in place as per its achievements across all decision making criteria (performance measures) defined by Nakamura's decision model.

4. CONCLUSION

An effective decision should be based on the achievement of objectives and therefore, criteria and objectives are to be used to measure how well the goals are achieved. For the purpose of this analysis, the Analytic Hierarchy Process of Saaty was used in assessing the achievements in applying an Enterprise Integration Programme across six Japanese companies. The following are drawn from the industry decision maker's remarks:

*advantages:

- an ability to make effective decisions in the face of complexity can be significantly increased using AHP and Expert Choice decision support software;
- Expert Choice only facilitates a decision process while a decision maker structures a problem, provides the judgement and makes the decision

*difficulties:

- a detailed knowledge of both AHP and Expert Choice is required in order to smoothly proceed with the analysis;
- underpinning mathematics used to capture the important aspects of the decision to be made must be known to decision makers in order to develop their confidence in using a decision support software for analysis of the process under study.

In addition, as per the literature consulted, the need for improved methods of estimation and aggregation of relative weights is still present.

The beauty of Nakamura's decision model and the study undertaken is that intangible costs are being taken into account. A shortcoming of this study is that it does not analytically consider the knowledge accumulation process of a decision maker. That is the subject of further study.

REFERENCES

Elzinga, *et al.* (1995). Business Process Management: Survey and Methodology. *IEEE Transaction on Engineering Management* **42**:2, 119–128.

Expert Choice Inc., (1993). *Expert Choice — User Manual*. Expert Choice Inc.

Nakamura, T. (1995). *Computer Integrated Manufacturing*. MPhil Thesis, University of Electro-Communications, Tokyo.

Nakamura, T., Lj Vlacic and Y Ogiwara, (1996). Multiattribute-based CIE/CIM Implementation Decision Model. *Computer Integrated Manufacturing Systems*, **9**:2, 73–89.

Saaty, T, (1990). *The Analytic Hierarchy Process: Planning, Priority Setting, Resource Allocation*. RWS Publications.

Shim, J.P. (1989). Bibliographical Research on the Analytic Hiearchy Process (AHP). *Socio-Economic Plan Sci,* **23**:3, 161–167.

Vlacic, Lj. (1996). Multicriteria-based Decision Making Models for Computer Integrated Enterprise. In: *Modelling and Methodologies for Enterprise Integration* (P. Bernus and N. Nemes Eds), 103–112. Chapman & Hall.

Williams, T.J. (Ed.), (1994). *A Guide to Master Planning and Implementation for Enterprise Integration Programs*. Report No. 157, Purdue Laboratory for Applied Industrial Control, Purdue University.

Zahedi, F. (1986). The Analytic Hierarchy Process — A Survey of the Method and its Applications. *Interfaces* **16**:4, 96-108.

AN OPTIMAL CONTROL CHART FOR NON-NORMAL PROCESSES

Emmanuel DUCLOS, Maurice PILLET

ODS Europe, 12 rue pré paillard 74940 Annecy le Vieux
LLP/CESALP - BP 806 74016 Annecy Cedex - FRANCE
☎ : (+33) 50.66.60.80 Fax : (+33) 50.66.60.20
Email : duclos@esia.univ-savoie.fr

Abstract : In this paper we present the application of a control chart for non-normal processes. This chart, which looks like an $\overline{X}/S$ control chart is built with a least-squares L-estimator, which can replace the arithmetic mean and standard deviation usually calculated for Shewhart charts. This estimator has the property to provide a minimum variance estimation of the process position and scattering. This, disregarding data distribution. We focused our attention on « multi-generators » processes, like screw-machines or multi-die holder for injection molding, these processes have the property to generate non-normally distributed pieces.

Keywords : Least-squares estimation, Non-Gaussian processes, Optimal estimation, Process control, Statistics

1. INTRODUCTION

When piloting a process with S.P.C. control charts ($\overline{X} / R$) we usually assume that the mean is normally distributed. In spite of it robustness, this assumption is far from always being satisfied. In fact, either the $\overline{X}$ distribution is close to a normal distribution or the sample is large enough to satisfy the assumption of normality of the mean. However, large samples are prohibited for economic reasons. Disregarding this assumption can cause problems when the hypothesis of normality is not valid.
First of all, statistical error, type I and II, correspond no longer with those defined for a normal distribution when control limits are placed at $\pm 3\sigma$ from the target (Shilling, 1976).
The second point concerns precision of estimations. Indeed, we can show that the mean is not the optimal estimator in term of variance when the population is non-normal. It means that we can find an estimator without bias which can provide better performances than the average.

So in order to solve these problems, we propose in this article the use of a control chart (L chart) build with a minimum variance estimator whose performances have been compared to those of the average in term of variance and distribution shape.
We will study this estimator in the case of data incoming from a « Multi -generator » process.

2. MULTI-GENERATOR PROCESSES

Consider a process which consists of several elementary machines. Among these processes, we can quote for example injection presses.
Each criterion produced by an elementary machine is distributed among an elementary characteristic which can be normal or non-normal according to the criterion studied. Then, the global distribution or population is said to be a « mix of probability distributions » which is unlikely to give à normal population.

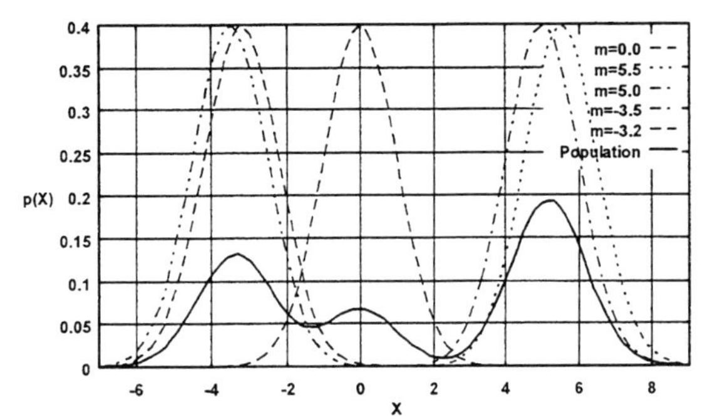

Fig.1. Non-normal population resulting of a mix of elementary distributions.

To illustrate this article we have chosen the following mix of distributions (Figure 1).
Assuming that each elementary distribution has the same probability $(\pi = 1/6)$, the population is defined by :
$f_p(X) = \pi \cdot f_{N1}(X) + \pi \cdot f_{N2}(X) + ... + \pi \cdot f_{N6}(X)$

Even when a Multi-generator process is under control, observations are non-normally distributed. Because of its physical characteristics, it is sometimes impossible to make it normal or economically unsuitable.

3. PRINCIPAL OF AN L-ESTIMATOR

If a population is normally distributed, we can prove that the variance of the arithmetic mean reaches the Frechet limit. So, the arithmetic mean is said to be the optimal estimator in terms of variance. Unfortunately, this is not the case when the population is non-normally distributed.
Therefore, we have been interested in an estimator based on order statistics known as L-estimators, which have the capability to take into account the population shape, through their coefficients (David, 1981).
The L-estimator proposed by E.H. Lloyd (1952), based on a Least-Squares algorithm, requires no hypothesis on the population shape whereas the method of maximum likelihood supposes a normal parent.

3.1 L-Estimators and L-Statistics

Consider a sample $(x_1, x_2, ..., x_n)$ of n independent observations sampled at a time k.
$x_{(1)}, x_{(2)}, ..., x_{(n)}$ are ordered observations of **Xk** such as $x_{(1)} \leq x_{(2)} \leq \cdots \leq x_{(n)}$.
$x_{(n)}$ is called n ordered statistics of the ordered vector **Xk**.
The linear combination of Xk's ordered statistics with a vector Cj of real coefficient is an L-statistic or L-estimator.

The estimation of θ parameter at a time k is given by :

$$\hat{\theta}_k = {}^tC \cdot X_k = \sum_{i=1}^{n} C_i \cdot x_{(i)} \quad (1)$$

The average is a particular case of L-estimator since all coefficients are equal : $C_i = \frac{1}{n} \; \forall i \in [1;n]$. The range is another particular L-estimator for a scale parameter : $C_1 = -1$, $C_n = 1$ and $C_i = 0 \;\; \forall i \neq 1, i \neq n$
The problem is then, to calculate the L-estimator's coefficients in order to obtain the expected performances.

3.2 The Least Square L-estimator

Choice of μ_p and σ_p parameters. The population's parameters of location and scale which can be estimated by the least-squares L-estimator are not necessarily the mean and standard deviation of the population. However, we show that they are the most appropriate parameters, if we want to maximise the process capability indice. (Pillet, 1997).
The purpose of S.P.C. is to improve production quality and therefore to minimise production cost. Taguchi defined the loss function (2) which represents the cost of non-quality :

$$L = K\left(\sigma^2 + \left(\overline{X} - \text{Target}\right)^2\right) \quad (2)$$

$$Cpm = \frac{IT}{6\sqrt{\sigma^2 + (\overline{X} - \text{Target})^2}} \quad (3)$$

Where K is a constant, $\overline{X}$ is the average of the population and σ its standard deviation.
We notice that minimising Taguchi's loss function is equivalent to maximising the capability indice Cpm (3).
So to maximise the Cpm indice we have to minimise the population standard deviation and the quadratique error between sample mean and the process target. In consequence, we will consider $\overline{X}$ and σ respectively as localisation μ_p and scale σ_p parameters of the population.

Generalised Least-Squares. In this paragraph we describe the construction of the least-squares L-estimator. We also expose some basic results concerning the Generalised Least Square theory.

Suppose the following multiple linear model $\boldsymbol{X} = \boldsymbol{Wb} + \boldsymbol{u}$ where :

- **X** is the variable of interest (sample n observations of the process).
- **u** is a random vector modelling noise on Y (common causes).

- **b** is the vector of parameters that we expect to estimate (localisation and scale of the process).
- **W** is a matrix including non random variables.

The following conditions are supposed to be satisfied : $E(u) = \vec{0}$ *et* $V(u) = \sigma^2 \cdot I_n$

Estimating parameter b by the least-squares consists of minimising the u random variable influence.
The Ordinary Least Square estimator is defined by :

$$\hat{b} = \left(W'W\right)^{-1} \cdot W'Y$$

Application on an ordered sample
Assume $(x_{(1)}, x_{(2)}, \ldots, x_{(n)})$ are ordered observations of a vector $\mathbf{X}_k$ where $x_{(1)} \leq x_{(2)} \leq \cdots \leq x_{(n)}$
$U_{(r)}$ is a standardized variable :

$$U_{(r)} = \frac{\left(x_{(r)} - \mu_p\right)}{\sigma_p}$$

and moments of the order statistic $U_{(r)}$ are :

$$E\left[U_{(r)}\right] = \alpha_r \quad Var\left[U_{(r)}\right] = \omega_{rr} \quad Cov\left[U_{(r)}, U_{(s)}\right] = \omega_{rs}$$

Lets resume now the problem exposed by Lloyd. The moment of order statistics X(r) can be written under vectorial form :

$$E\left[X(r)\right] = \mu_p . e + \sigma_p \alpha$$

where α is a vector of a_r and **e** is a unit vector

$$X = \begin{pmatrix} x(1) \\ \vdots \\ x(n) \end{pmatrix} \quad e = \begin{pmatrix} 1 \\ \vdots \\ 1 \end{pmatrix} \quad a = \begin{pmatrix} \alpha_1 \\ \vdots \\ \alpha_n \end{pmatrix}$$

The preceding equation can be written : $E[X] = A \cdot \theta$

where $\theta = \begin{pmatrix} \mu \\ \sigma \end{pmatrix}$ *and* $A = \begin{bmatrix} 1 & \alpha_1 \\ \vdots & \vdots \\ 1 & \alpha_n \end{bmatrix}$

The variance-covariance matrix of X is : $\sigma^2 \cdot \Omega$
where Ω is an (nxn) matrix of ω_{rs} elements.
Aitken (1935) proved that such a problem could be solved by applying a least-squares algorithm on ordered statistics.
Since $Var(u) = Var(X) = \sigma^2 \Omega$, the model is general. The solution of this model derives from the ordinary least-squares. In fact, it is noticeable that Ω is a positive matrix, so a M(n,n) regular matrix exists such as: $M' M = \Omega^{-1}$

A Generalised Least Square Estimation of θ is defined as (4) :

$$\hat{\theta} = \begin{pmatrix} \hat{\mu} \\ \hat{\sigma} \end{pmatrix} = \left(A' \cdot \Omega^{-1} \cdot A\right)^{-1} \cdot A' \cdot \Omega^{-1} \cdot X \qquad (4)$$

Variance of both estimators of location and scale is (5) :

$$Var\left[\hat{\mu}\right] = \sigma_p^{\,2} \cdot \frac{\alpha' \cdot \Omega^{-1} \cdot \alpha}{\det\left(A'\Omega^{-1}A\right)}$$
$$Var\left[\hat{\sigma}\right] = \sigma_p^{\,2} \cdot \frac{e' \cdot \Omega^{-1} \cdot e}{\det\left(A'\Omega^{-1}A\right)} \qquad (5)$$

This theoretical description of the Least-Squares L-estimator brings to light, that this estimator can't be used without knowing moments of the ordered observations. This can cause some problems of application. The next paragraph will present an application of this estimator for Statistical Control of a multi-generator process.

4. L-CHARTS PERFORMANCES

4.1 Calculation of the variance-covariance matrix.

To be able to appreciate improvement brought by the L-estimator compared to the arithmetic mean, we achieved computer simulations. In order to determine the L-estimator coefficients, it is necessary to calculate the matrix of variances-covariances of standardized ordered statistics. These calculations were made with random data generated according to a known distribution. Each coefficient of the matrix is calculated with the relation (6)

$$\omega_{ij} = \frac{1}{m-1} \sum_m \left[\left(x_{(i)} - \bar{x}_{(i)}\right)\left(x_{(j)} - \bar{x}_{(j)}\right)\right] \qquad (6)$$

Where m is the number of samples.
Coefficients of variance-covariance matrix can also be calculated from the theoretical expression :

$$Cov\left[X_{r:n} X_{s:n}\right] = \int_{-\infty}^{\infty} \int_{-\infty}^{y} \left(x - \mu_{r:n}\right)\left(y - \mu_{s:n}\right) \cdot f_{rs}(x, y) \cdot dx dy$$

(7)

with

$$f_{rs}(x, y) = \frac{n!}{(r-1)!(s-r-1)!(n-s)!} \cdot F^{r-1}(x) \cdot f(x)\left(F(y) - F(x)\right)^{s-r-1} \cdot p(y)\left[1 - P(y)\right]^{n-s}$$

This expression necessitates powerful means of calculation and even more, if the sample size is large and the distribution is non-normal. Further more, most practical applications deal with unknown populations, so that, this calculus can't be achieved. Although theoretical calculus was made to reach a great precision, computer simulations were preferred because of their similarity with the practical situation. Since processes are not necessarily well known and operating conditions are always evolving, building a model of the population's distribution does not seem realistic.

4.2 Computer Simulations

Since a control chart is a set of two estimations : Punctual estimation and confidence interval estimation, we studied the L-estimator performances in terms of variance and shape. Since the method is non parametric, performances of the L-estimator are different according to the distribution of the population.
The more different from a normal law the distribution is, the more efficient the L-estimator. In consequence we studied a case where the population is significantly non-normal (figure 1) to appreciate the L-estimator efficiency compared to the mean.

Variance and bias of estimations. Results in table 2, show that the variance of the L-estimator of location is always much lower than the average. The relative efficiency of the L-estimator compared to the mean is (8) :

$$eff = Var[\hat{\mu}]/Var[\bar{X}] \quad (8)$$

To provide the same performances in term of variance as the L-estimator when n=4, the average requires a sample size n=7. The benefits of using an L-estimator can be either the sample size reduction, or the improvement of commandability when sample size is maintained and the variance of the estimation is reduced.
In opposition, the L-estimator for the scale parameter brings no improvement compared to the standard deviation, in terms of variance.
However, it gives a non biased estimation of the population dispersion even when the population is non-normal (Figure 3, Table 2). Whereas the empiric standard deviation Sn, is a biased estimator when sample size is small. In fact, the coefficients c4 usually used to correct this bias is unsuitable when the population is non-normal.
A relative bias for estimators of a scale parameter can be defined as (9) :

$$b_{\hat{\sigma}} = (\hat{\sigma} - \sigma_p)/\sigma_p \quad (9)$$

Of course these values are specific to this example, but similar results can be found for other significantly non-normal populations.

Symmetry of distributions. The distribution shape was studied to determine how to apply confidence interval tests for control charts.
The setting of limits on $\bar{X}$ control charts are based on the assumption of normality, justified by the central limit theorem.

Table 1 Variance of both location estimators and relative efficiency of the L-estimator compared to the arithmetic mean

n	3	4	5	6	7	8	9	10
$Var[\hat{\mu}]$	4.29	2.45	1.52	1.01	0.72	0.55	0.44	0.36
$Var[\bar{X}]$	5.32	3.99	3.19	2.66	2.28	2.00	1.77	1.60
eff	0.80	0.62	0.48	0.38	0.32	0.28	0.25	0.23

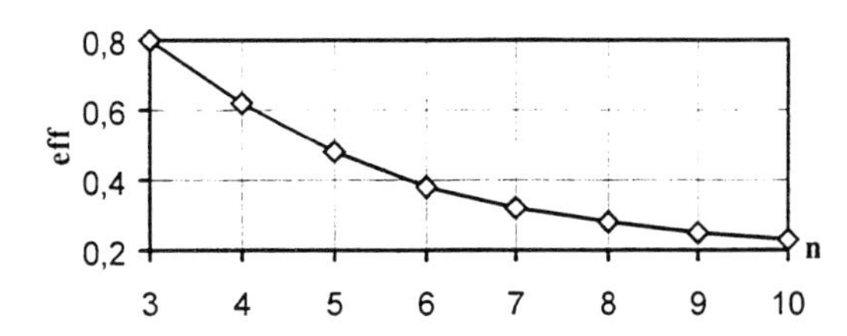

Fig. 2. Relative efficiency of the L-estimator compared to the arithmetic mean

Table 2 Relative bias of the L-estimator of dispersion and standard deviation

n	3	4	5	6	7	8
$b_{\hat{\sigma}}$	-9.2 %	-5.2%	-3.4%	-2.4%	-1.7%	-1.4%
b_{Sn}	0.1%	0.03%	0.06%	0.03%	0.03%	0.01%

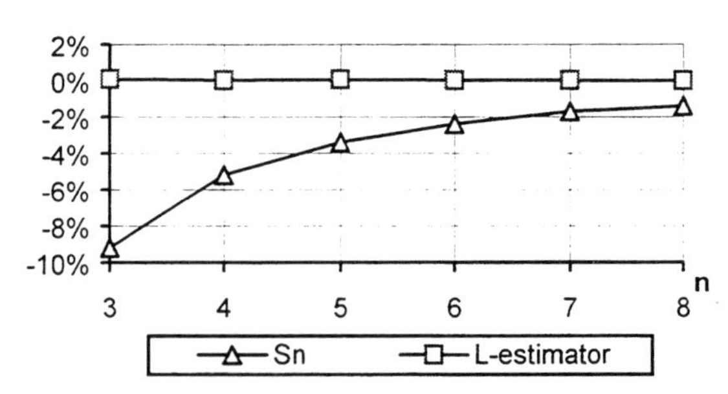

Fig. 3. Relative bias of the L-estimator of dispersion.

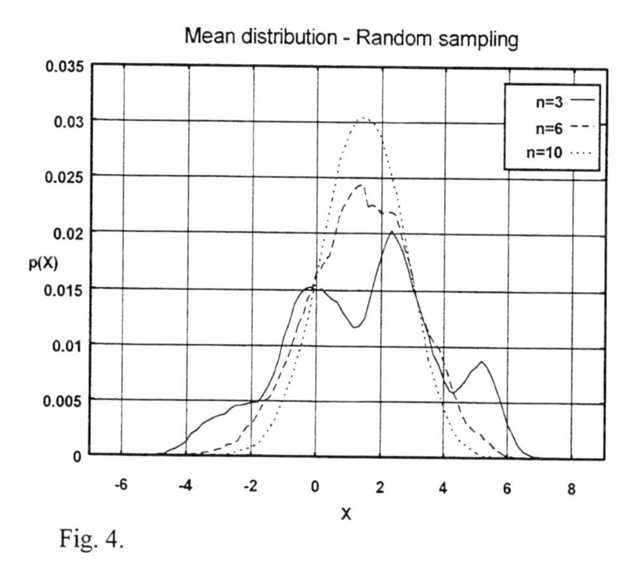

Fig. 4.

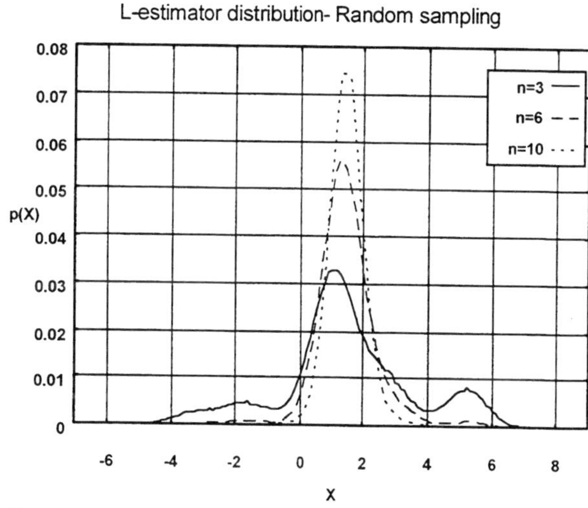

Fig. 5.

The theorem essentially states, that under general conditions, the distribution of sample means approaches normality. Nevertheless (Burr, 1967) showed that for significantly non-normal distributions, sample mean was far from being normal. Yourstone (1992) proposed recently modified control limits to keep risks close to 0.27% in the case of skewed population.

Figure 4 underlines the fact that when the sample size is small, normality is far from being satisfied. Coefficients of Kurtosis approaches 3 by inferior value as n increases. It means that the distribution has heavier tails than the normal law.

On the contrary we notice that the distribution of the L-estimator doesn't converge to a normal law (Figure 5), but a narrower one. Coefficients of Kurtosis are actually around 6. Since coefficients of the Means-Squares L-estimator are non zero for extreme values, this estimator doesn't reject outliers. As a result distribution tails are as heavy as the average ones, hence control limits are kept at $3\sigma/\sqrt{n}$ to minimise type I false alarms.

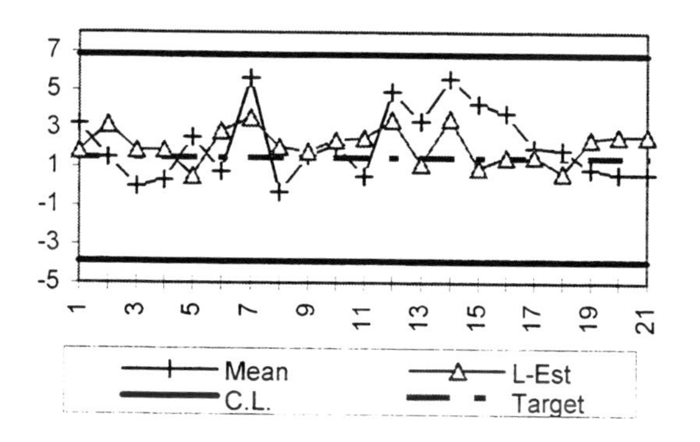

Fig. 6. L-chart of location compared to an $\overline{X}$ chart.

5. EXAMPLE OF APPLICATION

5.1 The L-chart

The interest for such an estimator is evident for an industrial application since the sample size can be reduced without loss of efficiency compared to the mean. The application of this chart requires a preliminary run to calculate the variance-covariance matrix. However this run isn't restricting as it was established by simulation that 40 samples were sufficient to calculate Ω precisely. Such a preliminary run is then short enough for an industrial application. This procedure is equivalent to using a preliminary control chart to determinate the mean and the standard deviation of the process.

When using standard control charts, the process which is under control, is always supposed to be stationary. Statistical Process Control aim is to keep the process under control, which means evolution of the process position or dispersion. In consequence, each time an out of control state is detected, the model (matrix Ω and vector α) has to be computed again to take into account the evolution of the process.

5.2 L-chart for Start-up processes

In order to reduce to a minimum the period of reference and then control the process with the only values μ_p and σ_p fixed by the user's knowledge, we propose a start-up procedure for the L-chart by exponentially weighted moving average (EWMA) of the L-estimator's coefficients.

Coefficients of the estimator ($\mathbf{C}_{EWMA}$) evolve from an initial value ($\mathbf{C}_{initial}$) to the optimal coefficients ($\mathbf{C}_{L-estimator}$) by using relationship (10).

$$C_{EWMA} = \frac{k}{k_{\max}} \cdot C_{L-Estimator} + (1 - \frac{k}{k_{\max}}) \cdot C_{initial} \quad (10)$$

Where k is the number of a samples and k_{max} is the number of samples required to calculate the matrix of variance-covariance with precision. The coefficient k_{max} should be larger than 40.

Coefficients of the vector $C_{initial}$ depend on the parameter to estimate. In order to estimate the parameter of location μ_p, coefficient of $C_{initial}$ are equal to the mean (1/n). As a result, the estimator will provide the mean of samples at the beginning of the run. On the contrary, as k increases, more weight is given to the coefficients of the L-estimator $\mathbf{C}_{L-estimator}$. which are more and more reliable. In order to estimate the parameter of scale σ_p, the vector $C_{initial}$ defined as : $C_{initial}(1)$= -1/d2, $C_{initial}(1)$= 1/d2 and $C_{initial}(i)$= 0 $\forall i \neq 1, i \neq n$ gives an estimation of population's standard deviation with R/d2.

Thanks to this step, variance of estimation for the location parameter, decreases from σ_p^2/n to $\sigma_{L-estimator}^2$.

6. CONCLUSION

Through this short example, we have shown that it is interesting to replace the traditional estimators of Statistical Process Control (the average and the standard deviation), by the Least-Squares L-estimator. This estimator has the advantage of providing non biased estimations with minimum variance. These two characteristics are essential to minimise the cost of non quality. In fact, it is shown that Taguchi's loss function is minimised when observations are centred on the target and their variance is minimal. In addition, the application of this L-estimator is not limited to Shewhart control charts. It can easily be extended to CUSUM charts.

On the other hand, the use of the Least-Squares L-estimator is facilitated by its systematic method of calculation whatever the distribution of the population. In order to make this method transparent for a manufacturer, we have proposed a procedure to launch production without any preliminary run.

Finally, for the construction of L charts, we have systematically placed control limits to $\text{Target} \pm 3\sigma/\sqrt{n}$. This specification allows us to reduce α risks (as compared to the average), however it seems unsatisfactory because, there are unquantifiable risks on both sides of the distribution. L-estimators are non non-normally distributed in spite of the fact that they are asymptotically normally distributed. We are presently working on a systematic method of calculation for control limits that would make L charts more efficient in term of Average Run Length (A.R.L.).

REFERENCES

Aitken, A. C. « *On least squares and linear combinations of observations.* » Proc. Roy. Soc. Edimb. 55,42, 1935

Burr, I. W. « *The effect of Non-Normality on Constants for $\overline{X}$ and R Charts* » Industrial Quality Control, Vol. 3, No. 11, May 1967, pp563-568.

David, H. A.« *Order Statistics* », New York, Willey, 1981

Downton, F. « *Least Square Estimates* », Annals of Mathematical Statistics, Vol 25, pp 303-316, 1954

Duclos, E. « *Optimisation de la Maîtrise Statistique des procédés par une méthode de filtrage d'ordre* », Revue de Statistiques Appliquées (RSA), N° XLIV Vol 2, p 61-79 - 1996

Lloyde, E. H. « *Least Square Estimation of Location and Scale Parameters using Order Statistics* », Biometrika, Vol 39, pp 88-95 - 1952

Pillet, M., Rochon, S., Duclos, E « *SPC - Generalisation of capability index Cpm. Case of unilateral tolerances* » Quality Engineering, 1997.

Shilling, E.G, Nelson, P. R. « *Effect of Non-Normality on the Control Limits of $\overline{X}$ Charts* », Journal of Quality Technology Vol 8, N° 4, October 1976.

Yourstone, S. « *Non-normality and design of Control Charts for Average* », Decision Science, Vol 23, pp 1099-1113 - 1992

DESIGN OF POST-PROCESS QUALITY-CONTROL FOR THE TURNING-PROCESS

O. Sawodny, G. Goch

Universität Ulm
Abt. Meß-, Regel- und Mikrotechnik

Abstract: For a quality control in cutting manufacturing processes, a method is presented for the implementation of closed quality-control loops at turning process. In order to integrate the method in common manufacturing systems, the control strategy is designed as a post-process measuring system. As an important condition for the control method, additional sensors and actors for the machine tools should be avoided. To design the control, a model has been developed, which describes the most important internal influences on the manufacturing process based on the known rules of the turning technologies. Using this model, different designs for the control are presented.

Keywords: Quality control, Manufacturing process, Modelling, Parameter estimation, Control design

1. INTRODUCTION

In the last years, the control of the dimensional quality characteristics in cutting manufacturing systems became more and more important. But till today, there are only a few approaches in industrial practice, which control these characteristics in a closed loop structure. These systems mostly assume an in-line-measurement of process parameters like the cutting forces or tool geometry (e.g. Harder *et al.*, 1995, Rao, *et al.*, 1995, Lundholm, 1991). This leads to a decreasing reliability of the rather complex machines, which are running under rough operating conditions. So, the industrial acceptance of these machines is farely low. In addition, a global quality strategy concerning the whole manufacturing system is only realizable, if all machines are upgraded with such an in-line control system, which causes very high investment costs. Therefore, the machine independency is another important condition for a convenient quality control. In the following, a method is presented, which is integrable in common manufacturing systems. For the turning process the modelling is shown. Different designs for the quality control meeting the above shown conditions are discussed based on measurement results.

2. CONDITIONS FOR THE QUALITY CONTROL

For the integration of a quality control in common manufacturing systems, no additional sensors and actors should be necessary. So, only the parameters, which can be changed by the NC-control of the machine, are used for the process inputs. The output variables are the dimensional quality characteristics (e.g. the diameter). Common measurement devices are used, which mainly operate off-line, sometimes even in separate measuring rooms. The control inputs are the given tolerances for the desired workpiece. As the control is based on a post-process-measurement system, the correction of process inputs are evaluated in the step of the produced workpieces (or samples). For a broad use in a manufacturing system, the modelling of the cutting process ensures the machine indepency of the control.

3. MODELLING OF THE TURNING PROCESS

Following the idea of a quality control, a model for the manufacturing process is derived from a formalized procedure (fig.1).

First, a detailed process analysis leads to the most important process disturbances. In the next step, a relationship between these disturbances and the affected geometrical deviations is formulated. The deviations are classified and each class is described by the equations of the manufacturing technologies, aiming at a description, which connects the geometrical deviations with internal process parameters. Lastly, the intended control of the manufacturing task defines the parameters, which are process inputs, output variables, disturbance variables and internal process parameters. The resulting control structure is the basis for the subsequent control design.

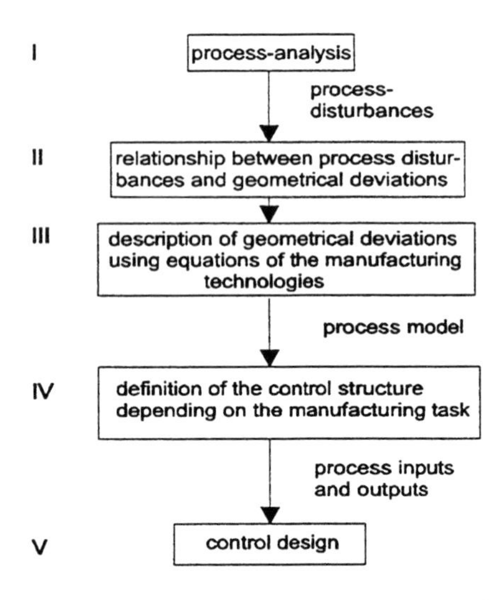

Fig. 1: Formalized modelling procedure for the quality control.

3.1 Process analysis for the turning process and relationship to the geometrical deviations

A detailed process analysis leads to the possible process disturbances of table 1. The process errors are investigated concerning their effect on the process. Next, their relationship to the affected geometrical deviations (or surface roughness) is determined.

3.2 Modelling of the geometrical deviations

The most important process errors concerning the geometry are summarized in an offset and flexional deviation. The flexional deviation is mainly caused by the passive force, the component of the cutting force, which is nearly rectangular to the surface of the workpiece (Tönshoff 1995). Different approaches for the description of the passive force are given in the literature (e.g. Meyer 1964). For the design of the control, the following equations are used (Sawodny, 1996) distinguishing two cases in dependence of the depth of cut a (1). R ist the radius of the cutting edge, κ the the cutting edge angle. These nonlinear equations connect the process inputs depth of cut a and forward feed f with the resulting passive force under ideal process conditions. In this case the parameter values of $1\text{-}x$ and $k_{a1.1}$ are known. Process disturbances effecting a changing passive force are reflected by changing of these parameter values.

$$F_p = \frac{1}{2-x} k_{a11} R f^{1-x} \left(\sqrt{\frac{a}{R}\left(2-\frac{a}{R}\right)}\right)^{2-x} \quad ;\text{if } a \le R_\kappa$$

$$F_p = k_{a11} f^{1-x} \left(\frac{1}{2-x} R(\sin\kappa)^{2-x} + \frac{\cos\kappa}{(\sin\kappa)^x} \cdot (a-R_\kappa)\right); \text{ if } a > R_\kappa \text{ where: } R_\kappa = R(1-\cos\kappa) \qquad (1)$$

Table 1: Process errors, effect on the process, and geometrical deviations

Process error	*effect on the process*
NC-control errors, spindle errors, elasticity of the machine	deviation of nominal and real path → **offset deviation**
chattering	waviness, form of the workpiece surface (not considered), **effect on surface roughness**
eccentricity of clamping	changing depth of cut, chattering (not considered)
wear of tool	increasing cutting forces → **flexion deviation**, wear of cutting edge → **offset deviation, effect on surface roughness**
temperature	microstructural change, **effect on surface roughness** continous material expansion → **offset deviation,**
cutting forces	flexion of the workpiece → **flexion deviation**
changing geometry of the cutting edge	changing cutting forces → **flexion deviation, effect on surface roughness**
tool material, workpiece material, cooling	changing cutting forces → **flexion deviation**

The passive force results in the flexion of the clamped workpiece (fig. 2). The geometrical deviation due to the flexion is described by the differential equation

$$\frac{d^2 \Delta x_{flex}}{dz^2} = \frac{F_p(z_{force} - z)}{EJ(z)} \qquad (2)$$

Equation (2) is solved numerically, depending on the workpiece geometry, which affects the geometrical moment of inertia by the nominal radius at a location z along the workpiece axis. The offset deviation is a simple constant for each workpiece section.

The surface roughness is described by the following theoretical equation (Degner, 1985):

$$R_z = \frac{f^2}{c_0 R} \;;\quad c_0 \approx 8 \qquad (3)$$

Changing process conditions are summarized in the parameter c_0, which is under ideal conditions identical to 8.

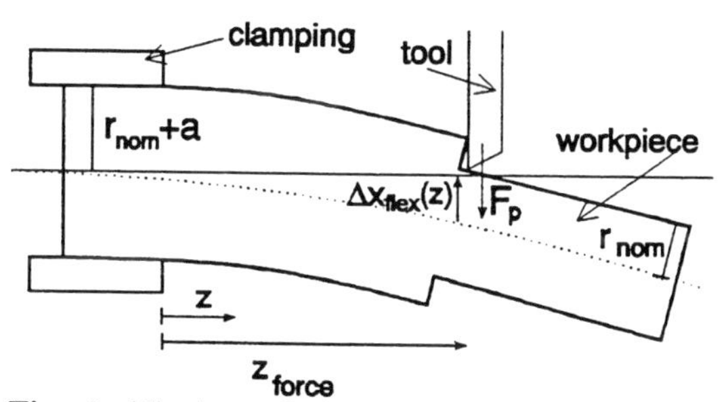

Fig. 2: Flexion of the workpiece in a one-sided clamping.

The result of this analysis is a static nonlinear model of the process. Its dynamic behaviour is taken into account by a description, which depends on the workpiece geometry and the updating of the process inputs following the step of the produced workpieces.

4. LINEARIZED DESIGN FOR THE CONTROL OF GEOMETRY AND SURFACE

For the manufacturing of shafts, a control structure will be developed. The geometry of the shafts is divided into basic geometry elements (cylindrical, conical and spherical areas), because each section can be tolerated differently. The control of a single workpiece section is separated from the other workpiece areas.

4.1 Control structure

The measured diameters along the workpiece axis of the considered workpiece sections will be (concerning the geometry) overlapped by a flexional and offset deviation (see 3.). The offset deviation can be compensated directly by its inverse value, which will be evaluated by a linear regression algorithm using the knowledge of the flexional line. In the case of a cylindrical workpiece section, the flexion line from $z=0$ to $z=z_0$ according to (2) is a function:

$$\Delta x_{mess} = t \cdot z_k^3 + \Delta x_{offset} \qquad (4)$$

The subtraction of the regression parameter Δx_{offset} results in a measured flexional deviation $\Delta x_{flexmeas}$. The nominal value for the flexion $\Delta x_{flexnom}$ will be tuned to the given tolerance in the desired workpiece section. So the variance as the controller input is the difference of these values (fig. 4). The control amplification is calculated by the inverse process amplification compensating the disturbances by reducing or increasing the passive force in the production step i due to the value ΔF_p.

$$\Delta F_{p,i} = \frac{3E\pi r^4}{4z_k^3}(\Delta x_{flexnom,i} - \Delta x_{flexmeas,i}) \qquad (5)$$

The passive force law (1), linearized in the working point, leads to the adjusting values Δa and Δf of the process inputs.

$$\Delta a_i = \frac{1}{\left.\frac{\partial F_p}{\partial a}\right|_{a=ai}}\left(\Delta F_{p,i} - \left.\frac{\partial F_p}{\partial f}\right|_{f=fi} \Delta f_i\right) \qquad (6)$$

Δf_i is determined by the result of the surface roughness control loop. The variance of this control loop is the difference between measured surface roughness R_{zmeas} and the nominal surface roughness R_{znom}. The control amplification is again the inverse process amplification , so that Δf_i will be

$$\Delta f_i = \frac{1}{\left.\frac{\partial R_z}{\partial f}\right|_{f=fi}}(\Delta R_{znom,i} - \Delta R_{zmeas,i}) \qquad (7)$$

Then, the new process inputs for the next production step will be

$$\begin{aligned} f_{i+1} &= f_i + \Delta f_i \\ a_{i+1} &= a_i + \Delta a_i \end{aligned} \qquad (8)$$

The working point is continously actualized from one production step to the next in order to improve the behaviour of the control in general. Fig. 4 shows its structure, as explained in the section above.

4.2 Experimental results

Experiments are carried out at the manufacturing of shafts with a nominal diameter of 17 mm and a length of 100 mm. The diameter tolerance of 55 μm leads to an allowed passive force of 74.5 N under ideal process conditions. The calculation of the starting values for the process inputs is (according to a nominal value of 10 μm of the surface roughness) a forward feed f_0 of 0.18mm and a depth of cut a_0 of 0.11 mm. Fig. 5 shows the surface roughness R_{zmeas} and the offset deviation Δx_{offset} for the first 5 produced workpieces. Because of process deviations, the nominal value of these output parameters is reached after the fourth workpiece. The controller adjusts the workpiece contour to the given ideal contour, as shown in fig. 6 at the example of the measured radii along the workpiece axis z for the first and fifth workpiece. The process inputs a and f tend to stable values (fig. 7).

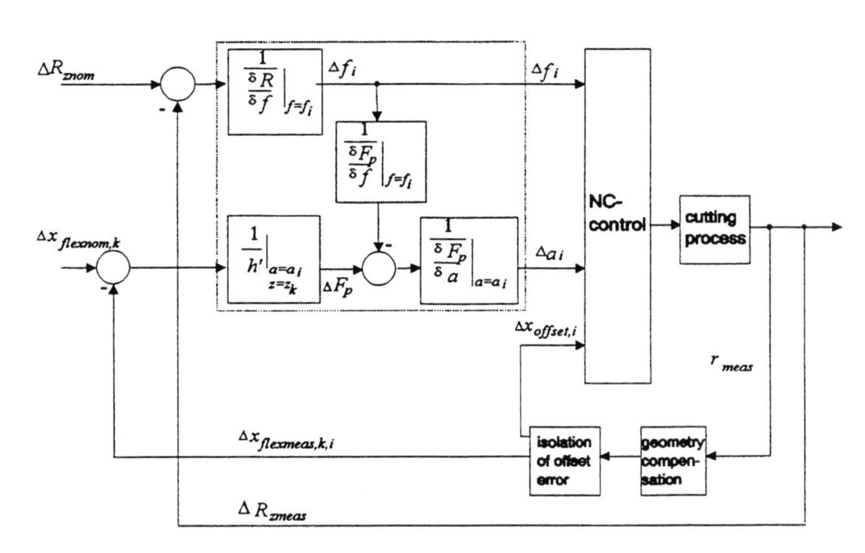

Fig. 4: Structure of the linearized geometry and surface roughness control.

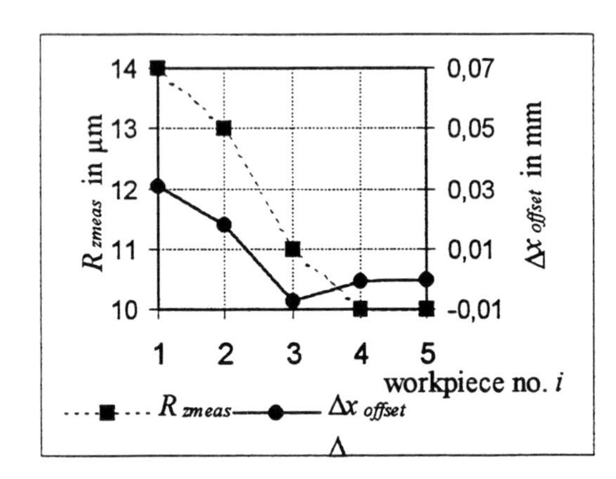

Fig. 5: Measured surface roughness R_{zmeas} and offset deviation Δx_{offset} for the first 5 workpieces.

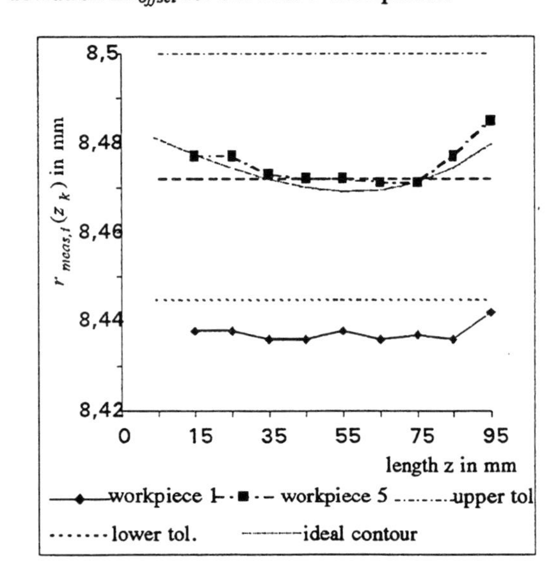

Fig. 6: Measured contour r_{meas} of the first and fifth workpiece

4.3 Discussion of the linear approach

In the above example, the actual process conditions deviate only a little from their ideal values. But as experiments show, in case of larger deviations, especially of the force parameters $k_{a1.1}$ and $1\text{-}x$, the linear approach tends to large oscillations of the process inputs.

These effects can be reproduced during simulations. In fig. 8 the force parameter $k_{a1.1}$ increases to 200 %. The controller is not able to adjust the contour and the process inputs a and f are alternating between their limits, which are given by the manufacturing technologies.

This is a very important aspect, since in cutting manufacturing systems such deviations are quite possible. Lastly one can summarize that the robustness of the linear approach is not sufficient for the use in cutting manufacturing systems.

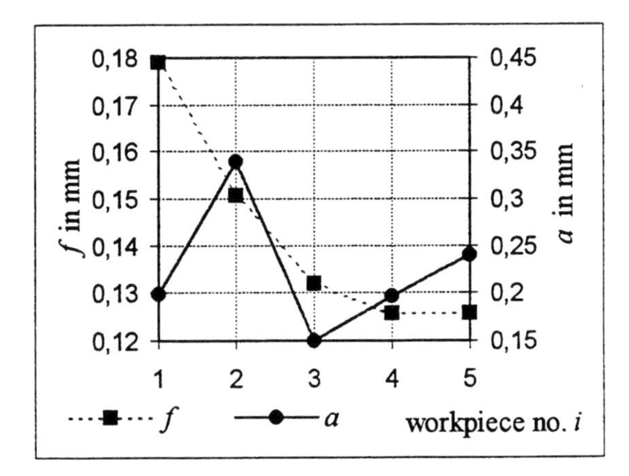

Fig. 7: Process inputs a and f for the production steps 1 to 5.

5. CONTROL BY ESTIMATION OF INTERNAL PROCESS PARAMETERS

5.1 Control sturcture

To improve the robustness, a controller was designed, which estimates varying process parameters. The expected geometry of the workpiece is overlapped by an offset and flexional deviation and a third deviation caused by the periodical surface roughness profile according to

$$\Delta x_{Rz} = R_{zTKM}\left|\sin\frac{\pi}{f}z_k + \varphi_0\right| \tag{9}$$

R_{zTKM} is leveled out due to the measurement on a CMM with a probing ball diameter r_{TKM}. Contrarily R_{zmeas} is measured with a fine stylus instrument. The differential equation (2) is solved once for a desired workpiece geometry assuming a defined passive force F_{pnorm}. In relation to the set tolerance, an allowed flexional deviation $\Delta x_{flexall}$ and subsequently a allowed passive force F_{pall} is calculated. Then, an estimated flexional deviation can be described by:

$$\Delta x_{flex} = \frac{F_{peff}}{F_{pall}} \Delta x_{all} \tag{10}$$

where F_{peff} is the assumed effective passive force. According to (1), Δx_{flex} is a function linear to the parameter $k_{a1.1}$ and nonlinear to the parameter $1\text{-}x$.

$$\Delta x_{flex} = k_{a1.1} g(a, f, 1-x) \tag{11}$$

The measured radius at the z-coordinate z_k is then

$$r_{meas}(z_k) = r_{nom}(z_k) + \Delta x_{flex}(z_k) + \Delta x_{offset} + \Delta x_{Rz}(z_k) \tag{12}$$

With the equations (9) to (12), a quadratic objective function for the estimation of the process parameters Δx_{offset}, $k_{a1.1}$, $1\text{-}x$ and φ_0 can be formulated:

$$Q(k_{a1.1}, 1-x, \Delta x_{offset}, \varphi_0) = \sum_{k=1}^{n} (r_{meas}(z_k) - r_{nom,k}(z_k) - k_{a1.1} g(a, f, 1-x, z_k) - \Delta x_{offset} - R_{zTKM} \left| \sin(\frac{\pi}{f} z_k + \varphi_0) \right|)^2 \tag{13}$$

The objective function Q requires n radii $r_{meas}(z_k)$ measured with a CMM at the positions z_k in the considered workpiece section.

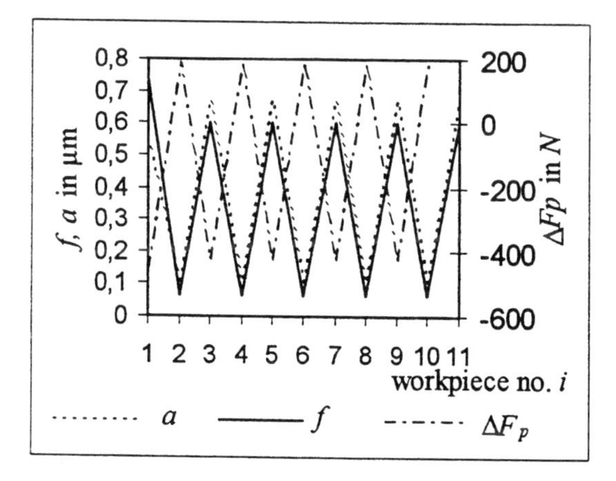

Fig. 8: Oscillations of the process inputs during simulations for an increased force coefficient $k_{a1.1}$ up to 200 %.

As (13) is a nonlinear optimization problem, in most cases the result is only a local minimum due to the noise of the measured data and the dependence on the starting values. To improve the optimization results, a sequential quadratic programming method is used (Sachs *et.al.* 1994, Großmann *et.al.* 1993). The possible ranges of parameter values are formulated as constraints (secondary conditions) according to the theory and technical conditions.

$$\begin{aligned} &k_{a1.1} \geq 0 \\ &0 \leq 1-x \leq 1 \\ &-1 \leq \Delta x_{offset} \leq 1 \end{aligned} \tag{14}$$

In addition, the linear subproblem is separated from the nonlinear parameter problem. The subproblem is solved by the direct solution.

$$\min\{Q(k_{a1.1}, 1-x, \Delta x_{offset}, \varphi_0)\} = \min_{1-x,\varphi_0} \left\{ \min_{k_{a1.1}, \Delta x_{offse}} Q(k_{a1.1}, 1-x, \Delta x_{offset}, \varphi_0) \right\} = \min_{1-x,\varphi_0} \{\tilde{Q}(1-x, \varphi_0)\} \tag{15}$$

Since (13) is overparameterized concerning the parameter $1\text{-}x$, the problem is modified according to (16)

$$u_i = \sum_{k=1}^{n} (\Delta x_{meas,i}(z_k) - k_{a1.1} g(a_i, f_i, 1-x) \frac{\Delta x_{flexall}}{F_{pall}} + \Delta x_{offset,i} + R_{zTKM,i} \left| \sin \frac{\pi}{f_i} z_k + \varphi_{0,i} \right|$$

$$Q(k_{a1.1}, 1-x, \Delta x_{offset,1..j}, \varphi_{0,1..j}) = \sum_{i=1}^{j} u_i \rightarrow \min. \tag{16}$$

Using this approach all interesting process parameters can be estimated after the second workpiece. The estimated values of the preceeding production step are used for the calculation of the following set of process inputs, based on the nonlinear equations (1) and (3) and adjusting the control parameters to their nominal values. The tolerance leads to an allowed flexion and thus to an allowed passive force. This passive force is the nominal value, which has to be achieved in order to receive the defined ideal contour of the considered workpiece section. The controller determines with the actual process parameters $1\text{-}x$, $k_{a1.1}$ the values for the process inputs a and f, which will adjust this passive force. The offset deviation is compensated directly. The structure of the control is shown in fig. 9.

5.2 Experimental results

Again, with the control explained above, shafts with a length of 100 mm are produced. The geometry is divided into 3 sections: a cylinder (z=0mm to z=40mm; diameter 10mm), a cone (z=40mm to 70mm) and a cylinder (z=70mm to 100mm, diameter 7mm). In this experiment the section from 70mm to 100mm is considered.

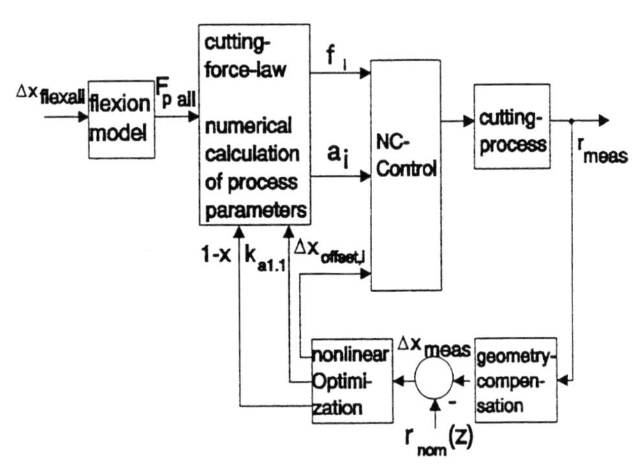

Fig. 9: Structure of the control by estimation of internal process parameters.

Fig. 10 shows the calculated deviation between the estimated and allowed passive force ΔF_p ,which continuously tend to 0. Its effect on the geometry is very important for the discussion of the control behaviour. The calculated form error of about less than 10 μm in the considered workpiece section is a satisfying result for a one-sided clamping (fig. 11). More experiments show a good behaviour in case of large deviations of the process inputs as well.

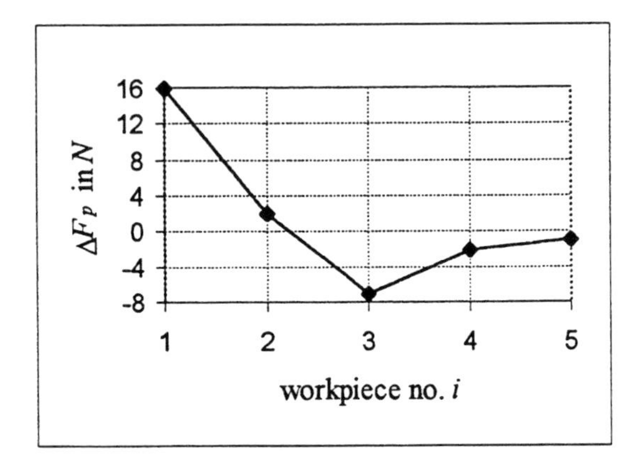

Fig. 11: Calulated difference between allowed and estimated passive force for the workpieces 1 to 5.

6. CONCLUSION

The shown control method was designed for the use in common manufacturing systems. Therefore, the investment costs for an upgrading with such a quality control should be as low as possible. So, common post process measuring devices and NC-machines are assumed, which control the quality in the step of the produced workpieces or samples (in case of slowly changing process parameters SPC-methods can easily be adapted).

After a model for the turning process was derived, two different control designs were presented. A linearized control, which is not able to adjust the workpiece gemoetry to the ideal contour even for large process disturbances, and a second control based on a nonlinear estimation of internal process parameters. This control shows a enhanced robustness. Also for widely varying process parameters, the control is able to guarantee the claimed tolerances.

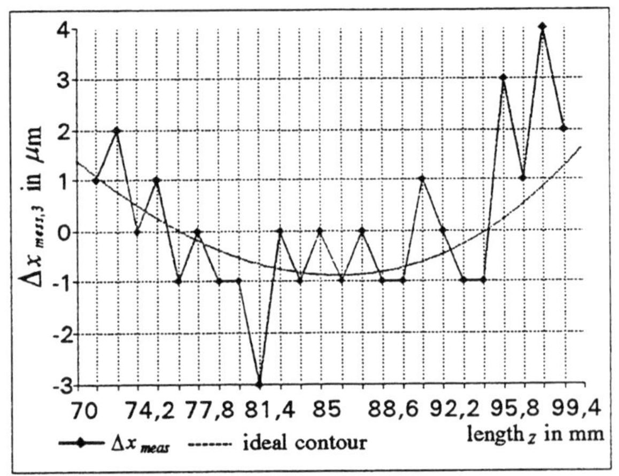

Fig. 11: Measured deviation from the ideal contour at the fifth workpiece.

REFERENCES

Großmann Ch., Terno J. (1993): Numerik der Optimierung, Teubner Verlag, Stuttgart

Harder L., Isaksson A.J. (1995). Robust PI-control of cutting forces in turning. In: Proceedings of the 31th International MATADOR conference, Macmillan press, Lonodon, p. 261-266

Lundholm, Th. (1991). A Flexible Real-Time Adaptive Control System for Turning. In: Annals of the CIRP, Paris, Vol. 40/1, p.441-444

Meyer K.F.(1964): Der Einfluß der Werkzeuggeometrie und des Werkstoffes auf die Vorschub- und Rückkräfte des Drehens, Industrie-Anzeiger, **86**, 835-844

Rao, B.C., Gao R.X. (1995). Integrated force measurement for on-line cutting geometry inspection, In: IEEE transactions on instrumentation and Measurement, vol. 44, no. 5, p.977-980

Sachs E.W. (1994): Control Applications of Reduced SQP Methods; in: Bulirsch, R., Kraft, D. (Ed.): Computational Control, Birkhäuser Verlag, Basel, 89-104

Sawodny O. (1996) Modellbasierte Qualitätsregelung in der spanenden Fertigung am Beispiel der Drehbearbeitung, VDI-Verlag, Düsseldorf

Tönshoff H.-K.(1995): Spanen, Springer Verlag, Berlin

Zhang G.M., Hwang T.W.: (1995) Mathematical modelling of the uncertainty for improving quality in machining operations, p.500-505, Proceedings 2nd International Symposium on uncertainty modelling and analysis, IEEE Computer Society Press, Los Alamitos, 1995

A ROBUST CONTROL DESIGN FOR MINERAL PROCESSING PLANT

S. Yahmedi*, A. Pomerleau and D. Hodouin****

* *Université d'Annaba, BP 12, Algérie*
** *Université Laval, Québec, G1K 7P4, Canada*

Abstract: Modern controllers are based on the use of mathematical models. However the models are always obtained through a reduction in the complexity of reality. Consequently, their ability to properly represent the general behavior of the processes is very limited. Therefore, it is advantageous to analyze the problem resulting from model uncertainty in the control of mineral processing plant. This problem has often been ignored in theoretical studies and in practical process control. This paper will first show an H infinity (H_∞) method on how to achieve the benefit of feedback in the face of uncertainties. Then it will present grinding process example which illustrates the use of robust control to provide satisfactory performance despite of system uncertainties in mineral processing plant.

Keywords: H infinity control, Robust stability and performance.

1. INTRODUCTION

The standard modern approach to process control consists in constructing a mathematical model of the process and then using explicitly this model in the controller. However, there are two major problems with this approach: first, the model is only a simplified representation of the process which is generally much more complex; second, the process behaviour continuously changes. For these two reasons there is inevitably a mismatch between the plant and the model. Such model uncertainties are responsible for the degradation of the controller. Hence, the first step in a robust control study is to quantify these uncertainties. For that purpose, a phenomenological simulator is used and by varying the operating conditions: the ore grindability, the hydrocyclone tuning and the operating levels, the models for five operating conditions different from the nominal conditions are obtained and their multiplicative uncertainties are also determined. In the next step an H_∞ controller for a grinding circuit is found.

2. PRELIMINARIES

It is necessary to recall the basic required performances of a control loop in the frequency domain. Fig.1 shows the classical structure of a control loop with the main components: the controller (transfer matrix K(s)), the process (transfer matrix G(s)), the multiplicative model uncertainty at the process output $\Delta_m(s)$, the set-point r, the loop's error e and finally the manipulated variable u and the output y. Let G'(s) the transfer matrix of the true plant, all perturbed regimes, then the following relation can be written:

$$G'(s) = [I+\Delta_m(s)]G(s) \qquad (1)$$

The largest singular value of $\Delta_m(s)$ is obtained from relation (1):

$$\sigma_{max}[\Delta_m(s)] = \sigma_{max}([G'(s)-G(s)]G^{-1}(s)) \qquad (2)$$

Relation (2) is used to quantify the multiplicative models uncertainties.

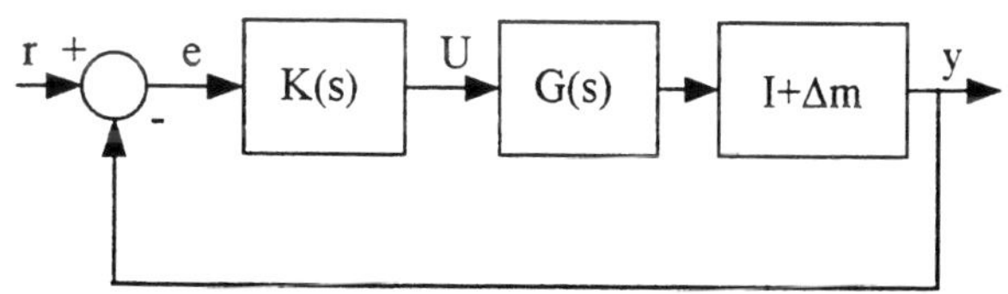

Fig.1 : Feeback configuration with multiplicative uncertainties

2.1 Robust stability

Assume that the nominal feedback system G(s) (i.e with $\Delta_m(s) = 0$) is stable, then the true feedback system G'(s) is stable if the following inequality holds (Doyle and Stein, 1981):

$$\sigma_{max}[T(s)] < 1/\sigma_{max}[W_t(s)] \quad (3)$$

Where T(s) is the nominal closed loop transfer matrix given by:

$$T(s) = G(s)K(s)[I+G(s)K(s)]^{-1} \quad (4)$$

And $W_t(s)$ is a stability specification matrix such as:

$$\sigma_{max}[\Delta_m(s)] \leq \sigma_{max}[W_t(s)] \quad (5)$$

Then $\sigma_{max}[T(s)]$, the largest singular value of the nominal closed loop transfer matrix is a reliable indicator of the robust stability of the feedback system. Relation (3) is the robustness condition of the feedback system.

2.2 Robust performances

Let $W_p(s)$ a performance specification matrix, weighting matrix, then the robust performances of all perturbed regimes G'(s) are satisfied if the following inequality holds (Doyle and Stein, 1981) and (Safonov and Chiang, 1988):

$$\sigma_{max}[S(s)] \leq 1/\sigma_{max}[W_p(s)] \quad (6)$$

Where S(s) is the sensitivity matrix given by:

$$S(s) = [I+G(s)K(s)]^{-1} \quad (7)$$

In fact, the largest singular value of the sensitivity matrix $\sigma_{max}(s)$ is also an indicator of the sensitivity of the system response to a change of the plant character.

In conclusion, the inequalities (3) and (6) represent the robustness conditions and must be satisfied to obtain a robust controller.

3. H_∞ CONTROLLER SYNTHESIS

The optimal control theory using the H_∞ norm in the frequency domain has known important developements during the past decade. also the H_∞ controller synthesis is easily realized using the solutions of H_∞ control problem in the state space (Doyle et al., 1989) and (Maciejowski, 1989). From the augmented state space representation showed in fig.2, the transfer function T_{y1u1} relating y to u is given by:

$$T_{y1u1} = \begin{bmatrix} W_p(s) \\ W_t(s) \end{bmatrix} \quad (8)$$

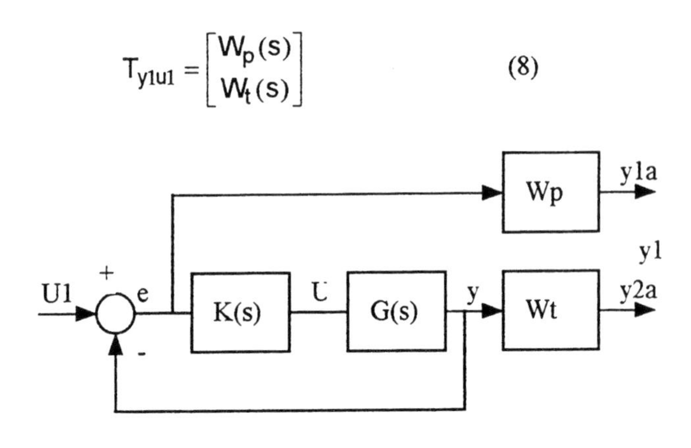

Fig.2 : Augmented system configuration.

Then the standard problem of H_∞ control theory is that of obtaining all controllers which make the H_∞ norm of T_{y1u1} less than one, i.e:

$$\left\| T_{y1u1} \right\| < 1 \quad (9)$$

Where $\|.\|$ is the H_∞ norm defined as:

$$\|.\| = \sup(\sigma_{max}[.]) \quad (10)$$

The use of H_∞ method (Doyle, 1983) and matlab program (Safonov and Chiang, 1988) and (Chiang and Safonov, 1988) gives a robust H_∞ controller (Yahmedi, 1993). The performance specification matrix $W_p(s)$ is adjusted successively by a " γ" parameter. In others words the performance specification is considered of the form: $\gamma * W_p(s)$, where γ is a real number. The following organigram in fig.3 gives the design method:

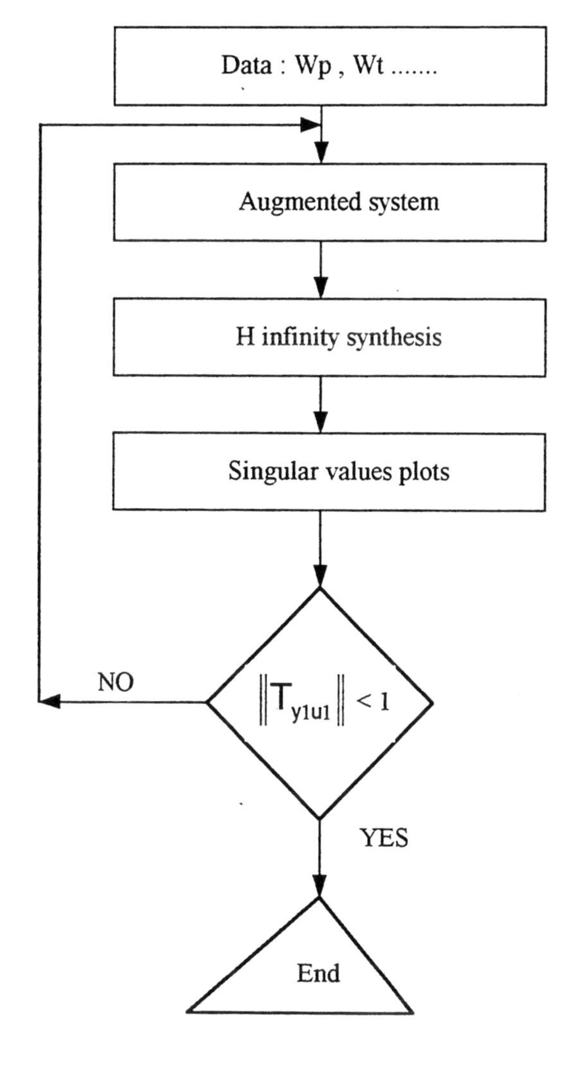

Fig.3 : Organigram of H_∞ synthesis

4. APPLICATION: GRINDING CIRCUIT

The representation of grinding circuit is given in fig.4 where the manipulated variables u_1 and u_2 are respectively the fresh ore feed rate and the water feed addition to the cyclone pump sump. The final controlled variables are the ball mill throughput y_1 and the product fineness y_2. A dynamic process model was developed with the aid of step responses. The transfer function of the nominal regime of the process is given by:

$$G(s) = \begin{vmatrix} G_{11}(s) & G_{12}(s) \\ G_{21}(s) & G_{22}(s) \end{vmatrix} \tag{11}$$

Where:

$$G_{11}(s) = 1.53e^{-180s}/(1+998.5s)$$

$$G_{12}(s) = (.49e^{-30s}-.25e^{-570s})/(1+1110s)$$

$$G_{21}(s) = -2.25(1-348.8s)e^{-60s}/(1+634.1s)^2$$

$$G_{22}(s) = .2(1+2488s)e^{-60s}/(1+630s)$$

This regime is called A_2 and the operating conditions are u_1 = 70(t/h), u_2 = 96.18(t/h), y_1 = 175(t/h) and y_2=68%. The models for five operating conditions different from the nominal conditions (perturbed regimes) are also obtained with the aid of step responses from a phenomenological simulator of grinding circuit.

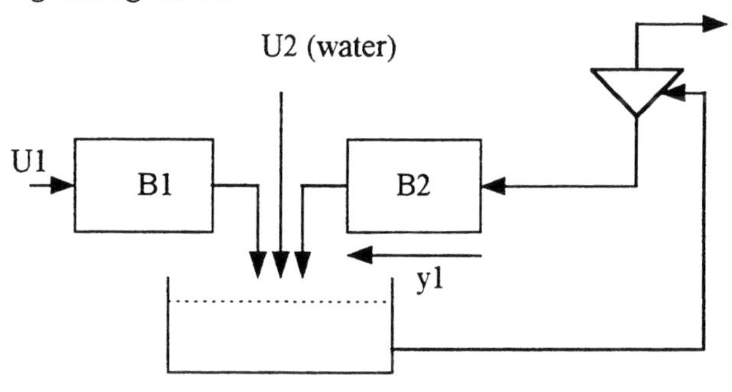

Fig.4 : Grinding circuit.

4.1 Evaluation of multiplicative uncertainties $\Delta_m(s)$

The largest singular values of the multiplicative uncertainties $\Delta_m(s)$ are determined from relation (2). The result is given in fig.5, where it is verified that the norms of these uncertainties are less than one at low frequencies and increase at high frequencies (Doyle and Stein, 1981).

4.2 Robustness conditions

using relation (5) and the result given in fig.5, the stability specification matrix $W_t(s)$ is represented as:

$$W_t(s) = \begin{vmatrix} 0.8(1+50s) & 0 \\ 0 & 0.8(1+50s) \end{vmatrix} \tag{12}$$

Then, the condition for robust stability is given by inequality (3). The performance specifications for all possible plants, perturbed regimes, are defined such these regimes have the same response time that the nominal regime A_2. Then, the performance specification matrix $W_p(s)$ is given by:

$$W_p(s) = \begin{vmatrix} (1+1000s)/(1000s) & 0 \\ 0 & (1+1000s)/(1000s) \end{vmatrix} \tag{13}$$

The condition for robust performance is given by relation (6). Finally, the robustness conditions for grinding circuit are represented in fig.6.

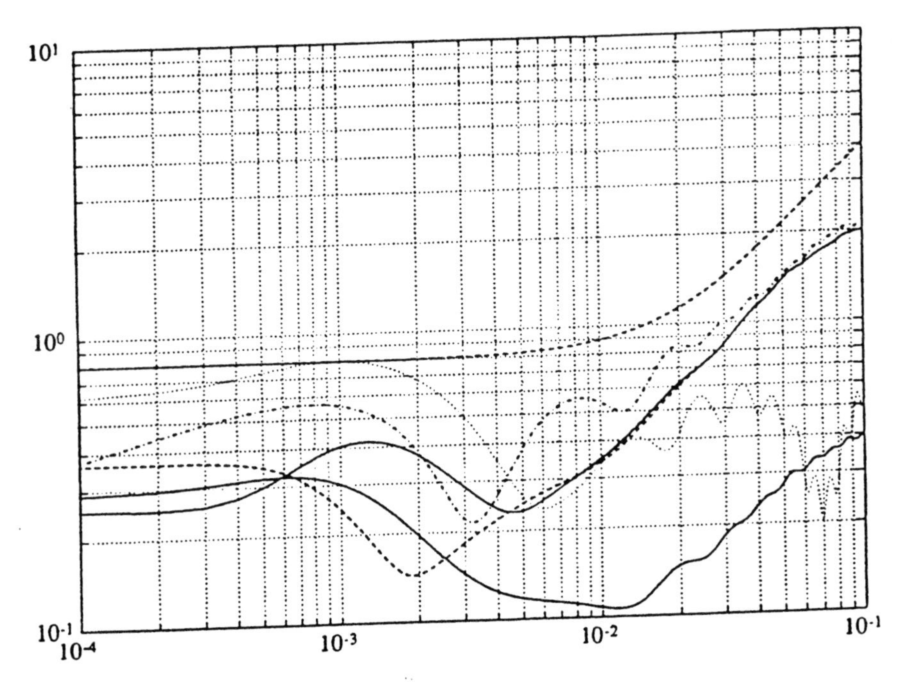

fig.5 multiplicative uncertainties $\Delta_m(s)$

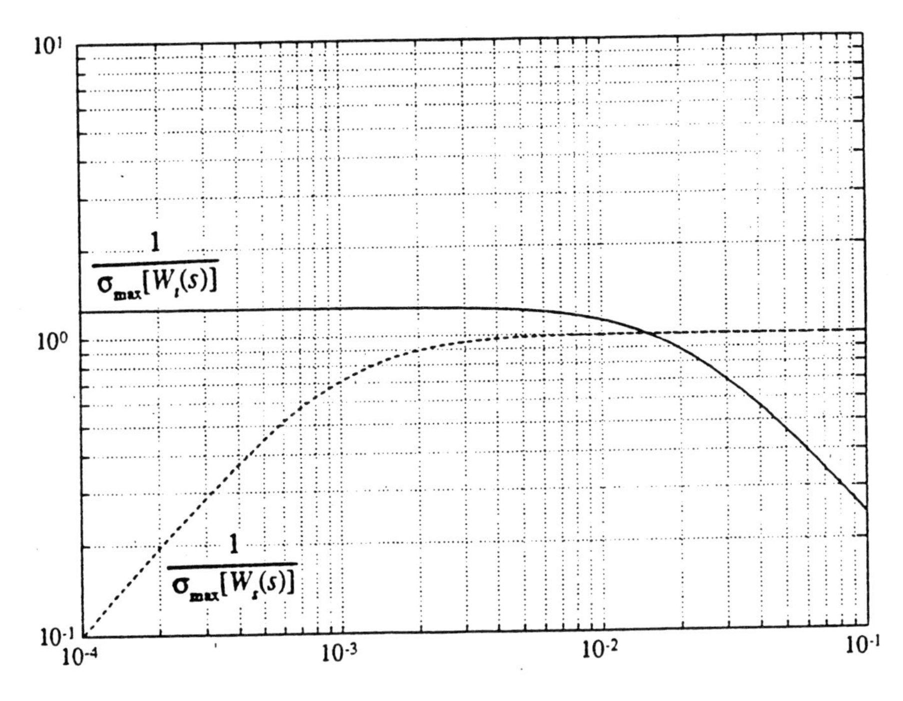

fig.6. robustness conditions

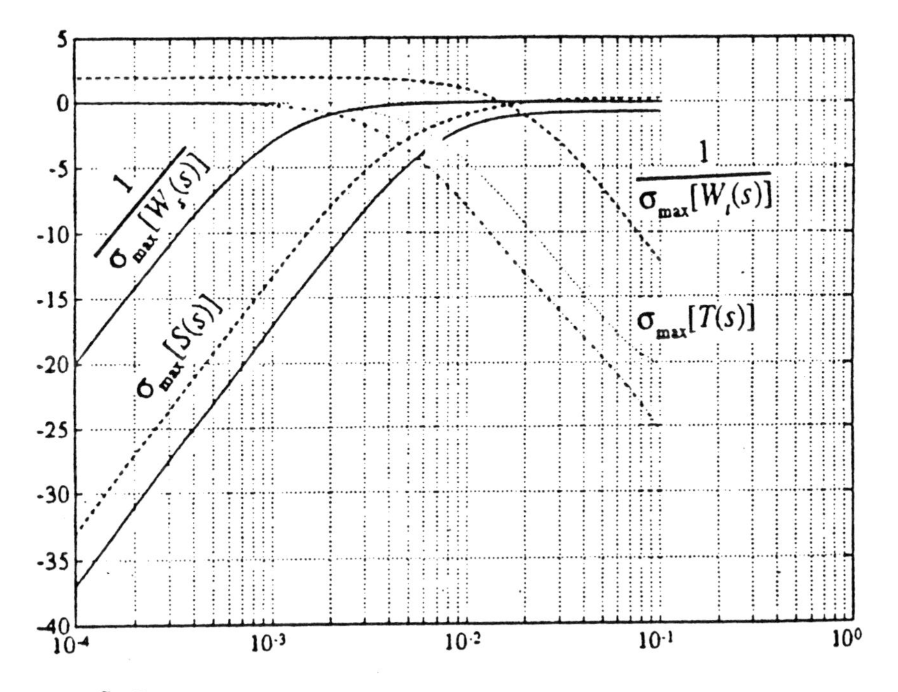

fig.7 frequency results

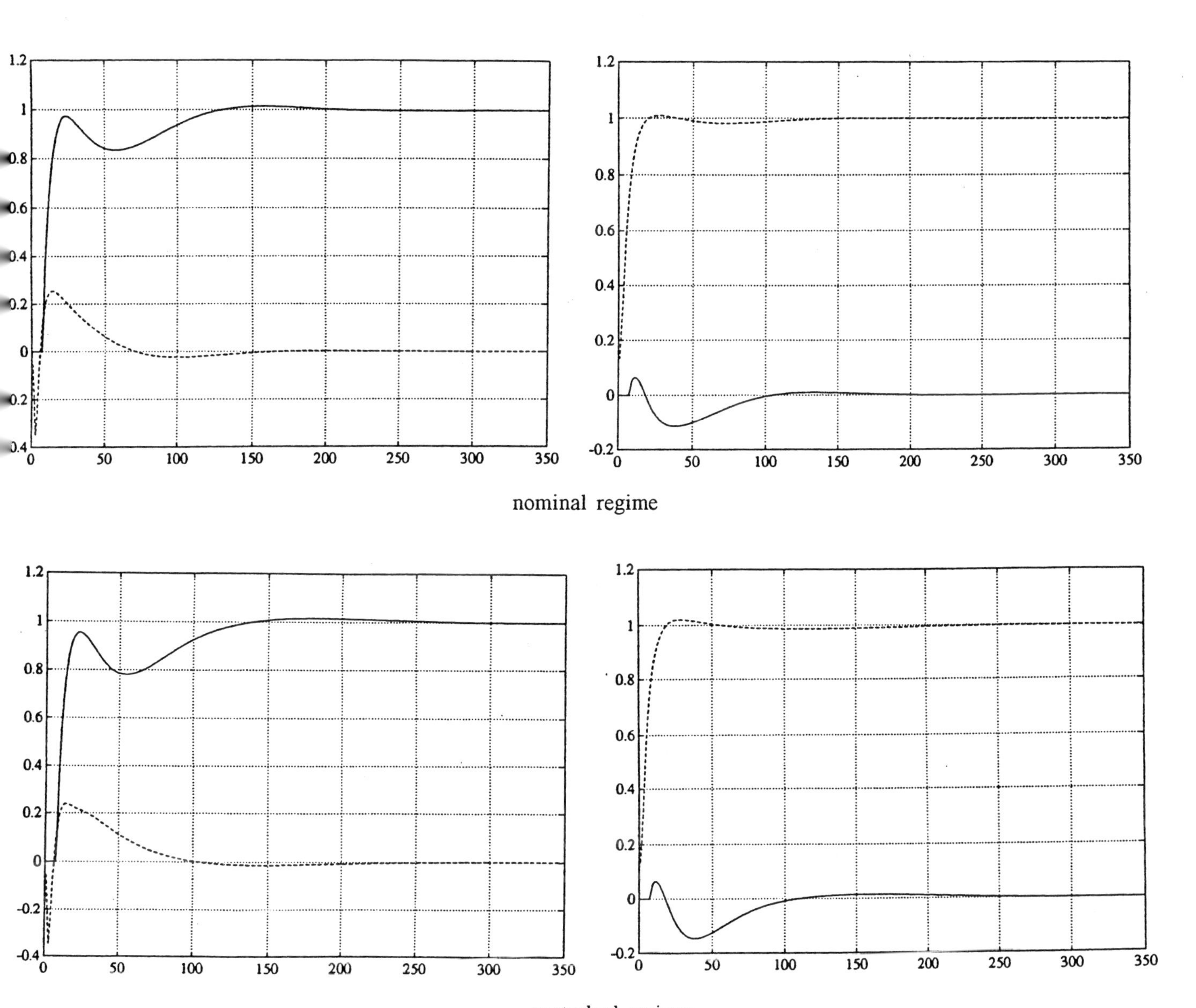

fig.8 step responses,

4.3 H_∞ controller for grinding circuit

The H_∞ synthesis for grinding circuit consists of finding a controller K(s) such that the small gain theorem given by relation (9) be satisfied and the conditions (3) and (6) for robust stability and performance are also verified. A simplified model of nominal regime A_2 is used in the design. This model is defined by:

$$G_s = \begin{bmatrix} 1.53/(1+1000s) & 0.24/(1+1000s) \\ 0.2(1+2500s)/(1+630s) & -2.25(1-350s)/(1+630s)^2 \end{bmatrix}$$

The controller is obtained using the logiciel H_2H_{inf} (Yahmedi, 1993):

$$K(s) = C(sI-A)^{-1}B+D \qquad (14)$$

where

$$A = [A1\ A2\ A3]\ ;\ C = [C1\ C2]$$

With:

$$A1 = \begin{matrix} -.1705 & -.0264 & -.0005 & -.0001 \\ -.0025 & -.003 & 4.5844e\text{-}6 & -2.6336e\text{-}6 \\ 1 & 0 & 0 & 0 \\ 0 & 1 & 0 & 0 \\ 0 & 0 & 1 & 0 \\ 0 & 0 & 0 & 1 \\ 2.0916e\text{-}19 & 2.58e\text{-}20 & 8.2274e\text{-}22 & 1.0172e\text{-}22 \\ 2.1757e\text{-}19 & 1.085e\text{-}19 & 1.897e\text{-}24 & 4.2315e\text{-}22 \\ -.1663 & -.0264 & -.0005 & -.0001 \\ -.0025 & .0012 & 4.5844e\text{-}6 & 3.0605e\text{-}6 \\ .0015 & .0002 & 4.8571e\text{-}6 & 7.619e\text{-}7 \\ -1.6187e\text{-}5 & -2.6651e\text{-}6 & -4.6403e\text{-}8 & -9.0315e\text{-}9 \end{matrix}$$

$$A2 = \begin{matrix} -4.2397e\text{-}7 & -6.6448e\text{-}8 & .6919 \\ 7.0947e\text{-}9 & -6.4452e\text{-}10 & .0016 \\ 0 & 0 & 0 \\ 0 & 0 & 0 \\ 0 & 0 & 0 \\ 0 & 0 & 0 \\ -1.9631e\text{-}26 & 1.0006e\text{-}25 & -1e\text{-}5 \\ -8.2564e\text{-}25 & -2.0647e\text{-}25 & -2.8736e\text{-}19 \\ -4.2145e\text{-}7 & -6.6448e\text{-}8 & .6919 \\ 7.0947e\text{-}9 & 1.875e\text{-}9 & .0016 \\ 3.8549e\text{-}9 & 6.0469e\text{-}10 & 0 \\ -3.8214e\text{-}11 & -7.6208e\text{-}12 & .0013 \end{matrix}$$

$$A3 = \begin{matrix} .1318 & 0 & 0 & 0 & 0 \\ .0067 & 0 & 0 & 0 & 0 \\ 0 & 0 & 0 & 0 & 0 \\ 0 & 0 & 0 & 0 & 0 \\ 0 & 0 & 0 & 0 & 0 \\ 0 & 0 & 0 & 0 & 0 \\ -1.0568e\text{-}35 & 0 & 0 & 0 & 0 \\ -1e\text{-}5 & 0 & 0 & 0 & 0 \\ .1318 & -1 & 0 & 0 & 0 \\ .0067 & 0 & -1 & 0 & 0 \\ 0 & 0 & 0 & -.02 & 0 \\ .0053 & 0 & 0 & 0 & -.02 \end{matrix}$$

$$B^T = \begin{matrix} 0 & 0 & 0 & 0 & 0 & 0 & 1 & -2.382e\text{-}25 & 0 & 0 & 0 & 0 \\ 0 & .1134 & 0 & 0 & 0 & 0 & -3.9408e\text{-}39 & 1 & 0 & .1134 & 0 & .09 \end{matrix}$$

$$C1 = \begin{matrix} -.1663 & -.0264 & -.0005 & -.0001 & -4.2145e\text{-}7 \\ -.0025 & .0012 & 4.5844e\text{-}6 & 3.0605e\text{-}6 & 7.0947e\text{-}9 \end{matrix}$$

$$C2 = \begin{matrix} -6.6448e\text{-}8 & .6919 & .1318 & 0 & 0 & 0 & 0 \\ 1.875e\text{-}9 & .0016 & .0067 & 0 & 0 & 0 & 0 \end{matrix}$$

$$D = \begin{vmatrix} 0 & 0 \\ 0 & 0.1134 \end{vmatrix}$$

The results in frequency domain are given in fig.7, where it is showed that the robustness conditions are not violated because, for multivariable systems, the stability is guaranted if the largest singular value of closed loop transfer matrix ($\sigma_{max}[T(s)]$) is lower than the upper bound of the largest singular value of the model uncertainties ($1/\sigma_{max}[W_t(s)]$). The same idea is used for the robust performance criterion. In fig.8, the results in temporal domain are given; the stability of all regimes and a good performance which means small interactions and a fast response time, are observed.

5. CONCLUSIONS

In conclusion, the following remarks can be made:

- H_∞ method can be successfully applied to mineral process.
- The dynamic behaviour of mineral processing plant is generally difficult to model, it could be modelised.
- The theory behind the robust control tools is simplified to be easily transmitted to mineral processing students and engineers.

6. REFERENCES

Chiang, R.Y. and M.G. Safonov (1988). *Robust-Control Toolbox*, The Mathworks, A Tutorial. South Natick,MA.

Doyle, J.C. and G. Stein (1981). Multivariable Feedback Design: Concepts for a Classical Modern Synthesis. *IEEE. Trans. Automat. Control*, Vol. AC-26,N^{0}1, pp.4-16.

Doyle, J.C (1983). Synthesis of Robust Controllers and Filters. in Proc. *IEEE Conf. Decision Control* San Antonio, TX, pp. 109-114.

Doyle, J.C., K. Glover, P.P. Khargonekar and B.A. Francis (1989). State space solutions to standard H_2 and H_∞ Control Problems. *IEEE Trans. Automat. Contr.*, Vol.34, N^{0}8, pp. 831-847.

Maciejowski, J.M (1989). *Multivariable Feedback Design*. Addison-Wesley, New York.

Safonov, M.G. and R.Y. Chiang (1988). CACSD using the State Space L_∞ Theory- A Design Example. *IEEE. Trans. Automat. Control.*, AC 33, N^{0}5, pp. 477-479.

Yahmedi, S. (1993). Mise en Oeuvre d'Outils Algorithmiques Permettant l'Etude de la Robustesse de la Stabilité et des Performances des Systèmes Multivariables Bouclés, PhD *Thesis* (French Text), Université Laval, Québec, Canada.

MODULAR CONTROL SYSTEM FOR FMS FOR PRISMATIC WORKPIECES

Mahbubur Rahman[1]
Jouko Heikkala[2]

[1]*Researcher, Production Technology Laboratory, University of Oulu,
PL 444, 90571 Oulu, Finland, E-mail: mrahman@me.oulu.fi*
[2]*Laboratory Manager, Production Technology Laboratory, University of Oulu,
PL 444, 90571 Oulu, Finland, E-mail: jouko.heikkala@oulu.fi*

Abstract: In a modular approach of controlling a flexible manufacturing system a large control program was broken down into smaller program units in a way that they can work independently of each other. By this modular approach an interlinked information flow between shop floor level and management level could be obtained via data transfer by FTP protocol. This modular approach increases the effectiveness of information flow within FMS, thus increases productivity. This up-down information flow makes it easier for the management to control FMS cells just as a transparent production system. In this paper, the job order and other necessary information transfer from upper level to FMS cell level and from cell level to upper level has been described. The structure of the modular control system of FMS is discussed and it has been shown how the FMS control system effects on production planning, process planning and tool management.

Key words: Flexible Manufacturing Systems, Productivity, Production Control System.

1. INTRODUCTION

The design work of the flexible manufacturing system in Oulu University Production Technology Laboratory (OUPTL) started in 1985. After that it has been developed and expanded continuously. For example the whole control software and hardware have been renewed several times. The system was carried out as a pilot-plan, and it has been used as a versatile research and testing environment for more than ten years. It has also been used for manufacturing different prismatic parts, e.g. pump housings, gearboxes etc. (Pylkkänen, J. 1984).

Formerly the control system (OS QNX version 2) for the FMS was such that a single control module was controlling the whole system. In the current version of FMS control system (OS QNX version 4.1) we have developed a modular approach where the large program has been broken down into smaller modules. These modules are communicating with each other in a transparent way to control the whole system in conjunction with other subsystems like production planning and control (PPC). The main advantage of modular control system is that it is easier to reroute the production plan due to sudden change of any system variables.

If different processes are rigidly integrated into special purposes in highly productive production system such as transfer line for large batch production, then neither modular development, nor flexible operation is possible. So a flexible manufacturing system is such a system which is capable to adjust the system with varying market demands by re-programmable production planning and control system. The modular approach could be applied to the whole integrated system. In this paper, for brevity, a small portion of the whole integrated system has been described with a focus on the FM-system in the Oulu university Production Technology Laboratory (abbreviated here OUPTL).

1.1 Definition of FMS

A flexible manufacturing system (FMS) may be defined as a system dealing with high level distributed data processing and automated material flow using computer controlled machines, assembly cells, industrial robots, inspection machines and so on, together with a computer integrated materials handling and storage systems (Ranky, P. 1983). FMS can have a variety of interconnections with different materials handling systems (Robots, AGVs, Machining centers etc.) and they have to communicate with data processing networks for successful integration of the system. They can also enable the feedback of data to the upper level of the control hierarchy, thus providing a facility for further analysis of performance or for real time fault recovery.

In this paper, a manufacturing system is considered to be a system which integrates different modules and requires a properly defined input to create the expected output. Input may be raw materials and/or data (e.g. CNC part programs, tooling data etc.) which have to be processed using various auxiliary components of the system such as tools, fixtures and clamping devices and sensors and their feedback data. The output may also be data/or material which can be processed on units (often called FMS cells) of the manufacturing system. FMS can also be thought as a distributed management information system linking together intelligent subsystems (known as nodes) of machining, welding, washing, flame cutting, assembly etc. and materials handling and storage modules. This modular system approach to FMS can be applied not only while designing and operating machining system, but also in other computer integrated projects, where machines work under computer control in an integrated manner. Such an application includes flexible assembly using industrial robots and other assembly devices.

2. MODULAR CONTROL SYSTEM

Formulation of a complex control system into numerous small sequences are called modules or control modules. Such formulation of computer code is referred to as modular programming. By this modular approach in FMS, different modules talk and communicate with each other by network topology yielding each module's status in certain time interval or yielding some command for performing some specific task (Warnecke, H.J. and R. Steinhilper 1985). Modules are divided into two categories: local and global; local modules can access local work station and global modules can access any resources based on it's permission via network.

Different modules can communicate synchronously/
Different modules can communicate synchronously/
asynchronously with each other. Any module sends a message in the form of data which is understandable to the recipient module (Quantum Software, 1992). In the following figure module A sends a message to module B and B understands the message and does some specific task and then replies about it's own status. Modules A or B could be a storage control module, a tooling database control module or any work station control module. These modules could communicate with each other in response to user input thus providing output to the user. All the individual modules are linked by using LAN.

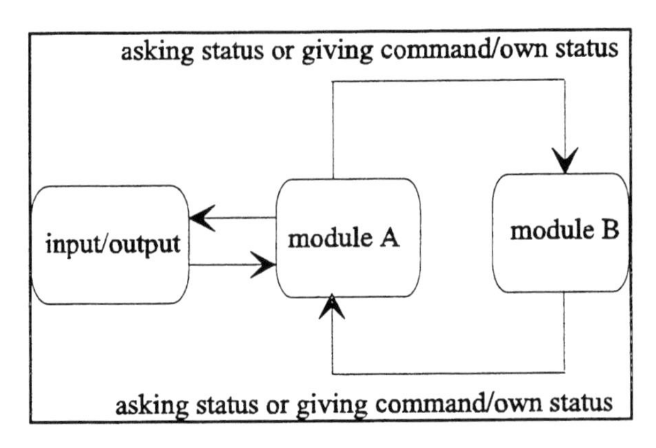

Fig. 1. Communication between modules A and B

Expanding this modular approach to the factory information management system, materials handling and storage and computer aided design (CAD), a whole computer integrated manufacturing system could be obtained taking advantage of modularity.

2.1 Advantages of a modular control system

Several advantages of modular approach can be listed:

- Reprogramming of the production planning and control system is easier.
- Easy to adapt the FMS with sudden market changes.
- Easy to adapt with future expansion.
- The intelligent, self correcting system (i.e. machine equipped with sensory feedback systems) increases the overall reliability of production.
- Modules are easier to comprehend.
- Different modules can be developed by different people.
- Debugging and testing of the system could be done in more orderly fashion.
- Documentation can be made easily understood.
- Modifications may be localized.
- Frequently used tasks can be programmed into modules that are stored in libraries and used by several programs.
- Different modules can be located in different networks with different operating systems in the workstation computers.
- Subsequent error checking in the whole production system.
- The expenses are smaller because of higher rate of return.

2.2 *Components of FMS*

For a successful implementation and realization an FMS should have the following components as a total system:

- Distributed database management system for up and down information flow.
- Integrated CAD/CAM system and part program preparation.
- Distributed tool database.
- Database for clamping devices and fixtures.
- Processing units such as machine tools, washing machines, industrial robots, automated warehouses, automatically guided vehicles etc. and their working modules.

3. NETWORKING FOR THE FMS AT OUPTL

The operating system for the control of the FMS is QNX - a real-time, network based multi-tasking operating system in which all work stations (nodes) are in the same level of hierarchy. This allows several tasks to be carried out simultaneously, which is necessary when controlling a manufacturing system consisting of several workstations and a Local Area Network and provides a mechanism for sharing database/files and peripheral devices among several interconnected computers. Several modules can be run in one workstation computer depending on it's capacity and power. The workstation computer keeps also a local database for its own use. This database can be used by this workstation and as well as by other workstation computers if necessary. All workstations have access to network database (NDBMS).

Two types of networks are being used in the FMS Laboratory. The logical network is an ethernet network which is connected to global network via Oulu University LAN. The sub-network is a local area network used only for the FMS inter-module communication. Logical sub-network is driven by Arcnet -network adapters. Currently there are five nodes working as shown in the following diagram. Nodes 4 and 1 are connected to both networks.

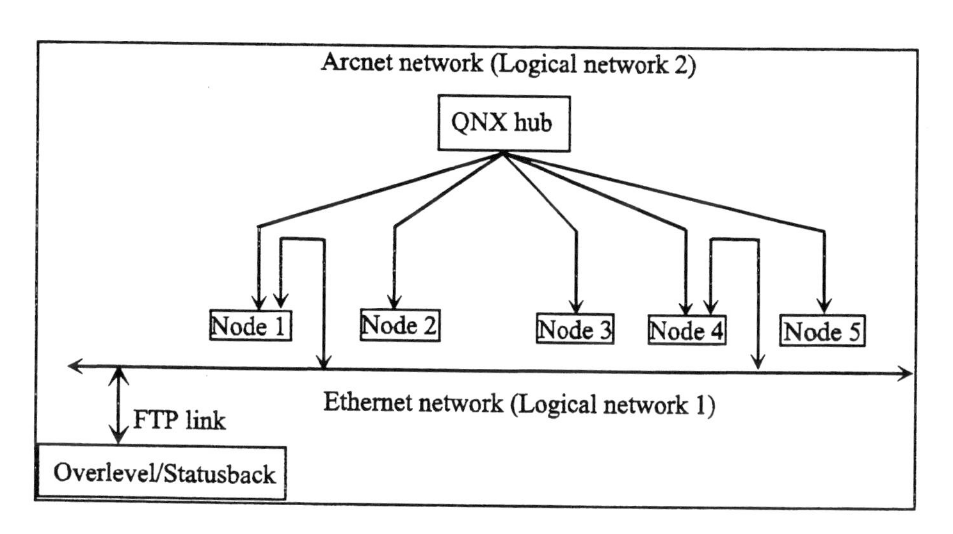

Fig. 2. QNX-based network at the FMS at OUPTL

3.1 *Application of a modular control system*

The production planning and control (PPC) module is running in MS-Windows-environment. This module is supported by other modules like storage and material control modules, tool management module, part program control module etc. by distributed database system. One of the advantages is subsequent error checking in the production system. For example if some part program is not available, then PPC can not assign a particular job to the FMS cell but can reassign or reprogram it based on other modules' status.

Any working module can be located on anywhere on the network with necessary software support. In the FMS the necessary job order comes from the production planning and control department via FTP link. This job order is the key information for production. This coming information is used in the FMS to produce a certain product. After production the necessary information goes back to PPC via FTP. The system's operator's work is confined within the QNX environment. Figure 2 shows how the information is coming from other department to QNX environment for running the FMS. By this up-down information flow we can update the database for performance analysis and fault recovery of the FMS cell.

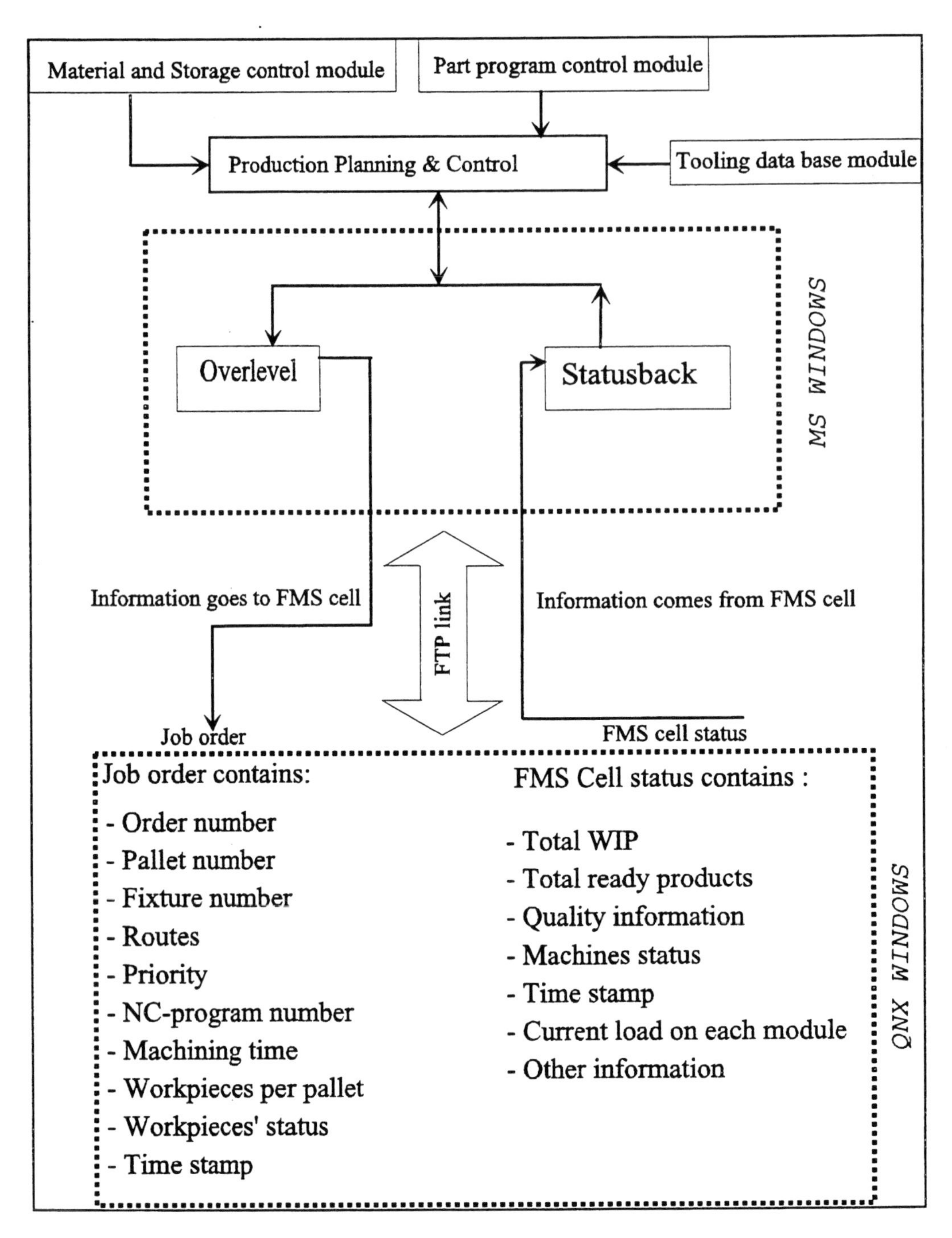

Fig. 3. Total information flow in the FMS at the OUPTL

In a QNX environment, the automatic production is carried out based on modular approach. For example if a machining center workstation is ready to take new pallet it asks for a new pallet from the transport module. Transport module has its own database and knows who is telling to fetch pallet. In the same way the material and storage control modules, the tool management module, the part program module, loading/unloading station control modules etc. work.

3.2 Overlevel and Statusback

These two modules are running under MS Windows. The Overlevel conducts the QNX windows via FTP (File Transfer Protocol) with necessary information to run the FMS. The information can contain pallet number, fixture number, priority, routes etc. The Statusback module informs the production planner and controller about the current status of all equipment of the FMS. If some equipment is out of order, production planner can reroute the job with different routes.

The statusback information can be current equipment's status, time elapsed for processing a pallet, ready product information etc. These status information is coming from QNX to MS-windows via FTP link. The feedback information is used to recover fault and to take corrective action thus improving system performance. In turn, these information is used by other departments to satisfy other goals of the whole business system. Here is shown a typical job order from the Overlevel to FMS.

```
message {
 sender = overlevel
 receiver = FMS-link
 timestamp = 823459993
 check = yes
 priority = 1
}
jobqueue {
 routes {
  order="1"
  routepoint="6"
 }
 routes {
  order="2"
  routepoint="5"
 }
 workpiecepallet="2"
 machiningtime="1234"
 deliverydate="12.3.96"
 priority="5"
 ncprogram="6000"
 fixturenumber="12"
 palletnumber="12"
 jobnumber="1234"
 datestamp="4.2.1996"
}
```

Statusback sends information to the upper level to report the activity of the cell in the following way.

```
message {
 sender = FMS Operator
 receiver = overlevel
 timestamp = 823432100
 checkup = YES
 priority= 1
}

work queue {
Job Number = "1234"
Product Data {
        Time Stamp = "Sun Feb 02 13:08:20 1997"
        Elapsed Time = "324566"
        Station 1 Time = "3243"
        Station 2 Time = "4803"
        Station n Time = "1234"
        Product Name = "name "
        Lot Produced = "9"
        Pallet Number = "12"
        Status = Ready/In process
        Workpieces per pallet = "2"
        Quality = "Excellent"
  }
}
Machine Status {
        Time Stamp = "Sun Feb 02 13:08:20 1997"
        AGV Status = "ok" /* AGV is busy or idle
or out of order. */
        Assigned job = "23"  /* AGV is busy 23
min. */
        Machining center = "ok"
        Assigned job = "29"
        Washing machine = "ok"
        Assigned job = "59"
        Robot Status = "ok"
  }
```

4. CONCLUSIONS

In this paper the application of a modular control system for FMS has been shown. The whole production database could be made transparent to all other subsystems of the total computer integrated manufacturing system. For example the whole tool database could be made available to the CAD subsystem so that the product designer could design any product according to the tooling capacity. In this same way other subsystems of the management information system could work. All these approach could be tested in real life by taking the advantage of computer technology with realization of greater computer integrated manufacturing.

Acknowledgments- RM would like to acknowledge supporting from the Production Technology Laboratory at the University of Oulu.

REFERENCES

Pylkkänen, J. (1984). Finnish FMS exceeds expectations. In: *The FMS Magazine*, April 1984, pp. 82-85.

Quantum Software Systems Ltd. (1992). QNX System Architecture, 123 pages.

Ranky, P. G. (1983). *The design and operation of FMS Flexible Manufacturing systems* 345 pp, IFS (Publications) Ltd, UK.

Warnecke, H. -J and R. Steinhilper (1985) *Flexible Manufacturing systems*, pp. 143-151, IFS (Publications) Ltd, UK.

DESIGN METHOD ON OPTIMAL FLEXIBLE MANUFACTURING SYSTEMS

Ferreiro Garcia, R.,Pardo Martínez, X.C., Vidal Paz, J. & Coego Botana, J.

Dept. Electrónica e Sistemas. Universidade da Coruña
Facultade de Informática. Campus de Elviña, s/n. 15071. A Coruña. Spain.
E-mail: {ferreiro, pardo}@des.fi.udc.es

Abstract: This paper presents a methodology for designing optimal flexible process sequencing using SFC (sequential function charts) based approach. In this approach SFC is used as a unified framework for representing the optimal process sequence by determining the optimum condition for transition between two macrosteps, completing the flexible manufacturing cycle at the minimum cost. Dynamic programming has been used to find the optimal trajectory, in which a strong correspondence with its sequential function chart is stated.

Keywords: Dynamic Programming, Flexible Manufacturing Systems, Optimal Systems, Optimal Conditional Branching Transition, Sequential Control.

1. INTRODUCTION

It is known that there are two major knowledge representation schemes in process planing systems: a rule-based approach and a frame-based approach. In particular, a rule-based approach has been extensively used for representing process planning knowledge. There exists a strong correspondence between a rule-based and a SFC approaches (Kyung-Huy and Moo-Yung, 1995). A frame based approach can be described on a SFC basis by defining its properties by means of the mathematical description of the function chart. The association of a set of conditions for transition or rules, to an optimisation problem under a performance criteria or quality index to be satisfied, constitute an approach to the problem of optimal flexible or adaptive manufacturing system (Gershwin, 1994), under a SFC description, which means a powerful tool to be used on a real time computer control implementation.

The mathematical model of the function chart is derived from that of Petri nets (IEC 848, 1988). A function chart is a directed graph defined as a quadruplet:

$$[X,\ T, L, X_0]$$

where:

X = $(x_1,...x_m)$ is a finite, non empty, set of steps;

T = $(t_1,...t_n)$ is a finite, non empty, set of transitions;

X and T represent the nodes of the graph;

L = $(l_1,...l_p)$ is a finite, non empty, set of directed links, linking either a step to a transition or a transition to a step. There are two types of links: parallel branches and optional branches. The system can only be described by means of both links. The type of link depends on the characteristic of the system stage to be processed, that is, parallel or optional branching.

X_0 **which belongs to X** is the set of initial steps. These steps are activated at the beginning of the process and determine the initial situation.

Moreover, the graph is interpreted, meaning that:

with the steps, commands or actions are associated;
with each transition, a logic transition condition is associated.

Steps are represented by labelled squares, transitions by dashes. In addition to the static representation, the graph also has a dynamic aspect defined by evolution rules.

Given the SFC description of a problem, the second step is how to optimise such implementation. Several methods have previously been used to solve the minimum transition cost problem (time or safety optimisation criteria, between others, are performance criteria), for example, calculus of variation, Pontryagins maximum principle and dynamic programming. A disadvantage of the two first methods is that they may find local minima. Dynamic programming does not have these problems, but the necessary problem discretisation limits the routes to a finite number of gridpoints and the computational burden may become prohibitive if a fine grid is needed. Fortunately, this is not the case for the given optimization problem, so dynamic programming has been chosen.

2. PROBLEM STATEMENT

The ability of SFC to describe a global approach to the process sequencing, justify its application as a powerful tool not only in the problem description but in its real-time implementation. In addition, there are high level contexts in which a process may be splitted in order to simplify its global description. A macrostep $Mx = (x_i, t_i, L_i)$ is a high level context where steps linked by transitions are embedded. A global description consists in linking series of macrosteps.

In this way, this work is based on the consequences of the following propositions, which are trivial contributions:

a) A flexible manufacturing system (FMS) is synthesised by linking macrosteps with only conditional branches.

b) There is a total correspondence between an optimal path of a state space with an optimal conditional branch linking two macrosteps in a SFC description.

c) The application of the Principle of Optimality determines univocally the optimal path (the conditions for transition, CFT) that are the conditional branches linking macrosteps of an OFMS.

d) An optimal flexible manufacturing system (OFMS), is synthesised if, and only if,

1) macrosteps are linked with conditional branches.

2) the principle of optimality holds for deciding the transition between macrosteps.

The triviality of such assumptions don't need demonstration here, being clearly illustrated if figure 1, in which two concepts are associated: the state space, optimised by applying the principle of optimality, with an OFMS defined on a SFC basis by macrosteps.

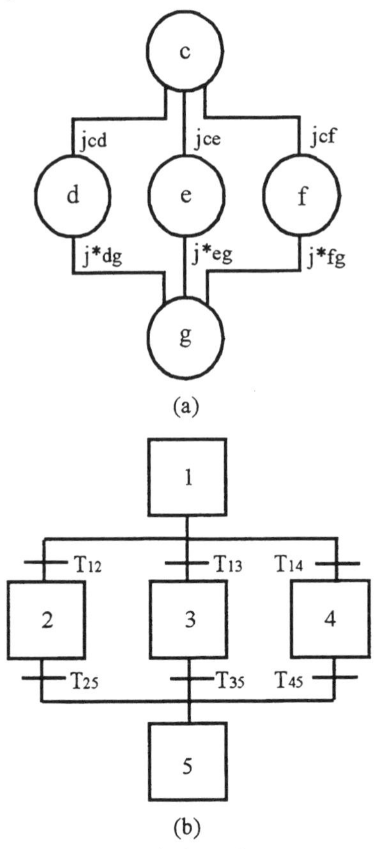

Fig. 1. (a) Process consisting of numerous paths from state c to g. (b) SFC description of same process.

The problem of OFMS belongs to a class of highly nonlinear multivariable system with many constraints, which represent some of the most difficult systems to control optimality. Dynamic programming may provide the best approach, avoiding the troublesome questions of existence and uniqueness, conditional upon the existence of at least one feasible control sequence. As a result, the direct-search method guarantees that the absolute minimum cost is achieved.

Dynamic programming provides a computational technique to apply the Principle of Optimality (Bellman, 1957) to sequences of decisions which define an optimal trajectory (Bersekas, 1987). Consider the process shown in figure 1 (a), at a current state c and optimal paths to the goal state g resulting from the admissible decisions taken at c (White, 1992). The principle of optimality states that if segment c-d is the initial segment of the optimal path from c to g , then d-g is the terminal segment of the optimal path. If it is applied the same principle to paths c-e and c-f then the decisions of paths c-d, c-e and c-f are the only admissible candidates for achieving the optimal trajectory from state c to g. The optimal trajectory from state c to g can be determined by comparing the minimum cost J of the admissible decisions as follows:

$$J^*_{c,d,g} = J_{c,d} + J^*_{d,g}$$
$$J^*_{c,e,g} = J_{c,e} + J^*_{e,g} \qquad (1)$$
$$J^*_{c,f,g} = J_{c,f} + J^*_{f,g}$$

where $J^*_{c,d,g}$ is defined as the minimum cost for going from c to g via state d.

Extending this principle to a multi-stage decision process the three equations above could be replaced with the more general form:

$$J^*_{t,t+1,N} = J_{t,t+1} + J^*_{t+1,N} \qquad (2)$$

and the optimal decision at state t, $U^*(t)$ is the one which achieves

$$J^*_{t,N} = min(J^*_{t,t+1,N}) \qquad (3)$$

for all admissible decisions at state U(t) to reach an adjacent state at t+1. These two equations, (2) and (3) form the basis for dynamic programming.

In the same way, the association of optimal decisions to the conditions for transition, in a SFC description of the same problem, it is achieved straightforward the SFC description to be applied on an OFMS.

The optimum trajectory is a sequence of optimum segments. There is a strong correspondence between the set of optimum segments of an optimal trajectory and optimal conditions for transition linking macrosteps by conditional branches in a OFMS.

The mathematical model of the OFMS is derived from that of SFC. Such a function chart is a directed graph defined as a triplet:

[Mx, T, L]

where:

Mx = $(\mathbf{m_1,..m_m})$ is a finite, non empty, set of macrosteps;

T = $(\mathbf{t_1,..t_n})$ is a finite, non empty, set of transitions, where the set of uniquely transitions to be enabled will be defined at state t by the optimal decision U(t), that is the one stated by equation (3).

Mx and T represent the nodes of the graph;

L = $(\mathbf{l_1,..l_p})$ is a finite, non empty, set of directed links, linking either a macrostep to a transition or a transition to a macrostep. There are only one type of link: optional branches. The system can only be described by means of such links.

3. APPLICATION TO AN OPTIMAL FLEXIBLE MANUFACTURING PROBLEM

In order to illustrate the optimization problem, the flexible manufacturing system described in figure 2 is presented as an application example. This system consits of a process dedicated to manufacture and deliver or store two kinds of product (Pr1, Pr2) in two different sizes (Si1, Si2). Its state space trajectories are processed to find the optimum path by applying the principle of optimality so that the optimum trajectory is the sequence of conditions for transition on the optimum conditional branches which links macrosteps in a SFC description, as shown in figure 3. The admissible trajectories are the possible combinatorial different ways to move from beginning to the end; in this problem there are only $2^2 = 4$ different candidates trajectories to be computed under the principle of optimality.

The objective function to be minimised is the cost of both production and stock of the four products, keeping the plant running. The cost of the possible trajectories in this problem are:

Si1+Pr1+St1 = J1
Si2+Pr1+St2 = J2
Si1+Pr2+St3 = J3
Si2+Pr2+St4 = J4

The trajectory selected is the one who satisfy the minimum cost given as min(Ji)= min (Sii+Pri+Sti). Assuming the following values for all trajectories,

2 + 10 + 20 = 32
3 + 10 + 50 = 63
2 + 15 + 10 = 27
3 + 15 + 30 = 48

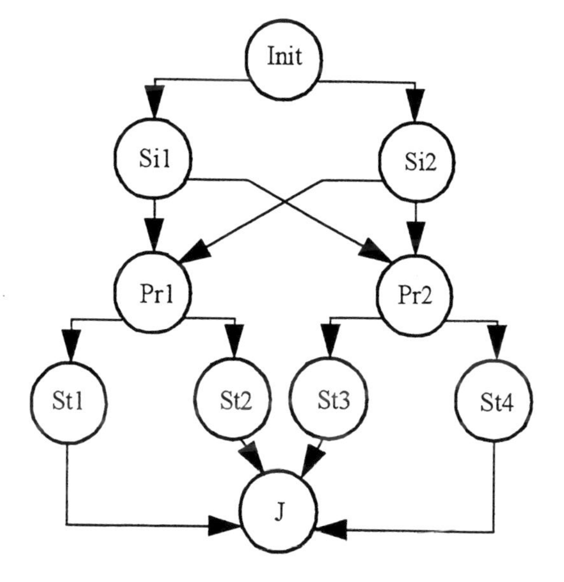

Fig. 2. State space of a FMS

As the minimum cost is 27, it belongs to the trajectory Si1+Pr2+St3. So that, next manufacturing cicle is decided as the conditional branches whose optimum transitions are T01, T14, T45 linking the macrosteps by conditional branches. This cycle will be repeated until the cost of this trajectory ceases to be the minimum cost. Then, another trajectory will be selected, the one that satisfies the minimum principle of optimality. The SFC is illustrated in figure 3.

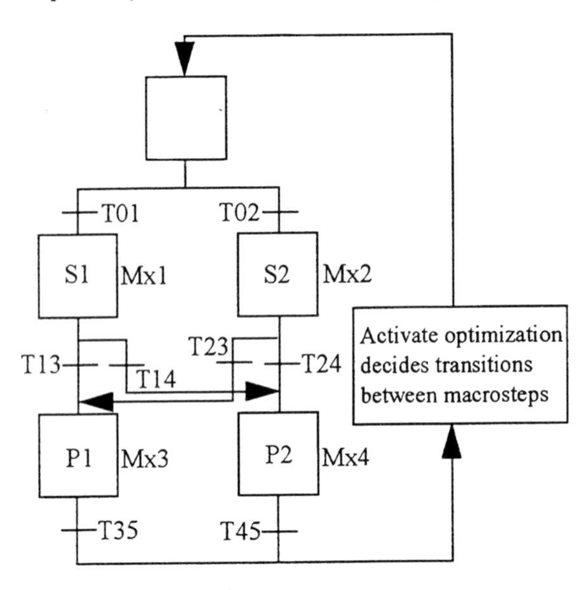

Fig. 3. SFC implementation of an optimal flexible manufacturing project

4. CONCLUSIONS

A strategy to implement optimal flexible manufacturing systems has been introduced. The application of the principle of optimality satisfies the requirements of searching the optimal set of conditions for transition in a OFMS network. Following the optimal trajectory in sequential order, it will be found there only the optimum conditions for transition to be enabled. The application of this method is simple and clear. There is not ambiguity in the decision making strategy.

REFERENCES

Bellman R.E. (1957). *Dynamic Programming.* (Priceton University Press). N. J. USA.

Bertsekas, D.P. (1987) *Dynamic programming: Deterministic and Stochastic Models.* Prentice-Hall, Englewood Cliffs. USA.

Gershwin, S.B. (1994) *Manufacturing Systems Engineering.* Prentice-Hall. USA.

IEC publication 848 (1988) *Preparation of function charts for control systems.* (Atar S.A.(1 ed.), Geneva, Switzerland

Kyung-Huy, L. and Moo-Young, J. (1995) *Flexible process sequencing using Petri Net Theory.* (Computers ind. Engng.) Vol. 28, No.2, pp. 279-290.

White, D. (1992). *Handbook of Intelligent Control. Neural, Fuzzy and Adapptive approaches.* (Ed. David A. White, Donald A. Sofge) Chap. 6. pp-185-214. Van Nostrand Reinhold. New York. USA.

ADVANCES IN MANUFACTURING ALGEBRA: DISCRETE-EVENT DYNAMIC MODELS OF PRODUCTION PROCESSES

Donati F., Canuto E., Vallauri M.

Dipartimento di Automatica e Informatica, Politecnico di Torino, Corso Duca degli Abruzzi 24, I-10129 Torino, Italy

Abstract: The paper presents a further original development of the Manufacturing Algebra aiming at describing the dynamics of the production processess taking place in factories. Starting from the definition of events and event sequences, a theory of discrete-event dynamic systems is developed which is appropriate for modelling the time and space evolution of discrete (or by part) production processes, at any level of detail. The result is a set of state equations modellling the state evolution of the storage and production units of a factory under the action of a real-time production control.

Keywords: Discrete-event dynamic systems, Manufacturing systems, Production systems.

1. INTRODUCTION

The concepts and results presented in this paper have been developed within the ESPRIT Basic Research HIMAC 8141. The basic elements of the Manufacturing Algebra have been already introduced in previous publications. In this paper, the attention is drawn towards the application of the discrete-event dynamic systems to production process modelling. The discrete-event approach hereafter presented is original and has been expressly studied to meet the modelling requirements of the discrete production systems.

It lies outside the paper to summarize the results which have been already achieved in the development of the Manufacturing Algebra. The reader is therefore referred to other papers, and specifically to the companion paper presented at the same Workshop (Canuto et al., 1997), where one can find a summary of the whole HIMAC work and a complete bibliography. In this paper, due to space constraints, the authors will limit themselves to recall those concepts of the Manufacturing Algebra which are thought to be strictly necessary for understanding the discrete-event dynamic modelling applied to production processes.
In the second section, still before dealing with the results of the Manufacturing Algebra, the basic concepts of event and of event sequence are introduced. They will be then used for modelling the production processes and for writing the relevant discrete-event state equations. In the third section the concept and the elements of the product modelling, which are preliminary to production modelling, are shortly presented. Finally in sections 4 and 5, first the mathematical elements of the Manufacturing Algebra to be used in production process description will be introduced and then the discrete-event dynamic models will be formulated.

2. TIME, EVENTS AND EVENT SEQUENCES

2.1 Basic definitions

Time, fact and event. Given a real variable $t\in\mathcal{T}\subseteq\mathcal{R}$. (where $\mathcal{R}$ denotes the set of the real numbers) called *time* and given a generic set Ξ of elements $\xi\in\Xi$ called *facts*, the following Cartesian product is defined $\mathcal{E}=\mathcal{T}\times\Xi$. Their elements $e=(t,\xi)\in\mathcal{E}$ are called *events*. Any event is therefore described by the pair t = *"the occurrence time instant of the event"*, ξ = *"the fact associated to the event"*.

Addition of events. Two different events $e_1=(t_1,\xi_1)$, $e_2=(t_2,\xi_2)$ are said to be simultaneous, when they have the same occurrence time, i.e. when it holds

$t_1=t_2$. The occurrence of simultaneous events is only admitted in the case when different facts belonging to the same set are mutually *compatible*. Different facts occurring simultaneously should be considered as a single fact. Therefore the occurrence of simultaneous events will be only allowed as far as, in the set Ξ of the facts, an addition operation is defined and the sum of two or more facts always belong to the original set Ξ. In this case the addition can be extended to the set of simultaneous events such that, given three simultaneous events $e_1=(t,\xi_1)$, $e_2=(t,\xi_2)$, $e_3=(t,\xi_3)$, it holds $e_3=e_1+e_2$ if $\xi_3=\xi_1+\xi_2$. When the set of facts Ξ includes the null element θ, the event $e=(t,\theta)$ is called the *null event*, meaning that no fact is occurring at time t.

Event sequences. An event sequence σ is a finite set of events belonging to the same event set $\mathcal{E}$. The elements of the sequence can be ordered with respect to their occurrence times, i.e.

$$\sigma=\{e_1=(t_1,\xi_1),\ldots,e_i=(t_i,\xi_i),\ldots,e_n=(t_n,\xi_n)\} \quad \text{with the constraint } t_{i+1}>t_i \tag{1}$$

By denoting with $\mathcal{T}_\sigma\subset\mathcal{T}\subseteq\mathcal{R}$ the set $\{t_1,\ldots,t_i,\ldots,t_n\}$ of the occurrence times of an event sequence σ, the sequence itself may be defined as a function $\sigma:\mathcal{T}_\sigma\to\Xi$ which is everywhere defined in $\mathcal{T}_\sigma$ and possesses a single value in Ξ. Therefore an event sequence will be indicated either with $\sigma=\{(t_i,\xi_i)\}$, $i\in[1,n]$ or equivalently as a function value $\xi(t_i)$, $t_i\in\mathcal{T}_\sigma$.

The set of all the (finite) sequences which can be constructed over the event set $\mathcal{E}=\mathcal{T}\times\Xi$ will be indicated with $\Sigma(\mathcal{E})$.

2.2 *Operations between event sequences.*

Over the set $\Sigma(\mathcal{E})$ of the event sequences, the following operations are defined.

Restriction of a sequence. The restriction of an event sequence σ to a time interval $t_i>t$ is defined as the operation $R[\sigma,t_i>t]$ whose result is a subsequence $\sigma'\in\Sigma(\mathcal{E})$ of σ such to include only the events e_i occurring at times $t_i>t$. The restriction operation will be indicated with $\sigma'=R[\sigma,t_i>t]$ or simply with $\sigma'=\sigma(t)$.

Addition of sequences belonging to the same set. A pair of sequences $\sigma_1,\sigma_2\in\Sigma(\mathcal{E})$, $\mathcal{E}=\mathcal{T}\times\Xi$, is said to be summable if and only if the addition operation is defined in their set of facts Ξ. On the contrary, if no addition operation is defined in the set of facts, a pair of sequences is summable only if they do not include simultaneous events. When a pair of sequences is summable, their sum $\sigma_3=\sigma_1+\sigma_2$ is defined as the new sequence whose events are the union of the events of the two addenda.

Addition of two sequences defined in the same time interval, but with facts belonging to different sets, although not incompatible. Let us consider now two sets of events $\mathcal{E}_1=\mathcal{T}\times\Xi_1$ and $\mathcal{E}_2=\mathcal{T}\times\Xi_2$, having different sets of facts Ξ_1 and Ξ_2. Assume further that the facts belonging to the different sets are compatible. In other words, we allow the simultaneous occurrence of two different events belonging respectively to $\mathcal{E}_1$ and $\mathcal{E}_2$. This assumption is essential to make possible the addition of two event sequences belonging respectively to $\Sigma(\mathcal{E}_1)$ and $\Sigma(\mathcal{E}_2)$.

Given now a pair of sequences $\sigma_1\in\Sigma(\mathcal{E}_1)$ and $\sigma_2\in\Sigma(\mathcal{E}_2)$, their sum $\sigma=\sigma_1+\sigma_2$, $\sigma\in\Sigma(\mathcal{E})$ is defined as follows.

1) Unless both sets of facts Ξ_1 and Ξ_2 do not possess already the null element θ, equip each of them with a null element, and denote the resulting sets with Ξ_{10} and Ξ_{20}.
2) Build a new set of facts $\Xi=\Xi_{10}\times\Xi_{20}$ as the Cartesian product of the two original sets equipped with the null element and derive from it a new event set $\mathcal{E}=\mathcal{T}\times\Xi$.
3) Transform the pair of sequences σ_1 and σ_2 into new sequences belonging to the set $\Sigma(\mathcal{E})$, paying attention to the following remarks. The ordered pair of facts (ξ_1,θ), being θ the null element of Ξ_{20} and $\xi_1\in\Xi_{10}$, is equivalent to the simple fact $\xi_1\in\Xi_1$. The pair (θ,ξ_2), such that $\xi_2\in\Xi_{20}$ and $\theta\in\Xi_{10}$, is equivalent to the simple fact $\xi_2\in\Xi_2$, as well.
4) Having brought the pair of sequences to belong to the same set, their addition is obtained with the rules previously defined.

3. THE PRODUCT MODELLING

Product modelling is developed outside space and time, being only concerned with the manufacturing operations and the different types of objects which may be required by a production process. The goal is that of describing the content of a product in terms of a predefined set of objects (raw materials. semi-finished products, finished products and reusable tools) and a set of admissible and interconnected manufacturing operations.

It is always necessary to put product modelling before production process modelling, because the latter one has the goal of manufacturing finished products according to the description provided by the product model. The interaction between product and production modelling is so tight that pairs of elements exist in both modelling approaches which share the same name, although being mathematically different. As a matter of fact, the elements of the production modelling describe the implementation by the production process itself of the concepts expressed by the product model elements. Therefore, for such pairs of elements, the same notations will be maintained, with the only care of starring the product modelling elements. The latter ones will be also indicated as the *reference elements*. The following sets are defined.

The object set. Objects are mathematical elements o describing material parts (raw materials, semi-finished products, finished products) and reusable parts (fixtures, machines, tools, resources, workers, etc.) which are involved in the production process. The set or universe of all the objects is denoted by $\mathcal{O}$. It is assumed to be a numerable set.

The object type set. An object type k is a mathematical element introduced to describe technological similarities between the different objects employed and produced by a production system. The finite set of the elements k, the *object type set*, having cardinality n_k is denoted by $\mathcal{K}$. An *object type* can also be defined as a value of the function $type{:}O\rightarrow\mathcal{K}$, written as $k=type(o)$, assigning one and only one element $k\in\mathcal{K}$ to each object $o\in O$.

Within product modelling, the analysis is limited to distinguish between the object types in a static context. Therefore the following equivalence relation holds.

Type equivalence. Two elements o_1, o_2 belonging to $\mathcal{O}$ are said to be equivalent with respect to the object type, if $type(o_1)=type(o_2)=k\in\mathcal{K}$. The type-equivalence partitions the universe O into a finite set of equivalence classes O_k, each class being the numerable subset of the elements $o\in\mathcal{O}$ of type k.

The quantity vector space. Consider now a subset $O\subset\mathcal{O}$ and denote the object quantity of type k, i.e. the cardinality of the objects of O belonging to the class O_k, as $q(k)$. The quantity vector q^* is defined as the vector of all the components $q(k)$ and belongs to the vector space $\mathcal{Q}^*$ of dimension n_i. The natural basis of the vector space $\mathcal{Q}^*$ will correspond one-to-one to the finite set $\mathcal{K}$ of the object types. The subsets $O\subset\mathcal{O}$ possessing the same quantity vector q^* are said to be type-equivalent and make up a family $[O]$ of type-equivalent sets.

Manufacturing operation (MO). An MO is defined by the ordered pair

$$A^*=(u^*,y^*),\ u^*,y^*\in\mathcal{Q}^*,\ A^*\in\mathcal{Q}^*\times\mathcal{Q}^* \quad (2)$$

where u^* and y^* denote respectively the *input quantity vector* (the vector of the consumed quantities) and the *output quantity vector* (the vector of the product quantities) of A^*.
Over the set of the MOs, four operations are defined: the addition, meaning that two MOs are performed in parallel, the multiplication, meaning that two MOs are performed in series, the scalar multiplication and the raising to a power. By employing such operations, one can easily build algebraic expressions of MOs such to describe an MO as a function of other MOs. Specifically, having established a basis made by elementary MOs, one can generate the numerable set of all the complex MOs which can be written as an algebraic expression of elementary MOs belonging to the basis. A detailed presentation of this chapter of the Manufacturing Algebra is outside the scope of this paper and the reader is referred to (Canuto et al., 1995, 1996)

4. THE PRODUCTION MODELLING

The production process develops itself along space and time. Time t is defined as a real variable; space instead is defined as the countable and finite set of the factory places where the objects can be located.

4.1 The storage units

The storage unit (SU) set. An SU describes an available *space* in the production system or factory, where the objects can be located. The finite set of the elements s, the *storage unit (SU) set*, having cardinality n_s, will be indicated with $\mathcal{S}$. An SU can be also defined as a value of the function $loc{:}\mathcal{O}\rightarrow\mathcal{S}$, written as $s=loc(o)$, assigning one and only one element $s\in\mathcal{S}$ to each object $o\in\mathcal{O}$.

Type-location equivalence. At time t, two elements o_1, o_2 belonging to $\mathcal{O}$ are said to be equivalent with respect to object type and location if

$$type(o_1)=type(o_2)=k\in\mathcal{K} \text{ and } loc(o_1)=loc(o_2)=s\in\mathcal{S}. \quad (3)$$

The type-location equivalence partitions the universe $\mathcal{O}$, at time t, into a finite set of equivalence classes O_{ks} each class being the numerable subset of the elements $o\in\mathcal{O}$ of type $k\in\mathcal{K}$ and located in $s\in\mathcal{S}$.

The quantity vector space. Consider now a subset $O\subset\mathcal{O}$ and denote the object quantity of type k and located in s, i.e. the cardinality of the objects of O belonging to the class O_{ks} as $q(k,s)$. The quantity vector q is defined as the two-index vector of all the components $q(k,s)$ and therefore belongs to the vector space $\mathcal{Q}$ of dimension $n_k\times n_s$. The natural basis of the vector space $\mathcal{Q}$ will correspond one-to-one to the cartesian product $\mathcal{K}\times\mathcal{S}$. An object quantity q is always referred to a time instant t.

The SU state variables. The state of an SU at a

given time t is defined as the quantity of the objects of each type which are stored in that SU. The state of the set of the SUs is then defined as a quantity vector and denoted as $x(t) \in \mathcal{Q}$.

The (object) delivery/drawing event set. Consider now the facts *"delivery/drawing of objects"*. They are described by quantity vectors $q \in \mathcal{Q}$ whose components $q(k,s)$ represent an object quantity of type $k \in \mathcal{K}$, delivered to, if $q(k,s)>0$, or drawn from, if $q(k,s)<0$, the an SU $s \in \mathcal{S}$. Clearly, in the set of facts $\mathcal{Q}$, the addition and the null element are defined. Those events belonging to the Cartesian product $\mathcal{E}_q = \mathcal{T} \times \mathcal{Q}$ will be indicated as (object) delivery/drawing events and the set $\mathcal{E}_q$ as the (object) delivery/drawing event set.

4.2 The production units

The production unit (PU) set. A PU is a mathematical element capable of performing manufacturing operations. In the factory there exists a finite set $\mathcal{P}$ of n_p production units. Each production unit $p \in \mathcal{P}$ is programmed to perform a specific and finite subset $\mathcal{M}_p \subset \mathcal{M}$ of the manufacturings $A_m^* = (u_m^*, y_m^*)$, already defined in terms of product modelling. The actuation of one of the programmed manufacturings is commanded by the control unit having in charge the management of that production unit. A PU can only actuate the admissible MOs in series, one at a time. The PU operations are regulated by a boolean variable z_p, which can assume only two values: *waiting* or *working*. A production unit p can receive the command of actuating one MO out of its programmed set $\mathcal{M}_p$, only when it is in *waiting* state. Then it switches to *working* state and there remains for a time interval strictly depending on the actuated MO. Note that such an interval is a variable which may be different from the time instants at which delivery of output objects has been requested.

The PU state variables. The state of a PU is defined by its boolean variable z_p. The state of the PU set $\mathcal{P}$ is then defined by the vector z of dimension n_p having components z_p, $p \in [1,n_p]$.

The PU (state) transition event set. To each PU p belonging the PU set $\mathcal{P}$, the following set of *facts* H_p is associated. A generic element h_p of this set can assume three values:

1) h_p=*"begin"*, indicating the transition of the p-th PU from waiting to working state;
2) h_p=*"end"*, indicating the transition of the p-th PU from working to waiting state;
3) h_p=*"null"*, meaning no transition.

In the fact set H_p the addition operation is not defined, which, as already said, implies that different facts cannot occur simultaneously. Let us introduce the following event sets:

1) The event set of the state transitions of the p-th PU, defined as the Cartesian product $\mathcal{E}_{hp} = \mathcal{T} \times H_p$.
2) The event set of the state transitions of the production unit set $\mathcal{P}$, defined by the Cartesian product $\mathcal{E}_h = \mathcal{T} \times H_1 \times ... H_p \times ... H_{n_p} = \mathcal{T} \times \mathcal{H}$, where $\mathcal{H}$ is the set of the vectors h of dimension n_p and components h_p.

4.3 The control units

The Control Unit (CU). It is a mathematical element which manages a set of PUs. It works according to suitable control algorithms and sends to each controlled PU the MO actuation commands, after having verified the occurrence of appropriate feasibility conditions. That is, a PU should be in its *waiting* state and the input objects required by the commanded MO should be available in the input storage units. Control algorithms are not treated in this note, whose aim is to model the system dynamic behaviour under CU commands.

The PU command event set. Consider now the set $\mathcal{P}$ of all the PU commanded by the control unit. To each PU p of this set, a new set of facts, denoted by U_p, is associated; it lists all the admissible commands of that PU. The generic element $u_p \in U_p$ can assume values from zero to m_p, the number of programmed MOs on the p-th PU. The value $u_p=0$ (the null element) means *"no command"*, while $u_p=m$ ($0<m \leq m_p$) expresses the actuation command of the m-th programmed MO. In the set U_p, the addition operation is not defined, which, as already said, implies that different facts cannot occur simultaneously. Then, we introduce the following event sets:

1) The event set of the MO actuating commands of the p-th PU, defined as the Cartesian product $\mathcal{E}_{up} = \mathcal{T} \times U_p$.
2) The event set of the commands of the production unit set $\mathcal{P}$, defined by the Cartesian product $\mathcal{E}_h = \mathcal{T} \times U_1 \times ... U_p \times ... U_{n_p} = \mathcal{T} \times \mathcal{U}$, where $\mathcal{U}$ is the set of the vectors u having dimension n_p and components u_p.

4.4 Programmed and actuated MOs

A PU can be programmed to perform MOs which are already defined in terms of product modelling (reference MOs), that is by a pair $A_m^* = (u_m^*, y_m^*)$ of input and output quantity vectors, with the quantities partitioned between the different object types.

Programming a production unit to perform a manufacturing operation $A_m^* = (u_m^*, y_m^*)$ means to define:

- from which input storage unit and at which time instant each consumed object listed in the input vector u_m^* has to be drawn,
- to which output storage unit and at which time

instant each produced object listed in the output vector y_m^* has to be delivered,

- at which time instant the production unit has to switch back to waiting state.

A programmed MO can be launched when a PU is in waiting state. Then all the programmed activities are actuated. Note that any MO actuation shall not respect in a precise manner the programmed activities. Therefore any actuated MO will be considered a different mathematical element from the programmed MO. Moreover, both the programmed and actuated MOs are different mathematical elements from the reference MO, from which they were derived.

A *programmed MO* is a *relocable* event sequence $\sigma'_q=\{(t_{oi},\boldsymbol{q}_i)\}$, $i\in[1,n]$ of delivery/drawing events $\boldsymbol{e}_{qo}=(t_o,\boldsymbol{q})\in\mathcal{E}_{qo}=\mathcal{T}_o\times\mathcal{Q}$ where $\mathcal{T}_o$ is the relocable time interval in which the MO is programmed. *Relocable* means that the time instant t_o at which the programmed MO has to be commanded is set equal to zero. The programmed MOs are then mathematical elements belonging to the sequence set $\Sigma(\mathcal{E}_{qo})$.

Any given reference MO $A_m^*=(\boldsymbol{u}_m^*,\boldsymbol{y}_m^*)$ can be programmed on a specific PU by creating a *relocable* event sequence $\sigma'_q=\{(t_{oi},\boldsymbol{q}_i)\}$, $i\in[1,n]$ of delivery/drawing events, which is related to the reference MO by the following condition. The drawn and delivered object quantities may be expressed, respectively, by the quantity vectors $\boldsymbol{u}_m^*$ and $\boldsymbol{y}_m^*$ when the drawn or delivered objects of the same type are assumed to be equivalent, i.e. no attention is paid to time and place (storage unit) of their drawing or delivery. Accordingly, the set of the reference MOs generates a partition in the set of the programmed MOs; in other words, each reference MO corresponds to an equivalent class in the set of the programmed MOs.

An *actuated* MO is an event sequence $\sigma_q=\{(t_i,\boldsymbol{q}_i)\}$, of delivery/drawing events $\boldsymbol{e}_q=(t,\boldsymbol{q})\in\mathcal{E}_q=\mathcal{T}\times\mathcal{Q}$, where $\mathcal{T}$ is the actual time interval in which the production process takes place. The actuated MOs are then mathematical elements belonging to the sequence set $\Sigma(\mathcal{E}_q)$.

Given a programmed MO described by a relocable event sequence $\sigma'_q=\{(t_{oi},\boldsymbol{q}_i)\}$, $i\in[1,n]$ belonging to the sequence set $\Sigma(\mathcal{E}_{qo})$, it can be actuated by a command dispatched to the PU at a time τ. The corresponding actuated MO is then described by an event sequence $\sigma_q=\{(t_i,\boldsymbol{q}_i)\}$, $i\in[1,n]$ where the event times t_i are specific functions $t_i(t_{oi},\tau)$ depending on the actuation model of the programmed MOs. In the simplest case, i.e. the linear time-invariant determistic model, it holds $t_i=t_{oi}+\tau$. More generally, as in the case of a stochastic model, t_i will be a random function of t_{oi} and τ.

A further event sequence has to be associated to each programmed or actuated MO. It is the event sequence describing the PU state transitions from *waiting* to *working* and back to *waiting*. For each PU p and for each programmed MO m of the same PU, the following two event sequences are defined.

- A programmed sequence

(4) $\sigma'_{hp}=\{e_b=(0,h=\text{"begin"}), e_e=(T_{pm},h=\text{"end"})\}$

  where T_{pm} is the programmed interval during which the p-th PU remains busy to perform the m-th MO.

- An actuated sequence

(5) $\sigma_{hp}=\{e_b=(t_b,h=\text{"begin"}), e_e=(t_e,h=\text{"end"})\}$

  where t_b is the starting time of the MO and t_e is the, possibly stochastic, function of t_b and T_{pm}.

5. THE DISCRETE-EVENT STATE EQUATIONS OF THE PRODUCTION PROCESS

Let us consider a factory made up by a set of SUs and a set of PUs commanded by a single CU. The following assumptions are made:

1) Each PU has been programmed to perform an its own set of MOs.
2) The control unit knows the state of the storage units, i..e the quantity of the stocked objects, and the state (waiting/working) of all the PUs under control.

As a consequence all the MO actuating commands which are dispatched to the different PUs can be assumed to be always *feasible*, i.e. to be consistent with the stock of objects in the SUs and with the operative state of the PUs.

The *input* variables to the production system are the commands dispatched by the CU; They take the form of an event sequence $\sigma_u=(t_{ui},\boldsymbol{u})$, $i\in[1,n_u]$, where the *fact* $\boldsymbol{u}\in\mathcal{U}$ is a the n_p-sized vector listing the commands which are dispatched in an asynchronous way to the PUs. Let us recall that u_p is the p-th component of such a vector, and it may assume values from 0 to m_p, the latter symbol being the total number of MOs programmed on the p-th PU.

The state of the production system is defined by a list of different variables, as follows:

- The state of the MOs under actuation. At time t, it is described by the *restriction* to the interval $t_i>t$ of the sum of the delivery/drawing sequences which have been actuated by all the MOs already commanded before t. In other words, the state is the set of the events belonging to those sequences, but which have not yet occurred. In the following, the event sequence resulting from the above restriction will be indicated by $\Sigma_q(t)$; it expresses the state of the MOs under actuation.

- The transition state of all the PUs, imposed by the MOs under actuation. At time t, it is described by the *restriction* to the interval $t_i>t$ of the sum of the transition sequences which have been actuated by all the MOs commanded before t. In other words, the state is the set of the events belonging to those sequences, but which have not yet occurred. In the following, the event sequence resulting from the above restriction will be indicated by $\Sigma_h(t)$; it expresses the transition state of all the PUs which are still committed to perform the already actuated manufacturing operations.
- The state of the factory PUs, which, as already said, is described by the vector of boolean variables $z(t)$.
- The state of the factory SUs, which, as already said, is described by the vector $\boldsymbol{x}(t)\in\mathcal{Q}$ of the stocked objects.

An input event $\boldsymbol{e}_u=(t_{ui},\boldsymbol{u})$ occurring at time t_u, and such that the component u_p of $\boldsymbol{u}$ be not null, forces the p-th PU to start the programmed MO specified by the value u_p. Starting an MO on a PU will actuate the following two event sequences:

1) The delivery/drawing sequence $\sigma_q=\{(t_i,\boldsymbol{q}_i)\}$, $i\in[1,n]$ of the actuated MO.
2) The state transition sequence of the commanded PU $\sigma_{hp}=\{\boldsymbol{e}_b=(t_b,h=\text{"begin"}),\ \boldsymbol{e}_e=(t_e,h=\text{"end"})\}$.

The sums of all the different sequences of the above two kinds, generated by the same input event $\boldsymbol{e}_u=(t_{ui},\boldsymbol{u})$, will be denoted respectively with $F_q[\boldsymbol{e}_u=(t_{ui},\boldsymbol{u})]$ and $F_h[\boldsymbol{e}_u=(t_{ui},\boldsymbol{u})]$.

At the occurrence time t_{ui} of an input event, the following state equations arise:

$$\text{(6)}\qquad \begin{aligned}\Sigma_q(t_{ui+}) &= \Sigma_q(t_{ui}) + F_q[\boldsymbol{e}_u=(t_{ui},\boldsymbol{u})]\\ \Sigma_h(t_{ui+}) &= \Sigma_h(t_{ui}) + F_h[\boldsymbol{e}_u=(t_{ui},\boldsymbol{u})]\end{aligned}$$

Having denoted with $\boldsymbol{e}_q=(t_{qi},\boldsymbol{q}_i)$ the first event of the sequence $\Sigma_q(t)$ $(t_{qi}\geq t)$, the occurrence of the event $\boldsymbol{e}_q=(t_{qi},\boldsymbol{q}_i)$ will force an updating of the state of the SUs and of the state of the MOs under actuation as it follows:

$$\text{(7)}\qquad \begin{aligned}\boldsymbol{x}(t_{qi+}) &= \boldsymbol{x}(t) + \boldsymbol{q}_i\\ \Sigma_q(t_{qi+}) &= R[\Sigma_q(t),\ t_{qj}\geq t_{qi}]\end{aligned}$$

Having denoted with $\boldsymbol{e}_h=(t_{hi},\boldsymbol{h}_i)$ the first event of the sequence $\Sigma_h(t)$ $(t_{hi}\geq t)$, the occurrence of the event $\boldsymbol{e}_h$ will force an updating of the state z of the PUs, specifically of the component z_p such that the component h_{pi} of the vector $\boldsymbol{h}_i$ be not null. The same event $\boldsymbol{e}_h$ will force also the updating of the transition state of the PUs, as follows.

$$\text{(8)}\qquad \begin{aligned}z_p(t_{hi+}) &= \text{"waiting" if } (h_{pi}=\text{"end"})\\ z_p(t_{hi+}) &= \text{"working" if } (h_{pi}=\text{"begin"})\\ \Sigma_h(t_{ti+}) &= R[\Sigma_h(t),\ t_{hj}\geq t_{hi}]\end{aligned}$$

The above state equations look from one hand very compact and simple, but capable of describing the factory dynamics at any detail level, including a detailed numerical simulator and the simplified models driving the control algorithms of the production control units.

6. CONCLUSIONS

The introduction of the concept of event=(time,fact) and event sequences in the Manufacturing Algebra has allowed first to carefully define programming on and actuation by a PU of the reference MOs describing the factory products; then to define in terms of event sequences the states of the factory dynamic elements, SUs and PUs, and the command schedule of a CU. The occurrence of a command event or one of the programmed activities (drawing, delivery, end) of a PU modifies the state as described by a compact set of discrete-event state equations. Those equations are being used to study and synthesize production control strategies for CUs organized in a hierarchical way.

ACKNOWLEDGMENTS

HIMAC 8141, started in 1994, was funded within the ESPRIT III Basic Research programme. It was coordinated by EICAS Automazione spa, Torino.

REFERENCES

Canuto E., Donati F. and Vallauri M. (1995). An algebra for modelling manufacturing processes. In *Proc. 3rd IEEE Mediterranean Symp. on New Directions in Control & Automation, Vol.II*, p.438-445. Limassol (Cyprus)

Canuto E., Donati F. and Vallauri M. (1996). Advances in manufacturing algebra: foundations of production planning. In *Proc. 1st HIMAC Workshop, A new mathematical approach to manufacturing engineering*, (Vallauri M. ed.), p.19-31. CELID, Torino.

Canuto E., Christodoulou M., Chu C., Donati F., Gaganis V., Janusz B., Proth J.M., Reithofer W. and Vallauri M. (1997). The ESPRIT basic research HIMAC: hierarchical management and control in manufacturing systems. To be presented to *1st IFAC Workshop on Manufacturing Systems: Manufacturing Modelling, Management and Control*. Vienna (Austria).

MANUFACTURING ENTERPRISE MODELING WITH PERA AND CIMOSA

K. Kosanke[1], F.B. Vernadat[2], T.J. Williams[3]

[1]CIMOSA Association e.V., Stockholmerstr. 7, D-71034 Böblingen, Germany
[2]LGIPM, University of Metz, Ile du Saulcy, F-57012 Metz, France
[3]Purdue University, Potter Center, West Lafayette, IN 47907-1293 USA

Abstract: Both CIMOSA and PERA are architectures developed for integrated manufacturing systems analysis, design and implementation. CIMOSA is an open systems architecture for Enterprise Integration which includes a powerful language for enterprise modeling and is aimed at model-based integration. The Purdue Enterprise Reference Architecture (PERA) is based on a detailed and pragmatic methodology covering the whole life cycle of an industrial project from inception to operation and even system disposal. This paper presents how the PERA methodology can be applied using the CIMOSA constructs to engineer a production planning and control system. PERA provides a clear indication of the role and place of humans in the architecture to be defined or specified while CIMOSA provides a consistent set of constructs to capture system details at all modeling levels.

Keywords: Enterprise Integration, Enterprise Engineering, Modeling, CIMOSA, PERA

1. INTRODUCTION

Enterprise modeling and integration techniques have great potential for improving engineering of modern manufacturing enterprises (Bernus and Nemes, 1996; Vernadat, 1996). Manufacturing systems are complex entities which fall under the category of discrete event dynamic systems and must be engineered like any other complex systems, i.e. using a structured approach covering their entire life cycle. It is therefore essential to develop sound approaches for *Enterprise Engineering* spanning system definition, requirements definition, design specification and implementation (Ladet and Vernadat, 1995; Williams, 1994).

The aim of this paper is to present such an approach obtained by combining concepts from CIMOSA and PERA, two advanced architectures and methodologies proposed for integrated industrial systems engineering. Basic knowledge both on CIMOSA and PERA is assumed on the part of the reader of this paper.

2. CIMOSA

CIMOSA (AMICE, 1993; CIMOSA Association, 1996) is an Open Systems Architecture for CIM and Enterprise Integration (EI) developed as a series of ESPRIT Projects (EP 688, 5288 and 7110) between 1986 and 1994 with the support of EU DG III.

Its aim is to provide the manufacturing industry with (1) an Enterprise Modeling Framework (EMF), which can accurately represent business operations, support their analysis and design, and lead to executable enterprise models; (2) an Integrating Infrastructure (IIS), used to support application and business integration as well as execution of the implementation model to control and monitor enterprise operations; and (3) major steps for a methodology describing the System Life Cycle (SLC).

CIMOSA provides a Reference Architecture (known as the CIMOSA cube) from which particular enterprise architectures can be derived. This Reference Architecture and the associated enterprise modeling framework are based on a set of modeling constructs, or generic building blocks, which altogether form the CIMOSA modeling language. The language is based on an event-driven process-based model centered on the concept of business processes and enterprise activities (Vernadat, 1993). Essential modeling constructs and their relationships are indicated by Fig. 1 for each modeling view. Definition of constructs can be found in (Heulluy and Vernadat, 1997).

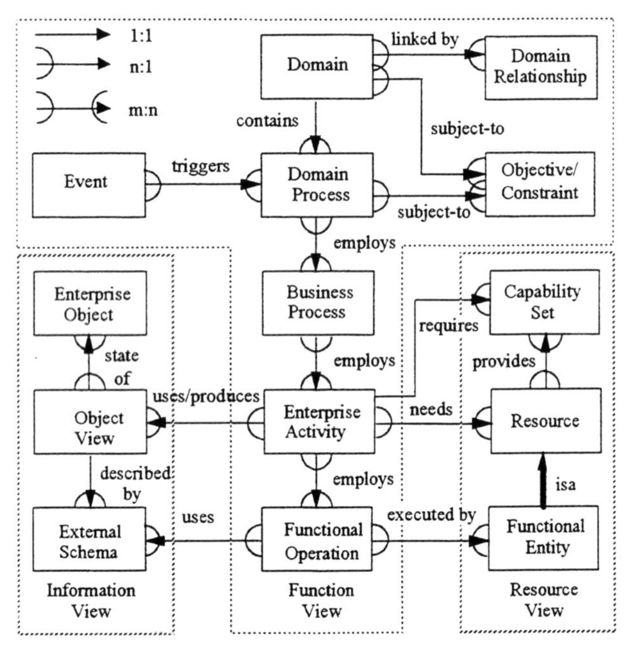

Fig. 1. Relationships among essential CIMOSA constructs

3. PERA

The Purdue Enterprise Reference Architecture (PERA) has been designed to assist industry in its effort to develop and implement integrated manufacturing systems. It is a complete and largely documented methodology to define, design, construct, install, and operate an integrated enterprise system or large automation project. It has been developed at Purdue University in collaboration with a consortium of industrial companies (Williams, 1992; 1994).

The methodology is not usually applied to the whole enterprise in one shot. It is intended to be applied to a so-called Enterprise Business Entity (EBE), whatever its size may be. Be the definition of the business entities obtained using a top-down or bottom-up approach makes no difference to PERA. However, once the business entity has been selected, PERA adopts a top-down approach from mission definition to complete system specification or detailed implementation description according to the level of analysis required by the business user.

PERA, as an enterprise system design methodology, is based on the following essential principles:

1. An overall detailed *Master Plan* for the desired enterprise integration migration, stating the TO-BE state, is absolutely necessary before attempting to implement any integration program.
2. An extensive and detailed *Instructional Manual* is necessary to guide and simplify the preparation of the Master Plan in a particular company.
3. Included in the Master Plan is an *Enterprise Integration Program Proposal*, i.e. a prioritized set of integrated projects, each within the resources of the company involved, whose ultimate completion will assure the success of the finally desired operational integration of the enterprise.
4. A reference architecture is necessary to provide the framework for the development and use of the Instructional Manual, the resulting Master Plan, and the ultimately implemented Integration Program Proposal. This is the role of the Purdue Enterprise Reference Architecture.
5. Included in the reference architecture could be example material providing much of the detail necessary in developing the Master Plan. This information is compiled in PERA in the form of *The Purdue CIM Reference Model*.

PERA is defined as an architecture for the development of program life history of an enterprise (or part of it) as opposed to architectures for the physical system (i.e. control or communication systems). It provides the necessary guidelines and a process to establish the Master Plan and Integration Program Proposal. It is documented by means of several books and it is, therefore, not the intention of this paper to describe it in full details. Only the general principles of the methodology are reminded.

The overall structure of the PERA methodology or PERA skeleton is presented in Fig. 2. The methodology is composed of a number of phases which are listed on the left-hand side of Fig. 2, starting from the top. Each phase is made of a number of tasks. A renovation or disposal phase is added to the end.

The PERA skeleton (Fig. 2) starts with the definition of the relevant EBE and then diverges into two branches in the definition phase: one for the information systems requirements and one for the manufacturing equipment requirements. This structure is then divided into three main blocks for each task of the successive phases: the left-hand side block corresponds to the customer product and service or *Manufacturing Equipment Architecture*, the right-hand side block corresponds to the information and control or *Information Systems Architecture*, and the central block mediating the two others corresponds to the *Human and Organizational Architecture*. The latter defines those activities solely performed by humans or mostly performed by humans assisted by machines or computer systems.

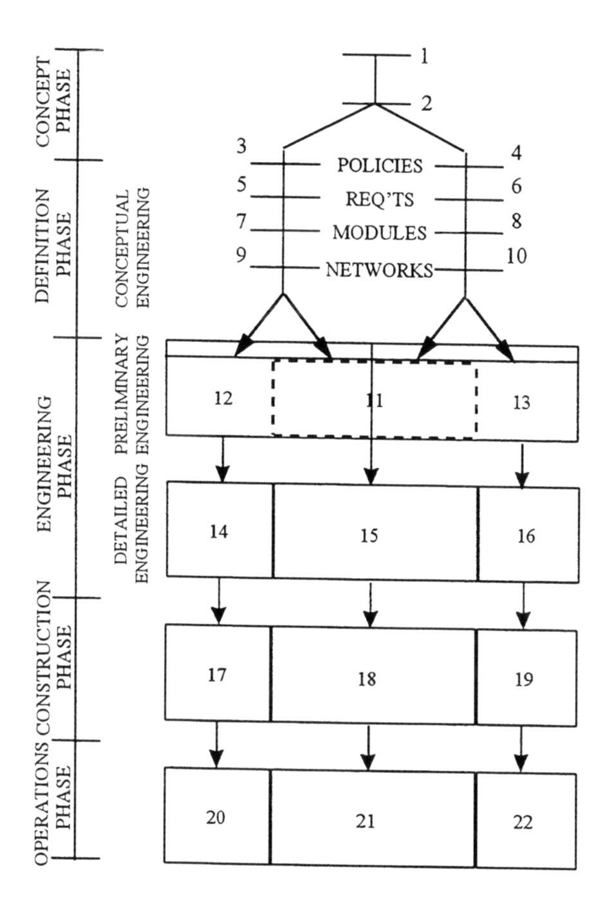

Fig. 2. Skeleton of the PERA methodology

Beside the detailed systematization of all steps for integration program implementation proposal, the originality of the PERA approach mostly relies on its definition of the place of humans in the enterprise, described in the Human and Organizational Architecture. Especially, concepts of *humanizability* and *automatability* have been introduced. Indeed, the frontier between man and machine activities is not clear-cut. This is true both for information system-based activities and manufacturing-based activities. Therefore, there exists a zone on both sides of a line materializing the extent of automation corresponding to automatability and humanizability. The extent of automation actually defines the boundary between the Human and Organization Architecture and the Information Systems Architecture on one hand, and between the Human and Organization Architecture and the Manufacturing Equipment Architecture on the other hand. The humanizability line shows the extent to which humans can be used to actually implement the tasks and functions of the integrated system. The automatability line does the same to show the limits of technology in achieving automation.

Starting with a given Enterprise Business Entity (EBE), the PERA methodology proceeds as follows (referring to numbers in Fig. 2):

(1) Task 1: Identification of EBE.

(2) Task 2: Description of the management's mission, vision and values for the EBE plus any further operational philosophies or mandated actions concerning it, such as choice of processes, vendor selection, etc.

(3) Task 3: Definition of the operational policies and goals related to the Customer Product and Service or Manufacturing goals and objectives (present or proposed production entity including product, process, and operational requirements).

(4) Task 4: Definition of the operational policies related to the Information goals and objectives (control capabilities, degrees of performance of processes and equipment, adherence to regulations and laws, etc.).

(5) Task 5: Definition of requirements to be fulfilled in carrying out the Customer Product and Service or Manufacturing related policies of the company (physical or operation production requirements).

(6) Task 6: Definition of requirements to be fulfilled in carrying out the Information policies of the enterprise (planning, scheduling, control, and data management requirements).

(7) Task 7: Definition of the sets of tasks, function modules, and macrofunction modules required to carry out the requirements of the Manufacturing or Customer Product or Service mission of the enterprise.

(8) Task 8: Definition of the sets of tasks, function modules, and macrofunction modules required to carry out the requirements of the Information or Mission Support side of the functional analysis.

(9) Task 9: Definition of process flow diagrams showing the connectivity of the tasks, function modules, and macrofunctions of the Manufacturing or Customer Product and Service processes involved (operations functional networks).

(10) Task 10: Definition of the connectivity diagrams of the tasks, function modules, and macrofunction modules of the Information or Mission Support (e.g. data flow diagrams or related modeling techniques).

(11) Task 11: It concerns the functional design of the Human and Organizational Architecture (skill level determination, ergonomy of work centers, etc.). Establishment of the Automatability, Humanizability and Extent of Automation Lines.

(12) Task 12: It concerns the functional design of the Manufacturing Equipment Architecture (e.g. process equipment to be required, control systems capabilities required, plant layout and design optimization, etc.).

(13) Task 13: It concerns the functional design of the Information Systems Architecture (functional analysis of software developments, database schemata, screen layouts, man-machine interfaces, computer hardware architectures, computer networks, etc.).

(14) Task 14: It concerns the detailed design of components, processes, and equipment of the Manufacturing Equipment Architecture. It covers de-

tailed equipment design and selection from vendors as well as final plant layout design.

(15) Task 15: It concerns the detailed design of the task assignments, personnel skills development, training courses, and organizational planning of the Human and Organizational Architecture.

(16) Task 16: It concerns the detailed design of the equipment and software of the Information Systems Architecture. It covers such tasks as computer equipment selection, software system configuration, and computer systems layout.

(17) Task 17: It deals with the construction, checkout, and commissioning of the equipment and processes of the Manufacturing Equipment Architecture. Machines are bought, installed, and verified. NC and robot programs are tested.

(18) Task 18: It concerns staffing, implementation of organizational development training courses, and on-line skill practice for the Human and Organizational Architecture.

(19) Task 19: It deals with the construction, checkout, and commissioning of the equipment and software of the Information Systems Architecture. Computer machines and off-the-shelf software packages are purchased and installed and built-in applications are installed and tested.

(20) Task 20: It deals with continued improvement of process and equipment operating conditions to increase quality and productivity and to reduce costs involved for the Manufacturing Equipment Architecture. Activities involved may deal with techniques such as Statistical Quality Control, Statistical Process Control, Total Quality Management and other related techniques.

(21) Task 21: It concerns continued organizational development and skill and human relations development training of the Human and Organizational Architecture.

(22) Task 22: It deals with the operation of the Information and Control System of the Information Systems Architecture including its continued improvement.

Using the PERA approach, a Master Plan can be developed for any part of the enterprise requiring integration. PERA suggests that the Master Plan compares the present state of the business enterprise (the AS-IS state) with the desired future state (the TO-BE state) to characterize the migration path to be followed between them.

It is recommended to first identify the EBE to be analyzed and then to immediately state the TO-BE policies of the desired system before doing the analysis of the AS-IS state to avoid any biais. The TO-BE system is described (using the procedure mentioned above) in terms of TO-BE physical, TO-BE information, and TO-BE human architectures. The same is done for the AS-IS system and then the migration path is derived. The whole analysis must be complemented by a costs/benefits analysis.

To date, PERA is certainly the most complete and sophisticated methodology for assisting business users in developing an integration implementation program. It is supported by a detailed documentation with procedures manuals and guides and provides examples of program implementation proposals.

4. APPLICATION

4.1 The manufacturing scenario

Let us consider a SME named CODEC Ltd. producing special heavy gear-boxes of different sizes and complexity for large industrial equipment on demand. Gear-boxes are ordered by unit or by small batches (less than 10 units). The product design and engineering is done within the enterprise according to customer specifications. The enterprise comprises only one plant having under the same roof the administrative department, the product design and engineering department, the production planning and control department, the manufacturing department, and the shipping and receiving department.

The enterprise has adopted a just-in-time (JIT) approach with a partner company located in its vicinity for procurements of raw materials (blank parts, bars, sheet metal plates). However, it works on stocks for purchased items (such as oil barils, bearings, bolts, screws, and other manufactured items). Stocks are minimized as much as possible on the basis of planned orders and forecasts on future demand.

Customer orders are collected every day to produce so-called 'factory orders'. Production planning and control is performed by means of an off-the-shelf MRP system. The package supports master production planning, material requirements planning, inventory control, capacity planning, and production plant monitoring. It is essentially used to track production data on a day-to-day basis, to control inventories, and to produce daily manufacturing plans. Manufacturing orders to be executed by the manufacturing facility are produced by a dedicated package based on a dynamic scheduling algorithm taking into account the current status and available capacity of the shop-floor.

Figure 3 presents the part of the enterprise (enterprise business entity) which will be the focus of the case study. It is made of three subsequent core processes: Customer Order Processing, Production Planning and Control and Shop Floor Operations.

4.2 Case study

The problem is to organize the three core processes into an integrated process chain.

To perform Tasks 1 to 4 of PERA, the domain, domain relationship and objective/constraint constructs of CIMOSA are used. The domain construct is used to define the EBE. It comprises the three core processes listed as domain processes. Relationships with other parts of the enterprise can be documented in the form of domain relationships. Business goals and policies (qualitative or quantitative) are documented using the objective/constraint construct.

To perform Tasks 5 to 10 (Conceptual Engineering) of PERA, CIMOSA proposes to define event-process chains (EPC's). To do that, the user must define events triggering domain processes and then decompose domain processes into business processes (or sub-processes) and enterprise activities (basic process steps). In this case, it is decided that the arrival of a customer order triggers the Customer Order Processing process which produces a Factory Order. The creation of a Factory Order triggers the Production Planning & Control process producing a list of Manufacturing Orders. The release of a set of Manufacturing Orders then triggers the Shop Floor Operations process. Thus, the event, domain process, business process and enterprise activity constructs of CIMOSA are used. For instance, the Receive Customer Order event and the Customer Order Processing process can be described as follows (where START indicates the beginning of the process, FINISH represents its end, and ES means 'ending status of'):

EVENT
- Identifier: EV-1
- Name: Receive Customer Order
- Description: Receiving customer order by phone
- Triggers: DP-1 / Customer Order Processing
- Source: external
- Object View: OV-1 / Customer Order

DOMAIN PROCESS
- Identifier: DP-1
- Name: Customer Order Processing
- Events: EV-1 / Receive Customer Order
- Objectives: To receive and process customer orders
- Constraints: All orders received during the day must be processed before the end of the day
- Process Behavior:
  - WHEN (START WITH EV-1) DO Get Order
  - WHEN (ES(Get Order) = Confirmed) DO Update Order Database
  - WHEN (ES(Get Order) = Rejected) DO FINISH
  - WHEN (ES(Update Order Database) = Done) DO Generate Factory Order
  - WHEN (ES(Generate Factory Order) = OK) DO FINISH
- Comprises: EA-1 / Get Order, EA-2 / Update Order Database, EA-3 / Generate Factory Order

At this point, it is decided that Customer Order Processing will be performed by a sales person using a PC-based package, that Production Planning & Control will be performed by the MRP package and a in-house built dynamic scheduling program both running on a super-mini computer, that raw material procurement orders will be automatically sent via an EDI system, and that Shop Floor Operations will be managed by a control system to be developed by a software company according to CODEC's specifications.

To perform Task 11, CIMOSA provides the capability set construct to define human skills and abilities required by tasks executed by humans. Similarly, the capability set construct is also used in Tasks 12 and 13 to state capabilities required on the part of machines and software applications necessary to perform tasks of the Manufacturing Equipment Architecture and the Information Systems Architecture. These tasks are defined in terms enterprise activities of domain or business processes. At this stage, inputs and outputs of each enterprise activity must be defined in terms of object views, another CIMOSA construct used to describe enterprise object states (an activity is defined as a timed function transforming an input state into an output state). An object view is defined by a list of properties of enterprise objects. For instance, the Get Order enterprise activity and Customer Order object view are as follows:

ENTERPRISE ACTIVITY
- Identifier: EA-1
- Name: Get Order
- Objective: To receive customer order by phone
- Function Input: Customer File, Order File
- Control Input: Customer Order
- Resource Input: Sales Person, Sales Support Package
- Function Output: Order File
- Control Output:
- Resource Output:
- Activity Behavior: {}
- Required Capabilities: CS-1 / Order Proc. Cap.
- Ending Status: Confirmed, Rejected
- Used By: DP-1 / Customer Order Processing

OBJECT VIEW
- Identifier: OV-1
- Name: Customer Order
- Leading Object: Customer Order
- Related Objects:
- Properties:
  - Order #: STRING [10]
  - Date: date
  - Customer: STRING [20]
  - Due date: date
  - Product #: STRING [8]
  - Product name: STRING [20]
  - Quantity: INTEGER
  - Net Price: MONEY ($)

Tasks 14 to 16 of PERA concern detailed system design. To support them, CIMOSA provides design specification constructs. These include specified event, specified domain process, specified business process, and specified enterprise activity for all tasks. The term 'specified' means that the same constructs than before are used but they are more detailed (i.e. they get additional attributes to be defined). For instance, enterprise activities must be specified in terms of their functional operations (i.e. basic primitives that constitute their elementary steps defined by a verb and a list of input/output parameters). In addition, time must be added to events (date of occurrence) and activities (duration). Also, resource input of activities must be defined in terms of functional entities or active resources. A resource construct is thus provided which can be specialized into human, machine or application functional entities with their relevant set of provided capabilities. For Task 15, CIMOSA provides two constructs: the organization unit and the organization cell to define authority and responsibility on activities and resources previously defined. In Task 16, all object views must be specified in terms of external schemata (using the entity-relationship model or an object-oriented model), thus providing the conceptual database schema of the underlying information system.

For Tasks 17 to 19 of the installation phase, CIMOSA provides basically the same constructs than for the design phase but all process triggering conditions and activity pre- and post-conditions must be defined as well as exception handling mechanisms. In addition, constructs are provided to describe the internal schema of the information systems (using the SQL language). Constructs of the resource view are used to describe all manufacturing and information resources and their installation. This also applies to plant layout, logistic systems, and computer network configuration.

Because the CIMOSA enterprise modeling framework has been designed for analysis and design of integrated manufacturing environments, no constructs are provided for the operations phase.

5. CONCLUSION

The aim of the paper was to provide an indication on how CIMOSA and PERA can be jointly applied for industrial systems engineering. Because both CIMOSA and PERA require full books for their complete treatment, it was only possible in this paper to give an indication of where and when CIMOSA constructs can be used in the PERA methodology.

REFERENCES

AMICE (1993). *CIMOSA: Open System Architecture for CIM, 2nd extended and revised version.* Springer-Verlag, Berlin.

Bernus, P. and L. Nemes, eds. (1996). *Modelling and Methodologies for Enterprise Integration.* Chapman & Hall, London.

CIMOSA Association (1996). CIMOSA Formal Reference Base. CIMOSA Association e.V., Stockholmerstr. 7, D-70731 Böblingen, Germany, April.

Heulluy, B. and F.B. Vernadat (1997). The CIMOSA Enterprise Modelling Ontology. Proc. Manufacturing Systems: Manufacturing Modelling, Management and Control (MIM'97), Vienna, Austria, 5-7 February 1997. These proceedings.

Ladet, P. and F.B. Vernadat, eds. (1995). *Integrated Manufacturing Systems Engineering.* Chapman & Hall, London.

Vernadat, F. (1993) CIMOSA: Enterprise modelling and enterprise integration using a process-based approach. In : *Information Infrastructure Systems for Manufacturing* (H. Yoshikawa and J. Goossenaerts, eds.), pp. 65-84. North-Holland, Amsterdam.

Vernadat, F.B. (1996) *Enterprise Modeling and Integration: Principles and Application.* Chapman & Hall, London.

Williams, T.J. (1992) *The Purdue Enterprise Reference Architecture.* Instrument Society of America, Research Triangle Park, North Carolina.

Williams, T.J. (1994). The Purdue Enterprise Reference Architecture. *Computers in Industry*, **24**(2-3), 141-158.

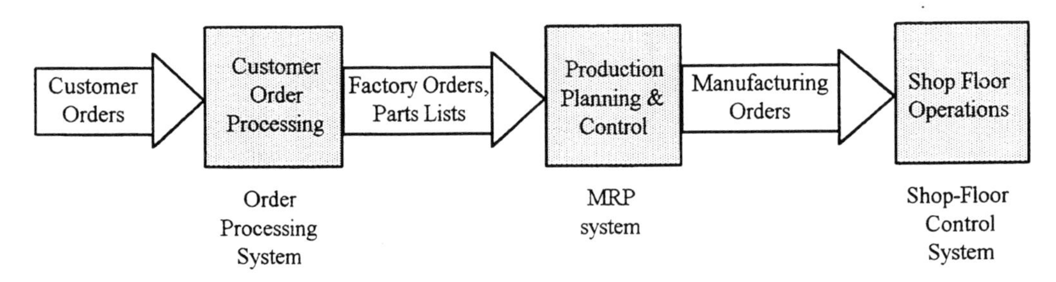

Fig. 3. Simplified scenario for customer order acceptance and execution.

THE CIMOSA ENTERPRISE MODELLING ONTOLOGY

B. Heulluy, F.B. Vernadat

LGIPM, University of Metz/ENIM, Ile du Saulcy, F-57012 Metz, France

Abstract: To model and evaluate business processes, enterprise modelling tools and techniques are required. It is therefore necessary that all concepts and relationships underlying enterprise modelling languages be precisely defined. This is the role of ontologies. In this paper, we present an ontology for enterprise modelling based on constructs identified in the CEN ENV 12 204 (Constructs for Enterprise Modelling) and the CIMOSA modelling language. The language addresses functional, information, resource and organisational aspects of an enterprise and includes the time dimension. The ontology formally defines such concepts as activity, process, event, resource, resource capabilities, enterprise object, object view (or object state), organisation unit and organisation cell. A notion of cost can also be defined in relation to the concepts of activity and resource.

Keywords: Enterprise Integration, Enterprise modelling, CIMOSA, Enterprise ontology

I. INTRODUCTION

To accurately model and reengineer business processes, share experience with business partners and integrate business applications, business users require widely accepted enterprise modelling methods.

Enterprise modelling (EM) is the process of abstracting a reality for the purpose of common understanding, design, (qualitative and quantitative) analysis, simulation, control or simply representation of a part of an enterprise. An enterprise model represents a consensus, a common view, shared by a group of users within an enterprise. An enterprise model is always biased by its finality, i.e. the angle from which the reality under scrutiny is abstracted. Enterprise models are either expressed by means of some notations or using formal languages. A modelling language is a set of constructs used to describe the building blocks forming the enterprise model.

Various notations and languages have been proposed for modelling the different aspects of an enterprise (e.g. CIMOSA, ER models, EXPRESS, GRAI, IDEF0/SADT, IDEF1x, IDEF3, Merise, M*, NIAM, Petri nets or SA/RT, to name a few) and even global frameworks combining modelling tools and methodologies for enterprise modelling have been proposed (ARIS, CIMOSA, GIM, PERA). All these have been surveyed in a book by Vernadat (1996).

CIMOSA is an open systems architecture for CIM and enterprise integration (AMICE, 1993). CIMOSA has been developed by the ESPRIT Consortium AMICE as an pre-normative initiative for integration in manufacturing and serves as a reference architecture in the field. It has contributed a lot to enterprise modelling by providing a comprehensive Enterprise Modelling Framework (known as the CIMOSA cube) defining three modelling levels (requirements, design and implementation), four essential modelling views (function, information, resource and organisation) and three levels of genericity (generic, partial, particular). The generic layer contains a set of predefined modelling constructs, called Generic Building Blocks (GBB's) and their sub-types or Building Block Types (BBT's), from which partial and particular enterprise models can be instantiated. This set of constructs has been partly the basis for the development of the European pre-norm CEN ENV 12 204 - Constructs for Enterprise Modelling (CEN, 1995).

This aim of this paper is to present an ontology for enterprise modelling using CIMOSA constructs.

II. ENTERPRISE MODELLING ONTOLOGIES

2.1 Enterprise Ontologies

An ontology is defined in computer science as a conceptualisation of knowledge for a given domain of discourse (Gruber, 1992). It consists of a formal representation of the kinds of things or concepts (both physical or conceptual), their associated properties and the relationships that hold among them as captured by the terminology used in this domain. Ontologies are important for EM and CIM for the following reasons:

(1) There is a need for a widely accepted *enterprise modelling language* to develop EM tools. Constructs of this language must be formally defined in a neutral, i.e. system independent, way and their semantics precisely specified. These constructs must be based on an agreed upon EM terminology and domain theory.

(2) There is a need to develop standard, reusable CIM *(partial) reference models*. These models (for function, information, resource or organisation) could be exchanged by tools and reused by business users in their own architecture. This would reduce the cost of reinventing the wheel any time a conceptual model is reengineered.

(3) The use of ontologies is perceived as the basis for *knowledge sharing* among intelligent agents (Neches *et al.*, 1991).

Various projects for the development of ontologies for enterprise modelling are currently under way. The most notable ones are: (1) the pioneering work in the TOVE project at University of Toronto proposing an ontology for activities, resources, cost and quality (Fox and Grüninger, 1994) and (2) the Enterprise Ontology developed at Edinburgh (Ushold *et al.*, 1995) having some similarities with CIMOSA. A method for acquiring CIM ontologies has been recently proposed as part of the IDEF suite of methods known as IDEF5 (Benjamin *et al.*, 1995).

2.2 Ontology Description Method

An ontology description method is made of two parts: *terminology definition* and *concept specification*. Terminology definition consists in identifying all relevant concepts and their relationships (including synonyms and homonyms) for a given domain. This is represented in the form of a lattice or semantic network connecting concepts with links. Concept specification consists in defining formally each concept used (properties and semantic rules).

The notation for concept specification used in this paper is based on algebraic specifications, and especially order-sorted algebras (Ehrig and Mahr, 1985). This method has been used in a previous work for defining a formal production management language as reported by Hilger and Proth (1990). It consists in describing each construct of the EM language in terms of so-called *sorts* defined by:

- its sort name
- a set of typed variables
- a set of operations on variables
- a set of axioms defining the semantic rules
- a set of relationships defining semantic or relational links of this concept with other concepts of the ontology

A sort defines a set of similar elements characterised by variables (structural properties or derived properties) and subject to operations. Operations are defined by their signature. Variables and operations can be used in axioms defining the semantic rules to be verified by the concept and expressed in first order logic or logic propositions. We have added a set of relationships to the sort to formalise the terminology lattice.

Three kinds of links can be used to relate concepts to one another:

- *semantic relationships* including the generalisation hierarchy (or is-a link with property inheritance), the aggregation hierarchy (or part-of link) and the instantiation link (to relate an instance to its sort);
- *user-defined relationships* which represent domain-specific links. These are binary links characterised by their name and cardinalities, i.e. a pair (min, max) defining the number of sort occurrences involved in the link;
- *terminology relationships* used to relate concepts in a terminological sense, i.e. using types of links such as 'is-equivalent-to', 'is-related-to', 'may-be-related-to', 'is-synonym-to', etc.

The advantage of using order-sorted algebras as a neutral means of defining ontologies is that specifications expressed can be formally verified using rewriting systems such as OBJ and REVE (Ehrig and Mahr, 1985).

III. THE CIMOSA ONTOLOGY

CIMOSA includes an enterprise modelling language based on an original event-driven process-based model. It emphasises the concept of business process to model event-driven process chains (EPC's) where each process step is an enterprise activity (Vernadat, 1993).

The ontology proposed formally defines such concepts as time, activity, process, event, enterprise object, object view (or object state), resource, resource capabilities, organisation unit and organisation cell to cover functional, information, resource and organisation views of CIMOSA. These concepts correspond to generic CIMOSA constructs or GBB's and are defined by description templates in the CIMOSA Technical Baseline (CIMOSA Association, 1996). Figure 1 presents the semantic network defining relationships among these concepts.

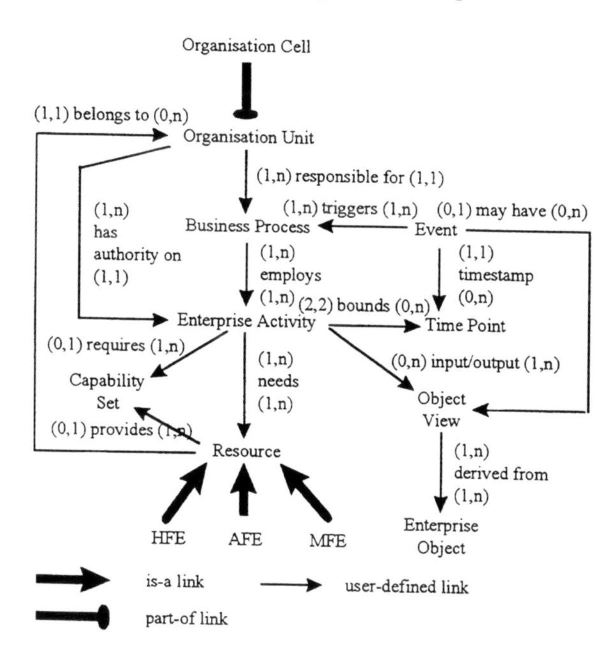

Fig. 1. Ontology concepts and their relationships

3.1 Ontology of time

We use the ontology of time defined in (Grüninger and Fox, 1994). Time is assumed to be a continuous line on which time points and time intervals can be defined as sorts using the order relation '<' (respect. '≤') over time points with the intended interpretation that $t<t'$ (respect. $t\leq t'$) iff t is strictly earlier (respect. earlier) than t'. Uncertainty on a time point is defined by the following predicate where *min* and *max* are two real numbers:

$$time\_point(t,min,max) \Leftrightarrow min \leq t \leq max$$

We also need two functions $SP(t)$ and $EP(t)$ to denote the start and end points of an interval t. Using these relations and the order relations, other axioms on time intervals can be defined such as:

$(\forall t,t',p_1,p_2,p'_1,p'_2,p''_1,p''_2,p'''_1,p'''_2)\ during\ (t,t') \Leftrightarrow$
$time\_point(SP(t),p_1,p_2) \wedge time\_point(EP(t),p'_1,p'_2) \wedge$
$time\_point(SP(t'),p''_1,p''_2) \wedge time\_point(EP(t'),p'''_1,p'''_2)\ /$
$p''_1 \leq p_1 \wedge p'_2 \leq p'''_2$

$(\forall t,t',p_1,p_2,p'_1,p'_2)\ strictly\_before\ (t,t') \Leftrightarrow$
$time\_point(EP(t),p_1,p_2) \wedge time\_point(SP(t'),p'_1,p'_2)\ /\ p_2 < p'_1$

In the following, we assume that INTEGER, REAL, BOOLEAN and PROPERTY are sorts which have been previously defined and will be imported to respectively represent integer and real numbers, Boolean values, object attributes (a name or label and a data type). {.} denotes a set defined in extension. Sets are subject to usual operators defined in set theory. $P(A)$ represents the power set of set A.

3.2 Enterprise object

Enterprise objects are elements of a sort representing entities subject to processing or handling within an enterprise and creating a flow within the enterprise. From an information viewpoint, they are characterised by their name, abstraction mechanisms if any (is-a or part-of) and their list of properties.

SORT Enterprise_Object
IMPORT PROPERTY
VARIABLES
  Isa: *P*(Enterprise_Object)
  Partof: *P*(Enterprise_Object)
  Properties: *P*(PROPERTY)
AXIOMS
  $(\forall O \in$ Enterprise_Object$)$
    $\neg\exists O_1,O_2 \in$ Isa $(O)\ /\ O_1 = O_2$
  $(\forall O\in$ Enterprise_Object $\wedge \exists O' \in$ Enterprise_Object$)$
    $O' =$ Isa $(O) \Rightarrow O \neq O'$
  $(\forall O \in$ Enterprise_Object$)$ Properties $(O) \neq \varnothing$

3.3 Object view

Object views represent enterprise object states, i.e. manifestations of objects at certain instants in time. Properties of one or several enterprise objects can be combined to form an object view. The nature of an object view is either physical (for material flows) or informational (for information flows).

SORT Object_View
IMPORT Enterprise_Object, PROPERTY
VARIABLES
  Leading_Object: Enterprise_Object
  Related_Objects: *P*(Enterprise_Object)
  Nature: (physical, informational)
  Properties: *P*(PROPERTY)
AXIOMS
  $(\forall OV \in$ Object_View$)$ Properties $(OV) \neq \varnothing$
  $(\forall OV \in$ Object_View, $\forall$p $\in$ Properties (OV)
    $\exists O \in$ Enterprise_Object$)$ p $\in$ Properties (O)

3.4 Event

Events are facts (solicited or not) indicating a change in the state of the system. An event happens at a certain point in time. It may be related to an enterprise object or carry information in the form of an object view. It has a source of generation of occurrences of the event (either a resource, an enterprise activity or a source external to the enterprise).

SORT Event
IMPORT Object_View, Enterprise_Activity, Resource,
  BOOLEAN, REAL
VARIABLES
  Source: Resource $\cup$ Enterprise_Activity $\cup$ {external}
  ObjectView: Object_View
  TimeStamp: REAL
OPERATIONS
  CreateEvent: NIL $\rightarrow$ Event

Active: Event → BOOLEAN
AXIOMS
$(e \in \text{Event})$ *CreateEvent* $(e) \Rightarrow$ TimeStamp $(e) = SP(e)$
$(\forall\ e \in \text{Event})$ *Active* $(e) \Rightarrow$ *defined* (TimeStamp(e))

3.5 Business process

Business processes represent the flow of control (or workflow) within an enterprise. They are triggered by one or more events (conjunctive clause) and proceeds by executing activities employed. An event may be used to trigger several processes. Business processes are characterised by their behaviour or flow of control, defined as a partially ordered set of process steps (either business processes or enterprise activities), and their non-empty set of ending statuses, i.e. 0-argument predicates defined as labels and identifying all possible termination statuses of a process. Let us assume that LABEL has been defined as the sort of all possible ending statuses.

SORT Business_Process
IMPORT Event, Enterprise_Activity, LABEL, BOOLEAN
VARIABLES
Triggered_By: *P*(Events)
Ending_Statuses: *P*(LABEL)
Behaviour: δ (*P*(Enterprise_Activity))
Used_By: *P*(Business_Process)
OPERATIONS
Start: Business_Process → BOOLEAN
Finish: Business_Process → BOOLEAN
Ending_Status: Business_Process → LABEL
AXIOMS
$(\forall\ p \in$Business_Process) Triggered_By $(p) \neq \varnothing \wedge$
Ending_Statuses(p) $\neq \varnothing \wedge$ *defined* (Behaviour (p))
$\wedge\ p \notin$ Used_By (p)
$(\forall\ p \in$Business_Process) *Start* $(p) \Leftrightarrow$
Active (e), $\forall\ e \in$ Triggered_By (p)
$(\forall\ p \in$Business_Process) *Finish* $(p) \Rightarrow$
defined (Ending_Status (p)) $\wedge$
Ending_Status (p) $\in$ Ending_Statuses (p)
$(\forall\ p \in$Business_Process, $\forall\ es \in$ Ending_Statuses (p))
$\exists!$ a $\in$ Enterprise_Activity /
$es \in$ Ending_Statuses (a) $\wedge\ p \in$ Used_By (a)

The behaviour δ of a business process p is defined by behavioural rules of the form:

WHEN (condition) DO action

where *condition* represents logical conjunctive conditions associating events and/or ending statuses of activities and sub-processes and *action* defines the next step(s) to be followed by the process when the conditions evaluate to true. A syntax for a process behaviour grammar suitable to describe structured and semi-structured processes has been presented in (Vernadat and Zelm, 1996). They are summarised in the Appendix of the paper for the sake of conciseness.

3.6 Enterprise activity

Enterprise activities are the locus of action, i.e. they transform an input state into an output state using resources and time within the course of a process. An activity ontology for enterprise integration has been presented in (Grüninger and Fox, 1994).

In CIMOSA and the ENV 12 204, activities transform function, control and resource inputs into function, control and resource outputs using some transformation function (algorithm or script or scenario) and producing a certain ending status. The distinction between the three types of inputs and outputs makes it possible to differentiate the control flow from the information flow and the material flow. Time is also associated to activities in the form of a duration.

SORT Enterprise_Activity
IMPORT Event, Object_View, Resource, LABEL, REAL
VARIABLES
Funtion_Input: *P*(Object_View)
Control_Input: *P*(Object_View)
Resource_Input: *P*(Resource)
Funtion_Output: *P*(Object_View)
Control_Output: LABEL ∪ *P*(Event)
Resource_Output: Object_View
Ending_Statuses: *P*(LABEL)
Minimum_Duration: REAL
Maximum_Duration: REAL
Required_Capabilities: Capability_Set
Function: Algorithm
Used_By: *P*(Business_Process)
OPERATIONS
Start: Enterprise_Activity → BOOLEAN
Finish: Enterprise_Activity → BOOLEAN
Duration: Enterprise_Activity → REAL
Ending_Status: Enterprise_Activity → LABEL
AXIOMS
$(\forall\ a \in$ Enterprise_Activity) Function_Input $(a) \cup$
Control_Input $(a) \cup$ Function_Output $(a) \neq \varnothing \wedge$
Function_Input $(a) \cap$ Control_Input $(a) = \varnothing \wedge$
Resource_Input $(a) \neq \varnothing \wedge$
Control_Output $(a) \neq \varnothing \wedge$
Minimum-Duration $(a) \leq$ Maximum_Duration (a)
$(\forall\ a \in$ Enterprise_Activity) *Start* $(a) \Leftrightarrow$
defined (preconditions (Algorithm (a)))
$(\forall\ a \in$ Enterprise_Activity) *Finish* $(a) \Rightarrow$
defined (Ending_Status (a)) $\wedge$
Ending_Status $(a) \in$ Ending_Statuses $(a) \wedge$
Ending_Status $(a) \in$ Control_Output (a)
$(\forall$a $\in$ Enterprise_Activity)
Duration $(a) = EP(a) - SP(a)$

3.7 Resource

Resources are enterprise objects (identified by one of their object views) playing the role of a doer in the execution of an activity (Ushold *et al.*, 1995). In CIMOSA, active resources, i.e. system agents or actors, are called functional entities. They are characterised by their set of offered capabilities and their set of executable functional operations. Each functional operation is a basic primitive that the resource can perform upon request. To do this, a call is made to the resource in the form of the operation name and a list of parameters corresponding to the operation arguments. Functional entities can be specialised into three fundamental subclasses using the generalisation hierarchy: *human functional entities* (HFE) for human operators, *machine functional entities* (MFE) for autonomous machines and hardware devices and *application functional entities* (AFE) for software applications (Fig. 1). Each of these sorts inherits properties and axioms of the resource sort and may add its own properties and axioms as illustrated below. Resources can be made of resources

SORT Resource
IMPORT Object_View, Functional_Operation
VARIABLES
  Entity: Object_View
  Provided_Capabilities: Capability_Set
  Operations: *P*(Functional_Operation)
  Comprises: *P*(Resource)
AXIOMS
  $(\forall\ r \in$ Resource) Operations $(a) \neq \varnothing$

SORT HFE
Isa: Resource
IMPORT Competency
VARIABLES
  Competencies: *P*(Competency)

SORT MFE
Isa: Resource
IMPORT Position, REAL
VARIABLES
  Location: Position
  Capacity: REAL

SORT AFE
Isa: Resource
IMPORT Site
VARIABLES
  Host_Site: Site

3.8 Capability set

Capabilities define the abilities required to execute a given task (i.e. an activity) or provided by a functional entity. They are expressed as functional requirements (e.g. to be able to drill holes with diameters between 2 and 13 mm) or technical constraints and limitations (e.g. hole depth < 5 cm for steel-made parts) expressed in the form of a value, a range of values, a list of values or some descriptive text. In CIMOSA, capabilities are grouped to form consistent sets of technical abilities called capability sets. For HFE's, capabilities refer to skills and competencies. Capability sets are very useful to define resource profiles required for performing certain activities.

SORT Capability_Set
IMPORT Resource, Capability
VARIABLES
  Capabilities: *P*(Capability)
AXIOMS
  $(\forall\ cs \in$ Capability_Set) Capabilities $(cs) \neq \varnothing$

3.9 Organisation unit

Organisation units are organisation view constructs used to define responsibility and authority of individuals or units of decision-making in the enterprise. Thus, any organisation unit must equate to a functional entity, usually a HFE. Each functional unit must belong to an organisation cell, indicating its position in the organisation hierarchy.

SORT Organisation_Unit
Partof: Organisation_Cell
IMPORT AFE, HFE, Organisation_Cell, Capability, Text
VARIABLES
  Functional_Entity: AFE $\cup$ HFE
  Job_Description: Text
  Skill_Profile: *P*(Capability)
  Responsibilities: Text
  Authorities: Text
AXIOMS
  $(\forall\ u \in$ Organisation_Unit)
    defined (Job_Description (u)) $\wedge$
    defined (Responsibilities (u)) $\wedge$
    defined (Authorities (u))
  $(\forall\ u \in$ Organisation_Unit) $\exists!\ c \in$ Organisation_Cell /
    Partof $(u) = c$

3.10 Organisation cell

Organisation cells are constructs of the organisation view used to define decision centres into decision levels and build the organisation hierarchy of the enterprise (corporate level, plant/division level, shop/department level, cell level, work-centre level, equipment level). They are used to define responsibility and authority on various parts of the enterprise, i.e. other constructs of the model and especially business processes, enterprise activities, enterprise objects and their object views and resources. Each organisation cell is placed under the authority of a manager which must be a HFE.

SORT Organisation_Cell
IMPORT Event, Business_Process, Enterprise_Activity,
  Enterprise_Object, Object_View, HFE, Resource,
  Capability_Set
VARIABLES
  Manager: HFE
  Process_Responsibility: *P*(Business_Proccess) $\cup$
    P(Enterprise_Activity) $\cup$*P*(Event)
  Information_Responsibility: *P*(Enterprise_Object) $\cup$
    P(Object_View)
  Resource_Responsibility: *P*(Resource) $\cup$
    P(Capability_Set)
  Organisation_Level: {corporate, plant, shop, cell,
    work-centre, equipment}
  Comprises: *P*(Organisation_Unit) $\cup$
    P(Organisation_Cell)
AXIOMS
  $(\forall\ c \in$ Organisation_Cell,$\forall\ v \in$ Comprises (c))
    $\exists\ u \in$ Organisation_Unit $\cup$ Organisation_Cell /
    $v = u$

IV. CONCLUSION

Ontologies are useful means to formally describe relevant knowledge of a given domain of discourse. In this paper, a method for expressing a domain ontology has been introduced and an ontology for enterprise modelling has been presented on the basis of CIMOSA and CEN ENV 12 204 constructs for enterprise modelling.

REFERENCES

AMICE (1993) *CIMOSA: Open System Architecture for CIM, 2nd extended and revised version.* Springer-Verlag, Berlin.

Benjamin, P.C., C.P. Menzel, R.J. Mayer and N. Padmanaban (1995) Toward a method for acquiring CIM ontologies. *Int. J. of Computer-Integrated Manufacturing,* **8**(3): 225-234.

CEN (1995) ENV 12 204: Computer-Integrated Manufacturing. Constructs for Enterprise Modelling, CEN/CENELEC, Brussels, December.

CIMOSA Association (1996) CIMOSA Technical Baseline. CIMOSA Association e.V., Stockholmerstr. 7, D-70731 Böblingen, Germany.

Ehrig, H-D. and B. Mahr (1985) *Fundamentals of Algebraic Specifications.* Springer-Verlag, Berlin.

Fox, M.S. and M. Grüninger (1994) Ontologies for enterprise integration. Proc. Second Int. Conf. on Cooperative Information Systems, Toronto, May. pp. 82-89.

Gruber, T. (1992) Ontolingua: A Mechanism to Support Portable Ontologies. Res. Rep. KSL-91-66, Stanford Knowledge Systems Lab., Final Version, Stanford University, Stanford, CA.

Grüninger, M. and M.S. Fox (1994) An activity ontology for enterprise modelling. Proc. Third IEEE Workshop on Enabling Technologies: Infrastructures for Collaborative Enterprises (WET ICE'94), Morgantown, West Virginia.

Hilger, J. and J-M. Proth (1990) A production Management Language (PML). Proc. IFAC Congress'90, Tallin, Estonia.

Neches, R. *et al.* (1991) Enabling technology for knowledge sharing. *AI Magazine,* **12**(3): 36-56.

Uschold, M., M. King, S. Moralee and Y. Zorgios (1995) The Enterprise Ontology. Res., Rep. Artificial Intelligence Applications Institute, Edingburgh, Scotland, June.

Vernadat, F.B. (1993) CIMOSA: Enterprise modelling and enterprise integration using a process-based approach. In : *Information Infrastructure Systems for Manufacturing* (H. Yoshikawa and J. Goossenaerts, eds.), pp. 65-84. North-Holland, Amsterdam.

Vernadat, F.B. (1996) *Enterprise Modeling and Integration: Principles and Application.* Chapman & Hall, London.

Vernadat, F. and M. Zelm. Business process modelling methods for enterprise integration: IDEF3 and CIMOSA. Proc. Fourth Int. Conf. on Control, Automation, Robotics and Vision (ICARCV'96), Singapore, 3-6 December 1996. pp. 1585-1589.

APPENDIX

The following rules are used in CIMOSA to describe a process workflow:

- Process triggering rules: They are used to start a process by means of events.

WHEN (START WITH event-i) DO EF1

In this case, the process starts with process step EF1 any time that an occurrence of event-i occurs.

- Forced sequential rules: They are used when a process step EF2 must follow another step EF1 whatever the ending status (given by the function ES) of EF1 is. The reserved word 'any' is used in this case (not an ending status).

WHEN (ES(EF1) = any) DO EF2

- Conditional sequential rules: They are used to represent branching conditions in a flow of control.

WHEN (ES(EF1) = end_stat_1) DO EF2
WHEN (ES(EF1) = end_stat_2) DO EF3
WHEN (ES(EF1) = end_stat_3) DO EF4

Spawning rules: They are used to represent the parallel execution of process steps in a flow of control. Two types of spawning rules can be defined:

(a) Asynchronous spawning: When EF1 is finished with status 'value', EF2, EF3 and EF4 are requested to start as soon as they are ready (& is the parallel operator).

WHEN (ES(EF1) = value) DO EF2 & EF3 & EF4

(b) Synchronous spawning: When EF1 is finished with status 'value', EF2, EF3 and EF4 are all requested to start exactly at the same time (SYNC indicates the synchronisation).

WHEN (ES(EF1) = value) DO
SYNC (EF2 & EF3 & EF4)

- Rendez-vous rules: They are used to synchronise the end of spawning rules.

WHEN (ES(EF2) = value_2 AND ES(EF3) = value_3 AND ES(EF4) = value_4) DO EF5

- Loop rules: They are used to execute the same process step(s) several times as long as a loop condition is true:

WHEN (ES(EF1) = loop_value) DO EF1

- Process termination rules: They are used to indicate the end of the process.

WHEN (ES(EF1) = end_stat_x
AND ES(EF2) = end_stat_y) DO FINISH

Using these rules, a process behaviour is said to be consistent if FINISH can be reached from all STARTs and all process steps used in the set of rules belong to at least one path from START to FINISH.

CIMOSA AND ITS APPLICATION IN AN ISO 9000 PROCESS MODEL

M. Zelm and K. Kosanke,

CIMOSA Association e.V., Aachen/Germany

Abstract: Methodologies in enterprise integration and business re-engineering as well as for integrated quality management can be represented very effectively through enterprise modelling. CIMOSA provides a process oriented modelling methodology enabling representation of the modelling methodology and the identification of the enterprise information needed in the different tasks. The paper describes a process model of the ISO 9000 methodology for implementation, usage and maintenance of quality management systems. Application of the model provides an improved understanding of the ISO 9000 methodology and a very comprehensive and easily maintainable quality documentation.

Keywords: CIMOSA, Enterprise Modelling, ISO 9000, Methodology, Process Model, Quality Management.

1. INTRODUCTION

Process oriented modelling can enhance significantly the representation and thereby the usability of general methodologies. Both the information needed for the different tasks and the results of the tasks can be explicitly identified in a process model of the methodology, thereby guiding its users. CIMOSA provides a process oriented modelling methodology covering the enterprise system and enabling the collection and representation of enterprise information (CIMOSA Association e.V. 1996, Kosanke, *et al.*, 1996). CIMOSA allows to model the ISO 9000 methodology in terms of business processes which specify the requirements for a quality management system.

Following a brief presentation of the CIMOSA concepts and modelling methodology the paper describes a process model of the implementation of an ISO 9000 quality management system starting with the establishment of the relevant enterprise domains and the identification of the global information exchange between the domains. The model covers both introduction and use of ISO 9000 standards (DIN EN ISO 9001, 1994; Gumpp/ Wallisch, 1995).

Process modelling of QM systems has been investigated elsewhere (Mertins, *et al.*, 1996). Here a process model of ISO 9000 was developed with the object-oriented 'Integrated Enterprise Modelling' (IEM) methodology. IEM describes functions and information in one integrated model of the enterprise. The model representation is based on the object classes 'Product', 'Order' and 'Resource' and can handle additional quality specific attributes.

2. CIMOSA PROCESS BASED ENTERPRISE MODELLING AND ENGINEERING

The CIMOSA modelling framework supports process oriented modelling of enterprise operations, providing different views (function, information, resource and organisation) on the model of a particular enterprise. The concept of views allows the business user to work with a subset of the model rather than with the complete model thereby reducing the model complexity for his particular area of interest. CIMOSA model engineering is using the concept of enterprise domains, representing their functionality by a set of domain processes, business processes and enterprise activities. Domain processes communicate with each other via events and process results. The connecting control flow between enterprise activities is represented by a set of behavioural rules. An enterprise activity has inputs and outputs of function, control and resource. Inputs/outputs are views on enterprise (information) objects.

3. THE ISO 9000 PROCESS MODEL

The implementation of an ISO 9000 based quality management system has to be supported by a methodology which guides the implementation team. In the following a CIMOSA based process model for implementation, use and maintenance of a quality management system complying with the ISO 9000 standard is developed. The quality management relevant documentation will serve for audits, certifications and as a proof that the enterprise is well under control. Whenever relevant changes in the enterprise occur which may impact the operation and/or the quality system, it is necessary to make a re-certification. This leads to a new version of the quality handbook.

3.1 Model Overview

Figure 1 shows the relations between quality management, enterprise management and operation. The processes occurring in the quality management system are described in the domain DM2 'Quality Management'. Domain DM1 'Management' is concerned with the management processes. Domain DM3 contains the operational enterprise processes. Modelling focus is on the introduction, usage and maintenance of an ISO 9000 compatible quality management system and its interaction with the other enterprise processes. Fig 1 also shows the information exchanged between the domains.

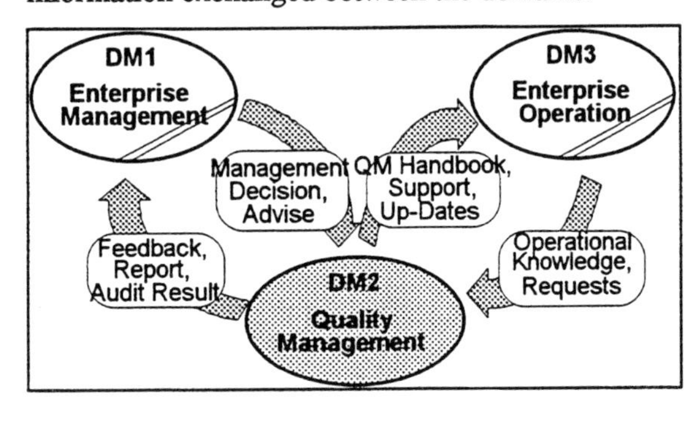

Fig. 1: ISO 9000 Process Model Overview

The overall process scenario starts with a management decision to introduce an ISO 9000 compliant quality management system. In the domain 'Quality Management', two processes take place to introduce ISO 9000: At first all current enterprise procedures and supporting documents - specifications for the processes manufacturing, test, production control - are collected, evaluated and registered. Then, the implementation team defines the Quality Management Elements which are required by the ISO 9000 standard and at the same time looks at the specific characteristics of the enterprise which are represented in documents, files, procedures, processes - generally speaking the enterprise knowledge. As a result of this task, a documentation of the enterprise processes and the QME's relevant for this particular enterprise is generated and documented in the quality handbook.

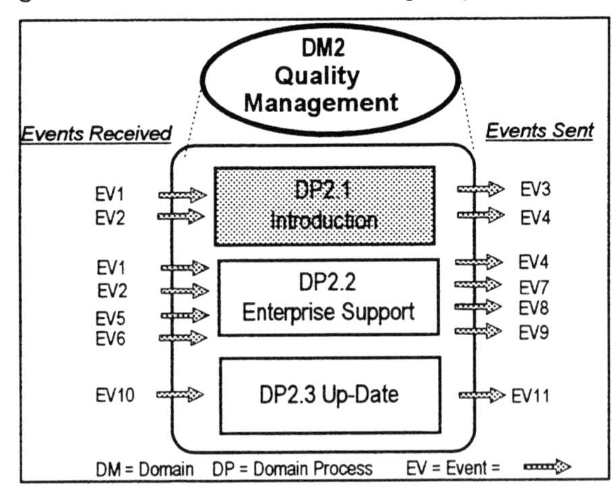

Fig 2: Domain 'Quality Management'

	List of Events Received	List of Events Sent
DP2.1	EV1 = Management Decision EV2 = Operational Knowledge	EV3 = QM Handbook EV4 = Report
DP2.2	EV1 = Management Decision EV2 = Operational Knowledge EV5 = Feedback from Operation EV6 = Support Request	EV4 = Report EV7 = Support EV8 = Document EV9 = Audit Result
DP2.3	EV10 = Change Request	EV11 = QM Handbook Up-Date

Table 1: Events of Domain 'Quality Management'

The processes to develop and to use a quality handbook are modelled in the domain processes DP2.1 - DP2.3 shown in Figure 2. For the list of events which are exchanged between domains see Table 1.

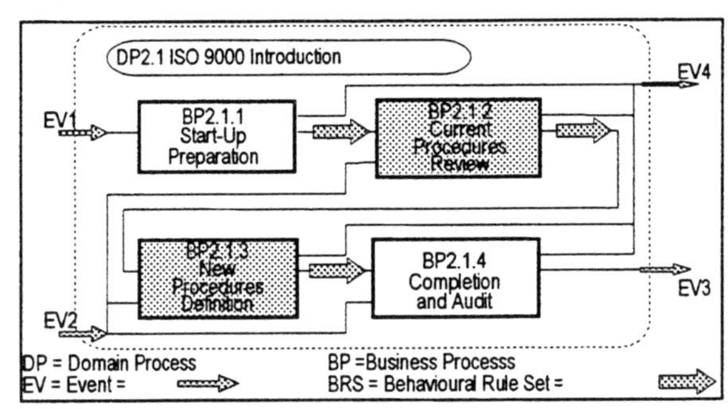

Fig. 3: Breakdown of Domain Process DP2.1 'Introduction

The next level of details for the domain process DP 2.1 'Introduction' is shown in Fig 3. The introduction of ISO 9000 is triggered by a management decision and modelled as a series of business processes starting with establishing the QM strategies and installing the implementation team. In the next process, 'Current Procedures Review' the current enterprise procedures and supporting documents are collected, evaluated and registered.

The following process 'New Procedures Definition' is concerned with the definition of the new quality management procedures as they apply to the particular enterprise. The individual process steps are modelled as activities with the main inputs being operational knowledge in terms of the relevant subset list of QM procedures. The output of each activity is a formal document, a QME fulfilling the ISO 9000 standard requirements. The aggregated output of the process 'New Procedures Definition' forms the major part of the quality related documentation of the enterprise, the QM Handbook. The process 'Completion and Audit' completes the methodology of ISO 9000 Introduction

3.2 Decomposition of Business Processes

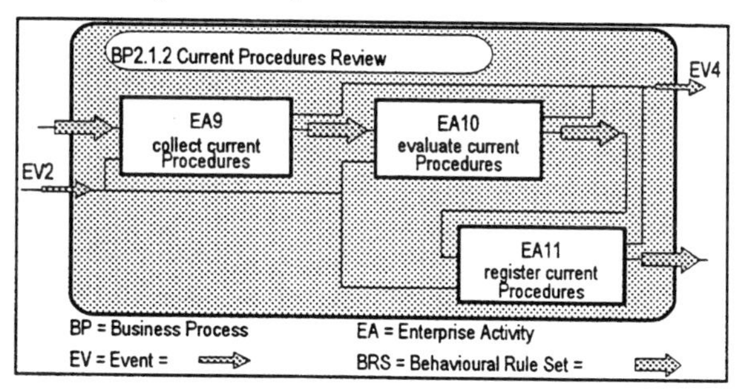

Figure 4: Business Processes BP2.1.2 'Current Procedures Review'

The decomposition of business processes leads to the identification of enterprise activities. Figures 4 and 5 show the contents of the processes 'Current Procedures Review' and 'New Procedures Definition' respectively. These processes consist of activities EA9 to EA29 and the connecting flow of control. The process steps (enterprise activities) of the 'Current Procedures Review' process describe the tasks to collect all quality relevant current enterprise procedures, to evaluate and to register them. The process 'New Procedures Definition' is concerned with selecting quality elements from the list of the ISO 9000 standard, defining the QME's relevant for the particular enterprise and adjusting them to the quality objectives of the enterprise. The so defined QME's are the outputs of the activities and contribute to the build up of the ISO 9000 quality documentation of the enterprise, the quality handbook. Details of selected activities, their inputs, outputs and functionality are provided in Table 2.

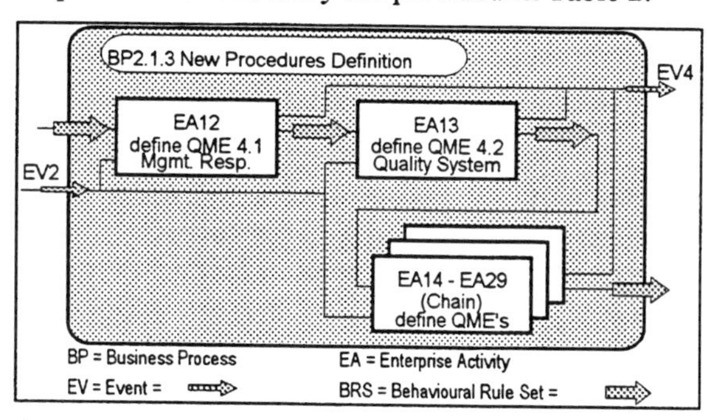

Figure 5: Business BP2.1.3 'New Procedures Definition'

This table provides examples of two enterprise activities with their inputs, outputs and functionality. EA11 'register current procedures' describes the last activity of the process 'Current Procedures Review'. The activity produces a list of quality relevant procedures needed to establish a new documentation compliant with ISO 9000 and an enterprise organisation chart, required in part II of the QM handbook.

Table 2: Inputs, Outputs and Functionality of selected Enterprise Activities

EA-ID/Name		
Input	**Functionality**	**Output/Result**
EA11/register current procedures (part of the process 'Current Procedures Review')		
- set of relevant enterprise procedures and supporting documents.	registration of QM relevant enterprise procedures and supporting documents.	- list of QM relevant enterprise procedures and supporting documents - QM plan - enterprise organisation chart (part II of QM Handbook).
EA20/define QME 4.9 Process Control (part of process 'New Procedures Definition')		
- relevant subset of list of QM relevant enterprise procedures and supporting documents - list of current processes	identification and planning of production, installation and servicing processes which directly affect quality establishment, documentation and maintenance of relevant procedures.	- process control procedure (QME 4.9 in part IV of QM Handbook).

EA 20 'define QME 4.9 Process Control' describes one activity of the process 'New Procedures Definition', namely to identify and - if not available- plan the relevant procedures to insure appropriate control of the production, installation and servicing processes of the enterprise. The activity produces the process control procedure(QME 4.9), which is an essential portion of the QM handbook.

The other activities to introduce ISO 9000 are similarly structured: They use the relevant current procedures as input and produce a new procedure - a QME of the QM handbook - as output. The entire model has been implemented on a modelling tool (Gaches, 1996).

3.4 Model Structure - Information Analysis

Information and information flow in the enterprise are analysed by considering all information objects and identifying their origin/source (where produced) and their usage/destination (where used). An example of information objects related to management, to operation and to the QM handbook is shown in Table 3.

Summarising and structuring all information of the enterprise results in the semantic information model as shown in Fig 6. It explains the relationships between the information objects of the two domains management and operation and the parts of the quality handbook. The information model is used to visualise and to understand the relationships between enterprise objects. This representation could also serve as a prerequisite or input for the design and structuring of an enterprise data base.

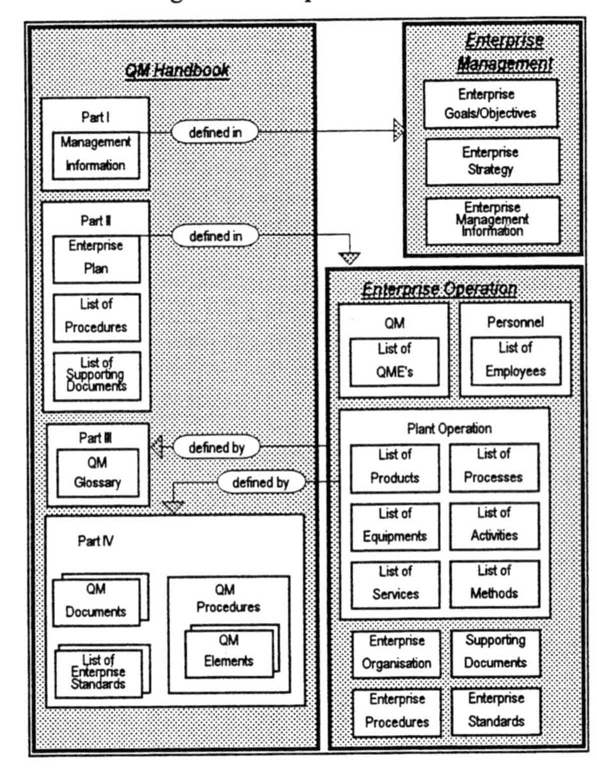

Fig 6: Semantic information model

Table 3: Information Objects used and produced (Examples)

Information Object	Where Used	Where Produced
Management related		
enterprise strategies and goals	BP2.1.1	DM1 Management
management commitment	BP 2.1.1	-"-
management statement on quality policy	BP 2.1.3 (EA12)	-"-
Operation related		
enterprise organisation chart	BP2.1.1 BP2.1.2 (EA10)	DM3 Operation
current enterprise procedures and supporting documents	BP2.1.2 (EA10)	-"-
list of current processes	BP2.1.2	-"-
list of current customer-supplied products	BP2.1.2	-"-
QM Handbook		
QM strategy document (part I of QM handbook).	BP2.1.1	DM1 Management
QM plan (part II of QM handbook).	BP2.1.1	BP2.1.2 (EA11)
list of QM relevant enterprise procedures and supporting documents (part II of QM handbook).	BP2.1.3 (EA12 -EA31)	BP2.1.2 (EA11)
set of new procedures (QME's in part IV of QM handbook).	BP2.1.4	BP2.1.3 (EA12 - EA31)
list of rel. enterprise standards (part IV of QM handbook)	BP2.1.1.	BP2.1.1

Table 4: Process-QME-Matrix

QME	4.1	4.2	4.3	4.5	4.6	4.7	4.8	4.9-4.14	4.15	others tbd
DP1 'Order Handling'	x	x								
EAop1 'enter Cust. Order'			x							
EAop2 'control Production'							x			
EAop3 'purchase Parts'			x		x	x				
EAop4 'produce Product'				x				x		
EAop5 'store Product'						x			x	

4. AN EXAMPLE

Applying the methodology has to take into account the enterprise operation and its relevant processes, information, resources and organisational aspects.

Relations between the operational processes and quality management have to be made explicit and easily maintainable. An enterprise model representing the enterprise operation is the easiest way of documenting and maintaining these relations. This is illustrated through an example of an enterprise modelled with one domain process 'Order Handling', which consists of five enterprise activities (see Figure 7).

A customer order arrives, is registered in the order entry department and sent to production control. Production control starts production and requests purchasing to order parts or material. Following production the product is sent to the storage area, ready for shipment to the customer.

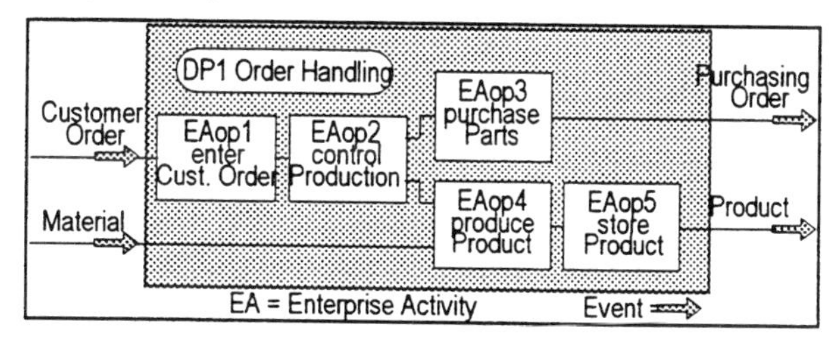

Fig. 7: Domain Process 'Order Handling

Fig. 7 shows the process model, consisting of five activities and their flow of control. Each activity is described with a modelling template. The template contains the responsibility for this activity as one attribute. In addition, one or more QME may be required to control a process. QME's are also modelled in terms of attributes in the modelling template. In order to determine the ISO 9000 quality criteria required for the overall process, the relations between the processes or activities and the required quality management elements have to be established. This is done by introducing a matrix table of enterprise processes and QME's (see Table 4).

The list of enterprise processes and activities is shown on the vertical axis of the matrix. The horizontal axis shows the Quality Management Elements. The matrix indicates which QME's are required to control a process or an activity. A QME is modelled as one entry (attribute) in the modelling template of an enterprise activity or domain and business process.

In Fig 8 the template of the enterprise activity EAop4 'produce Product' is shown. It indicates the responsibility of a person for the design and for the operation of the activity. In addition, the template contains the QME's which refer to this activity. QME's control the activity through the control input. Both appear as attributes of the template in the model of the activity

5. DOMAIN PROCESSES 'ENTERPRISE SUPPORT' AND 'UP-DATE'

The process 'Enterprise Support' describes the *utilisation* of the quality management system to support the enterprise in all quality matters, to run internal or external certifications or to conduct audits. As shown in Fig 9 the process employs three sub-processes 'Certification', 'Audit' and 'General Support'.

Header	
Name:	Produce Product
Identifier:	[EAop4]
Design **Responsibility:**	**A. Meier**
Operation **Responsibility:**	**B. Müller**
Body	
OBJECTIVES: Meet cost and quality target	
CONSTRAINTS: Capacity limits: no hiring	
DESCRIPTION:	
Describes the current operation of the enterprise to produce products	
INPUTS:	
Function Input:	Material
Control Input:	**QME 4.9-4.14**
Resource Input:	Operation Equipment
OUTPUTS:	
Function Output:	Final Product
Control Output:	

Figure 8: Example of a modelling template

All three sub-processes need operational knowledge - new enterprise procedures, documents, test reports - as inputs. The process 'Certification' describes the activities to certify a new operational process, a new production line or new equipment. This process is triggered by a management decision (for List of Events see Table 1) and produces a certification document, which may be presented to the enterprise management or to an external certification agency to receive a certificate. The process 'Audit' describes the tasks of conducting an audit- installing the audit team and auditing operation documents and data for compliance with the specified requirements. The process is triggered by a management decision and produces an audit report. The process 'General Support' describes any form of support of the quality organisation given to the enterprise. Examples are quality training and education of employees, or the introduction of new statistical techniques. The process is started when a support request arrives from the enterprise operation and produces relevant documents in a report.

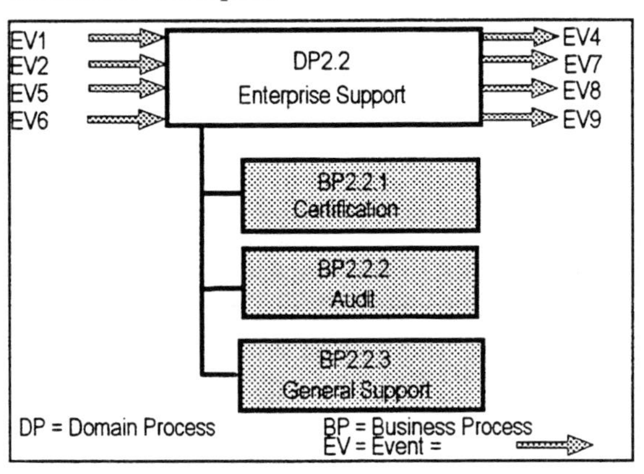

Fig. 9: Support Processes

The domain process 'Up-Date' which is not further decomposed here, is concerned with the modification or an update of the quality management system. The change process itself or the *method* to modify the quality management system is modelled. The process starts with a change request. The quality management system must be analysed including potential impacts from the enterprise operation and re-defined. The output of the process is an updated QM handbook. Examples of change requests could be new rules for workers safety or new environmental guidelines issued by governmental agencies. The CIMOSA structuring concept of domains enables easy extension of the enterprise model.

6. SUMMARY

CIMOSA provides a process model of the ISO 9000 methodology thereby supporting introduction, usage and update of a quality management system. The CIMOSA methodology offers several benefits: user oriented representation of the methodology, identification of all required and created information, easy maintenance of quality documentation, use of international standards in enterprise modelling and the concept of model views to reduce model complexity. The structuring principle of domains helps to reduce complexity even further. New legislative or environmental requirements (ISO 14000 series) can be easily implemented in the overall enterprise model and the QM handbook.

The model can be customised to specific requirements and can fulfil the needs of fast inter-organisational changes of virtual enterprises. The presented model illustrates implementation, operational use and maintenance of a quality management system in a particular company. The application installed on a modelling tool (Gaches, 1996) provides as output a comprehensive electronic documentation; the Quality Management Handbook in hypertext format.

More work is required for a full implementation of an integrated, process oriented enterprise model covering all aspects of enterprise management, quality, environment and operation. CIMOSA modelling will provide a valuable methodology to achieve this goal. Another pre-requisite will be the availability of powerful, user oriented process modelling tools. It seems very obvious that complex, methodologies can be much better represented by a process model. In addition, the hypertext documentation further enhances the usability and maintainability compared with the still widely used paper documentation.

REFERENCES:

CIMOSA Association e.V. (1996). *CIMOSA Open System Architecture for CIM, Technical Baseline.* Vers. 3.2, Private Publication.

DIN EN ISO 9001 (1994), *Quality systems - Model for quality assurance in design/development, production, installation and servicing.*

Gumpp G.B. /F. Wallisch (1995). *ISO 9000 entschlüsselt.* mi verlag moderne industrie.

Gaches R.(1996) CimTool Version 2.7, Product Description.

Kosanke, K. F,B. Vernadat and M. Zelm: (1996). *CIMOSA: Industrial Applications of Enterprise Modelling.* ASI'96, Toulouse

Mertins, K. M. Schwermer and F.-W. Jäkel (1996). *Confidence in a network of co-operating companies - QM documentation for certification.* DIISM96, Eindhoven

COOPERATIVE ENTERPRISE PLATFORM

Francisco Edeneziano Dantas Pereira [1]
João Maurício Rosário [2]
Marcius Fabius Henriques de Carvalho[1]
Mauro Ferreira Koyama[1]
Olga Fernanda Nabuco Araujo[1]

(1) Fundação Centro Tecnológico para Informática - CTI Rodovia SP-65, Km 143.6
13089-500 Campinas - SP, Brazil
e-mail: dantas@ia.cti.br
(2) Departamento de Projeto Mecânico, Faculdade de Engenharia Mecânica,
Universidade Estadual de Campinas - Unicamp
C.P. 6122 13083-970 Campinas - SP, Brazil

Abstract: This article presents a cooperative enterprise environment to carry out research, teaching, training and development on integration techniques and technologies to be transferred to small and medium enterprises (SMEs) in order to narrow the gap of them for innovation. The environment is the result of the PIPEFA project: Industrial Platform for Research, Teaching, and Training in Automation involving CTI and Unicamp.

Keywords: Cooperative enterprises, Inter-networking enterprises, enterprise integration, system integration, system automation.

1. INTRODUCTION

To increase their competitiveness and to survive in rapid changing market the companies are making efforts not only toward manufacturing automation and integration; they also are putting emphasis on total manufacturing business systems.

Total manufacturing business considers integration between value-adding activities along the value chain. The objective of integration should not be restricted to a single plant or production facility, but it must include chains or networks of production, physical distribution, customers and suppliers.

Therefore, integration ranges a set of enterprises actuating as a group of cooperating processes making arise a new concept *Cooperative Enterprise*: a dynamic set of enterprises working or acting together for a common purpose or benefit to reach and sustain a competitive advantage in the global market.

The *Cooperative Enterprise* concept establishes a new paradigm in which companies go to a cooperative relationship from short term to long term and from confrontational to cooperative partnerships (Bessant, 1994). Named as extended enterprise concept (Browne, 1995), it means that core product functionalities are provided separately by different companies that come together for the purpose of providing a customer-defined product/service. The organizational (generic, sequential and reciprocal interdependencies between activities) are also considered (Migliarese, 1996).

To apply this new concept intra-enterprise and inter-enterprise networking are required to support and to make the cooperation viable. This cooperation ranges management, planning and operational activities into the Cooperative Enterprise environment.

In order to have an effective cooperation, as for example, on co-designing a new product, involving different partners, the increasing interdependence between all the related activities must be considered. The objective of this article is to present the efforts that have been doing during the PIPEFA project to study and develop techniques that allow an effective enterprise cooperation.

The following sections describe the efforts developed to build a platform which aims to accomplish the presented concepts. Section 2 presents PIPEFA framework. PIPEFA composition is described in section 3. Section 4 explains PIPEFA organization. Section 5 shows the system behavior. Finally, section 6 concludes the article.

2. PIPEFA FRAMEWORK

The PIPEFA framework is shown on the Fig. 1. It outlines the interrelationship of the several cooperative enterprise partners. The framework depicts the PIPEFA environment, based on the overall structure of any enterprise (CIMOSA Technical Baseline, 1994). The environment is stratified on the following three plans: decision, information and physical.

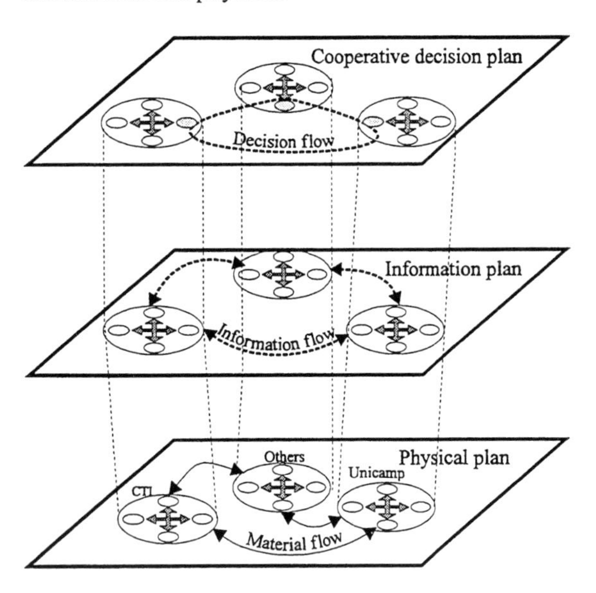

Fig. 1. PIPEFA framework.

The Cooperative and decision plan considers dynamic and holistic production management concepts for the whole cooperative enterprise. The coordination of the decision making process takes place in order to integrate globally distributed production processes of value-adding logistic chains. The decision considering the best for the Cooperative Enterprise is termed cooperative decision. It must regard the particularities, constraints and objectives of both: individual partners and the Cooperative Enterprise. The prime (fundamental) element of cooperative decision is the global coordination related to involved enterprises' available resources. This is only possible by the integration of information.

The Information plan concerns with individual partners and cooperative enterprise information. The coordination of information exchanges for the Cooperative Enterprise must regard local and shared information needs. It must support the cooperative decision making, promote the enterprise integration via information sharing and assist groups in eliminating no-value adding activities in manufacturing and processing of products. It enables partners and the Cooperative Enterprise to be more customer responsive. This plan is contemplated in the PIPEFA project as an information technology platform. The purpose of the platform is to transform a highly distributed heterogeneous environment into one which looks centralized (information can be transparently accessed wherever it is stored) and homogeneous (standard languages are used for data communications, data presentation, data access, machine access etc. (Kosanke, 1955).

The Physical Plan is composed of physical systems which transform raw material, assembly parts, transport and storage products. Material flows integrate physical systems. In order to satisfy time based competition these systems should be flexible enough to produce an increasingly product variety on a smaller lot size. They should also operate in accordance with manufacturing strategies defined in conformity with enterprise global strategy. The interactions - parts exchanges - between the plants should be guided by a logistic system. This system considers decentralized production management approach, as for example DECOR (Kuhlmann, 1996).

3. PIPEFA COMPOSITION

In order to reach a globally distributed manufacturing network, in accordance with Cooperative Enterprise paradigm, several requirements must be met. PIPEFA project considers physical (connectivity), application (interoperability) and business integration levels (ESPRIT, 1989), involving remote sites.

The first guideline adopted to meet the mentioned requirements is to use open systems concepts by partners participating in the project.

The PIPEFA project is a joint effort involving the Automation Institute of the Technological Center for Informatics Foundation - CTI and Laboratory of Integrated Automation and Robotics, Faculty of Mechanical Engineering, State University of Campinas - Unicamp.

The project started in 1995 and has the collaboration of the *Laboratoire d'Ingéniere Intégrée des Systèmes Industriels* (LIISI), Toulon, France. Its first purpose is to obtain and develop concepts, techniques and technologies on enterprise integration and then to transfer these knowledge to SMEs in order to support them on the current market requirements. In this sense the project develops a Platform which allows research, teaching and training on cooperative enterprise.

The Platform creates a cooperative enterprise environment implementing an inter-networking of partners which interact on close collaboration supported by Internet. It is being gradually implemented with low operation costs incorporating the most recent techniques and technologies in integrating infrastructure for manufacturing systems. To be successful, the platform allows the incorporation of new partners and is capable of being reproduced by entities (enterprises, institutions etc.) interested on it.

3.1 Technical Rationale

The platform has been created as an enterprise engineering process based on system life cycle concept (ESPRIT, 1989). In the same sense three levels of integration: physical, application and business integration (ESPRIT, 1989) are also considered to be accomplished by all the partners. Business integration refers to the coordination of operational and control processes in order to satisfy business objectives. Application integration is concerned with the usage of information technology to provide interoperation between manufacturing resources (Zwegers, 1995).

This project has chosen "The Reference Model for Computer Integrated Manufacturing (CIM)" (Williams, 1989) as a reference model. Based on the model partners can map their interest areas and expertise in the project and also to derive particular models. The model has six hierarchical levels, nine plant functional entities and eight external entities. The levels are: Equipment, Station, Cell, Section, Facility and Enterprise The functional entities are: Order Processing, Production Scheduling (PS), Production Control, Raw Materials Control, Procurement, Quality Assurance, Product Inventory Control, Product Cost, Accounting and Product Shipping Administration. The external entities are: Marketing & Sales, Corporate R. D. & E, Suppliers, Vendors, Customers, Transport Companies, Accounting and Purchasing.

4. PIPEFA ORGANIZATION

In order to get together CTI and Unicamp expertise, the functional entities Production Scheduling and Production Control have been chosen.

The Production Control is installed at Unicamp and Production Scheduling at CTI. These institutions are geographically separated and interlinked by Internet.

4.1 Production Scheduling

It determines production scheduling specifying what, when and how much parts and products must be produced on a period of time and send the correspondent production orders to shop floor. Performed by production planners aided by a decision support system (DSS), the scheduling takes place an on-line cooperative process, involving both CTI and Unicamp. On the DSS, demand is simulated using forecasting models considering known and predicted orders (Carvalho, 1996). The system is executed every time a new sale order or shop floor status information has arrived. The DSS current behavioral model is based on Ward & Mellor approach (Ward and Mellor, 1985) as shown on the Fig. 2. Its data model is based on Entity-Relationship approach.

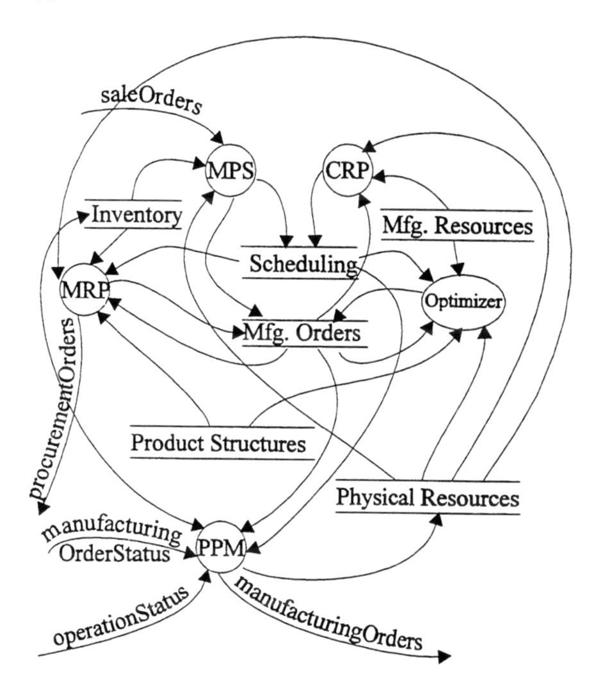

Fig. 2. DSS behavioral model.

The DSS is composed of the ensuing modules: Master Production Schedule (MPS), Material Requirements Planning (MRP), Capacity Requirements Planning (CRP), and an Optimizer. It also includes configuration and actualization functions which provide needed information from others functional and external entities. The DSS fits Section level and the function Production Scheduling (Williams, 1989).

The mentioned modules are being implemented in two software: Prodcon (Passos *et al.*, 1996) and Retra (Optimizer) (Carvalho and Silva, 1996).

4.2 Manufacturing shop floor

The manufacturing shop floor is an assembly line which operates on stocks for purchased items and searches to limit them on the basis of planned orders and predicted demand. It meets the levels Cell, Station and Equipment accomplishing the Production Control function (Williams, 1989).

It is being modeled based on Ward & Mellor approach as shown on the Fig. 3 and Entity-relationship data model. Also, models based on Petri nets (Peterson, 1981) and GRAFCET IEC 848 are being used to describe and solve synchronism problems on shop floor.

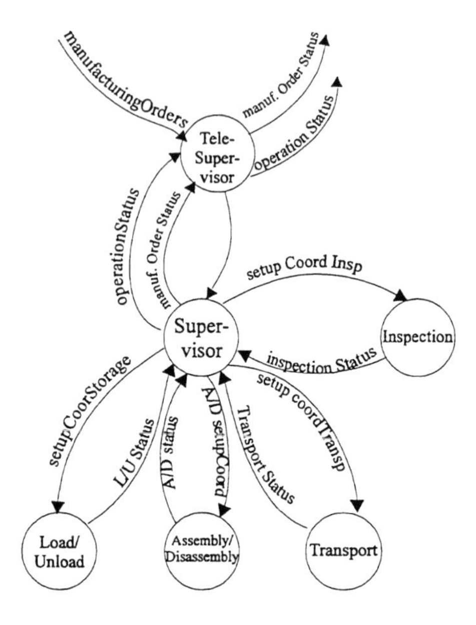

Fig. 3. Shop floor behavioral model.

The shop floor operates on a semi-automated mode and uses LEGO (TM LEGO Group) parts (cubes and boards) as raw material. It can produce a mix of twenty six products with different lot size. It also enables to explore environmental benign production concept (Browne, 1995) taking life cycle product concept (ESPRIT, 1989) into account.

This functionality is carried out by the following operations: load/unload, assembly/disassembly, transport and inspection. These operations take place on four work centers interlinked by a transport system (conveyor), a Cell Supervisor and a tele-supervisor (technological resources).

Two work centers transform parts' shape and characteristics performing assembly and disassembly operations. The third work center verifies whether the finished good satisfies or not quality standards. The last one executes load and unload operations. The transport system consists on a controlled conveyor.

4.3 Information Technology (IT) platform

Information systems contain organized, complete, updated and shareable information representing a common way through which heterogeneous systems (federated) can talk to each other (Rabelo, 1995). These systems support enterprise integration.

In this sense PIPEFA's IT platform supports shop floor and PS integration, enables coordination and knowledge sharing by work teams. It allows information exchanges and provides groupware services (electronic mail and teleconferencing). Finally, it makes shop floor images available via Internet and will support close collaboration to any project partner, as shown on Fig. 4.

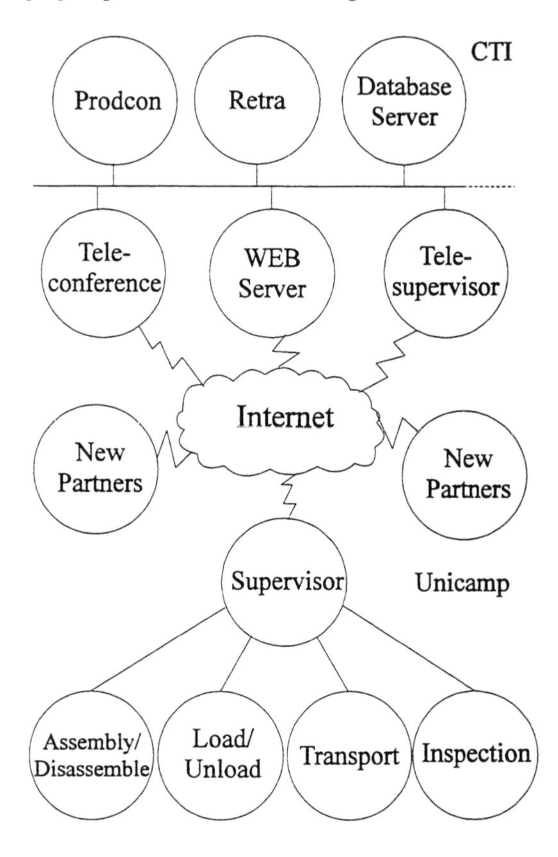

Fig. 4. PIPEFA Information Technology platform.

At CTI, the platform includes stations running PS programs, Prodcon (Santos, 1996) and Retra (Carvalho, 1996), on a client-server mode. A PS general database is composed of three modules: Basic, Production Planning Monitoring (PPM) and Planning. Basic module provides information coming from entities as Marketing & Sales, Corporate R. D. & E. (work centers, product structures, manufacturing processes, clients, vendors, suppliers and sales forecast). PPM keeps tracking shop floor information. The Planning module contains MPS and MRP algorithms. The Retra optimization function is being integrated on it. This program processes information in order to optimize production systems providing a better production schedule in terms of the available resources and finally, sends a new production schedule to the database.

The database management system is being developed using SQLWindows for implementing the clients and SQLBase for building the server. The server runs on an Unix station and the clients run over Windows 95 stations.

Tele-supervisor provides a Production Control window running at CTI. It works together, via Internet, with a supervisor executing at Unicamp. Tele-supervisor provides man-machine interface, shows pictographically shop floor status information and supports decision making referred to the shop floor operation. It has a proprietary database with selected image of the shop floor operation database. This module exchanges information with database through SQL primitives. On this way, tele-supervisor reduces reaction time referred to production system.

Supervisor runs in real-time, but tele-supervisor does not support real-time control. Both are connected via Internet. Tele-supervisor has the same resources as Supervisor does (man-machine interface with pictographic images from shop floor) running over Wizcon 5 SCADA system in Windows NT platform.

At Unicamp, a Supervisor System (Cell Controller) runs in OS/2 supporting Shop Floor real-time monitoring and control. This system interacts with tele-supervisor via Internet and exchanges information with PLCs trough serial lines (RS-232 and RS-485). Nowadays, the PLCs exchange data with the supervisor using proprietary protocols, but soon, equipment providing standard protocols, as for example, MMS (Manufacturing Message Specification) will be applied. All the actuators and sensors are controlled by PLCs. Fieldbus solution based on PROFIBUS standard DIN 19245 and international standards as proposed by IEC 65C are being considered.

In parallel to these exchanges, a teleconferencing is being implemented in order to provide a service that fits decision making based on consensus. This teleconferencing uses CU-SeeMe software running over Windows 95.

5. SYSTEM BEHAVIOR

Production process starts with Production Scheduling activities. MPS receives sale orders and estimated sales forecasting, verifies inventory and physical resources and generates factory orders. MRP explodes these orders into detailed manufacturing orders. Retra uses both orders, physical resources, product structure, manufacturing processes and time intervals to provide optimized solutions. These solutions show on time and delayed quantities to be met and machine utilization per period. CRP module makes loading machines balancing and elaborates a schedule per machine.

Based on all these information, production planners consolidate factory orders. They store them in the PS database and send them to shop floor via PPM.

Cell level is being implemented by Supervisor and Tele-supervisor modules. These modules monitor the shop floor, in an automatic mode,. Supervisor runs at Unicamp and the latter at CTI. Tele-supervisor receives through PPM module, factory and manufacturing orders and sends them to the Supervisor. It also receives resources status, raw material availability, finished and rejected products etc. from the Supervisor and sends them to PPM. These information exchanges enable to alter production plans and allow to establish new goals. Tele-supervisor maintains a shop floor selected data mirror and carries out the interface between Production and Cell levels. It also can provide information to Manufacturing Engineering and other functional and external entities.

Supervisor monitors, controls and coordinates tasks sequencing performed by the stations, on real-time. It provides man-machine interface, collects process data, keeps aware about each part in the stations and status information from the stations. These information are maintained in a local database named Operational database.

The functions inspection, assembly/disassembly, loading/unloading and transport are carried out by five stations under Supervisor control. At this level information are processed by commands, as for example: read, write, start, stop, load, upload etc. Stations report to Supervisor status information from the controlled equipment and themselves.

6. CONCLUSION

This article has presented PIPEFA project: a joint effort to build a research and teach environment for Cooperative Enterprise support. The project involves CTI and Unicamp. It also has the collaboration of LIISI, France. The infrastructure allows the participation of new partners.

The levels Equipment, Station, Cell and Section are contemplated by the environment. The three first levels are located at Unicamp. The two latter are located at CTI. The Cell level is implemented on distributed mode. It comprises a Supervisor and a tele-supervisor modules using the Internet support. This support enables to implement the Cooperative Enterprise concepts.

The environment purpose is to support teaching, training and development on enterprise integration concepts. The current infrastructure is acting as a catalyst between the teams, promoting cooperation and diffusion of integration concepts into academic and industrial environment.

REFERENCES

Browne, J., P. J. Sackett and J. C. Wortmann (1995). Future manufacturing systems - Towards the extended enterprise. *Computers in Industry*, Vol. 25, Num. 3, pp. 235-254.

Carvalho, M. F. and O. S. Silva Filho (1996). Two-Stage Strategic Manufacturing Planning System: A Practical View. Proceedings of the 12th CARS & FOF'96, England. Pp. 574-580. Edited by Raj Gill and Chanan S. Syan. London, England.

CIMOSA Technical Baseline (1994). CIMOSA Association e. V. , Stockholmerstr. 7, D-70731 Böblingen, Germany.

ESPRIT Consortium AMICE, 1989. *Open Systems Architecture for CIM*, **Vol.1**, Springer-Verlag, Berlin.

Kosanke, K (1995) CIMOSA - Overview and status. *Computers in Industry*, Vol. 27, Num. 2 pp. 101-109.

Kuhlmann, T., Lamping, R. Almeida, L. T., J. Hynyen, C. Lischke, F. Gomes Lopes, M. Ganzer, K. Focke (1996). The DECOR - Toolbox. ESPRIT Conference on Integration in Manufacturing (IiM), Galway, Ireland.

Migliarese, P. and E. Paolucci. (1995). Improved communications and collaborations among tasks induced by Groupware. *Decision Support Systems*, Vol. 14, pp. 237-250.

Passos, C. S., E. Teixeira, M. Rodrigues, R. Haddad, K. Bottini, E. Cavalcante and C. Cavalcante (1996). SIGEP: Um Sistema de Planejamento e Controle da Produção da Nova Geração. Anais do *I Simpósio de Automática Aplicada*, pp. 73-78. SBA, Brazil.

Peterson, J. L. (1981). *Petri Net Theory and the Modeling of Systems*, Prentice-Hall, Englewood Cliffs, NJ.

Rabelo, J. R. and L. M. Camarinha-Matos (1994). Control and Dynamic Scheduling in Virtual Organization of Production Resources. *IFIP WG5.7 Working Conference on Evaluation of Production Management Methods*, pp. 141-150. IFIP Transactions, B-19.

Ward, P. T. and J. Mellor (1985). Structured Development for Real-Time Systems. **Vols. 1 and 2**. Yourdon Press, New York.

Williams, T. J. (1989). *Reference Model for Computer Integrated Manufacturing, A Description from the Viewpoint of Industrial Automation.* Instrument Society of America, Research Triangle Park, NC, USA.

Zweger, A. J. R and T. A. Gransier (1995). Managing re-engineering with the CIMOSA architectural framework. *Computers in Industry*, Vol. 27, Num. 2, pp. 143-153.

LEGO is trademark of LEGO Group, SQLWindows and SQLBase are marketed by Gupta, Windows 95 and Windows NT are trademark of Microsoft, Wizcon 5 is trademark of PCSoft International, CU-SeeMe is sold by White Pine and OS/2 is trademark of International Business Machine.

IMPLEMENTATION OF A DYNAMIC CONTROL SYSTEM FOR SCARA ROBOT

Jang M. Lee[1], M.C. Lee[2], K. Son[2], M.H. Lee[2] and S.H. Han[3]

1 Dept. of Electronics Eng., Pusan National University, Pusan 609-735, Korea
2 Dept. of Mechanical Eng., Pusan National University
3 Dept. of Mechanical Eng., Kyungnam University

Abstract: A control system for the SCARA robot is designed for the implementation of a dynamic control algorithm. This study focuses on the use of DSPs in the design of joint controllers and interfaces in between the host controller and four joint controllers and in between the joint controllers and four servo drives. To demonstrate the performance and efficiency of the system, a robust dynamic algorithm is implemented for a faster and more precise control. Experimental data for the proposed algorithm are presented and compared with the results obtained from the PID algorithm.

Keywords: Dynamic control system, PID, Robustness, DSP

1. INTRODUCTION

Robots are being more frequently utilized in factories to alleviate workers in laborious and time-consuming tasks. Currently, noticeable improvement in speed and precision makes robots more popular tools in manufacturing processes where highly robust and reliable operations are required. Many researches have been performed through the computer simulation to find a suitable control algorithm for tasks specified at a high level. As a result, many control algorithms have been developed (Lewis et al, 1993; Spong and Ortega, 1994; Song et al, 1994; Shyu et al, 1996), however, few are implemented in the real system. The main difficulty lies in the extended computing time for complex algorithms whose control period is longer than those of simpler PID algorithms (Rocco, 1996; Kelly and Salgado, 1994). Practically, the prolonged period severely deteriorates the control performance of the system.
To overcome this difficulty, we designed an efficient control system using four TMS320C50 digital signal processors (DSPs) for a SCARA robot. There are four major parts in this system: a host controller, four joint controllers, four servo drives and the mechanical body of SCARA robot.

The control system introduced here can perform a lot of calculations required for high-speed control algorithms in real time, since DSPs are used as main processors of the joint controllers to reduce the computation time. There are two types of DSPs: fixed point types and floating point ones. A fixed point DSP, TMS320C50, is selected for our design, as it provides the maximum value of performance/cost for the controller at this moment (Lin et al, 1987). A joint controller is assigned to each joint motor to meet the case that independent joint control is required for the control algorithm and that heavy computations are required for each joint control. In our design, data representing interactions among joint motions are saved and released from the global memory through a 16 bits data bus. A host computer (PC 586) is used for the trajectory planning and high level task planning of given tasks as well as the graphic display of task performance. The host computer has serial communication channels to each controller to provide auxiliary communication path in urgent cases. For example, when a fatal error happened to a joint controller, the controller can not send messages to the host computer through the 16 bits data bus shared by all of the joint controllers. In this case, a hardware-supported signal will notice the fatal error of the controller to the host computer through the additional serial communication channel.

To demonstrate that the designed system is suitable for the application of dynamic control algorithms involving lots of calculations, experiments are conducted with a SCARA robot and the experimental results are presented.

2. ARCHITECTURE OF THE CONTROLLER

2.1 System Overview

An overall system block diagram of the suggested controller is shown in Fig. 1. The system is composed of a host computer (PC 586), four joint controllers, four servo drives, a mother board, actuators, and two power supplies. The mechanical body and servo drives of a SCARA robot (made by Samsung Electronics Company) are being utilized as the control system. Through the mother board (refer to Fig. 1), the host computer sends position commands to the joint controllers in every 16 msec. With this data transmission it receives actual position and velocity values from the controllers to monitor the control performance in real time. The communication between the host computer and the four controllers is mainly performed through the 16 bits data bus connected to the global memory which is partitioned and used by the memory-map method. Reset and start signals are sent from the host computer to the individual joint controllers and status signals of the joint controllers are sent back to the host computer through the I/O map I/O interface.

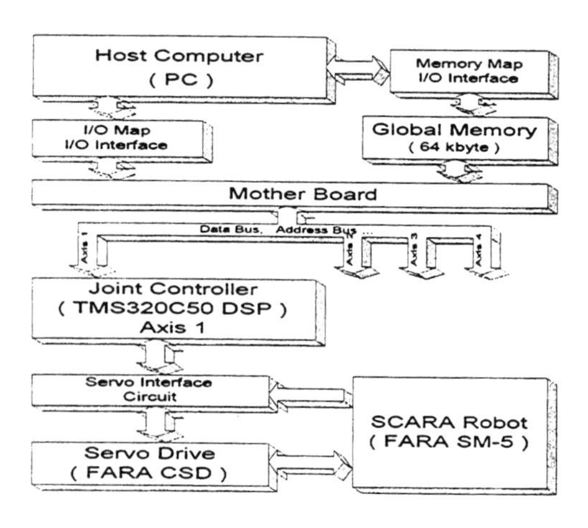

Fig. 1. Schematic diagram overall system

2.2 Joint Controllers

Joint controllers are designed using fixed-point DSPs (TMS320C50) as shown in Fig. 2. A DSP can perform a single-cycle fixed-point instruction every 50 nsec (20 MIPS) when a 40 MHz crystal is used as the clock oscillator. It can address 64k times 16 bits of program and data areas individually, and it has 64k I/O capability. In the DSP, 9k times 16 bits of single-cycle program/data RAM and 2k times 16 bits of single-cycle boot ROM are imbedded inherently. There are also several 32 bits registers that can perform 16 bits multiplications without repetition. Each joint controller uses 32k times 16 bits SRAM as the program memory whose access time is 20 nsec. It also uses 32k times 16 bits SRAM as the data memory individually and shares 32k times 16 bits SRAM as the global memory with others. The memory architecture is determined to remove the memory access conflicts among the DSPs.

The communication path in between each joint controller and the corresponding servo drive (Pillay and Krishnan, 1989; Maeno and Kobata, 1972; Zubek et al, 1975) is supported by a 34 pin connector as shown in Table 1. A 12 bits encoder residing in the servo drive generates the Z phase pulse for the indication of crossing the reference point on the disk of the encoder, and A and B phase pulses for the indication of position and the direction of revolution. The Z phase pulse is noticed to the joint controller by the interrupt, and the A and B phase pulses are fed to the 16 bits up/down counter which will be accessed by the joint controller in every 1 msec. A servo alarm signal, three alarm codes and a limit sensor signal are also checked by the joint

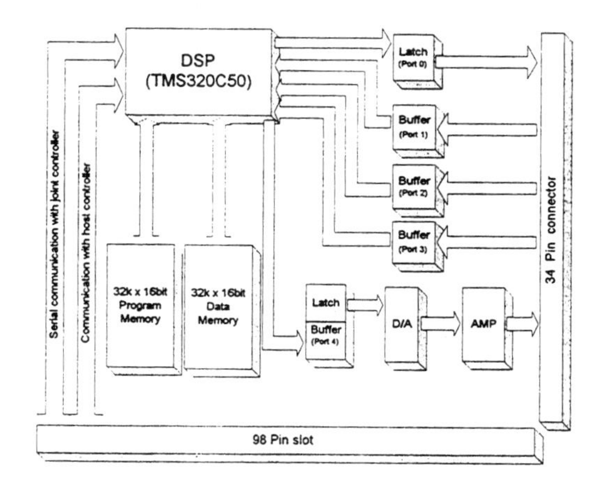

Fig. 2. Schematic diagram of joint controller

controller; a servo on, a servo off, a reset signal and a current command are sent to the servo drive from the joint controller also in every 1 msec. Servo setting DIP switches are utilized for the setting of the servo drive mode. For example, servo command can be velocity value or current value depending on the setting of the DIP. switches. The composition of the 34 pin connector and the I/O port address of the DSP are summarized in Table 1.

The serial communication network is established among the joint controllers, in addition to the parallel communication path through the global memory, to guarantee the emergency communications among them. Through this serial communication network, all controllers are interconnected. All the communication with the host computer is supported by a 98 pin connector which is plugged onto the mother board.

Table 1. Communication signals between a joint controller and a servo drive

Address	Bit	Signal	In/Out (DSP)
50H	1	Servo Reset	Output
50H	1	Servo On/Off	Output
51H	3	Limit Sensor Signal	Input
51H	4	Servo Alarm	Input
52H	16	Feedback Position	Input
	7	Reserved, Ground	Input/Output
54H	Analog	Current Command	Output
INT3	1	Z-Pulse Interrupt	Input

2.3 Mother Board

The mother board configuration is shown in Fig. 3. The board is composed of the global memory, the global memory arbitration logic, serial communication arbitration circuits, EPROM, and several buffers. The global memory consists of 32k times 16 bits SRAM (access time: 20 nsec), and it is mainly used for sending to joint controllers the position commands which are generated by the host computer according to the trajectory planning. Joint controllers can use this global memory to send control results back to the host computer, and they can also use it to save and share the coupling variables among themselves.

The global memory arbitration logic resolves possible collisions which may happen when the global memory is accessed by more than one of the host computer and joint controllers at the same time. In the arbitration logic circuits, the order of priority is set by the hardware: the first one is the host computer, and the joint

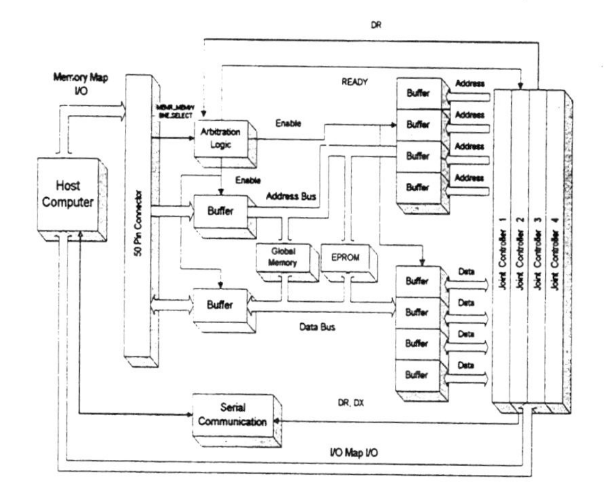

Fig. 3. Schematic diagram of the mother board

controllers follow from lower one to higher one. A lower joint controller needs high priority since coupling effects (which can be accessed through the global memory) in a lower joint are more serious than those in the higher joints. Note that the resource allocation, specifically global memory sharing, can be effectively done by closely monitoring coupling variables among joint controllers, even though this effectiveness of the global memory highly depends upon a control algorithm. For example, to make position commands - to be sent to the joint controllers from the host computer - not to cause collisions, the transmission sequence can be predetermined and kept consistently. The serial communication arbitration circuit arbitrates the RS-232C communication (the serial communication channels between the host computer and joint controllers) using the multiplexer and demultiplexer. The channel allocation is governed by the host computer and the connection is activated by the polling method. The EPROM is used for the initial down-loading of programs for joint controllers when the power switch is set on.

3. CONTROL FLOW OF JOINT CONTROLLERS

The DSP at a joint controller may down-load the main program from the EPROM when power is on. In this design, we down-load the main program from the host computer to the joint controllers directly with the selection signal to the joint controllers through the I/O map I/O interface. Note that the TMS320C50 can be used as two modes: a micro-computer mode and a microprocessor mode, which can be selected by the bias of MP/MC pin of the DSP. In this study, the micro-computer mode is selected to use the boot loader existing inherently in the DSP. Each DSP waits until all DSPs are down-loaded and a start signal from the host

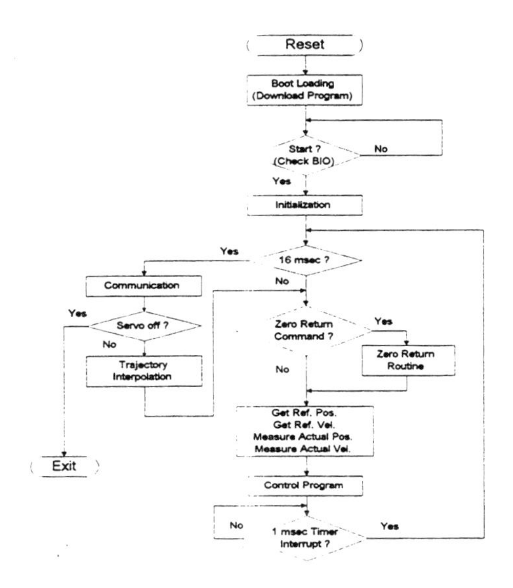

Fig. 4. Control flowchart of joint controller

computer comes into the BIO pin of the DSP. After this step, the control flow comes into the normal processing mode as shown in Fig. 4. At the initialization stage of operation, all registers are initialized first and then the initialization of timers for sampling is followed. After the initialization processes, the joint controllers start to perform the down-loaded program using the position and velocity data if necessary. The desired position command is received from the host computer in every 16 msec and it is interpolated to 16 segments. Before getting into the servo control loop (16 times of servo control loop in every 1 msec), the joint controller checks whether the command is for the home position (zero return) or not. And then, the controller generates the reference position and velocity values; it measures the actual position and velocity values. With these values, the control algorithm is performed and the current command is sent to the servo drive every in 1 msec.

4. ROBUST INVERSE DYNAMIC ALGORITHM

Most of industrial robots are using the PID algorithm which is well developed for the control of linear systems (Rocco, 1996; Kelly and Salgado, 1994). When the robots are operating at a slow speed, the algorithm shows relatively good performance. However, when the robots are operating at a high speed, the inertia, centrifugal, Coriolis and gravitational terms are activated and the control performance suddenly drops. To overcome these nonlinear and coupling effects and to make the controller robust against disturbance, we combined the inverse dynamic algorithm and the PID algorithm. Fig. 5 shows the control block diagram of the proposed algorithm.

Generally, the dynamics of the n-link manipulator can be represented as follows:

$$H(q)\ddot{q} + B(q,\dot{q})\dot{q} + G(q) = \tau, \qquad (1)$$

where $H(q)$ is the $n \times n$ inertia matrix, $\ddot{q}$ is the $n \times 1$ generalized acceleration vector, $B(q,\dot{q})$ represents the centrifugal and Coriolis force terms, $\dot{q}$ is the $n \times 1$ velocity vector, q is the $n \times 1$ configuration vector, $G(q)$ represents the gravitational force vector, and τ is the $n \times 1$ torque vector.

The manipulator dynamics can be considered as the disturbance for the individual actuator. Assuming the manipulator dynamics are obtained precisely, the disturbance d to the actuators are equal to the joint torques τ. That is,

$$d = r\tau \qquad (2)$$

where $r_{n\times n}$ is the diagonal gear ratio matrix of the n joints, and d is the $n \times 1$ disturbance vector. Therefore, the joint controller computes the disturbance value using the measured values of q, $\dot{q}$, and the desired value of $\ddot{q}^d$. The computed disturbance vector can be represented as follows:

$$\text{dequi.} = H(q)\,\ddot{q}^d + B(q,\dot{q})\,\dot{q} + G(q). \qquad (3)$$

When the acceleration sensors are used in the control system, $\ddot{q}^d$ will be replaced by $\ddot{q}$ to make the disturbance compensation more accurate. Here, we assumed $\ddot{q}^d = \ddot{q}$. Now, the control block diagram of the proposed algorithm is shown in Fig. 5.

In Fig. 5, q_i^d represents the desired joint angle, r_{ii} represents the gear ratio, K_P, K_D, and K_I represents the PID gains, respectively, J_m represents the inertia moment of the i-th actuator, q_{mi}^a represents the actual angle of the i-th actuator, and q_i^a represents the actual angle of the i-th joint.

$B_{eff.}$ is the equivalent friction coefficients of the

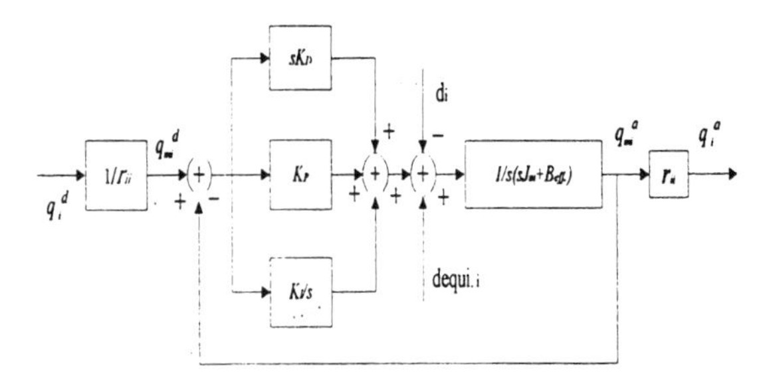

Fig. 5. Block diagram of the proposed control algorithm

i-th actuator and it is given as (Spong and Vidyasgar, 1989),

$$B_{eff.} = B_m + K_b K_T / R, \quad (4)$$

where B_m is the friction coefficient, K_b is the back e.m.f constant, K_T is the torque constant, and R is the resistance of the i-th actuator windings. With the computed torque feedback of dequi., the system becomes linear. And, the characteristic equation of the system becomes

$$J_m S^3 + (B_{eff.} + K_D)S^2 + K_P S + K_I = 0. \quad (5)$$

This third order linear system can be designed according to the desired overshoot, response time requirements. To make the system stable, we have the constraint as (Spong and Vidyasgar, 1989)

$$\frac{(B_{eff.} + K_T K_D / R)K_P}{J} > K_I. \quad (6)$$

With this constraint and the desired damping and overshoot conditions, we can determine the optimal values of K_I, K_P, and K_D. In this research, we obtained the optimal values of the PID gains to reduce the tracking errors, experimentally.

5. PERFORMANCE EVALUATION

To check the performance of the desired controller system, we applied the PID algorithm and the robust inverse dynamic algorithm to the SCARA robot. As it is shown in Fig. 6, the end-effector of the SCARA robot is initially located at A (610, 0, 408). The goal point is set to B (433, 400, 378) in the three dimensional space. As shown in Fig. 6, the end-effector of the SCARA robot moved from A to B point along the straight line with the prespecified velocity profile shown as Fig. 7, and came back to A point without any prespecified path. The trajectory planning is done based upon the B-Spline method which generates a continuous trajectory in the cartesian space. Fig. 8 shows the trajectory tracking error of the SCARA robot under the PID algorithm; Fig. 9 shows the trajectory tracking error when the robust inverse dynamic algorithm is applied to the SCARA robot. As we notice through the comparison of the two figures, the robust inverse dynamic algorithm shows better dynamic performance and worse

static performance than the PID algorithm. We started with the well-tuned PID gains, and the same gains are used for the robust dynamic algorithm. It is easily predicted that more gain tunings for the robust inverse dynamic algorithm will give better results than the current ones. It should be noted that the ultimate goal of our experiment is showing that the controller system is working correctly for a given algorithm.

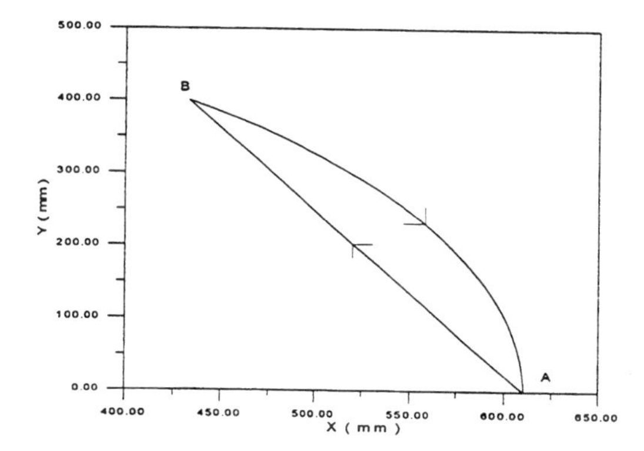

Fig. 6. Motion trajectory projected on the x-y plane

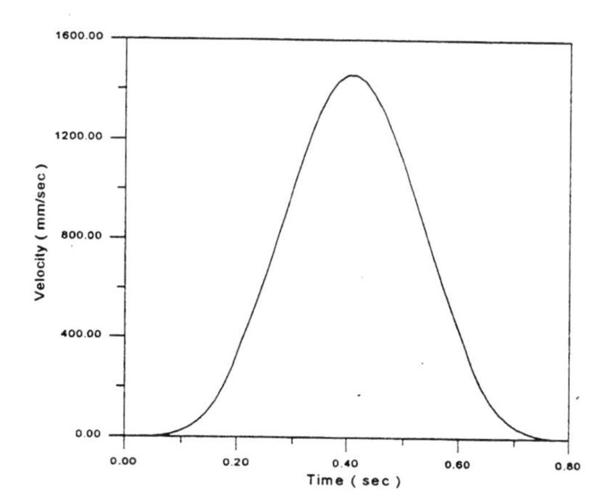

Fig. 7. Velocity profile of the trajectory

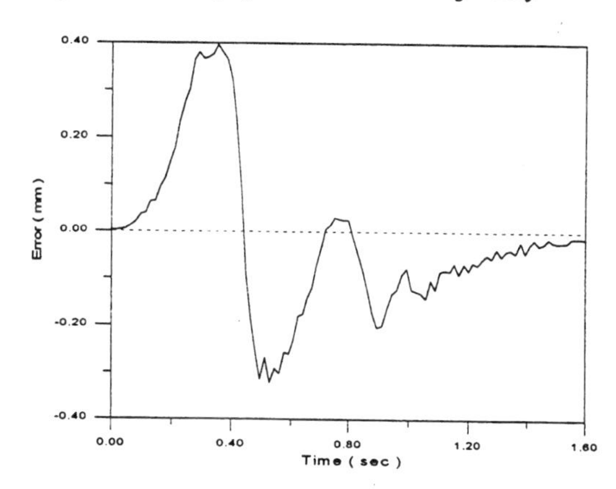

Fig. 8. Trajectory following error with the PID algorithm

To evaluate the controller system performance in the sense of speed, we calculated and measured the time spent for each control algorithm and the servo drive, respectively. The major point of this measurement is whether the designed system is suitable for the implementation of complex dynamic algorithms or not. Fig. 10 (a) shows the time spent for PID algorithm; Fig. 10 (b) shows the time spent for the proposed robust inverse dynamic algorithm, as well as the time spent for the servo drive, within a servo control loop.

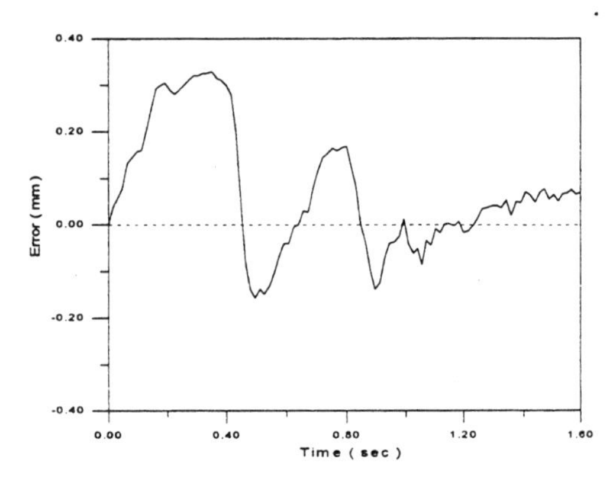

Fig. 9. Trajectory following error with the robust inverse dynamic algorithm

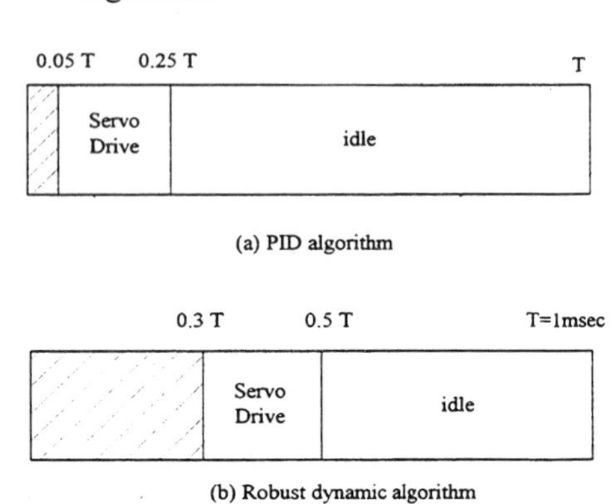

Fig. 10. Controller speed analysis

As seen from Fig. 10, the computing power of the joint controller is good enough for the robust dynamic algorithm. Therefore, we claim that there is still enough room for the implementation of learning algorithms which are supposed to be not suitable for real-time control. The gain tunings of the robust inverse dynamic algorithm are under the implementation using fuzzy rules where the desired velocity, acceleration and z-directional displacement are used for input variables.

6. CONCLUSION

We have proposed a fast control system for the SCARA robot, which is designed using DSPs to implement complex control algorithms requiring many calculations in real time. Four controller boards are used for control of each joint motion independently and quickly, and they are all connected to the host computer through the local bus and the global memory. In this system, both of the parallel and serial communication networks are provided between the host computer and joint controllers to broaden the application fields. To demonstrate the capability and correctness of the designed system, we implemented a PID control algorithm and a robust inverse dynamic control algorithm in this system and demonstrated the performance of a trajectory following task. It is shown that the designed controller can be used to implement different dynamic control algorithms in real time. Our results encourage us to implement various control algorithms in the SCARA robot so that we plan to develop and implement a high-speed and high-precision control algorithm that utilizes full capability of this system. The self-tuning of the PID gains by an intelligent algorithm, with the robust inverse dynamic control, is left for our future research.

7. REFERENCES

F.L. Lewis, C.T. Abdallak, and D.M. Dawson(1993), *Control of Robot Manipulators*, Macmillan Publishing Company.

M.W. Spong, and R. Ortega(1994), *On Adaptive Inverse Dynamics Control of Rigid Robots*, IEEE Trans., AC-39, pp. 1866-1871.

Y.D. Song, T.L. Mitchell, and H.Y. Lai(1994), *Control of a Class of Nonlinear Uncertain Systems via Compensated Inverse Dynamics Approach*, IEEE Trans., AC-39, pp. 1866-1871.

K.K. Shyu, P.H. Chu and L.J. Shang(1996), *Control of rigid robot manipulators via combination of adaptive sliding mode control and compensated inverse dynamics approach*, IEE Proc.-Control Theory Appl., Vol. 143, No. 3, May.

P. Rocco(1996), *Stability of PID Control for Industrial Robot Arms*, IEEE Trans., Vol. 12, No. 4, August.

R. Kelly and R. Salgado(1994), *PD Control with Computed Feedforward of Robot Manipulators: A Design Procedure*, IEEE Trans., Vol. 10, No. 4, August.

M.W. Spong and M. Vidyasgar(1989), *Robot Dynamics and Control*, John Wiley & Sons, Inc.

K.S. Lin, G.A. Frantz, and R. Simar, Jr.(1987), *The TMS320 Family of Digital Signal Processors*, Proceedings of the IEEE, Vol. 75, No. 9, Sep.

P. Pillay and R. Krishnan(1989), *Modeling, Simulation, and Analysis of Permanent-Magnet Motor Drives, Part II: The Brushless DC Motor Drive*, IEEE Trans. Ind. Appl., Vol. 25, No. 2, Mar.

T. Maeno and M. Kobata(1972), *AC Commutatorless and Brushless Motor*, IEEE Trans. Power Appl. Syst., Vol. PAS-91, pp. 1476-1484, Jul.

J. Zubek, A. Abbondanti, and C. J. Nordby(1975), *Pulsewidth Modulated Inverter Motor Drives with Improved Modulation*, IEEE Trans. Ind. Appl., Vol. 1A-11, pp. 695-703, Nov.

IMPLEMENTATION OF A REAL-TIME ADAPTIVE CONTROLLER FOR ROBOTIC MANIPULATOR USING DSP'S

**S. H. Han[1], K. Son[2], M. H. Lee[2], M. C. Lee[2],
J. M. Lee[3], and J. O. Kim[4]**

1 Dept. of Mechanical Engineering, Kyungnam Univ., Masan, Korea
2 Dept. of Mechanical Engineering, Pusan National Univ., Pusan, Korea
3 Dept. of Electronics Engineering, Pusan National Univ., Pusan, Korea
4 Samsung Electronics Co. Ltd., Suwon, Korea

Abstract : Real-time implementation of an adaptive controller for the robotic manipulator is presented in this paper. Digital signal processors are used in implementing real time adaptive control algorithms to provide an enhanced motion for robotic manipulators. In the proposed scheme, adaptation laws are derived from the improved Lyapunov second stability analysis based on the model reference adaptive control theory. The adaptive controller consists of an adaptive feedforward controller and feedback controller and time-varying auxiliary control elements to the nominal operating point. The proposed control scheme is simple in structure, fast in computation, and suitable for real-time control. Moreover, this scheme does not require any accurate dynamic modeling, nor values of manipulator parameters and payload. Performance of the adaptive controller is illustrated by simulation and experimental results for a SCARA robot.

Keywords: Adaptive controller, manipulators, digital signal processor, real-time, robust control

1. INTRODUCTION

ROBOTIC manipulators consist of independent joint controllers which control joint angles separately through simple position servo loops (Ortega and Spong, 1989). This basic control system enables a manipulator to perform simple positioning tasks such as in the pick-and-place operation. However, joint controllers are severely limited in precise tracking of fast trajectories and in sustaining desirable dynamic performance for variations of payload and parameter uncertainties. Design of a high performance controller for robotic manipulators has been an active topic of research(Ortega and Spong, 1989; Tomei, 1989).

The dynamics of manipulators are highly coupled and nonlinear. In many servo control applications the linear control scheme proves unsatisfactory, therefore, need for nonlinear techniques seems increasing. Today there are many advanced techniques that are suitable for servo control of a large class of nonlinear systems including robotic manipulators(Sadegh and Horowitz, 1990; Bortoff, 1994; Slotine and Li, 1987). Since the pioneering work of Dubowsky and DesForges (Sadegh and Horowitz, 1990), interest in adaptive control of robot manipulators has been growing steadily (Ortega and Spong, 1989; Tomei, 1989; Sadegh and Horowitz, 1990; Bortoff, 1994). This growth is largely due to the fact that adaptive control theory is particularly well-suited to robotic manipulators whose dynamic model is highly complex and may contain unknown parameters. However, implementation of these algorithms generally involves intensive numerical computations.

Digital signal processors(DSP's) are special purpose microprocessors that are particularly powerful for intensive numerical computations involving sums and products of variables (Ahmed, 1991). Digital version of most advanced control algorithms can be defined as sums and products of measured variables, thus can naturally be implemented by DSP's. In addition, DSP's are as fast in computation as most 32-bit microprocessors and yet at a fraction of their prices(Bortoff, 1994; Ahmed, 1991). These features make them a viable computational tool for digital implementation of advanced controllers.

In order to develop a digital servo controller one must carefully consider the effect of the sample and hold operation, the sampling frequency, the computational delay, and that of the quantization error on the stability of a closed-loop system(Ahmed, 1991; Parks, 1986; Dubowsky and DesForges, 1979; Choi, *et al*, 1986; Gavel and Hsia, 1987). Moreover, one must also consider the effect of disturbances on the transient variation of the tracking error as well as its steady-state value.

This paper presents a new approach to the design of an adaptive control system using DSP's, TMS320C30, for robotic manipulators to achieve trajectory tracking by the joint angles.

This paper is organized as follows: in Section 2, the dynamic modeling of robotic manipulator and the adaptive control algorithm are derived. Adaptation laws are derived based on the model reference adaptive control theory using the improved Lyapunov second method. Section 3 represents simulation and experimental results is obtained for a SCARA robot. Finally, Section 4 discusses findings and draws some conclusions.

2. ADAPTIVE CONTROL

2.1 Dynamic Modeling

Let us consider a nonredundant-joint robotic manipulator in which the $n \times 1$ joint torque vector $\tau(t)$ is related to the $n \times 1$ joint angle vector $q(t)$ by the following nonlinear dynamic equation of motion:

$$D(q)\ddot{q} + N(q, \dot{q}) + G(q) = \tau(t) \quad (1)$$

where $D(q)$ is the $n \times n$ symmetric positive-definite inertia matrix, $N(q, \dot{q})$ is the $n \times 1$ Coriolis and centrifugal torque vector, and $G(q)$ is the $n \times 1$ gravitational load vector.

Equation (1) describes the manipulator dynamics without any payload. Let the $n \times 1$ vector X represent the end-effector position and orientation coordinates in a fixed task related Cartesian frame of reference.

In order to consider payload on the manipulator dynamics, suppose that the manipulator end-effector is firmly grasping a payload represented by point mass ΔP. For the payload to move with acceleration $\ddot{X}(t)$ in the gravity field, the end-effector must apply the $n \times 1$ force vector $T(t)$ given by

$$T(t) = \Delta P[\ddot{X}(t) + g] \quad (2)$$

where g is the $n \times 1$ gravitational acceleration vector. The end-effector requires additional joint torque

$$\tau_f(t) = J(q)^T T(t) \quad (3)$$

where $J(q) = [\partial \lambda(q)/\partial q]$ is the $n \times n$ Jacobian matrix of the manipulator. Hence, the total joint torque vector can be obtained by combining equations (1) and (3) as

$$\tau(t) = J(q)^T T(t) + D(q)\ddot{q} + N(q, \dot{q}) + G(q) \quad (4)$$

Substituting equations (2) and (3) into equation (4) yields

$$\Delta P\, J(q)^T [J(q)\ddot{q} + \dot{J}(q, \dot{q})\dot{q} + g] + D(q)\ddot{q} + N(q, \dot{q}) + G(q) = \tau(t) \quad (5)$$

Equation (5) shows the explicitly effect of payload ΔP on the manipulator dynamics.

This equation (5) can be define as the implicit nonlinear dynamic equation

$$D^*(\Delta P, q, \dot{q})\ddot{q} + N^*(\Delta P, q, \dot{q})\dot{q} + G^*(\Delta P, q, \dot{q}) = \tau(t) \quad (6)$$

2.2 Adaptation Law

In order to cope with changes in operating point, gains are varied with the change in external working condition. This yields the adaptive control law

$$\tau(t) = [\ P_A(t)\,\ddot{q}_r(t) + P_B(t)\,\dot{q}_r(t) + P_C(t)\,q_r(t)] + [\ P_P(t)E(t) + P_V(t)\,\dot{E}(t) + P_I(t)] \tag{7}$$

where $P_A(t)$, $P_B(t)$, $P_C(t)$ are feed forward time-varying adaptive gains, and $P_P(t)$ and $P_V(t)$ are feedback adaptive gains. And $P_I(t)$ denotes the time varying control signal corresponding to the nominal operating point term, generated by a feedback controller driven by position backing error $E(t)$.

Fig. 1 represents the block diagram of adaptive control system for robotic manipulator.

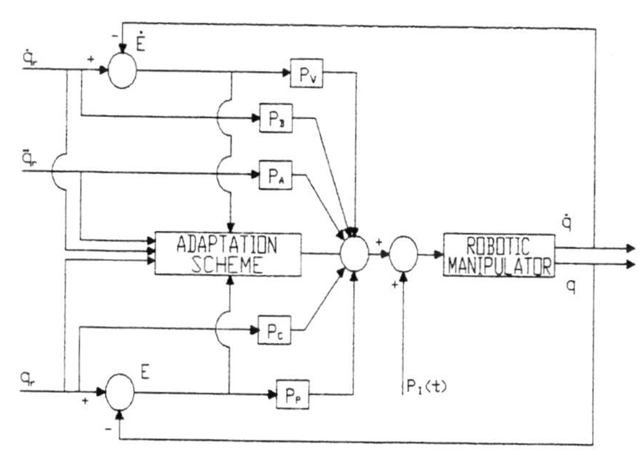

Fig. 1. The block diagram of adaptive control scheme for robotic manipulator.

On applying adaptive control law (7) to the nonlinear robot dynamic equation (6) the error differential equation can be obtained as

$$D^*\ddot{E}(t) + (N^* + P_V)\dot{E}(t) + (G^* + P_P)E(t) = P_I(t) + (D^* - P_A)\ddot{q}_r(t) + (N^* - P_B)\dot{q}_r(t) + (G^* - P_C)q_r(t) \tag{8}$$

Defining the $2n\times 1$ position-velocity error vector $\varepsilon(t) = [\ E(t),\ \dot{E}(t)]^T$, equation (8) can be written in the statespace form

$$\dot{\varepsilon}(t) = \begin{pmatrix} 0 & I_n \\ w_1 & w_2 \end{pmatrix} \varepsilon(t) + \begin{pmatrix} 0 \\ w_3 \end{pmatrix} q_r(t) + \begin{pmatrix} 0 \\ w_4 \end{pmatrix} \dot{q}_r(t) + \begin{pmatrix} 0 \\ w_5 \end{pmatrix} \ddot{q}_r(t) + \begin{pmatrix} 0 \\ w_6 \end{pmatrix} \tag{9}$$

where

$$\begin{aligned} w_1 &= [\ D^*]^{-1} [\ G^* + P_P] \\ w_2 &= [\ D^*]^{-1} [\ N^* + P_V] \\ w_3 &= [\ D^*]^{-1} [\ G^* - P_C] \\ w_4 &= [\ D^*]^{-1} [\ N^* - P_B] \\ w_5 &= [\ D^*]^{-1} [\ D^* - P_A] \\ w_6 &= -[\ D^*]^{-1} [\ P_I] \end{aligned}$$

The adaptation laws are now derived by ensuring the stability of error dynamics. To this end, let us define a scalar positive definite Lyapunov function as

$$\begin{aligned} L = {} & \varepsilon^T R \varepsilon + trace\{[\ Q_1]^T K_1 [\ Q_1]\} \\ & + trace\{[\ Q_2]^T K_2 [\ Q_2]\} \\ & + trace\{[\ Q_3]^T K_3 [\ Q_3]\} \\ & + trace\{[\ Q_4]^T K_4 [\ Q_4]\} \\ & + trace\{[\ Q_5]^T K_5 [\ Q_5]\} \\ & + [\ Q_6 K_6 Q_6]^T \end{aligned} \tag{10}$$

where $Q_1 = w_1 - S_1 - w_1^*$, $Q_2 = w_2 - S_2 - w_2^*$, $Q_3 = w_3 - w_3^*$, $Q_4 = w_4 - w_4^*$, $Q_5 = w_5 - w_5^*$, and $Q_6 = w_6 - w_6^*$. R is the solution of the Lyapunov equation for the reference model, $K_1, \ldots, K_6$ are arbitrary symmetric positive definite constant $n\times n$ matrices, and matrices $w_1^*, \ldots, w_6^*$ are functions of time which will be specified later.

From the stability analysis by the Lyapunov second method, required adaptive controller gains are obtained as

$$P_P(t) = p_1 [\ P_{p1}E + P_{p2}\dot{E}]\,[\ E]^T + p_2 \int_0^t [\ P_{p1}E + P_{p2}\dot{E}]\,[\ E]^T dt \tag{11-a}$$

$$P_V(t) = v_1 [\ P_{v1}E + P_{v2}\dot{E}]\,[\ \dot{E}]^T + v_2 \int_0^t [\ P_{v1}E + P_{v2}\dot{E}]\,[\ \dot{E}]^T dt \tag{11-b}$$

$$P_C(t) = c_1 [\ P_{c1}E + P_{c2}\dot{E}]\,[\ q_r]^T + c_2 \int_0^t [\ P_{c1}E + P_{c2}\dot{E}]\,[\ q_r]^T dt \tag{11-c}$$

$$P_B(t) = b_1 [\ P_{b1}E + P_{b2}\dot{E}]\,[\ \dot{q}_r]^T + b_2 \int_0^t [\ P_{b1}E + P_{b2}\dot{E}]\,[\ \dot{q}_r]^T dt \tag{11-d}$$

$$P_A(t) = a_1 [\ P_{a1}E + P_{a2}\dot{E}]\,[\ \ddot{q}_r]^T + a_2 \int_0^t [\ P_{a1}E + P_{a2}\dot{E}]\,[\ \ddot{q}_r]^T dt \tag{11-e}$$

$$P_I(t) = \lambda_2 [\ P_{i2}E] + \lambda_1 \int_0^t [\ P_{i1}E]^T dt \tag{11-f}$$

where $[\ \lambda_1, p_{p1}, p_{v1}, p_{c1}, p_{b1}, p_{a1}]$ and $[\ \lambda_2, p_{p2}, p_{v2}, p_{c2}, p_{b2}, p_{a2}]$ are positive and nonnegative scalar adaptation gains.

3. EXPERIMENT AND RESULTS

This section represents DSP's-based control results of the position and velocity control for a SCARA robot with four joint as shown in Fig. 2, and discusses the advantages of using DSP's for robotic motion control.

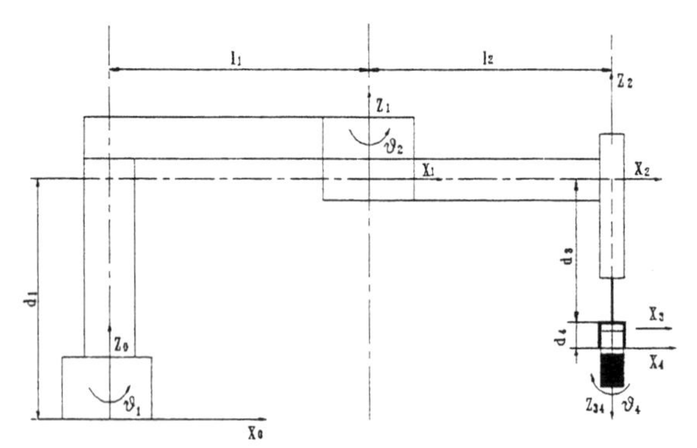

Fig. 2. Link coordinate systems of a SCARA robot.

A set of experiments for the proposed adaptive controller was performed for four joints of the SCARA robot. To implement the proposed adaptive controller, we used our own TMS320C30 assembler software developed.

Also, a TMS320C30 emulator was used in experimental set-up as can be seen form Fig. 3. The TMS320C3x emulator is an application development tool which is based on the TI's TMS320C30 floating point DSP chip with an instruction cycle time 50ns. At each joint, a harmonic drive was used to transfer power from the motor, which has a resolver attached to its shaft for sensing angular velocity with a resolution of 8096 pulses/rev.

Fig. 4 represents the main hardware structure of control system of a SCARA robot.

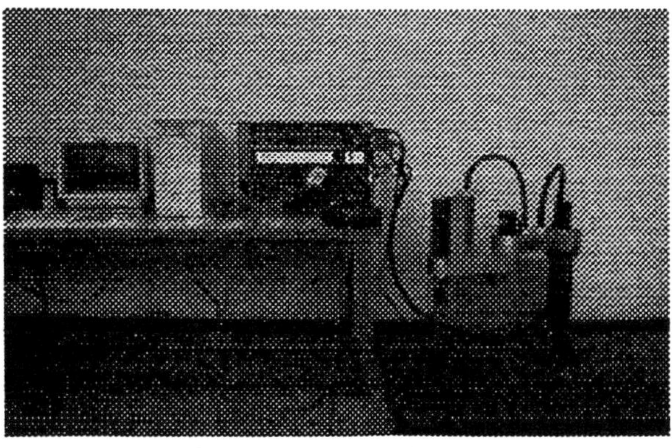

Fig. 3. Experimental set-up.

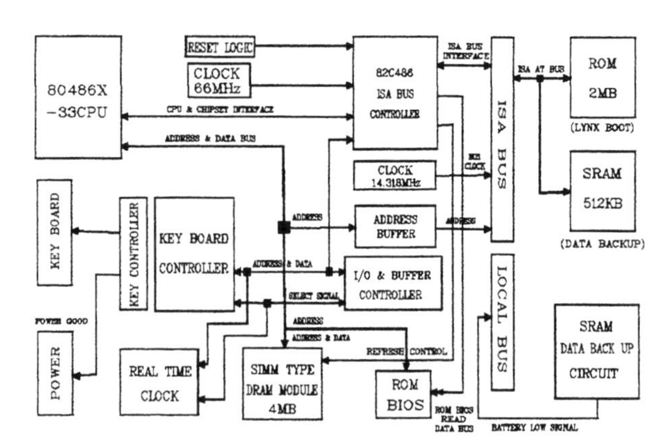

Fig. 4. The main board block diagram of hardware structure of SCARA robot.

The performance evaluation of proposed adaptive controller was performed in the joint space and cartesian space.

In the joint space, experiment was carried out to evaluate the position and velocity control performance of the four joints for variation of payloads. Fig. 5 shows the results of the position and velocity tracking control for the first joint with 3.5 kg payload. In the joint space, as can be seen from results, the DSP-based adaptive controller shows extremely good tracking performance even with the added external disturbance. Fig. 6 shows the experimental results of the position and velocity tracking performance for the second joint with 3.5 kg payload in the joint space.

From experiment results, proposed adaptive controller shows very good control performance in the test for trajectory tracking of the velocity and position in the joint space.

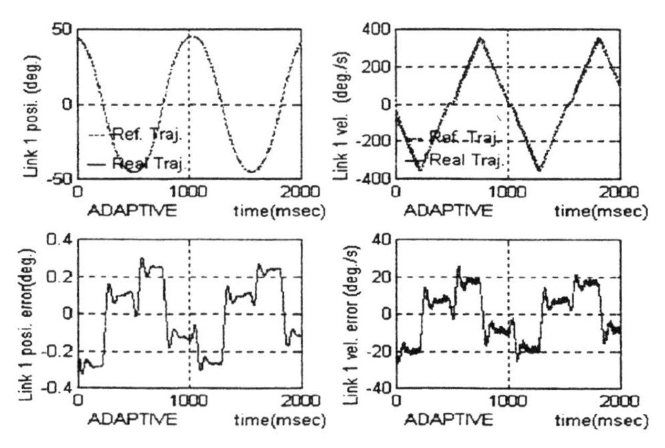

Fig. 5. Experimental results for the position and velocity tracking at the first joint with 3.5 kg payload.

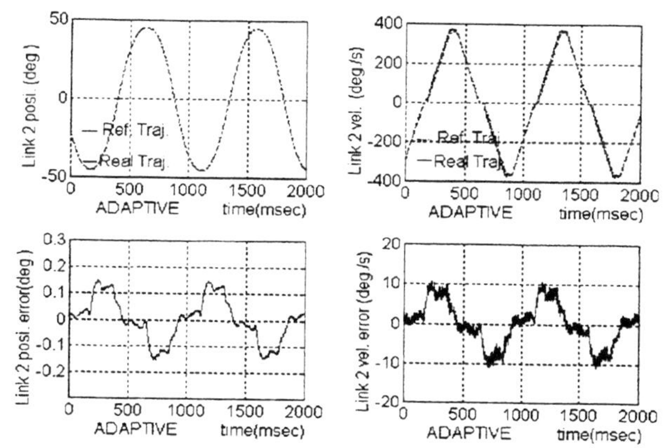

Fig. 6. Experimental results for the position and velocity tracking at the second joint with 3.5 kg payload.

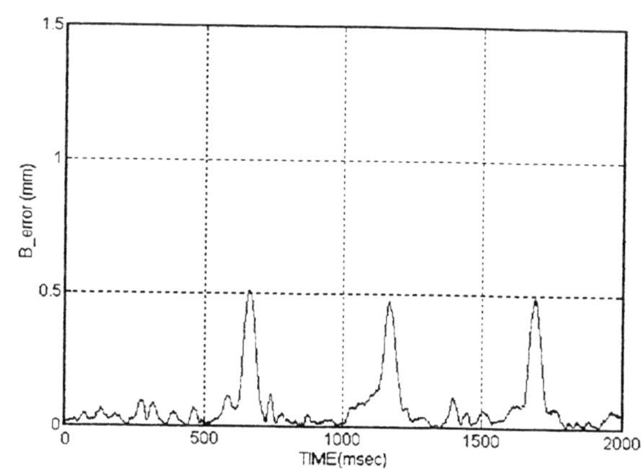

Fig. 8. Experimental result of the adaptive controller for tracking of B shaped reference trajectory with 3.5 kg payload.

In the cartesian space, the adaptive controller was evaluated in a peg-in-hole task, and in a tracking task of B shaped reference trajectory.

Fig. 7 represents the B shaped reference trajectory in the cartesian space. Fig. 8 represents the experimental results of adaptive controller for the B shaped reference trajectory with 3.5 kg payload and maximum velocity (2.2 m/s) in the cartesian space. Fig. 9 shows the experimental results of PID controller for the B shaped reference trajectory with 3.5 kg payload. Fig. 10 represents the kinematic configuration of peg-in-hole task in the cartesian space. In Fig. 10, each length of link1 and link2 is 350㎜ and 260㎜ respectively.

Table. 1 represents the experimental results for the peg-in-hole tasks with 3.5 kg payload and maximum velocity(2.2 m/s) during 8 hours running time in the cartesian space.

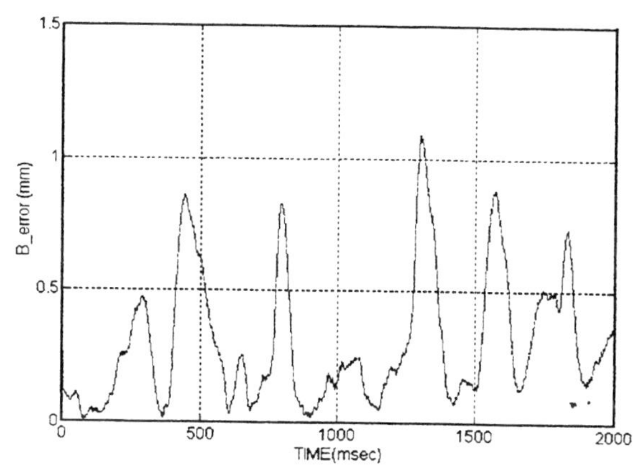

Fig. 9. Experimental result of PID controller for tracking of B shaped reference trajectory with 3.5 kg payload.

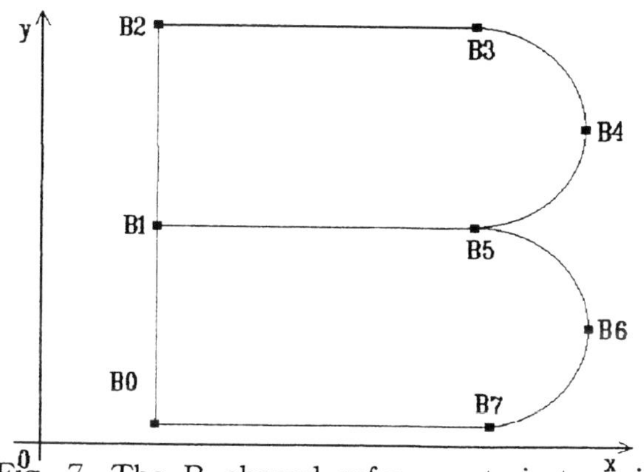

Fig. 7. The B shaped reference trajectory in the cartesian space.

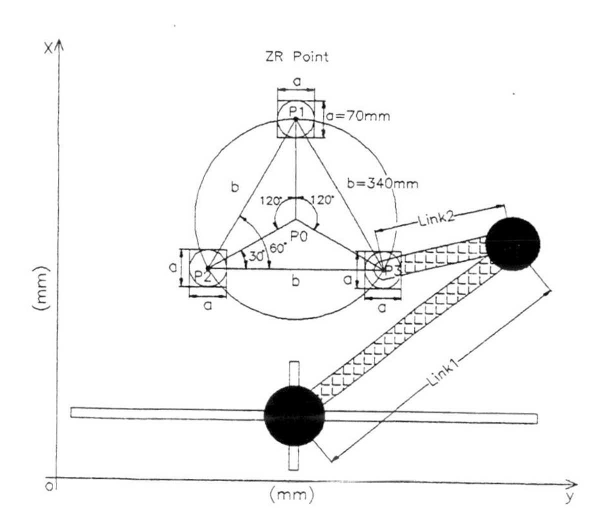

Fig. 10. Kinematic configuration for peg-in-hole task in the cartesian space.

Table. 1 Comparision of the failure rate between the adaptive controller and PID controller in the peg-in-hole task.

Task speed	80.00 (%)	100 (%)
Failure (%) of adaptive controller	0.008 (%)	0.012 (%)
Failure (%) of PID controller	0.015 (%)	0.038 (%)

As can be seen from the experimental results(Table. 1), the adaptive controller shows the better control performance and reliability than existing PID controller in the higher speed.

4. CONCLUSIONS

A new adaptive digital control scheme is described in this paper using the TMS320C30 chips for robotic manipulators. The adaptation laws are derived from the model reference adaptive theory using the improved direct Lyapunov method. The simulation and experimental results show that the proposed DSPs-adaptive controller is robust to the payload variation, inertia parameter uncertainty, and change of reference trajectory. This adaptive controller has been found to be suitable to the real-time control of robot system. A novel feature of the proposed scheme is the utilization of an adaptive feedforward controller, an adaptive feedback controller, and a PI type time-varying control signal to the nominal operating point which result in improved tracking performance.

Another attractive feature of this control scheme is that, to generate the control action, it neither requires a complex mathematical model of the manipulator dynamics nor any knowledge of the manipulator parameters and payload.

Control scheme uses only the information contained in the actual and reference trajectories which are directly available. Futhermore, the adaptation laws generate the controller gains by means of simple arithmetic operations. Hence, the calculation control action is extremely simple and fast. These features are suitable for implementation of on-line real time control for robotic manipulators with a high sampling rate, particularly when all physical parameters of the manipulator cannot be measured accurately and the mass of the payload can vary substantially.

REFERENCES

D. Gavel and T.C. Hsia(1987), *Decentralized Adaptive Control of Robot Manipulator,* In Proceeding of the 1987 IEEE Conference on Robotics and Automation, Raleigh, NC.

I. Ahmed(1991), *Digital Control Applications with the TMS320 Family, Selected Application Notes,* Texas Instruments Inc.

J. J. E. Slotine and W. Li(1987), *Adaptive Manipulator Control - A Case Study,* pp. 1392-1400, Proc. IEEE Conf. Robotics and Automation.

N. Sadegh and R. Horowitz(1990), *An Exponentially Stable Adaptive Control Law for Robot Manipulators,* IEEE Trans. Robotics and Automation, Vol. 9, No. 4.

P.C.V. Parks(1986), *Lyapunov Redesign of Model Reference Adaptive Control System,* pp. 362-267, IEEE Trans. Auto. Contr., Vol. AC-11, No. 3.

P. Tomei(1991), *Adaptive PD Controller for Robot Manipulators,* IEEE Trans. Robotics and Automation, Vol. 7, No. 4.

R. Ortega and M. W. Spong(1989), *Adaptive Motion Control of Rigid Robots: A Tutorial,* pp. 877-888, Automatica, Vol. 25.

S. A. Bortoff(1994), *Advanced Nonlinear Robotic Control Using Digital Signal Processing,* IEEE Trans. Indust. Elect., Vol. 41, No. 1.

S. Dubowsky and D.T. DesForges(1979), *The Application of Model Reference Adaptative Control to the Robot Manipulators,* pp. 193-200, ASME J. Dyn. Syst., Meas., Contr., Vol. 101.

Y.K. Choi, M.J. Chang, and Z. Bien(1986), *An Adaptive Control Scheme for Robot Manipulators,* pp. 1185-1191, IEEE Trans. Auto. Contr., Vol. 44, No. 4.

THE HIMAC PROJECT: HIERARCHICAL MANAGEMENT AND CONTROL IN MANUFACTURING SYSTEMS

Canuto E.[1], Christodoulou M.[2], Chu C.[3], Donati F.[1], Gaganis V.[2], Janusz B.[4], Proth J.M.[3], Reithofer W.[4], Vallauri M.[1]

[1] *Dipartimento di Automatica e Informatica, Politecnico di Torino, Corso Duca degli Abruzzi 24, I-10129 Torino, Italy*
[2] *Dept. of Electrical & Computer Engineering, Laboratory of Automation, Technical University of Crete, GR-73100 Chania (Crete), Greece*
[3] *INRIA Lorraine, Technopôle Metz 2000, Rue Marconi 4, F-57070 Metz, France*
[4] *Institute for Real-time Computer Systems and Robotics, University of Karlsruhe, Kaiserstrasse 12, D-76128 Karlsruhe, Germany*

Abstract: The paper presents the major features of the ESPRIT Basic Research HIMAC, proposed to the EC after a thorough investigation of the state of the art of manufacturing systems control. HIMAC has followed a new approach, by developing a specific mathematics, the Manufacturing Algebra (MA), for modelling and controlling the production processes of discrete manufacturing systems. As a key feature, the MA models can be bottom-up aggregated starting from the very detailed levels used in simulation, thus making possible a coherent model hierarchy at the base of the design and realization of hierarchical control strategies and architectures. The formulation of the original mathematical approach and its validation criteria are outlined in the paper. For a deeper understanding, the interested reader can take advantage of the list of references given at the end.

Keywords: Manufacturing systems, Production control, Hierarchical control.

1. INTRODUCTION

The Basic Research HIMAC, where the acronym stands for HIerarchical MAnagement and Control of manufacturing systems, started in the spring of 1994 within the third ESPRIT programme. The HIMAC research does not include practical implementations or realization of prototypes: it aims simply at showing and validating through simulations, feasibility and merits of the original concepts developed as major part of the programme. The ultimate objective was the design of a closed-loop modular and hierarchical control structure between shop-floor level and higher decision levels: a control structure such to be easily implemented by low-cost and distributed IT and applicable to any type of discrete manufacturing systems.

Due to its complexity, production of discrete manufacturing systems is usually managed in a hierarchical way, i.e. decisions are taken and progressively detailed top-down, from strategic and tactical plans looking at several months down to daily/weekly shop-floor orders and finally to the real-time dispatching of machine operations. Disturbances and random events are at least tentatively attenuated bottom-up, but it is a common experience that discrepancies between plans and actual production easily propagate and drift bottom-up, forcing frequent recalculations of the higher-level plans which reveal themselves all but not optimal.

What seemed missing and justified HIMAC was a unifying methodology for the design of the various levels of the control hierarchy, based on a specific mathematical formulation of the discrete production processes. The mathematical models should have been capable of describing the production phenomena in a way suitable at any decision level, by progressively and in a coherent way aggregating the very detailed models used in simulation runs.

2. MODELS OF MANUFACTURING SYSTEMS

HIMAC modelling procedure is axiom-based, i.e. although the authors were keeping their eyes on the manufacturing technology, they meant to formulate a self-consistent mathematics, the Manufacturing Algebra (MA), with axioms and theorems. The MA employs, to denote its elements, the same terms cu-used in manufacturing engineering, but the Algebra elements should not be confused with the physical reality, which behind the same name hides something complex and changing from case to case. The modelling body of the MA consists of two parts, the *Product* and the *Production Modelling*, which are intimately connected since the latter part describes the implementation in time and space of the former elements. To this end, the same symbols and acronyms were used, with the care of starring the Product Modelling elements and of referring them as the *reference* ones.

Modelling problem have been tackled and solved in two ways:

1) Definition of five mathematical elements - objects, manufacturing operations, stores, production units, control units - apt to describe the dynamic phenomena of production processes as the effect of the decisions taken at any level.
2) Rules for simplifying the detailed models and creating aggregate elements having the properties of the five basic elements. The higher-level decisions will be taken from simplified models inferred from the detailed ones. The key concept is the aggregation of production units, which at low level describe shop-floor stations and at a higher level may describe a set of such stations managed by a single control unit.

2.1 Product modelling

Product modelling aims at describing the set of products that the factory management plans to produce. A product will be described as the result of a series of manufacturing operations which, starting from raw materials, makes the end-products by assembling semifinished and components. Product modelling is not concerned with *when*, *where* and by *whom* products will be made, but just with *what* and *how*. To this end, only two mathematical elements need to be defined: the *objects*, their types and their quantities, and the *manufacturing operations* together with their algebra. For a deeper understanding, refer to (Canuto & al., 1991, 1993b, 1994b, 1995a, 1996a and 1996c).

Objects, types and quantities. All the parts which can arise or die during production are collected in a numerable set $\mathcal{O}$, called the *universe of the objects*. Objects which can be each other replaced during production are collected into equivalence classes, called *object types*. The key assumption is that the set $\mathcal{K}$ of the object types is *finite*. Each object type can be therefore indicated by a finite integer $k=1,...,n_k$. Product modelling is only concerned with object types and their quantities $q^*(k)$. The ordered list of all the object quantities is collected in a vector $\boldsymbol{q}^*$ belonging to the quantity vector space $\mathcal{Q}^*$. Note that the quantity vectors might be very sparse, since most of their entries might be equal to zero.

Manufacturing operations and their algebra. A (reference) manufacturing operation (or shortly MO) is defined as an ordered pair $A^*=(\boldsymbol{u}^*,\boldsymbol{y}^*)\in\mathcal{Q}^*\times\mathcal{Q}^*$ of quantity vectors defined over a set $\mathcal{K}$ of object types. The vector $\boldsymbol{u}^*$ is called the *input vector* and the vector $\boldsymbol{y}^*$ the *output vector*. Accordingly, the object types having a non zero quantity in $\boldsymbol{u}^*$ will be the input objects and those having a non zero quantity in $\boldsymbol{y}^*$ will be the output objects. The reference MOs are mathematical entities which can be algebraically composed by addition (or parallel) and multiplication (or series) (Canuto & al., 1993b). Multiplication is useful to built series of MOs which are each other interconnected, i.e. exchange their input/output objects. Starting from a finite set $\mathcal{M}$ of independent MOs (called the elementary MOs), the algebra allows to build a numerable set of complex MOs, each one. defined by an algebraic expression of elementary MOs.

Product model. Using the above concepts, a list of factory products can be described by a finite and independent set $\mathcal{M}$ of elementary MOs. The analysis of the input and output objects of the various MOs partitions the set $\mathcal{K}$ into four disjoint subsets: the non-used objects, the raw materials occurring only in input, the finished products occurring only in output and the semifinished, i.e those objects which are exchanged between different MOs and may be either consumed or released (reusable objects) at the process end. Given now a pair $\{\mathcal{K},\mathcal{M}\}$, each finished product in $\mathcal{K}$ can be uniquely described by a list of the MOs belonging to $\mathcal{M}$, called the Bill-of-Manufacturing Operations (BOMO) of that product. Note that all the MOs of a BOMO are each other interconnected in tree-like form. Then, a BOMO contains all the MOs that the factory has to implement in order to yield the relevant finished products.

2.2 Production modelling

Production modelling aims at describing *when*, *where* and *who*, in a production plant, will implement the different BOMOs describing the planned products. To this end, first the concept of object quantity was revised, by introducing space and time coordinates; then the asynchronous time variation of object quantities and other variables was modelled with an algebra of event and event sequences. Finally, three dynamic elements, the storage units (SU), the production units (PU) and the control units (CU), together with their state variables and equations, were introduced to describe the time evolution of production processes. For a deeper understanding, refer to (Canuto & al., 1992, 1993a, 1994a, 1995b, 1995c and 1996b).

Quantity vectors and events. A first modelling issue concerned the fact that, during a production process, objects of the same type, but located in different factory spaces and at different times, cannot be considered each other replaceable. Therefore, a new finite set $\mathcal{S}$ of equivalence classes, the *locations s*, was introduced in the universe $\mathcal{O}$, and a quantity vector $\boldsymbol{q}(t)\in\mathcal{Q}$ was defined as a two-index vector, having components denoted by $q(k,s;t)$ and always referred to a time instant t. A second aspect concerned the asynchronous born and death of hundreds, thousands of different objects during production processes. They were formulated by introducing in MA the concepts of event and event sequence. An event $\boldsymbol{e}$ is a pair (time t, fact ξ), where the time t is a real number and the fact ξ is an element of a set. For instance a drawing/delivery event is a pair (t,q) where $q \in \mathcal{Q}$ is a quantity vector, listing all the quantities of different object types to be drawn from or delivered to the locations of the production plant. A finite list of events built over the same set of facts is defined as an event sequence. Event and event sequences can be summable with certain restrictions.

Storage units. A storage unit is defined as one of the factory locations $s \in \mathcal{S}$ and by the state variables $x(k,s;t)$, listing the quantities of the objects stocked in the location s at time t. The two-index quantity vector $x(t) \in \mathcal{Q}$, listing all the components $x(k,s;t)$, is the state vector of all the factory SUs. A *factory* is defined as a network of SUs and production units.

Production units. They are defined as dynamic elements which at given dates are commanded by a CU to actuate an MO by transforming the input objects into output objects. The input objects are drawn at a certain time from their input SUs and the output objects are delivered after a certain interval to their output SUs. Any PU can actuate only one MO at a time, which takes a finite and variable interval, during which the PU can not actuate other MOs (*working* state). At the end, the PU becomes free (*waiting* state) and it can actuate other commands. The set $\mathcal{P}$ of the factory PUs is finite and each PU is denoted by an index p. The concept of PU is very general and encompasses manual, automatic working stations and transport units. Each PU is programmed to actuate only a finite set of admissible MOs, selected within the set $\mathcal{M}$ of the reference MOs defining the product BOMOs. It is a task of the relevant CU to command its own PUs to actuate since a time t one of their programmed MOs.

Programmed and actuated MOs. When a reference MO has to be programmed on a specific PU, one has to *schedule* when and from which SU the different input objects will be drawn and similarly when and to which SU the output objects will be delivered. Using the concept of event sequence, a programmed MO was defined as a sequence of drawing and delivery events. Of course, programming the same reference MO on different PUs will bring to different event sequences. The event sequences of a programmed MO are said to be *relocable*, since their real times of occurrence are not known. Only when a programmed MO is commanded to a PU to be actuated since a time τ, the real occurrence time t of a relocable drawing/delivery event (t_0,q) will become known, as a (possibly stochastic) function $t=t(t_0,\tau)$ of the relocable time t_0 and of the MO starting time τ. At the end, three different MO models have been defined in MA: the reference model, out of space and time, the programmed model, listing the expected events requested to fulfill the MO on a specific PU, and the actuated model, describing how the event sequence will actually occur.

Discrete-event state equations of a factory. The model of the above elements have lead to a set of state equations capable of describing the asynchronous evolution of any factory, i.e. of a network of SUs and PUs, receiving their commands from a CU. They are presented in the companion paper (Donati & al., 1997). The basic concepts are the *input and state variables*. An input is an event sequence of the commands dispatched by the CU the the factory PUs and asking the actuation of a list of programmed MOs. The state variables are the event sub-sequences of the actuated MOs which at the current time t have not yet been occurred. As far as a new command is received by a waiting PU, a new MO is actuated and new event sequences are added to the state variables; as far as an event of the actuated MOs occurs the state variables are updated. Note that the state variables of a factory are not just the quantities of the objects stocked in the factory SUs, but also the so-called work-in-progress, i.e. all the programmed events of the already actuated MOs, which did not yet occur.

3. HIERARCHICAL CONTROL

The main objective of HIMAC was a methodology for designing and validating closed-loop production control systems organized in a hierarchical way. Hierarchy is the common way in industry and in social life to manage and control very complex, uncertain and highly variable systems. In a hierarchical control:

1) the dynamic models and the information from the real system to be managed become simpler and simpler as far as one goes towards higher control levels;
1) the decisions (commands) become more and more detailed in time and space as far as one goes down in the hierarchy.

A specific aggregation theory was developed in HIMAC to made applicable such concepts in the design of production control systems, taking into account that the factory dynamic models are discrete-event state equations. The key concept is the aggregation of the PUs managed by the same CU into a new aggregate PU.

3.1 Aggregation theory

Aggregation theory is a chapter of the Manufacturing Algebra which was developed to simplify in a coherent way product and production models. The key concepts are MO and PU aggregation.

Aggregation of MOs. Given the BOMO of a finished product, an aggregate MO is defined as a subset of the elementary MOs defining the BOMO such that they are each other interconnected. Hence an aggregate MO can be seen as a sub-tree of a BOMO. Using aggregate MOs, any BOMO can be simplified in terms of input-output objects and MOs. At the end of the aggregation process, any BOMO can be aggregated into a single MO, having as input objects only the raw materials and as output object its finished product. An aggregate MO will denoted with MO_k, where $k \geq 1$.

Aggregation of PUs. MO aggregation becomes useful when it is employed in the creation of aggregate PUs, capable of actuating a finite set of aggregate

MOs. It should be however pointed out that, once a factory has been designed, the only aggregation process of interest can not start from the MOs,, but only from the PUs themselves. In principle any set of PUs can be aggregated; in practice the only aggregations of interest are suggested by the factory design and the management customs. An aggregate PU is defined as a set of elementary PUs managed by the same control unit. Then the main issue is to build the dynamic model of an aggregate PU, using the elements of the Manufacturing Algebra. If such a problem would have a solution, one could build, starting from the shop-floor PUs, a hierarchy of aggregated PUs having at the top the whole factory described as a single PU (the highest-level PU). HIMAC has developed a methodology to solve this problem. An aggregate PU will be denoted by PU_k.

The most significant step is the construction of the aggregate MOs which are to be programmed on and actuated by aggregate PUs. The aim is to construct suitable *bundles* of aggregate MOs derived from the MOs already programmed on the elementary PUs; they should encounter the interest of the management by making efficient use of production capacity and by favouring enterprise policies. For such reasons, this step can not be fully automatic, but in the hands of production managers, who have to validate and certify the programmed MOs of their aggregate PUs (think to a workshop, to a department, to the whole factory). HIMAC methodology offers a formulation of this step, which more or less corresponds to the common practice of lot sizing and product mixing. Of course each *MO bundle* will have a programmed and actuation model, obtained by scheduling the relevant elementary MOs.

3.2 Control architecture

HIMAC control architecture has been conceived to be hierarchical, modular (the control functions and hence the software of any CU have to be tightly the same), customizable to any discrete manufacturing enterprise and implementable on low cost IT. The key concepts concern the functionalities of the Control Units and how the different CUs are configured and exchange the real-time data.

Control units. A CU is mathematical element which describes the entity (f.i. a computer node visible and maneuverable by a manager), which has in charge a subset of the factory PUs. It receives production orders from the upper-level production manager and has the task of synchronizing the operations (MOs) of the PUs in charge, in order to punctually fulfill the received order. Note that in HIMAC, the orders coming from an upper-level, say *k+1*, are formulated as programmed MO_{k+1} of the aggregated PU_{k+1} defined by the CU_k itself and all the PU_k which are in charge of CU_k. The main functions of a CU_k are:

1) Disaggregation of the received MO_{k+1} into the list of the programmed MO_k to be dispatched to the different PU_k.
2) Record keeping of the semifinished objects stocked in internal SUs and of the PU work-in-progress: this is made by real-time updating the discrete-event state equations of the PUs.
3) Feasibility checks on the pending MO_k and dispatching of feasible MO_k to the waiting PUs (dispatching may be constrained by precedence rules established by scheduling).
4) Updating of the pending list and transmission to the upper level CU_{k+1} of the occurred events (objects delivery/drawing) of all the production orders MO_{k+1} still not completed.

The second and fourth functions are observer functions of the control theory. Instead the first and third functions require specific control algorithms.

3.3 Control algorithms

The production control algorithms are usually subdivided into three classes: planning, scheduling and real-time. In HIMAC formulation, the first stage (from day to weeks) of medium-term production plans is the input to the highest-level CU and hence to HIMAC control hierarchy. The outputs of production plans are usually quantities of finished products, which are further exploded (MRP) into orders of semifinished products for the different factory departments and shops (Chu & al., 1996a). Only the shop orders are then converted by routing and scheduling into machine operations. Note that HIMAC suggests, instead, of formulating the production orders at any level in terms of pre-defined (aggregate) MOs, whose models and properties had been already validated.

The second class of algorithms concerns scheduling. Due to their mathematical complexity, myriad of heuristic algorithms have been proposed in the literature (Chu & al., 1996b, Proth & al., 1996). In HIMAC scheduling algorithms are mainly seen as a tool for programming on a PU the event sequences of any non trivial MO (especially aggregate MOs and their bundles).

The third class is that of the real-time algorithms, used to select the feasible MOs to be dispatched to the waiting PUs. The HIMAC real-time algorithms are still under investigation and test, but they will be very simple, like push-pull strategies, due to their closed-loop nature. Closed-loop algorithms, based on Neural Networks, have been also studied.

3.4 Neural networks

HIMAC was supporting also a research effort in the field of Neural Networks (NN) aiming at studying and assessing alternative control algorithms. After preliminary studies (see Christodoulou & al., 1996a), the research focused on the dynamic modelling (Christodoulou & al., 1996b, Rovithakis & al., 1995) and on the scheduling of small cells including different machines capable of processing different part types (Kosmatopoulos & al., 1995). The control objective is to reach the demanded production rate starting from empty internal buffers.

The cell state equations were adapted from

continuous-time dynamic NN models. The control input was defined as the machine working rate, which lead to model nonlinearities approximated by high-order sigmoid functions. The control input is synthesized from an adaptive closed-loop function of the buffer states. To transform the continuous-time control into a sequence of asynchronous dispatching times, the controller output is sampled at appropriate time instants and the working rate converted into a working time interval for the waiting machines. In case of multi-product machines, a priority criterion selects the part to be dispatched.

4. PERFORMANCE ASSESSMENT

To assess HIMAC theory, two test cases of increasing complexity and taken from existing plants have been studied and are under simulation on a powerful simulation test-bed, which was ad hoc implemented with the SIMPLE++ package. For further details, please refer to (Janusz & al., 1995, 1996, Reithofer & al., 1996).

4.1 Test case 1

The first case was derived from an assembly plant installed in a university laboratory. The plant automatically assembled sixteen different variants of a robot gripper, shown in Figure 1.

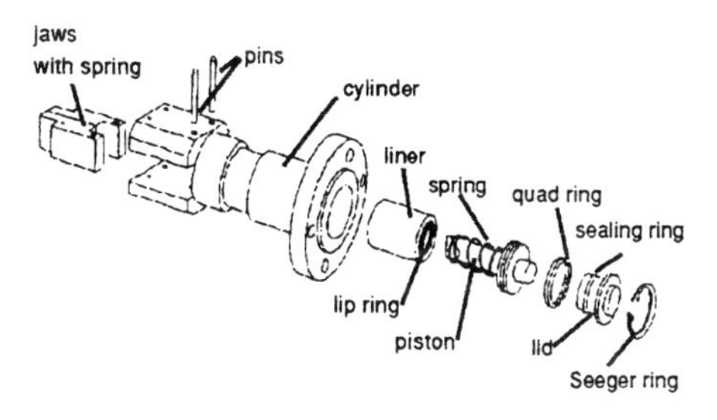

Fig. 1. Exploded view of the robot gripper.

The assembly plant is made by eight stations connected by a conveyor belt (see Figure 2).

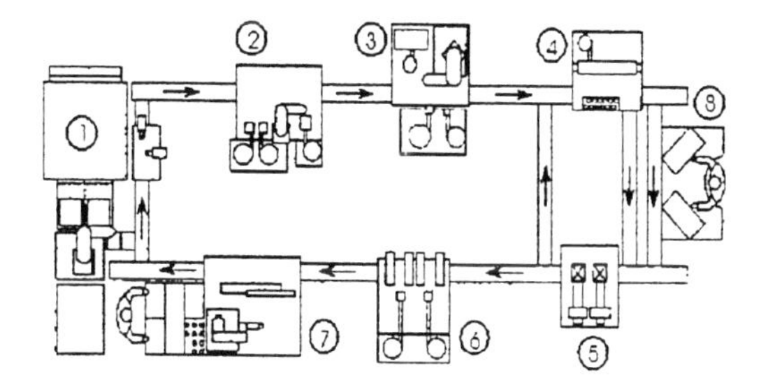

Fig. 2. Layout of the gripper assembly plant

The main parts of each gripper are transported by an AGV to the first station, where a robot places them on a pallet, and the desired gripper variant is coded on a read/write memory. Each pallet passes through each station where robots perform the assembling operations and are capable of recovering a few errors. In case of not recoverable errors, the pallet is coded as defective and sent to a manual station for error removal. Gripper quality and functionality are inspected in the final station, before the gripper is removed from the pallet. The pallet is then sent back to the first station ready to be equipped again.

Working times are around 60s and transport times around 20s. A detailed model of the plant was built on the HIMAC testbed. The control strategies, under development, are based on a simplified model of the plant. This case is suitable to test a single-layer control system made up of a single CU coordinating all the eight (or less) assembly stations.

4.2 Test case 2

The second case is the Machine Tool Division of the oldest industrial German enterprise. The division employs about 120 people and manufactures universal machining centres. Usually, only simple milling machines, like that in Figure 3, are made regularly in lots of five items per year. All the other centres are produced on order. To simplify manufacturing, all the centres are assembled from standard modules, which can be partly customized.

Fig. 3. A standard milling machine produced by the Machine Tool Division selected as a test case.

The layout of the production plant is shown in Figure 4, together with the main part-flows.

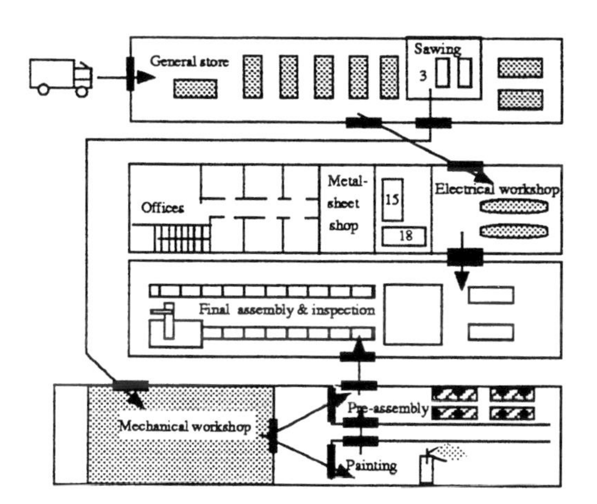

Fig. 4. Layout of the Machine Tool Division.

The module production starts in pre-production

shops, then passes to the mechanical shop and finally to the pre-assembly shop. The modules are then assembled together and electrically connected in the final assembly and inspection shops. Electrical parts and harness are prepared in the electrical shop.

The most critical and complex shop to be modelled and managed was the mechanical shop, which is committed to different productions: the mechanical working of the module parts of the on-order machining centres and sub-contracted orders in lots of hundreds of equal parts (in total about six hundred orders per month). The shop includes automatic machining centres and manual machine tools (MT).

First the BOMOs were studied (f.i. the mechanical shop may require from four to twenty different MOs). Then the production model was developed. The most critical issue was the definition of the PUs. Three alternatives were analyzed: worker, machine, machine+worker. For several reasons PU=worker was assumed to model manual MTs and assembling stations; PU=machine was instead assumed to model automatic machining centres. In the former case the MTs are modelled as reusable objects: vice versa workers are reusable in the latter case. The third model was never used. Design and test of a two-level hierarchical control system are under development.

ACKNOWLEDGMENTS

HIMAC 8141 was funded by the ESPRIT III Basic Research programme. It has been coordinated by EICAS Automazione spa, Torino.

REFERENCES

Canuto E., Donati F. & Vallauri M. (1991). Factory modelling and production control. In *Proc. 10th IASTED Int. Symp. Mod. Ident. & Contr.* (Hamza M. ed.), p.475-478. Acta Press, Anaheim.

Canuto E., Donati F. & Vallauri M. (1992). Manufacturing plant dynamics. In *Proc. 11th IASTED Int. Symp. on Mod. Ident. & Contr.* (Hamza M. ed.), p.498-502. Acta Press, Zurich.

Canuto E., Donati F. & Vallauri M. (1993a). Factory modelling for production control. In *Proc. Int. Conf. on Ind. Eng. and Production Management*, p.737-746. FUCAM, Mons (Belgium).

Canuto E., Donati F. and Vallauri M. (1993b). Factory Modelling and Production Control", *Int. J. of Modelling and Simulation*, **13** (4), p.162-166.

Canuto E., Donati F. & Vallauri M. (1994a). Modelling manufacturing dynamics for production control. In *Proc. 2nd IEEE Mediterranean Symp. on New Directions in Control & Automation*, p.142-149. Chania (Greece).

Canuto E., Donati F. & Vallauri M. (1994b). A new approach to modelling manufacturing systems. In *Proc. 27th ISATA, Int. Conf on Lean-Agile Man.*, p.317-324. ISATA, Aachen (Germany).

Canuto E., Donati F. & Vallauri M. (1995a). An algebra for modelling manufacturing processes. In *Proc. 3rd IEEE Med. Symp. on New Directions in C. & A., Vol.II*, p.438-445. Limassol (Cyprus)

Canuto E., Donati F. & Vallauri M. (1995b). An approach to factory dynamics based on manufacturing algebra. In *Proc. 3rd IEEE Med. Symp. on New Directions in C. & A., Vol.II*, p.446-453. Limassol (Cyprus).

Canuto E. (1995c). A control-oriented dynamic model of discrete manufacturing systems. In *Integrated Man. Systems Eng.* (Ladet P. & Vernadat F. eds.), p.174-188. Chapman & Hall, London.

Canuto E., Donati F. & Vallauri M. (1996a). Advances in manufacturing algebra: foundations of production planning. In *Proc. 1st HIMAC W/S, A new math. approach to manufacturing eng.*, (Vallauri M. ed.), p.19-31. CELID, Torino.

Canuto E., Donati F. & Vallauri M. (1996b). Factory models based on manufacturing algebra. In *Proc. 1st HIMAC W/S, A new math. approach to manufacturing eng.*, (Vallauri M. ed.), p.33-40. CELID, Torino.

Canuto E., Donati F. & Vallauri M. (1996c). Production planning using Manufacturing Algebra. In *Proc. 4th IEEE Med. Symp. on New Directions in C. & A.*, p.177-182. Chania (Greece).

Christodoulou M. & Gaganis V. (1996a). Determining manufacturing cell topology using neural networks. In *Proc. 4th IEEE Med. Symp. on New Directions in C. & A.*, p.737-742. Chania (Greece).

Christodoulou M. & Lambrou L. (1996b). Manufacturing algebra interface to neural networks. In *Proc. 1st HIMAC W/S, A new math. approach to manufacturing eng.* (Vallauri M. ed.), p.63-81. CELID, Torino.

Chu C., Proth J.-M., Sauer N. & Mehra A. (1996a). Hierarchical production management systems. In *Proc. 1st HIMAC W/S, A new math. approach to manufacturing eng.* (Vallauri M. ed.), p.55-61. CELID, Torino.

Chu C. (1996b). Production scheduling and control: a unified framework. In *Proc. 4th IEEE Med. Symp. on New Directions in C. & A.*, p.381-385. Chania (Greece).

Donati F., Canuto E. & Vallauri M. (1997). Advances in Manufacturing Algebra: discrete-event dynamic models of production processes. In *Preprints 1st IFAC W/S Man. Systems: MIM'97*, p. 461-467. Vienna (Austria).

Janusz B., Reithofer W. & Raczkowsky J. (1995). Simulation for hierarchical management and control in manufacturing systems. In *Proc. 3rd IEEE Med. Symp. on New Directions in C. & A., Vol.II*, p.430-437. Limassol (Cyprus).

Janusz B., Reithofer W. & Raczkowsky J. (1996). Interfaces of simulation to algebra based in manufacturing systems. In *Proc. SCSC '96*. Ottawa (Canada).

Kosmatopoulos E. & Christodoulou M. (1995). A new model used for suboptimal solutions in control of manufacturing processes. In *Proc. 3rd IEEE Med. Symp. on New Directions in C. & A., Vol.II*, p.423-429. Limassol (Cyprus).

Proth J.-M. & Sauer N. (1996). Sensibility analysis for scheduling manufacturing systems. In *Proc. 4th IEEE Med. Symp. on New Directions in C. & A.*, p.743-747. Chania (Greece).

Reithofer W., Janusz B. & Raczkowsky J. (1996). HIMAC validation by simulation test cases. In In *Proc. 1st HIMAC W/S, A new math. approach to manufacturing eng.*, (Vallauri M. ed.), p.41-54. CELID, Torino.

Rovithakis G. & Christodoulou M. (1995). On modelling the factory dynamics. In *Proc. 3rd IEEE Med. Symp. on New Directions in C. & A., Vol.II*, p.408-422. Limassol (Cyprus).

PROTOTYPE FOR DECENTRALIZED COMPUTER ASSISTED BUDGETING FOR LARGE NETWORKED ENTERPRISES

Johannes Grobe*

* *Department of Informatics in Mechanical Engineering (HDZ/IMA), University of Technology (RWTH) Aachen, Germany*

Abstract: Large enterprises like hospitals are highly complex systems in teams of dynamic interrelations between people, the organization they work in and the technology they work with. Such complex systems must constantly develop and redesign themselves to meet all requirements of their task. This paper deals with organizational and technical development and the reorganization of large hospitals. The development of some technical components of the information system for these hospitals is described in some more detail. This example shows the close interrelation of technological change and organizational development: the hospital wards operate on the basis of decentral revenues and costs considerations. Revenues and costs should be allocated over the decentral units and decentralized data capture and processing of the diagnosis and treatment of any patient. Decentral performance however is to be measured and accounted for every single ward similar to profit centres in industry.

Keywords: Hospitals, Decentralization, Profit-Centre, Budgeting, Re-Designing Technology Organizational Development

1. INTRODUCTION

Large enterprises like hospitals are economically challenged especially due to their complex structures. They are frequently controlled by central data administration.

On the one hand, there is hardly any awareness of costs which has resulted in too little transparency of profits and losses, on the other hand, the hierarchical differences between doctors and nurses often hinder economic acting.

Therefore profit-centre structures have been introduced in a university hospital, in order to transform new economic ideas into action. Structural and strategic differences have, thus, been introduced in the way that profit and loss accounts are decentrally organized. Each accounting system involves the profits and costs caused by only one ward or a group of related wards.

Within the framework of the research project "Organizational Development in Hospitals - Introduction of Profit-Centre-Structures" financed by the German Government the HDZ/IMA University Aachen has developed a profit-centre concept which is supported by a prototype-software suitable for the different requirements of each ward.

The university hospital to be discussed here consists of 1500 beds, more than 4000 employees and approximately 3000 medical students. The Centre of Neurology and Neurosurgery will be considered in some more detail and in this report. It consists of several different wards. It comprises 143 beds.

All wards, serving units and laboratories are located in one building. The wards are assigned to the centre´s two departments "Neurosurgery" and "Neurology".

To a large extent important plans and decisions occur within this centre aiming at a cost-oriented and economic acting as in any economic enterprise.

Thus, sufficient information (cost-income balances, pre-calculations/pre-estimating etc.) has to be provided for a clear view of the centre´s own profit-cost situation. Decentral support systems are needed processing relevant information and comparing profits and costs to determine the success of the organization unit within the respective period.

In the centre´s wards a computer-supported profit-loss accounting system that integrates nursing and medical tasks as well as administrative tasks has been implemented: this is the main content of the research project described here.

Each profit-centre unit, called "profit-loss unit" (EKE) is managed by a management team consisting of a leading senior physician, a senior nursing officer and an employee from the centre´s administration. This management group works together with the director of the centre (Fig. 1).

All actions of other service units within the centre are calculated as to how they contribute to the economic performance of these wards. In the some way, all 'business' of the centre with any other centre' or outside service unit is calculated as to how it contributes to the performance of the wards. In this way, the wards are the only entities of the centre which transmit the centre´s economic performances to the public (outside world)

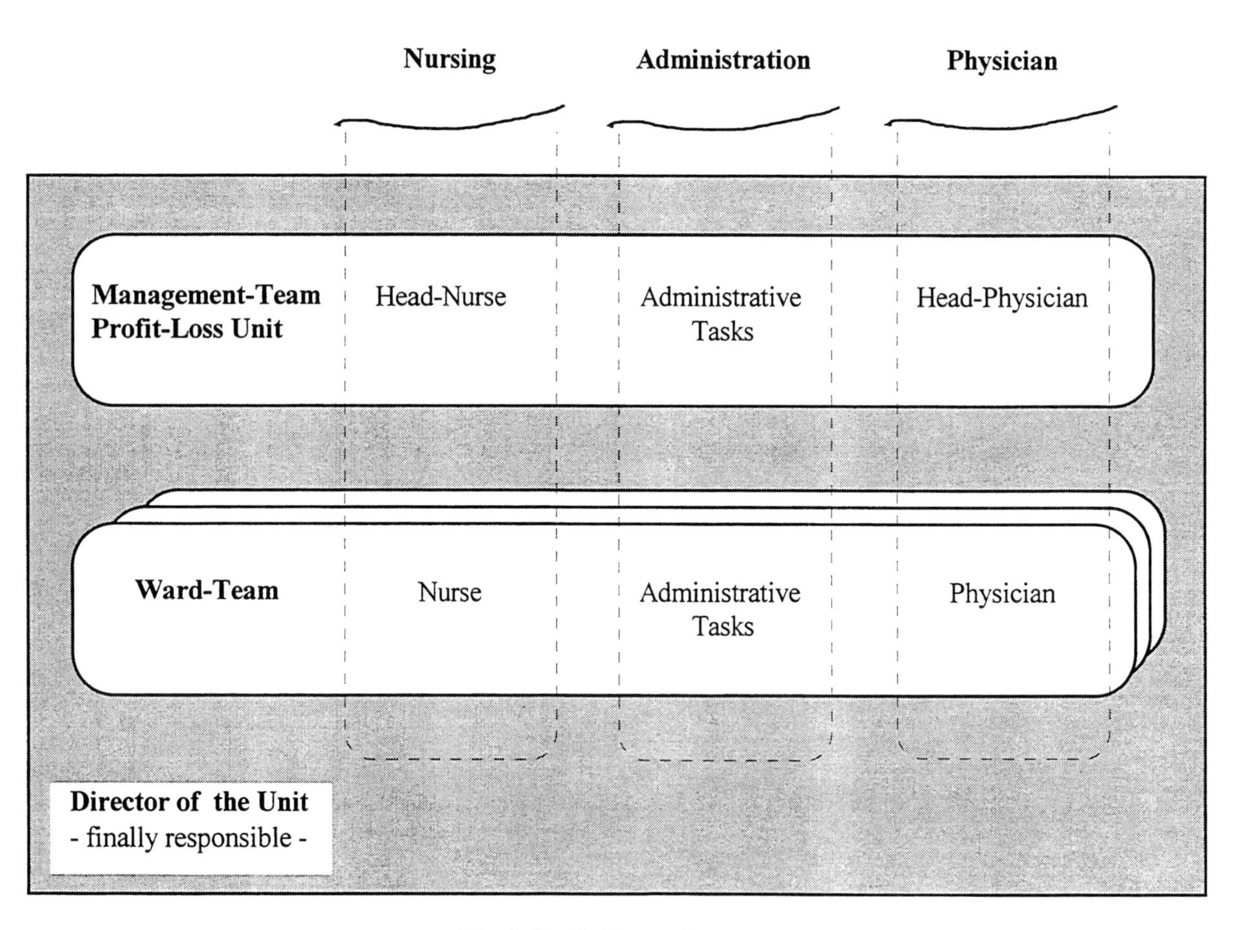

Fig. 1: Profit-Centre Structure

2. THE TECHNICAL SYSTEM

The newly developed profit-loss accounting system guarantees a cause-oriented classification of all informal expenditures and profits in the ward: costs and profits of the different departments, e.g. the operating theatre, EEG/EGG or physiotherapy, which have caused these costs are allocated to the respective wards.

In order to make the new system work, previous administrative rules became obsolete, the cost-splitting ratios of the hospital at large needed modernization to clearly show and demonstrate the real profit-loss situation with all its interconnections, for any of the wards, at any time.

According to the method developed by the

HDZ/IMA those clearing methods that require long, confusing data and figure lists are to be excluded.

A specification of result-oriented ratios to be allocated on the respective cost bearers has turned out to be unavoidable in the centre "Neurosurgery", which is already organized relatively autonomously.

Figure 2 shows the different departments of the centre, e.g. the intensive care unit, the physiotherapy unit in which the wards which are listed as single accounts to be balanced with each other according to respective cost-splitting rules/orders. The cost allocation structure combines centrally determined splitting factors and decentrally elaborated result/capacity features. Thus a direct data transfer from the central computer to the computer in the pilot ward has been initiated.

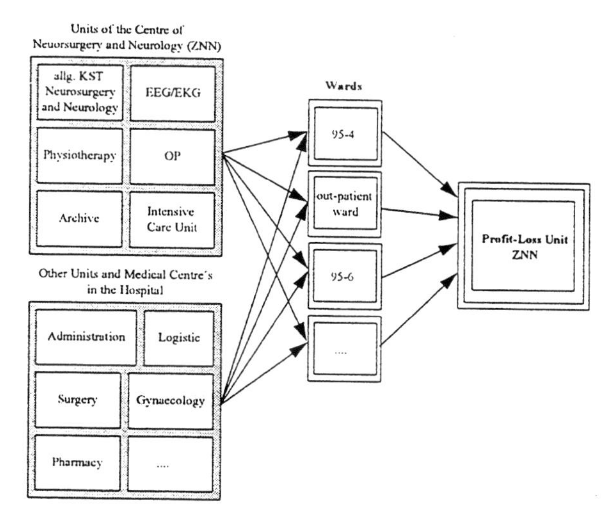

Fig. 2: Cost Allocation Structure

For the decentralized data capture and processing it is necessary the whole prodecure for accounting ?? with regard the medical treatment of any individual patient: the ward is expected to record the diagnosis, the treatment suggested by the physician and the treatment performed. The physician and nurse use specific forms, which are created in this process. They consist of one or two sheets of paper listing the mean diagnosis and common treatment to one of the wards. The list are typed normal language, each item is characterized by barcode. The physician marks the items corresponding to the treatment of the patient. Subsequently he hands the filled forms to the administrative nurse or the ward. The nurse scans these forms in order to the data of the diagnosis and the treatment of the patient into the ward's PC. This data is correlated to the costs depending whether they accrue from the ward or from an outside service.

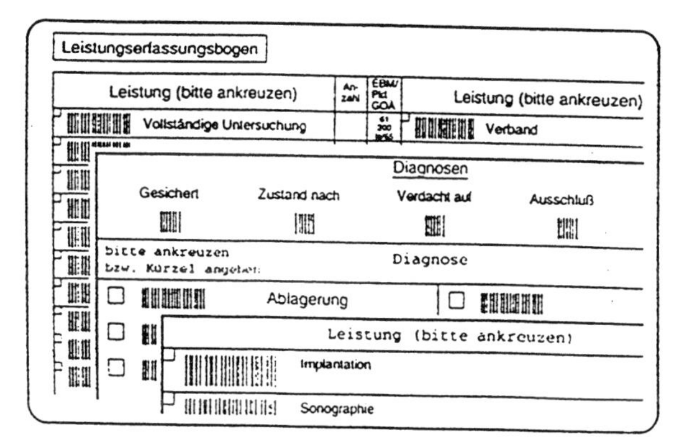
Leistungserfassungsbogen
Leistung (bitte ankreuzen)
Leistung (bitte ankreuzen)
Vollständige Untersuchung
Verband
Diagnosen
Gesichert
Zustand nach
Verdacht auf
Ausschluß
bitte ankreuzen bzw. Kürzel angeben
Diagnose
Ablagerung
Leistung (bitte ankreuzen)
Implantation
Sonographie

Fig. 3: Forms for diagnosis and treatment

This computer system needs to allow to record and process decentrally the diagnosis and treatment data through the PC. Only the frame data's networked and transmitted to the central hospital computer invoicing of patients and insurance companies.

The combination of there two approaches gives a specific structure to this large hospital: all actions within the ward are recorded, processed and budgeted within the wards where they take place. The overwhole budgeting of the large hospital is based on the client-supplier concept of the wards as autonomous profit-centres. This concept is only possible through implementing networked PC's in the wards and the corresponding decentralized software. In several aspects this concept is contradictory to the implementation of large, centralized computer systems as they are presently taking place.

3. THE USER INTERFACE FOR PROFIT-LOSS ACCOUNTING

Despite the large number of different functions of the implemented programs the user interface has been designed to be user-friendly.

The division of "choice of calculated month", "input of data", "monthly profit-loss-sheet" and "result-and cost-splitting settlement" as the main functions of the interface make time the profit-loss developments of the preceding month for the first (Fig. 4). The user can thus pre-calculate the current month and compare the pre-calculated profit and loss data with the current bookings of the financial accounts department.

The user interface corresponds to the users' wishes: all necessary relevant data appear at the display when the user clicks the input key.

In the right top half of the display appears the profit-loss comparison of the pre-calculation of the ward and a current booking.

A sort of "speedometer" shows profit and loss in either a green or red range. The users can thereby get an overall picture of the current economic situation.

The particularily revised settlement rules for the system are documented in the computer and can, if necessary, be fetched with the respective key by the user.

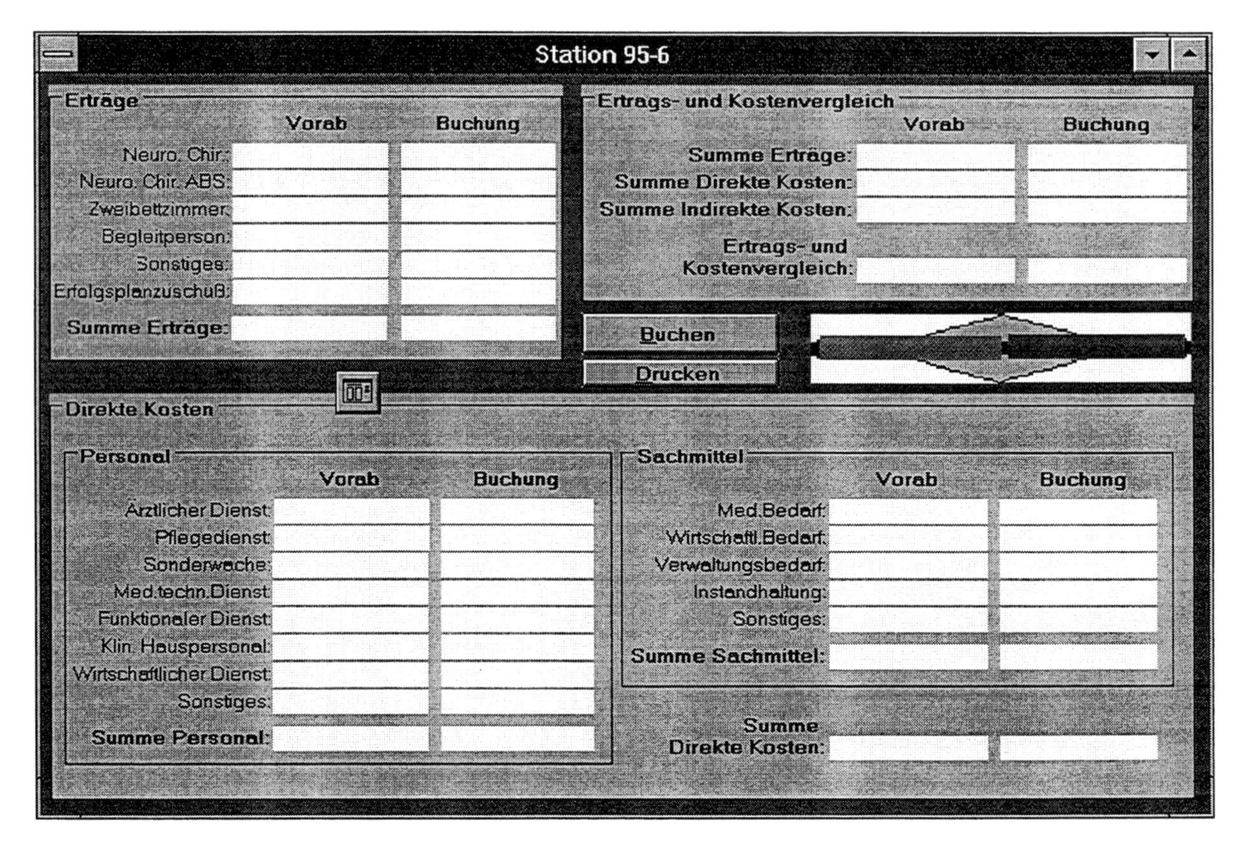

Fig. 4: User Interface

4. EVALUATION

A new cost-centre-structure and changed settlement rules have, already after a short period of time, guaranteed a high transparency of profits and costs.

Profits and costs become transparent and the ward members are given the opportunity to determine and even "plan" their own requirements in the ward. On these conditions they can react flexibly on temporary changes.

The requirements on the prototype are developed in co-operation with the staff members. The prototype has been tested in the pilot ward, improved later on and integrated in the whole process. This iterative proceeding also includies the specific requirements of the users. The staff member are shown the application of the system accept a computer in the ward. Hereby fear of contact should be prevented.

Moreover, the software for the profit-loss-units, is particularlly focused on the criteria "human", "organization" and "technology". That implies that the technology does not determine the system because the technology has been developed to meet the requirements of each ward. It is a technology that its users can apply without any further experiences in the field of computers.

Interactions and cross linkage of technical and organizational spheres have also been taken into account in order to integrate the system in higher levels, like in the entire centre.

The developed decentralized accounting system has finally turned out to be an appropriate means to support the work in the ward of the centre of Neurosurgery.

The users' requirements on the system showed that the software had to be oriented on the ward-specific equipment. At the same time it had to be flexible and sometimes modified to be suitable for further wards.

The integration of the system in the entire system of the whole hospital is thereby guaranteed. With regard to the multitude of data (25000 bookings per month) the decentral computer system is an excellent support system.

The Prototype for decentralized computer assisted budgeting means for the ward members to assume not only medical but also economic responsibility for their own field of action.

5. CONCLUSION

The outcomes of these two examples and several simular projects implemented by the HDZ/IMA - have shown that it is possible to change large complex organizations like hospitals or industrial companies. In order be successful it helps to have the following available:

- an analysis and innovation strategy which is based on a client-supplier concept of the wards as autonomous profit-centres,
- an introduction of project work: various innovation tasks were solved by teams working on certain projects. These teams consisted of members from different sections of the hospital.
- The aim 'economic responsibility' could finally have been translated into action - verified by experiences of staff members of the centre 'Neurosurgery' - thanks to the decentral, partially autonomous profit-loss units.

Economic thinking supported by user-friendly profit-loss-system encourages capacities and economic efficiency in the ward, the centre and maybe even in the whole hospital.

That comes in useful especially to the ones having a legitimate claim to the fulfillment of the hospital's mission - we are talking about the patients!

REFERENCES

Henning, K.; Isenhardt, I.; Grobe, J. and Steinhagen, U. (1996),; Organisationsentwicklung in Groß-kliniken - dargestellt am Beispiel der Einführung von Ertragskosteneinheiten, in: Zeitschrift für Chirurgie, Hüttgen Verlag Heidelberg

Isenhardt, I. (1994); Komplexitätsorientierte Gestaltungsprinzipien für Organisationen - dargestellt an Fallstudien zu Reorganisationsprozessen in einem Großkrankenhaus, Aachener Reihe Mensch und Technik, Aachen

Isenhardt, I.; Grobe, J.; Selbstorganisierte Teamarbeit und Gruppenlernen im Betrieb - Beitrag zur BMBF - Expertenkonferenz „Selbstgesteuertes lebenslanges Lernen?“, 06. - 07. Dezember 1996. Bonn

Isenhardt, I.; Grobe, J. and Steinhagen de Sánchez, U. (1995); Organizational Development in Hospitals - Exemplified by the Introduction of Profit-Centre Structures, IFAC Symposium September 27.-28. 1995, Berlin

IMPLEMENTATION OF COMMUNICATING SEQUENTIAL PROCESSES FOR DISTRIBUTED ROBOT SYSTEM ARCHITECTURES

Gen'ichi Yasuda* and Keihachiro Tachibana**

**Department of Mechanical Engineering, Nagasaki Institute of Applied Science,
536 Aba-machi, Nagasaki 851-01, Japan
**Department of Information Machines and Interfaces, Hiroshima City University,
151-5 Ozuka, Numata, Asaminami-ku, Hiroshima 731-31, Japan*

Abstract: This paper describes an implementation procedure of distributed control software for robot-based manufacturing systems. Robots and other devices such as conveyors and manufacturing machines are defined as object-oriented communicating sequential processes. A modular and hierarchical approach is adopted to define a set of Petri net type diagrams, which represent concurrent control processes for such devices, and asynchronous and synchronous interactions in global process interaction nets. Operational specifications are directly transformed to executable codes in a parallel programming language, such as occam. Implementations of control software on a multiprocessing environment and transputer networks are comparatively evaluated.

Keywords: Multiprocessing systems; parallelism; Petri nets; object-oriented programming; software development; industrial robot systems.

1. INTRODUCTION

Robot-based manufacturing systems, where individual robots pursue its own goals in such a way that some sorts of cooperation and conflict avoidance among related robots are performed in real time, can be viewed as complex discrete event systems of asynchronous, concurrent processes with distributed data and decentralized control. In such multirobot applications, run time synchronization of actions without rigorous scheduling should be achieved by direct communications among robots. Thus, distributed autonomous control architectures are desirable for effective real-time control performance, flexibility of system configuration and reduction of software development time (Hatvany, 1985). In this respect, object-oriented bottom-up approaches seem to be natural for such complex systems (Bucci and Vicario, 1992).

For the specification of flexible manufacturing cells, Petri nets have been utilized as a language suitable for purposes of verification, analysis and simulation. However, ordinary or extended Petri net modelling of complex manufacturing systems involves a highly large number of places and transitions. Petri net based interpreted or compiled program codes are inefficient because they do not exploit the parallelism of Petri nets. Owing to an increasing degree of operational integration of robots and the complexity of real-time control software, suitable techniques and tools for the specification and validation of requirements and for their systematic transformation to executable codes are required.

In this paper, robot systems are specified as a collection of communicating sequential processes, and multiprocessing control structures and their implementations on parallel distributed processing architectures are presented. Global Petri net modelling is described to specify the discrete event dynamics of robots and other industrial devices, and their asynchronous and synchronous interactions. Control structures of the data flows in control systems are defined as a set of protocols for interprocess interactions to perform cooperative tasks. A modular and hierarchical approach for task specifications is adopted toward the integration of more complex robotic manufacturing systems, such that some important structural properties for manufacturing systems, such as boundedness, liveness and reversibility, are maintained in the implementation of control systems. The operational specifications using Petri nets are independent of the implementation mechanisms, and directly transformed to executable codes in a general parallel programming language, occam, the language designed for Inmos transputers.

2. SPECIFICATIONS OF ROBOT SYSTEMS

The specification procedure first defines physical objects in manufacturing systems as state machines, a subclass of ordinary Petri nets. The basic idea behind the distributed autonomous realization of robot systems is the decomposition of the overall control system into a set of autonomous control processes or state machines which communicate asynchronously or synchronously among themselves.

2.1 Object-oriented Petri Net Modelling of Basic Robot Process

To model basic robot processes in manufacturing systems, strongly connected Petri nets, where tokens circulate according to the firing rules associated with transitions, with communication interfaces are introduced. The firing consists in removing the tokens from the input places and adding tokens to the output places after performing the associated actions. In the proposed specification procedure, the firing of a transition corresponds to the beginning or end of a robotic or machine task, or a communication operation, and at the transition firing some instruction in the program code is executed. Thus, a procedure which executes a task can be associated with the transition corresponding to the beginning and the place between the two successive transitions. Hence, a place represents a state of executing a task or waiting for some message by asynchronous communication which can be a precondition or postcondition of a transition.

State machines, a Petri net subclass where each transition has exactly one input place and one output place, model autonomous control processes of robots and other machines. State machines, like real single robot processes, do not include synchronization of concurrent processes. A modular, bottom-up approach is adopted, although an ordinary Petri net can be decomposed into a set of state machines (Breant and Paviot-Adet, 1992). The discrete event dynamics of a robot or machine can be defined in terms of state machines at various aggregate levels. Fig.1 shows a global Petri net representation of a robot process; the highest level control structure using global working and free states of the robot. Each robot performs an independent task when command from the coordinator is for independent action. On the other hand, when command is for cooperation, the robot perform a cooperative task synchronously with the other robot or machine. Additional pieces of information such as timing requirements and priorities can be associated with mutually exclusive transitions in charge of task request. Each robot process or autonomous control process can be viewed as an object which controls the data flow according to the current state and incoming messages.

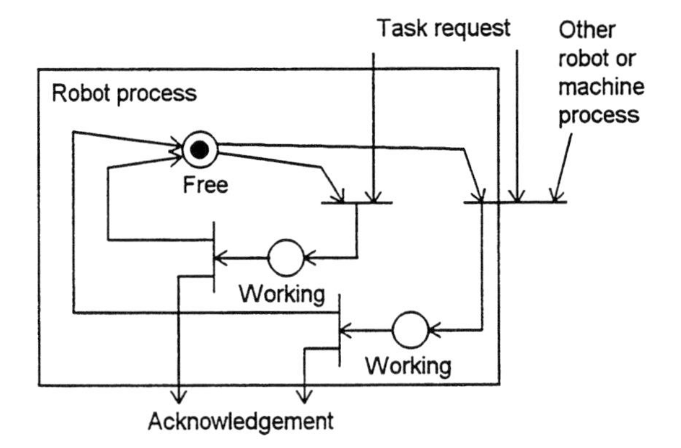

Fig.1. Global Petri net representation of robot process.

2.2 Interaction Mechanisms between Autonomous Processes

The modular approach affords to build complex validated Petri nets by specifying how autonomous control processes will interact with each other. For discrete event systems, two types of interactions among processes, asynchronous and synchronous ones, are distinguished. After the specifications of autonomous control processes, a process interaction net which describes interactions among these processes as asynchronous and synchronous communications, is specified (Breant and Paviot-Adet, 1992). Ordinary and asynchronous interaction is modelled by a place which does not belong to any autonomous control process. On the other hand, synchronous interaction is modelled by a shared transition as shown in Fig.2. Local data exchange within one process, or state machine, can also be specified by a place belonging to the process.

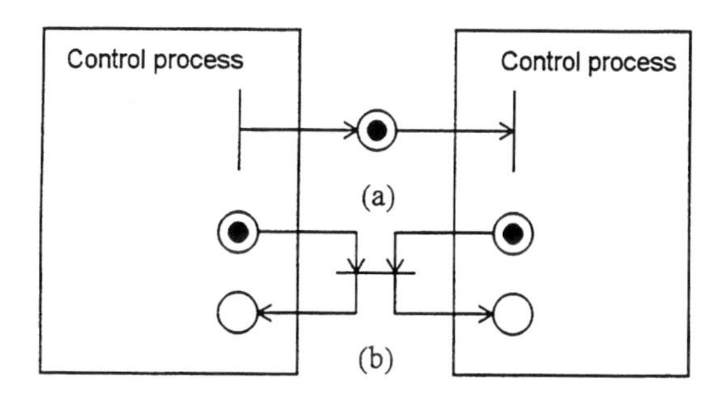

Fig.2. Schematic of (a) asynchronous and (b) synchronous interactions.

A process interaction net (Breant and Paviot-Adet, 1992) of a typical robot system composed of two cooperating robots is shown in Fig.3. The coordinator process and the two robot processes are autonomous control processes, that is, state machines. The places and the transition model asynchronous and synchronous interactions between the associated processes, respectively. The two places are in charge of control functions for each robot to perform its own tasks exclusively or cooperatively with the other robot. The transition has one precondition place and one postcondition place which models the monitoring process. The functions of these places and transition are managed by the dedicated asynchronous and synchronous interaction control processes.

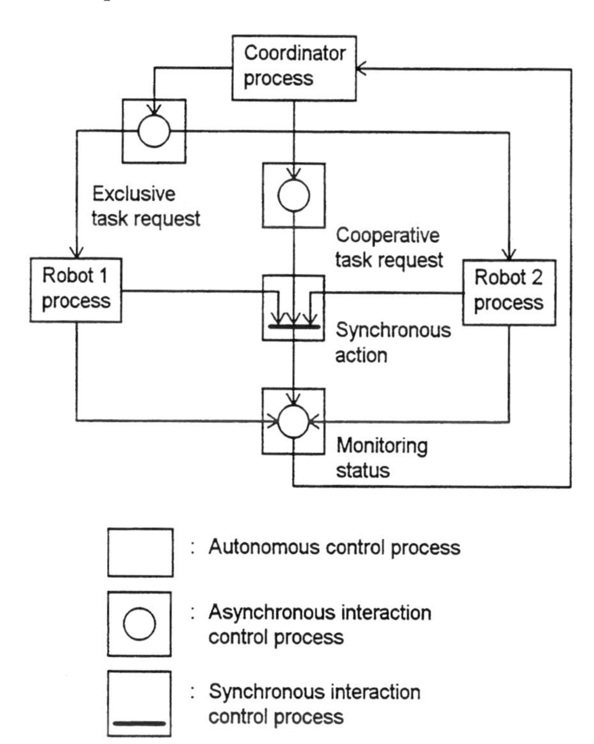

Fig.3. Example of process interaction net of two cooperating robots.

System coordination is performed based on the transition firing of the Petri net model. For example, a detailed Petri net representation of a loading task from a conveyor by a robot is shown in Fig.4. The coordinator process sends messages of request to the autonomous control processes, that is, the robot process and the conveyor process, for the execution of specified tasks. The robot and conveyor processes perform their tasks concurrently with mutual communications, which are managed by the dedicated asynchronous interaction control process. When the tasks are successfully completed, messages of acknowledgement or status are sent to the coordinator process in return through the monitoring process. Messages for exception handling and error recovery are sent from the monitoring process to the coordinator process if any error arises during task execution. Cooperation between two robots is represented by similar control structures (Yasuda and Tachibana, 1994).

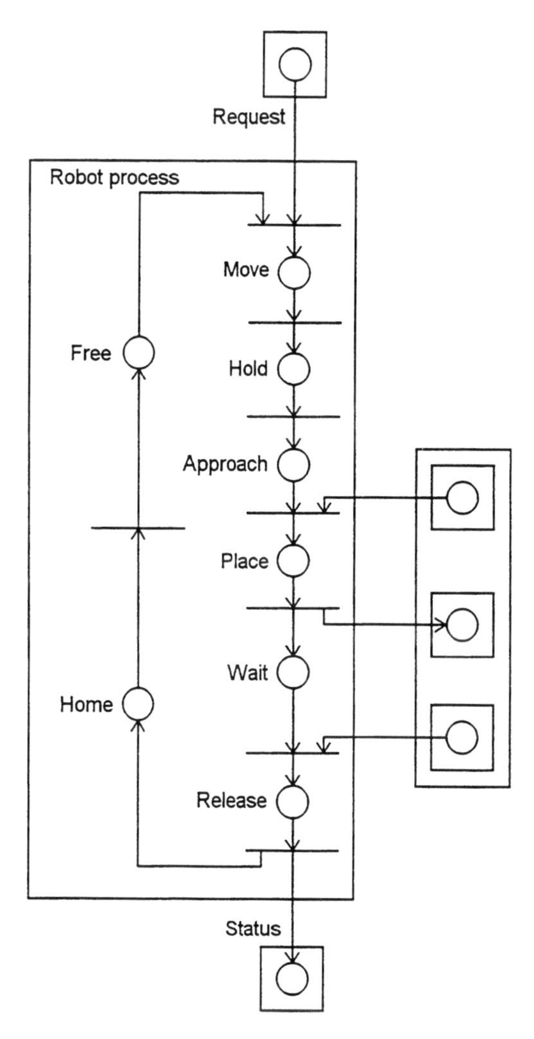

Fig.4. Petri net representation of robot process for loading task to conveyor.

3. MAPPING OF SPECIFICATIONS ONTO OCCAM CODE

The Inmos transputer with its associated parallel programming language, occam, was selected as the target architecture owing to the sound formal theory behind the model of parallelism and synchronous communication (Hoare, 1985). The correspondence between Petri nets and a subclass of occam constructs is adopted. The nature of the mapping between specifications and occam processes is such that state reachability trees are analysed empirically using corresponding Petri nets (Tyrell and Holding, 1986) to show the desired properties for robot-based manufacturing systems. The details of the above control processes are described below.

3.1 Autonomous Control Process

Autonomous control processes are essentially sequential processes as state machines although they may include control flows with decisions implemented through IF and WHILE constructs in occam, which are also represented using corresponding Petri nets. Each place of a state machine is associated with a state of the sequential process, and the marking by a single token indicates the current state of the process. A simple sequential process is mapped onto occam code using an IF construct, where each case of the IF construct corresponds to one place or state of the Petri net representing the sequential process. Fig.5 shows an example occam program of the robot process, where there are three cases corresponding to the places of the Petri net model in Fig.3. An independent or cooperative task is selected alternatively by an ALT construct. After completion of each requested task, the process moves the robot to its home position.

Autonomous control processes can also include decisions which perform conflict resolution or choice between alternatives. Conflict resolution is transformed to guarded alternative processes; an ALT construct with guards such as input, Boolean and timing expressions in occam.

```
PROC robot.process( )
  SEQ
    . . . initialize process
    WHILE continue.flag
      IF
        state = AUTO
          SEQ
            . . .
        state = SYNC
          SEQ
            . . .
        state = HOME
          SEQ
            move.to.home( )
            ALT
              auto.task.request ? auto.ready
                state := AUTO
              SKIP
                state := SYNC
        . . .
    . . . end process
:
```

Fig.5. Example occam program of robot process.

3.2 Asynchronous Interaction Control Process

Asynchronous interaction is represented as a place and managed by the dedicated asynchronous interaction control process. Token passing between two autonomous control processes is implemented through asynchronous communication. When a transition is fired, one process sends a message and a token to the asynchronous interaction control process in charge of the postcondition place. The other process sends a message to the asynchronous interaction control process to evaluate the precondition place and to try to receive a token. Thus, the asynchronous interaction control process updates the token counter as a postcondition or precondition place when firing of any connected transition occurs. Fig.6 shows an example program of the asynchronous interaction control process for token passing from the coordinator process to the synchronous interaction control process in Fig.3. Acceptance of a cancelling message from the control process can be included to restore a token when precondition evaluation fails.

```
PROC async.control( )
  WHILE control.flag
    ALT
      . . . receive request from coordinator
        SEQ
          . . . evaluate token counter
          . . . send response to coordinator
      . . . receive request from synchro.control
        SEQ
          . . . evaluate token counter
          . . . send response to synchro.control
:
```

Fig.6. Example occam program of asynchronous interaction control process.

The monitoring process is the special asynchronous interaction control process and intermediates asynchronous communications regarding system status. In case of mutual exclusion, a token which means completion of the task by one robot process is sent to the place which the monitoring process manages. The coordinator process communicates with the monitoring process in order to generate the next command. The monitoring process also manages messages from external sensors to detect exceptional situations such as operation failure and machine stoppage, sends emergency messages to the coordinator process and other autonomous control processes, and coordinates some error recovery process.

3.3 Synchronous Interaction Control Process

Synchronous interaction or rendezvous among autonomous control processes is represented as one transition and managed by the dedicated synchronous interaction control process. An example program for cooperating robots is shown in Fig.7. When every robot is ready to perform rendezvous, the procedure for synchronous action is called. The procedure includes an appropriate cooperative task among associated robots. If any robot is not ready for rendezvous, the interaction control process informs the ready robot process that the rendezvous can not be performed, then the process changes its state. Timed rendezvous can be implemented on the robot processes and the synchronous interaction control process.

```
PROC synchro.control( )
  WHILE control.flag
    SEQ
      . . . initialize
        WHILE continue.flag
          ALT
            . . . receive request from robot1
            . . . receive request from robot2
        SEQ
          . . . execute synchronous action
          . . . send a token in postcondition place
          . . . send messages to robot1 and robot2
:
```

Fig.7. Example occam program of synchronous interaction control process.

3.4 Hierarchical Specifications of Robot Process

Generally, for real robot-based manufacturing systems, Petri nets generate large, unmanageable nets. Based on the hierarchical approach of the proposed procedure, Petri nets are specified and translated into occam codes by stepwise refinements from the highest global control level to the lowest local control level. A global and simple Petri net model is first chosen which describes the aggregate system structure and satisfies the boundedness, liveness and reversibility. Then, at each step of specification, some parts of the Petri net, transitions or places, are substituted by a subnet in a manner which maintains the structural properties. This procedure enables information hiding, because the detailed behavior of each robot or machine process can be easily extracted from the net.

Fig.8 shows that the robotic task specified within the robot process is substituted by the detailed specifications of the robot motion process, the robot controller interface process and the sensor process. The motion process for trajectory generation and its real-time modification has the interface part, that is, places and transitions for asynchronous communication with the global robot process. Overall control software is implemented using simple asynchronous communications with messages of request and acknowledgement between two associated autonomous control processes.

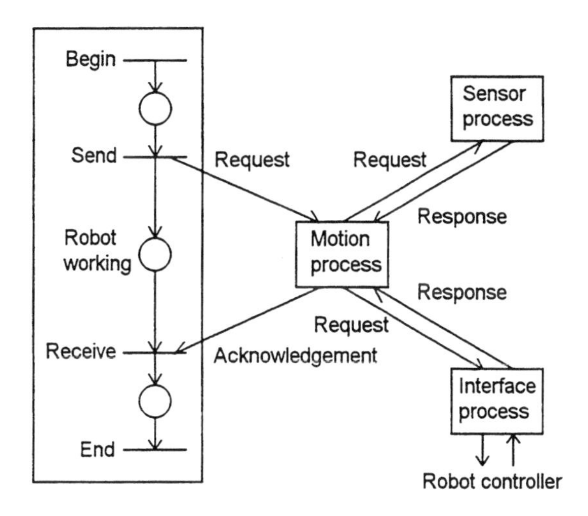

Fig.8. Hierarchical specification of robot process.

4. IMPLEMENTATION AND PERFORMANCE

4.1 A Multiprocessing Control System

To perform sensor-based robot control and cooperative multirobot control, a multiprocessing control system using a general real-time multitasking operating system based on object-oriented programming on a single microcomputer has been implemented (Yasuda and Tachibana, 1994; Yasuda, 1996). The preemptive scheduling and 10~150 ms time slicing with priorities are incorporated into the rescheduler's program. Context switching is successfully realized under the restrictions caused by the non-reentrant MS-DOS.

The multiprocessing control system has four types of processes: the coordinator process, the robot motion process, the robot controller interface process and the sensor process. Each process communicates with other processes via mailboxes. Each robot motion process handles a separate robot arm, executing high-level motion commands sequentially from its own mailbox. If the queue is empty, it suspends itself and waits for a new command. For the proposed procedure, the autonomous control processes as well as asynchronous and synchronous interaction control processes were successfully simulated and proved to have freedom from deadlock.

Currently, several examples of robot systems have been developed using the proposed specification and implementation procedure. Robotic handling operations synchronized with a conveyor were successfully implemented. For vision-based robot positioning, multitasking processing is effective owing to cooperation between a robot and its vision sensor system in comparison with sequential processing; the robot can reach, grasp and start moving the workpiece while the vision system is busy making a decision from its extracted feature parameters. As an example of cooperative multirobot control, pick and place operations, in which the two robots perform synchronous communication with each other for exchange of a workpiece, were also successfully implemented and experimentally validated.

However, such control systems using conventional processors cannot handle the full complexity of the implementation of parallel distributed control structures due to the following reasons:

(1) in addition to preemptive or event driven scheduling, round robin scheduling, where the time slice is required to be shorter than the time required for processing and communication of one motion command, about 100 ms using NEC PC-9801FA, is essentially needed for two robots to move simultaneously;

(2) for smooth robot motion control, the sampling interval largely depends on the time required for processing of one motion command; and

(3) executable code is inefficient and the safety cannot be logically proven, because multiprocessing control software is written as polling loops or with interrupt routines.

Owing to these disadvantages, control system implementation using a multitasking operating system on a conventional processor lowers the performance, flexibility and extensibility.

4.2 Transputer Networks

For distributed implementation of Petri net based control structures on transputer networks with true parallel processing and message passing primitives effectively handled in hardware, a bidirectional communication link between transputers or a pair of channels within one transputer, is assigned to each directed arc in the process interaction net. The overall control software can be implemented as a parallel construct of the control processes on the transputer network, and does not need any operating system software for managing the transputer and the whole transputer network.

The control system can be implemented on different hardware and software structures. Hierarchical control structures are classified according to autonomy of low level control modules. Real-time control performance of distributed autonomous control systems is evaluated with respect to throughput, response time and synchronization between transputer modules for simultaneous or coordinated actions in response to user command.

The experimental results showed that the proposed procedure generates correct codes. Program code in occam is very efficient because of its small instruction size. Advantages of the transputer network for robot system architectures are confirmed with respect to real-time control performance, flexibility of system configuration and software development time in comparison with conventional approaches using commercial real-time multitasking operating systems.

5. CONCLUSIONS

The specification and implementation procedure of communicating sequential processes for distributed robot system architectures in robot-based manufacturing systems has been described. Distributed implementations of Petri net based control structures are illustrated aiming at high performance parallel processing of low level control modules on transputer networks. The procedure supports rapid prototyping and reusability, since a communicating sequential process or control process already described using a Petri net can be kept as a software module in a library and instantiated repeatedly to link it with other processes for real robot system applications.

REFERENCES

Breant, F. and E. Paviot-Adet (1992). OCCAM prototyping from hierarchical Petri nets. In: *Transputers'92*, 189-209, IOS Press, Amsterdam.

Bucci, G. and E. Vicario (1992). Rapid prototyping through communicating Petri nets. *Proc. 3rd IEEE Int. Workshop on Rapid System Prototyping*, 58-75.

Hatvany, J. (1985). Intelligence and cooperation in heterarchical manufacturing systems. *Robotics and Computer Integrated Manufacturing*, **2**, 101-104.

Hoare, C.A.R. (1985). *Communicating Sequential Processes*. Prentice-Hall International, London.

Tyrell, A.M. and D.L. Holding (1986). Design of reliable software in distributed systems using the conversation scheme. *IEEE Trans.*, **SE-12**, 921-928.

Yasuda, G. and K. Tachibana (1994). A parallel distributed control architecture for multiple robot systems using a network of microcomputers. *Computers & Industrial Engineering*, **27**, 63-66.

Yasuda, G. (1996). An object-oriented network environment for computer vision based multirobot system architectures. *Proc. 20th Int. Conf. on Computers & Industrial Engineering*, 1199-1202.

MANUFACTURING SYSTEM IN THE MICRO DOMAIN: MICROROBOTICS

Gábor Felső

Department of Process Control, Technical University of Budapest
Műegyetem rkp. 9., Budapest, 1111 Hungary

Abstract: Microrobotics combines the manufacturing technology of robotics with activities on small-sized pieces with a great accuracy. The goal of this paper is to investigate the capabilities of several vision robot sensors in the field of microrobotics. Two methods are discussed in details: the out-of-focus and the shadow detection principles. The combination of these methods provides a sensor structure, which can be used to determine the position of a micromanipulator observed through a microscope.

Keywords: microsystems, robot vision, manipulation, mobile robots, image sensors.

1. INTRODUCTION

In the last decade there was high demand for developing mobile robots, which can move in industrial environment, can transport materials between different workstations and can perform several tasks on them. Nevertheless, the miniaturizing of lots of parts of a machine generated demand on development of automatic equipment for handling, transporting and manufacturing of small-sized pieces.

Microrobotics combines the manufacturing technology of robotics with activities on small-sized pieces with a great accuracy. The applications of microrobotics are handling and manufacturing of miniature objects, testing and repairing of microchips and unreachable areas of macro-object (e.g. inner side of a pipeline). In addition, microrobots could be great tools in the surgery of cells as well.

In the project a working-cell is developed, where the microrobot moves, controlled by a vision sensor system, in order to perform transport and assembly operations in the range of cm till μm. Therefore, the paper of our research in the field of microrobotics combines mobile robot technics and the object of vision based control technology, respectively. The main goal of the research is to develop a sensor system for the microrobot to make it intelligent for performing tasks accurately.

This paper presents the moving principle of the microrobot and its manipulator, which is based on the inverse piezoelectric effect[1]. The movement of the microrobot in the transport domain is perceived by a laser based camera vision system. The manipulation sensor system controls the position and the orientation of the manipulator of the microrobot by observing it through a camera connected to a microscope. All of the manipulating and manufacturing operations occur in the field of sight of the microscope. This distributed sensor system layout enables a great area in transport and a high-precision in manufacturing, respectively.

2. THE MICROROBOT AND ITS MOVING PRINCIPLE

The goal of the paper is to investigate the capabilities of several robot sensors in the field of microrobotics. This paper focusses first of all on a certain microrobot structure, a piezoelectric driven microrobot (Fatikow and Rembold, 1993; Magnussen, *et al.*, 1995), but the investigation could be wider (Fatikow, *et al.*, 1994; Dario *et al.*, 1992). The body of the robot is standing on three

[1] The microrobot PROHAM / MINIMAN was originally developed by the Institute for Real-Time Computer Systems and Robotics, University of Karlsruhe. The research is sponsored by the Volkswagen Foundation.

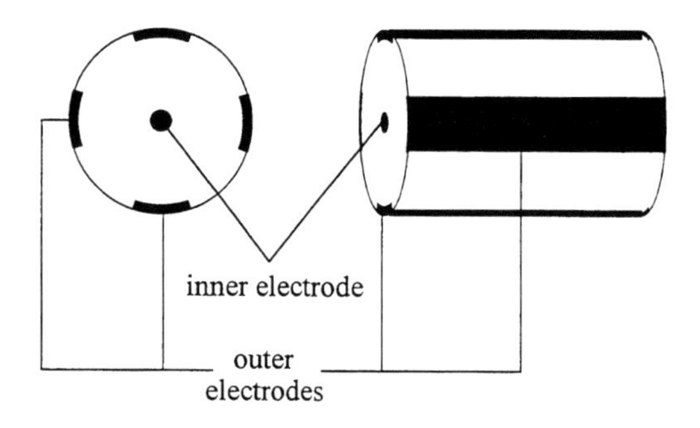

Fig. 1. Layout of electrodes on a piezoelectric leg

piezoelectric legs. Every leg is equipped with four longitudinal electrodes on the cylindrical superficies and one in its axis, as shown in Fig. 1. Depending on the voltage between the inner and the outer electrodes, the legs can be lengthened, contracted and bent. If every leg in a certain sequence is contracted, bent, lengthened and then re-bent, the microrobot will step about 3 μm. The repetition of this process will lead to a continuous movement of the robot.

The manipulator of the robot is placed by a spherical metal on the lower part of the body of the microrobot, hold by a magnet and supported by three piezoelectric legs, which are identical with the above mentioned ones. As the stepping of the legs of the body led to the moving of the microrobot, the stepping of the supporting legs of the manipulator on the surface of the metal ball lead to a rotation of the ball and, therefore, to a rotation of the endeffector (manipulator).

However, one step of the body against the gravitation is carried out by its slight tilt, the continuous stepping of the robot will decrease the vertical rocking of the body.

3. MICROMANIPULATION WORKING CELL

The microrobot is designed to carry out various tasks in a micromanipulation working cell (Sulzmann and Jacot, 1995). The moving principle of the robot enables to transfer microobjects on a relatively large distance. In addition, the transportation area of the robot is only bounded by the power and control supply, practically infinite. To increase the accuracy of micromanipulations they can be made in the field of sight of a microscope. Therefore, the working area of the microrobot is divided into two domains, which are functionally different: transportation and manipulation domain.

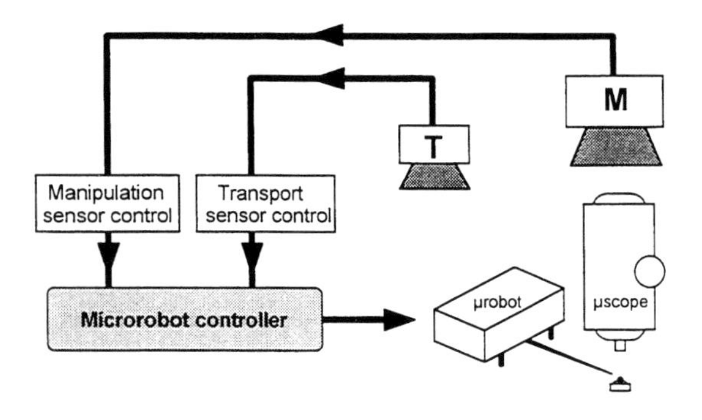

Fig. 2. Functional layout of the micromanipulation working cell

The transportation tasks of the microrobot are moving microobjects and assemblies from one part of the working place to the other. During the transportation procedure the moving speed should be high without an accurate positioning compared to the micromanipulation accuracy. The micro-manipulations consist of accurate positioning and performing microoperations in the micro domain. Therefore, the high speed of moving is not required here. The working cell can be seen in Fig. 2.

Table 1. Properties of the transport and the manipulation domains

	Transport domain	**Manipulation domain**
Size of area	large	small
Accuracy of positioning	low	high
Moving speed	high	low

The most important features of the domains can be found in Table 1. Below this paper deals with the properties of sensors in the manipulation domain.

4. VISION SENSORS

The control of the movement of the microrobot claims accurate information about the 3D position and the orientation of the robot body itself and of its manipulator. Based on the functional difference of the two working domains of the microrobot, different sensor structures are required to control the movements of the microrobot. Below the general requirements of a sensor structure will be discussed.

4.1 What is to be measured?

The application aim of such kind of microrobots determines that it should be put into a wide variety of environments: it should be installable into a new place fast and inexpensive, the environment may be more or less unknown. Therefore, the sensor structure should be compact, transportable and independent from any specific environment. Furthermore, there are a lot of restrictions in the micro domain that many sensor structures are not applicable. The contact sensors require to measure very small quantities of force, power, etc. (Shimoyama, 1995), which can be performed only by high precision measuring units. In addition, sensors, which physically touch the surface of the object to be manipulated are excluded from our area of interest, because a lot of application does not enable any contact with the microobject.

To illustrate this problem let us see Fig. 3. Here the manipulator of the microrobot is a needle, which has to drop into a biological structure without to touch it. It comes out that besides the 2D placing the vertical positioning is also significant.

Therefore, the above mentioned requirements can be satisfied best by non-contact sensors, first of all by vision sensors. In the following some important characteristics of commonly used and newly suggested procedures of 3D vision sensors will be introduced.

The vision sensors can be divided into two categories (Levi and Vajta, 1987). The passive systems do not interfere with the environment to be measured. These are the stereo vision systems and the indirect mono procedures: passive focusing, texture processing, etc.

The main feature of the active systems is that they interfere with the environment to be measured. These systems "illuminate" the environment and the object to be measured. The reflected or refracted signals are captured and analyzed. The commonly used active vision sensors are the following:

- active focusing
- Fresnel-diffraction
- holographic interferometry
- polarized illumination
- Moire-technics
- radar or lidar systems
- active triangulation

The radar (or lidar) systems measure the time of the reflection of the radiated signal. The time can be measured in a direct way (time-of-flight) or in an indirect way (amplitude- or frequency-modulation).

The active triangulation requires a structured illumination, which can be a point, a strip or a complex lighting and a detector unit. The range-finding is based on trigonometric relations determined by the geometry of the arrangement of the measurement unit and the objects to be measured.

Though most of the above mentioned sensor structures are known, not all of them are used in the field of robotics, especially in microrobotics yet. Our goal is to investigate the capabilities of sensor systems in order to manage the information collection about the position and orientation of the manipulator of the microrobot.

4.2 Sensors in the manipulation domain

The micromanipulations require a sensor structure to control the positioning of the microrobot in the μm domain, where two basic methods are to be taken into account: interferometry methods and the use of a microscope. The solution between this two methods is mainly influenced by the technological requirements. The interferometry procedures provide relative position information about the manipulator of the microrobot. In addition, this method makes the processing of the image of the object to be manipulated more difficult, and it is quite expensive. On the other hand, the procedure can provide accurate position data and is selected under other technological circumstances. The second method, the use of a microscope equipped with a camera is an inexpensive, commonly used image acquisition procedure. Information about the 2D position of the manipulator in the field of sight of the microscope can be collected quite easily. This problem can be solved by commonly used image processing tools, therefore, this task will be not investigated in more details. The more interesting problem is how to measure the depth position of the manipulator. To solve this problem two methods and a combination of them will be discussed.

4.3 Shadow detection

The sensor system, called shadow detection, consists of a camera (in our case on the tube of the microscope) and an image processing unit. The object to be measured is e.g. a needle, which is in front of the background (object plane). The area of interest is illuminated by a parallel light source, the direction of which is known and the camera makes a picture from a direction, which is different from that of the illumination. Then, a shadow can be seen both on the background (object plane) and on the image.

The 3D position of the end of the manipulator can be determined by the intersection of two lines. One of

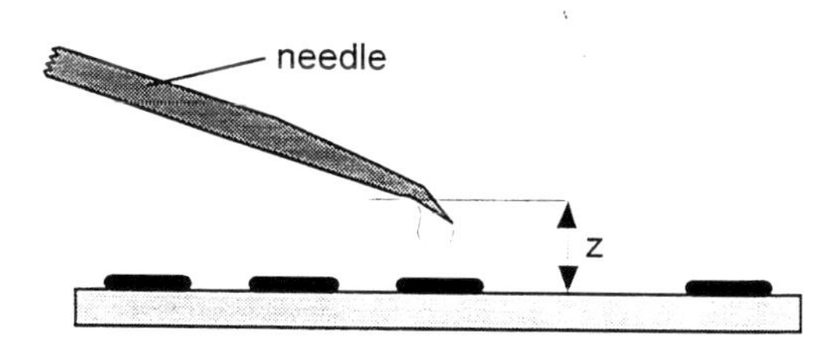

Fig. 3. Positioning with a needle

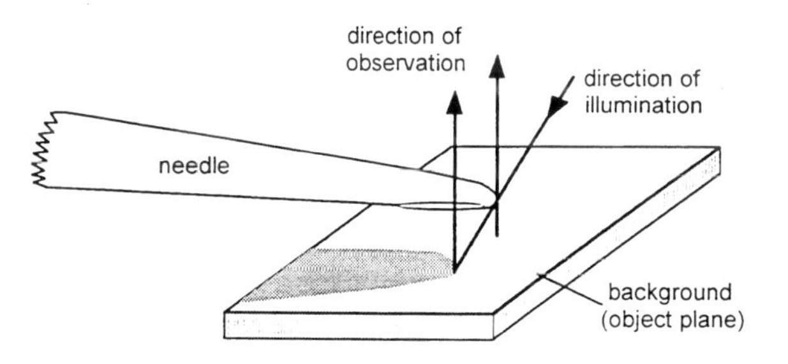

Fig. 4. Detecting of the shadow of a micromanipulator

the lines is determined by the planar position of the manipulator in the 2D scene and by the direction of observation. The second one is determined by the planar position of the shadow of the end of the manipulator on the background and by the direction of illumination. The manipulator can be found on the intersection of the two lines as shown in Fig. 4.

The advantage of the method is that no additional equipment is required to measure the depth position of the manipulator. The method is based mainly on image processing procedures. Nevertheless, the application domain of this method is limited. First, the segmentation of the shadow of the manipulator and the texture on the background is not obvious. On the other hand, the farer the manipulator is from the object plane the more blurred the intensity of the shadow is on the background. This is caused mainly by the diffraction of the light on the edges of the manipulator. Therefore, this method can be used first of all in small distances from the object plane. Another requirement is that the illumination should be directed so that the manipulator does not cover its shadow observed through the microscope.

4.4 *Out-of-focus sensor system*

The out-of-focus principle means the following procedure. The image acquisition is carried out by a single camera. The focal length is set up to the background objects, therefore these object are in focus. The manipulator, the depth of which is to be measured, is in front of the background, out of focus (see Fig. 5). The fact that the manipulator is out-of-focus blurs the image and it generally makes the image processing more difficult. In our case the advantage of this effect will be exploited.

The transition of the intensity of the image points of the background and of the manipulator is of primary importance. The farer the object is from the object plane the more blurred the transition is. In ideal circumstances (white background, black object) the transition curve of the intensity should be linear. Moreover, the extension of the blur of the object is connected to the distance from the object plane by the following equation:

$$y = \frac{af}{t-f} \cdot \frac{|z|}{t-z} \qquad (1)$$

The above mentioned out-of-focus procedure provides an easy method to determine the depth of an object from the object plane. The disadvantage of this method is that the determination of the transition on a complex background structure is not obvious, because the image of the in-focus objects and the blur of the out-of-focus objects overlap. The method can be improved by investigating several transitions of an extended object on the image. The change of the

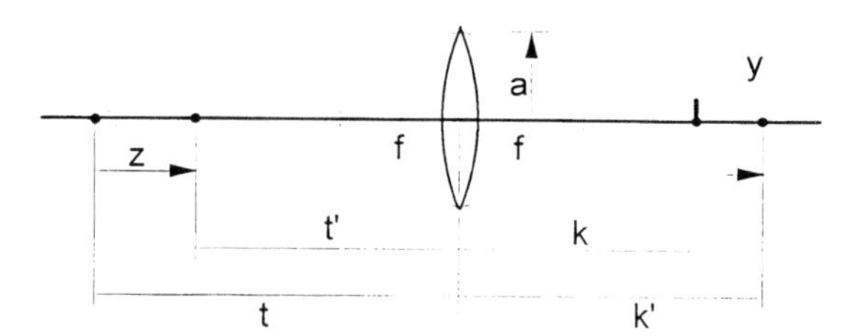

Fig. 5. Out-of-focus principle

transitions is in strong correlation with the slope of the object itself. In this case the lack of information in certain domain on the image does not result in the defect of the measurement. If the depth of a certain point of the object is known the depth of all object points can be determined by its slope.

Because of the diffraction of light, the image of a point does not become a single point on the image plane, but a spot of a finite size. That is why the method is not applicable in very small distances. A comparison between the theoretical results given by the equation (1) and the experimental ones can be found in Fig. 6. (The parameters are: f=1 mm, a=2 mm.)

4.5 *Combination of the two sensor systems*

It can be seen from the previous sections that the sensors with the out-of-focus and the shadow detection methods are applicable in different micro domains. The position of a micromanipulator can be determined by the out-of-focus method at depth distances of above 10 μm. The procedure of shadow detection, based on our experiences, provides a technics for position determination at distances within 20 μm. The combination of the two methods seems to be obvious. The combined position detection does not require a double sensor system, because both of them are based on the same camera sensor. The image processing algorithm should decide which image processing method should be used. This can be solved by the following way. The draft position detection is made by the out-of-focus method. This provides information about the depth of the micromanipulator. If the distance between the

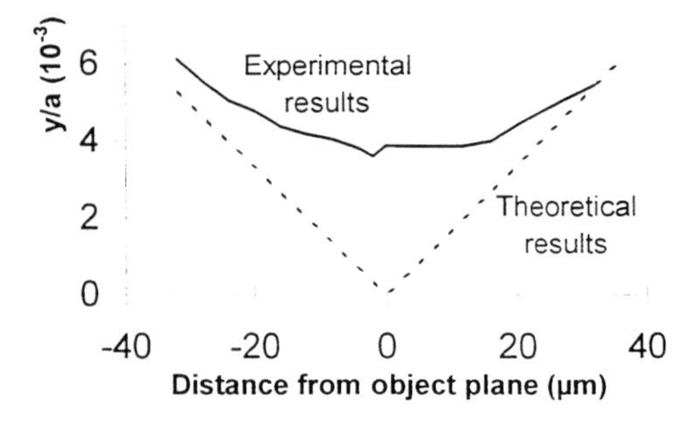

Fig. 6. Theoretical and experimental results of the out-of-focus principle

end of the manipulator and the object plane is equal to or less than about 10-20 μm, the image will be analyzed according to the shadow detection. This method is suitable to increase the accuracy of the position determination in this order of magnitude. Based on the experiences, the method of shadow detection could identify distances of several micrometers depending on the arrangement of the manipulator and the direction of observation.

5. CONCLUSION

This paper presents a combined sensor system used for determining the position of a micromanipulator in the micro domain. This sensor system is an extension of the commonly used image acquisition and processing sensors, where the image is produced by a camera looking through a microscope. The experimental results have shown that despite the requirements in the arrangement of the micromanipulator and the sensor system these can be fulfilled and the methods can be deployed in a wide area of application.

The main advantage of the methods is that the sensor is quite inexpensive and does not require special optical elements or additional sensors for the position determination. Based on the development and experiences described here the sensor structure will be realized and tested.

ACKNOWLEDGEMENTS

The author would like to thank the members of the Institute for Real-Time Computer Systems and Robotics, University of Karlsruhe and personally Prof. U. Rembold for their support in the project.

REFERENCES

Dario, P., Valleggi, R., Carrozza, M. C., Montesi, M. C. and Cocco, M. (1992). Microactuators for microrobots: a critical survey. In: *Journal of Micromechanics and Microengineering*, **2**, pp. 141-157. IOP Publishing, Bristol

Fatikow, S. and Rembold, U. (1993). Principles of Micro Actuators and their Applications. In: *Proc. of the IAPR Workshop on Micromachine Technologies and Systems*, pp. 108-117. Tokyo

Fatikow, S., Rembold, U. and Wöhlke, G. (1994). A survey of the present state of microsystem technology. In: *Journal of Design and Manufacturing* **4**, pp. 293-306.

Levi P. and Vajta L. (1987). Sensoren für Roboter. In: *Robotersysteme*, **3**, pp. 1-15. Springer-Verlag.

Magnussen, B., Fatikow, S. and Rembold, U. (1995). Actuation in Microsystems: Problem Field Overview and Practical Examples of the Piezoelectric Robot for Handling of Microobjects. In: *INRIA/IEEE Conf. on Emerging Technologies and Factory Automation*, Paris

Shimoyama, I. (1995). Scaling in Microrobots. In: *IEEE Proceedings Robotics and Automation*, pp. 208-211.

Sulzmann A. and Jacot J. (1995). 3D computer graphics based interface to real microscopic worlds for μ-robot telemanipulation and position control, pp. 286-290.

SIMULATION AND EXPLOITATION OF THE DYNAMIC BOND GRAPH MODEL OF THE RTX ROBOT

M.L. KOUBAA * **M. AMARA ****

* *LAboratoire de Recherche en Automatique, Ecole Nationale d'Ingénieurs de Tunis, BP 37 Le Belvédère Tunis,TUNISIE, Tel : (216) (1) 874.700 Telex : (216) (1) 872.729*
** *CEntre de REcherche en Productique, Ecole Superieure des Sciences et des Techniques de Tunis, TUNISIE, Tel : (216) (1) 392.559 Telex : (216) (1) 391.166*

Abstract : In this paper, we will develop, step by step, a systematic procedure to succeed finally to the bond graph model of the RTX robot. The simulation with a causal bond graph will allow us to calculate the evolution of the geometric, kinematics and dynamic sizes the long of the trajectory imposed by specification conditions. This allows to predict the dynamic behaviour that must have the robot in its operational environment The dimensionnement of the energy source of every articulation, in term of instantaneous useful power, is necessary to could carry away the mechanical structure and respect consequently the specification conditions . All the mentioned treatments, will be based on some conventional sizes such as strength, velocity, torque...

Keywords : Bond graph, Model, robot, simulation, specifications.

1. INTRODUCTION

The description of a system involves two distinct parts: a description of the nature of the primitive system elements and an enumeration of the topological structure of interconnections between the elements. The bond graph is a graphic representation of physical systems which belongs to the class of representations type networks. Often this graphic representation concerned initially a specific domain but they were then formalised mathematically under shape of graphs (Amara, 1991).

For the construction of dynamic model of a polyarticulate system, we are going to adopt the systematic method of modelisation in bond graph developed by BOS (1986). The model of every subsystem is then developed to obtain a complete model, constituted by bond graph elements. The polyarticulate mechanical system (case of RTX robot) is described first like the assembly of words bond graph correspondent to every segment and of junctions' structures which represent the articulations; some supplementary bond graph words represent the actuators as well as the exterior actions.

2. MODELISATION OF POLYARTICULATE SYSTEM

2.1 Word bond graph of RTX robot

To develop the word bond graph of RTX robot, the first step consists in drawing the words which represent the solids. Every solid is represented by a word. On the figure 1 which represents the mechanic structure of robot, we projected respectively : a word bond graph called "SOLID 0" which represents the column of linear guidance , a word called "SOLID 1" which represents the shoulder, a word called "SOLID 2" which represents the superior arm, a word called "SOLID 3" which represents the fore-arm, the words

called "SOLID 4", "SOLID 5" and "SOLID 6" which represent the wrist.
The second step gives models, by bonds and junctions, for the links between the different solids. In the studied case, we distinguish 6 links :

- A prismatic link following the Z axis called P1;
- Three rotational links around Z axis, named respectively R2, R3, R4;
- A rotational link around Y and a final link around X,
 called respectively R5 and R6.

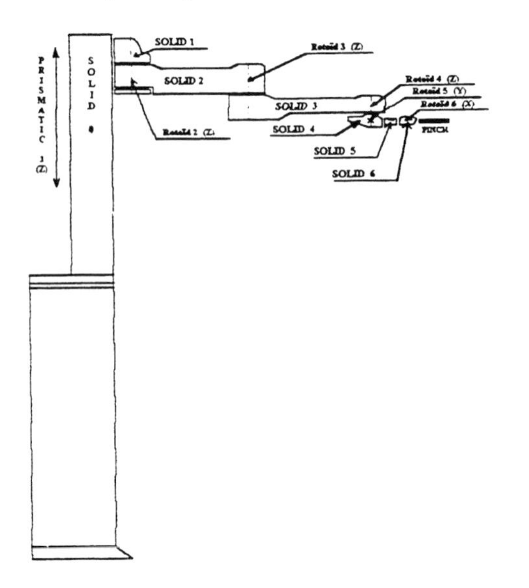

Fig. 1. Different solids of RTX robot.

A kinematics representation of robot is the following :

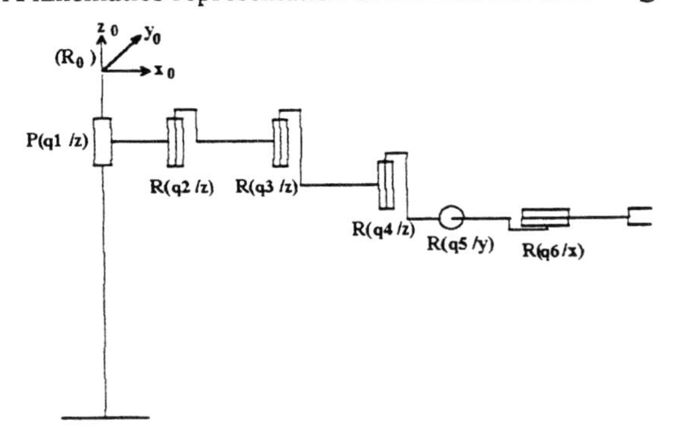

Fig. 2. Kinematics diagram of RTX robot.

From kinematics diagram, we decompose the robot in different solids. In every solid, we introduce two points of attachment, one with the upstream solid and the second with the downstream solid , and an other point which represent the centre of mass :
O_{i1}: point of connection with the solid B_{i-1}
 i.= 1, 2,..., 6.
O_{i0}: Centre of mass of solid B_i , i = 0, 1,..., 6.
O_{i2}: point of connection with the solid B_{i+1}
 i.= 0, 2,..., 6.

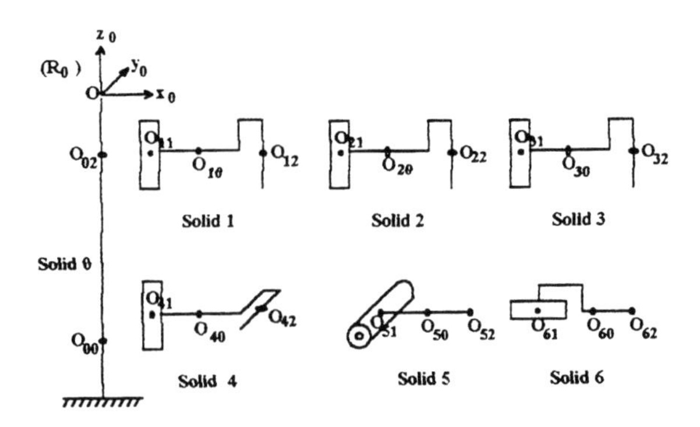

Fig. 3. Points definitions of different solids of RTX robot.

The robot is decomposed in different solids represented by words. Between the different words of the solids, we introduce the corresponding words of different articulations (Koubaa, 1995). The figure 4 represents the partial word bond graph of RTX robot.

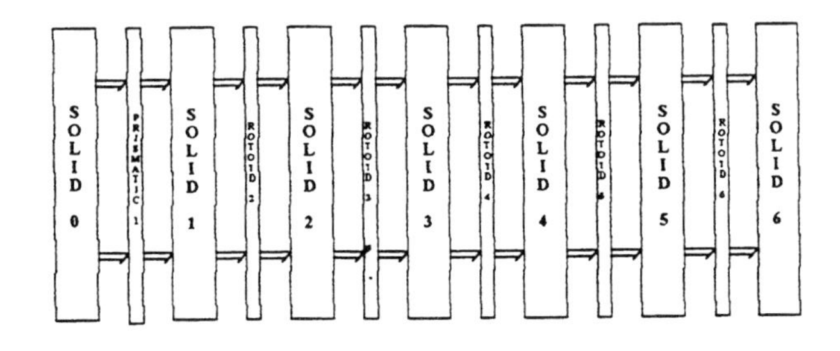

Fig. 4. Partial word bond graph of RTX robot.

2.2 Complete multi bond graph of RTX robot

The word bond graph of the figure 4 is developed in a complete multi bond graph of robot, in replacing every word "SOLID" and every word articulation "PRISMATIC" or "ROTOÎD", by their models bond graph (Koubaa, 1995; Koubaa, et al ,1996).

The robot includes six degrees of active liberty. The actuators are placed in the level of every couple of points of attachment ($O_{(i-1)2}$, O_{i1}) with i = 1,..., 6. all these limit conditions will be translated by the addition, on the bond graph sources of effort and of external flow :

- Sources of no void flow associate to the prismatic link and to the different rotational links;
- Two sources of void flow in the level of the column of linear guidance which is embedded, the limit conditions impose void speeds of translation and rotation in the levels of the two bonds of the left extremity of the prismatic link ;
- Two sources of efforts in the level of the terminal organ, then the limit conditions which impose to the bond corresponding to the rotation a void moment of strength, and to the bond corresponding to the translation a no void strength. This last opposes the weight of the whole arm.

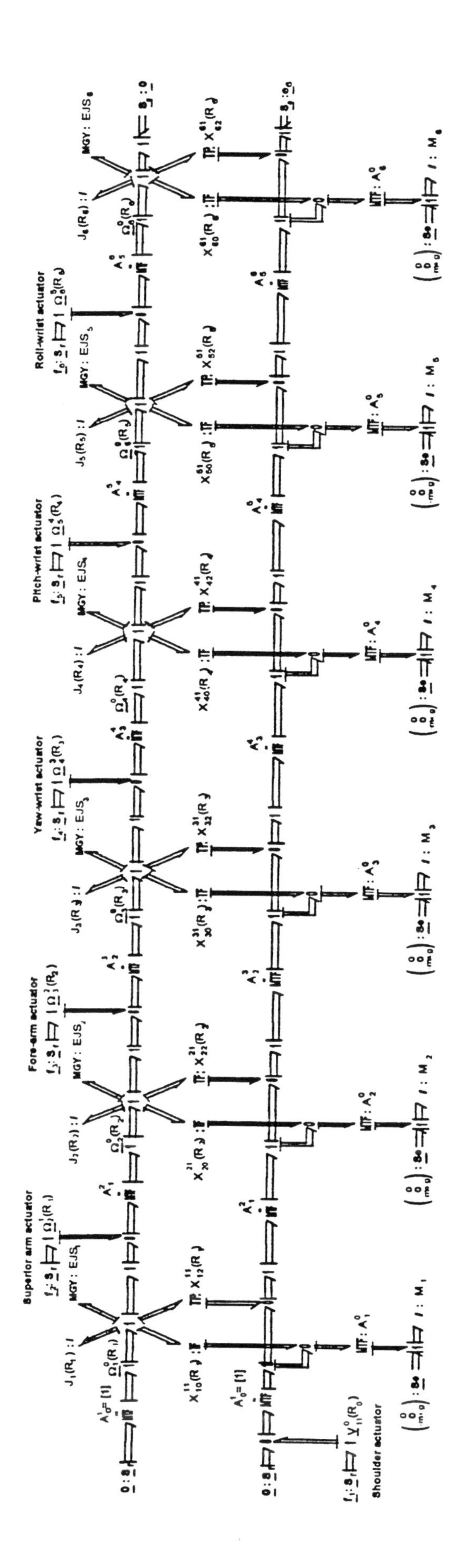

Fig. 5. Affectation of a derivative causality on the bond graph of RTX robot.

2.3 Causal multi bond graph of RTX robot

The implantation of the dynamic bond graph model of the robot on the simulation software SIMULINK, necessitates in first place a causal analysis which is a graphic procedure applied on the bond graph of the whole model.

Every element of bond graph, including junctions, represents one or several mathematical relations between the effort variables (strength, torque,...) and flow variables (speed,...).These relations have not any calculating structure (not causal = acausal)

In bond graph the affectation of calculating structure is called assignation of the causality and permits to defining the relations of causes to effects. We choose to assign a derivative causality. It permits, indeed, to translate the specifications under the shape of a law of given speed and calculating the law of effort.

From these conditions, we propagate the causality in all the structure with respect to the own rule of every element. The bond graph affected with a derivative causality is given by the figure 5.

3- IMPLANTATION OF THE MODEL ON SIMULINK

The model of RTX robot determined by causal bond graph is then implanted on SIMULINK. The simulation consists in pick up the entrances of chosen actuators with a view to observing and studying its dynamic behaviour. The specification conditions were expressed under the shape of kinematics constraints translated by a law of speed. The actuators must be capable to communicate for the different articulations a law of speed. This condition will be translated by the assignment of a flow causality in the level of all the active articulations.

The law of speed imposed in the level of the actuator M1 which assures the translation following Z is given by the following figure :

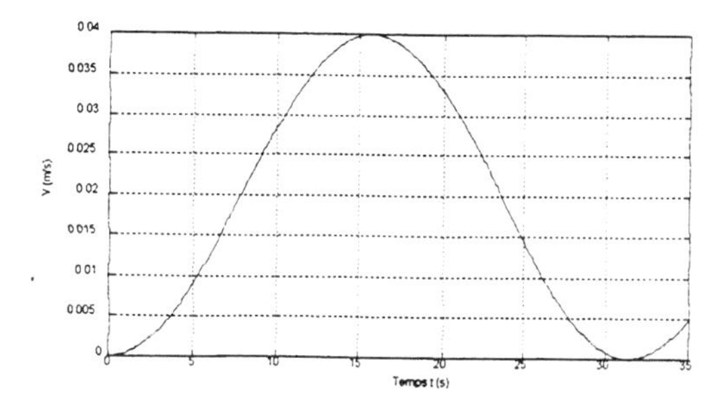

Fig. 6. Law of speed imposed in the level of the actuator M1.

The law of speed imposed in the level of actuators M2, M3, M4, M5 and M6 is given by the following figure:

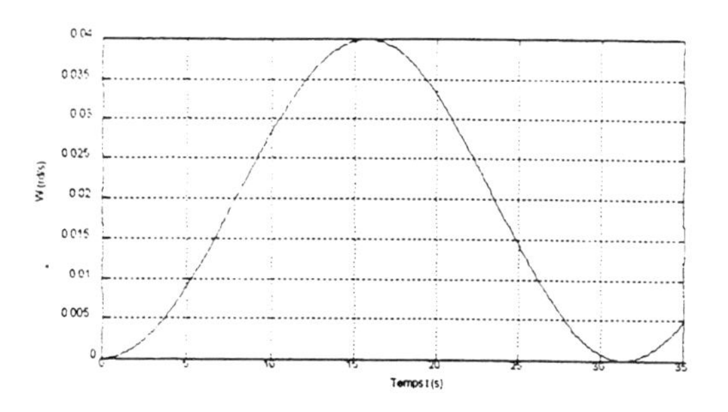

Fig. 7. Law of speed imposed in the level of the actuators M2, M3, M4, M5 and M6.

The local treatment of specifications translated under the shape of a speed law permit sequentially :

1- To calculate at every instant, the evolution of positions and the distances travelled by the extremity of different segments.

2- To calculate the angular speeds and the linear speeds at the extremity of every solid.

3- To determine the dynamic characteristics of movements of robot and notably the strength and the torque in the different actuators

The knowledge of these efforts which characterise the dynamic constraints, allow us to determine for every actuator the useful power Pu(t).

For the phase of exploitation of simulator, we chose in a first step to inject, for the actuator M1, the law of speed represented by the figure 6 and nullify the speeds in the level of other actuators. We applied a law "A.sin^2(t)" with maximal amplitude 0.04 m /s and of pulsation 0,1 rd/s during 35 seconds. The length of simulation is chosen so as to the law of speed imposed travel a period. The results of simulation allowed to have the curves represented by the figures 8, 9 and 10. Note than all the angular speeds are void, the robot does only a movement of translation.

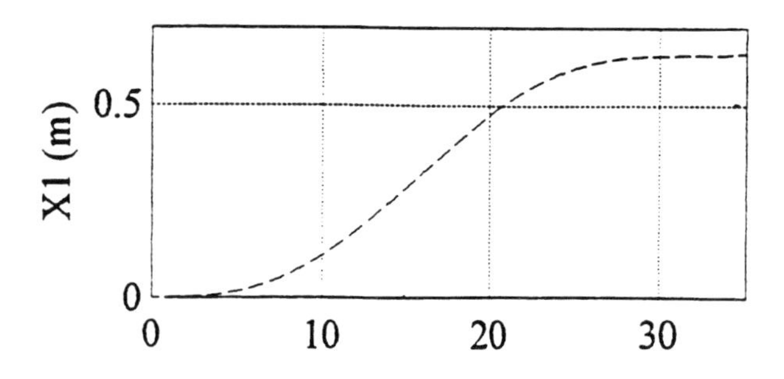

Fig. 8. Position of the linear guidance.

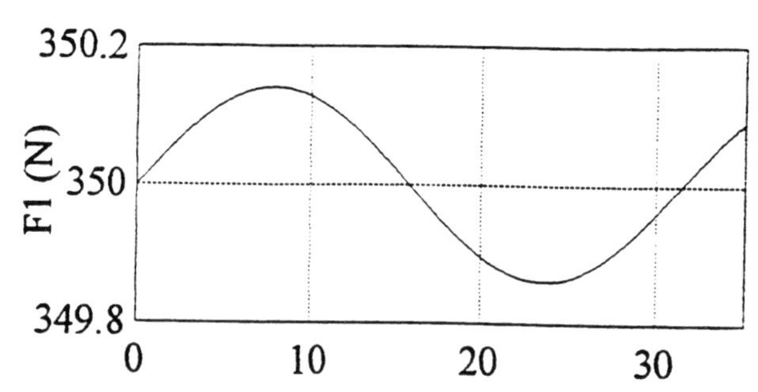

Fig. 9. Strength developed by the M1.

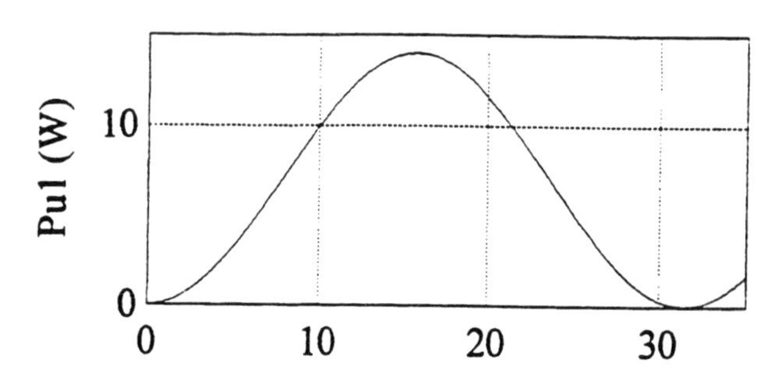

Fig. 10. Power which must be supplied by the actuator 1.

In a second step, we nullify the translation speed and apply, for actuators which ensure the rotation of articulations, the law of speed represented by the figure 7 during 35 seconds.

This law of speed is of the shape "A.sin^2(t)", its maximal amplitude is 0.04 rd/s and its pulsation is 0,1 rd/s. The result of simulations allowed to have the different geometric, kinematics and dynamic characteristics. represented by the figures 11 to 27

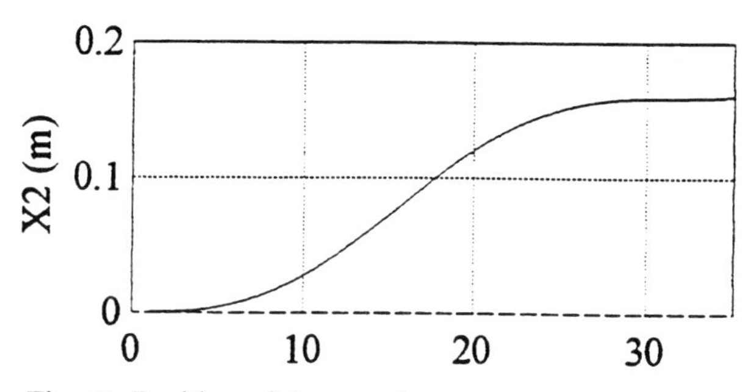

Fig. 11. Position of the superior arm.

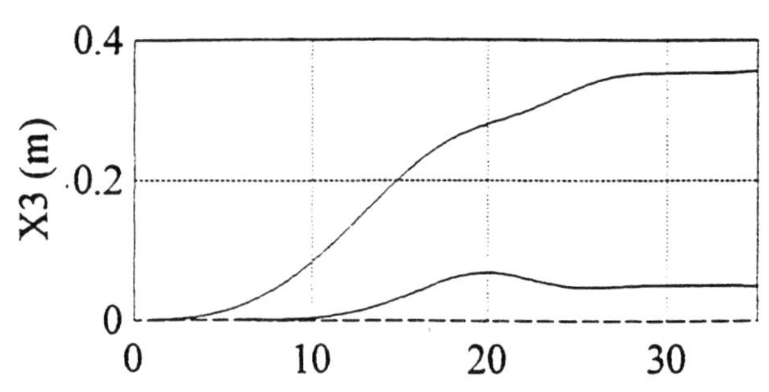

Fig. 12. Position of the fore-arm.

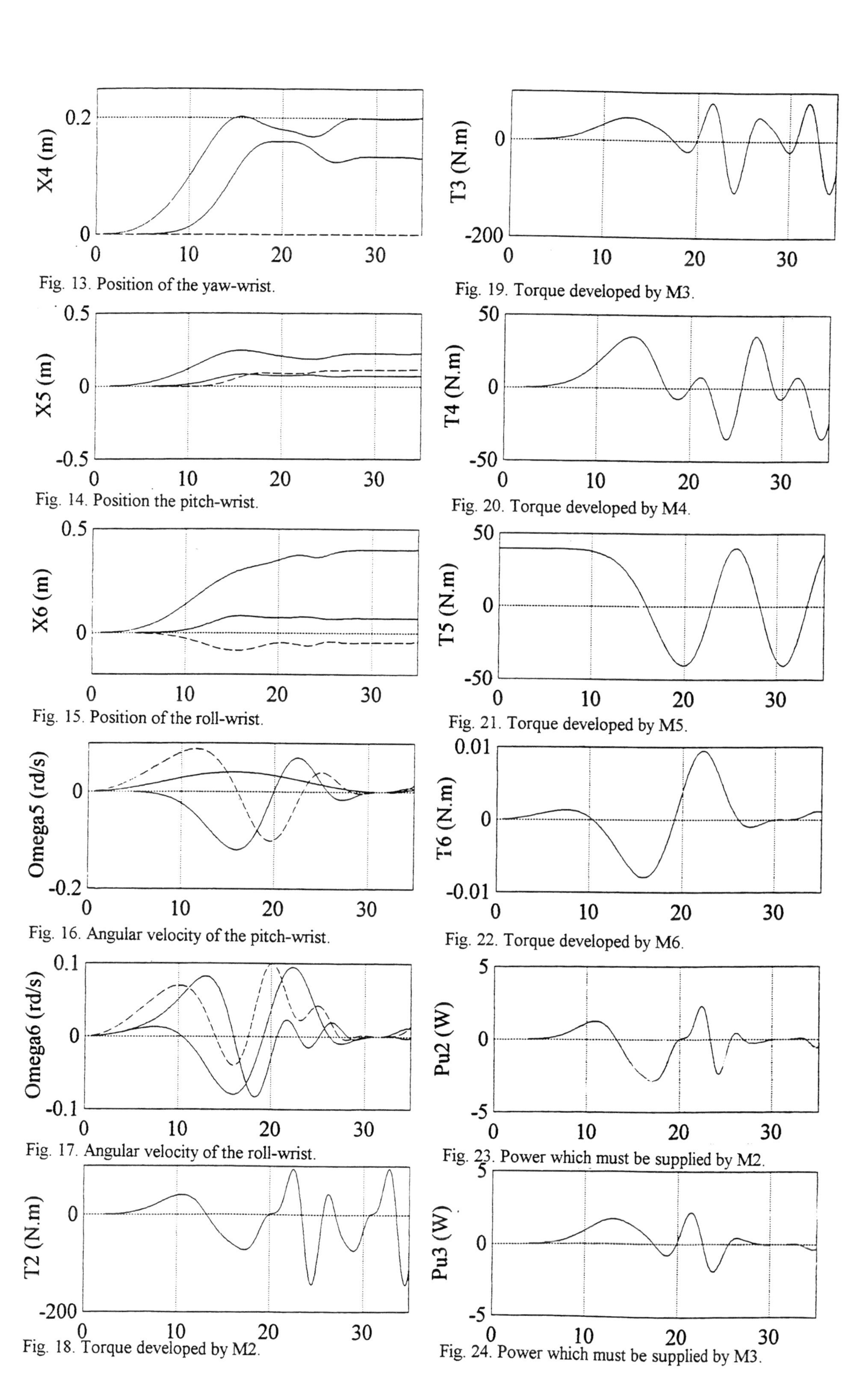

Fig. 13. Position of the yaw-wrist.

Fig. 14. Position the pitch-wrist.

Fig. 15. Position of the roll-wrist.

Fig. 16. Angular velocity of the pitch-wrist.

Fig. 17. Angular velocity of the roll-wrist.

Fig. 18. Torque developed by M2.

Fig. 19. Torque developed by M3.

Fig. 20. Torque developed by M4.

Fig. 21. Torque developed by M5.

Fig. 22. Torque developed by M6.

Fig. 23. Power which must be supplied by M2.

Fig. 24. Power which must be supplied by M3.

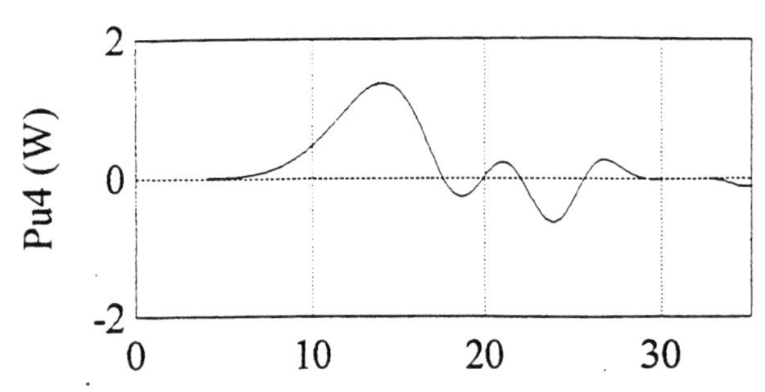

Fig. 25. Power which must be supplied by M4.

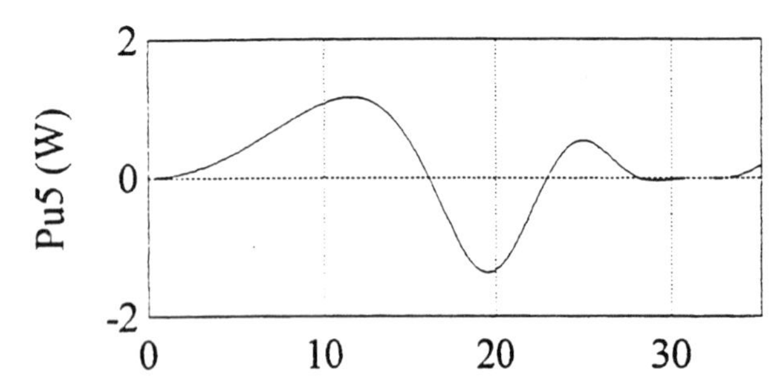

Fig. 26. Power which must be supplied by M5.

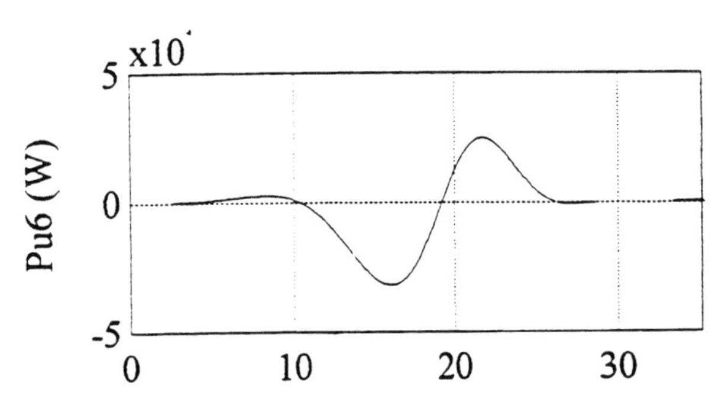

Fig. 27. Power which must be supplied by M6.

4- CONCLUSION

In order to study the dynamic performances of RTX robot and particularly those of its actuators, we developed in the first part the three-dimensional dynamic model of its mechanical structure.

Of methodical view point, we applied the systematic procedure of modelisation in bond graph of polyarticulate systems developed by BOS. This procedure stake in work step by step to lead finally to acausal bond graph model. The affectation of a derivative causality for the bond graph allowed to solicit the mechanic structure of robot by sources of speed which corresponds at the specifications.

The second part has for objective the quantification of dynamic behaviour of RTX robot so the transition to the phase of numeric simulation was achieved. The bond graph model was implanted on the software Simulink.

In this sense, specifications translated under the shape of a law of speed imposed in the level of active articulations was fixed. This specifications impose the minimal kinematics and dynamic performances that have to respect the actuators systems and consequently the robot.

The numeric exploitation of this model leads to the calculation of the law of variation of all the "conventional" variables that is speed, torque, strength,.. in the level of every point of robot and particularly in the level of every actuator system and its corresponding articulation.

In evidence, this phase provides a help for the quantitative study and the dimensionnement of components to verify the adequacy of the set of actuators of robot.

REFERENCES

Amara, M. (1991). *Contribution à l'étude des aspects énergetiques en robotique mobile*, Thèse, pp. 94-130, Institue National des Sciences Appliquées de Lyon, France.

BOS, A.M. (1986). *Modelling multibody system in terms of multibond graphs with application to motorcycle*, Ph.D. Thesis, pp. 33-64, Netherland.

Koubaa, M.L. (1995). *Contribution à l'étude des systèmes mécatroniques par les formalismes bond graph et scattering - Application au robot six axes RTX,* DEA Automatique, pp. 88-120, Ecole Superieure des Sciences et Techniques de Tunis.

Koubaa, M.L., M. Amara and M. Betemps (1996). exploitation du formalisme scattering pour la representation 2D d'un segment du robot RTX, JTEA'96 **Vol 1,** pp. 235-243, Nabeul Tunisia.

Koubaa, M.L., M. Amara and M. Betemps (1997). contribution for the study of mecatronic systems by the scattering and bond graph formalisms, Accepted paper in IFAC-CIS'97, Belfort France.

A HEURISTIC APPROACH FOR PROCESS MODELLING

Luca Settineri

Dipartimento di Sistemi di Produzione ed Economia dell'Azienda - Politecnico di Torino
C.so Duca degli Abruzzi, 24 - 10129 TORINO - ITALY

Abstract: In this paper a computationally efficient approach to stochastic modelling of widely-intended stationary signals using ARMA models is presented. The estimator minimises a modified version of the likelihood function using only linear techniques. The proposed approach is therefore easy to implement, it does not require explicit statistical information of the second order and gives very accurate estimates with low computational load. Furthermore, an iterative version of the algorithm is proposed, to be used for the in-process implementation, and the modelling strategies are discussed. The validity of the proposed approach is tested with numerical simulations, in comparison with other heuristic schemes.

Keywords: Process modelling, ARMA models, Linearization, Heuristics, Simulation.

1. INTRODUCTION

It is common practice today to monitor the time history of properly chosen variables, useful to characterise operating conditions and possible failures of industrial plants. Therefore a wide range of signal analysis techniques have been developed, allowing for the extraction of useful information from the measured signals, many of which reported in the literature (Pandit and Wu, 1983; Braun, 1984; Braun, 1986, Tansel and Mclaughlin, 1993).

In the 80's parametric signal analysis methods have gained a big popularity (Braun, 1986; Kay and Marple, 1981) with respect to the non-parametric classical methods in the frequency domain (Wu, *et al.*, 1980; Kay and Marple, 1981; Romberg, *et al.*, 1984). Auto-Regressive (AR) models are commonly used, due to the availability of AR modelling methods efficient and easy to implement and to the difficulties that the estimation of the parameters of more complex models, like the Auto-Regressive Moving Average (ARMA) involves (Pandit and Wu, 1983; Kay and Marple, 1981, Tansel and Mclaughlin, 1993, Lombardo *et al.*, 1997); nevertheless, AR models are in general not adequate in effectively describing the behaviour of signals.

The most classical and direct approach to ARMA process parameters estimation is based on the principle of the Maximum Likelihood (ML). According to this approach a likelihood function is formulated, which is maximised by means of a suitable procedure (Box and Jenkins, 1976). The resulting estimator is characterised by useful properties like consistency, efficiency, asymptotic normality, but the technique is complex and computationally heavy, because of the strong non-linearity of the likelihood function, requiring an iterative maximisation algorithm. Furthermore, the possible presence of several local maximums can lead to completely wrong results (Åström and Söderström, 1975). Therefore the need of alternative heuristic techniques arises, in order to obtain good parameter estimation with limited computational load.

A commonly used heuristic approach is based on the extended Yule-Walker equations, and operates the estimation of the auto covariance function of the signal with a relatively high number of iterations, along with efficiency strongly dependent on the kind of process considered (Kay and Marple, 1981). An alternative procedure has been proposed, known with the name of two-stadium least-squares method this difficulty at the expense of a bigger computational load (Mayne and Firoozan, 1982).

In this paper an heuristic ARMA modelling approach based on the principle of the maximum likelihood is proposed, based only on linear computational techniques. The algorithm allows to obtain accurate estimations with low computational load. The problem of discrete ARMA modelling and the proposed approach are presented in section 2. An iterative version of the algorithm is presented in section 3. In section 5 the results of the simulation tests along with the comparison with existing algorithms are presented.

2. THE ESTIMATION ALGORITHM

A stationary signal in the discrete time domain $[x[t]]$ is defined an ARMA process of order (n,m) if it can be represented with a stochastic difference equation of the following form (Pandit and Wu, 1983):

$$x[t]=\sum_{i=1}^{n}\phi_i^o\cdot x[t-i]=w[t]+\sum_{i=1}^{m}\vartheta_i^o\cdot w[t-i] \quad (1)$$

where $\{\phi_i^o\}_{i=1}^{n}$ and $\{\vartheta_i^o\}_{i=1}^{m}$ represent the parameters of the Auto-Regressive part (AR) and of the Moving Average part (MA), respectively, and $\{w[t]\}$ a white noise with the following characteristics:

$$E\{w[t]\}=0; E\{w[t]\cdot w[s]\}=\delta_{t,s}\cdot\sigma_w^2 \quad (2)$$

where $E\{\bullet\}$ is the expected value, $\delta_{t,s}$, the Kronecker delta and σ_w^2 the variance of $\{w[t]\}$. On the AR and MA polynomial:

$$\Phi^o(z^{-1})\underline{\Delta}1+\sum_{i=1}^{n}\phi_i^o\cdot z^{-i};\Theta^o(z^{-1})\underline{\Delta}1+\sum_{i=1}^{m}\vartheta_i^o\cdot z^{-i} \quad (3)$$

the hypothesis of being prime and minimum phase is made. Given the set of observations $\{x[t]\}_{t=1}^{N}$, the estimation of an ARMA model with the method of the maximum likelihood is translated in the problem of minimising the following function:

$$Q_N(e)=\frac{1}{2N}\cdot\|e[t]\|_{l_2}^2=\frac{1}{2N}\cdot\sum_{t=1}^{N}e^2[t] \quad (4)$$

where $\{e[t]\}$ represents the residual defined from the model equation:

$$\Phi(z^{-1})\cdot X(z^{-1})=\Theta(z^{-1})\cdot E(z^{-1}) \quad (5)$$

The signal $\{e[t]\}$ can be thought as the one-step-ahead prediction error sequence for the model (5), implying that the optimal ARMA model is the one giving the minimum prediction error. The problem of the ML method is the non-linear dependence of $Q_N(e)$ from the MA polynomial $\Theta(z^{-1})$, that can be verified by rewriting eq. (4) in the following form:

$$Q_N(e)=\frac{1}{2N}\cdot\frac{1}{2\pi j}\oint_{|z|=1}\left|\frac{\Phi(z^{-1})}{\Theta(z^{-1})}\right|^2\cdot\left|X(z^{-1})\right|^2\cdot\frac{dz}{z} \quad (6)$$

where j denotes the imaginary unit. Minimising $Q_N(e)$ is therefore computationally complex and requires non-linear techniques. This difficulty can be overcome by defining the following function (Settineri, 1993):

$$Q'_{Ni}\underline{\Delta}\frac{1}{2N}\cdot\frac{1}{2\pi j}\oint_{|z|=1}\left|\frac{\Phi_i(z^{-1})}{\Theta_i(z^{-1})}\right|^2\cdot\left|\frac{\Theta_i(z^{-1})}{\bar{\Theta}_i(z^{-1})}\right|^2\cdot\left|X(z^{-1})\right|^2\cdot\frac{dz}{z} \quad (7)$$

where i is an iteration index.

Under suitable hypotheses:

$$Q_N(e)\cong Q'_{Ni} \quad (8)$$

implying that the minimising Q'_{Ni} should lead to an approximated sub-optimal estimator of $\Theta_i(z^{-1})$. Q'_{Ni} can also be seen as a quadratic function of the residual sequence modified as follows:

$$E'_i(z^{-1})\underline{\Delta}\frac{\Theta_i(z^{-1})}{\bar{\Theta}_{i-1}(z^{-1})}\cdot E(z^{-1}) \quad (9)$$

and its minimisation can be obtained by defining the filtered sequence:

$$\tilde{X}_{i-1}(z^{-1})\underline{\Delta}\frac{X(z^{-1})}{\bar{\Theta}_{i-1}(z^{-1})} \quad (10)$$

and by observing that:

$$Q'_{Ni}=\frac{1}{2N}\cdot\frac{1}{2\pi j}\oint_{|z|=1}\left|\Phi_i(z^{-1})\right|^2\left|\tilde{X}_{i-1}(z^{-1})\right|^2\frac{dz}{z} \quad (11)$$

is a quadratic function of the AR polynomial coefficients and can be minimised with the linear least-squares method. By combining the (5), (9) and (10), in the:

$$\tilde{x}_{i-1}[t]=\tilde{\xi}_{i-1}^T[t]\cdot\phi_i+e_i^r[t] \quad (12)$$

where:

$$\tilde{\xi}_{i-1}[t]\underline{\Delta}[-\tilde{x}_{i-1}[t-1]\ldots\tilde{x}_{i-1}[t-n]]^T \quad (13)$$

$$\phi_i\underline{\Delta}[\phi_{1i}\quad\phi_{2i}\quad\ldots\quad\phi_{ni}]^T \quad (14)$$

it follows that the AR polynomial parameters estimator:

$$\bar{\phi}\underline{\Delta}\arg\min Q'_{Ni}=\arg\min\frac{1}{2N}\cdot\sum_{t=n+1}^{N}(e_i^r[t])^2 \quad (15)$$

where "arg min" denotes the minimising argument, can be expressed by:

$$\bar{\phi}_i=\left(\sum_{t=n+1}^{N}\tilde{\xi}_{i-1}[t]\cdot\tilde{\xi}_{i-1}^T[t]\right)^{-1}\cdot\left(\sum_{t=n+1}^{N}\tilde{\xi}_{i-1}[t]\cdot\tilde{x}_{i-1}[t]\right) \quad (16)$$

After the AR polynomial parameters estimation has been operated, the MA polynomial parameters can be obtained from the expressions:

$$\bar{\theta}_{ij}=\bar{I}_j+\sum_{k=1}^{j-1}\bar{\vartheta}_{ki}\bar{I}_{j-k}+\bar{\phi}_{ij}\quad(j=1,\ldots,m) \quad (17)$$

where the first index denotes the MA parameter and $\{I_j\}$ the inverse function of the ARMA process defined as (Pandit and Wu, 1983):

$$\frac{\Theta(z^{-1})}{\Phi(z^{-1})}\underline{\Delta}\frac{1}{I(z^{-1})}\underline{\Delta}\frac{1}{1-I_1\cdot z^{-1}-I_2\cdot z^{-2}-\ldots} \quad (18)$$

The first l terms are estimated, by considering an AR model of the following form:

$$x[t]-\sum_{j=1}^{l} I_j \cdot x[t-j] = e_{AR}[t] \quad (19)$$

The validity of this approximation is due to the Wold decomposition theorem. The parameters vector:

$$I \triangleq [I_1 \quad I_2 \quad \cdots \quad I_l]^T \quad (20)$$

is then estimated as:

$$\bar{I} = \arg\min \frac{1}{2N} \cdot \|e_{AR}[t]\|_{l_2}^2 = \arg\min \frac{1}{2N} \cdot \sum_{t=l+1}^{N} (e_{AR}[t])^2 \quad (21)$$

which gives:

$$\bar{I} = \left(\sum_{t=l+1}^{N} \xi_{AR}[t] \cdot \xi_{AR}^T[t]\right)^{-1} \cdot \left(\sum_{t=l+1}^{N} \xi_{AR}[t] \cdot x[t]\right) \quad (22)$$

with:

$$\xi_{AR}[t] \triangleq [x[t-1] \quad \cdots \quad x[t-l]]^T \quad (23)$$

An appropriate order p of the AR model can be evaluated through feasible criteria, like the Akaike criterion, or by evaluating the residuals correlation $\{e_{AR}[t]\}$ (Söderström, 1977); furthermore, the estimator (22) is consistent and, for Gaussian signals, asymptotically convergent (Graupe, *et al.*, 1975).

To complete the algorithm, it is necessary an MA initial parameters estimator. This estimator can be obtained by using the following ARMA (n,m) process inverse function property (Pandit and Wu, 1983):

$$I_j + \sum_{k=1}^{m} \theta_k \cdot I_{j-k} = 0 \qquad \text{for } j > \max(n,m) \quad (24)$$

From (24) a linear system is obtained for the estimation of the inverse function $\{\bar{I}_j\}_{j=1}^{r}$. The system is resolved for the initial estimations of the MA parameters $\{\bar{\theta}_{k0}\}_{k=1}^{m}$. The integer r is to be selected so that the following condition is satisfied:

$$r \geq \max(n,m)+m \quad (25)$$

The proposed algorithm can therefore be formally sketched as follows:

Step 1: Inverse function $\{\bar{I}_j\}_{j=1}^{l}$ estimation using eq. (22). The value of l is to be properly selected, by satisfying the following condition:

$$l \geq r \geq \max(n,m)+m \quad (26)$$

where n,m are the orders of the ARMA model.

Step 2: Initial estimations of the MA parameters $\{\bar{\theta}_{k0}\}_{k=1}^{m}$ through the solution of eq. (25).

Step 3: Computing of $\{\tilde{x}_{i-1}[t]\}$ by filtering the signal $\{x[t]\}$ with the MA polynomial $\bar{\Theta}_{i-1}(z^{-1})$ (eq. 10).

Step 4: Estimation of the parameters vector ϕ_i of the AR polynomial by solving eq. (16).

Step 5: Estimation of the parameters vector θ_i of the MA polynomial by solving eq. (17).

In Fig. 1 is represented a diagram of the algorithm, where the information flow is represented by the thick line and the signal flow by the thin line.

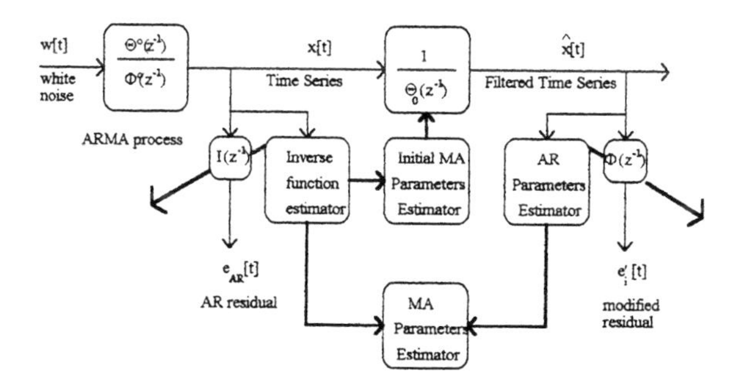

Fig.1. Flow diagram of the proposed approach.

The following observations should be made:

a) Going through the steps from 1 to 5 only one time is enough in most cases.

b) The computational load of the proposed algorithm is equivalent to two least squares operations of p and n orders respectively, 1 linear system of order m, an iterative filtering of order m and a set of equations requiring $\frac{1}{2} \cdot m \cdot (m-1)$ multiplication. The number of required iterations for the ML approach is impossible to determine *a priori*.

c) The required computational load for the solution of eq. (25) can be reduced from (m^3) multiplication operations, required by the standard algorithms, to (m^2) operations, thanks to the Toeplitz structure of the square matrix (see section 3).

3. THE ITERATIVE ESTIMATION ALGORITHM

When the parameters of a process model evolving in time must be estimated in-process, an iterative version of the proposed algorithm is needed, presented in this section. The sub-optimal iterative algorithm ML consists of the following steps:

Step 1: Estimation of the inverse function $I[t]$.
An iterative version of eq. (22) is used. To make the estimator able to track time-variant signals, expression (21) is modified as follows:

$$\bar{I}[t] \triangleq \arg\min \sum_{i=1}^{t} \mu[t,i] \cdot (e_{AR}[i])^2 \quad (27)$$

where:

$$\mu[t,i] \triangleq \prod_{j=i+1}^{t} \lambda[j] \qquad \mu[t,t] \triangleq 1 \quad (28)$$

with $\{\lambda[t]\}$ decaying factor introduced to improve the tracking capabilities of the algorithms. The iterative estimator defined in the (18), assumes therefore the following form:

$$\bar{I}[t] = \bar{I}[t-1] + \Gamma[t] \cdot \left(x[t] - \xi_{AR}^T[t] \cdot \bar{I}[t-1]\right) \quad [l \times 1] \quad (29a)$$

$$\Gamma[t]=\frac{Z[t-1]\cdot\xi_{AR}[t]}{\lambda[t]+\xi_{AR}^{T}[t]\cdot Z[t-1]\cdot\xi_{AR}[t]}\quad[l\times 1]\qquad(29b)$$

$$Z[t]=\frac{1}{\lambda[t]}\cdot\left[Z[t-1]-\frac{Z[t-1]\cdot\xi_{AR}[t]\xi_{AR}^{T}[t]\cdot Z[t-1]}{\lambda[t]+\xi_{AR}^{T}[t]\cdot Z[t-1]\cdot\xi_{AR}[t]}\right]$$
$$[l\times l]\qquad(29c)$$

where the terms in brackets represent the dimensions of vector and matrixes, and the decaying factor $\{\lambda[t]\}$ is upgraded every iteration by means of the following expressions:

$$\lambda[t]=\begin{cases}\lambda_0\cdot\lambda[t-1]+(1-\lambda_0) & (t<t_0)\\ \lambda & (t\geq t_0)\end{cases}\qquad(29d)$$

where the constants λ_0 e $\lambda[0]$ are properly selected barely less than one. To initialise the (29) it can be posed:
$\bar{\phi}[0]=\bar{0}, Z[0]=\alpha\cdot I_l, (\alpha>>0)$.

Step 2: Estimation of the MA polynomial initial parameters, in vector $\theta_0[t]$. This step requires the solution of eq. (25), that to this purpose can be written in a short form:

$$\tilde{N}_m[t]\cdot\theta_0[t]=\tilde{f}_m[t]\qquad(30)$$

with obvious definitions of the symbols. The solution of this equation with standard techniques requires (m^3) multiplications. Since $\tilde{N}_m[t]$ is characterised by a Toeplitz structure, non-Hermitian, the computational load can be reduced to $3(m^2)$ multiplications (Zohar, 1974). This is made dividing both terms of the (30) by $\bar{P}_{r-m}$ and introducing the notation:

$$N_m\underline{\underline{\Delta}}\begin{pmatrix}\gamma_0 & \gamma_{-1} & \gamma_{-2} & K & \gamma_{-m+1}\\ \gamma_1 & \gamma_0 & \gamma_{-1} & K & \gamma_{-m+2}\\ M & M & M & O & M\\ \gamma_{m-1} & \gamma_{m-2} & \gamma_{m-3} & K & \gamma_0\end{pmatrix}\qquad(\gamma_0\underline{\underline{\Delta}}1)$$
$$(31)$$

$$f_m\underline{\underline{\Delta}}[\delta_1\quad\delta_2\quad K\quad\delta_m]^T\qquad(32)$$

where the symbols N_i e $f_i(1\leq i\leq m)$ are used to denote respectively the sub matrix top-left of the i-th order of N_m and the vector made by the i elements to the left of f_m^T. The system:

$$N_i\cdot\bar{\theta}_{0i}=f_i\qquad(33)$$

where:

$$\bar{\theta}_{0i}\underline{\underline{\Delta}}[\bar{\theta}_{01}\quad\bar{\theta}_{02}\quad K\quad\bar{\theta}_{0i}]^T\quad(1\leq i\leq m)\qquad(34)$$

is therefore solved iteratively.

Step 3: Iterative filtering: According to the (10), at every iteration a filtering operation of the following form is to be made:

$$\tilde{x}[k]\underline{\underline{\Delta}}x[k]-\sum_{j=1}^{m}\bar{\theta}_{0j}\cdot\tilde{x}_t[k-j]\qquad(35)$$

which requires a number of multiplications superior to t. To overpass this difficulty different techniques can be used. The one used in the simulation is here presented:

At the time instant t to evaluate the required filtered values $\{\tilde{x}_t[k]\}_{k=t-n}^{t}$ using the (35) and the approximation:

$$\tilde{x}_t[k]\cong\tilde{x}_{t-1}[k]\quad(k=t-n-m,.......,t-n-1)\qquad(36)$$

The computational load of this approach is of $n\times m+m$ multiplications.

Step 4: Estimation of the AR parameters vector $\phi[t]$: the estimator (15) is modified as follows (see step 1):

$$\bar{\phi}[t]\arg\min\sum_{i=1}^{t}\mu[t,i]\cdot(e'_t[i])^2\qquad(37)$$

where $\mu[t,i]$ is defined from (28) and:

$$E'_t(z^{-1})\underline{\underline{\Delta}}\Phi_t(z^{-1})\cdot\tilde{X}_t(z^{-1})\qquad(38)$$

Under the hypothesis of slowly-varying process, such that:

$$\tilde{x}_t[t-k]\cong\tilde{x}_{t-1}[t-k]\qquad(k=1,2,...,n)\qquad(39)$$

the estimator (37) can be put in iterative form:

$$\bar{\phi}[t]=\bar{\phi}[t-1]+\tilde{\Gamma}[t]\cdot\left(\tilde{x}_t[t]-\tilde{\xi}_t^T[t]\cdot\bar{\phi}[t-1]\right)\quad[n\times 1]$$
$$(40a)$$

$$\tilde{\Gamma}[t]=\frac{\tilde{Z}[t-1]\cdot\tilde{\xi}_t[t]}{\lambda[t]+\tilde{\xi}_t^T[t]\cdot\tilde{Z}[t-1]\cdot\tilde{\xi}_t[t]}\quad[n\times 1]\qquad(40b)$$

$$\tilde{Z}[t]=\frac{1}{\lambda[t]}\cdot\left(\tilde{Z}[t-1]-\frac{\tilde{Z}[t-1]\cdot\tilde{\xi}_t[t]\cdot\tilde{\xi}_t^T[t]\cdot\tilde{Z}[t-1]}{\lambda[t]+\tilde{\xi}_t^T[t]\cdot\tilde{Z}[t-1]\cdot\tilde{\xi}_t[t]}\right)$$
$$[n\times n]\qquad(40c)$$

where $\lambda(t)$ is upgraded according to the (29d).

Step 5: Estimation of the MA polynomial parameters vector $\theta[t]$: this step requires the solution of the (17), that will be operated in a completely analogous way to how it is operated by the non-iterative algorithm.

The total computational load of the proposed ARMA modelling iterative algorithm, when the first filtering method is used, is equivalent to $3.5m^2+10(l+n)x.5m+m*n+8$ multiplications/ divisions per iteration.

4. MODEL IDENTIFICATION METHODS

In most applications the order of the appropriate ARMA model it is not known a priori and it is to be evaluated on the base of the available data. In the present paper a systematic method has been used, consisting in evaluating the fitting to the experimental data of ARMA $(2k,2k-1)$, k=1,2,3,.... models, until an adequate representation ARMA $(2k^*,2k^*-1)$ is reached. This can be furthermore compared with the representation ARMA $(2k-1^*,2k^*-1)$, before the definitive choice is made (Pandit and Wu, 1983; Lombardo *et al., 1997*). The adequacy of a particular model can be verified by using an F-test and/or the information Akaike criterion (AIC).

5. SIMULATION RESULTS

In this paragraph the performances of both versions of the algorithm, iterative and off-line, have been evaluated. The signals that have been modelled by an ARMA model are the result of the measurement of the *Fx* and *Fy* components of the cutting force in lathe turning operations in continuous cutting.

The identification procedure of the model described in paragraph 4 has led to the choice of an ARMA(6,5) model, whose parameters are reported in table 1.

The proposed algorithm has been compared with others algorithm: Graupe's (Pandit and Wu, 1983) and Mayne and Firoozan's (1982). As algorithm performance index the sum of squares of the residuals has been used, defined as:

$$RSS = \sum_{t=1}^{L} e^2[t] \tag{41}$$

where $\{e[t]\}$ is given by equation (5). The values of RSS relative to the parameters reported in table 1 reveal the bigger estimation precision of the proposed heuristic approach with respect to the methods of Mayne e Firoozan and to the one of Graupe, with a lower computational load.

Table 1. Parameters estimation results of an ARMA (6,5) model of a cutting force signal, carried on with the following algorithms: 1) Proposed algorithm; 2) Mayne e Firoozan algorithm; 3) Groupe algorithm.

Sample dimension: 1000; Sampling frequency: 10.000 Hz.

	ϕ_1	ϕ_2	ϕ_3	ϕ_4	ϕ_5	ϕ_6	ϑ_1	ϑ_2	ϑ_3	ϑ_4	ϑ_5	RSS
1)	-1.40	0.06	0.53	0.25	-0.69	0.28	-0.14	0.13	-0.16	-0.41	0.17	4.66
2)	-1.07	-0.16	0.29	0.37	-0.63	0.23	0.16	0.29	-0.13	-0.37	-0.09	4.91
3)	-1.04	-0.23	0.40	0.34	-0.59	0.14	0.22	0.26	-0.11	-0.44	0.00	5.15

Fig. 2 reports the power spectra of the real signal, and of the model ARMA(8,7) whose parameters have been estimated with the proposed algorithm. In table 2 the parameter estimation results of an ARMA(3,2) process are presented, operated with the iterative version of the proposed algorithm. In table 3 the estimation results of an ARMA (3,2) process with time-variant parameters are presented.

The time evolution of some of the parameters is shown in Fig. 3, where the real values of the parameters are shown in dotted lines. Constant and time variant parameters are considered.

Some other results have been obtained and, based on these, can be made the general observation that the iterative algorithm seems to offer an acceptable precision and a good tracking capability both with time invariant and with time variant signals. The transient performances have improved by the use of iteration (29d), while the converging rapidity depends on the initialisation constant α.

Table 2 Simulation results with the iterative version of the algorithm.

	Process parameters	Estimated values (iterations needed to convergence)		
		zero noise conventional	5% noise conventional	10% noise conventional
ϕ_1	-0.534	-0.535 (72)	-0.535 (67)	-0.536 (41)
ϕ_2	0.645	0.645 (93)	0.643 (86)	0.643 (69)
ϕ_3	0.154	0.155 (108)	0.155 (86)	0.157 (53)
ϑ_0	0.000	0.000 (22)	0.001 (15)	0.001 (20)
ϑ_1	-0.877	-0.878 (112)	-0.878 (84)	-0.875 (59)
ϑ_2	-0.587	-0.587 (98)	-0.587 (83)	-0.586 (68)

Sample dimension: 2000; Equivalent AR model order: 14; q=10; Initialisation: $\alpha = 10^{10}$; Forgetting factor: $\lambda = 1$

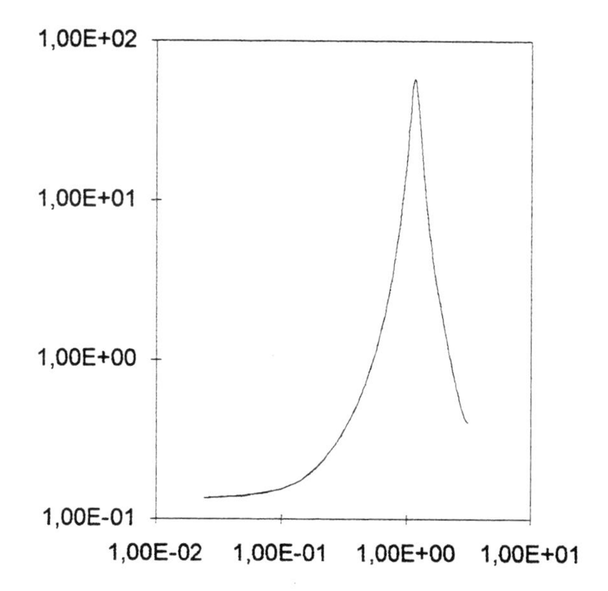

Fig. 2. Power spectra comparison between the real process and the process with the conventional algorithm estimated parameters.

Table 3 Estimated parameters of a machining process with time-variant parameters. Initialisation: $\alpha = 10^{10}$.

	Process parameters before/after variation	Estimated values before/after the variation (iterations to convergence)		
		no noise	5% noise	10% noise
ϕ_1	-0.434/-0.477	-0.434 (98)/-0.477 (82)	-0.433 (88)/ -0.475 (54)	-0.435 (84)/ -0.475 (57)
ϕ_2	0.785/0.432	0.785 (100)/ 0.432 (74)	0.784 (65)/ 0.431 (32)	0.789 (175)/ 0.432 (4)
ϕ_3	0.064/0.0235	0.064 (125)/ 0.0235 (64)	0.065(105)/ 0.0235 (84).	0.065 (103)/ 0.0237 (71)
ϑ_0	0.000/0.000	0.000 (24)/ 0.000 (13)	-0.001 (17)/ 0.001 (12)	-0.001 (17)/ -0.001 (8)
ϑ_1	-0.365/-0.402	-0.365(106)/-0.402 (67)	-0.366 (84)/ -0.404 (63)	-0.367 (48)/ -0.405 (58)
ϑ_2	-0.345/-0.759	-0.345(109)/-0.759 (63)	-0.343 (96)/ -0.755 (57)	-0.343 (100)/ -0.76 (88)

6. CONCLUSION

An algorithm has been presented, able to operate the estimation of the parameters of ARMA models. The proposed algorithm operates the minimisation of a modified version of the likelihood function and it is based only on linear optimisation techniques, much less complex then the non-linear techniques required by the maximum likelihood method.

The advantages of the proposed algorithm include:

- Implementation facility;
- It does not require explicit second order statistical information;
- High estimation precision with a computational load much lower of the ML method.

Furthermore, an iterative version of the algorithm has been presented as well as the implementation strategies. The effectiveness and precision of the proposed approach have been demonstrated with the use of numerical simulation and comparison with other sub-optimal schemes.

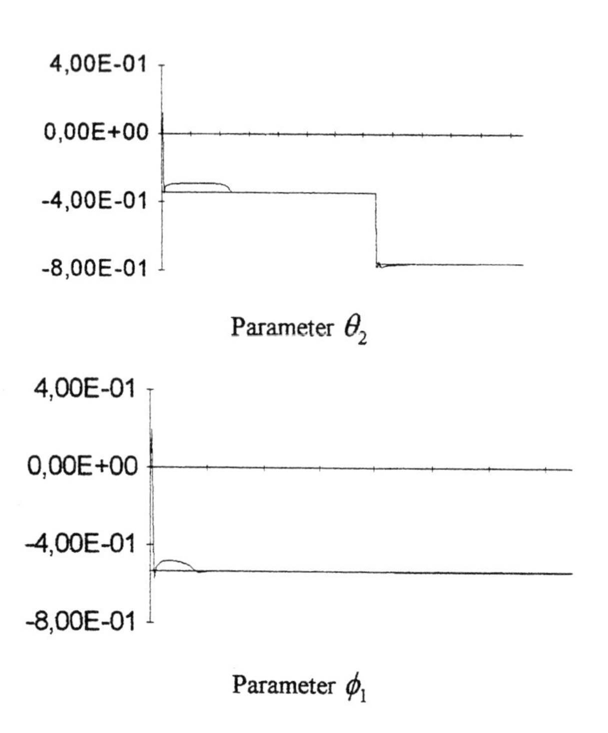

Fig. 3. Time behaviour of parameters θ_2 and ϕ_1, estimated with the iterative version of the algorithm.

REFERENCES

Åström, K.J. and T. Söderström (1974). Uniqueness of the Maximum Likelihood Estimates of the Parameters of an ARMA Model. *IEEE Trans. on Automatic Control* **Vol. 19**, pp. 769-773.

Box, G.E.P. and G.M. Jenkins (1976). *Time Series Analisys Forecasting and Control* (Holden-Day).

Braun, S. (1984). MSA-Mechanical Signature Analysis. *ASME Journal of Vibration, Acoustics, Stress, and Reliability in Des.* **Vol. 106**, pp. 1-3.

Braun, S., (ed.) (1986). *Mech. Signature Analysis: Theory and Applications* (Academic Press).

Graupe, D., D.J. Krause and J.B. Moore (1975). Identification of Auto-Regressive Moving-Average Parameters of Time Series. *IEEE Trans. on Automatic Control* **Vol. 20**, pp.104-107.

Kashyap, R.L. and R.E. Nashburg (1974). Parameter Estimation in Multivariable Stochastic Difference Equations. *IEEE Trans. on Automatic Control* **Vol. 19**, pp. 787-797.

Kay, S.M. and S.L. Marple Jr. (1981). Spectrum Analysis-A Modern Perspective. *Proceeedings of IEEE* **Vol. 69, No.11**, pp. 1380-1419.

Lombardo, A., A. Masnata and L.M. Settineri (1997). In Process Tool Failure Detection by Means of AR Models. accepted for publication on the *Int. J. of Advanced Manufact. Technol.*.

Mayne, D.Q. and F. Firoozan (1982). Linear Identification of ARMA Processes. *Automatica* **Vol. 18**, pp. 461-466.

Pandit, S.M. and S.M. Wu (1983). *Time Series and System Anal. with Appl.* (John Wiley and Sons).

Romberg, T. M., A.G. Cassar and R.W. Harris (1984). A Comparison of Traditional Fourier and Maximum Entropy Spectral Methods for Vibration Analysis. *ASME Journal of Vibration, Acoustics, Stress, and Reliability in Design* **Vol. 106** pp. 36-39.

Settineri, L.M. (1993). Un Algoritmo per l'Identificazione ed il Controllo Adattativo di un Processo di Lavorazione per Asportazione di Truciolo. *Proc of the I AITEM Conf.*, pp.43-45 (in Italian).

Söderström, T. (1977). On Model Structure Testing in System Identification. *Int. J. Control* **Vol. 26, No.1**, pp.1-18.

Tansel, I.N. and C. Mclaughlin (1993). Detection of Tool Breakage in Milling Operation - the Time Series Approach. *Int. Journal of Machine Tools and Manufacturing,* **Vol. 33, No.4**, pp.531-544.

Wu, S.M., T.H. Tobinan Jr. and M.C. Cho (1980). Signature Analysis for Mechanical Systems via the Dynamic Data System (DDS) Monitoring Technique. *ASME Journal of Mechanical Design* **Vol. 102**, pp. 217-221.

Zohar, S. (1974). The Solution of a Toeplitz Set of Linear Equations. *Journal of the Association for Computing Machinery* **Vol. 21**, pp. 272-276.

CONCEPTUAL MODELLING IN GENERIC PROTOTYPING APPROACH

Gabriel Neagu

Research Institute for Informatics
8-10, Averescu Av., 71316 Bucharest
Tel:+40-1-222 3778; Fax:+40-1-312 8539;
E_mail: gneagu@u3.ici.ro

Abstract: The generic prototyping approach (GPA) has been conceived according to some major orientations in complex system engineering field, with the emphases on manufacturing systems and their decision support capabilities. In the framework provided by this approach, the conceptual modelling is a central activity in the object oriented analysis and design of the future generic system. The paper presents the place of the conceptual model in the GPA development cycle, the specific structure of this construct and its role as the major result of the requirement definition phase. Then, the adopted model structure is argued and the solutions proposed for the implementation of the object, control and user-interface components are detailed, putting the stress on the decision control Petri nets formalism for the behaviour modelling. A case study of a job shop control system has been chosen to illustrate the approach.

Keywords: requirements analysis, object oriented modelling, dynamic models, agents, events, Petri-nets, decision support systems, blackboard architecture, production control

1. INTRODUCTION

The **Generic Prototyping Approach (GPA)** belongs to the class of modelling approaches, having in common the emphasis put on the abstract description of the future system as a determinant feature of the development process (Neagu, 1994b). The approach has been intended to comply with some major orientations in the complex system engineering regarding: the end-user involvement into the system development process starting with the early requirement definition phase; the open system architecture as a prerequisite for the gradual improvement of system functionality and performances; the object-oriented analysis and design with their major benefits regarding the natural matching between the reality and the model, the abstraction power, the structuring flexibility and the interaction flexibility, the robustness of OO design with respect to the later changes in system requirements. A specific emphases has been put on integration issues (Neagu, 1994c).

The *prototype* is an work product, with a limited functionality as compared with the target system (TS). It facilitates answering questions about the target system based on the end-user's involvement as partner of the development team, which is responsible for the final result. On the other side, the *generic model* facilitates the transition of the responsibility for the TS development towards the end-user. GPA is a combined approach, using the *generic prototype (GP)* as its central concept. GP inherits features specific to both the prototype concept (as being a preliminary version for the future system) and the generic system concept (due to its deliverable character as a final product, with host system (HS) capabilities for TS generation).

The *GPA life-cycle* includes prototyping, integration and instantiation processes, defined in the conceptual framework provided by the instantiation and derivation processes of *CIMOSA* methodology (Kosanke and Vernadat, 1996). The prototyping and integration processes deal with the *GP development*, while the instantiation process refers to the *GP tuning* and *maintenance*.

The paper is devoted to the **Conceptual Model (CM)** as the major result of the *Requirement Definition* phase. It consist of *object*, *control* and *user-interface* sub-models. The first one is similar to the object model of the *OMT* method (Rumbaugh *et al.*, 1991). The second sub-model describes the behaviour of the system and corresponds to the dynamic OMT model. The user-interface sub-model aims at emphasizing the importance of this component for system architecture design and the necessity to tackle this aspect as earlier as possible in the system development life-cycle. *Section 2* of the paper outlines the GPA life-cycle. RD activities preceding the CM development are described in *section 3*. The *fourth section* is devoted to the presentation of the CM sub-models, with the stress on the behaviour modelling solution, developed according to the multi-agent paradigm (Agha, 1986). A case study of a job shop control system has been selected to illustrate the approach. The CASE support for the conceptual modelling is tackled in *section 5*. The *final section* presents some concluding remarks underlining the role of CM in the GP approach.

2. GPA PROCESSES

The GPA development cycle is presented in Fig. 1. The *prototyping process* has a diagonal orientation as reported to the CIMOSA instantiation and derivation processes, with an emphasis put on the OO approach for the requirements definition (RD) and design specification (DS) phases. It starts with the RD phase, which is placed at the generic level of the CIMOSA cube. The DS phase delivers the GS architecture (GS/A) and the GP detailed design specifications (GP/DS). The *architecture* defines a structured taxonomy of classes of objects involved in the GS construction, as presented in table 1.

Table 1. Object taxonomy in GPA architecture

GS/A layers	Types of objects			
User interface	Menus	Options	Forms	Reports
Services	Resources utilization and administration			
Resources	Structural	Procedural	Cognitive	

During the third phase - implementation (I), *particular prototypes (PPs)* are developed as GP building blocks, in order to validate the implementation solutions for prioritary GP/DS modules. Structural and cognitive resources are implemented at partial level of CIMOSA cube, considering their integration capabilities. The back-forward arrows in Fig. 1a illustrate the opportunity of an iterative refinement of the results generated by this process.

The *integration process* aims at implementing the GP, based on a relevant set of PPs. Further, the GS implementation phase may be tackled, using a set of GPs developed according to the same GS/CM and GS/A. A careful analysis of the opportunity of this phase, based on the cost/performance criteria, is strongly recommended, that explains its optional character in Fig. 1.

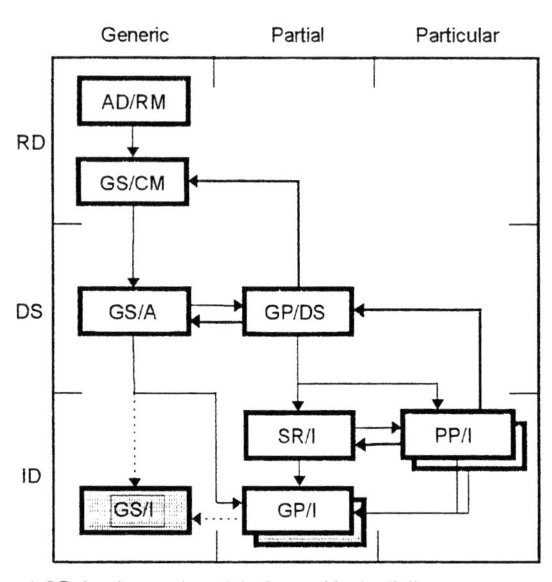

a) GP development: prototyping and instantiation processes

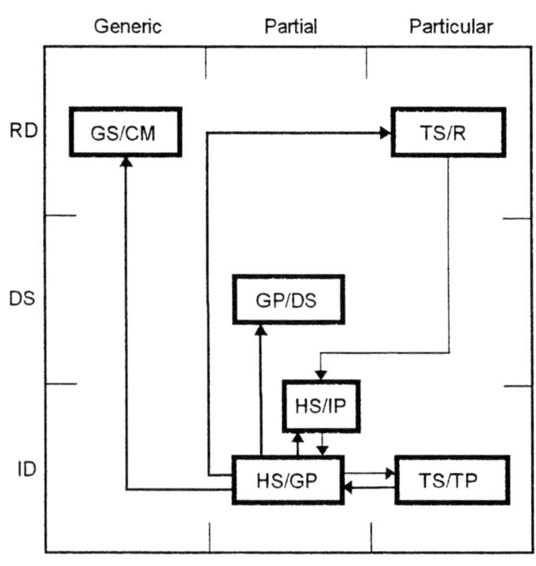

b) GP tuning and maintenance: instantiation process

AD - Aplication Domain
GS - Generic System
GP - Generic Prototype
PP - Particular Prototype
SR - Structural Resources
HS - Host System
TS - Target System
TP - Target Prototype
RM - Reference Model
CM - Conceptual Model
SA - System Architecture
DS - Detailed Specification
I - Implementation
R - Requirements
IP - Instantiation Parameters

Fig. 1. Processes of the GPA development cycle

The *instantiation process* deals with concrete target systems (TSs) generation using GP as a host system (HS). The process is driven by the instantiation parameters (HS/IP) expressing the TS requirements (TS/R). The result of the instantiation process should be also considered a prototype in its own rights: the target prototype (TP). As usual, it is the subject of some supplementary developments during its implementation, in order to solve specific TS requirements, which are not covered by GP. Therefore, TS is the result of two consecutive processes: GP instantiation and TP implementation. Based on the TS evaluation as compared with the initial TS requirements, this sequence may be iteratively re-activated according to feed-back links in Fig. 1b. At the same time, this evaluation provides the maintenance support for GP and, consequently, for the conceptual model used for its development.

3. PREPARING CM DEVELOPMENT

During the phase of the prototyping process, the conceptual modelling activity is preceded by the analysis of the given application domain (AD), which the future GPs belong to. It starts with the AD business architecture (AD/BA) development, using *domain* and *business process* concepts of the CIMOSA function view. Then, the AD/BA construct is further detailed through the functional decomposition operating with the CIMOSA *activity* concept. The *IDEF 0* method (Bravoco & Yadav, 1985) is used in this regard and the reference model for the given application domain (AD/RM) is generated. A list of system entities is also provided as a primary input for the conceptual modelling.

4. CM STRUCTURE

4.1. The Object sub-model

The object component of CM (CM/O) is similar to the corresponding sub-model of the OMT method. It reflects the static view of the system describing links, associations, generalization and inheritance relations between classes of objects. Fig.2 illustrates this component with the *manufacturing profile* section of the object sub-model developed for a job shop control system.

4.2. The Control sub model

The control component (CM/C) is devoted to the dynamic view of the system which is modeled as a population of agents interacting each other. This sub-model is composed of the diagrams describing their behaviour, which are developed according to the formalism of *Decision Control Petri Nets* (Neagu, 1994a).

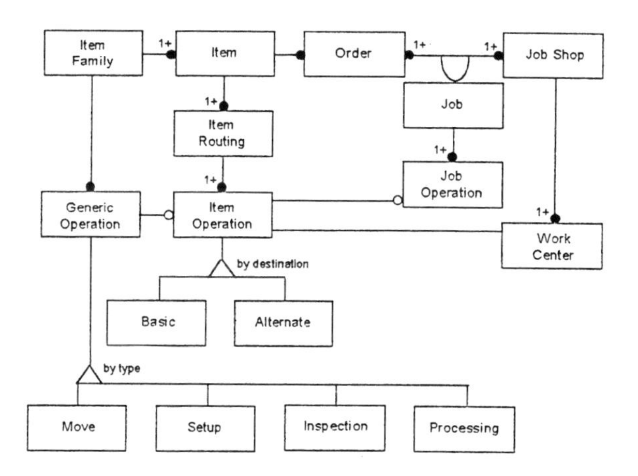

Fig.2. The object sub-model for *manufacturing profile*

This formalism has been defined as an extension of the control PN model (Mercier des Rochettes, 1988) to meet the requirements specific to DSS development.

$$DCPN = (CdPN, E, OP, f_K, f_T, R_D, T_D, f_D)$$

where: $CdPN$ is a colored PN; E is a finit set of events; $f_K : K \rightarrow E$ is a token function (with K the finite set of tokens moving in the net); OP is a finite set of process execution and control operators; and $f_T : T \rightarrow OP$ is a transition function (with T the finite set of transitions). These elements define a control PN. R_D is a finite set of decision resources (belonging to problem solving knowledge), for the identified token, transition or place conflicting situations; T_D is a finite subset of decision transitions ($T_D \subset T$) related with these conflicting situations; $f_D : T_D \rightarrow R_D$ is a function associating to each decision transition the decision resources required to solve corresponding conflicting situations. Specific graphical representations are used for decision transitions, depending on the related conflicting situation.

According to this formalism, the transition from one state to another is enabled by a *stimulus* - a set of events linked by logical operators. The event activating the stimulus is called the *control event*. All the other events in the stimulus structure are *conditions events*. When a stimulus is activated, the transition is fired and the associated operators are executed; consequently, new events are generated as candidates for future stimuli.

In the framework of the PN formalism the conflicting situations that could arise during a transition are characterized as *token, transition* or *place* selection conflicts. To cope with these situations the DCPN model proposes the class of *decisional transitions*. The set of associated operators includes in this case specific decisional resources (algorithms, rules), destined for solving given conflicts. An example of behaviour diagram is presented in Fig.3: the *job* agent belonging to the shop floor control system mentioned before. The list of states (positions) is given in the table 2.

Table 2. List of positions for the *job1* agent

P_1	waiting for the first operation
P_2	in transit for the next operation
P_3	waiting in the input queue the work center for the current operation
P_{3a}	waiting on the AGV to enter the input queue (in case it is full)
P_4	with machine allocated for the current operation
P_5	waiting for the available robot (if required for the current operation)
P_6	execution of the current operation
P_7	being unload after the current operation
P_8	waiting for an available AGV
P_{8a}	same, in case of the last operation
P_9	waiting for the allocated AGV
P_{9a}	same, in case of the last operation
P_{10}	re-allocated on the same machine for the next operation
P_{11}	finished

For objects with a more complex behaviour, additional agents may be defined. For the above example a second job agent has been defined, which is acting during the generic state P_7 state of the first agent. Depending on the evolution of the agent 2, at the output of state P_7 the agent 1 may switch to one of six possible states.

There are two decision transitions in this diagram: t_1 for the selection of an available AGV to move the job to the work center for next operation, and t_6 for the selection of an available machine to execute the current operation. In fact t_8 corresponds also to a decision transition, but the selection of the output state is solved by the second job agent.

In order to cope with the interaction and cooperation between agents the *co-ordination level of CM/C* has been defined. It is based on the discrimination of events depending on their influence on the agent population. *Agent, object* and *system* classes of events are proposed in this regard. Agent events correspond to local state changes and have no influence on the others agents. Object events are significant also for the other agents of the given object, while the influence of system events goes beyond the boundaries of the source object. Obviously, object and system events are *active (stimulus type)* events, while agent events are *passive* ones.

To solve this coordination level of CM/C the *blackboard* paradigm has been considered, with the event oriented control strategy (Nii, 1994). According to this strategy, one or more knowledge sources (agents) are activated on the basis of the current state of the blackboard. As a result of this activation one or more changes, called events, are generated to the blackboard state.

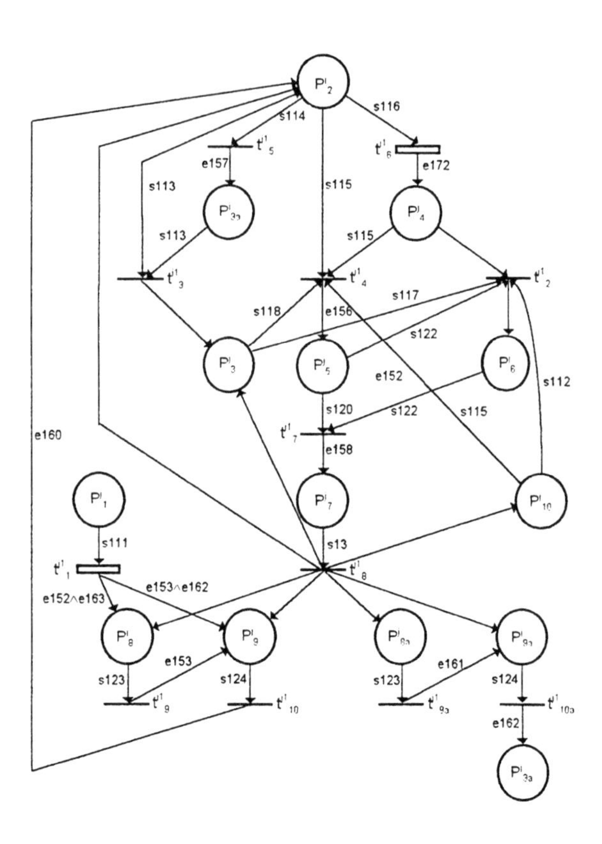

Fig.3. The control sub-model for the *job 1* agent

These changes are new solution candidates or modifications to the existing solution. A new event is selected in the current context and knowledge sources matching its significance are activated. This CM/C level is described in more detail in the paper (Neagu, 1995), as a support for a multi-agent simulation controller in manufacturing systems.

4.3. The user-interface sub-model

The CM/U sub-model has been approached according with the general principle stating that the user interface should be treated apart from the application itself in terms of both system architecture and development methodology (Barthes, 1995). It means that this component of the future system should be validated and freeze as early as possible during the development cycle. This specificity comes from the particular consistence and relevance of the interface for the negotiations between user and developer.

In the GPA framework the user interface is defined as a separate layer in the overall architecture (see table 1). The functional analysis and decomposition included in the AD/RM construct (see section 3) provides the support for defining menus and options. Further, the information about classes of objects, their attributes and relations specific to the CM/O sub-model is used for a draft specification of different forms, messages and reports. Obviously, the emphases is put on the structural view of the interface, using the object model formalism. Its dynamic view usually complies with the standard solutions specific to the technological environment selected for development and implementation phases of the future system. Depending on the facilities provided by this environment, the development of an working version for the user interface is recommended. It is important not only for the validation of the interface. Playing the role of an early, evaluation prototype of the future GS, this empty shell may provide a valuable support to delimit and establish the development priorities for different GPs.

5. CASE SUPPORT

OO analysis and design activities are widely assisted by CASE tools. For GPA implementation the *System Architect* software has been selected (PS&S, 1995). This solution provides support for both OMT and IDEF0 methods used during the requirement analysis phase.

As much as the dynamic modelling is concerned, the compatibility between state diagrams and PN formalisms has facilitated the mapping of GPA/CM requirements on the facilities provided by *System Architect* in this regard. The only exception has been provided by the decision transitions, special matching rules being defined to catch their peculiarity. The decision resources of each decision transition have been considered as *actions* associated with different types of events defined in the OMT method, depending on the nature of conflicting situation tackled by the given transition (Rumbaugh *et al.*, 1991, section 5.5.2). Therefore, in case of position selection conflicts, the decision support is associated with the *exit* event of the source state. For transition selection conflicts this support is activated by the event firing the transition, while for token selection conflicts decisional resources are attached to the *entry* event of the destination state. The principle adopted in defining this solution is to activate the decision support as close as possible to the moment the decision should be made, in order to take advantage of the freshest context.

CONCLUSIONS

In the GPA framework, the conceptual model is relevant for major system engineering orientations underling this approach. The CM structure reflects the attention paid to the system behaviour and to its interaction with the user, which is relevant for highly dynamic manufacturing control systems, including DSS. A specific solution has been proposed for the dynamic sub-model. As compared with the OMT configuration of OO models - as a reference point of view in this field - the functional sub-model is missing in case of GPA/CM because this aspect is reflected in the reference model of the application domain developed in a classical manner. From this point of view, GPA may be considered a specific method as reported to the taxonomy of OO analysis and design methods proposed in (Filip and Neagu, 1995). The methods of this kind try to reconcile the OO approach with the classical approaches in this field, being considered more pragmatic as soon as they encourage the former experience of the development team.

The user-interface sub-model is promoted as a CM component, considering its relevance for the end-user's validation of RD results, as well as its importance for the future GP structuring.

The importance of CM construct is twofold: on the one hand, it reflects the genericity of the future system, as an aggregated and consistent expression system requirements. On the other hand, together with the system architecture, it provides the support for the step-wise development and validation of generic prototype(s) for the given application domain. From both points of view, CM may be considered by itself an important result of the development process, encapsulating domain specific as well as system engineering expertise. Together with a set of compatible class libraries implementing the resource level of GPA architecture and developed for different computing platforms, this system modelling solution could be considered as one of the most reasonable goals of the endeavors to fully valorize the OO capabilities in the complex system engineering.

REFERENCES

Agha, G.A. (1986). Actors: a model of concurrent computation in distributed systems. *MIT Press*.

Barthes, J.-P. (1995). Man-machines interfacesin simulation and knowledge-based systems. *2nd Workshop of SiE-WG*, June, Brussels.

Bravoco, R.R. and S.B. Yadav (1985). A methodology to model the functional structures of an organization. *Computers in Industry*, **6**, 345-361.

Filip, F.G. and G. Neagu (1995). Object oriented approach to software engineering for CIMIA. *Studies in Informatics and Control*, **4**(1), 47-58.

Kosanke, K. and F. Vernadat (1996). CIMOSA: Open System Architecture for CIM. An example for Specification and Statement of Requirements for GERAM. *CIMOSA Association Report*, v 2.0, June, Böblingen, Germany.

Neagu, G. (1994a). Modelling approaches at the manufacturing shop level. *PhD report*, Politehnica University, Bucharest (in Romanian).

Neagu, G. (1994b). Generic modelling vs. prototyping: an object oriented approach to the decision support at the shop floor level. In: *Manufacturing Research and Technology 22: Advances in Manufacturing Systems-Design, Modeling and Analysis* (R.S. Sodhi,Ed.), 19-26, Elsevier Science.

Neagu, G (1994c). Integration levels in Generic Prototyping Approach. In: *Preprints, IFAC Conference on Integrated Systems Engineering* (G. Johannsen, Ed.), 477-482, Pergamon, UK.

Neagu, G. (1995). A multi-agent model for job-shop scheduling. In: *Preprints, IFAC Workshop on Intelligent Manufacturing Systems* (Th. Borangiu and I. Dumitrache, Eds), 275-280, Bucharest.

Nii, H.P. (1994). Blackboard systems at the architecture level. In: *Expert Systems with Applications*, 7, 43-54.

PS&S (1995). *System Architect. User Guide & Reference Manual.* Popkin Software & Systems Incorporated, USA.

Rumbaugh, J. *et al.* (1991). *Object-Oriented Modelling and Design*. Prentice Hall.

DEVELOPING AN OBJECT-ORIENTED REFERENCE MODEL FOR MANUFACTURING

Ricardo Chalmeta*; Theodore J. Williams; Francisco Lario***; Lorenzo Ros*****

*Grupo IRIS. *Departamento de Informatica. Universidad Jaume I. Campus Penyeta Roja s/n. Castellón. SPAIN. E-mail: rchalmet@inf.uji.es Fax: INT + 34-64-34 58 48*
** *Institute for Interdisciplinary Engineering Studies. Schools of Engineering. Purdue University. West Lafayette. Indiana. USA. E-mail: tjwill@ecn.purdue.edu*
*** *Grupo GIP. Departamento de Organizacion de Empresas. Universidad Politecnica de Valencia.Valencia. SPAIN. Tel.: INT + 34-6-387 76 80. Fax: INT + 34-6-387 76 89*

Abstract: A fundamental early step in the *design* process for a *Manufacturing System* is to develop a **model** which describes adequately the proposed system, in order to be able to study and to evaluate the impact of the design decisions on the system performance, before its construction. Among the different modeling methodologies applicable, the Object-Oriented Methodology is very useful due to its capability to represent all aspects involved in a complex Manufacturing System, including its dynamic dependencies. However, no such models exist at present in the literature. This paper shows such a detailed Object-Oriented Reference Model for Manufacturing System. This new model can be used to develop and to simulate particular manufacturing object-oriented models.

Keywords: Design, Manufacturing System, Object Models, Dynamic Models, Simulation.

1. INTRODUCTION[1]

The *Design Process* for a *Manufacturing System* requires that the designer have a clear knowledge of the overall system in order to be able to evaluate and to decide among different design alternatives involved in developing the system. However, the overall complexity of a Manufacturing System, especially when its design decisions are taken by multidiscipline groups with different needs and methods, requires the use of a **model** that describes the proposed system, with the required detail level, in a formal, concise and suitable way, in order to: (1) make the system understanding **easier**, (2) direct the **analysis, design and construction** of the system, integrating the four views (functional, informational, decisional and physical) that compose it (Amice, 1988) and (3) to be able to study the future impact of the design **decisions** on the **system performance**, before its construction, from the changes produced in the model.

[1] This research has been supported by the CICYT inside the TAP Project Number 95-0880.

2. MODEL REQUIREMENTS

In order to use a model in the design of a Manufacturing System, the model requires the following features:

1. **Static features** that describes the Manufacturing System from its four views:
   - *physical view*: represents the resources to carry out the activities
   - *informational view*: models all information needed to take and to execute decisions.
   - *organizational view*: describes who and how the decision of the system are taken.
   - *functional view:* describes the system activities.
2. **Dynamic features** that represent the process behavior.

3. **Execution features**, that is the possibility to bind the dynamic model with a computer language, in order to create a simulation computer program, to analyze the system performance.

A criterion which has been applied for many years is to use different modeling techniques for the different views. Therefore, a *variety of modeling techniques* exist and have been applied in the industrial enterprise field. For instance, general methods have been adapted to model manufacturing systems (i. e. structured analysis and design techniques) and specific tools have been developed for functional modeling (i.e. IDEFO), for informational modeling (data and process oriented methods) and for dynamic modeling (i.e. PetriNets).

The connections between the different views are essential to understand and to help in the Manufacturing System design. However, the above techniques cover different manufacturing views but none of them can model the manufacturing system as a whole. Also no one method is able to group all the partial models generated using these different methodologies in a coherent way. Therefore, the techniques used up to now are not yet suitable to generate a **Manufacturing System Integrated Model.**

3. OBJECT-ORIENTED METHODOLOGY

To fulfill the above need, the ***Object-Oriented Methodology* (OOM)** offers the features to properly carry out this process not only for its general applicability but also because there are many software tools supporting this methodology (Taylor, 1995)] The **OOM** develops models that contain all the necessaries features to assist in the design process of a manufacturing system. The resulting model can represent all informational, organizational, technological and functional aspects involved in a complex production system. It allows us to simulate them in order to analyze the dynamic behavior of the manufacturing system and it enables to develop informational computer systems to support the functions and decisions. That is to say, the **OOM** integrates **static** system features (considering all system views), **dynamic** modeling and **simulation.**

The application of this modeling focus based on objects to manufacturing systems has been recent. Thus, it is difficult to find detailed reports of projects. For instance, (Mize and Prat, 1991; Gaspart, 1991; Quellenberg, 1994) can be mentioned. However, none of these references present the detailed level of the Model that is proposed in this paper.

4. DEVELOPING OBJECT-ORIENTED MODELS

Although different object-oriented modeling techniques (Yourdon, 1979; Booch, 1996; Rumbaugh, 1991) have been developed, no one method is widely accepted as a standard and none is able to solve all problems appearing in its industry application. For that reason, a new methodology, named **RCOO,** is used.

The procedure to apply **RCOO** (Chalmeta, 1997), is structured into **four stages**, (domain, structural, functional and dynamic) that are shown in detail in the section five of this paper. For those who have a background in OOM, the last three phases are very close to the Rumbaugh proposal but have some difference like the *graphic tools* or the use of *message flow diagrams* in the functional model, instead of the data flow diagrams (that is because the Rumbaugh methodology is more orientated to software design and RCOO is oriented to model general systems).

5. OBJECT-ORIENTED REFERENCE MODEL FOR MANUFACTURING

Although as it has been showed above, an Object-Oriented model of a manufacturing system can assist in the design process, *detailed examples of how to apply the object methodology to develop models of manufacturing systems* do *not yet exist in the literature* (Mertins, 1994).

For this reason, this paper presents a Object-Oriented Reference Model for a Manufacturing System, using the **RCOO** methodology. This model can be used by different users to develop a particular manufacturing object-oriented model of their system that can help to understand the manufacturing complexity and to analysis the system behavior and the impact of decision making through simulation of the model. The methodology used follows:

5.1 Domain

Before starting any modeling process it is necessary to define the model (system) *limits* showing the **internal objects** that are included in the model, the **external objects** and the **linking objects** which connect external and internal entities (Taylor, 1995). Domain definition is a first approximation to delimit what entities are going to be completely modeled (internal objects and interface objects) and what entities are only considered from the point of view of their influence on the system without internally analyzing them (external objects). Figure 1 shows such an example domain for the model.

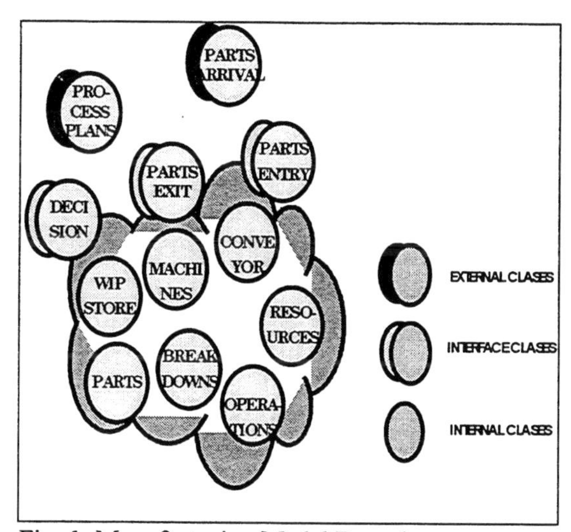

Fig. 1, Manufacturing Model Domain

In the following discussion, every one of sub-models which form the **Manufacturing System Reference Model** (Structural, Functional and Dynamic) are presented (see figure 2).

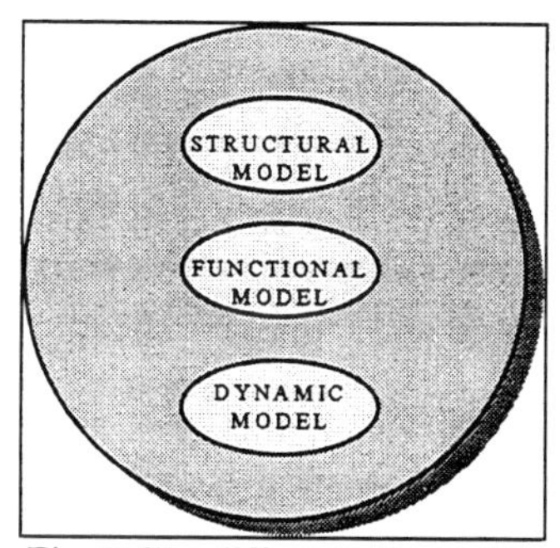

Fig. 2, The different sub-models that form the Reference Model

5.2. Structural model

Classes identification. Object-oriented modeling with the **RCOO** methodology allows one to structure a complex system in an hierarchical way through **composition** and **heritage** relations. Thus, in the maximum aggregation level only the main entities are shown. The selection of such entities must be done according to the system analysis focus and in the detail required. In a first approximation, the considered entities appear reflected in the following description of a manufacturing system:

"*A manufacturing system is considered to produce a set of final products through a set of elemental transformations (operations) of raw material (initial products) which are carried out in one or several machines. To regulate production, storage areas are used. These are physical spaces where products are placed temporarily. The movement or product flow into the system is carried out by transports. Finally, to execute an operation, auxiliary elements are eventually required, named resources*" (Proth, 1995). Therefore, the fundamental entities used in this model are **six object classes** as follow:

- **parts** (raw materials, WIP and final products)
- **stores** (elements of temporary pieces storage)
- **machines** (facilities to produce parts)
- **transports** (elements to carry out part movement)
- **resources** (material and human ones)
- **operations** (transformation or assembly)

Stores, ***machines***, ***transports*** and ***resources*** are classes which form the production system fundamental structure, that is, the physical structure.

Over this structure the **parts** flow is defined, and then, new classes are included in the model. Raw materials enter the system from the external class named ***parts arrival*** which acts as a source and is connected to the system through ***entry parts.*** Pieces leave the system through ***exit parts***. These classes allow the system interconnection with its environment of material suppliers, product customers, etc.

Operation is a class which acts as a link between the plant static structure and the material flows. A external class called ***process plan*** determines the operational sequence necessary to make a product and relates it with machines, pieces, resources and transports.

The Manufacturing System requires an element that carries out the system management, that is to say, that implement the decisional view. In the developed model, the decisions are represented by the ***decisional*** class. This class receives from its associated classes information about what kind of pieces are in the system, what operations can be made, the machines and resources available, etc. in order to take decisions according to some implemented rules.

Finally, it is necessary to consider an ***interruption*** class describing the system stops produced due to breakdowns, maintenance, personnel resting, etc.

In the OOM, the structural model has a different meaning than in a conventional modeling methodology, that would be more oriented to the physical architecture of the system. With the OOM methodology, the structural model collects all system classes, including those corresponding to the physical, decisional, functional and informational entities. This is an advance in order to develop a integrated model.

Once the main classes of the system are identified, successive activities will gradually give a more detailed model description.

Relationships among classes. Production system classes are not independent but a set of *Relationships* among them can be defined. These relationships can be classified in three categories: **Heritage**, **Association** and **Composition**.

Heritage. This is a feature of the OOM through which a class named **child** acquires the properties of other system classes named **father**. However, it is not necessary that father classes exist in the system. It can happen that several classes use similar attributes and methods. It is thus convenient to introduce a **virtual father class** defining the common features. In the model this circumstance arises with stores, transports and machines where the same procedure to assign and place pieces is used. It happens the same with transports, machines and resources where a similar protocol for interruption management is used.

Figure 3 shows a **hierarchical tree** of **classes** of the main manufacturing system model entities, which appear due to the heritage relationship. When a class is derived from another, an arrow is painted from the more specific class towards the more general one.

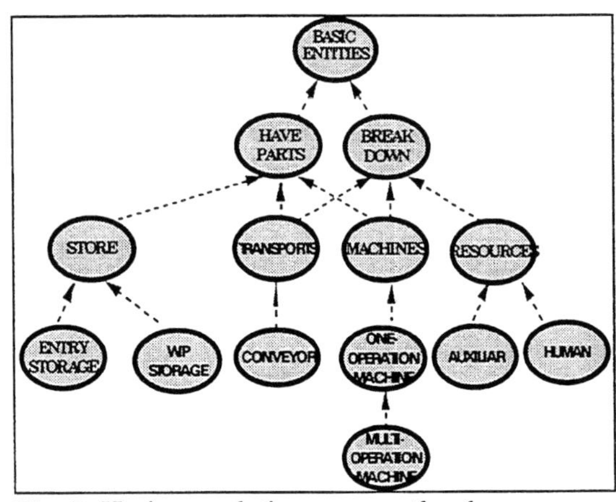

Fig. 3, Heritage relations among the classes.

Class hierarchy also introduces additional classes to allow one to complete the model and, at the same time, simplify it. The model is made more simple due to the common attributes and methods of the several classes which need to be specified only once in the super-class.

Association. The model classes are not isolated but they interact among themselves. This relationship is called **association** and is represented graphically by lines linking the related classes. Association adds much information to the model and offers a global view of it. Besides, it allows one to identify those attributes which link classes. Figure 4 shows the main associate relations between the classes.

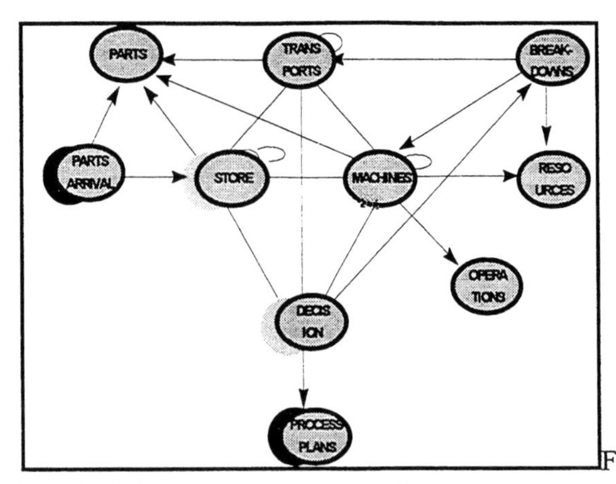

Fig. 4, Main associations among the classes

The first relation appears due to the **parts movement** through the system. The ***generator*** class is an entity in charge of introducing ***parts*** objects into the system (it model the Material Requirements System), assigning them to the ***entry parts store*** class. Then, parts pass through a net composed of ***stores***, ***transports*** and ***machines*** until they are final products. The flow of the parts in the system is managed by the ***decisional*** class, asking the ***process plans*** class and receiving information about the state in the plant of the different classes. The decision class take decisions about the pull and push parts rules for the stores and the machines, the parts inventory rule for the stores, the machine operations, and so on.

Another important link is due to the **interruptions**, that can affect the transports, machines and resources.

Composition. The composition relationship, also named **aggregation**, describes those objects that contain other objects. This kind of relationship is given many times in enterprises. For instance, a final product is made with several components, or a department is formed by employees. Formally, in the Object-Oriented Methodology, the composed object does not physically contain the component objects but refers to them by means of attributes. This allows the same object to be a component of more than one other object. Figure 5 show an example of a composition relationship in the manufacturing system.

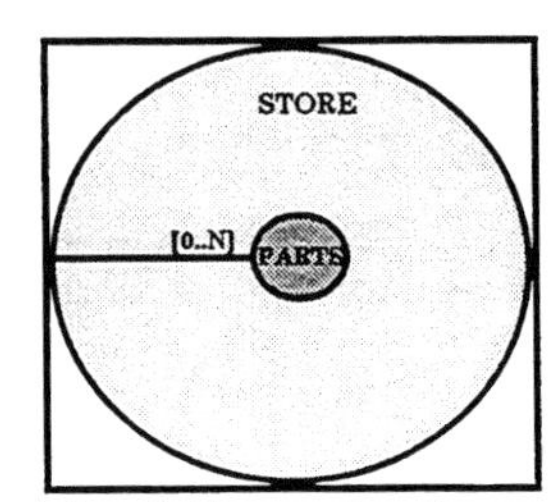

Fig. 5, Example of composition relation

Attributes. From the relations among classes, attributes of every entity are obtained. An attribute is any property or feature that all the objects of a same class have. Table 1 shows some characteristic attributes of the main classes.

Table 1. Some classes attributes

break down	machine	parts	operation
name	**name**	**name**	**name**
break time	**state**	**state**	**set-up time**
duration	**operation**	**type**	**duration**
state back	**efficiency**		**items**
	resources		
	location		

5.3. Functional model

Model objects interact among themselves creating a **communication flow** in which data transfer, control flow and service requirements are combined. The functional modeling phase consists of defining different **scenarios** that reflect the communication among classes. A scenario describes, in terms of *message flow diagrams* (MFD), interactions produced among system objects due to the execution of a specific process associated with a predetermined event collection.

Figure 6 presents a scenario corresponding to a PULL example, using a DFM. It is described below.

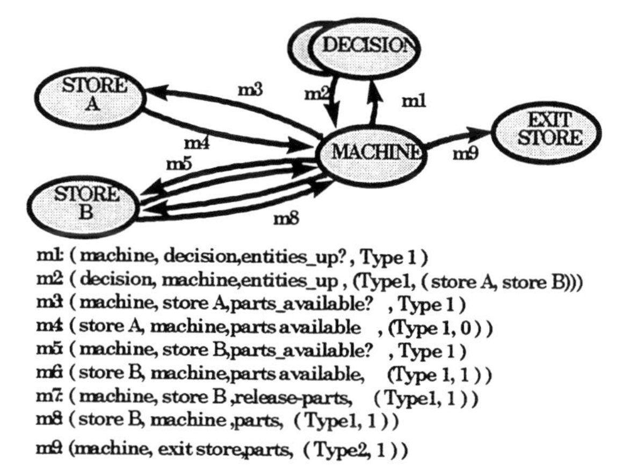

Fig. 6. Example of a PULL scenario

It is supposed that a system is composed of *two stores, a machine* and a *exit store*. Type 1 parts arrive to the machine coming from both stores. An operation is made on these parts and the type 2 resulting products are left in an exit store. Further, it is considered that as pull rule or a push one is by priority and a store A priority is higher than a store B priority.

The initial state of each entity is as follows. Store A has no Type 1 parts, store B has some Type 1 parts and the machine is processing a part. In this situation, the machine needs a new part Type 1.

Figure 6 shows the message sequence taking place. The machine asks the *decisional class* about entities that can send it Type 1 parts, through the message **entities_up?**. As a result, it receives the answer message **entities_up** saying that it must receive type 1 parts from store A and store B. First, it tries to obtain a part from store A (because this store is the higher priority one) through message **parts_available?**. In return, it receives the message **available_parts (0)** saying that store A has no available pieces. Then, the machine starts the same process with store B receiving the message **available_parts,** where store B notifies it has a type 1 part. Thus, the machine sends store B the message **release_parts** indicating for it to transfer the part. Then, the machine receives the message **parts** with the mentioned piece. Finally, when the machine has processed the Type 1 part, it sends the message **parts** to the exit store, sending it one Type 2 part.

Messages and Methods. Once the existing relations among objects in a system is understood, by means of the ***scenarios analysis***, the **messages** (elements used to communicate among objects) are determined and the **methods** which deal with them are described. Some of the methods of the machine class are:

- **Obtain entities up.** Send the message *entities up?*, asking for the entities that have the parts that are needed, and receive the message *entities up* with these entities

- **Choose entities up.** Following the pull rule, send to the entities the message *parts available?* And receive the message *parts available* with the number of parts.

- **Assign parts.** Send the message *release parts*, and receive the message *parts*, with the part.

If the objective were to model an information system, these attributes will be the data to introduce in the data base, and the methods will be the computer programs that transform them into information.

The importance of a graphic notation to understand the final model has led us to develop a diagram called **EAMM** (*Entity-Attribute-Method-Message*). It represents, in a graphical way, the fundamental mechanisms of a class. Examples of this diagram appears in (Chalmeta, 1997).

5.4. Dynamic model

Since the manufacturing system is not static but changes with time, it is necessary to represent its temporal behavior in order to have a total understanding of the system. In the **RCOO** Methodology, this task is carried out by means of modeling the **events** that happen in the system and the different states that these produce in a class. A **class state** is defined as the value of its attributes at a specific time. Thus, state evolution describes the temporal behavior of every class.

Table 2 shows the main events that can happen in a manufacturing system, and figure 7 represents a state diagram for the machine class. This diagram shows the different states of a machine, the events that produce the state changes and the value of some attributes. The dynamic behavior of all the entities of the model can be modeling following the same approach.

Table 2. Main events of the manufacturing systems

e1: Assigning a part	**e7:Notifying operation**
e2: Releasing a part	**e8: Assigning resources**
e3: Notifying parts to transport	**e9: An operation starts**
	e10: An operation ends
e4: A transport starts	**e11:Interruption starts**
e5: A transport ends	**e12: Interruption ends**
e6: Return to origin	**e13: Final product**

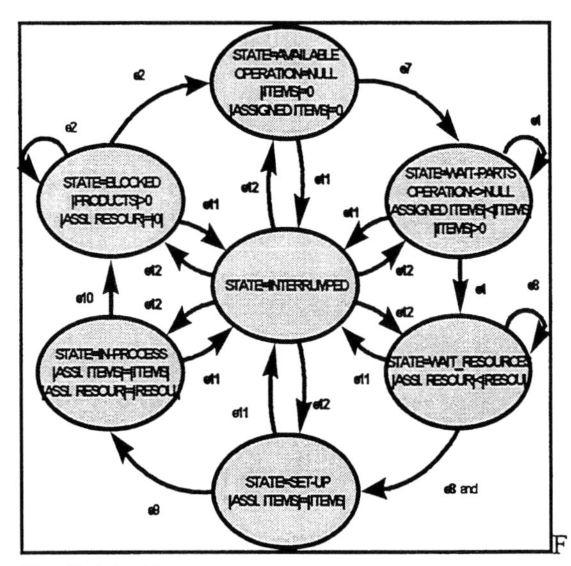

Fig. 7, Machine states

This dynamic model is an essential part of the Reference Model, because it introduce the concept of time, representing the dynamic behavior of the system and the state changes that happen in the different entities . So, this part is what really allows the **simulation** of the model.

6. CONCLUSIONS

To understand completely a manufacturing system and to analyze correctly its efficiency and performance before its construction is necessary the use of a integrated model. However, the modeling techniques used up to now were not able to model all requirements at the same time. Contrasting with these techniques, the **Object-Oriented Methodology** (OOM) allows one to model *all enterprise activity views* including their *dynamic dependencies.*

In this paper, a **Reference Model of a Manufacturing System** has been presented. This model is useful for an enterprise to develop a Particular Model (Amice, 1988) of its Manufacturing System. Then this particular model can be used for two applications. On the one hand, to help in the development of a Manufacturing Information System to support the management of the system. From this point of view, **attributes** are the *data* to incorporate to the data base and **methods** constitute the software program *algorithm.* On the other hand, the model incorporates the system dynamic features. So, with Object-Oriented Computer Technologies, it is possible to directly transform the model into an executable model, that can be simulated by a computer.

REFERENCES

Amice (1988). *"Open System Architecture for CIM".* Ed. Springer-Verlag

Booch, G. (1986) "Object-Oriented development". *IEEE Transactions on software Engineering.*

Chalmeta, R. (1997) *"Reference Architecture for Enterprise Integration".* Ed. Servicio Publicaciones Jaume I. SPAIN.

Gaspart, P., et al. (1991) "An Object Oriented Multi-Layered Approach to Multi-Resource Production Scheduling". *(FAIM'91).* Limerick.

Mertins K. et al. (1994). "Object-Oriented and Analysis of Business Processs". *European Workshop on Integrated Manufacturing Systems Engineering.* IMSE. (France).

Mize, J., and Pratt, D (1991). "A comprehensive Object Oriented Modeling Enviroment for Manufacturing Systems". *Proceedings of FAIM'91.* Limerick.

Proth, J. (1995) "*Les reseaux de Petri pour la conception et la gestion des systemes de produccion*". Ed. Masson. Francia. 1995

Quellenberg, T., and Schubert, J. (1994) "A Flow Management for On-line Processes in Production Systems". *IMSE'95.* Francia.

Rumbaugh, J (1991). "*Object-Oriented Modeling and Design*". Ed. Prentice Hall.

Taylor, D. (1995) "*Business Engineering with Object Technology*". Ed. John Wiley.

SIMULATION STUDIES OF PHYSICAL DISTRIBUTION SYSTEMS FOR AIRPORTS

Kenji Ozaki, Kohji Igakura, and Yoshitami Miki

Industrial Plant Engineering Division, Kawasaki Heavy Industries, Ltd.

Abstract: This paper is concerned with the simulation studies for three large-scale physical distribution systems for airports which are stochastic transient simulations with the arrival process of objectives. The authors present the idea to make simulation models separately for the arrival process and for the equipment motion. Then the antithetic variate method which is one kind of the variance reduction method, is examined especially how to apply the method for multivariable simulation experiments. Moreover, a guideline for the optimum number of vehicles in a large scale AGV system is also presented.

Keyword: Simulation, Discrete-event systems, Modelling, Queuing, Stochastic systems, Automated guided vehicle, Variance reduction method.

1 INTRODUCTION

Recently large-scale physical distribution systems have been planned for several airport projects. For these systems, simulation studies are indispensable to evaluate and confirm the system performance. In this paper, two examples of simulation studies for the baggage handling system and the cargo handling system in the Kansai International Airport (coded as KIX) and an example for a large-scale automated guided vehicle (AGV) system for an airport are presented.

The first example shows the simulation modeling and results for an automatic transportation and sorting system for passengers' baggage. The simulation has been made for around 10,000 passengers a day using passengers arrival time distribution, baggage number distribution, and other presumed information for stochastic distributions and also equipment dynamics with designed control logic. It gives us capacity evaluation results for this system in the form of baggage processing time histograms before it's operation.

The second example includes simulation modeling for ASRS with complicated control logic for elevating transfer vehicles. Also in this model, the process of loading cargo into containers by workers has been included. The result enables us peak-time capacity evaluation. Moreover, it gives us evaluation for future expansion of facilities.

The third example shows that the simulation study can give us the optimum vehicle number criterion for the transporting system with around 40 to 70 AGVs taking account of the interference among vehicles and their transporting routes. Also, the formula for a guideline is derived to estimate the optimum number of vehicles for a given route.

Lastly, it is shown that the above examples, which are stochastic transient simulations, require us to take a systematic variance reduction approach. This paper shows a study of adopting the antithetic variate (AV) method to reduce the variances of the simulation results. A technique to use the AV method for multiple stochastic variables is presented and evaluated with these examples and also with a transient M/M/1 queuing system.

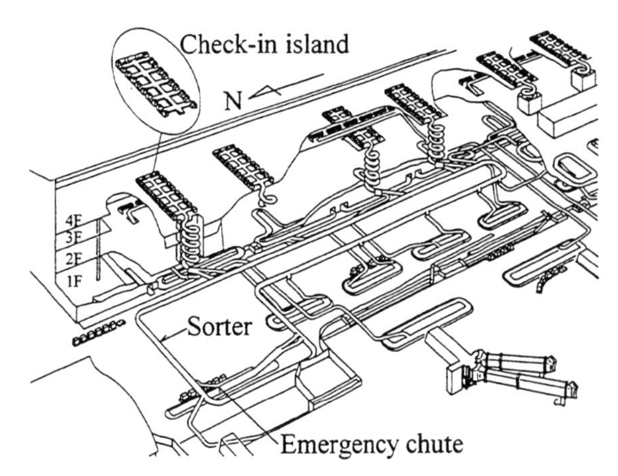

Fig. 1. Whole view of Baggage Handling System in KIX.

2 BAGGAGE HANDLING SYSTEM (BHS)

2.1 *Configuration of KIX BHS*

Figure 1 shows an overall view of this system. This system is symmetrical in the north-south direction and consists of four sub-systems: the domestic arrival, the international arrival, the domestic departure, and the international departure system. The international departure system is the most complicated one and the main target for this simulation study.

Passengers with baggage arrive at the international check-in counters located on the third floor of the terminal building. There are twenty check-in counters in each of eight check-in islands. At the check-in counter, an operator places the passenger's baggage on a weighing conveyor during the check-in procedure. After being weighed, the baggage is transferred to a queuing conveyor and at the same time, a dedicated segment on the following collecting conveyor, which we call "window", is reserved and the baggage waits for the reserved window to arrive. When the window arrives, the baggage is put onto the window and is transferred to a spiral conveyor. The spiral conveyor transfers the baggage from the third floor to the first underground floor, where there are sorting machines. Arriving at the underground floor, the baggage is put onto an empty tray of the sorter via an induction conveyor. The tray of the sorter rotates and discharges the baggage down to the make-up conveyor designated for the flight. The discharged baggage runs around on the make-up conveyor, waiting to be picked up by a worker and then it is loaded onto the aircraft container. An emergency discharging chute is also provided for wrongly transferred baggage and for safety when the make-up conveyor is fully occupied.

2.2 *Objectives of simulation*

The performance requirements for this system were given by the customer as follows.

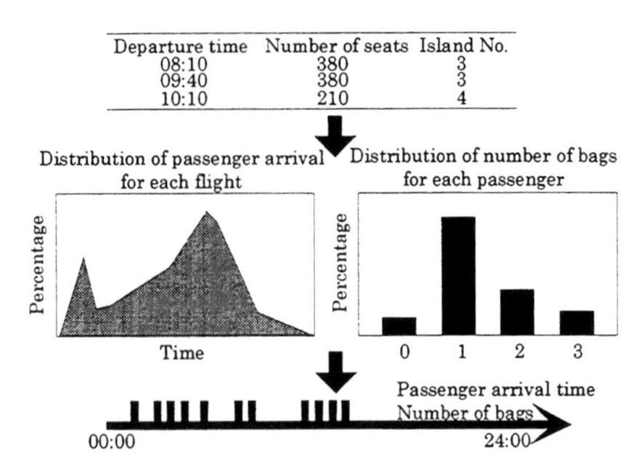

Fig. 2. Schematic of BHS arrival model.

–the capacity of one sorting machine is 3,000 pieces of baggage per hour and the capacity of the collecting conveyor is 980 pieces per hour.
–the transfer time from the weighing conveyor to the designated make-up conveyor is less than 11.5 minutes.
This system is composed of many devices and several branches and junctions. In order to obtain the precise figures of their performances as an entire system, therefore, we must take account of the imbalance of load and the difference in the timing of motion among several devices; they cannot be statically calculated. The authors also need to confirm the performance figures including the necessary number of buffers and also to confirm the control logic to satisfy the desired performances. As a consequence of this, the authors have carried out discrete system simulation with the following objectives:

–capacity evaluation of devices including their control logic for the sorter, collecting conveyor, make-up conveyors, and so on,
–confirming the transfer time of baggage to be less than a predetermined value, and
–checking the equality of availability of check-in counters.

2.3 *Outline of simulation model*

It is necessary to make a BHS model taking account of the arrival process of baggage in order to evaluate the performance through a day. Therefore, the authors have made two models separately. One is an "arrival model", which generates the arrival process of baggage as shown in Figure 2, and the other is a "main model", which represents all movements through the constituent devices. The arrival model calculates the passenger and baggage arrival process for one day according to a flight schedule and several stochastic distributions of: the number of passengers on each flight, the number of baggage brought in by each passenger, arrival time of passengers on each flights and so on. In the equipment model, the service process at the check-in counter is calculated using a random number stream. The equipment model

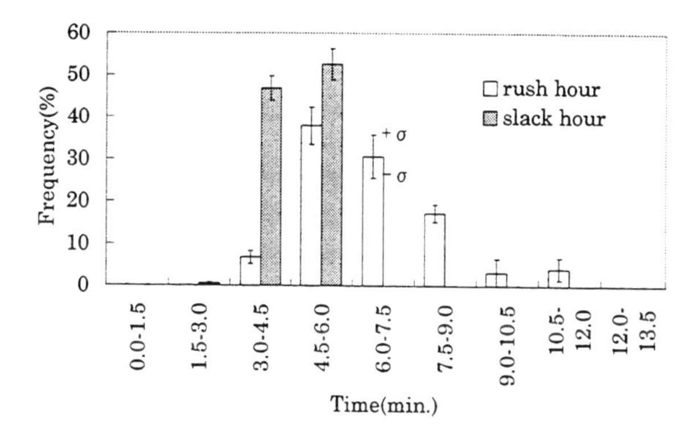

Fig. 3. Baggage processing time histogram.

represents weighing, collecting, spiral and induction conveyor motions, taking account of each control logic like "window" control as explained previously. After be sorted, the baggage is picked up and put into a container by the service process which is an Erlang distribution with a mean time of 6 sec.

The results of simulation experiments will now be explained. These results have been obtained, using the variance reduction technique mentioned in section 5.

2.4 *Simulation results*

Figure 3 shows the histograms of processing time for baggage, which is the duration from baggage check-in on the weighing conveyor to arrival at the make-up conveyor. The mean processing time for slack time was 5.8 min. and the transfer to the make-up conveyor was finished in 6.6 min. at the latest. Nevertheless, the processing time for rush time became longer than that for slack time because of the increase in waiting time at the queuing conveyor and at the induction conveyor. Moreover, when the make-up conveyor is full, the baggage on the sorter tray may run around again and it will cause an increase in the processing time. These results show that the total transfer time in this system was less than 11.5 min, satisfying the required performance mentioned previously. Figure 4 shows a change in the amount of baggage discharged to the emergency chute through a day. According to this result, the amount of baggage discharged to the emergency chute was larger than expected at the beginning stage of planning. By the previous control logic, two pieces of baggage on successive trays with the same destination failed to be discharged successively to the same outlet and the second one was discharged to the emergency chute. Therefore the control logic was modified by this simulation result. The result of this modification gives us satisfactory performance.

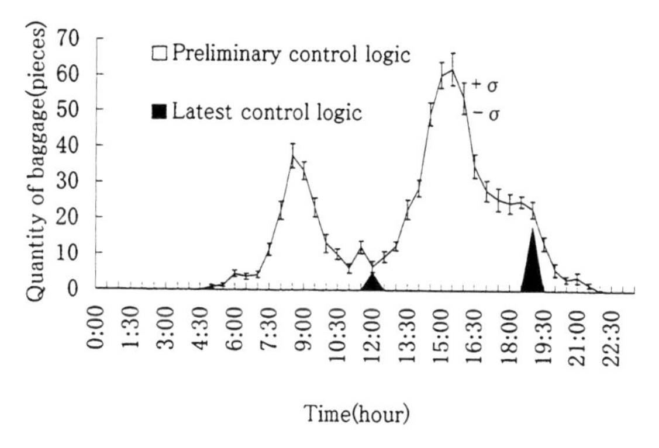

Fig. 4. Quantity of discharged baggage to emergency chute for one day.

3 CARGO HANDLING SYSTEM(CHS)

3.1 *Configuration of KIX CHS and its function*

Figure 5 shows an overall view of CHS in KIX. In this terminal, export goods received from forwarders are classified into three categories and stored in bulk storage. Export goods stored in bulk storage are retrieved at an appropriate time before the flight departure and made up on a pallet or in a container at one of twelve workstations. There are open spaces also to make up bulk goods in the case all workstations are fully occupied. Pallets or containers made up are called unit load devices "ULDs". ULDs are transferred to an ASRS, using two transfer vehicles (TVs) and then stored in the shelves with two elevating transfer vehicles (ETVs). This ASRS has an induction conveyor to transfer made-up ULDs coming directly from forwarders. From two to three hours before the flight departure time, the containers or pallets for the corresponding flight are retrieved through the lowest shelves at lamp side and carried to the airplane by dollies.

This ASRS is completely controlled by a computer and has high functions as follows;

–One ASRS shelf for ULDs can store one 20- foot or two 10-foot containers or pallets.

–Two elevating transfer vehicles (ETVs) move on the

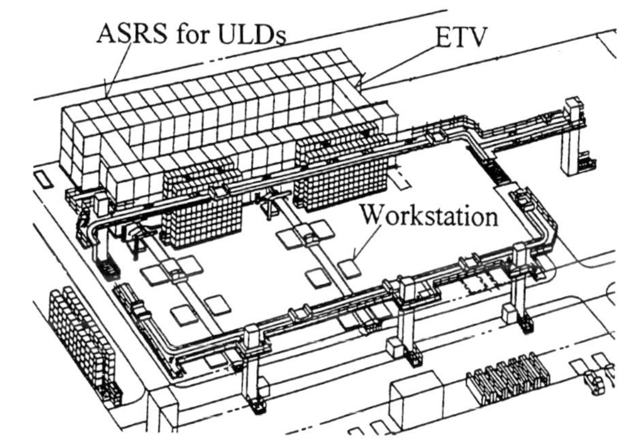

Fig. 5. Whole view of KIX Cargo Handling System (CHS).

same rails. Each ETV has an exclusive zone and a cooperation zone to store in and retrieve from.
–ASRS can store empty containers or transit cargoes, which can be retrieved to workstations at land side.

3.2 *Objectives of simulation*

The following performance figures were determined at the beginning stage of the project, taking account of the increase in cargo demand;

–handling capacity of cargoes should be 134,000t per year or more.
–time limit for receipt of export goods is 2 hours before the flight departure.
–minimum time from retrieval of stored ULD to flight departure is 30 minutes.

To confirm these figures, the authors have carried out simulations, paying attention to the following;

–performance of ASRS at peak time,
–activity rate of ETVs, and
–cycle time of each moving machine and each worker's operation.

To examine operation performances, the following figures were also considered:

–container or pallet build-up time,
–necessary open-space build-up area, and
–future expansion capability.

3.3 *Outline of simulation model*

The authors made two models for this CHS based on the same idea as that of BHS, one was a CHS equipment model and the other was an arrival model which generates the process of bulk material arrival at the build-up workstation and also ULD arrival directly from forwarders. For the equipment model, complicated control logic has been taken into account in order to simulate the precise movement of each device. The distribution of build-up time for bulk material is presumed to follow Erlang distributions with a mean time of 45 min. for a pallet and 30 min. for a container according to results actually measured at another airport. Moreover in order to study future expansion possibilities, several devices and shelves of the ASRS for expansion were also taken into account within the equipment model.

Table 1 Simulation result of CHS capacity and activity rate of ETV

	ETV1	ETV2
max. activity rate	71.8%	87.3%
mean activity rate	34.7%	24.0%

Table 2 Evaluation result for future expansion

Year	1994		2000		2000	
Future facility	no		no		yes	
Build-up efficiency	100%	200%	100%	200%	100%	200%
Necessary build-up space	8.2	3.6	14.2	9.8	15.5	11.1

3.4 *Simulation results*

Table 1 shows an example of the CHS simulation results. In the initial stage of this project, the authors estimated that the activity rate of ETV should be 78 % and that the quantity of ULDs handled should be 37 units in an hour by static calculation. By the simulation, the activity rate of ETVs was 77.1 % and the quantity was 32 units in an hour at peak-load time. Next, the authors carried out a simulation for checking the performance of the entire CHS with facilities for future expansion. Simulation experiments with or without facilities for future expansion and with various efficiencies of building-up work were carried out. The results of these experiments are shown in Table 2. This table shows the result of the comparison of the maximum number of necessary open-space areas between the year of opening (1994) and the year 2000. This result means that it is important for future operation by the client to consider not only an expansion of facilities but also how to improve the efficiency of building-up work.

4 LARGE-SCALE AGV SYSTEM

4.1 *Example of system*

As a third example, a simulation experiment will be introduced as one method for designing large-scale AGV transportation systems for transporting bulk cargo in boxes between the bulk storage ASRS and several points inside a cargo terminal. Figure 6 shows a schematic layout of the transporting route. This layout itself was determined according to results of several simulation experiments for some candidate layout designs. The vehicles run on rails having branches and junctions which facilitate efficient loading and unloading without much interference with other vehicles running on the main line. This system has also a special feature that the necessary number of vehicles becomes very large to meet capacity demand which amounts to 320 boxes per hour during the peak time. The optimum number of vehicles to meet the capacity demands has also been obtained by the simulation experiments.

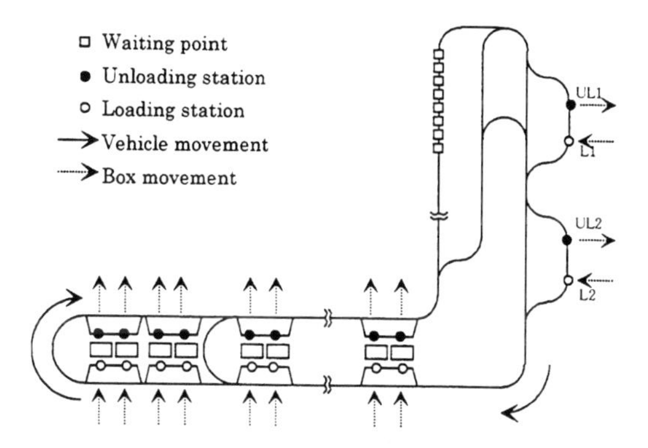

Fig. 6. Schematic layout of AGV system.

4.2 *Outline of model*

As illustrated in the figure, this AGV system has along the main transporting route, 18 loading stations and 18 unloading stations each having a dedicated branch route. When running to each station, the vehicle runs to the branch route in order to avoid interference with the other vehicles running on the main route. After the loading or unloading operation, the vehicle returns to the main route after waiting for the proper timing for merging if necessary. Moreover, the distance between two vehicles is checked. The latter one will be automatically decelerated when the distance is less than the predetermined value.

4.3 *Optimum number of vehicles*

The necessary number of vehicles fairly relates to the spending on equipment when the route and the number of branches and junctions are determined. On the other hand, too many vehicles cause a decrease in the total transporting capacity because of the interference among vehicles. It means that we have an optimum number of vehicles for the corresponding layout of the route.

As results of simulation experiments, Figure 7 shows the queue length at the most critical loading point versus the number of vehicles. We can find that the optimum number of vehicles is around 60. This figure also shows that there is an appropriate number of vehicles corresponding to the layout and length of the route, which we can intuitively recognize. On the other hand, we need many simulation experiments to find such characteristics, for example, in this case, the authors need to carry out 40 sets of experiments for accuracy at 5 points of the number of vehicles, i.e. 200 experiments in all. The following formula can be proposed to obtain the optimum number of vehicles, especially taking account of the interference length among vehicles;

$$N_{opt} \cong \frac{L_{main}}{l_{junc}} + N_{wait} \qquad (1)$$

where L_{main} is the length of main route, N_{wait} is sum of empty stand-by vehicles on loading branch, and l_{junc}

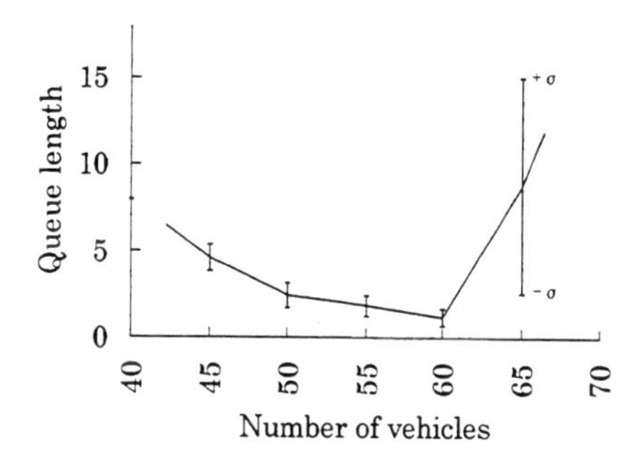

Fig.7. Queue length versus number of vehicles.

means the distance between two vehicles which will be come about at the most critical junction points when the latter vehicle will merge from stopping state.

In this example, L_{main}=828.9m, N_{wait}=20, L_{junc}=20.7m.

With this formula, we can obtain $N_{opt} \cong 60$ for this example and this figure is just same as the actual optimum number shown in figure 7. This formula give us an approximate optimum number of vehicles for preliminary design without numerous simulation studies.

5 APPLICATION OF VARIANCE REDUCTION METHOD

The three simulations mentioned earlier are transient stochastic simulations through one day according to the flight schedule. Normally, the coefficient of variance of an observed variable obtained by transient simulation becomes larger than the one for steady-state simulation (Law & Kelton 1991a). This fact requires us to carry out more experiments to obtain reliable results. The authors have examined the antithetic variates (AV) method with an extension to the multivariable system.

Generally, even though the objective system is highly complicated queuing system, we can intuitively select one primary set of arrival and service processes which are dominant in the results. Therefore, we will consider a case with two stochastic variables as the following, but we can easily extend this result to multivariable cases.

5.1 *Application technique for AV method*

Stochastic simulation using random numbers is described as follows.

$$Z = F(X, U(V)) \qquad (2)$$

where Z is one output from an experiment,

$$X=(x_1,x_2,\cdots,x_m) \qquad (3)$$

is a deterministic operative variable,

$$U=\{U1(v1),U2(v2),\cdots,Un(vn)\} \quad (4)$$

is an n-set of stochastic variables, v_i is a random number stream for simulating the i-th stochastic variable U_i.

In case $n=2$, $4p$ sets of simulation results can be obtained by using antithetical variable for each random number stream respectively as follows;

$$\begin{aligned} Z_{NN}^1 &= F(X,v_1^1,v_2^1),\cdots, & Z_{NN}^p &= F(X,v_1^p,v_2^p) \\ Z_{AN}^1 &= F(X,1-v_1^1,v_2^1),\cdots, & Z_{AN}^p &= F(X,1-v_1^p,v_2^p) \\ Z_{NA}^1 &= F(X,v_1^1,1-v_2^1),\cdots, & Z_{NA}^p &= F(X,v_1^p,1-v_2^p) \\ Z_{AA}^1 &= F(X,1-v_1^1,1-v_2^1),\cdots, & Z_{AA}^p &= F(X,1-v_1^p,1-v_2^p) \end{aligned} \quad (5)$$

Then we have two possibilities for estimation:

$$E_1[Z]=\bar{\xi}=\frac{1}{p}\sum_{i=1}^{p}\xi_i=\frac{1}{2p}\sum_{i=1}^{p}(Z_{NN}^i+Z_{AA}^i) \quad (6)$$

$$E_2[Z]=\bar{\varsigma}=\frac{1}{p}\sum_{i=1}^{p}\varsigma_i=\frac{1}{4p}\sum_{i=1}^{p}(Z_{NN}^i+Z_{AN}^i+Z_{NA}^i+Z_{AA}^i) \quad (7)$$

The authors' idea is that if a strong negative correlation between Z_{NN} and Z_{AA} can be expected beforehand, we may use $\bar{\xi}$ as the case 1 and on the other hand if not, we use $\bar{\varsigma}$ as case 2.

5.2 *Result of application to M/M/1 system*

Simulation experiments were made for the simplest M/M/1 queuing system with a mean arrival interval of 1 min. and a mean service time of 0.9 min. for the first 100 arrivals from the initial state. Figure 8 shows a comparison between the true value (Law and Kelton 1991b) and three estimated values for the mean waiting time in the queue for the first 100 arrivals. The values $\bar{\xi}$ and $\bar{\varsigma}$ estimated by the AV method have less deviation from the true value than δ, which is a simple mean obtained without using the AV method. We can also recognize that the estimate $\bar{\varsigma}$ has less variance around the true value than $\bar{\xi}$.

5.3 *Application to simulation*

This technique was applied to BHS and CHS simulations and the necessary number of experiments could be reduced as a result. In the case of BHS, the passenger arrival process and the service process for each flight are primary stochastic variables that affect the performance especially the waiting time for the entire system. In this case, it is difficult to assume a correlation between Z_{NN} and Z_{AA}, and so $\bar{\varsigma}$ was used.

On the other hand, in the case of CHS, the stochastic variables affecting the characteristics of the system are the distribution of build-up service time at workstations, which has two different distributions of build-up time for pallets and containers.

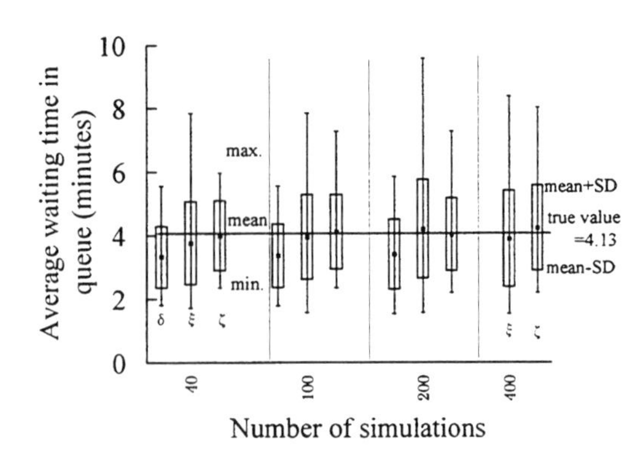

Fig. 8. Comparison of variances for transient M/M/1 system.

Because we can expect a negative correlation between Z_{NN} and Z_{AA} in these two service processes, $\bar{\xi}$ was used.

In both cases, we could satisfactorily obtain the several results after only 40 experiments even though these systems are quite large-scale objectives for simulation.

6 CONCLUDING REMARKS

In this paper, three examples of stochastic simulation studies for large-scale physical distribution systems have been presented especially in order to evaluate the overall system performances before their operations. Through these studies, systematic approaches to obtain accurate results with a minimum number of simulation experiments are recognized to be important especially for such large-scale stochastic simulations.

The authors have presented an application technique to use the antithetic variate method for multi-stochastic variable simulation to reduce the variance of the result. A guideline for the optimum number of vehicles has also been presented as a formula for an example of a large-scale AGV system. This guideline also enables a decrease in the number of simulation experiments and moreover provides us with preliminary criteria for determining an appropriate figure for the number without simulation experiments.

REFERENCES

Law and Kelton (1991 a, b). *Simulation Modeling & Analysis*, 525-527, 582-585, McGraw Hill

DESIGN OF FLEXIBLE MANUFACTURING SYSTEMS WITH REUSE IN A KB SIMULATION ENVIRONMENT

George L. Kovács, Sándor Kopácsi, Ildikó Kmecs

Computer and Automation Research Institute
Hungarian Academy of Sciences
H-1111 Budapest, Kende u. 13-17., Hungary
Phone: (36)(1)1811-143, Fax: (36)(1)166-7503
e-mail: gyorgy.kovacs@sztaki.hu

Abstract: In this paper the application of software reuse and of object-oriented methodologies will be introduced as assistance for the simulation based design of FMS.. This system has been implemented as a combination of a traditional simulation/animation language (SIMAN/Cinema) and an intelligent expert environment (G2). Reuse in SIMAN/Cinema and in G2 will be analyzed separately and together in general, and in the context of FMS to get acceptable solutions, to build up new simulation models using parts of earlier simulation models. The design methodology used for model development is based on the object-oriented Rational Rose CASE tool of the OOD methodology and on the use-case design of the OOSE methodology.

Keywords: Simulation, Modeling, Design, Flexible Manufacturing Systems, Software Reuse, Object-Oriented Design, SIMAN, G2

1. INTRODUCTION

Software developers often meet the problem of creating new components of an application that someone probably previously has already produced. Without having effective reuse tools usually it is more natural to create new components from scratch than to seek for useful elements in other programs and/or systems.

In the field of simulation of flexible manufacturing systems (FMS) this issue often occurs, when different systems with some similar features have to be managed. The components of different FMS and FMC (Flexible Manufacturing Cell) are the same type of machine tools, robots, transfer equipment, etc. In the relevant aspects they usually differ from each other only in their amounts and working parameters. This fact itself breeds the idea of reuse of FMS and FMC elements.

Simulation can be used for analyzing the behavior of an existing or would-be system. Simulation is a design assistance tool to check different design possibilities, ideas, etc. It is useful for calculating utilization statistics, finding bottlenecks, pointing out scheduling errors, and even for creating manufacturing schedules. Simulation is a very effective technique for dynamic analysis, but it is short of optimizing ability. A satisfactory solution can be obtained through modeling and simulating alternative approaches with iteration.

There are several elements defined during the analysis, design and implementation of a simulation model, as ideas, concepts, object classes, and lines of source code created, etc. that we should reuse in new applications. Usage of reuse methodology and practice will reduce our efforts in developing new simulation models to assist the design of new systems.

In this paper the reuse of FMS model components in new modeling applications will be discussed. The simulation system has been implemented as a combination of a traditional simulation/animation system SIMAN/Cinema (Pedgen, 1990) and an intelligent KB environment G2 (G2, 1995). The analysis, design and re-design phases of different FMS simulation models are discussed using reuse in all possible phases of the life-cycle. The design methodology for analysis and model design is based on the object-oriented methodologies OOD (Booch, 1991) and OOSE (Jacobson, 1992).

2. SOME IDEAS OF SOFTWARE DESIGN FOR/WITH REUSE

Reuse of software elements is becoming more and more important in the life-cycle of software products. However, it does not necessarily mean that only the code of a software product can be reused. One view is that reuse efforts should focus on code, as this work is more likely to have practical results (Frakes, 1990). Another opinion is that all the results and resources used in a project, including human expertise, should be reused (Basili, 1988). In this paper we concentrate on code reuse, but we keep in mind that all documents created during the perception-design-implementation-testing of a product, as ideas, methodologies, requirement specifications, design results, documentation, etc. could be reused in later projects.

Two kinds of software reuse can be distinguished: reuse across vertical vs. reuse across horizontal domains (Banker, 1993). In vertical (software) reuse the components are used in the area within a specific domain. In horizontal reuse the (software) components are used across different domains. As we are dealing with modeling and simulation aspects of FMS we deal with vertical reuse mostly.

One proposal for representing software components for reuse is based on the concept of program families. A program family is a set of software components derived from a common set of design decisions (Frakes, 1990). One possible problem with the family-based approach is the amount of design information that should be captured for large systems. A possible solution of this problem is that not all representation and design possibilities are stored during the analysis and design phase. General analysis and design decisions should be made in such a way that it would be easy to derive all the necessary variations of a system based on the stored information.

Chapter 3.1 and 3.2 will discuss the application of a CASE tool and of reuse during FMS model analysis and design, while chapters 4. and 5. deal with reuse during model implementation in G2 and in SIMAN/Cinema respectively.

Working with reuse generally means that previously designed and/or implemented elements are used again (and again) in later projects with or without changes. The effectiveness of reuse is increased when during the design and implementation phase of new (software) components it is kept in mind that these elements may be reused later on. If we take into account the requirements of reuse we are making analysis, design, coding, etc. for reuse, or simply it is design for reuse, while the application of reuse for analysis, design, implementation, etc. is the design with reuse.

3. MODELING AND SIMULATION AS DESIGN TOOLS OF FMS

The basic purpose of the simulation of a modeled FMS in our hybrid simulation and scheduling system is to compare different scheduling strategies in the modeled system which can be a system under design or an existing system. A remote goal is the intelligent control of real FMS that is based on the evaluation of simulation results and is done by execution of a 'best' schedule created after the appropriate scheduling experiments with the simulated model.

The other goals of the simulation system is to simulate different production tasks on a given FMS and finally to facilitate the evaluation and comparison of different FMS designs for the same tasks. This last target requires to build up several, new simulation models.

There are different tools and methods to build up a simulation model of an FMS, but all of them have a common property, i.e. they consist of basic elements. The programmer has to create the simulation model by putting these elements together. Using this method the developer often has to build certain sets of elements, that represent regularly appearing parts of different FMSs, like a general purpose manufacturing centre, a robot, a workpiece buffer, etc. Creating a repository of these sets of simulation elements where they are kept in reusable form seems to be obvious.

We use SIMAN/Cinema as a traditional simulation/animation tool and the G2 intelligent expert system environment to find the most effective solution of simulating and scheduling of FMS (see Kovács, 1994 and Kovács, 1996). Three G2 based expert system modules help the SIMAN simulation engine working together as a hybrid system. The Preparation Expert System creates the simulation model based on the user's instructions. The Advisor Expert System gives advises to the Simulation

Module in scheduling and quality assurance decision points. The Evaluation Expert System evaluates the results of the simulation and gives suggestions for modifications.

3.1 Analysis and design of FMS simulation models using object-oriented tools

The information about the state of the whole modeled FMS is necessary not only in G2 where updating is done continuously during the simulation run, but in the SIMAN/Cinema system, too for decision making during the simulation-scheduling program run.

Therefore we have decided to use the Rational Rose (RR) software analysis and design tool that is based on the Object-Oriented Design (OOD) of the Booch methodology (Booch, 1991). We extended Rational Rose with the use-case concept of the Object-Oriented Software Engineering (OOSE) of Jacobson (Jacobson, 1992). This combination results in a well structured and understandable description (use-case) besides the - sometimes hard to follow - object diagrams (domain object models). The model representation in such a tool is language independent. RR offers the possibility to directly transform the designed model into C++ source code. The C++ Generator produces the appropriate C++ class for each class in an RR model. For standard and user-defined operations the C++ Generator produces skeletal member functions that must be filled up with member function bodies. Any other object-oriented definition system, as G2 for example can be used in a similar way with restricted efforts.

The analysis and design of a multifunctional virtual model of FMS are done using the extended Booch methodology. This model contains a generic manufacturing component library consisting of templates of manufacturing elements to serve for interactive and flexible FMS model design and configuration. The generic representation of an FMS is given in the form of a class hierarchy with top classes presenting abstract levels of manufacturing elements and with the bottom classes presenting concrete manufacturing components. This representation of an FMS was developed reusing the representation of FMSs from previous projects. The reused elements contain the basic FMS concepts and the task sharing among different functional parts of the FMS model. During the use of the system a user can work with existing FMS models or she or he can put together a new model creating instances from the library of the ready-to-use classes by cloning them and filling out their attributes. It is also possible to create new elements by refining classes from higher abstraction levels.

The application of the use-case methodology is proceeded at two different levels. The definition of the simulation model elements means to build up the component library, or repository by means of use-case sequences using icons, text patterns, relations, lists, etc. On the analysis and design level the use-case sequences are used to collect the appropriate model elements from the repository, and then to connect them according to the requirements of the FMS to be modeled for simulation.

3.2 Reuse during modeling of flexible manufacturing systems for simulation

It is advantageous to abstract common FMS concepts and to create their representation in an abstract way. For instance, instead of designing milling, drilling, turning, and other manufacturing operations one by one, it is time- and labor-saving to design the manufacturing operation as such. Object-oriented CASE tools are very good in this effort. Such concepts can be easily modified or widened also in non object-oriented programming languages.

In Fig. 1. we can see a simple example of a class that is designed in Rational Rose. It represents a manufacturing operation - like milling, drilling, turning that are the concrete instances of the class manufacturing operation. An operation in the real manufacturing means the processing of the corresponding NC/CNC programs. A process plan of a workpiece is a sequence of such operations, possibly with alternatives.

This class has two attributes (Duration, EquipmentSpec) that are then implemented in both SIMAN and G2. Duration means the length of the operation in time and EquipmentSpec defines the manufacturing equipment on which the given operation should be performed. Tools, fixtures, and other relevant data are taken into account, too.

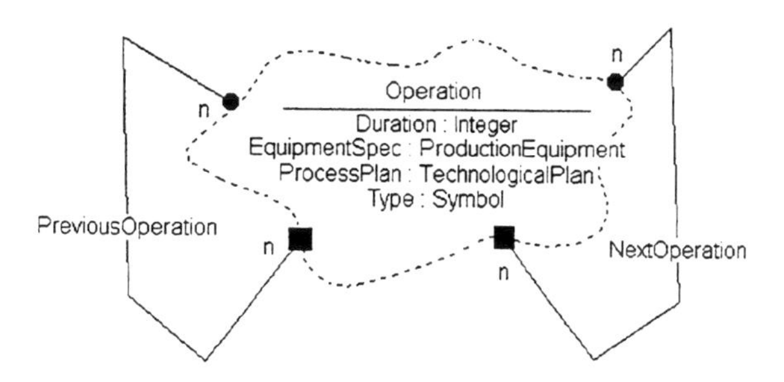

Fig. 1. Operation Class in Rational Rose

Two actions are designed for this class that are common for both SIMAN and G2: CreateRelation and DeleteRelation. By their means we can represent the fact that one operation follows another one in the process plan. This fact is then stored in the relation NextOperation with its opposite in

PreviousOperation. There is another attribute and there are two other actions defined that are implemented only in G2: ProcessPlan, InsertToList and RemoveFromList. ProcessPlan points to the process plan that contains the given operation.

In the FMS model developed in G2 process plans are allowed to have alternatives. Starting operations of a process plan are collected into a list. These two operations serve for inserting/removing operations into/from such a list.

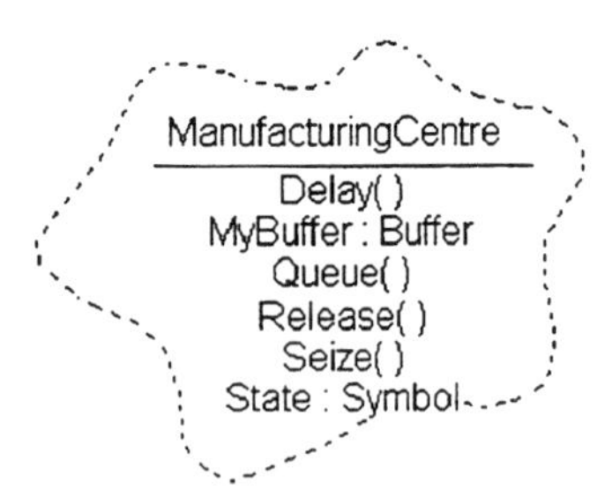

Fig. 2. Manufacturing Centre Class in Rational Rose

In Fig. 2. another example of a class (ManufacturingCentre) designed in Rational Rose can be seen. It has two attributes that are implemented in G2: State, inherited from ProductionEquipment class and MyBuffer denoting the buffer that serves the ManufacturingCentre. It has also four operations defined - three of them, Queue, Release and Seize are implemented in SIMAN and one of them, the Delay operation is implemented both in G2 and SIMAN. Delay operation is implemented in G2 in the method Perform_operation of the Manufacturing_centre class.

As an example a possible use-case description of the manufacturing centre is the following: The operator actor fixes a work-piece on the palette and then the palette is put on its place in the manufacturing centre, the programmer actor loads the appropriate CNC program into the controller of the machine tool and then pushes the start button of the machine. The manufacturing centre makes the appropriate metal-cutting procedures according to the CNC code and when it is done gives a ready signal. Then the operator actor removes the palette. The next use-case of the same machine-tool may star with fixing the palette, as fixing the work-piece can be done separately in advance. This description supposes that there is a palette belonging to the machine tool, and it is simplified as we do not deal with any error possibilities, etc.

All parts of a simulation model should be analyzed and designed in a similar way, based on the RR notation applying use-cases of OOSE reusing as much as possible from previous analysis and design models. As this paper is not dedicated to deal with practical procedures of reuse we do not deal with the appropriate searching, browsing, etc. systems and methods.

4. REUSE POSSIBILITIES OF FMS MODEL ELEMENTS IN G2

The range of appropriate expert system shells available on the market is rather wide. Our first experiments with intelligent systems were the application of the CS-PROLOG (Futó, 1987) language and then the CS-PROLOG based ALL-EX expert system shell (ALL-EX, 1991), both developed in Hungary in the late eighties. After having finished some successful prototype programs we got a chance to start to work with the G2 expert system environment (G2, 1995) which proved to be suitable for our research purposes. G2 is a real-time object-oriented rule-based intelligent environment with the possibility to implement the behavioral knowledge not only in rules, but also in the form of procedures, functions and formulas. G2 is used world-wide mainly for real-time control, monitoring and supervising of discrete and continuous processes in many industries. G2 is equipped with a standard interface to external processes which was a very useful base in interfacing G2 with SIMAN in our projects.

G2 has different types of reuse possibilities which will be given in the following without details:

- Built-in reuse possibilities of G2. The reuse of software elements in G2 is basically supported by its object-orientedness. A class is designed and implemented only once and its instances can be used any required times. The behavior of a class is described in methods in such a way that a class inherits generally defined behavior from higher classes and only the class-specific behavior has to be detailed in its own methods.

- Reuse of G2 modules. The other feature that increases the reusability in G2 is modularization. The code in G2 can be separated in a rather free way into modules that are organized in dependency hierarchy and may be saved and loaded separately. A module M1 is depending from a module M2 if it contains instances of a class defined in the module M2. It is then possible to have a system with the same FMS functionality, like scheduling, quality assurance, but with different FMS layout - a module containing the layout of FMS1 may be changed for the module containing the layout of FMS2.

- Reuse of G2 classes, rules, procedures and other items. There are G2 elements that can be reused vertically, e.g. in a later project that has common features with the current project. For instance, the production_equipment class can be reused with or without modifications in every FMS related application. In a second group such G2 elements can be found that are horizontally reusable, e.g. independently from the area of the current application. Two examples from this group: 1. There is a rule that first cleans up the screen and then starts the initialization procedure that is application dependent. This rule provides that starting the application from any state will give the same visual appearance. 2. There are special rules to control the rather complicated access rights in the G2 system.

- Reuse of G2 icons. During the design of a G2 class much effort is devoted to the iconic representation of the class. If a sophisticated representation is needed the development of an icon for the class may take much more time then designing other features of the class. It is reasonable to place the classes with complicated icon descriptions into a (software) repository.

5. FMS SIMULATION AND REUSE IN THE SIMAN /CINEMA ENVIRONMENT

SIMAN (Pegden, 1990) is a general purpose simulation language, but because of some features it is especially suitable for simulating manufacturing systems. In SIMAN the modeling of transportation systems (automated guided vehicles, conveyors, etc.) is principally simple. The workpieces are represented by entities, that control the flow of the simulation. Each entity carries its attributes, that can store all important parameters of the workpieces.

A simulation model consists of two parts: a Model Frame and an Experimental Frame. The Model Frame describes the logic of the analyzed system while the Experimental Frame contains its working parameters. Simulation models can be run with or without animation. Animation is feasible when the adequate animation layout has been created by a special software belonging to SIMAN called Cinema.

Using the SIMAN/Cinema system we examined several FMS and FMC of different size and type (a CIM pilot plant at the Technical University of Budapest, a Furniture Factory in Hungary, a workshop of the Korean GoldStar Cable Factory, etc.). These systems were rather different, but there were some common properties that led us to consider the analysis of the reuse possibilities of SIMAN programs for current and future systems to be simulated.

Making repository from and for SIMAN simulation systems four main components has to be taken into account. A SIMAN program itself consists of the above mentioned two frames (Model Frame and Experimental Frame), and one or more Cinema animation layouts may belong to them. The fourth group is the interface unit that creates the link between the simulation and expert part of the hybrid system.

- Reuse of SIMAN Model Frame Blocks. The basic elements of the Model Frame, the Blocks can be interpreted as commands to be executed every time when an entity arrives in a block. These blocks represent the atomic elements of the simulated (in our case the manufacturing) processes. Almost every imaginable process can be modeled using these components, however in most of the cases stereotype situations are typical. It occurs more frequently in the models of FMS, where identical groups of blocks can be found in most application. Blocks appearing regularly together can be combined and stored in a SIMAN software repository. A later time or another user doesn't have to create this set of blocks again, only their parameters have to be redefined. As a typical example for SIMAN Block reuse the sub-model of a manufacturing centre can be mentioned. It can be found in every FMS simulation model, however the developer has to build its model every time.

- Reuse of SIMAN Experimental Frame elements. After describing the logic of the system to be examined the developer has to define the parameters of the components given in the Model Frame. It means that he or she has to specify the capacities of the resources, express the lengths of the waiting queues, detail the parameters of the transport system and so on. These elements are reusable and should be linked to the components given in the Model Frame. If we consider the description of a manufacturing centre mentioned in the previous section, the developer has to define the capacity of the resources and the length of the waiting queue in front of the resource. These parameters should be given in the RESOURCES and QUEUES elements in the Experimental Frame.

As there is a close connection between the Elements of the Experimental Frame and the Blocks of the Model Frame, the repository should contain the links between them, as well.

- Reuse of Cinema animation layout. Before discussing this topic we have to distinguish between the static and dynamic elements of an animation layout. The static elements of a layout remain unchanged during the animation, while the changes in the form, color, etc. of dynamic components represent alteration of the states of the represented elements. The static components can be imported from CAD programs (as AutoCAD) to Cinema, while dynamic components should be created using the graphical editor of Cinema. The animation layout of the modeled system is different from case to case, therefore there are two ways to go. If we don't want to create realistic animation and a symbolic representation of the modeled system is satisfactory, then different systems can be animated using the same graphical objects. In this case we can use both static and dynamic components of the constructed layout. General icons for manufacturing centres, robots, temporary buffers, etc. can be defined and linked to the components of the Model Frame and the Experimental Frame in the repository, and can be reused in other applications.

In the other case, when we want realistic animation, we can use only some dynamic components, like the working place of a machine, that doesn't differ from each other significantly. The static components representing the real appearance of the modeled elements, should be created from case to case.

- Reuse of the C-language interface unit. The link between the simulation and expert modules of the hybrid simulation and scheduling system from the side of SIMAN is realized by a C-language interface unit compiled to the SIMAN program. Some routines browse the Experimental Frame to identify the name of the elements (resources, queues, etc.) in the system. Others send standardized messages using the previously obtained information to the expert modules, and further routines set the attributes of the simulation entities according to the messages from the expert modules. These routines are fairly general, therefore they can be used with minimal alteration in every application.

6. CONCLUSION

It is of basic importance to be able to simulate the sophisticated, very expensive manufacturing and assembly facilities before of running a new production task, or during the design phase prior to implementation of new facilities.

In this paper some ideas and a hybrid (expert combined with traditional) simulation environment were introduced to be able to assist in FMS design by building new FMS simulation models easier, faster and more reliably. Reuse at all levels of the software development and the application of the object-oriented design techniques help in approaching the above goal.

It can be concluded that there are many possibilities of reuse in all parts of the system. Reusable elements in the expert (G2) and in the traditional (SIMAN/Cinema) systems are separated, however they correspond to each other. In recent development some parts of the system should be developed and implemented in G2 and in SIMAN, as well, and reuse is done separately. It was very important to find the common conceptual roots of the different types of elements and to create the links among them.

The application of use-cases from OOSE and of the Rational Rose CASE tool leads us toward this direction. Components can be analyzed and defined using these tools and applying reuse, and they should be implemented either in G2 or in SIMAN or in both systems with limited further efforts.

REFERENCES

ALL-EX (1991), ALL-EX Reference Guide, Multilogic Computing Ltd, 1991.

Basili, V, Caldierra, G (1988): "Reusing existing software", Computer Science Technical Report Series UMIACS-TR 88-72, 1988.

Booch, G. (1991): Object-Oriented Design with Applications, Benjamin/Cummings, 1991.

Frakes, W.B., Gandel, P. B (1990): "Representing Reusable Software", Information and Software Technology, Vol. 32, No. 10, 1990, pp. 653-664.

Futó, I. (1987): "AI and Simulation on Prolog Basis", International Symposium on Artificial Intelligence, Expert Systems and Languages in Modelling and Simulation, Barcelona, Spain, June 2-4, 1987.

G2 Reference Manual (1995), G2 Version 4.0, Gensym Corporation, Cambridge, MA, USA, 1995.

Jacobson, I. Christerson, M. Jonsson, P. Övergaard, G. (1992): Object-Oriented Software Engineering, Addison-Wesley, 1992.

Kovács, G.L. (1996): "Changing Paradigms in Manufacturing Automation", Proc. of the 1996 IEEE ICRA Conf, Minneapolis, Minnesota, USA, 22-28 April, 1996, pp. 3343-3348.

Kovács, G. L., Mezgár, I., Kopácsi, S., Gavalcová, D., Nacsa, J. (1994): "Application of Artificial Intelligence to Problems in Advanced Manufacturing Systems", Computer Integrated Manufacturing Systems, Vol. 7, No. 3, 1994, pp. 153-160.

Pegden, C.D. , Shannon, R.E. Sadowski, R.P. (1990): "Introduction to Simulation Using SIMAN", McGraw- Hill, Inc. New York, 1990.

Manufacturing Systems: Modelling, Management
and Control, Vienna, Austria, 1997

SIMULATING A MANUFACTURING ENTERPRISE USING INTEGRATED PROCESS MODELS

Albert W. Chan* and Vince Thomson**

**National Research Council of Canada, Ottawa, Canada*
***McGill University, Montreal, Canada*

Abstract: A two-level process model of a large electronics manufacturer has been developed to allow the study of the management and operational issues that have an enterprise wide impact. An order fulfillment (business process) submodel considers the end-to-end processes from order entry to delivery to the customer. A shop floor (production process) submodel captures the flow of information and materials on the shop floor. The integrated model is being used to reduce total product cycle time while allowing the development of department specific solutions as well as the resolution of corporate wide constraints.

Keywords: business process engineering, enterprise modelling, simulation, process models

1. INTRODUCTION

In an era when the business and manufacturing environment is constantly changing, and when information processing and telecommunication power is growing fast, many manufacturing firms are adopting time-based competitive strategies to reduce their lead times -- both the time to bring new products to market, and the time to manufacture an existing product from raw materials. While the major focus is on looking for ways to speed up existing procedures, companies have demonstrated that other benefits and improvements can also be derived, simultaneously achieving low cost, high quality and rapid delivery. By working closely with networks of suppliers and subcontractors on the one hand, and customers on the other hand, competitive advantages can be gained by establishing special linkages with other partners in the supply chain and distribution chain. The objective is to reduce overall order fulfillment lead time through holistic solutions that optimize the best efforts of all activities associated with manufacturing quality, customer oriented products quickly.

In order to facilitate the study of enterprise wide issues and to formulate new manufacturing strategies to align with some of the new approaches, process modelling is an important and very useful technique. While certain issues such as order fulfillment require high level process modelling and redesign, other considerations such as production scheduling and line balancing call for more detailed studies of lower level processes. The study of all these operational issues simultaneously through modelling and simulation to determine enterprise wide solutions has not been possible due to the inadequacies of modelling and computer technology to handle the very different types of activities and the different levels of detail that are necessary. The research undertaken by the authors has demonstrated such a capability and is described below.

2. THE PROJECT

A large supplier of electronics equipment has been collaborating with McGill University and the National Research Council of Canada to develop a computer model/simulation of its manufacturing operations in order to assess the potential benefits of applying simulation technologies to the study of their manufacturing and business operations, as well as the immediate interest in results from the study of their manufacturing cycle.

The objective of the Manufacturing Enterprise Model was two fold: first, to provide insight into the issues and mechanisms for the reduction of overall manufacturing cycle time; secondly, to develop the techniques necessary to build very large enterprise models/simulations which included both manufacturing and business processes.

At the time of the project, the manufacturing company was implementing a strategy to reduce overall cycle time and to offer increased merchandise products. As part of this effort, the company reviewed its order fulfillment process, materials management, purchasing and supplier policies, as well as its manufacturing shop floor operations. The modelling/simulation approach complemented the ongoing effort, and was aimed at helping to highlight the areas of operation which could best assist in cycle time reduction, as well as those areas of operation which would subsequently be impacted by the reduction. The proposed simulation models were also to be used as the basis of further studies of selected business and fabrication processes.

To accomplish manufacturing cycle time reduction, the company was investigated from the stand point of the Order Fulfillment Process. This process included customer interaction, order entry, custom product configuration, trigger to shop floor, trigger to purchasing, material planning, production scheduling, distribution and delivery to customer. To complement the Order Fulfillment Process, a detailed model of Shop Floor Operations was also created which modelled fabrication from trigger to shop floor to placing finished goods into inventory.

3. A TWO-LEVEL PROCESS MODEL

3.1 Methodology

A logical model of the enterprise was made from the point of customer order entry to invoice payment. As stated above this model considered all the business and manufacturing processes involved in all the business units from marketing and sales through production to invoicing. The intent of the project was to use a state-of-the-art computer modelling/simulation system so that research efforts would be focused on solutions to manufacturing cycle time issues rather than creating modelling systems. No single system was found which had the necessary modelling and simulation capabilities. The various functions necessary for a Manufacturing Enterprise Model were considered and compared to the capabilities of the software systems available.

The capabilities thought necessary in a modelling/simulation system for the project were:

- the modelling and simulating of business and manufacturing processes,
- the tracking of materials and information items ('forms', data requirements, etc.),
- the provision of statistics on resource and time usage,
- the ability to handle the logic for complex routings of materials and work processes,
- tool sets for developing models and for creating reports on modelling scenarios,
- the ability to handle very large models, and
- a system which was simple to learn and use, i.e., team personnel would not have to be experts in modelling or computer technology.

The latter item was very important. Since the modelling of a big, complex (knowledge work as well as fabrication processes) manufacturing enterprise needed a large team with diverse skills, it could not be assumed that all would have extensive skills in computer modelling. As well, the company was interested in ongoing use of the model; an easy to use system was imperative. It should also be noted that the specific objective of studying methods for accomplishing the reduction in manufacturing cycle time throughout an enterprise did not, of itself, help in differentiating potential systems.

Two types of modelling systems were identified and reviewed: business process modellers and shop floor modellers, which were thought capable of modelling the necessary set of activities in a manufacturing enterprise. These were two somewhat mutually exclusive sets of activities from the point of view of existing computer modelling systems. Business process modellers had capabilities for defining and modelling networks of processes, tracking materials and different types of knowledge work information, as well as reporting statistics on modelled activities. They typically did not have sufficient logic capabilities to allow for the complex scheduling and routing which was necessary for the modelling of work orders and materials on a shop floor.

Since two modelling systems were now necessary, it was deemed desirable to run the models simultaneously during simulation, so that enterprise wide issues and processes which impacted on manufacturing cycle time and which covered both business and manufacturing processes could be studied together. Two additional characteristics of the modelling/simulation systems were now required: the capability of exchanging data necessary for both models and the ability to be externally synchronized with other cooperating software systems. The two systems which were chosen were FirstSTEP, a business process modelling/simulation system from Interfacing Technologies (Montreal), and ProModel, a manufacturing modelling/simulation system from ProModel Corp. (Orem, Utah).

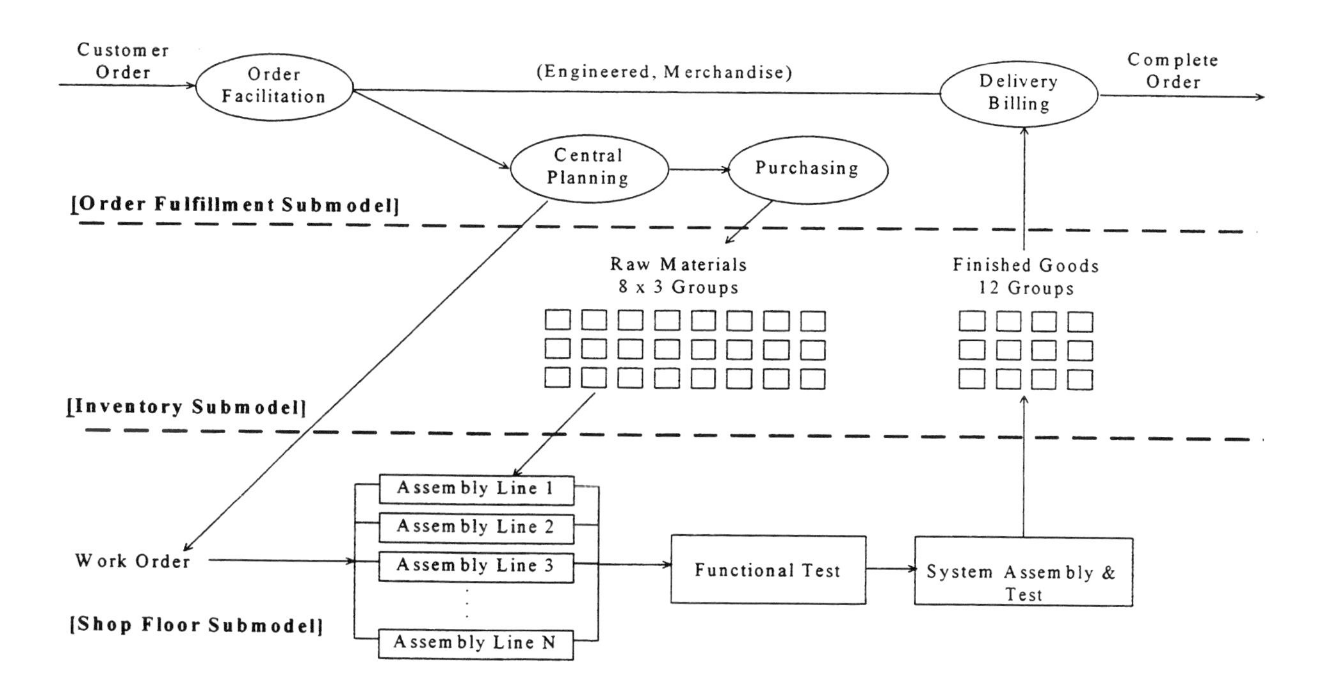

Fig. 1. An Overview of the Manufacturing Enterprise Model

3.2 Model Design

The Manufacturing Enterprise Model was composed of three submodels: an Order Fulfillment Submodel, an Inventory Submodel, and a Shop Floor Submodel.

The Order Fulfillment Submodel was composed of those business processes from receiving a customer order to invoicing of the customer. The Shop Floor Submodel was composed of those processes from reception of work orders from the production planning function to placing finished goods into inventory. The creation of a separate Inventory Submodel was necessary for two reasons: first, a lot of emphasis was placed on tracking raw materials, work-in-process (WIP), and finished goods; secondly, the Inventory Submodel was seen as the means for coordinating the Order Fulfillment and Shop Floor submodels.

A schematic of the Manufacturing Enterprise Model showing an overview of the Order Fulfillment, Inventory and Shop Floor submodels is shown in Figure 1. These three submodels are described in more detail below. It should be noted that the Inventory Submodel is a separate logical model, but in terms of implementation it is distributed between the Order Fulfillment (business process) Submodel and the Shop Floor (fabrication process) Submodel. The points of communication between the latter two submodels are the work orders which come from the central planning function, the raw materials which are controlled and purchased by the planning and purchasing functions, and which are drawn upon by the shop floor, and finished goods which are placed in inventory by the shop floor, and then, delivered to the customer as part of the order fulfillment function by distribution.

4. ORDER FULFILLMENT SUBMODEL

The Order Fulfillment Submodel starts when a customer places a specific order or a request for quotation, and ends when the customer receives the merchandise at its dock. An outline of the order fulfillment process can be seen in Figure 2.

From Figure 2, *Define Customer Solution* is where a request for quotation is received from the customer, a quotation is prepared and returned. The acceptance of customer orders which result from quotations is handled statistically. *Accept Order* identifies the orders coming in between merchandise orders (standard items), and engineered orders (customer specific, system configurations). The ratio of merchandise to engineered orders can be modified statistically to mimic order variation. In addition, the orders received can be either through an electronic data interchange (EDI) medium or not; this has time and cost implications for order transactions. Merchandise orders move directly to the *Deliver Products and Services* where items are retrieved from inventory. Engineered orders need specific *Product Configuration* for the ordering customer.

All orders are received by *Central Planning*. The coordinator in central planning receives the order specifications, and then, goes through the availability process, which consists of checking for the availab-

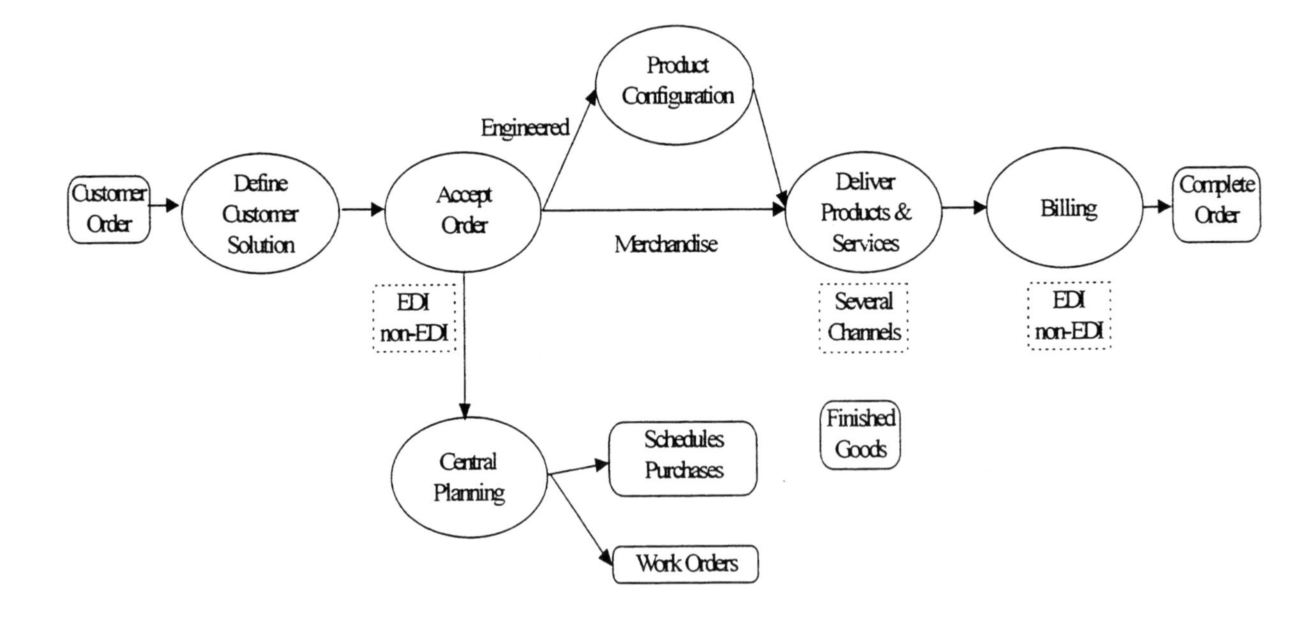

Fig. 2. An overview of the Order Fulfillment Submodel

ility of manufactured goods. As appropriate, receipts for purchased items and raw materials are scheduled along with work orders to the shop floor. The schedule of work orders is communicated to the Shop Floor Submodel, and triggers start of production. The Shop Floor Submodel in its turn communicates the resultant finished goods back to the Order Fulfillment Submodel at the *Deliver Products and Services* function. Finally, complete orders are delivered to the customer through one of the several distribution channels available and the customer is billed. Invoicing is dependent upon the various EDI and non-EDI billing paths.

The fabrication and distribution processes for orders are decoupled. The fabrication process replenishes the stock of finished goods according to the forecasted plan coming from central planning, while the distribution process depletes the finished goods inventory according to demand. Due to the various possibilities for 'paper work', distribution, and merchandise or engineered orders, there are many conceivable order transactions paths.

The Order Fulfillment Submodel, then, allows the measurement of:

- the effects of the shifting demand from engineered orders to merchandise orders,
- the cost for the various order transaction paths,
- reduction in manufacturing cycle time (customer order to delivery),
- resource utilization in each department,
- times for various activities under the various scenarios, and
- the effect of shortages in raw materials on customer's required date.

In addition, the simulation allows the determination of bottlenecks, resource imbalance, and activity costs.

5. SHOP FLOOR SUBMODEL

The Shop Floor Submodel includes the modelling of the fabrication lines, which produce printed circuit boards used in the production of electronics equipment. A typical line starts with Board Preparation and Screen Printing. A number of automated steps, including Surface Mount, Automatic Insertion, Mold Seal, and Connector Assembly follow. These are, then, followed by the Before-Wave, Wave and After-Wave operations for soldering, as required. In-Circuit Testing is performed next, followed by Mechanical Assembly. The boards are, then, passed through a series of testing stages, including Functional Test, System Assembly and Test, and Environmental Test, where they have to be tested in conjunction with boards of other types produced on other lines. The model ends with packing and Finished Goods inventory (Figure 3).

Whereas the fabrication of the various types of boards is carried out on each of the production lines, a number of the testing stages require the merging of boards from more than one line. Key features of the Shop Floor Submodel include the use of variable

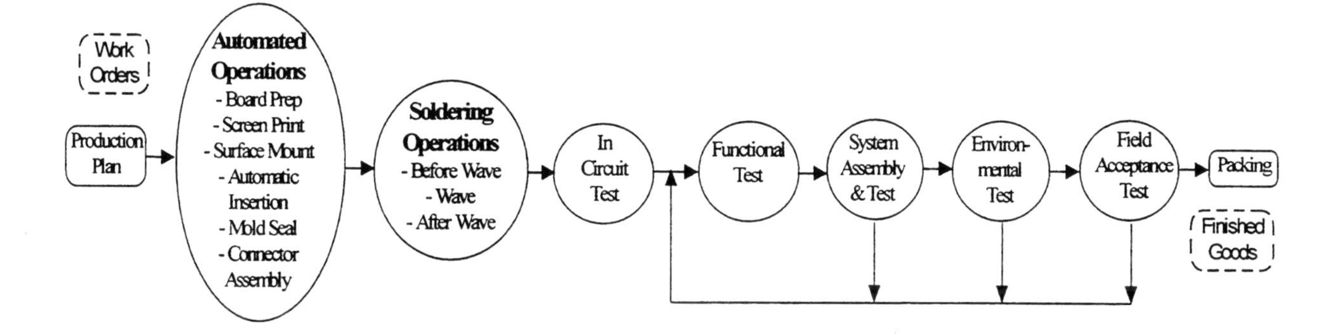

Fig. 3. An Overview of the Shop Floor Submodel

numbers of stations at several of the fabrication and testing stages, permitting the study of resource allocating issues. The yield rates at the various testing stages are modelled, and the occurrences of boards failing specific tests are treated by routing them to be repaired and then re-tested. The batching of boards into lots for both the regular and re-routed flows is modelled in detail. Quality issues and repair and re-test strategies can thus be studied using the model. The requirement of specific combinations of boards for environmental testing is also modelled, again allowing experimentation with various testing configurations.

One of the key measures of performance in a production system such as the one described is the overall product cycle time, i.e., the time a board stays in the system from the initial entry as a blank board to final packing as a finished product for shipment. Due to the requirements for multi-stage testing and for coordination with other types of boards for testing, the amount of time a board spends in the system can be significantly greater than the total amount of processing time it has to undergo. There is certainly room for improved synchronization, and thus, shorter cycle time. The impact of reduced cycle time on the cost of WIP can be significant, as some of the components used in the manufacture of these circuit boards can be very costly. The value of in-process inventory is another performance measure that is tracked by the Shop Floor Submodel, providing a basis to reduce the total value of WIP. The impact of production quality on overall cycle time and WIP cannot be overstated. The amount of time needed to repair a faulty circuit board, and the associated delay incurred in re-testing can be eliminated by improving the production quality in the first place. The yield rate at the various testing stages can be adjusted and used to quantify the significance of quality considerations. As well, the use of sampling for the purpose of testing, especially for the lengthy environmental tests, can be applied in the model.

In short, the Shop Floor Submodel is designed so that it can address issues such as equipment utilization, manpower utilization, capacity, materials management, quality, and overall manufacturing cycle time. It allows the measurement of:

- total product cycle time,
- value of work in process,
- draw down on raw material inventory, and
- quality of circuit boards and systems produced.

6. INVENTORY SUBMODEL

The Inventory Submodel is a conceptual model involving three different sets of programs: 1) the Order Fulfillment Submodel in FirstSTEP, 2) the Shop Floor Submodel in ProModel, and 3) a series of computer programs for grouping inventories, calculating the required number of parts for any given fabricated item, and determining the chance of stock outages. All inventory policies, namely, raw materials strategy, finished goods strategy, and re-ordering and safety stock rules, are incorporated in the Inventory Submodel. This planning data is used for modelling purchasing and fabrication activities which impact inventory.

In the Order Fulfillment Submodel, work orders are developed, raw materials are purchased and stocked, and finished goods are withdrawn and delivered to the customer. In the Shop Floor Submodel, work orders are received, raw materials are withdrawn from stocks and used, WIP is calculated, and finished goods are placed in inventory in the Order Fulfillment Submodel. The withdrawing of raw material and the calculation of WIP are done at two points in the Shop Floor Submodel. This reduces the amount of calculation necessary in the model, and is all that is necessary to obtain a good understanding of these inventories.

As indicated in Figure 1, raw materials are modelled as 8 groups each consisting of three classes (A, B, C) for a total of 24 inventory bins. Finished goods are modelled as 12 bins of items. Each group of items contains anywhere from 20 to 2000 different types of parts and a range of 1 to 100,000,000 parts used per year for each type. This grouping of parts (raw material and finished goods) is necessary in order to reduce the complexity of the model. It does, however, mean the necessity of calculating stock out probabilities for each inventory bin since bins are modelled as a collection of parts. This is done by using a Monte Carlo technique for each inventory bin taking into account the actual types and numbers of parts, and the inventory policy for the parts in each bin.

The Inventory Submodel allows the evaluation of:

- inventory policies,
- the effect of different manufacturing strategies on inventory,
- the effects of variation in order type, and
- resource utilization for inventory management.

7. INTEGRATING THE SUBMODELS

Both FirstSTEP and ProModel are simulation tools which provide capabilities for the linkage with other software tools. Spreadsheets are being used in the first instance for the exchange of data between the submodels; however, the integration is limited to the exchange of model data at the beginning and at the end of each simulation run. For example, using the Order Fulfillment Submodel and customer orders as input, a required build plan can be established for a certain production period. This is used to develop production schedules for each day in the period. Thus, the output of the Order Fulfillment Submodel will be a spreadsheet describing the build plan, and the input to the Shop Floor Submodel will be a spreadsheet describing the daily production schedule (work orders). Based on this schedule, the Shop Floor Submodel generates a finished goods inventory profile as output. This is in turn fed back to the Order Fulfillment Submodel to determine the implications on shipment and delivery to customers. The future will see both models running simultaneously and the dynamic exchange of model data.

Another key interface between the Order Fulfillment Submodel and the Shop Floor Submodel is raw material inventory. As the Shop Floor Submodel executes the build plan to produce circuit boards and switching equipment, raw material and components are used up, and raw material inventory will be drawn down. Depending on the reorder policy for a particular component, this may necessitate reordering from the corresponding supplier. Thus, the drawn down profiles of raw materials are an output from the Shop Floor Submodel, which become the input to the Order Fulfillment Submodel for the purpose of purchasing. The impact of various reordering policies and partnering relationships with suppliers can be assessed via the use of the integrated model.

The development and use of an integrated two-level process model has permitted the study of many manufacturing issues which involve several functions throughout an enterprise. In the future, while a single modelling system may permit simultaneous study of business and manufacturing activities at an operational level, the use of an integrated model built with a business process and production process modeller does work at present. The system described in this paper will allow the study of the reduction of manufacturing cycle time, the order fulfillment process, and inventory policies. All of these have enterprise wide implications with the need to coordinate the activities across many functions inside and outside the manufacturing company.

8. CONCLUSION

A two-level Manufacturing Enterprise Model was developed to model business and fabrication processes for an electronic equipment manufacturer. While the submodel at each level was developed using a different simulation tool, the integration of the submodels was achieved via the use of spreadsheets, which allowed the output of one model to be used as the input to another model. The next challenge will be to run the submodels simultaneously and have data between the submodels exchanged dynamically as required by the simulation.

The integrated process model provided a multi-level view of the order fulfillment process at a large manufacturer, permitted the study of cross department issues, and in particular, allowed research into the reduction of overall manufacturing cycle time. The investigation of large enterprise wide problems through modelling and simulation at the operational level will be greatly facilitated by the use of integrated process models.

Acknowledgement

The authors wish to thank the following who helped build the computer models: Rola Abdul-Nour, Venus Chan, Martijn Kamphuis, Abdulcadir Mohamud, Philippe Schick, and Ziad Yazbek. In addition, they wish to thank Dr. Ahmet Satir of Concordia University for his participation.

SIMULATION AND NOVEL CONTROL TECHNIQUES FOR RE-ENGINEERING PURPOSES

Gödri, I.; Monostori, L.

Computer and Automation Research Institute, Hungarian Academy of Sciences
Kende u. 13-17, H-1111 Budapest, Hungary, Tel.(+36 1)1665644, Fax: (+36 1) 16667503

Abstract: The high investment cost of manufacturing and assembly systems justifies the use of computer simulation support during their whole life cycle In addition to simulation, novel control techniques are also important tools not only in the every day operation of the systems, but also in the re-engineering process- Both approaches are addressed within the framework of a research project supported by the European Union, which constitutes the base of investigations presented in the paper.

Keywords: product design, manufacturing & assembly process and system planning, concurrent engineering

1. INTRODUCTION

Flexible manufacturing and assembly systems, through a careful combination of computer control, communications, manufacturing processes and related equipment, enable a section of the production oriented aspects of an organisation to respond rapidly and economically, in an integrated manner, to significant changes in its operating environment. Such systems typically comprise: process equipment, material handling, assembly equipment, a communication system and a sophisticated computer control system.

Advantages of such systems include:
- improved capital/equipment utilisation,
- reduced work-in-progress and set up,
- substantially reduced throughput times/lead times,
- reduced inventory and smaller batches, and
- reduced manpower.

The main trade-off in the decisional process, when a manufacturing or assembly system is chosen, is among *flexibility, cost efficiency* and *productivity.* This means that we want to get a compromise among flexibility maximisation, cost minimisation and lead time minimisation (Gödri and Monostori,1996).

The high investment cost of manufacturing and assembly systems justifies the use of computer simulation support both in the design and operation phases of the system:
- in the case of new investments in the *design phase,*
- in *normal operation* to maintain high system performance by predicting the system behaviour under any feasible production schedule among the alternatives prior to its implementation,
- in *re-configuration* of the system due to new products, or to machine malfunctions,
- in *re-engineering* of the whole system as a part of the whole enterprise (business process re-engineering).

The importance of the flexibility of a system as a competitive strategy is necessary (Stevenson, 1990). It is important to know that flexibility does not always offer the best choice in processing decision. Flexible systems or equipment are often more expensive and less efficient than less flexible alternatives. In certain situations, flexibility is not a necessity when products are in mature stages, requiring few design changes, and there is a high volume of output. In this instances specialised equipment are needed, with no need for flexibility. The consequence is: Flexibility should be treated with great care, applications should be matched with situations in which a need for flexibility clearly exists.

1.1 Criteria for production system design

We define a production system as consisting of design and execution subsystems. The manufacturing and assembly processes form the execution subsystem. Designing a production system three main criteria i.e.: technological, economical and strategic criteria have to be taken in consideration.

The strategic criteria will mostly influence the choice of the whole system. The technological and economical criteria will be subordinated to the strategic one, that dictates whether the system is planned either for a long or for a short working period.

The strategic criteria are:

- to design a system for a long or a short life cycle assessment (LCA) (Alting and Legharth, 1995),
- to forecast the life of the product family which is planned to be produced, as a result of a market research.

The technological criteria are such as:

- to respond quickly to the change of demand on the market,
- to shorten the average production time.

The economical criteria is:

- to optimise the production cost, the production system investment cost, with respect to the expected lifetime of the system.

These criteria will be approached with financial techniques. Complex evaluation criteria in system design are presented in Fig. 1. (Chan, 1996).

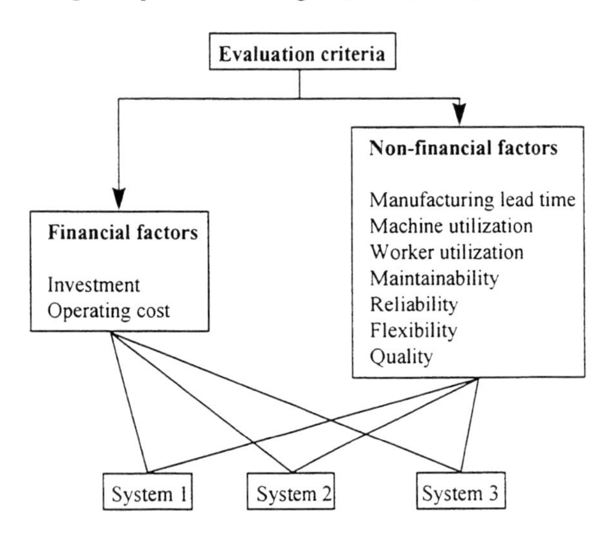

Fig. 1. Structure of evaluation algorithm

1.2 Performance measures of investment

The necessity to measure the performance of the investments is a straightforward requirement. There are three important measures of performance mentioned in (Thomson, 1993):

1. *Economy*, which means 'doing things cost effectively'. Resources should be managed at the lowest possible cost consistent with achieving quantity and quality targets.
2. *Efficiency*, which implies 'doing things right'. Resources should be deployed and utilised to maximise the returns from them.
   Economy and efficiency measures are essentially quantitative and objective.
3. *Effectiveness*, or 'doing the right things'. Resources should be allocated to those activities which satisfy the needs, expectations and priorities of the various stockholders in the business.
   Effectiveness relates to outcomes and need satisfaction, and consequently the measures are often qualitative and subjective.

The aim of this paper is to present the first results on applying simulation techniques for different phases of production systems′ life cycles from the design phase, through the normal operation phase, until the re-configuration, re-engineering phases.

In addition to *simulation, novel control techniques* are also important tools not only in the every day operation of the systems, but also in the re-engineering process. Both approaches are addressed within the framework of a research project supported by the European Union.

2. CASE STUDY

In this case study three types of assembly systems for a gas valve family are presented. These systems are to assemble 4 types of gas valves, each of them in 2-5 sizes. The systems are the following:

1. Manual assembly system, (Fig. 2.),
2. Flexible assembly system with 1 robot, (Fig. 3.),
3. Flexible assembly system with manipulators, (Fig. 4.)

The *manual assembly system* (Fig.2.)is served by 7-9 people. The first worker takes component nr.1 and keeps it under a blowing air jet and after that he/she combines it with component nr.2 taken from another container. The assembled two components are sent with the transporting belt to the next station.

Collar nr.3 and limitator collar nr.4 are put on the second component and the screw nr.5 is screwed on

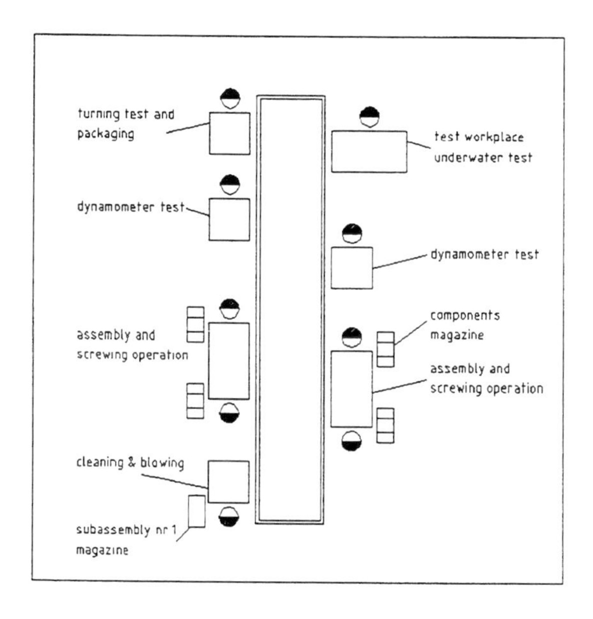

Fig. 2. Manual assembly system

as a final operation. These operations are executed by 4 people individually. In the next station one or two workers test the momentum of the screw with a dynamometer.
After that the gas valves are put on the transportation belt and carried till to the test workplace. Here the gas valves are tested under special circumstances. The first test is under water. The gas valve is put on the test bank. The valve is switched in off position. An air jet with 6 atm is introduced inside the gas valve when the gas valve is in water and it is monitored whether there is any air escapement.

In the *flexible assembly system with one robot* (Fig.3.) the inner parts are on the pallet magazine (2). There are 12 parts on a 3x4 arrangement. The external parts are on the pallet magazine (1). There are 12 parts on a 3x4 arrangement. The SCARA robot will put parts 1 on parts 2. Between the picking and leaving operation the robot will move the piece under a blowing jet in order to clean it from dust.

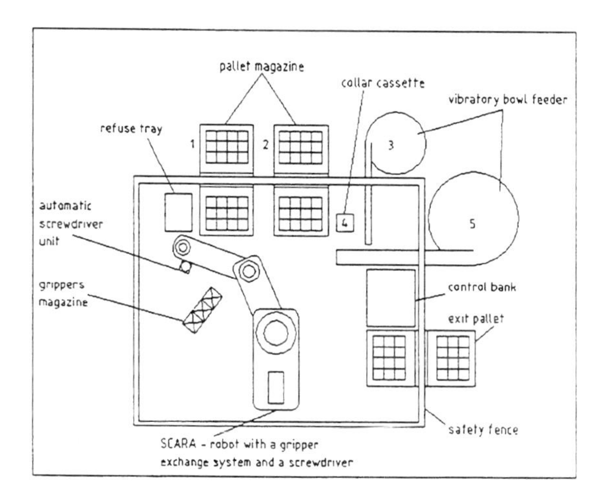

Fig.3. Flexible assembly system with one robot

Collar nr.3. fed by the vibratory bowl feeder nr.3 is picked up by the robot and it is put on the inner piece (2). Limitator collar nr.4. arranged in a vertical axis cassette is picked up by the robot and it is put on collar nr.3.. Screw nr.5. fed by the vibratory bowl feeder nr.5 is screwed on part 2 by the screwdriver. The momentum is limited from the robot's set up. When subassembly is finished the robot puts the gas valve on the control bank. The test is a relative move test between part 2 and part 1..

The *flexible assembly system with manipulators* (Fig. 4.) is composed by 5 stations/ workplaces. In the first station parts nr.2 arrive to the first station on a Europallet (1.2m x 0.8m). The portal robot from this station puts parts nr.2 on pallets from the transportation line. Each pallet can contain only one assembly. The pallets from station 1 are transported to station 2. In station 2 a similar portal manipulator picks up part nr.1 from the Europallet nr.1. The portal manipulator puts part nr.1 on part nr.2, bringing it under a blowing air jet in order to clean it from dust. After part nr.1 was put on part nr.2, the transportation line will bring the pallet to station nr.3.

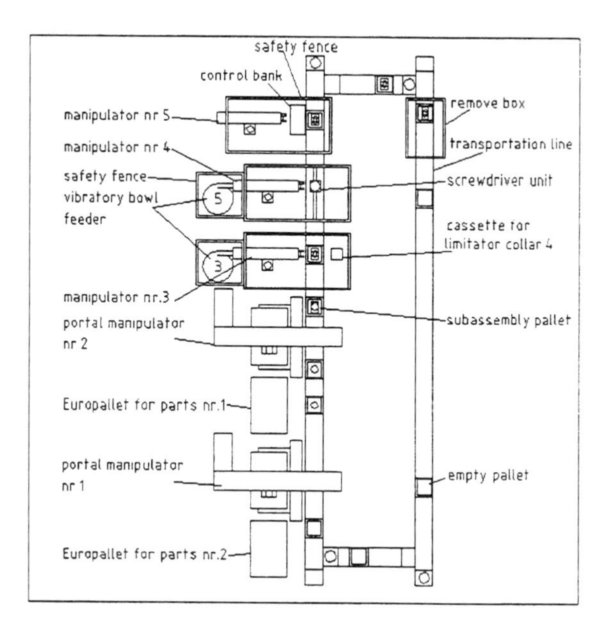

Fig.4. Flexible assembly system with manipulators

In station nr.3 the manipulator first time picks up collar nr.3 from the exit of vibratory trace connected with vibratory bowl feeder and puts it on part nr.2, second time it takes limitator collar nr.4 from the special cassette and puts it over collar nr.3. In the next step palette is in station nr.3. The manipulator takes the screw from the vibratory trace and puts it on part nr.2. In that moment the automatic screwdriver will insert/screw in the screw. When this operation is finished the palette will bring the assembled gas valve to the control bank. There, in station nr.6, a manipulator takes the gas valve and effectuates the tests described in the previous case.

Table 1 Attributes of the three systems

	Manual assembly system	Assembly system/ 1 robot	Assembly syst./ manipulators
General	7-9 workers, transp. belt	one robot, 1-2 workers	five manipulators, 2 workers
Investment	low investment, manual material handling	high investment, automatic	high investment, fully automated
Operating cost	7-9 workers' cost	1-2 workers' cost	2 workers' operating cost
Assembly lead time	average	high	low, very good
Machine utilization	average - transporting belt	high	high utilization
Worker utilization	most of time busy	average	under average
Maintainability	above average	highly sophisticated	sophisticated
Reliability	middle	highly reliable	highly reliable
Flexibility	high	high	reduced
Quality	average and variable	above average	very high

3. SIMULATION

The simulation of these systems was accomplished with the *Taylor II* simulation software. During the simulation of the three systems it was monitored how these systems react in different circumstances:

a, the first case, the so called 'ideal' case, it was supposed that neither of the systems have technical problems during their operation,

b, the second case, it was adopted that machine failure can occur periodically, following a negative exponential distribution,

c, the third case, it was additionally considered that the system produce waste products.

It was followed how the productivity varied in time for each system and for each situation. The variation of the average lead time to produce 10,000 products was also measured in the three different situations (Fig. 5.):

1. In the first case, the *small case*, batch sizes from the 10,000 products package were small, orders were under 800 products.
2. In the second case, the *medium case,* batch sizes were mixed with small and large orders together.
3. In the third case, the *large case,* were incorporated only large orders.

Productivity measures output relative to hours worked. In Fig. 6, the cycle time variation of the manual assembly system, is presented for different batch sizes. In Figures 7. and 8. the cycle time variation is presented for the flexible assembly system with robot and the flexible assembly system with manipulators, respectively.

4. EVALUATION OF THE THREE SYSTEMS

Each system had a specific setup-time during simulation. Manual assembly system had 600 s, assembly system with robot had 900 s and assembly system with manipulators had 1200 s.

It can be noticed in Fig. 6,7,8 that in each case as the order size/ batch size is increasing the cycle time of products is decreasing. The cycle time of each case has an order limit from where it becomes constant.

Regarding this case as an investment, studying Table 1 and following Fig.6,7,8 also having a concrete idea what is our strategic target, we can decide which system fit to our strategic, financial plan.

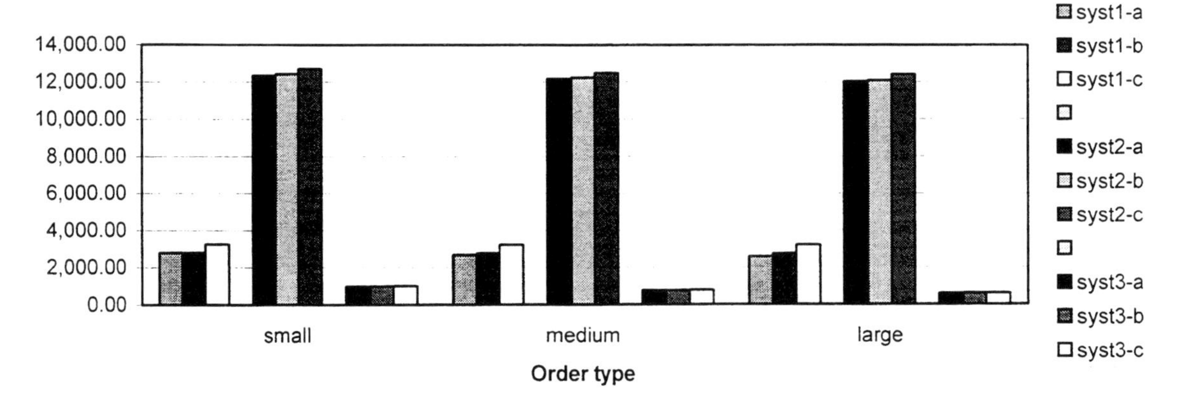

Fig. 5. Average lead time

System nr.1 - setup-time = 600 s

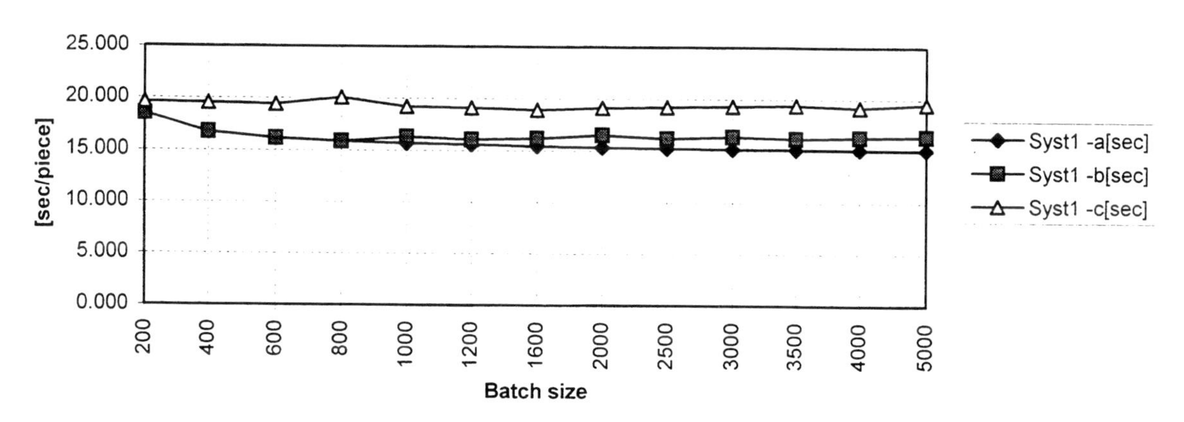

Fig. 6. Cycle time variation for manual assembly system

5. NOVEL CONTROL TECHNIQUES AS TOOL FOR ENHANCING SYSTEMS' FLEXIBILITY

Concerning *dynamic control and management of manufacturing processes*, the following new paradigms have come into the limelight:

- control and management of *holonic manufacturing systems* (systems consisting of autonomous, intelligent, flexible, distributed, co-operative agents, or *holons*),
- approaches for *reactive scheduling* of manufacturing systems (repairing/adjusting a complete but flawed schedule to keep it in-line with live system status)
- *monitoring* of production processes and their *proactive scheduling* (revision of a complete schedule which is going to be flawed at execution time, preventing an anticipated failure and minimising further performance deteriorations).

Given the limited space available, concerning the above issues, we can not go into details here, we refer to a plenary lecture of this Conference (Monostori et al., 1997).

CONCLUSIONS

The use of the simulation technique mostly for the design and operation phases was illustrated on a flexible assembly cell. Three types of assembly systems for a gas valve family were considered. These systems are to assemble 4 types of gas valves, each of them in 2-5 sizes. There were designed a manual assembly system, a flexible assembly system with one robot, and a flexible assembly system with manipulators.

Simulation based techniques were used for determining the productivity and robustness of the above solutions. The technical simulation made by the TAYLOR II package, is connected with an economic analysis of the different versions.

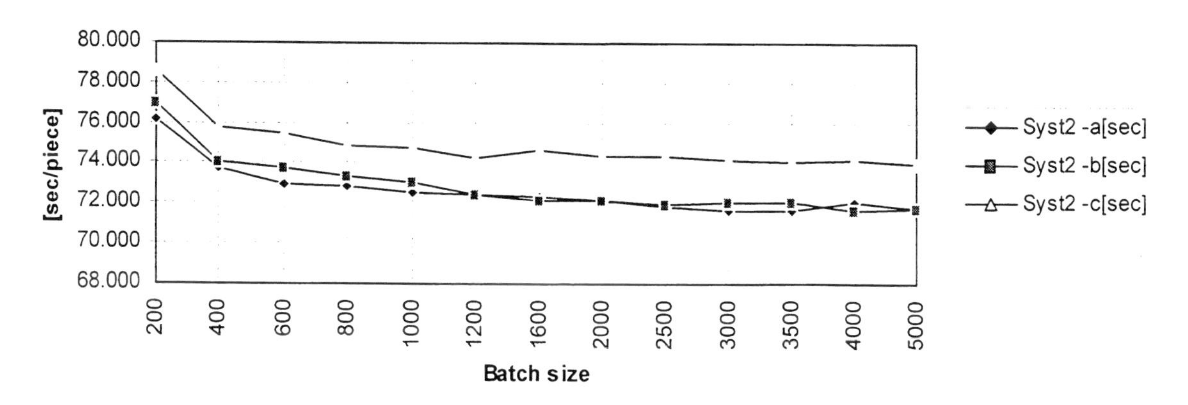

Fig. 7. Cycle time variation for flexible assembly system with one robot

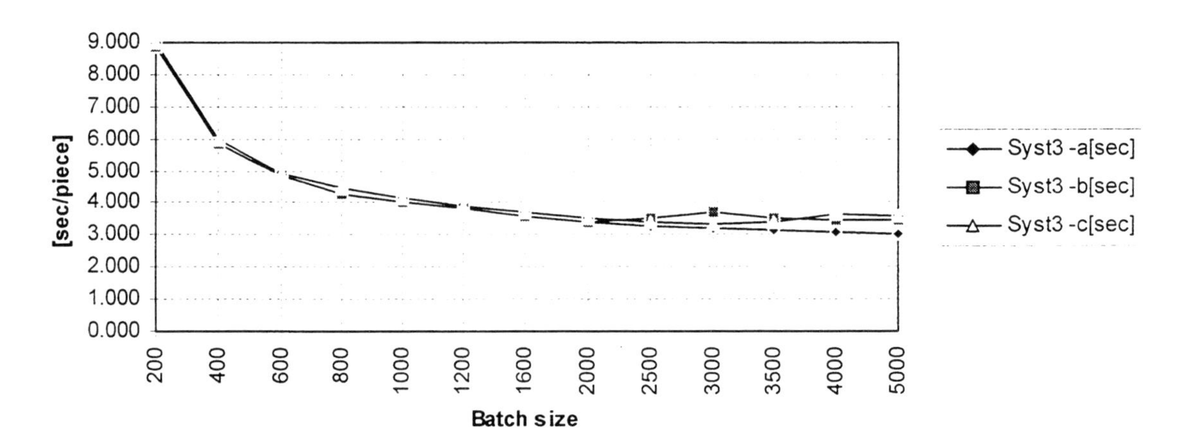

Fig. 8.Cycle time variation for flexible assembly system with manipulators

Final decisions are made taking into account both the technical and financial aspects, which will be the subject of a forthcoming paper.

ACKNOWLEDGEMENTS

This work was partially supported by the *National Research Foundation, Hungary,* Grant No. T 016512 and T016475. A part of the work was covered by the *PHARE TDQM Programme of the European Union,* Grant No. H 9305-02/1071 (REMADE)and by the *National Committee for Technological Development, Hungary* which promotes Hungarian participation in the *ESPRIT 21108 Working Group on Integration In Manufacturing and Beyond* (IIMB 21108).

REFERENCES

Alting, L. and Legharth, J.B. (1995). Life Cycle Engineering and Design, *Annals of the CIRP*, **Vol. 44**/2, pp.569-580.

Chan, T.S. Felix (1996). Design and evaluation of automated cellular manufacturing systems with simulation modelling and AHP approach: a case study, *Integrated Manufacturing Systems* **Vol.** 7/6, pp.39-52.

Gödri, I. and Monostori, L. (1996). An integrated approach to design, manufacturing and assembly/ diasassembly, *Proceedings of 7th International DAAAM Symposium,* Vienna, pp.125-126.

Monostori, L., Szelke, E. and Kádár, B. (1997). Management of changes and disturbances in manufacturing systems, *Preprints of the First IFAC Workshop on Manufacturing Systems: Manufacturing Modelling, Management and Control (MIM'97),* February 3-5, 1997, Vienna, Austria (invited plenary paper, in this issue)

Stevenson, J.William (1990). *Production & Operations Management,* IRWIN, Homewood, IL 60430, Boston, MA 02116, Third edition.

Thomson, J.L. (1993).*Strategic Management: Awarness and Change*, Chapmann & Hall, London.

MODELING METHODOLOGY FOR SYSTEMS WITH AMBIGUITY AND UNCERTAINTY

Naoki Imasaki, Satoshi Sekine, Hiroyuki Mizutani

Systems & Software Engineering Laboratory
Research & Development Center
Toshiba Corporation, Japan

Abstract: This paper discusses a modeling methodology for systems that contain ambiguity or uncertainty. Models built by the methodology are expected to handle the ambiguity and adapt to the uncertainty. The methodology is simple and versatile so that it can be applied to the description of various system architectures; *e.g.* multi-layered perceptron, fuzzy neural networks and so forth. A description language and a design tool that employ the method are also introduced.

Keywords: Modeling, Fuzzy logic, Neural networks, Software tools

1. INTRODUCTION

Any system, including manufacturing systems, has ambiguity or uncertainty. This fact always annoys engineers when they design models or controllers for systems.

If a system to be modeled contains any unignorable ambiguity, a solution is to introduce fuzzy logic(Dubois and Prade, 1980). The fuzzy logic is eligible for handling ambiguity in words or replaying a human's thinking way. If a system to be modeled is uncertain, learning from observed facts is effective to make the model accurate. A popular way to learn systems is to utilize neural networks. A combination of the fuzzy logic and the neural networks may realize an effective modeling of manufacturing systems that include human operators, a typical source of ambiguity and uncertainty. Generally speaking, if you have any knowledge on the behavior of the system to be modeled, it is a good idea to employ the fuzzy logic with which a frame is made when you design the model. The model would be able to learn more efficiently and become more accurate than using a simple neural network.

For the development of systems (models) that incorporate fuzzy logic and/or neural networks, this paper proposes a methodology called "bottom-up system construction." It has two steps to build a model:

STEP 1 making a frame,
STEP 2 filling the frame with contents.

The first step often utilizes fuzzy logic, and the second step is mainly done by the learning capabilities of neural networks.

In the first step, a frame of the model is assembled by collecting "components" necessary for the target system and connecting them each other. Fuzzy logic components realize the introduction of fuzzy logic to the model.

The methodology needs a component library. The library, which should be ready in advance, comprises enough amount of components in order to form the target system (model) with them. The essence of the methodology is that every component is designed to have a built-in tuning function as well as an intrinsic information-processing function.

By connecting the components, the model obtains a frame for the information-processing of the system to be modeled, and at the same time, this is an important feature of the methodology, it will be ready for the second step; *i.e.* the model automatically obtains the auto-tuning function. If the model gets reference data, it can begin to learn by itself. Furthermore, the model permits the replacement of its components anytime, when it still does not require any programming for the tuning function.

As for manufacturing systems, the methodology is applicable to the design of practical models by preparing components that perform various roles in the manufacturing systems. The manufacturing systems often include human operators that produce a lot of ambiguity and uncertainty. In some cases, the parts in which the human operators take part can be modeled by the fuzzy logic. The methodology may make it possible to perform a rapid modeling, and contribute to the management of the manufacturing systems.

In the next section, the modeling methodology is formulated first. Afterward, a class library and a design tool are introduced as an implementation of the methodology. Finally, how the methodology can be applied to the modeling of manufacturing systems will be discussed.

2. MODELING METHODOLOGY

This section shows how the modeling methodology is introduced and how the component library can be designed based on the methodology.

2.1 *DISASSEMBLY OF SYSTEM*

Suppose you have or can obtain reference sets $\{\bar{\boldsymbol{u}}, \bar{\boldsymbol{y}}\}$, where vector $\bar{\boldsymbol{u}}$ is r-dimensional and vector $\bar{\boldsymbol{y}}$ is m-dimensional, that represent the I/O characteristics of a system to be modeled, and some knowledge on the behavior or the configuration of the system. Let's define a model G as follows:

$$\boldsymbol{y} = \boldsymbol{G}(\boldsymbol{p}, \boldsymbol{u}), \qquad (1)$$

where the output $\boldsymbol{y}$ and the input $\boldsymbol{u}$ are r-dimensional and m-dimensional vector, respectively, $\boldsymbol{G}$ is a function that represents the I/O characteristics of the model. The n-dimensional vector $\boldsymbol{p}$ is an internal parameter set.

Consider a problem to design and build the model G. Let's introduce the *steepest descent method*(Amari, 1967) as the learning strategy. If an index function is defined as follows:

$$E = \frac{1}{2}\sum_{j=1}^{r}(y_j - \bar{y}_j)^2, \qquad (2)$$

you can solve the problem with the following modification rules:

$$\boldsymbol{p}(t+1) = \boldsymbol{p}(t) + \delta\boldsymbol{p}(t), \qquad (3a)$$

$$\delta\boldsymbol{p}(t) = -\varepsilon\frac{\partial E}{\partial \boldsymbol{p}} + \beta\delta\boldsymbol{p}(t-1), \qquad (3b)$$

where t is the iteration step, and ε and β are positive constants. If the model G is a multi-layered neural network, the equations (3) would be nothing but the backpropagation method.

The methodology, however, supposes the model G is composed of multiple sub-systems (not limited to *neurons*) based on the knowledge about the system to be modeled. The model would have a form of a feed-forward network of M sub-systems $g^1, \cdots, g^M$ (See Fig. 1). Each sub-system may be

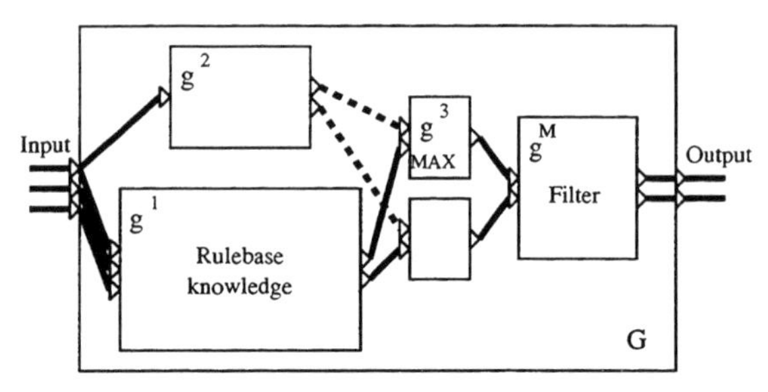

Fig. 1. System configuration as a network of sub-systems

a rulebase, a switch, a filter or any function that is a part of the whole system.

Let all sub-systems have the configuration shown in Fig. 2 so that the whole system G works properly. Each sub-system receives the input($\boldsymbol{x}$), executes an intrinsic signal processing procedure (Forward procedure), and produces the output($\boldsymbol{z}$). This process corresponds to each part of the knowledge about the system. Besides, each

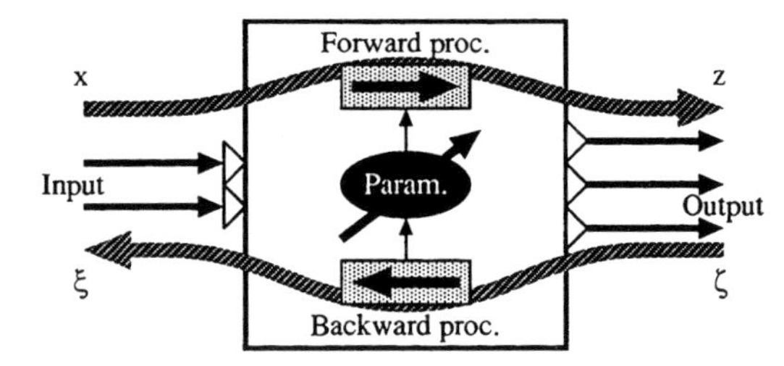

Fig. 2. Sub-system

sub-system receives an error-information($\boldsymbol{\zeta}$), executes a learning (modifying the internal parameter set) procedure (Backward procedure), and produces a new error-information($\boldsymbol{\xi}$) for the upper stream.

This function is formulated as follows. Let the I/O characteristics of a sub-system g^i be defined by the following equation.

$$\boldsymbol{z}^i = \boldsymbol{g}^i(\boldsymbol{p}^i, \boldsymbol{x}^i) \qquad (4)$$

The input is $\boldsymbol{x}^i = [x_1^i, \cdots, x_{m^i}^i]'$, and the output is $\boldsymbol{z}^i = [z_1^i, \cdots, z_{r^i}^i]'$. The internal parameter set is represented by $\boldsymbol{p}^i = [p_1^i, \cdots, p_{n^i}^i]'$, $\sum_{i=1}^{M} n^i = n$, which is a partial set of the internal parameter set $\boldsymbol{p}$ of the whole system G. The input x_j^i is given by Eq. (5).

$$x_j^i = \sum_{F_j^i} z_\ell^k \tag{5}$$

In Eq.(5), $\boldsymbol{F}_j^i$ is a set of originations, z_ℓ^k is the ℓ-th output of an origination g^k . If an origination is external, an element of the input $\boldsymbol{u}$ becomes z_ℓ^k. The output $\boldsymbol{y}$ is calculated in a similar way of Eq.(5).

The modification rule of the internal parameter set $\boldsymbol{p}^i$ of a sub-system g^i can be re-written as follows:

$$\boldsymbol{p}^i(t+1) = \boldsymbol{p}^i(t) + \delta\boldsymbol{p}^i(t), \tag{6a}$$

$$\delta\boldsymbol{p}^i(t) = -\varepsilon\frac{\partial \boldsymbol{g}^i}{\partial \boldsymbol{p}^i}\boldsymbol{\zeta}^i + \beta\delta\boldsymbol{p}^i(t-1), \tag{6b}$$

where $\boldsymbol{\zeta}^i = [\zeta_1^i, \cdots, \zeta_{r^i}^i]'$ is an error information, which is given as follows:

$$\zeta_j^i = \sum_{T_j} \xi_\ell^k. \tag{7}$$

In Eq.(7), $\boldsymbol{T}_j$ represents a set of destinations of the output z_j^i, and ξ_ℓ^k is the ℓ-th error-information of a destination g^k. If a destination is external, for instance it is y_j, an output of the system G, $\xi_\ell^k = y_j - \bar{y}_j$ is satisfied.

Note that the modification rule Eq.(6) for the sub-model g^i minimizes an index function:

$$E^i = \frac{1}{2}\sum_{j=1}^{r^i}(z_j^i - \bar{z}_j^i)^2, \tag{8}$$

based on the steepest decent method as well (Compare Eq.(3) and Eq.(6)). The error information $\boldsymbol{\zeta}^i$ is supposed to be nothing but the error vector $\boldsymbol{z}^i - \bar{\boldsymbol{z}}^i$ itself, and the reference output $\bar{\boldsymbol{z}}^i$ for g^i is created from the reference output $\bar{\boldsymbol{y}}$ for the whole system G by propagating the error backward as follows:

$$\boldsymbol{\xi}^i = \frac{\partial \boldsymbol{g}^i}{\partial \boldsymbol{x}^i}\boldsymbol{\zeta}^i, \tag{9}$$

where $\boldsymbol{\xi}^i = [\xi_1^i, \cdots, \xi_{m^i}^i]'$. In this way, it is shown that the modification of the network of the sub-system $\mathrm{g}^1, \cdots, \mathrm{g}^M$ is equivalent to that of the whole system G. The sub-systems that comprise the whole system G operate along the stream of signals in both of the forward procedure and the backward procedure. For example, suppose the system G consists of four sub-systems g^1, $\cdots$, g^4 as shown in Fig. 3. In the forward procedure, the sub-systems work in order of g^1, g^2, g^3, and g^4. On the other hand, the order of the operation becomes g^4, g^3, g^2, and g^1 in the backward procedure.

According to the above consideration, the learning function as well as the signal processing function

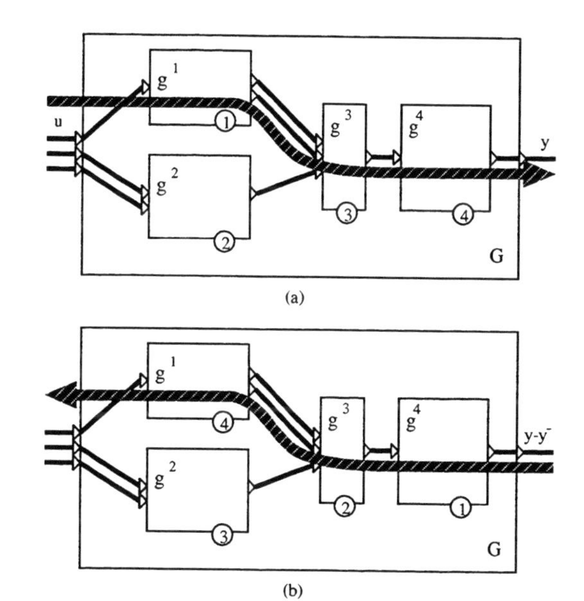

Fig. 3. Signal propagation

can be processed in each sub-system independently. That means if you prepare sub-systems which has functions as in Eq.(4),(6) and (9) as many as necessary, and assemble them into a network, then that is all of constructing a system with the learning capabilities.

2.2 *COMPONENT DESIGN*

Let us show some examples of components that are derived by the methodology.

2.2.1. *Proportional operator*

This component multiply the input by a constant and output the result.

Forward procedure:

$$z = K \cdot x, \text{ where } K \text{ is a constant} \tag{10}$$

Backward procedure:

$$K(t+1) = K(t) + \delta K(t) \tag{11a}$$

$$\delta K(t) = -\varepsilon x\zeta + \beta\delta K(t-1) \tag{11b}$$

$$\xi = K \cdot \zeta \tag{11c}$$

2.2.2. *MAX operator*

This component receives multiple inputs, selects the biggest value among them, and output it. The MIN operator can be designed in the same way.

Forward procedure:

$$z = \max_i x_i \tag{12}$$

Backward procedure: This component does not have any internal parameter.

$$\xi_i = \begin{cases} \zeta, & \text{if } z = x_i \\ 0, & \text{otherwise} \end{cases} \tag{13}$$

2.2.3. *Sigmoid function*

This component can be utilized as a neuron when a conventional neural network is required.

Forward procedure:

$$z = \frac{1}{1 + e^{-b(x-a)}}, \text{ where } a,b \text{ are constants} \quad (14)$$

Backward procedure:

$$a(t+1) = a(t) + \delta a(t) \quad (15a)$$

$$\delta a(t) = -\varepsilon\{-b \cdot z(1-z)\zeta\} + \beta\delta a(t-1) \quad (15b)$$

$$b(t+1) = b(t) + \delta b(t) \quad (15c)$$

$$\delta b(t) = -\varepsilon z(1-z)(x-a)\zeta + \beta\delta b(t-1) \quad (15d)$$

$$\xi = b \cdot z(1-z)\zeta \quad (15e)$$

2.2.4. *Frame component*

A network of components can be regarded as a component, too. This idea means the system structure can be hierarchical. The frame component manages the components in it.

2.2.5. *Other components*

You can design various and more complicated components such as membership function, neural networks and Gaussian function. When you design the backward procedure defined in Eq. (6), you sometimes may find you cannot differentiate the forward procedure function. In such cases, you should make some approximation.

2.3 *ADVANTAGES*

The methodology described above has some advantages:

- It allows a wide range of system configuration. By selecting and connecting components, you can make various kinds of system.
- It allows repetitive replacement of components. Because each component is independent, the other components are not affected by the replaced part.
- Users do not have to program anything. This feature realizes a rapid development.
- The built-up system always comes with the learning capabilities. The tuning of system is easy.

As for the system configuration, various combinations of the fuzzy logic and neural networks can be realized as the following.

- Fuzzy reasoning system
- Fuzzy-neuro system
- Neural network system
- Hybrid system of fuzzy reasoning system and neural networks

You can choose the most suitable configuration for the target system.

3. IMPLEMENTATION

3.1 *COMPONENT LIBRARY*

The implementation of the modeling methodology introduced above is suitable for the object oriented technology. We employ an object oriented language C++ in order to build the component library. For instance, the component library is implemented as a class library. The basic interface is defined in a primitive class, and every component is defined as a descendant class of the primitive class. The intrinsic signal processing functions and the learning functions are defined in the descendant classes.

Example 1 shows the essential part of the definition of the primitive class *EnBaseItem*.

Example 1.

```
class EnBaseItem {
protected:
  EnList<EnParameter> m_lLp;
  EnList<EnPort> m_lPort;
public:
  virtual void Forward();
  virtual void Backward();
  virtual void Modify();
};
```

The class has the internal parameter list (`m_lLp`) and a list of ports (signal terminals) which the object equips. The object can receive 3 kinds of messages: *Forward*, *Backward*, and *Modify*. When the object receives Forward message, it executes the forward procedure based on the internal parameter list and the input retrieved from the ports. Likewise when it receives Backward message, it executes the backward procedure. If the object receives Modify message, it updates the contents of the internal parameters by the results of the backward procedure. In the definition of descendants, message handling functions `Forward()`, `Backward()`, and `Modify()` are described based on Eq. (4), (6), and (9). Figure 4 shows a part of the class hierarchy.

When a user make a system with the component library, he/she defines an object that represents the target system. The class of the target system is often of the frame component, and in its constructor the frame is given multiple objects in

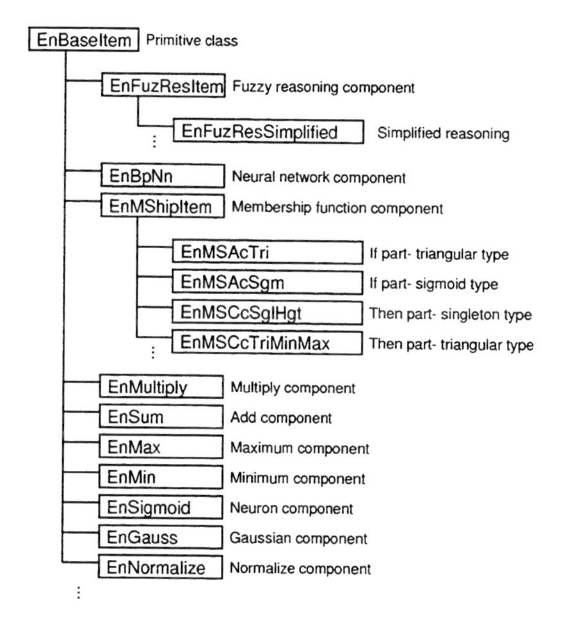

Fig. 4. Class hierarchy

order to realize the target function. Example 2 is a sample description.

Example 2.

```
// Instantiation of a fuzzy reasoning
// object. Contents of the object is
// generated in the constructor.
EnFuzResSimplified*
  pSys = new EnFuzResSimplified;

// Put the input
pSys->Input(inpVector);

// Forward procedure
pSys->Forward();

// Get the output
pSys->Output(outpVector);
```

3.2 *SYSTEM DESIGN TOOL*

To help users who are inexperienced in the C++ programming, a system design tool is provided.

Enfant™ is a design tool that provides a graphical user interface(GUI) module and a C++ class library EnCL(component library) for the design of systems that have the learning capabilities. Enfant™ hides the C++ class library from the users and provide a visual programming environment instead.

Enfant™ calls component objects defined in the class library *item*. Items work by receiving messages. If an item receive *Forward* message from somewhere, it executes the forward procedure. An item can be a network of items. That means item (any system in the world) can have a hierarchical structure. Messages are passed inside the network hierarchically.

If you want a system that have a complicated function, collect appropriate items from EnCL and connect those into a network in order to realize the function on a network editor provided by the GUI module. For example, if you want a fuzzy inference system, make a system like Fig. 5 (a). In the system, some *membership function*

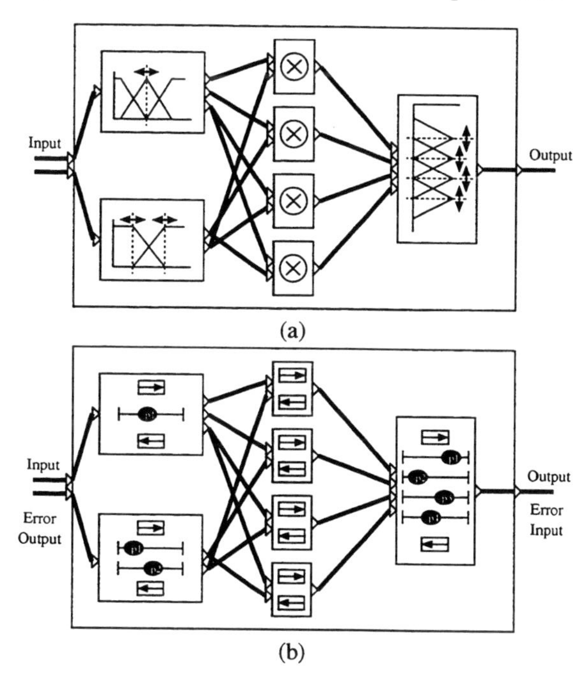

Fig. 5. Fuzzy inference system

items and *multiply* items, both kinds of item are provided by the class library, are connected each other.

The remarkable point is that the learning mechanism are automatically installed when the network is assembled. The users do not have to care about it. For instance, while you design a fuzzy reasoning system as in Fig. 5 (a), the GUI module regards the system as a learning system as in Fig. 5 (b). This is a contribution by the methodology. If you give a set of reference data to the system, it will tune the membership functions by itself.

Figure 6 shows a screen shot of Enfant™. For convenience, the GUI module provides a fuzzy rule editor and a membership function editor other than the network editor. If you input a rule set with the fuzzy rule editor, the GUI module generates a network for the fuzzy reasoning as in Fig. 5.

4. MANUFACTURING SYSTEMS

The methodology that this paper proposed can be applied to the modeling of manufacturing systems. Generally speaking, behavior of manufac-

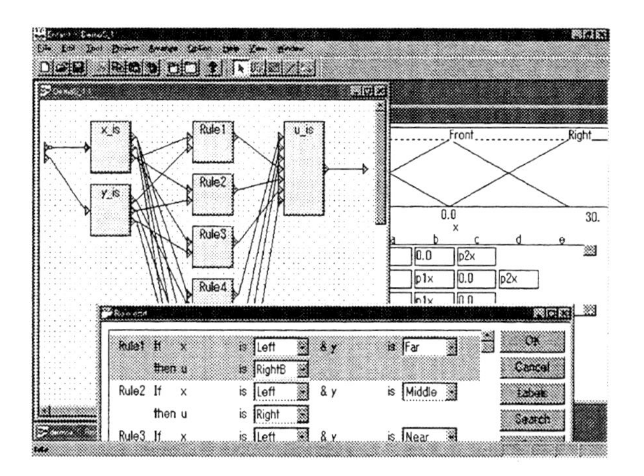

Fig. 6. GUI of Enfant

turing systems have ambiguity and uncertainty due to the following factors.

- Human operator
- Environment (temperature, humidity, maintenance, etc.)
- Dispersion of materials

Because the methodology allow you to join fuzzy reasoning system, neural networks, and whatever freely, complicated models can be created. See Fig. 7 for an example.

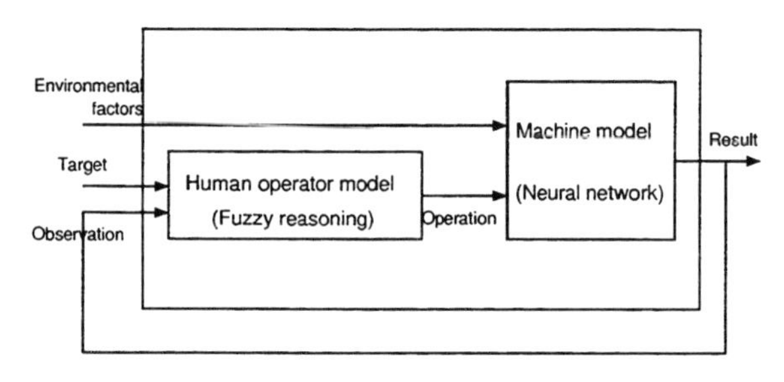

Fig. 7. Example

The fuzzy logic is suitable for modeling of a part that you have some knowledge about the behavior. On the other hand, if you do not have any knowledge with respect to a part in the system to be modeled, you had better employ the neural networks for the modeling of the part. You can employ multiple methodologies and the model still has learning capabilities.

5. CONCLUSION

This paper proposed a modeling methodology for systems with ambiguity and uncertainty. By components that have built-in learning mechanism, the model will have the learning capabilities without any programming. Designers can build models by selecting and connecting the components. The models can be realized in various configurations. An implementation of the methodology was also introduced.

REFERENCES

Amari, S. (1967). A theory of adaptive pattern classifiers. *IEEE Trans. on EC* (16), 279–307.

Dubois, D. and H. Prade (1980). *Fuzzy Sets and Systems*. Academic Press.

ON THE RECENT ADVANCES IN GRAFCET

J. Zaytoon, V. Carré-Ménétrier, M. Niclet and P. De Loor

Laboratoire d'Applications de la Microélectronique, Faculté des Sciences
B. P. 1039, 51687, Reims Cedex 2, France

Abstract: Grafcet is an international standard for specification and implementation of logic controllers in manufacturing systems. The objective of this paper is to review the current active research work that aims at providing the Grafcet both with a formal foundation to insure correctness and safety requirements, and with an integrated methodology that allows to link the design of the manufacturing systems with the development of their logic controllers. The contribution of the authors will be particularly emphasised.

Keywords: Logic controllers, Semantics, Methodology, Verification & Validation, Synthesis.

1. INTRODUCTION

Grafcet or function charts for control systems is an international standard used for the specification and the implementation of logic controllers in manufacturing systems (David, 1995 ; International Electrotechnic Commission, 1988). Throughout the twenty years that have passed since it was defined (AFCET, 1977), Grafcet is becoming widely used in the industry (Baracos, 1995) and in education (Kheir, *et al.*, 1996). The main contribution of Grafcet, which uses a Petri-net like formalism, is that it allows a clear modelling of inputs and outputs and of their relations. It also allows modelling of concurrency and synchronisation. In spite of these advantages, Grafcet has long been criticised because it is not supported by a formal foundation that allows to insure correctness and safety requirements on the one hand, and because it lacks adequate methodology that allows an efficient development of high quality models in the case of complex systems on the other.

The objective of this paper is to review the current active research works which have been undertaken recently to overcome the above problems. These works attempt to benefit from the recent advances in software engineering, in manufacturing-systems engineering, in discrete-event systems theory, and in formal methods for real-time systems. The contributions of the authors will be particularly emphasised.

The paper is decomposed into four parts, each of which treats one of the current research issues related to Grafcet.

2. SYNTACTIC AND SEMANTIC EXTENSIONS OF GRAFCET

Besides the definition of Macro-steps and forcing orders (Union Technique de l'Electricité, 1993), recent developments have provided hierarchical structure, abstraction levels and reuse facilities to Grafcet (Arzén, 1996 ; Duméry, *et al.*, 1996). These facilities allowed, for example, to extend Grafcet application scope to the control of batch systems.

Other extensions of Grafcet (Union Technique de l'Electricité, 1993), aiming to ensure synchrony, determinism and reactivity of the model, postulate two time scales for Grafcet. An internal one deals with state evolutions and an external one deals with stability. At the external time scale, all changes of the values of the logical expressions associated to the transitions are considered as soon as they occur. The whole of their consequential effects is perceived as occurring at the same time instant. This means that Grafcet reaction takes zero time in the external time scale. At the internal time scale, this reaction is carried out throughout a number of consecutive internal evolutions whose durations are as small as necessary. The consistency of this definition of the temporal behaviour of Grafcet has been proved by Frachet and Colombari (1993). However, the basic evolution rules of Grafcet and the above temporal behaviour are not sufficient to guarantee a unique interpretation of a given Grafcet (Lhoste, *et al.*, 1993). The semantics of Grafcet has therefore been completed by a semi-formal users guide called "Grafcet player" (Lhoste, *et al.*, 1997). This guide organises and explains the interactions between the basic evolution rules and the two time scales of the model. The resulting semi-formal semantics insure a deterministic interpretation of Grafcet and reinforce its synchronous and reactive nature.

3. DEVELOPMENT METHODOLOGIES

The existing complex relationships between a manufacturing system and its environment and the necessity of meeting the increasing quality and dependability constraints require a modelling framework which is able to link the design of these systems and the development of their logic controllers. Few methodologies have therefore been proposed to provide a framework for the development of Grafcet starting from a high level specification of the manufacturing system. Three of these methodologies (Chapurlat, *et al.*, 1993 ; Morel and Lhoste, 1993 ; Zaytoon, 1996) are functionally based, and two methodologies (Wilczynski and Wallace, 1992 ; Zaytoon and Villermain-Lecolier, 1997) are object oriented. The two methodologies proposed by the authors are briefly presented below.

3.1 A functionally based methodology

The methodology proposed by Zaytoon (1996) deals with the design of the control system according to three viewpoints :

1) functional specification to describe the organisation of the manufacturing activities and flows using SADT (Ross and Schoman, 1977). To distinguish the role of the different activities and production flows, without ambiguity, a temporal extension of SADT is proposed. This temporal extension is based on the use of the interval temporal logic (Allen, 1983).

2) high-level co-ordination and execution of SADT activities using Petri nets (Peterson, 1981). Mapping rules are provided to generate the Petri nets from the temporal extension of SADT. These Petri nets are used to ensure a modular development of the individual control tasks on the one hand, and to guarantee a number of safety and liveness properties and to prevent conflicts on the other hand.

3) specification of the local control tasks using Grafcets corresponding to the elementary sequences for individual machine. These local control tasks need only a partial view of the global system state and their execution is co-ordinated by the Petri net.

A co-operation framework is used to ensure a consistent shift from the external behaviour specification point of view to the specification and the implementation of local controllers tasks. This framework is based on the use of a generic Grafcet skeleton for each leaf activity of SADT on the one hand, and the use of a pre-established communication and synchronisation protocol between a Grafcet and the corresponding Petri sub-net on the other hand. In this way, the PLC designer can implement the local control tasks separately; the consistency of the global behaviour of the system can be guaranteed thanks to the formal verification possibilities of Petri-nets.

The application of the above methodology in a number of manufacturing systems has shown that it provides a useful and practical guide to the development of manufacturing systems when the requirements are well identified and the development cycle is relatively short.

3.2 An object oriented methodology

The methodology proposed by Zaytoon and Villermain-Lecolier (1997) is based on an extension of the object modelling technique "OMT" (Rumbaugh, *et al.*, 1991). This well-established software-engineering technique uses three complementary models (object, functional, and dynamic) which are well adapted to represent the different views of manufacturing systems.

The extension proposed is based on the definition of a generic functional layer on top of the three models of OMT. This layer uses mechanisms that are equivalent to those of the object model. These mechanisms are: functional generalisation, functional aggregation, sequential aggregation and iteration. The use of these mechanisms in both the object and the functional method of OMT allow to establish a generic functional scheme and a generic structural scheme for each specific manufacturing sector. Starting from such generic schemes, the following steps are used to generate the sequential controller for a given manufacturing system.

1) The generic functional scheme related to the appropriate manufacturing sector is instanciated, by graphical selection, to obtain a particular functional scheme corresponding to the sequence of processes required to carry out the target application. One or many objects of the generic structural scheme of the appropriate manufacturing sector must be associated with each of the processes of the lowest decomposition level of the particular functional scheme.

2) Transport and storage processes are introduced, in order to remove the discontinuities between the transformation processes of the particular functional scheme. This results in the definition of the functional architecture of the system. Each of the transport and storage processes must be associated with one or several objects which are instanciated from the generic structural scheme.

3) The dynamic model is automatically generated by means of dedicated translation rules. These rules go through the functional architecture and replace the processes related to an aggregation, a generalisation or an iteration, by a corresponding modular structure. The generated dynamic model corresponds to the specification of the control system at the co-ordination level, because it gives the co-ordination between the control tasks necessary to the fabrication of a part. These tasks correspond to the states of the lower decomposition level of the dynamic model. The objects controlled by these tasks are determined by the associations that have been established between the generic structural scheme and the functional architecture.

4) The control engineer should next develop the local controller for each control task by designing the partial Grafcet to be executed when the associated state in the dynamic model is active. Each partial Grafcet is used to control the operations and the interactions of the object(s) associated with its control task. The execution context of a partial Grafcet is given by the attributes of the associated objects. The values of input attributes of these objects condition the evolution of the partial Grafcet which in turn updates the output attributes according to its active actions.

This methodology is well adapted to manufacturing systems when the requirements are not well identified in advance or when the requirements are subject to rapid evolutions. It was validated using a number of real-sized systems belonging to the manufacturing sector of printed circuit boards. However for this method to be really efficient, considerable effort will have to be undertaken to establish certified libraries for the generic structural and functional shames for different manufacturing sectors.

4. VALIDATION OF GRAFCET

Active research work has been carried out recently to establish formal support for the verification (of internal consistency, stability, reinitialisation, deadlock-free) and the validation (of safety, liveness and timeliness properties) of Grafcet. The modelling power of Grafcet makes the validation of the model very hard to establish. In fact, once it has been modelled in terms of steps and transitions, a Grafcet model must be validated in terms of situations and evolutions between situations.

Most of the proposed approaches focus on the verification and the validation of the behaviour of Grafcet on its own without considering the controlled plant. These approaches are based on the use of synchronous languages (André and Gaffé, 1996 ; Marcé and Le Parc, 1993), time automata (Marcé, *et al.*, 1996), Petri nets (Aygalinc and Denat, 1993), Max+ algebra (Panetto, 1996), or state machines (Roussel and Lesage, 1996). Two other approaches are used to validate Grafcet together with the plant it controls. The first approach (Kristensen, *et al.*, 1995), which does not address the semantics of Grafcet properly, is based on the use of the TCCS (Timed Calculus of Communicating Systems) / TML (Timed Modal Logic) framework (Cerans, *et al.*, 1993). The second approach (De Loor, 1996), which will be presented below, is based on an extension of the TTM (Timed Transition Model) / RTTL (Real-Time Temporal Logic) framework (Ostroff and Wonham, 1990). In these two approaches, the integration of a model of the controlled plant is motivated by the fact that the plant represents the primary goal of the automation, whereas the control is only a means to achieve this goal and to ensure that the plant behaves safely and reliably. Therefore, the properties to be verified must be expressed in terms of the behaviour of the elements of the plant, and this requires the modelling of these elements.

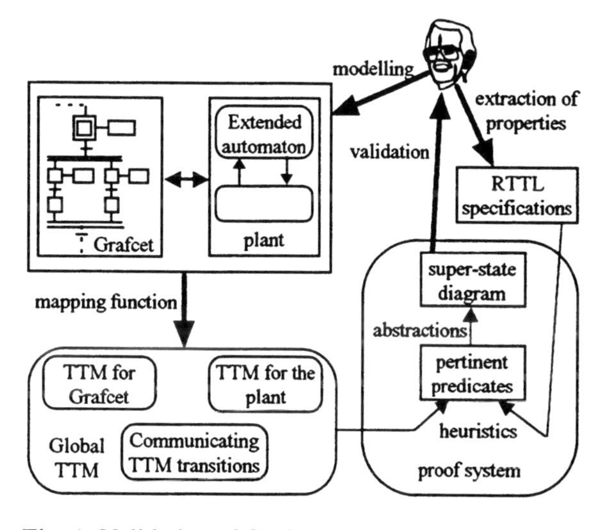

Fig. 1. Validation of Grafcet (De Loor, 1996)

The approach proposed by De Loor (1996) is illustrated by Fig. 1. In this approach, the control-system designer starts by modelling the controller using Grafcet, and the plant using extended automaton. A mapping function (Zaytoon, *et al.*, 1995) is used to translate the resulting model into an equivalent TTM, and a dedicated proof system (De Loor, *et al.*, 1997) allows to validate safety, liveness and timeliness properties, expressed in RTTL.

The mapping function insures a correct temporal behaviour of Grafcet in terms of synchronism, determinism and reactivity. It represents a formal definition of the temporal behaviour of Grafcet (Lhoste, *et al.,* 1997), in terms of the TTM. It also reflects the asynchronism and the non-determinism of the controlled plant. An adaptation of the semantics of the TTM is proposed to integrate synchronous and asynchronous behaviours within the same modelling framework.

The dedicated proof system is based on the verification of a concise super-state diagram, which focuses on certain evolutions of the system. These evolutions are given by pertinent predicates that preserve causality and certain temporal relationships between system evolutions related to the property. A pertinent predicate corresponds to an aggregation either of states of the plant, or of steps and transitions of Grafcet. Such a predicate is determined by heuristics which are based on the structure of the modelled system and on process-controller interactions. Safety properties can be proved as well as a class of liveness and timeliness properties. These properties are expressed using the RTTL.

5. SYNTHESIS OF GRAFCET

Two approaches use the supervisory control theory (Ramadge and Wonham, 1989) as a support for the synthesis of a supremal Grafcet that represents the minimal possible restriction of the behaviour of a given Grafcet, in a way to satisfy the given safety and liveness specifications and to assure non blocking behaviour of the closed loop system. The difficulty that faces these approaches is mainly related to the semantic difference between Grafcet model (based on conditions, events, logic operators, double time scale interpretation, synchronism, reactivity, possibility of simultaneous actions and simultaneous transition firings) on the one hand, and the formal model of the theory (based on asynchronism, particular events interpretation, controller-plant interaction) on the other.

The first approach (Charbonnier, *et al.*, 1995), which has been successfully applied to a real size industrial system, can only handle a sub-class of Grafcet in which the logical expressions of Grafcet transitions are limited to single events, representing the edges of input variables. This approach does not consider the constraints induced by the controlled process.

The other approach (Zaytoon, *et al.*, 1997a ; Zaytoon, *et al.*, 1997b) attempts to take into account all the features of Grafcet as defined by the International Electrotechnic Commission (1988). This approach (Fig. 2) is based on six steps which are described below.

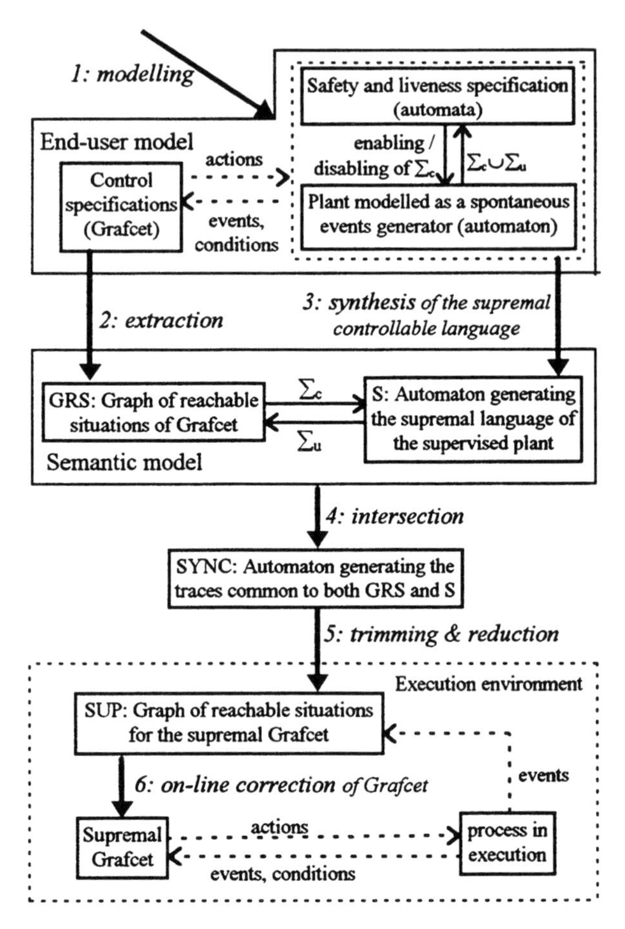

Fig. 2. Synthesis of a supremal Grafcet (Zaytoon, *et al.*, 1997a)

1) Modelling the plant behaviour and the safety and liveness user requirements, using automata. Then, modelling the required control behaviour using Grafcet. The automata representing the plant behaviour correspond to spontaneous event generators (Ramadge and Wonham, 1989). Controllable events Σ_c are associated to the activation and deactivation of Grafcet outputs. Uncontrollable events Σ_u are associated to the rising and falling edges of Grafcet inputs.

2) Extraction of the graph of reachable situations (GRS) of Grafcet according to the method given by Roussel and Lesage (1996). This method takes into account the possibilities of interpreted parallelism of Grafcet, the use of edges and step variables in the

receptivities of the transitions, and the reachability of a stable situation. The GRS, which is a uniform and completely specified Mealy machine, reflects the required semantics of Grafcet in terms of synchrony, reactivity and determinism (Lhoste, *et al.*, 1997).

3) The supremal language of the supervised plant is next obtained according to the classical synthesis algorithm (Wonham and Ramadge, 1987). The supervisor realisation which is used here is that of a discrete event system S, in which the enabling/ disabling action of the supervisor is implicit in the transition structure of S. Therefore, the transition structure of S corresponds to the maximum non-blocking allowable behaviour of the controlled plant with respect to the imposed safety and liveness specifications. The part of Grafcet (and consequently of GRS) that will be allowed to execute must be confined within the language generated by S.

4) The sequences of events which can be generated both by GRS and by S are extracted by intersection of the two corresponding languages. This language intersection is achieved by a dedicated synchronisation operator (Zaytoon, *et al.*, 1997a) that goes through the automata GRS and S to construct the automaton SYNC.

5) The fifth step is used to generate the maximum non-blocking execution model by trimming and reducing the automaton SYNC. The resulting automaton SUP represents the most permissive sub-set of the behaviour of Grafcet which is both non-blocking, and satisfies the supervisory specifications; that is, it gives the graph of reachable situations of the supremal Grafcet.

6) The automaton SUP corresponds to the required behaviour of the supremal Grafcet. During execution, this automaton is used to condition the execution of Grafcet outputs, in such a way as to confine the behaviour of Grafcet to the required supremal behaviour. This on-line correction of Grafcet is achieved by deactivating of some of the current outputs of Grafcet in a way that allows to avoid behaviours not included within SUP. The deactivated outputs can be highlighted so that the control designer can distinguish the control sequences which are maintained, inhibited or restricted by the on-line correction.

6. CONCLUSION

The four complementary aspects, that have been reviewed in this paper, provide solutions to specific problems related to the practice of Grafcet.

Syntactic and semantic extensions have been proposed in the literature to improve the modelling power of Grafcet, and to provide formal support that guarantees the synchrony, reactivity and determinism of Grafcet. Methodologies have also been developed to link the design of manufacturing systems and the development of their controllers given by Grafcet. This allows to derive the control specifications starting from high level functional and/or structural system models. Once these specifications are established, it is important to validate their consistency and their conformity to user requirements. Various formal methods have therefore been proposed for the validation of different types of safety, liveness and timeliness properties related to Grafcet. A more challenging problem is that of providing a formal framework for the automatic synthesis of an optimal Grafcet, starting from a given Grafcet and a number of user requirements. Such a framework has been recently established.

An important effort remains to be accomplished to develop an integrated framework that captures the benefits of the works carried out separately alongside each of these four areas.

REFERENCES

AFCET (1977). Normalisation de la représentation du cahier des charges d'un automatisme logique. *Final Report of AFCET Commission, J Automatique et informatique industrielle*, no. 61-62 (in Franch).

Allen, J.F. (1983). Maintaining knowledge about temporal intervals. *Com. ACM.*, **26**, 832-843.

André, C. and D. Gaffé (1996). Proving properties of Grafcet with synchronous tools. In: *Proceedings of IMACS/IEEE CESA'96 Multiconference*, 777-782. Lille, France.

Arzén, K.E. (1996). Grafcet for intelligent supervisory control applications, *Automatica*, **30**, 1513-1525.

Aygalinc, P. and J.P. Denat (1993). Validation of functional Grafcet models and performance evaluation of the associated system using Petri nets. *Automatic control production systems*, **27**, 81-93.

Baracos, P. (1995). Automation engineering in north America. In: *Proceedings of 2nd. Int Conf on indust automation*, 3-8, Nancy, France.

Cerans, K., J.C. Godskesen and K.G. Larsen (1993). Timed modal specifications - Theory and tools. In: *Lecture notes in computer science*, **697**, Springer, Berlin.

Chapurlat, V., T. Giaccone, G. Monneret, F. Pereyrol, F. Prunet and D. Simottel (1993). A structured model for specification of discrete events systems: The ACSY model. *Automatic control production systems*, **27**, 65-80.

Charbonnier, F., H. Alla and R. David (1995). The supervised control of discrete event dynamic systems: a new approach. In: *Proceedings of 34th. Conference on Decision and Control.* New-Orleans, USA.

David, R. (1995). Grafcet: A powerful tool for specification of logic controllers. *IEEE Trans Cont Sys Tech*, **3**, 253-268, 1995.

De Loor, P. (1996). *Du TTM/RTTL pour la validation des systèmes commandés par Grafcet.* PhD thesis, University of Reims, France (in French).

De Loor, P., J. Zaytoon and G. Villermain Lecolier (1997). Abstractions and heuristics for the validation of Grafcet controlled systems. *European Journal of Automation*, **31** (to appear).

Duméry, J.J, J.M. Faure, J.P. Frachet, S. Lampérière and F. Louni (1996). A tool for the structured modelling of discrete events systems behaviour : The Hypergrafcet. In: *Proceedings of IMACS/IEEE Multiconference CESA'96*, 519-524. Lille, France.

Frachet, J.P. and G. Colombari (1993). Elements for a semantics of the time in Grafcet and dynamic systems using non-standard analysis. *Automatic control production systems*, **27**, 107-125.

International Electrotechnic Commission (1988). *Preparation of function charts for control systems.* Publication 848.

Kheir, N.A., K.J. Aström, D. Auslander D., K.C. Cheok, G.F. Franklin, M. Masten and M. Rabins (1996). Control systems engineering education, *Automatica*, **32**, 147-166, 1996

Kristensen, C.H., J.H. Andersen and A. Skou (1995). Specification and automated verification of real-time behaviour - a case study. In: *Proceedings of 3rd. IFAC/IFIP Workshop AARTC*, 613-628. Ostend, Belgium.

Lhoste, P., H. Panetto and M. Roesch (1993). Grafcet: from syntax to semantics. *Automatic control production systems*, **27**, 95-105.

Lhoste, P., J.M. Faure, J.J. Lesage and J. Zaytoon (1997). Comportement temporel du Grafcet. *European Journal of Automation*, **31** (to appear, in French).

Marcé, L. and P. Le Parc (1993). Defining the semantics of languages for programmable controllers with synchronous processes. *Control Engineering Practice*, **1**, 79-84.

Marcé, L., D. L'Her and P. Le Parc (1996). Modelling and verification of temporized Grafcet. In: *Proceedings of IMACS/IEEE CESA'96 Multiconference*, 783-788. Lille.

Morel, G. and P. Lhoste (1993). Outline for discrete part manufacturing engineering. In: *Proceedings of 7th. annual European computer conference*, 146-155. Paris.

Ostroff, J.S. and W.M. Wonham (1990). A framework for real-time discrete event control. *IEEE Trans Autom Cont*, **35**, 386-397.

Panetto, H. (1996). Stability property of the Grafcet model with (Max,+) algebra. In: *Proceedings of IMACS/IEEE Multiconference CESA'96*, 771-776. Lille, France.

Peterson, J.L. (1981). *Petri Net Theory and the modelling of systems.* Prentice Hall, London.

Ramadge, P.J. and W.M. Wonham (1989). The control of discrete event systems. *Proceedings of the IEEE*, **77**, 81-98.

Ross, D.T., and K.E. Schoman (1977). Structured Analysis for requirements definition. *IEEE trans. S/W. Eng.*, **33**, 86-95.

Roussel, J.M. and J.J. Lesage (1996). Validation and verification of grafcets using state machine. In: *Proceedings of IMACS/IEEE CESA'96 Multiconference*, 765-770. Lille, France.

Rumbaugh, J., M. Blaha, W. Premerlani, F. Eddy, and W. Lorensen (1991). *Object-oriented modeling and design.* Prentice Hall, Englewood Cliffs, New Jersey.

Union Technique de l'électricité (1993). *Function charts Grafcet: Extension of basic principles.* document UTE C03-191.

Wilczynski, B.K. and B.K. Wallace (1992). OOPS in real-time control applications. In: *Object oriented software for manufacturing systems* (S. Adiga, (Ed.)), 194-229. Chapman & hall, London.

Wonham, W.M. and P.J. Ramadge (1987). On the supremal controllable sublanguage of a given language. *SIAM J. Control Optimization*, **25**, 637-659.

Zaytoon, J. (1996). Specification and design of logic controllers for automated manufacturing systems. *Robotics & CIM*, **12**, 353-366.

Zaytoon, J., P. De Loor and G. Villermain-Lecolier (1995). Using a real-time framework to verify the properties of GRAFCET. In: *Proceedings of 3rd. IFAC/IFIP Workshop AARTC'95*, 233-238. Ostend, Belgium.

Zaytoon, J. and G. Villermain-Lecolier (1997). Two methods for the engineering of manufacturing systems. *Control Engineering Practice*, **5** (to appear).

Zaytoon, J., C. Ndjab and J.M. Roussel (1997a). On the supremal controllable Grafcet of a given Grafcet. In: *Proceedings of 2nd IMACS MATHMOD Conference*. Vienna.

Zaytoon, J., C. Ndjab and V. Carré-Ménétrier (1997b). On the synthesis of Grafcet using the supervisory control theory. In: *Proceedings of IFAC Conference CIS'97*. Belfort, France.

CRITICAL PATH CONCEPT FOR MULTI-TOOL CUTTING PROCESSES OPTIMIZATION

J. Szadkowski

Technical University Lodz Branch in Bielsko-Biala, Department of Manufacturing Technology and Automation, ul. Willowa 2, 43-309 Bielsko-Biala, Poland

Abstract: in a cutting process an ordered set of pairs (an elementary cutting operation on an elementary machined surface, datum surfaces set) can be presented as a directed cycle free graph denoted as a K_L-graph. Some partitions of the K_L-graph lead to an assignment of machining stations to the partition classes of the elementary operations set. Problems similar to the (assembly) line balancing appear when some optimization tasks are formulated. Such tasks differ from the classical balancing problems because times of elementary operations are not constant but they are the functions of the cutting parameters. Since the tools actions may be mapped by a network (PERT methods) balancing and coordination constraints can be formulated as the conditions defining the critical path placement.

Keywords: cutting parameters optimization, line balancing problems.

1. INTRODUCTION

Discrete parts manufacturing technologies for high volumes and low variety require transfer machines and flow lines. Such manufacturing systems contain machining stations capable of using a large number of cutting tools. In designing and selecting new manufacturing technology the optimization objectives are usually formulated:
— cost minimization,
— output maximization.
The planning of machining process includes selection of machining stations, their tooling and set-ups, selection of cutting tools and calculation of cutting parameters, design of jigs and fixtures.
For machining operations the selection of tools and cutting parameters (speed, feed and depth or width of cut) is very important. In this paper a methodology has been proposed for solving the problems arising when a multi-tool machining process ought to be divided into parts assigned to various machining stations. This task has been linked with the cutting parameters optimization. As a base for the solution of the above mentioned problems the process modelling by use of graphs and networks is admitted as well as the concept of the critical path. The process features representation by topological models emphasises the structural approach, on the other hand the critical path introduces some parametrical optimization aspects. The influence of structural factors on parametrical ones belongs to problems concerning the manufacturing processes optimization, which are quite frequent. Furthermore, times of tools replacement ought to be taken into consideration. Balancing problems and the parametric optimization of the cutting processes can be treated as a complex optimization task. The mathematical model presented may be applied for transfer machines, multi-spindle automatics and for flexible automatic lines.

2. TOPOLOGICAL MODEL OF A WORKPIECE MACHINING PROCESS

If a workpiece is to be machined, a set of elementary machining operations (e.g. milling, drilling, boring etc) with proper cutting tools is defined. For elementary operations the precedence relation can be formulated. Surfaces in reference to which the position of other surfaces are defined are called datum surfaces. Datum surfaces define coordinate systems in which the position of the workpiece as a whole and positions of the separate surfaces are given. For surfaces to be machined the relations to reference (datum and location) surfaces are of great importance because they influence the economic and accuracy features of the manufacturing process.

The mapping λ of the set E of the elementary machining operations (corresponding to the workpiece surfaces machining operations) into the set L of the location surfaces is to be defined

$$\lambda : E \rightarrow L \tag{1}$$

where $E = \{e_1, e_2 .... e_i ... e_\alpha\}$, $L = \{L_1, L_2 ... L_j .. L_\beta\}$, e_i - the ith elementary operation, α - the number of elementary operations in the whole machining process, L_j - the jth set of location surfaces needed in the machining process, β- the number of locations in the process. Now pairs of the form (e_i, L_j) can be considered in which L_j is the set of location surfaces for the executing of the ith elementary operation. On the set of pairs (e_i, L_j) the precedence relation $\boldsymbol{R}$ is defined as a strict order relation (antireflexive and transitive, hence asymmetric). Thus the relation $\boldsymbol{R}$ possesses the following properties:

1) $(e_i, L_j)\ \boldsymbol{R}\ (e_i, L_j)$ does not hold for any (e,L),
2) if $(e_i, L_j)\ \boldsymbol{R}\ (e_k, L_l)$ and $(e_k, L_l)\ \boldsymbol{R}\ (e_m, L_n)$, then $(e_i, L_j)\ \boldsymbol{R}\ (e_m, L_n)$ holds,
3) if $(e_i, L_j)\ \boldsymbol{R}\ (e_k, L_l)$ holds, then $(e_k, L_l)\ \boldsymbol{R}\ (e_i, L_j)$ is impossible.

For the relation $\boldsymbol{R}$ a directed graph without circuits can be used as a model. An example is given in Fig. 1. The nodes of the graph represent the pairs (e_i, L_j), the arcs define the precedence constraints. The nodes are grouped in coherent classes in accord with the corresponding symbols L_j, (L_j-classes), such a graph is denoted as K_L -graph.

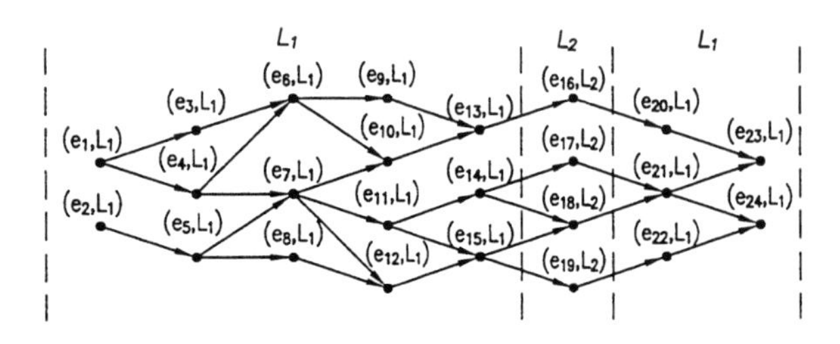

Fig.1. The K_L-graph for a workpiece: $E = \{e_1,...e_{24}\}$.

3. ASSIGNMENT OF ELEMENTARY OPERATIONS TO MACHINING STATIONS

The assignment of elementary operations to machining stations has to correspond with the partition of the K_L-graph on L_j-classes (it is the problem of the preserving of the location surfaces in the machining space of a machining station) but elementary operations from any of L_j-classes may bemachined in one or more than one machining substations belonging to the same transfer machine. To obtain more detailed partitions of the K_L-graph (containing the graph nodes S_{jn} (i.e. elementary operations) assigned to corresponding machining substations, $n = 1,2....\gamma$; γ - number of substations) one ought to observe the two following rules:

R1 — if two elementary operations e_i and e_k have to be assigned to the same machining substation all elementary operations mapped by the K_L-graph nodes belonging to all paths linking e_i and e_k must be included in the set of elementary operations appointed to this substation,

R2 — after the application of the rule 1 and after the elementary operations grouping for substations the elementary operations precedence relation must be preserved for elementary operations groups.

Fig. 2 contains the first L_1-partition class of the K_L-graph from the Fig. 1.

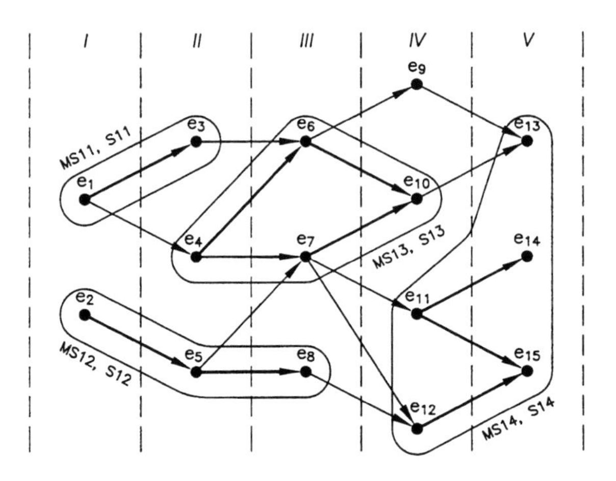

Fig. 2. Graph L_1 - the first partition class of the K_L-graph from the Fig.1 (I,...,V are niveaux obtained after the topological sorting).

In the Fig. 2 the grouping of elementary operations $e_1,...,e_{15}$ in subgroups (s_{11}, s_{12}, s_{13}, s_{14}) for the machining substations MS11, MS12, MS13, MS14 is shown. E.g. if the machining process necessitates it for operations e_4 and e_{10} to be executed in the same machining substation, operations e_6 and e_7 ought to be assigned to the same substation.

Adequate precedence relation for the elementary operations groups is presented in Fig. 3. From Fig. 3 an information can be deduced that a transfer machine using L_1 set of the location surfaces and having four machining substations may be designed in two substantial forms: with MS11, MS12, MS13, MS14 or MS12, MS11, MS13, MS14 lay-out of the substations.

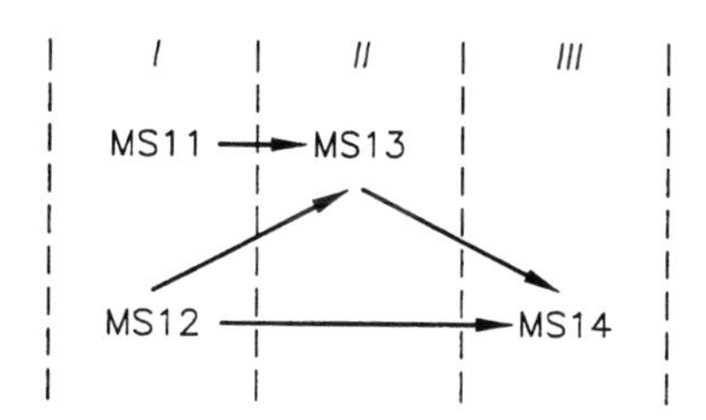

Fig. 3. Precedence relation for elementary operations groups.

Graphs in Fig. 2 and Fig. 3 are drawn as topologically sorted (devided on niveaux [2]). Such a presentation is more descriptive and provides possibility to further information concerning the structure of the machining process. *Note that the K_L-graph and its partitions are workpiece-oriented.*

4. TOPOLOGICAL MODEL OF TOOL ACTIONS

The problem of grouping elementary operations and assigning them to the machining (sub)stations would be a line balancing problem, if the times of elementary operations are not admitted as given values. Actually they are functions of cutting parameters being the decision variables. So the problem may be stated as a complex optimization task in which structural and parametric factors influence values of admitted optimization criteria. In order to explain the core of the considered mathematical model a question can be formulated: *what structure of a transfer line or a transfer machine is needed to enable the parallel executing of all α elementary operations*? A following statement brings the answer: the number of machining (sub)stations necessary for obtaining the simultaneous actions of all cutting tools executing the set of elementary operations belonging to a L_j-class of the K_L-graph partitions is equal to or greater than the number of the L_j-graph niveaux. So for the process presented by the graph from Fig. 2 a transfer machine in which all the 15 tools cut in parallel ought to include five or more machining substations. The minimal number of stations, i.e equal to the number of niveaux, can be designed only if collisions of the tools do not threaten.

The simultaneous cutting of all tools represents an interesting but rather special case; so further considerations take into account a more general case in which tools in any substation cut in a mixed mode (*sequential-parallel cutting*).

As shown above, the formulation of the mathematical model for the cutting process optimization includes some formalized steps; nevertheless the main tissue isof heuristic nature. Such is the transition from the L_j-graph to a graph (and then a network) mapping the cutting tools actions. The network is *machine-tool-oriented*. The basis for the network is laid by a graph L_j^* which can be treated as a form of the transformed L_j-graph.

The way in which the transformation is proceeded contains following steps:

1) For every elementary operations set S_{jn} assigned to corresponding machining substation MS_{jn} (n = 1,2,...,γ; γ - the number of machining substations) a subgraph SG_{jn} of the graph L_j is defined in which arcs mapping the precedence relation $\boldsymbol{R}$ between nodes $e_u \in S_{jn}$ are preserved,

2) The graphs SG_{jn} (for all n = 1,2...,γ) are put together as the parallel composed subgraphs of the graph L_j^* which contains additionally two nodes (O' or O") enabling to get a coherent structure of the graph L_j^* -Fig. 4.

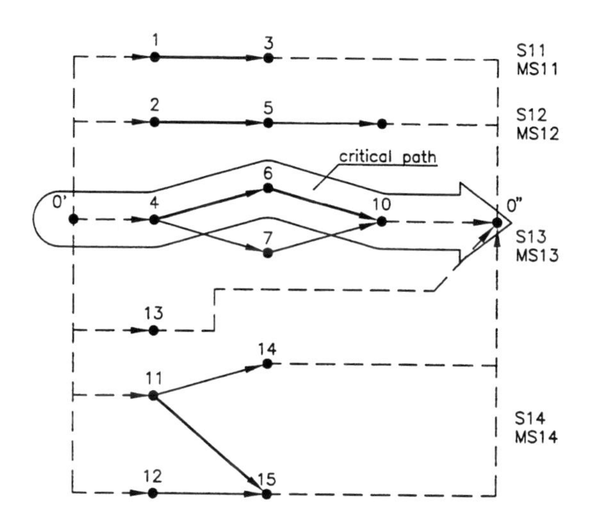

Fig. 4. Graph L_1^* for the graph L_1 from the Fig. 2.

5. CRITICAL PATH CONSTRAINTS FOR THE MACHINE-TOOL-ORIENTED TOOL ACTIONS GRAPH

The L_j^* graph offers the basis for a PERT-network in which machining times (as functions of cutting parameters) are assigned to the nodes (for the nodes O' and O" these values equal zero). *The main feature of the following optimization model consists in assuming that the critical path placement can be admitted as a decision of the designer.* Such decision ought to be supported by „critical path constraints" expressing the necessity for all other paths in the network to be shorter than the critical one. For instance, if in the network built on the basis of the graph L_1^* - Fig.4 - the critical path contains the nodes O'- 4 - 6 - 10 - 0" then

$$m_1 + m_3 \leq m_4 + m_6 + m_{10} \quad (2)$$

$$m_2 + m_5 + m_8 \leq m_4 + m_6 + m_{10} \quad (3)$$

$$m_7 \leq m_6 \quad (4)$$

$$m_{13} \leq m_4 + m_6 + m_{10} \quad (5)$$

$$m_{11} + m_{14} \leq m_4 + m_6 + m_{10} \quad (6)$$

$$m_{11} + m_{15} \leq m_4 + m_6 + m_{10} \quad (7)$$

$$m_{12} + m_{15} \leq m_4 + m_6 + m_{10} \quad (8)$$

where

$$m_i = f_i(v_i, f_i, a_i) \quad (9)$$

and v_i denotes the speed, f_i - the feed and a_i - the depth of cut for the ith tool (i = 1,2...,α). The functions (9) depend on the cutting tool design and the cutting method (e.g. turning, milling, etc) and are defined in handbooks - e.g. [1]. *Tools in multi-tools machining systems can be replaced individually or in blocks* [4], [5], [6]. A block contains tools replaced simultaneously, a preliminary study for the block replacement strategies optimization has been done in [5], [6]. The advantages of the block replacement in comparison with the individual one ought to be carefully checked and in the comparative analysis only the optimal block replacement strategies may be taken into consideration. To attain such optimal solution the base in the form of the optimal strategies for the individual replacement has to be built.

For the individual tool replacement case the production time and the production cost components (per one workpiece) which depend on cutting parameters can be expressed as follows

$$t = \sum_{i \in CP} m_i + \sum_{i=1}^{\alpha} m_i e_i r_i^{-1} \quad (10)$$

$$k = x \cdot t + \sum_{i=1}^{\alpha} m_i y_i r_i^{-1} \quad (11)$$

where: CP - the set of subscripts of the graph L_j^* nodes belonging to the critical path, e_i - the tool replacement time for the ith tool, r_i - tool life function for the ith tool, x - the machining station cost per time unit, y_i - the tool cost per tool life period for the ith tool. As tool life functions mostly Taylor

$$r_i = \frac{C_i}{v_i^{s_i} f_i^{u_i} a_i^{e_i}} \quad (12)$$

or Temcin

$$r_i = \frac{T_{G_i} C_i}{T_{G_i} v_i^{s_i} + C_i} \quad (13)$$

equations are used. Here C_i and T_i are constant values and s_i, u_i, e_i - exponents.

As a criterion for the two-criterion problems the function

$$W = \chi t + (1 - \chi) k \quad (14)$$

can be defined in which $0 \leq \chi \leq 1$. In such a way (when using the function (14) as the criterion) the Pareto optimum for a two-criterion problem can be obtained and the cost-time function (as put it Ravignani [3]) for the multi tool operation can be established. Here only a variation of the coefficient χ from 0 to 1 has to be performed.

To the criterion function (10) or (11) or (14) the critical path constraints like (2) - (8) must be attached as well as other typical constraints (for cutting forces, torques, powers, surfaces roughness e.t.c). Generally the computer application leads to the numerical values of the optimal cutting parameters which provide times for all machining (sub)stations.

For the block replacement case α tools are assigned to u blocks.

A tool replacement schedule (TRS) is determined by u integers: $(h_1, h_2, ..., h_k, ..., h_u) = \{h_k\}$ for which the integer a = least common multiple $\{h_k\}$ is computed. Then the vector of integers $\mathbf{A} = \{1, 2, ..., a_l, ..., a\}$ is defined and the matrix $\mathbf{H} = [h_{kl}]$ where:

$$h_{kl} = 1 \text{ if } a_l / h_k \text{ integer, } h_{kl} = 0 \text{ else} \quad (15)$$

The matrix **H** represents the structure of the TRS: columns of the matrix show which blocks are replaced in the subsequent phases of the schedule. Metric properties of the TRS are characterized by a number h - the quantitative parameter of the TRS expressed in time units or in numbers of workpieces machined so that the number of workpieces machined by the tools belonging to the kth block as long as the block is not replaced equals h_k workpieces. An example of the matrix **H** is given in Fig. 7.

The replacement time for the ith tool

$$e_i = {}^{\bullet}e_i + {}^{\circ}e_i, \quad i = 1, 2 ... \alpha \quad (16)$$

where ${}^{\bullet}e_i$ comprises preparatory activities and ${}^{\circ}e_i$ proper tool replacement activities. The assignment matrix $\mathbf{P} = [p_{ki}]$ is defined as

$p_{ki} = 1$ if the ith tool is assigned to the kth block

$p_{ki} = 0$ else

The proper replacement time for the kth block:

$$^{\circ}c_k(s) = f(s) \sum_{i=1}^{\alpha} p_{ki} \, ^{\circ}e_i \tag{17}$$

here the coefficient $0 < f(s) \leq 1$ describes the possibility of the replacement time diminishing by the application of s „servers" (workers or robots). The numbers $^{\circ}c_k(s)$, $k = 1,2...,u$ are the components of the row vector $^{\circ}C(s) = [^{\circ}c_k(s)]$, which contains the replacement times for all the phases of the TRS. Then the vector **C** of full replacement times for all the replacement phases can be formulated:

$$C = {}^{\bullet}e \, {}^{\bullet}H + {}^{\circ}C(s) H \tag{18}$$

$${}^{\bullet}H = [{}^{\bullet}h_l], \quad l = 1,2,...,a \tag{19}$$

$${}^{\bullet}h_l = \bigcup_l h_{kl} \tag{20}$$

where the last summation is done in the sense of the Boolean algebra. During the full replacement schedule time the ah workpieces are machined and for the tools replacement the time e_a is consumed

$$e_a = [C, 1] \tag{21}$$

where [...] denotes the scalar product of the vector **C** and the vector 1 whose all components equal one. The mean replacement time per one machined workpiece

$$e_1 = e_a / ah \tag{22}$$

For defining the mean tool cost per one machined workpiece y_1 the diagonal matrix $\hat{\mathbf{Y}}$ of tool costs y and the diagonal matrix $\hat{\mathbf{H}}_{-1}$ of inverse values h_k may be introduced

$$\hat{Y} = [y_i], \quad i = 1,2,...,\alpha \tag{23}$$

$$\hat{H}_{-1} = [1/h_k], \quad k = 1,2...\alpha \tag{24}$$

Now the mean tool cost is given by the equation

$$y_1 = h^{-1} \sum_{i=1}^{\alpha} \sum_{k=1}^{u} d_{ik} \tag{25}$$

$$[d_{ik}] = \hat{Y} \, P^T \, \hat{H}_{-1} \tag{26}$$

where $\mathbf{P}^{\mathrm{T}}$ is the transpose of the matrix **P**. The criteria have now the form

$$t = q + e_1 \tag{27}$$

$$k = xt + y_1 \tag{28}$$

and w is defined by (14).The element q in (28) is the „critical" component depending on the structure of the PERT-network in accord with former considerations. The criteria are assisted by the essential constraints

$$\hat{D} \, \hat{P} \geq h \, P^T \hat{H} \tag{29}$$

where $\hat{\mathbf{D}}$ is the diagonal matrix $\alpha \times \alpha$ of the tool life functions and $\hat{\mathbf{H}}$ is the matrix of values h_k. Other constraints are typical.

The analysis of the mathematical models (2) - (14) and (15) - (29) reveals that they are strongly influenced by structural factors. The structural problems connected with the tool blocks replacement strategies are:

— the tool blocks number u determination,

— the assignment of the tools to the blocks.

They can be solved by methods presented in [6].

Now an example illustrates the main features of the proposed method of the cutting process optimization. An automobile part is to be machined on a multi-spindle turning automatic. Twelve elementary operations: e_1, e_2,...e_i....e_{12}, (α = 12) ought to be executed and only one L-graph is needed (β = 1) - Fig. 5.

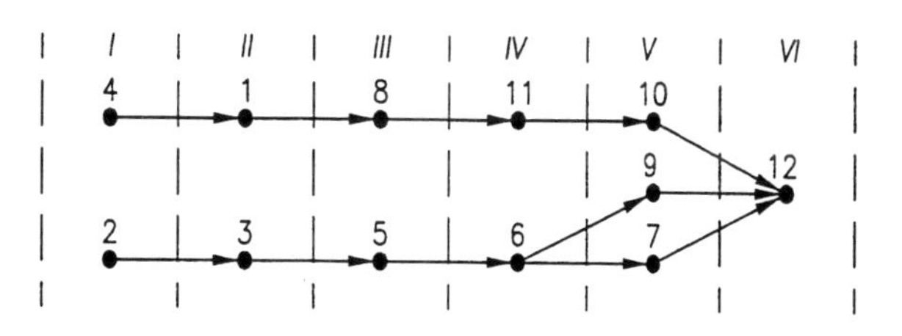

Fig. 5. L-graph for the machining of an automobile part.

Elementary operations are enlisted in Table 1 as well as other data for cutting parameters optimization.

The number of the L-graph niveaux equals here 6, so for the simultaneous cutting of all tools the six machining substations are needed and the L^*- graph assumes the very simple form - Fig. 6 - when an six-spindle automatic AS 67 (Gildemeister) is used (six machining substations).

Here the critical path constraints are of simple character tool

$$m_i <= m_{12}, \; i = 1,2,...,11$$

if the 12 th tool is admitted as the critical one. The existing software enables the computing of optimal cutting parameters (using equations (10), (11), (13), (14) and data from the data base) as well as coefficients contained in the 6th column of the

Table 1. These coefficients are the basis for the computing of the integers h_k (the number of the tool blocks u was admitted as equal 4) - the 7th column, as well as for the matrix **P** defining (the 8th column) - using the dynamic programming method presented in [6]. The optimal TRS is given in Fig. 7. It determines the values of all the cutting parameters.

Table 1. Elementary operations, tools, initial data and selected results for optimal cutting parameters computing

Operation	Cutting tool	Replacement time		
		e_i	$^{\bullet}e_i$	$^{\circ}e_i$
1	2	3	4	5
1	Twist drill ∅18	10	3	7
2	Form tool	5	3	2
3	Form tool	10	3	7
4	Twist drill ∅17.75	10	3	7
5	Form tool	10	3	7
6	Form tool	15	3	12
7	Form tool	10	3	7
8	Reamer	10	3	7
9	Form tool	18	3	15
10	Grooving tool	20	3	17
11	Reamer	20	3	17
12	Cut off tool	10	3	17

Operation	$(r_k/m_k)/(r_k/m_k)_{min}$	h_k	Transpose of the matrix **P**			
			h_1=1	h_2=10	h_3=15	h_4=30
1	6	7	8			
1	20	15	0	0	1	0
2	28	30	0	0	0	1
3	10	10	0	1	0	0
4	15	15	0	0	1	0
5	1	1	1	0	0	0
6	15	15	0	0	1	0
7	38	30	0	0	0	1
8	16	15	0	0	1	0
9	38	30	0	0	0	1
10	28	30	0	0	0	1
11	11	10	0	1	0	0
12	13	10	0	1	0	0

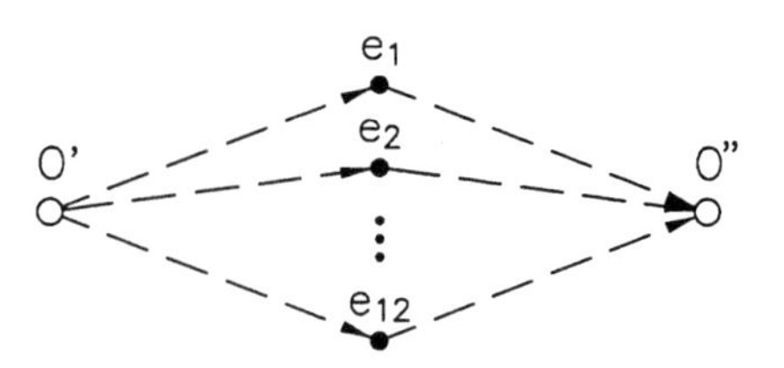

Fig. 6. L^*-graph for the machining of an automobile part.

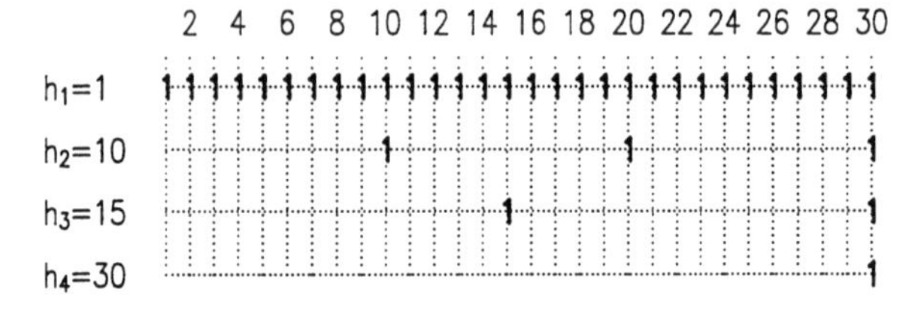

Fig. 7. The optimal tool replacement schedule (matrix **H**).

6. CONCLUDING REMARKS

Critical path concept applied to a complex optimization problem of multi-tool cutting provides the simple and objective approach giving as a solution not only optimal cutting parameters but also optimal balancing of elementary operations set.

REFERENCES

Armarego, E.J.A., B.H. Brown, B.H. (1969): *The Machining of Metals*. Prentice Hall Inc.

Deo, N. (1974): *Graph Theory with Applications to Engineering and Computer Science*. Prentice-Hall, Inc, N. Jersey.

Ravignani, G.L. (1974): Machining for a Fixed Demand. *Int. Journ. on Prod. Res. Nr 3*, p. 86-90.

Szadkowski, J. (1981): Projektowanie operacji wielonarzędziowych - warunki drogi krytycznej i polioptymalizacja warunków skrawania. *Postępy Technologii Maszyn i Urządzeń Nr 3-4*, p. 3-13.

Szadkowski, J. (1982): Projektowanie operacji wielonarzędziowych - polioptymalizacja warunków skrawania przy wymianie narzędzi blokami. *Postępy Technologii Maszyn i Urządzeń Nr 4*, p. 3-12.

Szadkowski, J. (1996): The Tool Blocks Replacement Strategies Optimization for Multi-Tool Machining. *Proc. of the IX th Int. Conf. on Tools*, Miskolc, p. 801-806.

MEASUREMENT OF THREE-DIMENSIONAL SHAPE OF INGOT AT FORGING PRESS MACHINE

Sun X.*, T. Ishimatsu*, T. Nomura*, D. Fujito*, K. Katsuki*, H. Imaizumi, M. Hirono****

**Mechanical Systems Engineering, Nagasaki University, Japan*
***Mitsubishi Nagasaki Machinery Mfg. Co. Ltd., Japan*

Abstract: In this paper, two methods are presented to measure 3D shape of red-hot ingot by visions. In the first method, contour analysis is used and a rather simple measuring system is configured. In the second method, the slit-ray projection is employed and the problem of fluctuation caused by gravity of target ingot is strongly considered. A technique of curve fitting is developed for reconstruction of cross-sections. Experiments show that both of contour analysis and slit-ray projection are applicable.

Keyword: red-hot ingot, 3D shape measurement, contour analysis, slit-ray projection, curve fitting

1 INTRODUCTION

3D vision is playing an increasingly important role in our world. It is such a versatile technique that it has been found applications in a very wide range of scientific disciplines and engineering. In the forging factory, shape measuring is one of the major reasons causing quality problem of forged products. If the shape of red-hot ingot can be measured during pressing and the results of measurement can be used to control the pressing procedure, the forging process can be fully automatic. However, the measuring work is still doing manually and since the radiant light and heat from red-hot ingot, the work is really hard and dangerous. The extremely inhospitable operating conditions make the measurement can only obtain diameters instead of the shape of red-hot ingot. Therefore, an automatic 3D shape measuring system for improving measuring accuracy and speed is highly desired in this field.

One method (Furuhasi, et al., 1991) was reported to measure an ingot using stereo vision technique. In this technique, finding the matching points between two images captured by two cameras is essential (Lim, 1988). However, because of the surface conditions of the red-hot ingot, it is difficult to find the reliable matching points. In this paper, two techniques are introduced for the 3D shape measurement of red-hot ingot. In the first one, contour analysis is used. It is described in section 2. The slit-ray projection is employed in the second one, and the detail is involved in section 3.

2 CONTOUR ANALYSIS

In this section, the method based on contour analysis is applied to measure 3D shape of cylindrical red-hot ingot rotated by a manipulator. During the rotation, contours information of the red-hot ingot are extracted from images captured by a CCD camera. By analyzing variations of the contour data during one rotation, cross-sectional data of the red-hot ingot are estimated and 3D shape of red-hot ingot is reconstructed.

Using contour analysis in the case of red-hot ingot, there exist at least three advantages: 1) The contour analysis does not use special projecting light source. Therefore, the affection of radiant light to the projecting light on the red-hot ingot can be avoided. 2) Since the red-hot ingot is luminous, the contours can be easily extracted. This is the reason the contour analysis is applicable in 3D measurement of red-hot ingot. 3) The system setting can be the most simple. In this method, one camera is enough to obtain sufficient contour information. Furthermore, simple set system is easy and economic to be put into practice in future. Figure 1 shows the configuration of measuring system.

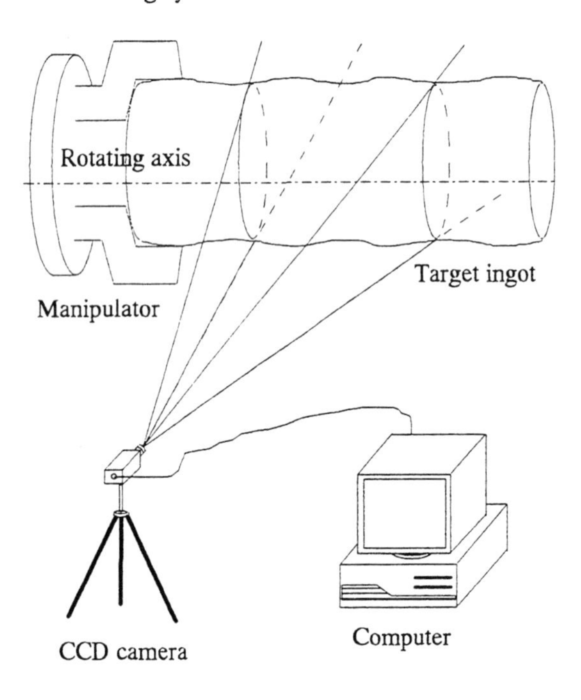

Fig. 1 Measuring system using contour analysis

2.1 Estimation of cross-sections

In order to reconstruct the global 3D profile of the target ingot, cross-sections of the target ingot at a set of planes S_d (d=1,2,...) are measured. Those planes are vertical to the rotating axis and placed with constant interval. Position of the plane S_d is specified by the position vector $\boldsymbol{v}_d$ as shown in Fig. 2, which starts from reference point O_c and goes along the rotating axis. The position vector $\boldsymbol{r}$ that used to determine O_c and the direction of vector $\boldsymbol{v}_d$ are obtained from a calibration technique (Sun, et al., 1995).

On the film plane, the projected contour is composed of many pixels. The position vectors of these pixels are denoted as $\boldsymbol{q}_i$ (i=1,2,3...). Vector $\boldsymbol{q}_i$ is tangent to the surface of the target ingot and ends at the film plane. The crossing point of the extrapolating line of vector $\boldsymbol{q}_i$ and the plane S_d is denoted by H_i. H_i is determined by position vector $\boldsymbol{p}_i$,

$$\boldsymbol{p}_i = \frac{|\boldsymbol{v}_d|^2}{\boldsymbol{q}_i \cdot \boldsymbol{v}_d} \boldsymbol{q}_i . \tag{1}$$

Connecting all points H_i (i=1,2...), a curve $M(\theta)$ on the plane S_d can be obtained. It should be noticed that cross-section of the target ingot on the plane S_d is enclosed by a set of curves $M(\theta_j)$ (j=1,2,...), which are tangent to the surface of target ingot. The set of curves $M(\theta_j)$ are determined at various rotating angles θ_j(j=1,2,...). Therefore, if the curves $M(\theta_j)$ (j=1,2,...) are connected by H_i (i=1,2,3...) at various rotating angles θ_j(j=1,2,...), the cross-section on the plane S_d can be estimated as a region enclosed by the set of curves.

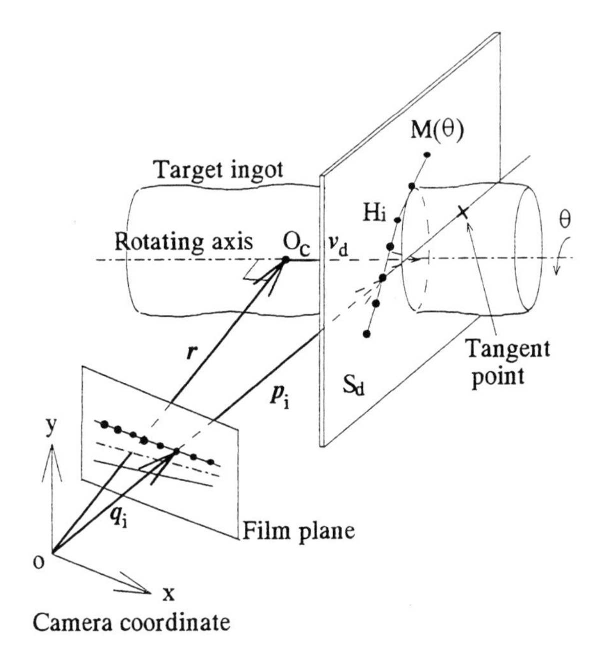

Fig. 2 Meaning of projected contour on film plane

A graphical approach is used to estimate the cross-section from these curves. Suppose the target ingot is rotated with step angle $\Delta\theta$ and a set of curves $M(\theta_j)$ (j=1,2,...) are obtained as shown in Fig. 3.

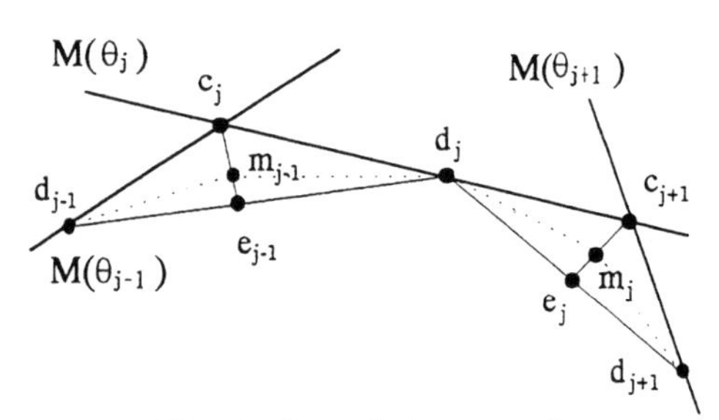

Fig. 3 Graphical approach

The approach is introduced as follows.
Step 1: Determine the cross-points c_j of every two curves $M(\theta_j)$ and $M(\theta_{j-1})$ (j=1,2,...).
Step 2: Determine the midpoints d_j between c_j and c_{j+1}.

Step 3: Determine the midpoints e_j between d_j and d_{j+1}.
Step 4: Determine the points m_j to divide the lines connecting e_j and c_{j+1} with the ratio k calculated by

$$k = \frac{\cos(\Delta\theta/2)[(\Delta\theta/2) - \sin(\Delta\theta/2)\cos(\Delta\theta/2)]}{\sin^3(\Delta\theta/2)}. \tag{2}$$

Suppose a circular cross-section is measured. Using this ratio k, the area of estimated polygon is equal to the circular one.
Step 5: The desired cross-section is estimated as a polygon obtained by connecting points d_j and $m_j (j=1,2...)$.
Estimating the cross-section on every plane S_d (d=1,2...), the global 3D profile of the target ingot is reconstructed.

2.2 Experiments of contour analysis

In order to test the method, two experiments were carried out at various temperatures.

Experiment 1: In the first experiment, a plastic bottle with irregular shape as target ingot as shown in Fig. 4 was measured at room temperature. The height of the bottle is 260.0mm.

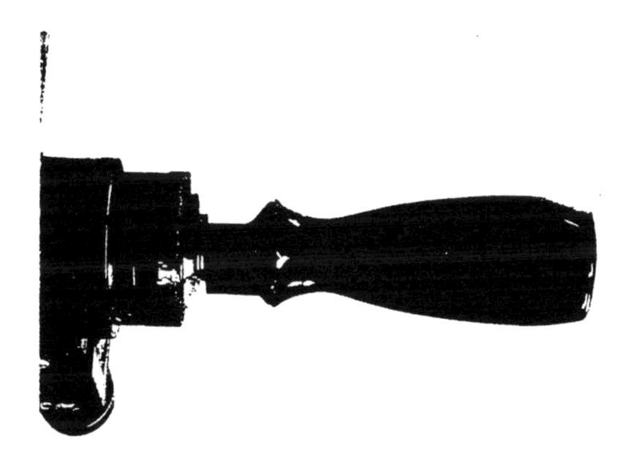

Fig. 4 Target ingot

The cross-sections were measured on 46 planes from the top to the bottom with 4.0mm interval. The computer obtained the contour data with step angle $2\pi/36$ during one rotation. The 3D profile of the bottle was reconstructed on the computer display as shown in Fig. 5. As an typical example, Fig. 6 shows a cross-section formed by a set of curves. Those data were obtained at position that is 184.0mm apart from the top of the bottle. The distances from the rotating axis to surface points at various step angles was measured to evaluate the accuracy of the algorithm. The variation of the distances was from 27.5mm to 48.3mm. The maximum difference of the measured data was 0.3mm.

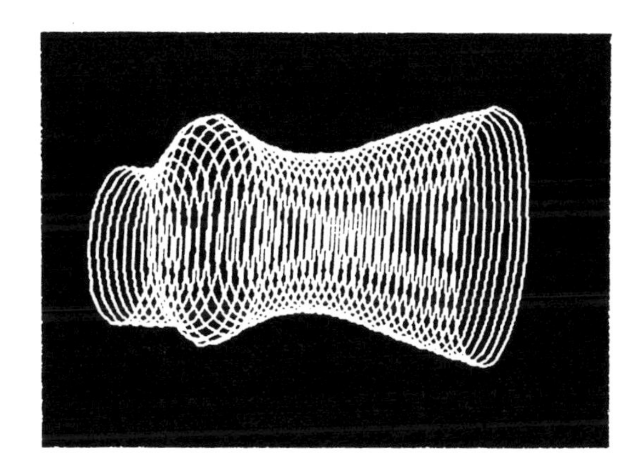

Fig. 5 Reconstructed 3D profile

Fig. 6 Cross-section formed by a set of curves

Experiment 2: For evaluating the applicability of contour analysis to red-hot ingots, the diameter of a cylindrical red-hot ingot with radius R=130.0mm was measured by the contour analysis at various temperature in the second experiment. Results are shown in Table 1. The results show the measuring accuracy is deteriorated as the increase of temperature. Main reason of this deterioration might come from the fluctuation of image caused by emitted light and heat from the red-hot ingot.

Table 1 Results of experiment 2

Temperature	Average diameter	Range of error
1100 ℃	131.4mm	-1.23%~1.85%
1000 ℃	130.5mm	-1.32%~1.79%
800 ℃	130.3mm	-0.86%~1.48%
Room	130.1mm	-0.61%~0.55%

3 SLIT-RAY PROJECTION

In previous section , a method of red-hot ingot using contour analysis is demonstrated. It must be noticed that the contour analysis can not give the 3D shape in detail such as concave region on the surface of the target ingots(Zheng, 1994). This drawback can be overcome by the technique using slit-ray projection in the case of measuring red-hot ingot. In forging manufacturing, a longer ingot often need to be measured. Causing by gravity of the longer red-hot ingot, the fluctuation occurs while it is rotated during the measurement. Because of the fluctuation, rotating axis becomes uncertainty. In this section, a measuring method using slit-ray projection for measuring red-hot ingot is proposed and how to deal with the problem of uncertain rotating axis of the red-hot ingot during the measurement is strongly considered.

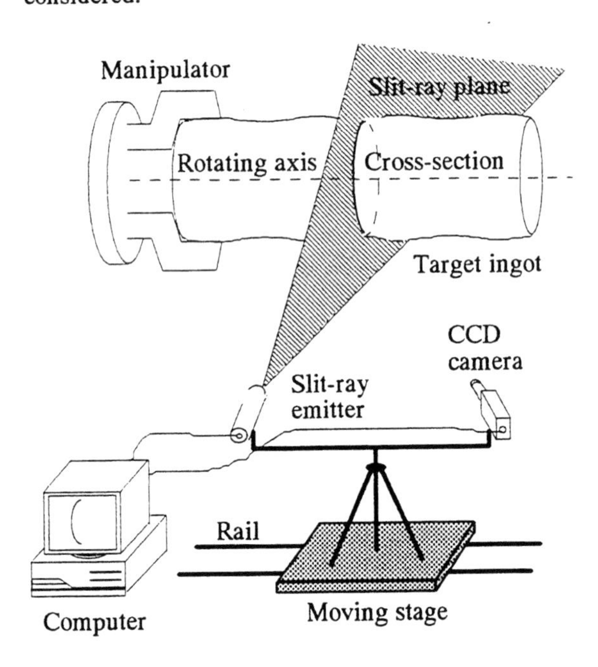

Fig. 7 System using slit-ray projection

Figure 7 shows the configuration of measuring system for completing the 3D measurement using slit-ray projection. In order to obtain 3D data, a cross-section is measured. During the measurement, red-hot ingot is rotated by a manipulator insuring all parts of a cross-section can viewed by a CCD camera. Based on the slit-ray projection method, a slit-ray is emitted onto the red-hot ingot. Images of the target ingot with projected slit-ray are captured by the CCD camera in one rotation. The relative position and posture of the slit-ray emitter and CCD camera are calibrated in advance to the measurement (Sun, et al., 1994). The slit-ray emitter and the CCD camera are fixed on a moving stage. In order to obtain the cross-section at different positions, the moving stage can be shifted along the target ingot. From the images of the red-hot ingot, a set of partial curves of cross-section are extracted by image processing techniques. The method used here finds overlapping portion between every two neighboring partial curves of the cross-section and connects them each other in the meaning of least squared difference. Connecting every partial curve obtained at various rotating angles, the cross-section can be reconstructed. After measuring every cross-section in various positions along the red-hot ingot, 3D shape is achieved.

3.1 Reconstruction of cross-section

Figure 8(a) shows a set of partial curves of the cross-section extracted from images in one rotation with 30 degree intervals. Because of the fluctuation of rotating axis, extracted partial curves of the cross-section need to be reconstructed. A curve fitting algorithm in the meaning of least squared difference is introduced to accomplish the reconstruction. In order to simplify the explanation about how to connect the partial curves of cross-section, two neighboring curves A and B are considered on coordinate system (X, Y) settled on the slit-ray plane. Suppose the difference of rotating angle of the two curves is $\Delta\theta$, curve A and curve B are rotated in different directions for compensating the affection of the rotating angle θ as shown in Fig. 8(b).

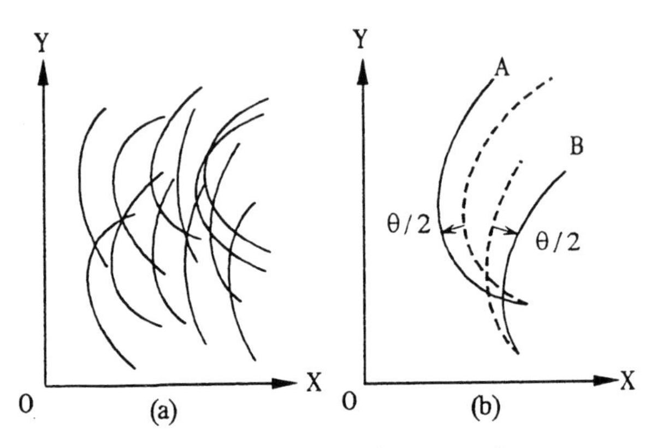

Fig. 8 Partial curves of cross-section

Since the overlapping portion on the two curves should be equal, curve B is shifted in the X and Y directions to fit curve A at optimum fitting position where difference of the overlapping portion becomes minimum in the meaning of least squared difference. In order to simplify the calculation to obtain the optimum fitting position, curve A and curve B are represented by a set of discrete points $i = 0,1 \cdots M \cdots N \cdots$ as shown in Fig. 9(a), whose coordinates are denoted by (X_{Ai}, ih) and (X_{Bi}, ih), where h is the interval of two points. Suppose the two curves overlap currently from point M to N, the number of overlapping points of the two curves is N-M+1. Firstly, the optimum shift $X_j (j = 1,2,\cdots)$ of curve B in X direction is calculated. The least squared difference between the two curves is evaluated by

$$J(\Delta X) = \frac{1}{N-M+1}\left[\sum_{i=M}^{N}(X_{Bi} - X_{Ai} + \Delta X)^2\right] \quad (3)$$

where ΔX is defined as shift of curve B in the X direction. $J(\Delta X)$ is minimized by

$$X_j = \frac{1}{N-M+1}\left[\sum_{i=M}^{N}(X_{Bi} - X_{Ai})\right] \quad (4)$$

and minimum $J_j(j=1,2,\cdots)$ is

$$J_j = \frac{1}{N-M+1}\left[\sum_{i=M}^{N}(X_{Bi}-X_{Ai})^2\right] - \frac{1}{(N-M+1)^2}\left[\sum_{i=M}^{N}(X_{Bi}-X_{Ai})\right]^2 \quad (5)$$

This means when the curve B is shifted with $X_j(j=1,2,\cdots)$ in X direction, the minimum $J_j(j=1,2\cdots)$ can be obtained under the constraint that the curve B is shifted with the displacement ih in the Y directionas shown in Fig. 9(b).

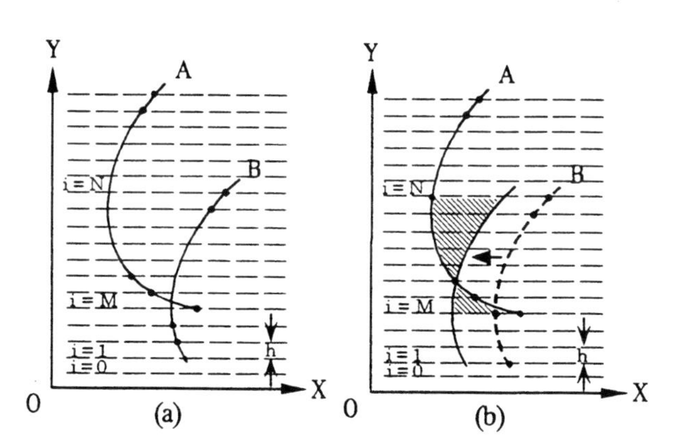

Fig. 9 Curves fitting

Secondly, considering the sampling effect of curve A and curve B, the optimum fitting position is obtained by interpolating the $J_j(j=1,2\cdots)$ and $X_j(j=1,2,\ldots)$. As shown in Fig. 10, suppose J_s is the smallest value among J_j and J_j are approximated by a quadratic function of Y determined by the three values J_{s-1}, J_s and J_{s+1}, the following equation is satisfied

$$J_i(Y) = \frac{(Y-sh)(Y-(s+1)h)}{2h^2}J_{s-1} - \frac{(Y-(s-1)h)(Y-(s+1)h)}{h^2}J_s + \frac{(Y-sh)(Y-(s-1)h)}{2h^2}J_{s+1}. \quad (6)$$

The value Y^o to minimize $J_j(Y)$ can be computed by

$$Y^o = \frac{(2s+1)hJ_{s-1} + 4shJ_s + (2s-1)hJ_{s+1}}{2(J_{s-1}+J_s+J_{s+1})} \quad (7)$$

Similarly, X_j are approximated by a quadratic function of Y determined by the three values X_{s-1}, X_s and X_{s+1} as shown in Fig. 11. X^o can be computed by

$$X^o = \frac{(Y^o-sh)(Y^o-(s+1)h)}{2h^2}X_{s-1} - \frac{(Y^o-(s-1)h)(Y^o-(s+1)h)}{h^2}X_s + \frac{(Y^o-sh)(Y^o-(s-1)h)}{2h^2}X_{s+1}. \quad (8)$$

Applying the above method to the all neighboring curves, all the curves can be connected as a closed one on the coordinate system (X, Y).

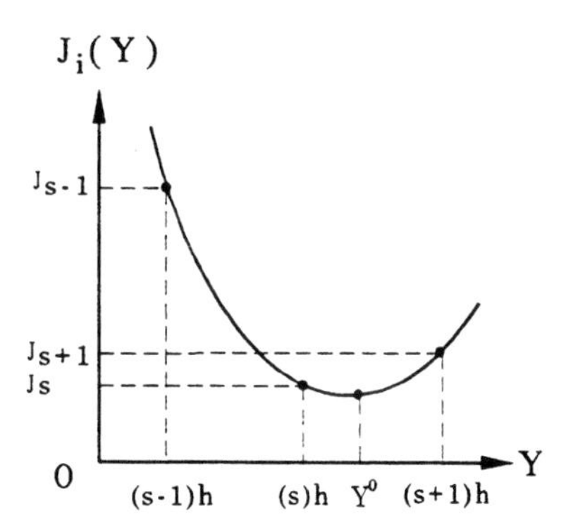

Fig. 10 Interpolation of J

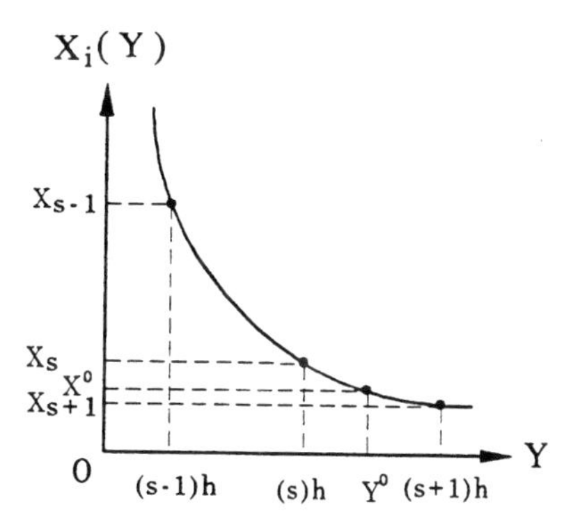

Fig. 11 Interpolation of X

3.2 Experiments of slit-ray projection

Two experiments were carried out in order to test this method. The camera and the slit-ray emitter were placed 2000.0mm away from the target ingots. The distance between the slit-ray emitter and the camera was 1000.0mm. The rotating axis was located arbitrarily apart from the gravity center of the target ingot. The partial curves of cross-sections were obtained 12 times during one rotation, that means the step angle was 30 degree.

Experiment 1: In the first experiment, the cross-section of an elliptic cylinder was measured at room temperature. Major axis of the ellipse was 240.0mm and minor axis was 186.0mm. Fig. 12 shows an example of reconstructed cross-sections on computer display. The maximum measuring error range was from -0.5% to +0.5% in 8 measurements.

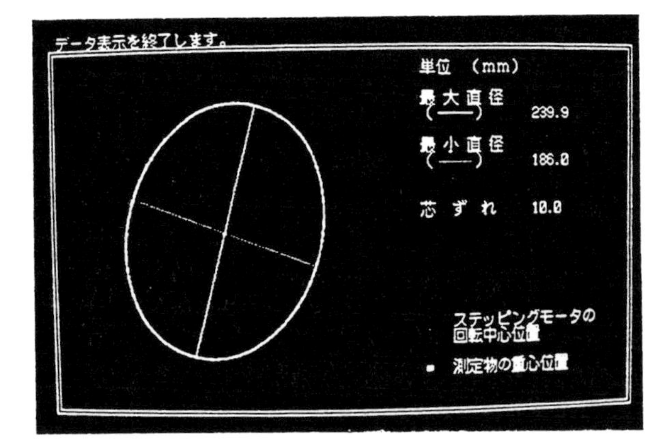

Fig. 12 Reconstructed cross-section

Experiment 2: In the second experiment, the target ingot was an iron cylinder whose diameter was 113.0mm. The cross-sections were measured in red-hot condition. Figures 13(a) and (b) show the error range of experimental results at temperature 800 °C and 1000 °C respectively. Comparison of Figs. 13(a) and (b) shows that the accuracy was deteriorated because of the increase of radiated light.

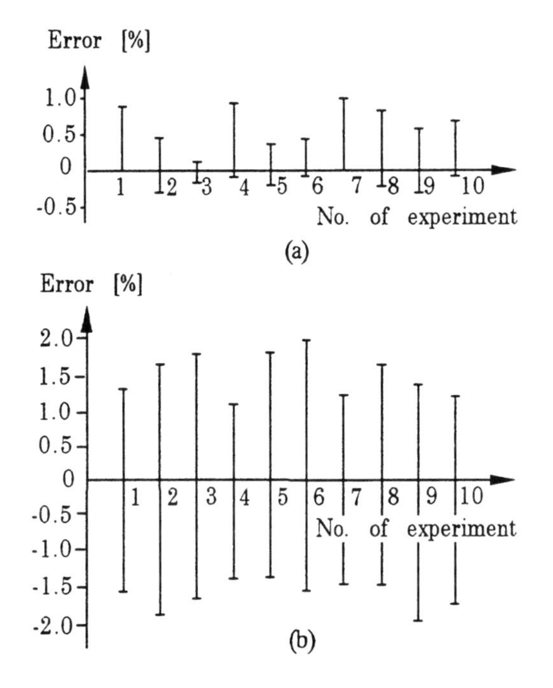

Fig. 13 Error range of experiments

4 CONCLUSIONS

Contour analysis: The 3D measurement of red-hot ingot is achieved using contour analysis. A feature of the method is that the system configuration is simple enough to be applied in the practical field. The system uses only one camera to capture the contour information of red-hot ingot and reliable data can be obtained even at higher temperature since no any special light source is involved in the system. Because of the drawback of contour analysis, this method can not give the shape in detail such as concave region on the surface of red-hot ingot. One assumption must be considered in this method that the fluctuation of red-hot ingot caused by bending is negligible. The experiments show that the contour analysis is an applicable way in the 3D shape measurement of red-hot ingot.

Slit-ray projection: In order to deal with the problem of the fluctuation cause by the bending of red-hot ingot, the measuring system using slit-ray projection is developed. Because uncertainty of rotating axis, a curve fitting technique in the meaning of least squared difference is used to reconstruct the partial curves of a cross-section into a complete one. In order to reduce the affection of radiant light from red-hot ingot to the projecting light source, a mercury lamp acting as slit-ray emitter and ultraviolet filter are used since the mercury lamp can give a broad spectrum light. This system is applicable and the algorithm is feasible. Experiments show that when the temperature is over 1000 °C, the accuracy of measurement were deteriorated by radiant light from red-hot ingot.

REFERENCES

Furuhashi, H., U. Ono, S. Hashimoto and M. Ooshima, 1991, One Fast Computing Technique of Image Registration, *Proc. 11th Pattern Measurement Division*, Yokohama, Japan

Lim, H. S., 1988, Curved Surface Reconstruction Using Stereo Correspondence, *Proc. Image Understanding Workshop*, 809-819.

Sun, X., T. Ishimatsu, H. Imaizumi, M. Hirono and T. Ochiai, 1994, Three-Dimensional Profile Measurement of the Ingot Pressed with Forging Machine, *Proc. First Asian Control Conference*, Tokyo, Japan, **vol. 3**, pp. 623-626.

Sun, X., T. Ishimatsu, H. Imaizumi, M. Hirono and T. Ochiai, 1995, A Three-Dimensional Measuring Technique Applied to Forging Press Machine, *Proc. 10th Korea Automatic Control Conference*, Seoul, Korea, pp. 138-141

Zheng J. Y., 1994, Acquiring 3D Models from Sequences of Contours, IEEE Transactions on Pattern Analysis and Machine Intelligence, **vol. 16(2)**, pp. 163-178

DEVELOPMENT OF A COMPUTATIONALLY EFFICIENT PROCESS SIMULATION FOR LASER DRILLING

Riccardo Cazzoli*, Luca Tomesani**

* Doctor in Mechanical Engineering, University of Bologna, Italy
**Assistant Professor, Department of Mechanical Construction Engineering
University of Bologna, Italy

Abstract: The development of a process planning aid for laser drilling operations is here presented. The system is made of two parts: first, an analytical model determines the shape of the material removing zone. Then, the distribution in time of the cavity geometrical extent, which corresponds to the melting isothermal, is used as an input in a finite element analysis, allowing a simplified solution for bulk material properties prediction. This hybrid approach, avoiding time-consuming full-FEM analyses, allows both to evaluate the heat affected zone extent and to predict the residual stress field at the end of the process. The proposed model has been experimentally veryfied by means of laser drilling experiences on low-carbon ASTM A366 steel and polimetil-metacrilate PMMA, evaluating the shape of the performed holes.

Keywords: Manufacturing processes, Process models, Process parameter estimation, Finite element analysis, Temperature distributions, Mechanical properties

1. INTRODUCTION

Laser drilling, due to its high material removal rate and operational flexibility, has already become a widely used tool in manufacturing of metal and non metal components, ranging from electrical components (electrical interconnect vias and component mounting holes in printed circuit boards, memory modules repairs) to thermo-fluid components (fuel injectors, combustors, nozzles and turbine blade bleed holes) to bio-medical applications (barrier filters, timed drug release devices).

The wide range of application opportunities of this technology gives interest to the forecast both of the geometrical characteristics of the hole to be performed, such as straightness, aspect ratio and dimensional accuracy, and of the bulk properties of the product material at the end of the process, such as heat-affected zone and residual stress state. This is especially true on difficult-to-work materials and on highly stressed components. Thus, precision-drilled holes are a really critical feature in manufacturing, in which tipical and less-tipical functional requirements must complain with process capability and economy.

On the other hand, the process complexity, coming from the different phenomena involved (phase change, plasma formation, surface reflection, convection, conduction) keeps from carry out an efficient process planning, which is very often performed in the form of a trial-and-error iteration. In laser drilling an effective planning and control depends on the availability of robust models which combine detailed description of the phisical phenomena with computational efficiency required for iterative solution of a multi-criteria optimization problem. Here, balancing factors such as material removal rate, hole taper, hole width, surface roughness and bulk material properties after drilling are to be considered.

At the present time, many analytical models for drilling operations are available to predict, under many simplifying assumptions, the final shape of the hole, the temperature field during the process and the material removal rate, but none of them can consider the heat affected zone extent and the residual stress field in the final product material.

On the other hand, numerical analyses of laser drilling are so computational intensive that their use in parameter optimization involved in process planning is impossible. Thus, the present paper describes an hybrid method approach to laser drilling processes, which integrates an analytical model for hole extent evaluation and a subsequent simplyfied FEM analysis, in order to predict both hole geometry and bulk material properties without renouncing to computational efficiency.

The analytical model used to determine 'a priori' the shape and extent of the growing cavity has been recently developed by Sheng and Wang (1996) for laser drilling processes, basing on that previously proposed by Sheng and Cai (1996) on laser cutting processes, in which a laser beam, with gaussian distribution and divergence along the focal lens axis, is considered. The model considers the dependence of material absorptivity from the angle of incidence between beam and cutting front, first determining the temperature distribution and then material removal, which corresponds to the melting isothermal.

The distribution in time of the cavity geometrical extent is then used as an input for the finite element code. This allows a fixed mesh analysis in which the solution, beeing already known at each step the melting isothermal, is less computational intensive than usual full-FEM analyses. To further increase computational efficiency, the analysis is uncoupled.

The numerical analysis estimates the HAZ extent as well as the residual stress state at the end of the process, thus beeing a useful tool in laser drilling process planning, which addresses computational efficiency and bulk material properties predictions.

The proposed model has been experimentally veryfied by means of laser drilling experiences on plain low-carbon steel and polimetil-metacrilate, evaluating the shape of the performed holes.

Further verification of the hybrid approach developed is in course, aiming at the heat affected zone extent evaluation.

2. ANALYTICAL MODELS FOR LASER DRILLING

2.1 Models review

Von Allmen (1976) developed a 1-D transient study in which the material is removed by means of evaporation and liquid expulsion, assuming that the melting front moves within the material at constant speed. Chan and Mazumder (1987) obtained an analytical solution in closed form by means of a 1-D stationary model, assuming evaporation and liquid expulsion, and solving the Stefan problem considering a monodimensional motion of the fluid phase. Minardi and Bishop (1993) solved at the finite differences a bidimensional transient model, evaluating the influence of the incident beam profile on the temperature profiles and on the spatial distribution of the phases. Ramanathan and Modest (1991) and Kar and Mazumder (1990) considered a finite thickness and integrated the bidimensional transient equation starting from an experimentally evaluated first attempt profile. Patel and Brewster (1991) used a 1-D transient model for the thermal change and a 2-D axisymmetric model to evaluate the assisting gas and molten layer condition, finding the penetration speed and the drilling time. Olson (1991) considered a single-impulse drilling, solving at the finite differences the conduction equation in 2-D and axisymmetrical condtions. He, *et al.* ((1993) numerically solved the axisymmetrical bidimensional transient problem, neglecting the latent heat and assuming an experimentally evaluated non-isothermal surface.

Concerning the vaporization kinetic models, the velocity and pressure fields at the liquid-vapor interface are found by Afanas'ev and Krokhin (1967), Anasimov (1968), Anasimov and Rakhmatulina (1973), Knight (1979).

Concerning the mechanical damage models, Steverding and Studel (1976) developed a monodimensional analysis of the laser-induced pressure waves shock and material fracture, Pirri (1973) modelled the formation of plasma on the material surface by means of gas dynamics and cylindrical impact wave, Pirri, *et al.* (1978) showed that two alternative regimes can develope within the system, depending on laser power intensity: laser supported combustion or laser detonation wave.

2.2 The Sheng and Wang model

The Sheng, Wang (1996) analytical model used for determining the shape and extention of the growing cavity considers a TEM_{00} gaussian laser beam, with divergence along the focal lens axis (Fig.1). The drilling front is divided into annular surface elements of depth d with orientation in the normal direction n.

The power balance at the surface element equals the sum of absorbed energy from the laser beam source and the reflected beam energy to the sum of conduction heat and phase change latent heat. The material surface absorptivity depends on the beam incidence angle and refractive indices through Fresnel's relation. The laser beam reflection on the hole profile is considered through a ray tracing method which simulates the beam path after the first incidence.

The reflected beam intensity at a particular surface element is the sum of contributions from all incident rays (see Fig.1). Only first order reflections are considered in the model to save computational effort.

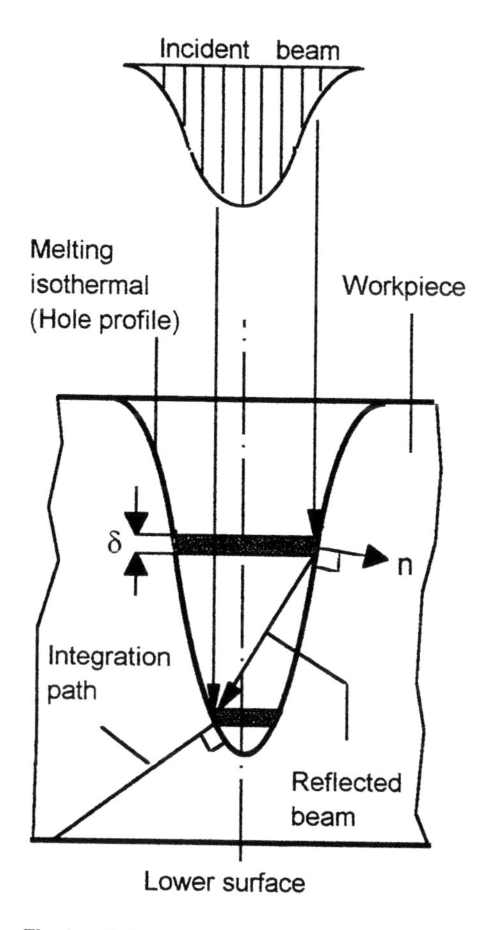

Fig.1 - Scheme of the Sheng and Wang model for laser drilling. The integration path is normal to the hole surface and considers direct beam and first reflection intensity.

The hole surface has been considered as perfectly smooth. In order to overcome the 2-D Stefan problem of heat conduction into the workpiece from a moving boundary, a simplified transient model is used by integrating in a local coordinate system the conductivity differential equation in the direction normal to the hole profile, thus determining the temperature distribution. The material removal then corresponds to the melting isothermal. Convection on the lower surface is also considered in solving the differential equation.

4. EXPERIMENTAL PROCEDURE FOR ANALYTICAL MODEL VALIDATION

The model was prooved to be effective by means of a set of experiences on different materials, such as polymers, steel and ceramics. Here only the PMMA experiment is reported. The model graphic output is shown in Fig.2 in a 1,4 [mm] thick sheet drilling by means of a 150 [W] power beam. The focal point is set 3,70 [mm] above the upper surface, as usual for these materials. The model gives the distribution of the hole shape at six different times. It can be noted that 7 [ms] are needed to reach the lower surface, while 20 [ms] are the complete drilling time.

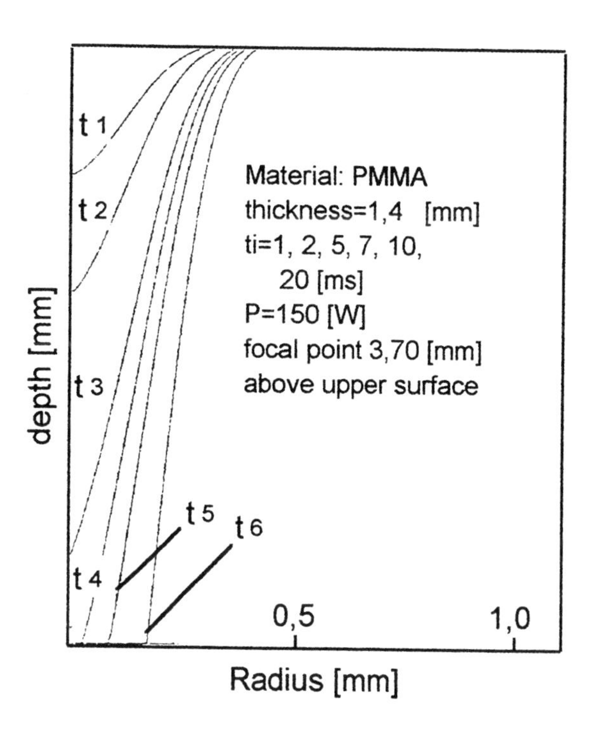

Fig.2 Prediction [1] of progressive hole extension in leser drilling of PMMA

In Fig.3 the experimental set-up for model validation is outlined. A circularly polarized laser beam is focussed on the sample workpiece, beeing cut by a constant-speed rotating alluminum disk with a radial slot. The fenditure allows the beam to shoot the workpiece for a definite period of time.

Thus, by regulating the angular speed of the disk, it is possible to exactly define the time of beam effectiveness and to perform partial drillings corresponding to the model output of Fig.2. Then, the hole shapes are measured and related to the model prediction, as shown in Fig.4.

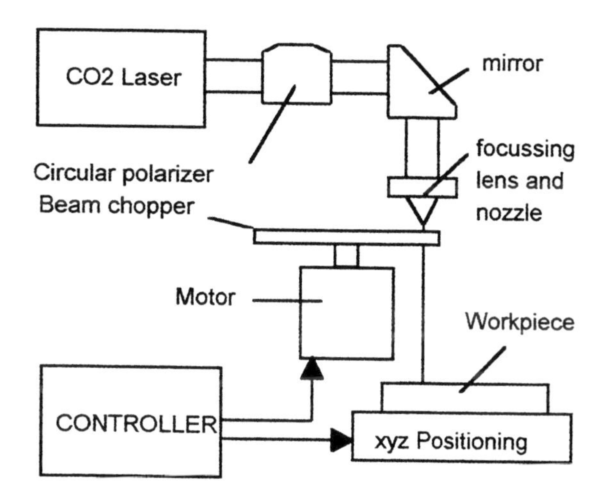

Fig.3 - Set-up for drilling experiences

Here, a slight overstimate of the time needed to complete the operation is outlined by the model, whith respect to the effective hole progression: 20 [ms] in spite of 10 [ms] measured on the workpiece. Nevertheless, the good agreement of predicted and measured final hole shapes allows to consider the Sheng and Wang model as a useful tool in drilling process planning.

In Fig.5 the analytical model output is shown for plain low-carbon steel processing. It is worth noting the greater drilling time and the hole straightness, much more than in PMMA drilling.

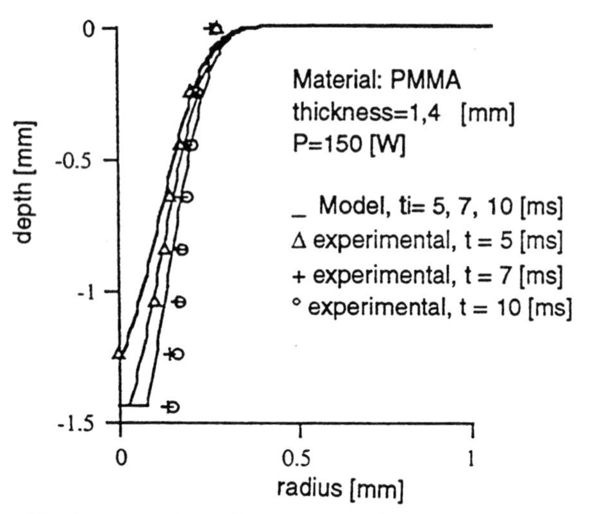

Fig.4 comparison between model prediction and experimental measures for PMMA

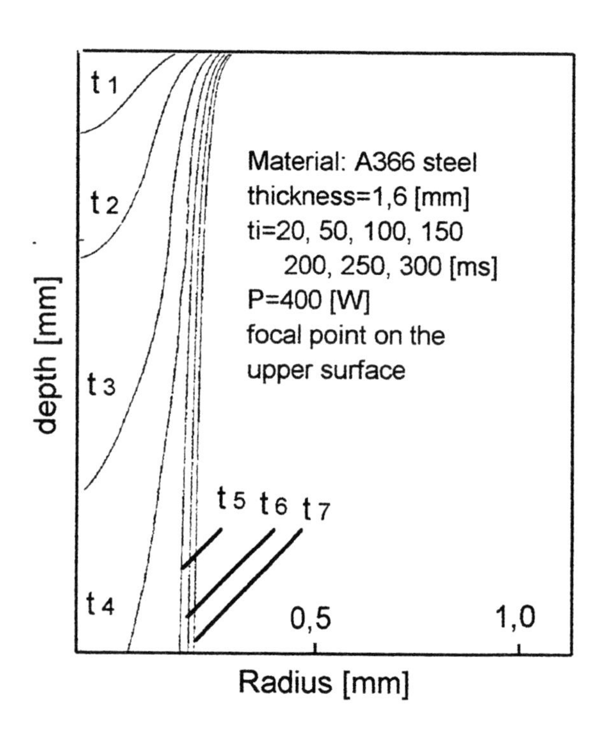

Fig.5 Prediction of progressive hole extension in laser drilling of A366 steel

4. FEM ANALYSIS

For the PMMA and for the low-carbon steel represented respectively in Figs. 2 and 5, a FEM analysis has been conducted by means of the commercial code ABAQUS at the Mechanical Engineering Department of the University of California, Berkeley.

The knowledge of the distribution in time of the cavity geometrical extent is an input to the finite element code. This allows a fixed mesh analysis in which the solution, beeing already known at each step the melting isothermal position, is less computational intensive than usual full-FEM analyses.

The simulation considered convection both on top and bottom surface, but neglected assisting gas convection on the inner hole surface, due to the partial balancing effect with plasma formation inside the hole, which was also neglected by the analytical model. Moreover, At high values of assisting gas pressure, convection phenomena related with the molten layer can be neglected, accordingly to Chan and Mazumder (1987) experimental verification. To further increase computational efficiency, FEM analysis is uncoupled; secondary changes in temperature due to the straining material are then neglected.

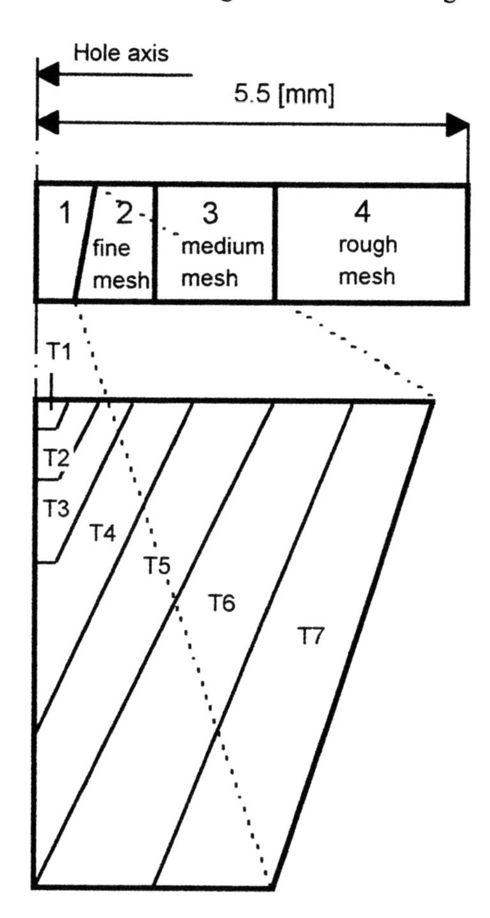

Fig.6 Mesh disposal and boundaries between sets of elements to be progressively removed, corresponding to the model prediction

Thus, the temperature field is found and a threshold value for HAZ extent evaluation has been set to 120 [°C] for PMMA and 400 [°C] for A366 steel, accordingly to Sheng and Joshi suggestion (1994).

The mesh geometry is divided in four zones (Fig.6). The first one corresponds to the hole extent at the end of the drilling process, as determined by the analytical model. This zone is further divided in sub-zones, corresponding to the evolution in time of the hole, which is also predicted by the model.

Zone 2 has the same mesh density of zone 1, but is not affected by material removal; zones 3 and 4, beeing at greater distance from the axis, have decreasing mesh densities. The complete mesh for the numerical analysis has 2161 nodes and 2048 axisymmetric elements.

The hole profile between time t_i and t_{i+1} is assumed to be fixed and equal to the hole extent at time t_i. The temperature field within the workpiece is then found by considering the hole surface at the melting temperature, until time t_{i+1} is reached, when the next hole profile is considered. As an example, in Fig.7 the temperature field is presented at the end of step 2 in A366 steel drilling, when hole extension is still that of time t_2 and the total cumulated time is t_3.

In Tab.1 and Tab.2 the complete output of the temperature analysis is presented, reporting current drilling depth, maximum radius and HAZ extent for the considered materials.

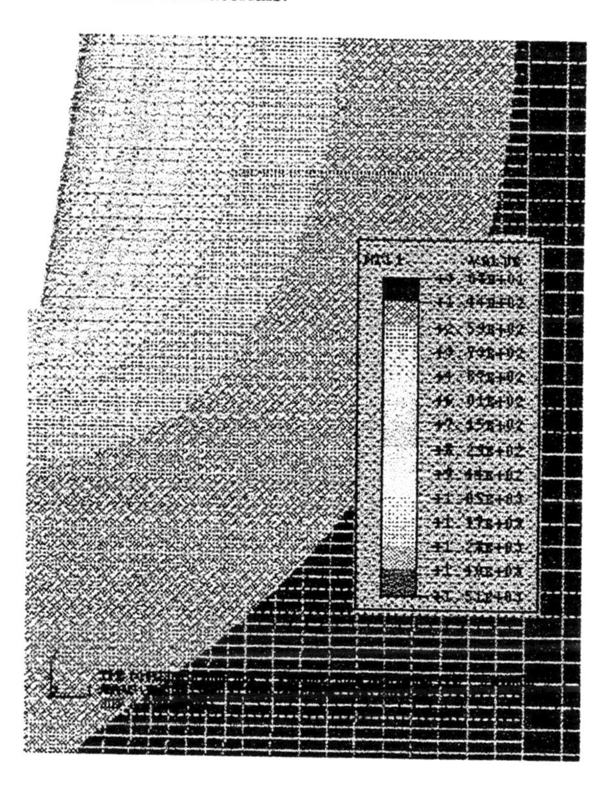

Fig.7 Temperature field in A366 steel drilling at the end of step 2 (drilling time = 100 [ms])

Tab.1 HAZ extent evaluation as determined by FEM analysis in PMMA drilling.

Time [s]	Drilling depth [mm]	Maximum radius [mm]	HAZ extent [mm]
heating			
0.002	0.175	0.10	0.027
0.004	0.351	0.15	0.030
0.008	0.527	0.20	0.034
0.015	1.030	0.30	0.038
0.020	1.404	0.40	0.043
0.040	1.404	0.459	0.055
0.050	1.404	0.518	0.061
cooling			
1.050	1.404	0.518	-------

Tab.2 HAZ extent evaluation as determined by FEM analysis in A366 steel drilling.

Time [s]	Drilling depth [mm]	Maximum radius [mm]	HAZ extent [mm]
heating			
0.020	0.194	0.06	0.171
0.050	0.595	0.09	0.305
0.100	1.116	0.12	0.384
0.120	1.335	0.18	0.448
0.150	1.587	0.24	0.526
0.200	1.587	0.295	0.576
0.250	1.587	0.35	0.632
cooling			
0.950	1.587	0.35	-------

Experimental verification of predicted HAZ extents, performed by means of optical methods, not here reported, gave good results for PMMA, but evidenced an HAZ understimate for steel.

After the thermal analysis is concluded, the subsequent stress analysis estimates the stress field during heating and after cooling. They were assumed elastic-perfectly plastic materials, with yield point and thermal properties dependent from temperature.

The circumferential stress fields for A366 steel are shown in Fig.8 at the end of step 5 and in Fig.9 after the cooling stage. It can be noted that compressive stresses are developed under the hole profile during heating, while intense residual tension stresses are present at the end of cooling.

6. CONCLUSIONS

A hybrid method for determining the heat-affected zone and the residual stress field in laser drilling operations is introduced by integrating an analytical solution for the hole extent geometry and a double-step numerical solution for the temperature and stresses distribution. The proposed approach has been experimentally veryfied by means of laser drilling experiences on low-carbon ASTM A366 steel and polimetil-metacrilate PMMA, evaluating the shape of the performed holes.

The computational effort required to develop a HAZ estimation can be reduced significantly compared to that for a three-dimensional finite-element model. The HAZ extent can be estimated through analysis of the transient extent of isotherms. Finally, this method is useful as a process planning aid for laser drilling, where fast computational time is essential in process parameters determination.

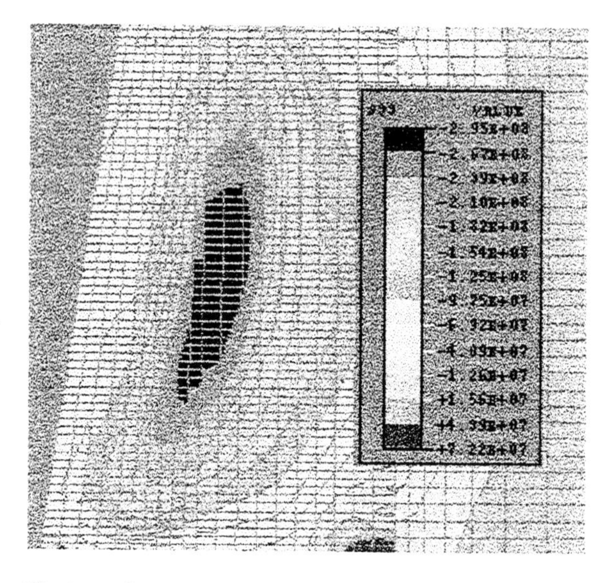

Fig.8 Compressive stresses are present under surface during heating (step 5).

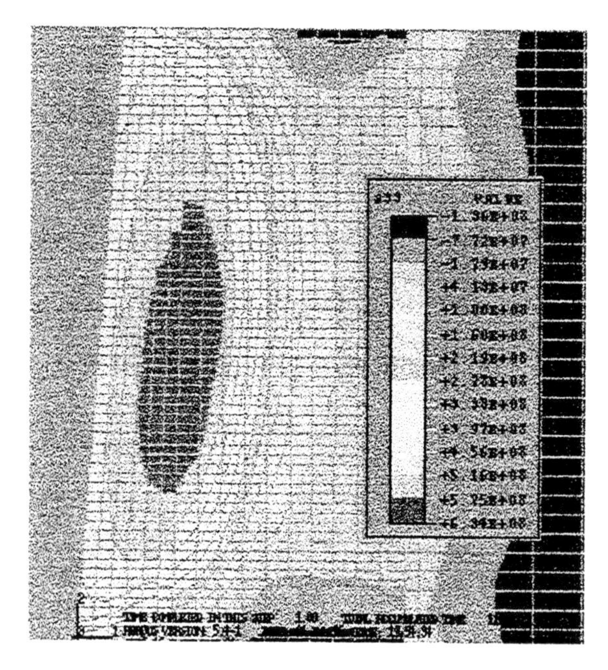

Fig.9 Tension stresses are present under surface at the end of the cooling stage (step 8).

ACKNOWLEDGEMENTS

The authors would like to thank MURST (Ministry of the University and of Scientific and Technological Research) for financial support.

REFERENCES

Afanas'ev, Y.V. and O.N. Krokhin (1967), In: *Sov. Phys.JETP*, **25**, pp.639-646

Anasimov, S.I. (1968), In: *Sov. Phys.JETP*, **27**, pp.182-189

Anasimov, S.I. and A.K. Rakhmatulina (1973), In: *Sov. Phys.JETP*, **37**, pp.441-448

Chan, C.L. and J. Mazumder (1987), One-dimensional steady state model for damage by vaporization and liquid. In:*J. Appl. Phys.*, **62** (12), pp.4579-4586

He, M., P.J. Bishop and A. Minardi (1993), Prediction of cavity shape and material removal rates using a two-dimensional axisymmetric heat conduction model for fibrous ceramic insulator and comparison with experiments. In: *J. Laser Appl.*, **5** (1), pp.33-40

Kar, A. and J. Mazumder (1990), Two-dimensional model for material damage due to melting and vaporization. In: *J. Appl. Phys.*, **68** (15), pp.3884-3891

Knight, J. (1979), In: AIAA Journal, **17**, pp.519-526

Minardi, A. and P.J. Bishop (1993), Two-dimensional transient temperature distribution whithin a metal. In: *J.Heat Transfer*, **110** (6), pp.1009-1011

Olson, J.W. (1991), Long pulsed laser drilling of composites. In: *J. Appl. Phys.*, **70** (12)

Patel R.S., Brewster M.Q., (1991), Gas assisted laser metal drilling. In: *J. Thermophysics*, **5** (1), pp.35-39

Pirri, A.N. (1973), Phys.Fluids, 16, pp.1435-1440

Pirri, A.N., R.G. Root and P.K.S. Wu (1978) In: *AIAA Journal*, **16**, pp.1296-1302

Ramanathan, S. and M.F. Modest (1991), CW laser drilling of composite ceramics. In: *ICALEO 91*, pp.305-326

Sheng, P. and V. Joshi (1994), Analysis of heat-affected zone formation for the laser cutting of stainless steel. In: *J. Mat. Proc. Techn.*, **53** (3-4), pp.879-892

Sheng, P. and L. Cai (1996), Analysis of laser evaporative and fusion cutting. In: *Transaction of the ASME, J. of Man. Sci. Eng.*, **118** (2), pp.225-234

Sheng, P. and S. Wang (1996), Planning model for long-pulsed laser drilling. In: *internal report*, Department of Mechanical Engineering, University of California, Berkeley, CA

Steverding, B. and H.P. Studel (1976), In: *J. Appl. Phys.*, **47**, pp.1940-1946

Von Allmen, M. (1976), Laser drilling velocity in metals. In: *J. Appl. Phys.*, **47**, pp.5460-5463

A REAL TIME COST MONITORING SYSTEM FOR BUSINESS PROCESS REENGINEERING

Julio Macedo

Institut Stratégies Industrielles, 229 Forest, Pincourt, P.Q., J7V8E8, Canada

R. Ruiz Usano and J. Framiñan Torres

School of Industrial Engineering, University of Sevilla
Av. Reina Mercedes, s/n, 41012 Sevilla, Spain

Abstract: The application of business process reengineering methodologies requires a real time monitoring system that warns when a business process is problematic. In addition, this monitoring system must be able to calculate the quantities of the activities that the products must use in order to improve the business proces efficiency. This paper implements this monitoring system using activity based costing theory and system dynamics software. The use of the suggested monitoring system is illustrated by applying it to reengineer culture media production in a Hospital laboratory.

Keywords: Business process engineering, Detector performance, Performance monitoring, Real time systems, Monitored control systems, Measuring elements.

1. INTRODUCTION

A business process is the set of manufacturing and management activities necessary to convert some resources into a product. In general, the product cost deviates from the desired one so that it is necessary to reengineer the business process. The output of reengineering is a set of improvements in the business process activities so that the desired product cost is reached. From this point of view, business process reengineering is a methodology to conceive the required improvements (Kubeck, 1995).

Current business process reengineering methodologies need a real time monitoring system that warns when the product cost deviates from its target cost. This is necessary to identify the problematic products and, lastly, the problematic businesses processes. In addition, the monitoring system must be able to calculate the quantities of the activities that the problematic products should use.

This is necessary to reduce the quantities of the nonvalue-add activities in order to reach the desired product cost (Elzinga et al., 1995). Surveys of business process reengineering tools (Klein, 1994) and activity based costing tools (Borden, 1994) do not report implementations of such monitoring system. This paper presents one possible implementation. In the second part of this paper, the structure of the suggested monitoring system is presented whereas in the third part, it is applied to reengineer the production of one type of culture media at Sainte Justine Hospital of Montreal city.

2. MONITORING SYSTEM STRUCTURE

An activity driver is a quantitative measure that characterizes a manufacturing or management activity required by a product. In addition, Activity Based Costing systems calculate the cost of a product by evaluating and summing the quantities of activity drivers utilized by the product during its flow through

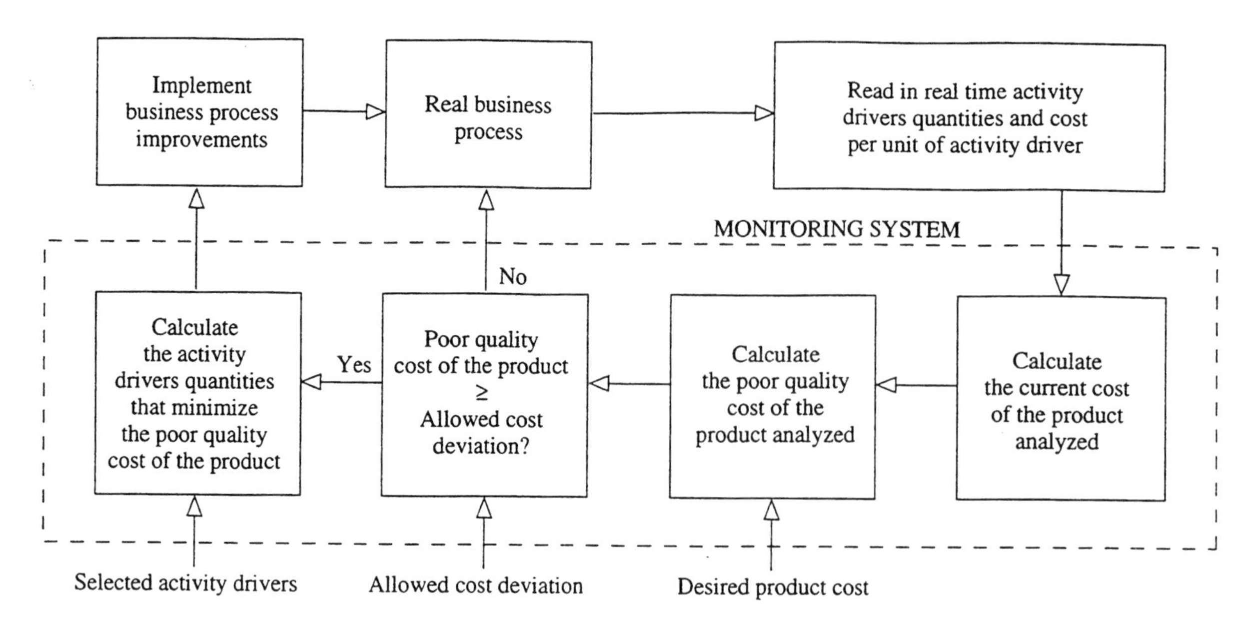

Fig. 1. Structure of the monitoring cost system. The desired product cost, selected activity drivers and allowed cost deviation must be given to the monitoring system.

its business process (Wiersema, 1995). Hence, the quantities of activity drivers and the costs per unit of activity driver can be used as starting points to reengineer a business process (Cooper and Kaplan, 1992). This is the foundation of the monitoring system proposed here.

The structure of the suggested monitoring system is illustrated in Figure 1. As shown, the monitoring system can be used as a real time cost calculator and warning system or as an optimizer of the quantities of activity drivers that minimize the poor quality cost of a product (Harrington, 1987). At this point it is important to note that the monitoring system is not tooled to select the activity drivers, so that it can support only part of a reengineering exercise. The identification of the activity drivers can be done using reference models developed by Macedo and Ruiz Usano (1995).

The equations of the suggested monitoring system are illustrated in Figure 2 and can be easily implemented using system dynamics software with subscript variables (Richardson and Pugh, 1981). As noted the formulae to calculate the total costs of the activity drivers are the ones currently used by activity based costing systems. However, the optimization module in dashed lines is new with respect to current activity based cost systems. Furthermore, the variables inside the hexagones are introduced in real time using an spreadsheet or a text file connected to the main program that solves the differential equations. At this point note in Figure 2 that the monitoring system progressivey cumulates the costs of the activity drivers quantities used by different batches of the product analyzed.

The monitoring system can be used as a a cost calculator and warning system or as a minimizer of the poor quality unit cost. When it is used as a warning system (in continuous lines in Figure 2), the values of the activity drivers and their unit costs are read in real time and summed to generate the poor quality unit cost. When the monitoring system is used as a minimizer (in dashed lines in Figure 2), some former activity drivers and costs per unit driver are selected as control variables and their values are optimized.

3. MONITORING SYSTEM APPLICATION

The suggested monitoring system was used as one of the tools in a reengineering exercise of culture media production at the laboratory of microbiology in Sainte Justine Hospital of Montreal city. The reengineering exercise was directed to identify the improvements required by this laboratory to lower the current cost of its produced culture media to reach the price offered by pharmaceutical factories.

Culture media is a gelatinuous substance for cultivating bacteria and viruses. There are hundreds of types of culture media mainly used for diagnostic of patient diseases in Hospitals. One of this culture media is named Columbia 5% Sheep Blood which is fabricated in large quantities (59 000 dishes of 100 x 15 mm per year) at the production unit of microbiology laboratory of Sainte Justine Hospital.

Sainte Justine production unit includes three full time employees that work in a room of 36 square meters with small sized machines. The manufacturing process of Columbia culture media is the following: Mix powder of culture media with water manually, put the liquified culture media in a small sterilizer, activate the sterilizer, connect the sterilizer to a small automatic distributor, activate the distributor that

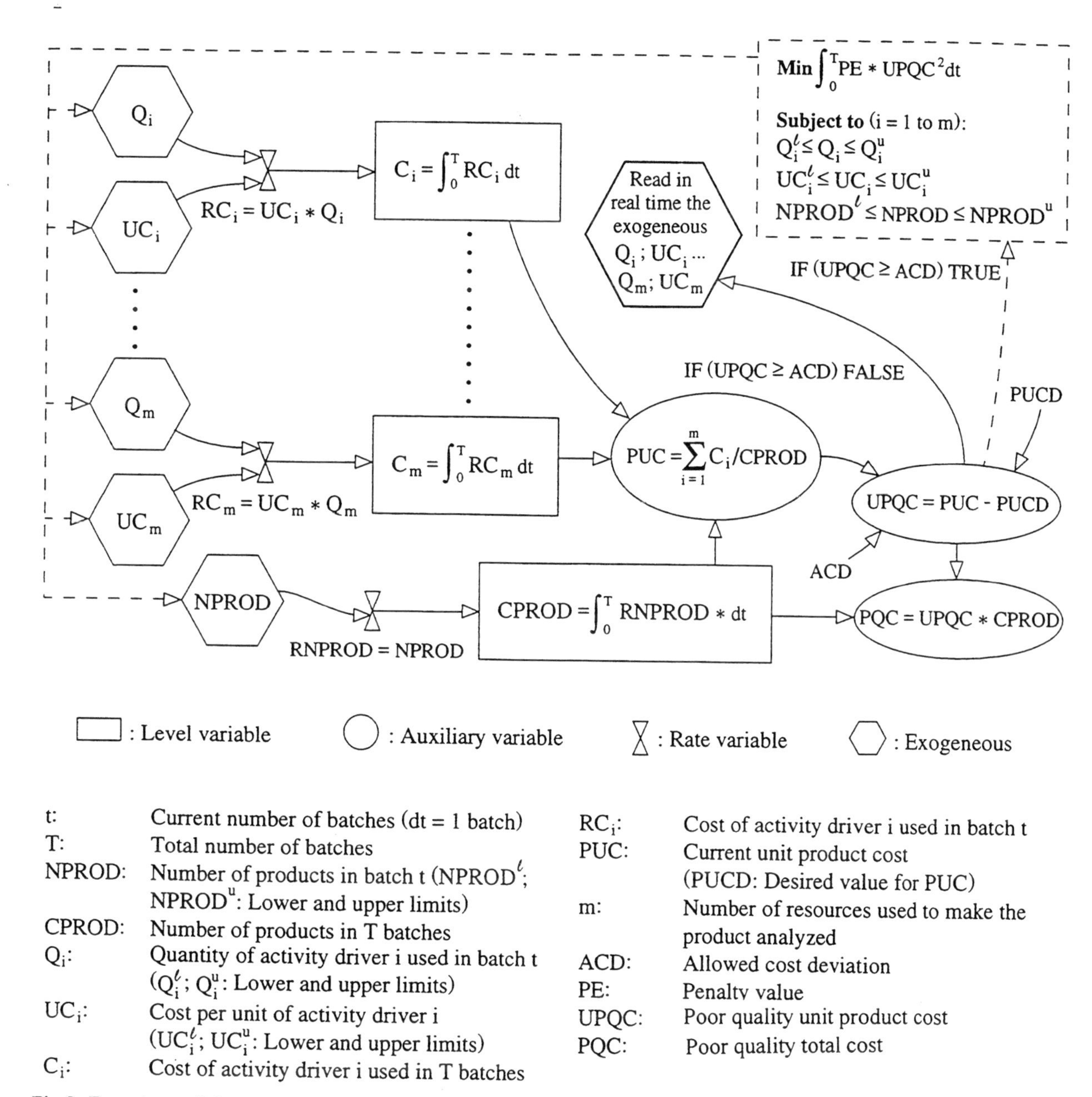

Fig.2. Equations of the monitoring cost system. When UPQC<ACD, the variables inside the hexagons are read in real time as indicated by continuous lines. When UPQC≥ACD, some variables inside the hexagons are selected and optimized as indicated in dashed lines.

pours the liquified culture media to plastic petri dishes of 100 x 15 mm, package the batch of dishes and transport it to a cool room located 11 meters out of the production unit. The culture media business process analyzed includes the preceding manufacturing process, and in addition, the quality control activities and the materials purchase activities.

The cost monitoring system was built by applying the equations of Figure 2 to the business process of Columbia culture media. The resultant minitoring system is illustrated in Figure 3, it was used in two modes: First, as a real time calculator of the cost to produce one 100 x 15 mm dish of Columbia media at Sainte Justine Hospital laboratory. Second, as an optimizer of the quantities of the activity drivers to make one dish of Columbia media at 0.21$/dish which is the average sales price offered by the pharmaceutical industry.

Before to use the monitoring system as a calculator cost, the costs per unit of activity driver were calculated using annual data available (Table 1). Then these unit costs (Table 2) were introduced in the equations of the monitoring system. Finally, the quantities of the activity drivers were progresively introduced in the monitoring system during the production of eight batches of Columbia media (Table 3). At the end of each batch the monitoring system calculated and displayed the cost to produce one 100 x 15 mm dish of Columbia (Figure 4). As noted, the current unit cost is 0.31$ which is 47% higher than the sales price of one Columbia dish (0.21 $). In addition, graphics displayed by the monitoring system indicated that the direct labor cost

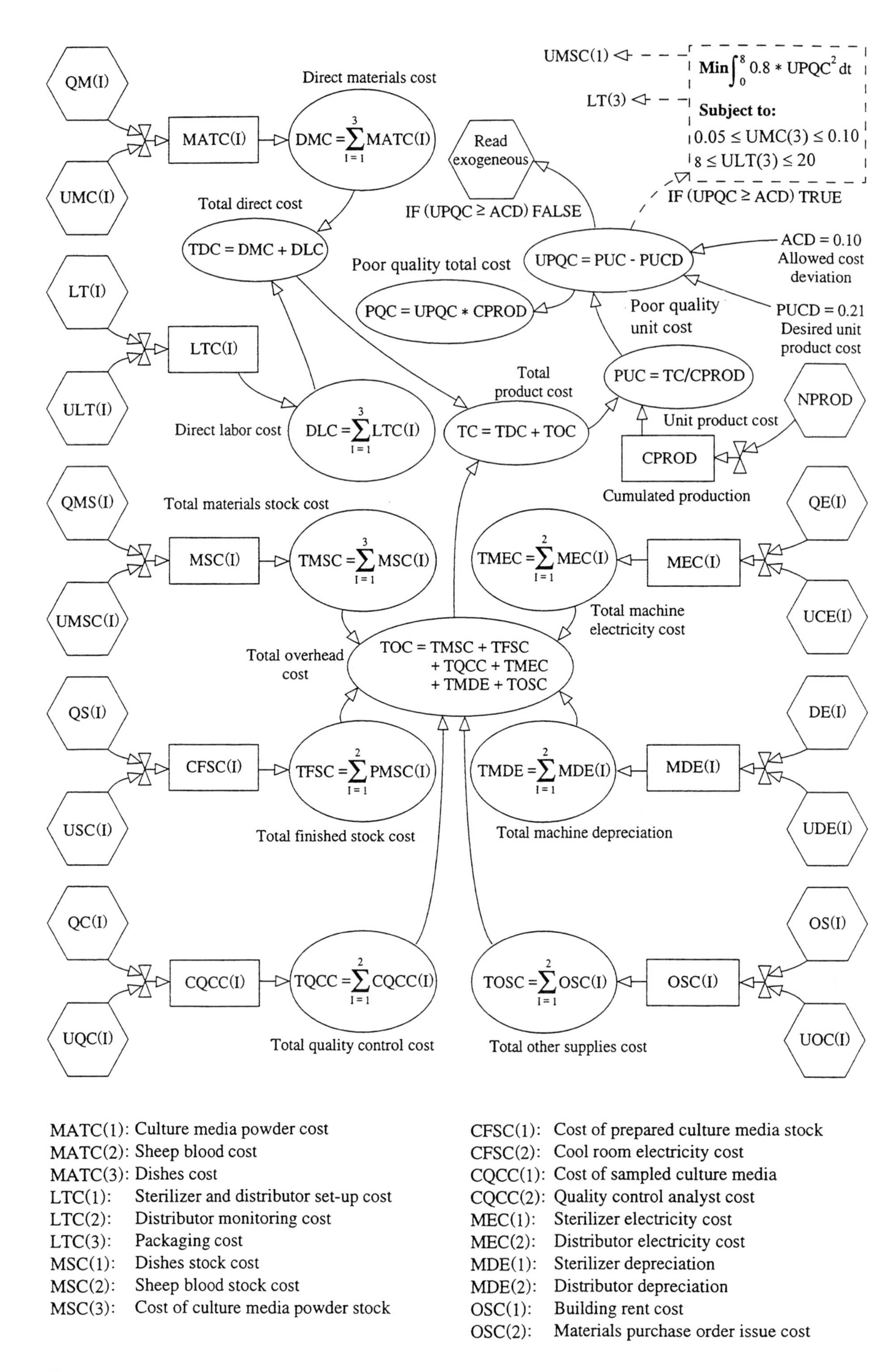

Fig. 3. Monitoring system of culture media cost at Sainte Justine Hospital.

Table 1. Calculation details of two costs per unit of activity driver: UMSC(3) and USC(2).

UMSC(3)	= Average number of powder bottles stocked per month * Price of a bottle * Interest rate per month * 12 months / Annual production in dishes
USC(2)	= KWh consumed per year * Price of one KWh / Annual volume of cool room available in m^3

Table 2 Cost per unit of activity driver read in real time during production of eight batches of Columbia culture media at Sainte Justine Hospital laboratory.

Cost per unit of activity driver	Activity driver	Total cost	Cost per unit of activity driver	Activity driver	Total cost
UMC(1) = 0.052$/gr	QM(1)	MATC(1)	UCE(1) = 0.1478$/hr	QE(1)	MEC(1)
UMC(2) = 0.065$/ml	QM(2)	MATC(2)	UCE(2) = 0.0336$/hr	QE(2)	MEC(2)
UMC(3) = 0.084$/dish	QM(3)	MATC(3)	UDE(1) = 0.0024$/dish	DE(1)	MDE(1)
ULT(1) = 17.11$/hr	LT(1)	LTC(1)	UDE(2) = 0.00096$/dish	DE(2)	MDE(2)
ULT(2) = 17.11$/hr	LT(2)	LTC(2)	UQC(1) = 0.012$/dish	QC(1)	CQCC(1)
ULT(3) = 17.11$/hr	LT(3)	LTC(3)	UQC(2) = 17.11$/hr	QC(2)	CQCC(2)
UMSC(1) = 0.00009$/dish	QMS(1)	MSC(1)	UOC(1) = 0.0.22$/hr	OS(1)	OSC(1)
UMSC(2) = 0.00003$/dish	QMS(2)	MSC(2)	UOC(2) = 0.00072$/hr	OS(2)	OSC(2)
UMSC(3) = 0.00009$/dish	QMS(3)	MSC(3)	USC(2) = 0.5107$/$m^3$	QS(2)	CFSC(1)
USC(1) = 0.00019$/dish	QS(1)	CFSC(1)			

Table 3 Activity drivers values read in real time during production of eight batches of Columbia culture media at Sainte Justine Hospital laboratory.

Activity drivers		Batch number							
		1	2	3	4	5	6	7	8
NPROD	: Batch size (dishes)	306	612	612	306	306	612	306	306
QS(2)	: Batch volume (m^3)	0.07	0.14	0.14	0.07	0.07	0.14	0.07	0.07
QM(1)	: Quantity of culture media powder (gr)	264	528	528	264	264	528	264	264
QM(2)	: Quantity of sheep blood (ml)	300	600	600	300	300	600	300	300
QM(3)	: Quantity of dishes (dishes)	306	612	612	306	306	612	306	306
LT(1)	: Labor time to set up sterilizer (hr)	0.42	0.42	0.42	0.42	0.42	0.42	0.42	0.42
LT(2)	: Labor time to monitor distributor (hr)	0.7	0.7	0.7	0.7	0.7	0.7	0.7	0.7
LT(3)	: Labor time to package (hr)	0.88	0.88	0.88	0.88	0.88	0.88	0.88	0.88
QE(1)	: Sterilizer working time (hr)	1	1	1	1	1	1	1	1
QE(2)	: Distributor working time (hr)	0.33	0.66	0.66	0.33	0.33	0.66	0.33	0.33
QC(2)	: Quality control analyst time (hr)	0.17	0.17	0.17	0.17	0.17	0.17	0.17	0.17

DE(1) = DE(2) = OS(1) = OS(2) = QS(1) = QC(1) = QMS(1) = QMS(2) = QMS(3) = NPROD

and the direct materials cost are the main contributors of the unit cost of Columbia media. Other graphics showed that the costs of culture media powder, sheep blood and plastic dishes represent 23%, 33% and 44% of the total direct materials cost. Finally, additional graphics showed that the sterilizer set-up time, the distributor monitoring time and the packaging time utilize 21%, 35% and 44% of the direct labor time. All this information allows to conclude that the production of Columbia culture media at Sainte Justine Hospital is a problematic process.

When the monitoring system was used as an optimizer, some former exogeneous variables (in hexagones in Figure 3) were chosen as control variables and used to build different optimization models with the structure indicated in Figure 3. The optimization model that generated the lower poor quality unit cost UPQC was the one inside the dashed box in Figure 3. As noted, this model minimizes the poor quality unit cost of Columbia media subject to two constraints. The first one limits the unit cost of a plastic dish UMC(3) to the range 0.05$ to 0.10$. The second one limits the cost of labor ULT(3) to the

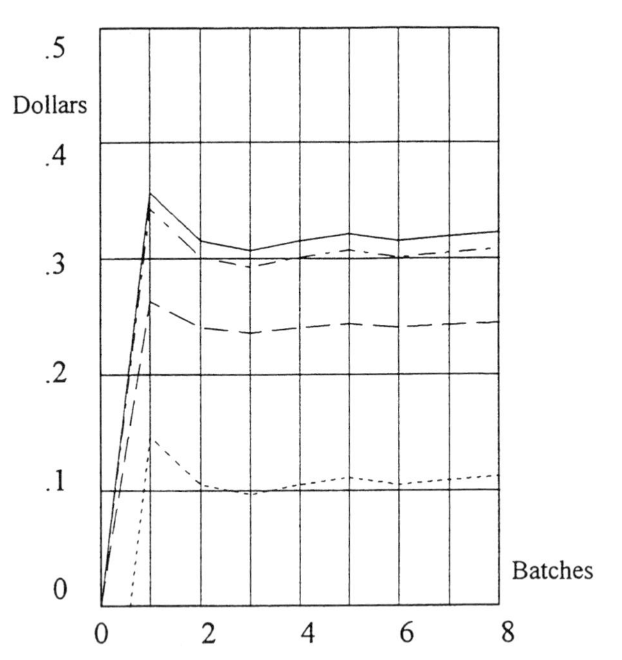

Fig 4. Current unit cost (___) and poor quality unit cost (.......) of Columbia media at Ste Justine Hospital. Expected unit cost if the following solutions are implemented: UMC(3)=0.05, ULT(3)=8 (-----); UMC(3)=0.07, ULT(3)=17.11 (-.-.-).

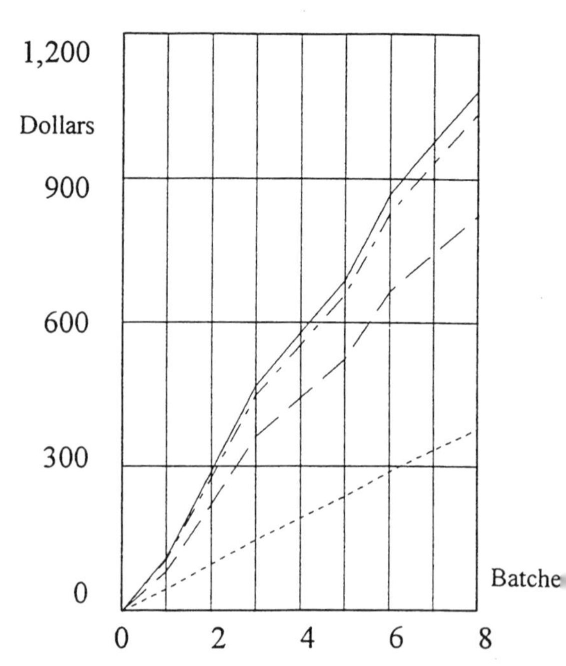

Fig 5. Current total cost (___) and poor quality total cost (........) of Columbia media at Ste Justine Hospital. Expected total cost if the following solutions are implemented: UMC(3)=0.05, ULT(3)=8 (-----); UMC(3)=0.07 and ULT(3)=17.11 (-.-.-).

range 8$/hr to 20$/hr. The solution of this model produced UMC(3)=0.05 $/dish and ULT(3)=8 $/hr. If these improvements are implemented, the product cost of one dish of Columbia media is expected to decrease dramatically as shown in Figure 4. These conclusions are confirmed by the corresponding total costs in Figure 5.

At this point, it is interesting to note that multiplying the poor quality unit cost of Columbia media (which is 0.1058$/dish in average, Figure 4) by 59 000 dishes/year produced at Sainte Justine Hospital gives 6 183$ which is the annual loss of Sainte Justine Hospital for making rather than buying Columbia media from industry.

One of the improvements above suggests to reduce the salary of Sainte Justine manufacturing operators from 17.11 $/hr to 8$/hr. In practice this is impossible due to Union Labor contracts. The other improvement suggests to pay 0.05$ per plastic dish in place of 0.084$. This second improvement can be only partially implemented at Sainte Justine because pharmaceutical industries which buy larger quantities of dishes than Sainte Justine pay a reduced price of 0.07$/dish. Hence, the strategy that can be implemented in reality is UMC(3)=0.07$/dish and ULT(3)=17.11$/hr. When these two numbers and the values of Tables 2 and 3 are introduced in the monitoring system this one displays the expected total costs of Figue 5. As noted, the new expected total cost for eight batches is 1085$ which is almost equal to the total current cost of 1038$. Hence, it is not interesting for Hospital Sainte Justine to partially implement the solutions above but to buy Columbia culture media from industry.

REFERENCES

Borden J.P. (1994). Activity based management software, *Journal of Cost Management*, Winter, 39-49.

Cooper R. and R. S. Kaplan (1992). Activity based management systems: Measuring the costs of resource usage, *Accounting Horizons*, Sept, 1-13.

Elzinga D. J., T. Horak, Ch. Lee and Ch. Bruner (1995). Business process management: Survey and methodology, *I.E.E.E. Transactions on Engineering Management*, **42** (2), 119-128.

Harrington J. (1987). *Poor Quality Cost.* Marcel Dekker, New York, NY.

Klein M. (1994). Reengineering methodologies and tools: A prescription for enhancing success, *Information Systems Management*, **11**(2), 30-35.

Kubeck L. C. (1995). *Techniques for Business Process Redesign.* John Wiley and sons, N. York.

Macedo J., R. Ruiz Usano (1995). Generating manufacturing system organizations using expert-neural models. In: *Intelligent Engineering through Artificial Neural Networks* (C. H. Dagli (Ed.)), p. 1001-1006. ASME Press, New York.

Richardson G., A. Pugh (1981). *Introduction to System Dynamics Modeling with Dynamo.* MIT Press, Cambridge, MA.

Wiersema W. (1995). *Activity Based Management.* AMACON, N. York.

USER-FRIENDLY MODELLING AND FORMAL VERIFICATION OF MANUFACTURING SYSTEMS

Mireille Larnac, Vincent Chapurlat and Janine Magnier

LGI2P - EMA/EERIE
Parc Scientifique Georges Besse
30000 Nîmes - France
Phone: +33 (0)466387025 - Fax: +33 (0)466387074
email: larnac,chapurla,magnier@eerie.fr

Abstract: This paper presents a new approach for modelling and formal analysis of discrete complex systems. The problematics addressed by this methodology concerns the analysis of complex systems, such as industrial production cells, or the organisation of production processes within an industrial plant. This approach is based on the collaboration of the Interpreted Sequential Machine (ISM) and a user-friendly description model (CANVAS Model).

Keywords: Modelling, Temporal logic, Formal verification, Industrial production systems, User interfaces

1. INTRODUCTION

The methodology presented in this paper addresses the problem of analysing the behaviour of complex systems, such as industrial production systems, or the organisation of the processes of a company. This approach is based on the use of two models: the CANVAS, a user-friendly description model of processes, and the Interpreted Sequential Machine (ISM) for the formal analysis of the behaviour of the system described by a set of CANVAS's.

The ISM model is a discrete state model and constitutes an extension of the classical Sequential Machine model as it handles any kind of data and allows us to describe the interactions of the environment of the state system. Moreover, it is possible to perform some verification on an ISM model. This process is based on expressing the behaviour of the ISM model into a formal system (temporal logic) which permits the formal proof of properties.

Nevertheless, using such a model requires some expertise; this is the reason why the CANVAS model has been designed to permit an easy description of a system by any user. The CANVAS models thus obtained are automatically translated into their equivalent ISM representation, and some verification can be handled. Obviously, the main problem due to the translation from CANVAS to ISM concerns the interpretation of the results of the verification process.

The ISM model is presented in the first part of this paper; the second one is dedicated to an overview of the proof of properties of ISM models using temporal logics. Then the CANVAS model concepts and the translation principles from an ISM to CANVAS are given. The conclusion contains some indications on a real industrial application, and the further developments of the methodology.

2. THE INTERPRETED SEQUENTIAL MACHINE

The Interpreted Sequential Machine (ISM) model is a discrete time state model. Basically, it is an extension of the Sequential Machine (Hartmanis and Stearns, 1966; Kohavi, 1978); the underlying concepts were adapted from the Extended Finite State Machine (Cheng and Krishnakumar, 1993).

The main characteristics of this model are:

- the core of the model is a state system (extension of state machine)
- the inputs and outputs can be of any type (Boolean, integer, real, data, event, etc.)
- the data which constitute the environment of the system and which influence the functioning of the sequential part of the system are represented separately

2.1 Structure of the ISM

As any sequential model, the state diagram, called Control Graph, is composed of an alternation of states and transitions. Informally, a transition between two states is activated if first an event appears on the inputs, and then if some conditions on the environment (data) are fulfilled; then the effects of this transition firing appear on the outputs and on the data of the environment which all change value. Structurally, the ISM model (Figure 1) owns inputs and outputs, and is made up of two parts:

- the Control Part (CP) contains the Control Graph and some necessary interpreters
- the Data Part (DP) is made up of the set of data which represent the environment of the system, and of the operations on these data

It is possible to partition the set I of inputs into two disjoint sets: the set I_C of the Control Part inputs and the set I_D of the Data Part inputs. Similarly, the set O of outputs is split up into Control Part outputs (O_C) and Data Part outputs (O_D).

The formal model of the ISM is thus the following:

ISM = < I, O, CP , DP > with:

- CP = <I_C, CII, **E**, **F**, CG, **U**, COI, **Z**, O_C> where:
  - I_C is the set of the Control Part inputs
  - **E** is the set of propositional input variables of the Control Graph
  - CII is the Control Input Interpreter. Its role is to evaluate some conditions on the Control Part inputs of I_C, and therefore to give some propositional (Boolean) values to the elements of **E**
  - **F** is the set of propositional enabling variables
  - CG is the Control Graph:

    CG = <**S**, **E**, **F**, **Z**, **U**, δ, λ, β> where:
    - **S** is the set of propositional symbolic state variables

      δ is the propositional transition function (next state): $\delta: \mathbf{S} \times \mathbf{E} \times \mathbf{F} \rightarrow \mathbf{S}$
    - λ is the propositional output function:

      $\lambda: \mathbf{S} \times \mathbf{E} \times \mathbf{F} \rightarrow \mathbf{Z}$
    - β is the propositional updating function

      $\beta: \mathbf{S} \times \mathbf{E} \times \mathbf{F} \rightarrow \mathbf{U}$
    - A transition t of CG is defined as follows:

      t: $(\mathbf{s_i}, \mathbf{e_j}, \mathbf{f_k}) \rightarrow\!- (\mathbf{s_l}, \mathbf{z_m}, \mathbf{u_n})$ where:

      $(\mathbf{s_i}, \mathbf{s_l}) \in \mathbf{S}^2$, $\mathbf{e_j} \in \mathbf{E}$, $\mathbf{f_k} \in \mathbf{F}$, $\mathbf{z_m} \in \mathbf{Z}$, $\mathbf{u_n} \in \mathbf{U}$

      $\mathbf{s_l} \in \delta(\mathbf{s_i}, \mathbf{e_j}, \mathbf{f_k})$, $\mathbf{z_m} \in \lambda(\mathbf{s_i}, \mathbf{e_j}, \mathbf{f_k})$ and

      $\mathbf{u_n} \in \beta(\mathbf{s_i}, \mathbf{e_j}, \mathbf{f_k})$
  - **U** is the set of propositional updating variables (assigned when β is evaluated)
  - **Z** is the set of propositional output variables of CG (assigned when λ is evaluated)
  - COI is the Control Output Interpreter. Its role is to give values to the Control Part outputs from the values of the variables of **Z**.
  - O_C is the set of the Control Part outputs
- DP = <I_D, D, P, EFI, UFI, **F**, **U**, O_D> where:
  - I_D is the set of the Data Part inputs
  - D is the set of internal variables
  - P is the set of parameters (fixed characteristics of the system, which never change value)
  - EFI is the Enabling Function Interpreter. Its role is to evaluate some conditions on the Data Part inputs, on the internal variables and on the parameters and therefore to give some propositional (Boolean) values to the elements of **F**
  - **F** is the set of propositional enabling variables
  - **U** is the set of propositional updating variables
  - UFI is the Updating Function Interpreter. It contains all the functions that are used for calculating the new values of the Data Part outputs and of the internal variables of D. these updating functions are activated by the variables of **U**
  - O_D is the set of the Data Part outputs

2.2 Behaviour of an ISM model

The behaviour of an ISM model is described by the dynamic evolution both of the Control Graph, and of the internal variables of D.
The behaviour of CG is expressed by the sequence of fired transitions.
The interpretation of the transition t of Figure 2 is:
if $\mathbf{s_i}$ is true ($\mathbf{s_i}$ denotes the current state of the Control Graph), then if $\mathbf{e_j}$ is true (the values of the inputs of CP verify the conditions which make $\mathbf{e_j}$ be true), then, if $\mathbf{f_k}$ is true (the value of the inputs of DP, of the internal variables in D and of the parameters of P verify the conditions which make $\mathbf{f_k}$ be true), then:

- at the same time, $\mathbf{z_m}$ and $\mathbf{u_n}$ become true until the next transition firing on CG (the functions associated with $\mathbf{z_m}$ in COI update the Control Part outputs, and the functions associated with $\mathbf{u_n}$ in UFI update both the internal variables and the Data Part outputs

- the next state will be $\mathbf{s_l}$ ($\mathbf{s_l}$ will be true, while $\mathbf{s_i}$ will become false).

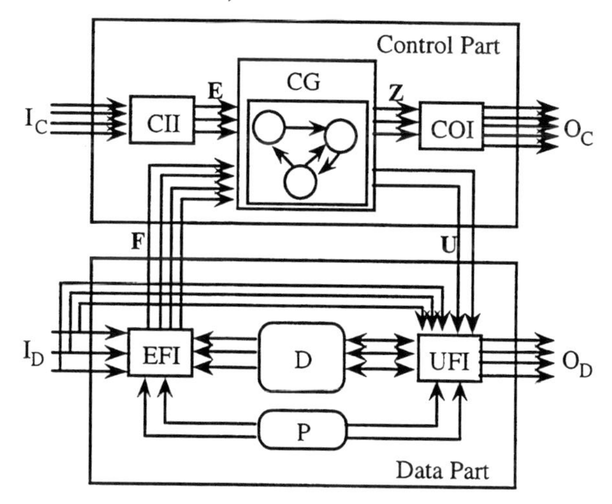

Fig. 1. Structure of the ISM model

A first model being described, it may be necessary in some cases (for example if one wants to perform some verification) to ensure that the Control Graph is non-ambiguous (deterministic) and/or completely specified (Larnac *et al.*, 1995). It is very important to note that these properties are studied on the couples (input/enabling variables) of each transition. These information appear in the Control Graph as propositional variables. However, these variables are not independent while they represent functions which are not exclusive. This means that verifying these properties requires to process some symbolic calculus on the interpretation of the propositional variables: the real inputs and internal data.

3. VERIFICATION OF THE ISM

A system being modeled by an ISM, it is then possible to implement some verification process. Once the behaviour of the model is perfectly defined (the functioning regarding time), the verification is based either on simulation or on proof of properties. The ISM handles both of them. The verification by properties proof is presented here.

3.1 Expression of the ISM behaviour in Temporal Logic

The proof of properties process requires a formal basis for reasoning. The ISM provides this support because of its clearly-defined semantics. The verification process consists in expressing properties of the system through the behaviour of the Control Graph, and then to formally prove that they are true. In order to perform this task, the behaviour of CG must be translated into an accurate formalism.

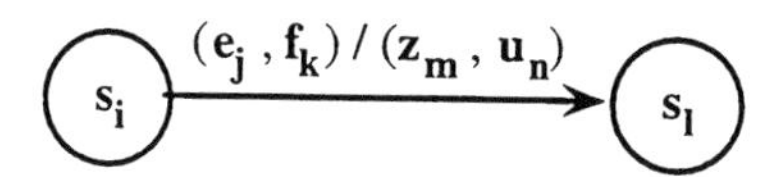

Fig. 2 - A transition of the Control Graph

Linear Time Temporal Logic (LTTL) has been chosen (Manna and Pnueli, 1982).

LTTL is an interpretation of a propositional modal logic in which time is discrete and which possesses the key property of being complete; this means that verifying the semantic validity of a formula comes down to syntactically verifying that this formula is a theorem.

The behaviour of the Control Graph of the ISM is expressed in temporal logic by a set of formulae, each of which being the translation of one transition of CG. Such a formula is called an *Elementary Valid Formula* (EVF). For example, the EVF of the transition of Figure 2, t: $(\mathbf{s_i}, \mathbf{e_j}, \mathbf{f_k}) \rightarrow (\mathbf{s_l}, \mathbf{z_m}, \mathbf{u_n})$ is the following:

$$\models \Box(s_i \wedge e_j \wedge f_k \supset \bigcirc s_l \wedge z_m \wedge u_n) \quad (1)$$

and its interpretation is that "it is always true ($\Box$ operator) that if $\mathbf{s_i}$ is true and if $\mathbf{e_j}$ is true and $\mathbf{f_k}$ is true, then, simultaneously, $\mathbf{z_m}$ and $\mathbf{u_n}$ are true, and $\mathbf{s_l}$ will be true at the next time ($\bigcirc$ operator)".

These formulae are symbolic, which means that there is no combinatorial explosion of their number. Moreover, the set of all the EVF gives an equivalent representation of the behaviour of the Control Graph and therefore constitutes the basis of all reasoning process for the proof of properties.

The analysis of a system which is modelled by an ISM may require a more global approach. This is the reason why the notion of Temporal Event (Et) has been defined as either a next state ($\bigcirc\mathbf{s_l}$), an output ($\mathbf{z_m}$) or an updating variable ($\mathbf{u_n}$). An Unified Valid Formula (UVF) then gathers all the conditions which involve a given Temporal Event Et.

$$\text{UVF(Et)} = \bigvee_{i,j,k:\ s_i \wedge e_j \wedge f_k \supset Et} (s_i \wedge e_j \wedge f_k) \quad (2)$$

The UVF is thus obtained by performing a logical OR of the right part of all the EVF which contain the Temporal Event Et in their left part. Note that it is also possible to extend the notion of temporal event to a n-future state, output or updating variable, as well as a sequence of states, outputs or updating variables.

3.2 Temporal Boolean Difference

The verification of most properties is obtained by application of the Temporal Boolean Difference (TBD) on some formula.

Definition. The Temporal Boolean Difference is a

temporal extension of the classical Boolean Difference (Kohavi, 1978). It was first defined (Magnier, 1990; Magnier *et al.*, 1994) for the formal verification of sequential machines.

The TBD of a formula A regarding a variable x is obtained by performing a logical exclusive-or between the logico-temporal formula A in which x equals 0 (or False) and the logico-temporal formula A in which x equals 1 (or True):

$$TBD(A,x) = \frac{\partial A}{\partial x} = A_{|x=0} \oplus A_{|x=1} \quad (3)$$

Interpretation. The resulting formula of TBD(A,x) contains all the conditions which make A change value (from True to False or False to True) when x changes value. It thus expresses the sensitivity of A with respect to x. So if TBD(A,x) equals False, A is totally independent of x.

3.3 Application to the verification of the ISM.

In order to verify properties of the ISM behaviour, the TBD is applied on Temporal Events regarding either a current state variable s_i, or a couple $(e_j \wedge f_k)$ of inputs/enabling variables of the Control Graph. This leads to the definition of the Difference Valid Formula (DVF):

$$DVF(Et,q) = \frac{\partial UVF(Et)}{\partial q} \quad (4)$$

where Et is a next state (Os_l), an output (z_m), or an updating variable (u_n), and q is a current state variable s_i, or a couple $(e_j \wedge f_k)$.

Note that this concept can be extended to both:

- Et being of a n-future state, output, or updating variable, or a sequence of states, outputs or updating variables,
- q being a n-future state, a n-future (input/enabling) couple of variables, a sequence of states, a sequence of (input/enabling) couples.

3.4 The verification process

The main properties that may be verified concern the states of the Control Graph, the output and/or the updating variables of this graph and sequence generation. In what follows, some properties are detailed (Vandermeulen *et al.*, 1995).

State properties. Many information linked to states are very interesting properties which help to better understand the behaviour of the system.

- Sink state: When the system reaches a sink state, while there exists no transition to leave it, the system is blocked. Note that some transition from a sink state to itself can exist. This means that the system will remain in this state when reached, but the data will go on evolving. Classically, verifying that s_i is a sink state consists in showing that all the EVF which contain s_i in their left part contain Os_i in their right part (any transition which leaves s_i goes back to s_i).
- Functional sink state: some states may not be structural sink states. But it is very interesting to highlight the conditions on the data which make a state become a sink state, when the conditions on the leaving transitions cannot be verified. For example, if s_i is the state which represents the optimal functioning of the system, it is very important to know all the possibilities (on the inputs and internal data) which make the system leave s_i. Searching the conditions which prevent the system to leave s_i comes down to study DVF(Os_i,s_i).
- Source state: The system can only leave a source state (or remain in it), but there exists no transition to reach it. A source state is often used as an initial state. Showing that s_i is a source state comes down to verify that either UVF(Os_i) is empty, or, if not, that the states which appear in UVF(Os_i) are s_i (the only way to reach s_i is through a transition which leaves s_i)
- Functional source state: it can be of great interest to know the conditions not to go back to a given state. The conditions which prevent the system to reach s_i are obtained from DVF(Os_i,s_j) for all $j \neq i$.
- Sensitivity of a state change regarding an input (and an associated enabling variable), a sequence of inputs, etc.

Output properties. Generally speaking, the only means for a user to get some information about a system consists in examining the outputs. It is therefore very important to be able to interpret these data, by calculating the DVF of one output variable (or sequence of output variables) regarding either a state, or a couple of (input/enabling) variables.

Sequence generation. Another interesting application of TBD is the generation of input sequences. The two main illustrations concern synchronizing and distinguishing sequences.

- A synchronizing sequence is an (input/enabling) variables sequence which, when applied to any state of CG will lead an unique n-future state. So if $(e \wedge f)^n$ is a synchronizing sequence, it means that there exists a state s_r such that:
  $\forall i$, UVF($O^n s_r$) $= s_i \wedge (e \wedge f)^n$
- A distinguishing sequence is an (input/enabling) variables sequence which, when applied to two different states of CG, produces two distinct output sequences.

4. THE CANVAS MODEL

A factory may be viewed as a complex organisation based on the interaction of human and material means, submitted to specific rules, strategies of evolution, etc. The modelling and the analysis of

hese kinds of structures are key problems. Indeed, it is difficult :
- to qualitatively or quantitatively analyse the consistency, the efficiency and the final impact of the entire structure of the factory behaviour on the product and on the production results,
- to analyse it, taking into account dynamical evolution of the full organisation, particularly when it is confronted to hazardous events,
- to model it if the user, who is in charge of this work, does not know all the characteristics of each level and mechanism used in the factory: decision level, management level or production level.

Therefore, it is necessary to develop methods, models and support tools able to help the user to clearly represent and decompose the system. He must specify all the entities which are involved into the factory behaviour, the different interactions between them, and his know-how.

In addition, the user may need to analyse the effects of a characteristic modification of an entity on the factory behaviour by searching limitations and weak points of the system.

A production, organisation or design process is composed of a succession of necessary steps to obtain the required result (product, rules, document, etc.).

For example, when describing a production system, one may consider the interactions of the set of production cells or machines and the human operators. At a higher level, one may consider the production design and validation levels which imply decision and management actions and other human responsibilities. Then some questions could be:
- Is the structure of the system coherent with respect to some given criterion?
- What are the execution rules of a given process?
- What is the added value of a particular step of a product or design process?
- What are the conditions in which two parallel steps may not be executed at the same time?
- Is Mr. X or Mrs. Y correctly employed?

Furthermore, it would be very interesting to know the consequences of the occurrence of an event (variation of a characteristic) on the behaviour of the factory. Finally, the modelling and analysis approaches may be powerful, user-friendly for a non specialist of the domain of complex models. The CANVAS model has thus be designed in order to fit all these requirements.

The CANVAS model allows us to describe low level models called sub-canvas. These sub-canvas are necessary to represent, decompose and manage an industrial complex system. For example, a factory behaviour and organisation may be described as a succession of steps, each one using actors, inputs and outputs. A sub-canvas is a structure containing a name and a set of fields.

Each field itself is described by a name and may contain typed data and/or references to other sub-canvas. A graphical chosen item named graphical form is then associated with each sub-canvas. A Generalisation relationship describes the possibility for a sub-canvas to inherit from another one. The Aggregation relation allows us to describe a sub-canvas which contains a set of other ones. Aggregation and Generalisation relationships allow us to decompose and to describe hierarchical levels in the system. Finally, a Use relation permits to describe dependencies between two sub-canvas.

The modelling process using the CANVAS model is based on three phases and associated sorts of users.

Modelling a complex system first consists in using the concepts of CANVAS to generate the set of sub-canvas, dedicated to the description of entities needed by the user to model the system. A sub-canvas contains fields which may represent static or dynamic characteristics. The user in charge of this phase is called administrator. He gives a set of sub-canvas skeleton completely defined for covering all aspects of the system modelling.

In the case of a real industrial project with Merlin Gerin Alès (Schneider Group), the sub-canvas are named Step, Actor, Function, Input and Output. A Step defines an activity at a given level of abstraction of the system. It is characterized by several fields such as actors (human and material) which are involved into this activity, inputs and outputs used for it and several functions describing how the step computes its outputs. Figure 4 constitutes a graphical representation of the components of the step.

A particular function, called Added Value, allows us to quantify qualitatively and/or quantitatively the objectives of the step. For example, the step called 'validation design' may evaluate added value 'quality of the validation', taking into account competence level of each actor and frequency of working group meetings. In the case of Merlin Gerin factory, about 20 added values have been defined.

The second phase consists in instanciating these sub-canvas in order to represent each entity in the real system. For example, it is necessary to define a production step, a scheduling command step, a technician actor, the sale service actor, the command input, etc. The user needed for this phase is called configurator. He specifies the possible contents of all the fields of each sub-canvas (name, definition domain and eventually unit), their hierarchy and other relations. For example, Mr.X is a technician who works in the maintenance team. These two factory components are actors even if Mr.X belongs to the maintenance team with particular abilities.

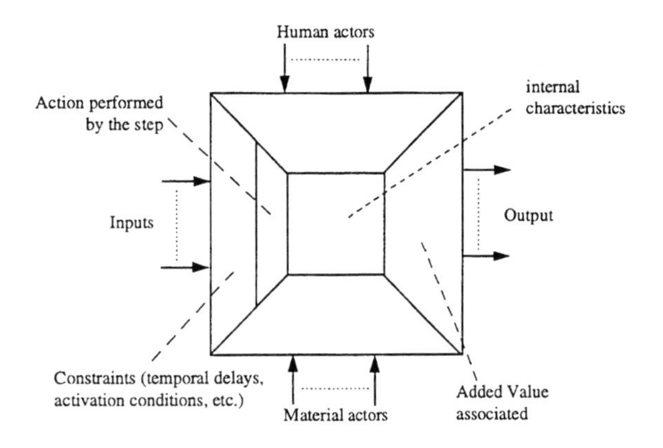

Fig. 4. the Step

Then, the final user, who is in charge of modelling the real system and the interactions between all sub-canvas, disposes of a specific components library of steps, actors, inputs, outputs and functions. This constitutes the third level of the use of the model. The components described in libraries are interconnected and specialized for describing a sub-system which may induce catastrophic situation. The obtained model is then composed of an interconnection of several Steps, Actors, Inputs and Outputs and uses specific functions and added values. The global modelling approach is then summarised in Figure 5.

CANVAS is a way to support the description of industrial complex systems but does not allow the verification of properties and the simulation of the system. The representation of the organisation of the factory with sub-canvas instances permits, in a first step, to empirically view some incoherences or inconsistencies at each level of description.

Furthermore, the CANVAS model of a system can be translated into an equivalent ISM model and therefore some relevant properties can be proven. For example, the behaviour of a step sub-canvas is immediately expressed by an ISM where:

- the different possible states of the step are defined by the different cases and associated actions; each of this state defines one of the Control Graph
- the inputs and outputs of the step are the inputs and outputs of the ISM and will make the CG evolve
- the actors are either inputs or internal variables of the Data Part of the ISM
- the internal characteristics are either internal variables or parameters
- the constraints constitute the enabling functions
- the actions are the updating and output functions

5. CONCLUSION

A new composite modelling and analysis approach based on the collaboration of CANVAS model and Interpreted Sequential Machine has been presented in this paper.

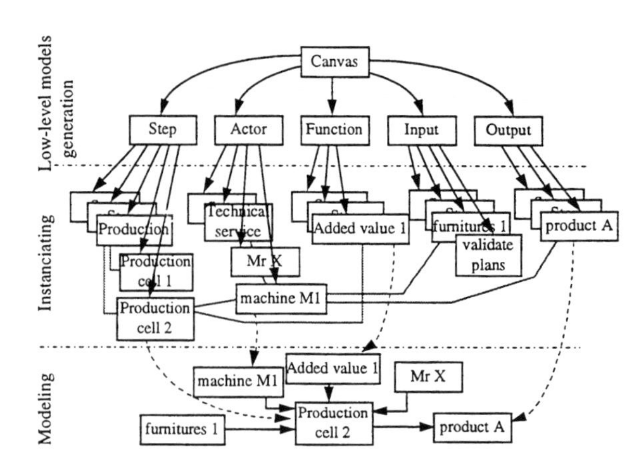

Fig. 5. The global modelling approach

It allows us to define low level models which are necessary for describing a particular industrial complex system. Interpreted Sequential Machine is then used for modelling behaviour of each entity and formally prove some temporal, structural and dynamic properties.

REFERENCES

Cheng, K.T. and Krishnakumar, A.S. (1993). Automatic functional test generation using the Extended Finite State Machine Model. *Procs. 30th ACM/IEEE Design Automation Conference.* USA

Hartmanis, J. and Stearns, R.E. (1966). *Algebraic Structure Theory of Sequential Machines.* Prentice Hall, Englewood Cliffs, N.J.

Kohavi, Z. (1978). *Switching and Finite Automata Theory.* Tata McGraw Hill, Computer Science Series

Larnac, M., Dray, G., Chapurlat, V. and Magnier J. (1995). Temporal and functional verification of a symbolic representation of complex systems. In: *Eurocast'95* (Lecture Notes in Computer Science), **10XX**, pp. nnn-nnn. Springer Verlag

Magnier, J. (1990). Représentation Symbolique et Vérification Formelle de Machines Séquentielles. *State Thesis.* University of Montpellier II

Magnier, J., Pearson, D. and Giambiasi, N. (1994). The Temporal Boolean Derivative applied to Verification of Sequential Machines. *Procs. European Simulation Symposium.* Istanbul, Turkey

Manna, Z. and Pnueli, A. (1982). How to cook a temporal proof system for your pet language. *Report n°STAN-CS-82-954 Department of Computer Sciences.* Stanford University

Vandermeulen, E., Donagan, H.A., Larnac, M. and Magnier, J. (1995). The Temporal Boolean Derivative applied to Verification of Extended Finite State Machine. *Computer and Mathematics with Application.* Vol. 30, n°2

INTEGRATED AND REACTIVE PART ROUTING POLICIES FOR MANUFACTURING SYSTEMS

Jean-Claude HENNET

L.A.A.S. - C.N.R.S., 7 avenue du Colonel Roche, 31077 Toulouse, FRANCE

Abstract: The considered manufacturing system provides several possible routes for processing the parts of a given product type. The proposed technique for implementing given loading ratios is through a cyclic valuated control Petri net. To take into account possible machine breakdowns, a supervision cyclic time-Petri net is superimposed onto the control net. In the one product-type case, the optimal routing probabilities are analytically obtained under product-form assumptions. The resulting control scheme combines the property of consistency with respect to optimal loading ratios, and automatic reactivity achieved without measuring the work-in-process on each route.

Keywords: Manufacturing systems, Optimal load flow, Load dispatching, Supervision, Petri-nets, Queueing network models

1. INTRODUCTION

In a job-shop manufacturing system, the existence of different possible routes for a given product-type may arise from the existence of several machines or production cells of the same type, from the use of flexible resources (robots, multi-purpose machines,...) or from tasks which can be achieved in different orders within a given job.

The use of cyclic schedules is appropriate for respecting production quota and average machine load balancing ratios. However, pure cyclic schedules poorly perform when demand and processing times fluctuate. In such conditions, a higher level of real-time reactivity can be achieved using state dependent policies, the current state of the system being characterized by work-in-process inventory levels or by waiting times.

The studied part routing policies are described by Petri net control graphs with control tokens and part tokens. The cyclic nature of the control graph and the arc valuations insure, on the average, the respect of prescribed routing ratios.

However, it is also important to avoid any blocking in the control network when one or several machines break down. It is proposed to superimpose onto the normal control network a supervision control network which gets activated only when the processing time of one manufacturing cell has become very low. It then causes a by-pass and a fast circulation of control tokens to the next route enabling place.

A structural analysis of the control and supervision graph provides conditions for its liveness, and allows for the determination of control arc valuations for which, on the average, the reference routing ratios are satisfied when there is no breakdown.

An efficient technique is proposed for optimizing the routing probabilities in the one product-type case.This paper provides explicit expressions of the optimal steady-state routing probabilities. The optimal mean values are computed under the simplifying assumptions of exponentially distributed input and service times, under a Bernoulli splitting of the input flow. The first step of the method determines which servers should be used and which ones should not be used by the policy. Then, the optimal routing parameters are obtained as the optimal solution of an unconstrained problem.

The performance of the proposed part routing control structure is finally evaluated by simulation, the manufacturing system and its control and supervision being described by a stochastic time-Petri net, with different values of processing time parameters. It is shown that the proposed part routing policy achieves a reasonable trade-off between decisional integration, through the respect of planned routing ratios, and efficient resource utilization, reacting to real-time changes of machine availability.

2. CONSTRUCTION OF A CYCLIC ROUTING POLICY WITH SUPERVISION

In terms of mean response time, superiority of dynamic load balancing policy versus static load balancing policy has been shown by many authors (Ephremides *et al.*, 1980), (Nelson and Philips, 1993). Furthermore, cyclic sequencing is a good intermediate stage in the design of a flexible loading device in which sequencing flexibility and routing flexibility can be introduced (Ohl *et al.*, 1993).

2.1 *Construction of a cyclic routing device*

The part routing device is represented by a Petri net. The part input node, labelled 0, is a free-choice place, linked by n arcs to the n route input transitions. On Fig.1, the firing of transition i corresponds to a part release on route i. A cyclic firing sequence for routes 1,...n is obtained by introducing n places for decision tokens and valuated arcs between these places and the route input transitions. The logics of such a cyclical loading device are represented on Fig.1.

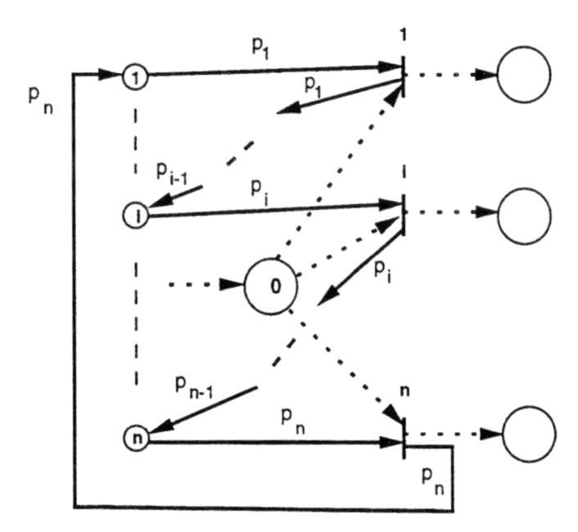

Fig. 1. A Petri net representation of the loading mechanism

To evaluate the average steady state routing ratios, an infinite feeding of tasks at place 0 can be assumed. Then, place 0 can be suppressed, and only the distribution of tokens determines the system evolution. The cyclic Petri net characterizing the loading device is represented by the following incidence matrix :

$$C = \begin{bmatrix} -p_1 & 0 & . & p_n \\ p_1 & -p_2 & . & 0 \\ . & . & . & . \\ 0 & . & . & -p_n \end{bmatrix} \qquad (1)$$

The invariant T-semiflot : $u \in \Re^n$ such that $Cu = 0$ is given by :

$$u_i = \frac{\prod_{j \neq i} p_j}{\sum_{k=1}^{n} [\prod_{j \neq k} p_j]} \qquad (2)$$

If the Petri net of Fig.1 is live, the invariant T-semiflot defined by the firing sequence u characterizes the cyclical stationary behaviour of the Petri net (Brams, 1983). To obtain the desired loading ratios $\alpha_1, ...\alpha_n$, it suffices to determine valuations $p_1, ..., p_n$ which satisfy the set of equations:

$$u_i = \alpha_i \text{ for i=1,...,n}$$

or equivalently, for $p_i > 0, \ i = 1, ..., n$,

$$\prod_{j \neq i} p_j = \alpha_i P \text{ with } P = \sum_{k=1}^{n} [\prod_{j \neq k} p_j]. \qquad (3)$$

From expression (2), it is clear that parameters p_i can be multiplied by any positive constant without changing the routing ratios. Furthermore, for the set of parameters $(p_1, ..., p_n)$ to be used as a set of valuations on the arcs of the control Petri net, it has to be approximated by a set of positive integers. Thus, if system (3) can be solved for a set of positive real parameters $(p_1, ..., p_n)$, the order of magnitude of parameters p_i has to be selected so as to achieve a trade-off between

- a sufficient precision in spite of the rounding to the nearest integers

- a "reasonable" number of tokens in the Petri net.

To solve the system of equations (3) in terms of real positive parameters, take the natural logarithm of each term of the equations to obtain the following set of linear equalities (with unconstrained real variables and P arbitrarily set to 1) :

$$Ay = b \text{ with } A = \begin{bmatrix} 0 & 1 & . & . & 1 \\ 1 & 0 & 1 & . & 1 \\ . & . & . & . & . \\ 1 & . & 1 & 0 & 1 \\ 1 & . & . & 1 & 0 \end{bmatrix} \qquad (4)$$

$$y^T = [y_1, ..., y_n] \text{ with } y_i = Log(p_i)$$
$$b^T = [y_1, ..., y_n] \text{ with } b_i = Log(\alpha_i).$$

Matrix A is square and regular for $n \geq 2$ and its inverse is explicitely given by :

$$A^{-1} = \frac{1}{n-1} \begin{bmatrix} -(n-2) & 1 & . & 1 \\ 1 & -(n-2) & . & 1 \\ . & . & . & 1 \\ 1 & . & 1 & -(n-2) \end{bmatrix}.$$

Solution of (4) is $y = A^{-1}b$, and the corresponding vector of real valuations, $p = [p_1, ..., p_n]^T$ has positive components. It is obtained by :

$$p_i = e^{y_i} \quad \forall i \in (1, ..., n). \tag{5}$$

For a normalized value of p_1 equal to π_1 ($\pi_1 = 1$ or $\pi_1 = 10$ for instance), the vector of integer valuations $\pi = [\pi_1, ..., \pi_n]$ is obtained as :

$$\pi = round(p). \tag{6}$$

where *round*(.) rounds the components of a positive real vector to the closest non-negative integers.

Now, the condition to be satisfied to obtain the desired routing ratios is liveness of the Petri net. Assuming that there is no blocking on any of the n routes, a necessary and sufficient condition for liveness can be obtained as an extension to a result by Ohl et al. (Ohl *et al.*, 1993) to the case of n routes. Vector $x = [1, \ldots, 1]^T$ is an invariant P-semiflot of the control Petri net : $x^T C = 0$. Therefore, if m_i denotes the number of tokens in place i, the total number of tokens in the control Petri net, $N = \sum_{i=1}^{n} m_i$, is a constant along the system evolution.

- if $N \leq \sum_{i=1}^{n} \pi_i - n$, all the transitions may be simultaneously disabled, for some particular markings,

- if $N > \sum_{i=1}^{n} \pi_i - n$, there is always at least one transition enabled, and since the control Petri net is an elementary cycle, all the transitions are live.

Therefore, the liveness condition is as follows :

The control Petri net is live if the number of tokens in the control Petri net, N, satisfies :

$$N \geq \sum_{i=1}^{n} \pi_i - n + 1. \tag{7}$$

Even for $N = \sum_{i=1}^{n} \pi_i - n + 1$, if $n > 2$, several transitions may be simultaneously enabled, and some decisions remain to be taken. And release flexibility increases with N, for $N > \sum_{i=1}^{n} \pi_i - n + 1$. This feature is basic for avoiding immediate general blocking when a failure occurs.

2.2 *Introducing reactivity through supervision*

Under condition (7), an increase of the number of tokens in the control graph provides some routing flexibility. However, supervision is needed to fastly react to strong throughput perturbations. In particular, if the breakdown of a machine occurs on route i, the throughput on this route considerably decreases or becomes null. The control tokens tend to accumulate in place i, and without a supervision mechanism, the whole system would become blocked.

The main idea of the proposed supervision network is to by-pass route i whenever the service time on this route gets greater than or equal to a threshold value θ_i. On figure 2, place i is now connected to place $i+1$ by two transitions :

- the normal route input transition, which requires the availability of a resource token to be enabled,

- the by-pass transition, which is a time-transition in the sense of Merlin (Merlin, 1976), and which is enabled only if at least p_i tokens have been present in place i for a time greater than or equal to θ_i.

From the token circulation viewpoint the result of the two firings are identical : p_i tokens have been transferred to place $i+1$, but in the latter case, the part has not be routed to route p_i tokens have been transferred to place $i+1$, but in the latter case, the part has not be routed to route i. It is now candidate for being routed to route $i+1$.

Clearly, the arc valuations of the supervision subnetwork are identical to the arc valuations of the supervision subnetwork. Structurally, the supervision network is thus live under condition (7). However, in the absence of a machine breakdown, the service time on route i should not be longer than θ_i, and thus the supervision transition is never fired in normal running conditions.

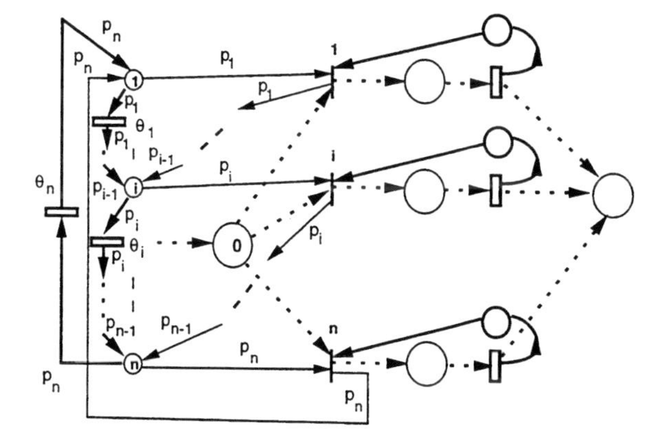

Fig. 2. Control device with Supervision

3. THE STEADY-STATE FLOW DISPATCHING POLICY

Consider an open queueing network which represents the parallel structure of the manufacturing system under a single type of products. To perform a mean value analysis of the network under steady state random input and service times, it is assumed that the input flow admits a Poisson distribution with mean arrival rate $\lambda > 0$. Upon its arrival, each part (of the given class) is routed to the queue of server i with probability α_i.

The *Bernoulli splitting* process does not change the memoryless property of the input, and under such an approximation, the controlled network admits a product-form (Baskett *et al.*, 1975). The optimal Bernoulli parameters are the optimal steady-state assignment probabilities for the various possible routings. In general, they can be computed by classical iterative algorithms (Frein *et al.*, 1988), (Smaili and Hennet, 1992). Here, explicit expressions of the optimal routing parameters are provided under the above product-form assumptions for the system represented on Fig.3.

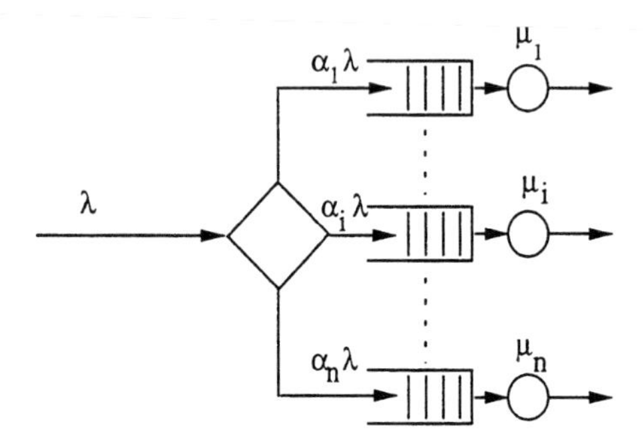

Fig. 3. A system of n parallel queues

The utilization rate of server i is defined by :

$$\rho_i = \frac{\alpha_i \lambda}{\mu_i}. \quad (8)$$

If $\rho_i < 1$ for $i = 1, ..., n$, the average value of the time spent in the system, T, (or mean response time) is (Kleinrock, 1976):

$$E(T) = \sum_{i=1}^{n} \frac{\alpha_i}{\mu_i - \alpha_i \lambda} \quad (9)$$

Without loss of generality, the n servers can be ordered in the decreasing order of their mean service rate:

$$\mu_1 \geq ... \geq \mu_n. \quad (10)$$

The constraints of the problem are related to the following requirements:

(1) The control policy has to be feasible. This condition is satisfied under the following constraints:

$$0 \leq \alpha_i \leq 1 \;\; \forall i \in (1, .., n) \; ; \quad (11)$$

$$\sum_{i=1}^{n} \alpha_i = 1 \quad (12)$$

(2) Stability of the network requires restrictions:

$$\frac{\alpha_i \lambda}{\mu_i} < 1 \text{ for } i = 1, ..., n. \quad (13)$$

the existence of admissible solutions to the steady state control problem is conditionned by the following result (Kleinrock, 1976):
A necessary and sufficient condition for the existence of a set of routing parameters ($\alpha_i; i = 1, ..., n$) satisfying constraints (11), (12), (13), is:

$$\lambda < \sum_{i=1}^{n} \mu_i \quad (14)$$

4. OPTIMIZATION OF ROUTING PARAMETERS

Assume that condition $\lambda < \sum_{i=1}^{n} \mu_i$ is satisfied. The optimization problem for criterion $E(T)$ takes the form:

$$\min_{\alpha_1, ..., \alpha_n} \sum_{i=1}^{n} \frac{\alpha_i}{(\mu_i - \alpha_i \lambda)} \quad (15)$$

subject to (11), (12), (13).

Constraints (11) and (13) can be replaced by:

$$0 \leq \alpha_i \leq \min(1, \frac{\mu_i}{\lambda}) \text{ for } i = 1, ..., n. \quad (16)$$

Constraints are linear and within its set of constraints, this problem is convex:

$$\frac{\partial^2 E(T)}{\partial \alpha_i^2} = \frac{2\mu_i \lambda}{[\mu_i - \alpha_i \lambda]^3} > 0 \text{ for } i = 1, \ldots n. \quad (17)$$

4.1 *The unconstrained optimality conditions*

Consider first the case when the optimum of the constrained problem is also the optimum of the unconstrained problem, namely, when the global minimum of (15) satisfies (12) and (16). In this case, the optimal solution is simply obtained by solving the first order optimality conditions of the unconstrained problem, that is, for $i = 1, ..., n$:

$$\frac{\partial E(T)}{\partial \alpha_i} = \frac{\mu_i}{[\mu_i - \alpha_i \lambda]^2} = 0 \quad (18)$$

Whenever the unconstrained optimum is identical to the constrained optimum, the optimal values of the parameters satisfy, for any $i \in (1, \ldots, n)$ and for any $j \in (1, \ldots, n)$,

$$(\mu_j - \alpha_j \lambda) = \sqrt{\mu_j} \frac{(\mu_i - \alpha_i \lambda)}{\sqrt{\mu_i}}. \quad (19)$$

Summing over j the 2 terms of this equation yields, for $i \in (1, \ldots, n)$,

$$\alpha_i = \frac{1}{\lambda}(\mu_i - \tau_n \sqrt{\mu_i}), \text{ with } \tau_n = \frac{(\sum_{j=1}^{n} \mu_j - \lambda)}{\sum_{j=1}^{n} \sqrt{\mu_j}}.$$

By construction, $\sum_{i=1}^{n} \alpha_i = 1$. The global minimum of (15) is feasible (and therefore optimal) for the constrained problem if and only if:

$$0 \leq \mu_i - \tau_n \sqrt{\mu_i} \leq \min(\mu_i, \lambda), \ i = 1, \dots n. \quad (20)$$

Stability condition (14) implies that τ_n is positive. Then, condition (20) can be replaced by:

$$0 \leq \mu_i - \tau_n \sqrt{\mu_i} \leq \lambda \text{ for } i = 1, \dots n \quad (21)$$

and the left part of these inequalities becomes equivalent to:

$$\mu_i \geq {\tau_n}^2 \text{ for } i = 1, \dots n. \quad (22)$$

4.2 *Solving the constrained problem*

If the left part of inequality (21) is violated for some $i_0; 0 < i_0 \leq n$, it is also violated for any $i \ ; i_0 \leq i \leq n$. Then, a restricted choice problem can be formulated, for which $\alpha_i = 0$ for $i = i_0, ..., n$. To show the relevance of the problem, define the *optimality parameter* associated to routings restricted to the m first routes:

$$\tau_m = \frac{(\sum_{i=1}^{m} \mu_i - \lambda)}{\sum_{i=1}^{m} \sqrt{\mu_m}}. \quad (23)$$

Under the convention $\mu_{n+1} = 0$, τ_m can be defined for $m = 1, \dots n+1$. It is not difficult to show that its evolution complies with the following lemmas:

Lemma 2
The evolution of τ_m for $m = 1, \dots n-1$, follows the rule:

If $\mu_{m+1} < \tau_m \sqrt{\mu_{m+1}}$ then, $\tau_{m+1} < \tau_m$.

If $\mu_{m+1} = \tau_m \sqrt{\mu_{m+1}}$ then, $\tau_{m+1} = \tau_m$.

If $\mu_{m+1} > \tau_m \sqrt{\mu_{m+1}}$ then, $\tau_{m+1} > \tau_m$.

Lemma 3
The evolution of the optimality parameter, *τ_m is increasing with m for $1 \leq m \leq m^*$.*
It takes its maximal value for m^* ($1 \leq m^* \leq n$), which is the smallest index such that:

$$\begin{cases} \sum_{i=1}^{m^*} \mu_i > \lambda, \\ \mu_{m^*+1} \leq \tau_{m^*}^2 \end{cases} . \quad (24)$$

Then, the value of τ_m monotonously decreases with m for $m^* < m \leq n+1$.

The constrained problem can then be solved using the following proposition:

Proposition 1
Under the feasibility condition $\lambda < \sum_{i=1}^{n} \mu_i$, consider the smallest index, m^*, with $1 \leq m^* \leq n$ satisfying conditions (24). Then, the optimal choice of routing parameters satisfies:

$$\begin{cases} \alpha_j = 0 \text{ for } m^* < j \leq n, \\ \alpha_i = \frac{1}{\lambda}(\mu_i - \tau_{m^*}\sqrt{\mu_i}) \text{ for } i = 1, ..., m^*. \end{cases} \quad (25)$$

Proof

- *Feasibility of the proposed policy :* The set of routing parameters defined in Proposition 1 satisfies:

$$\sum_{i=1}^{n} \alpha_i = \sum_{i=1}^{m^*} \alpha_i = \frac{1}{\lambda}(\sum_{i=1}^{m^*} \mu_i - \tau_m \sum_{i=1}^{m^*} \sqrt{\mu_i}) = 1.$$

If $m^* = 1$ satisfies (24), relations (25) define the set of feasible routing parameters : ($\alpha_1 = 1$, $\alpha_i = 0$ for $i = 2, \dots n$).

If $m^* \geq 2$, since $\mu_i \geq \mu_{m^*}$ for $i = 1, \dots m^*$, relation $\mu_{m^*} > \tau_{m^*}^2$ with $\tau_{m^*} > 0$, implies:

$$\mu_i - \tau_{m^*}\sqrt{\mu_i} > 0 \text{ for } i = 1, \dots, m^*. \quad (26)$$

Then, from (25), $\alpha_i > 0$ for $i = 1, \dots m^*$. The parameters of Proposition 1 (25), define a feasible solution to optimization problem (15).

- *Optimality among policies satisfying (24) and (25) :* Suppose the existence of an index k; $m^* < k \leq n$ also satisfying conditions (24). The selected routing parameters associated to the k routes problem is:

$$\alpha_j = 0 \text{ for } k < j \leq n \quad (27)$$

$$\alpha_i = \frac{1}{\lambda}(\mu_i - \tau_m \sqrt{\mu_i}), \text{ for } i = 1, ..., k. \quad (28)$$

Denote $J_k = \sum_{i=1}^{n} \frac{\alpha_i}{(\mu_i - \alpha_i \lambda)}$ with α_i defined by (27), (28). Replace α_i by its expression to obtain:

$$J_k = \frac{\sum_{i=1}^{k} \mu_i - \lambda}{\lambda \tau_k^2} - \frac{k}{\lambda}$$

If both m^* and k satisfy relation (24), use Lemma 3 to obtain $\tau_k \leq \tau_{m^*}$ and the majoration:

$$J_k \geq J_{m^*} + \frac{(k - m^*)\mu_k}{\lambda \tau_{m^*}^2} - \frac{(k - m^*)}{\lambda}$$

Then, $\mu_k \leq \mu_{m^*+1} \leq \tau_{m^*}^2$ implies:

$$J_k \geq J_{m^*}$$

The choice of the smallest value of m for which relations (24) are satisfied is therefore optimal in the class of policies considered in this paragraph.

- *Optimality relatively to any other policy :* By construction, the policy is optimal if $m^* = n$.

If $m^* < n$, the proposed solution is the best among those satisfying

$$\alpha_j = 0 \text{ for } m^* < j \leq n.$$

Under the convexity property of the problem, it now suffices to show that the criterion cannot decrease under any infinitesimal admissible move with $\delta\alpha_j > 0$ for any j; $m^* < j \leq n$, starting from solution (25).

$$\delta E(T) = \frac{1}{\mu_j}\delta\alpha_j + \sum_{i=1}^{m^*} \frac{\mu_i}{[\mu_i - \lambda]^2}\delta\alpha_i \quad (29)$$

under the admissibility constraint:

$$\delta\alpha_j + \sum_{i=1}^{m^*} \delta\alpha_i = 0 \quad (30)$$

Using relation $\frac{\mu_i}{[\mu_i-\lambda]^2} = \frac{1}{\tau_{m^*}^2}$, relation (29) takes the form:

$$\delta E(T) = \frac{1}{\mu_j}\delta\alpha_j + \frac{1}{\tau_{m^*}^2}\sum_{i=1}^{m^*} \delta\alpha_i \quad (31)$$

Conditions (24) imply:

$$\mu_j \leq \tau_{m^*}^2 \text{ for } j = m^* + 1, \ldots, n \quad (32)$$

Hence,

$$\delta E(T) = (\frac{1}{\mu_j} - \frac{1}{\tau_{m^*}^2})\delta\alpha_j > 0 \quad (33)$$

□

5. EVALUATION OF THE ROUTING CONTROL POLICY

The routing control policy has been evaluated by simulation, using the software MissRDP (IXI, 1996). Several results have been shown by simulation :

- In normal running conditions, the cyclic control scheme performs better than the Bernoulli dispatching rule, which randomly selects the route according to optimal steady-state routing probabilities.

- The proposed controlled scheme is outperformed by the shortest route policy in the one-product case. However, the shortest route policy requires the observation of the work-in process on each route. Furthermore, the shortest route policy is unable to satisfy loading ratios which do not correspond to a balanced loading of the different routes. In this sense, the proposed routing mechanism is better adapted to an integrated planning and scheduling framework in which the loading ratios are not locally determined, but rather imposed by the planning level.

- The supervision enabling times, $\theta_1, ..., \theta_n$ determine the rapidity of reaction to increases of processing times on routes. A careful choice of these time parameters is required to perform a good trade-off between a fast reactivity to perturbations (different possible running conditions of machines : fast, normal, slow, down) and a sufficient robustness of routing allocations in heavily perturbed conditions.

REFERENCES

Baskett, F., K.M. Chandy, R.R. Muntz and F.G. Palacios (1975). Open, closed and mixed networks of queues with different classes of customers. *J.ACM,* **vol.22, no.2,** *pp.248-260.*

Brams, G.W. (1983). *Réseaux de Petri : Théorie et Pratique.* Masson.

Ephremides, A., P. Varaiya and J. Walrand (1980). A simple dynamic routing problem. *IEEE Trans. Automatic Control,* **vol. 25, no. 4,** *pp.690-693.*

Frein, Y., Y. Dallery, J. Pierrat and R. David (1988). Optimisation du routage des pièces dans un atelier flexible par des méthodes analytiques. *RAIRO-APII,* **vol. 22,** *pp. 489-508.*

IXI, Company (1996). *MissRDP Software, under development.* 87/89 rue du Gouverneur Général Eboué, 92130 Issy les Moulineaux, France.

Kleinrock, L. (1976). *Queueing Systems, vol.1.* Wiley Interscience.

Merlin, P. (1976). Methodology for the design and implementation of communication protocols. *IEEE Trans. Communication,* **vol. 24, no. 6.**

Nelson, R.D. and T.K. Philips (1993). An approximation for the mean response time for the shortest queue routing with general interarrival and service times. *Performance Evaluation,* **vol. 17,** *pp.123-139.*

Ohl, H., E. Castelain and J.-C. Gentina (1993). State dependent release control in flexible manufacturing systems. *IEEE-SMC Conf., Le Touquet.*

Smaili, K. and J.C. Hennet (1992). Optimisation du routage des pièces dans un atelier flexible à contraintes de capacité locales. *RAIRO-APII,* **vol. 26,** *pp. 227-252.*

SOME PROBLEMS OF MODELLING THE COMPLEX MATERIAL FLOW SYSTEM

Đorđe N. Zrnić

Faculty of Mechanical Engineering University of Belgrade
11000 Beograd, 27. Marta 80, Yugoslavia

Abstract: The results of the modern design methods are objectivisation of the thinking process, which is important for the new system development. The aim of objectivisation of the design is obvious: it should enable a new approach in the problem solving and thinking of the designer, enriches the large quantity of new informations, which are important for the system designing. Particular attention is dedicated to application of the queuing theory and selection of the appropriate model. Some characteristics results are given for automated storage/retrieval system.

Keywords: System design, Stochastic modeling, Queuing theory, Deterministic behaviour, Database.

1. INTRODUCTION

The results of the modern design methods is objectivisation of the thinking process, which in the traditional way of designing, depends mostly on the designer. The aim of objectivization of design is obvious, it should enable the new approach in problem solving and enrich the designer thinking with a large quantity of new informations, facts and ideas, which are important for designing at the system level.

Designing is a complex activity whose success depends on correct utilization of mathematics and specialization, skill and experience. It should be noted that the probability for attaining a successful design by identification with only one of them is very small. The need of contemporary practice is to explain the designing essence which should be expressed in the aspect of standard procedure. In references it is frequently refereed to the designing sequence, methods and design contents, but nothing to the designing results. To find a reliable base for thinking, one should try to define the design not just on the base of the process moving, but on its results (Jones, 1982, Zrnić, 1993).

The designer should predict the final result of his design and to define the necessary criteria for attaining the needed results. The main difficulty for the designer is to predict on the base of the given data some future state, which originates only if the prognosis are exact.

The aim of this paper is to show some elements of developed procedure for modeling and designing the material handling systems and to point out relevant problems in modeling and designing. Because of the more complex systems, with different level of links between elements and with mutual influences which could be deterministic or defined by stochastic values, the particular problem presents the selection of appropriate model. The references give different types of queuing theory models, but there are no indications of adequate use in problem solving. To find the difference between the theory and practice, the research is conducted in industry to gather statistical data of behaviour of some processes. It is remarked that empirical law of relevant factors of material flow is possible to describe with Erlang, and normal distribution. Knowledge of these behaviour laws, guide to adequate description of material flow, the database formed used as a source of knowledge in systems modeling. The purpose of this work is to estimate possible errors by using queuing theory models and to predict some processes in simulations study. The designer is able to get more sophisticated results(Zrnić, 1993).

An example of the practical applications of the simulation study for input/output zone of the automated storage/retrieval system is given. It is analyzed influence of the input main flow and service time distribution of the storage machines.

2. COMPLEX MATERIAL HANDLING SYSTEMS

Designing of complex transportation systems, presents a challenge for the designer, from the viewpoint of selection and facilities layout, and appropriate corresponding software for automated guided systems. Because of the more complex systems with huge dimensions, with different level of links between elements (from very slack to very rigid), and with mutual influences which could be deterministic or defined by stochastic values, their analysis and solution selection for the given design task could be found only by the modeling procedure application. System approach is used as a way which gives the best results, and model as an investigating medium which contributes to complex reality observing.

The basic problems of these systems consist of: defining of the system (sub-systems), facilities layout, material flow and space efficiency. Facilities layout presents a particular problem of workshop design (complex technological requirements, required flexibility, frequency of transport requirements, etc.). Layout in designing of warehouses and transshipment systems has a relative simple form, but the particular problem is to find the solution of material flow.

The aim of this paper is to show some elements of the sophisticated procedure for transportation systems designing and to show the importance and some problems of modeling and designing from the type selection to the final solution. In the modeling process originates a sequence of particular problems, which should be specially noted (Zrnić, 1980, Zrnić, 1993):

A. System - surrounding, defining the system boundaries,
B. Sub-system determining (function linked for the subsystem or spaced to the several sub-systems),
C. Defining of elementary sub-systems with introducing the notion of the knot point (represents the bottle neck). Bottle necks could seriously menace the whole system performance. They appear when the incoming flow amounts to the capacity of serving and at that moment saturation appears (the level of congestion which makes the system disturbing). For the strongly less values of the level of congestion then it is predicted in the deterministic model, the saturation appears, for the case when the stochastic fluctuations in the mechanism of comings and servicing the requirements, are ignored.
D. Defining and selection of structure model.
E. Defining the criteria of solution selection.
F. System modeling - theory and practice.

The complexity of material handling systems requires particular methodology for model choice. Analytical models which are frequently used (models of queuing theory) for analysis of global solutions could estimate the systems performances. However, even with the simplification and decomposition of system, it is not always possible to adequately set the corresponding analytical model.

The particular problem presents the selection of the queuing theory model. References give different types of theoretical models but there are no indications of adequate use in problem solving and the difference between the theory and practice (Kleinrock, 1976, Cooper, 1981 etc.). To analyse this problem we have to start whit the classification scheme, Kendall, Lee, Taha, given as: (x/y/z):(u/v/w), where the symbols are:
x - arrival or interarrival distribution
y - service time distribution
z - number of service channels C
u - service discipline
v - maximum number in the system
w - size of the population

The following codes are used for x and y:
M - Poisson or equivalently exponential distribution
GI - general independent distribution
G - general distribution
D - deterministic process
E_K - Erlangian distribution with k phases
HE_k - hyper exponential distribution with k phases.

The symbols z,v and w are replaced by numerical designation. The symbol u is replaced by the following:

FCFS - first come - first served
LCFS - last come - first served
SIRO - service in random order
SPT - shortest processing time
GD - general service discipline
HELPF - full help between the servers
HELPS - partial help between the servers.

Additionally, a superscript is attached to the first or the second symbol if bulk arrivals or service is used ($M^{(b)}/M/C$ ...or $M/M^{(b)}/C$ ...). This designation is commonly used in the queuing literature.
The review of the most used queuing models given in the references is given below:

Poissons models:

(M/M/C):(FCFS/N/∞)	b=const.; b≠const.
(M/M/C):(FCFS/∞/∞)	(M/M/1):(FCFS/∞/∞)
(M/M/C):(GD/N/∞)	(M/M/1):(GD/N/∞)
(M/M/C):(GD/∞/∞)	(M/M/1):(GD/∞/∞)
(M/M/C):(PRI/∞/∞)	(M/M/C):(HELPF/N/∞)
(M/M/C):(GD/K/K)	(M/M/C):(HELPF/∞/∞)
(M/M/C):(GD/K/K)	(M/M/C):(HELPS/N/∞)
($M^{(b)}$/M/C):(GDFS/∞/∞)	(M/M/C):(HELPS/∞/∞)
(M/$M^{(b)}$/C):(GDFS/∞/∞)	

Non-Poissons models:

(M/G/1):(GD/∞/∞)	(GI/M/1):(GD/∞/∞)
(M/E_K/1):(GD/∞/∞)	(E_K/M/1):(GD/∞/∞)
(M/G/1):(PRI/∞/∞)	(D/M/1):(GD/∞/∞)

Specific non-Poissons models:

(M/E_2/2):(GD/2/∞)	(E_2/M/2):(GD/N/∞)
(M/HE_2/1):(GD/∞/∞)	(HE_2/M/1):(GD/N/∞)
(M/HE_2/2):(GD/2/∞)	(HE_2/M/2):(GD/N/∞)
(E_2/ME/1):(GD/N/∞)	(HE_2/E_2/1):(GD/1/∞)

The particular problem presents the classification of real processes (deterministic/stochastic), and the corresponding model application. Stochastic influences can be divided in the two categories: environment and system influences. Environment influences refer to properties of the system surrounding which affect the system and its performance. The stochastic influences in the system (process, layout, transportation paths, crossings, flow dynamic, sinhronizations, etc.), are possible to take into consideration and control during the modeling procedure.

From the viewpoint of designing practice, deterministic approach is more simple, but the possibility of its application is limited. Stochastic approach more realistic describes process and its dynamic character. Knowledge of these behavior laws, guide to adequate description of material flow, is used as a source of knowledge in system modelling (Zrnić, 1980).

It is created the database, on the basis of conducted researches, which is consisting of noticed laws which are presented in the material flow systems. The data were gathered by recording of some systems which enter the scope of industry in Belgrade. It is selected 15 representative systems and the gathering of data has been done at more than 100 entireties (workshops, manufacturing lines, warehouses, etc.). Those recordings were encompassed the work of the system, processes, the work of some transporting devices, etc. The data gathered by recording have been statistically prepared, it was also done the testing, and the hypotheses about belonging of data to corresponding theoretical distributions were created (Zrnić and Petrović, 1994).

The results which are systematize could be divided in two basic groups:

- system as a entirety (manufacturing, assembling, warehousing, etc.), and

- subsystem of material flow (transportation devices, unit loads (TU), etc.).

The creation of the database was shown that exist general laws of behaviour of some manufacturing and warehousing systems and processes. It was concluded: it is possible to describe the empirical law of relevant factors of material flow with theoretical distribution (Erlang, E_k, k=1,2,...∞, and normal distribution). The noticed laws of behaviour have for the aim the growing of quality of the designing process, and especially new systems when all relevant data about presented processes are not accessible to the designer. Those given laws have for the aim to give the general character of behaviour of systems or devices, so during the process of modelling a designer is approaching in an easier way to the field of real solutions and can to estimate the possible mistake during usage some models of the queuing theory.

Some examples of the real processes behaviour from the research study are given:
1) Behaviour of work of forklift truck in dependence on moving distance, (Fig. 1)
2) Work of storage machine, (Fig. 2)
3) Work of storage machine - manual order picking, (Fig. 3)
4) Process of manual paletisation and depaletisation, (Fig. 4).
5) Input in the manufacturing system, (Fig. 5), etc.

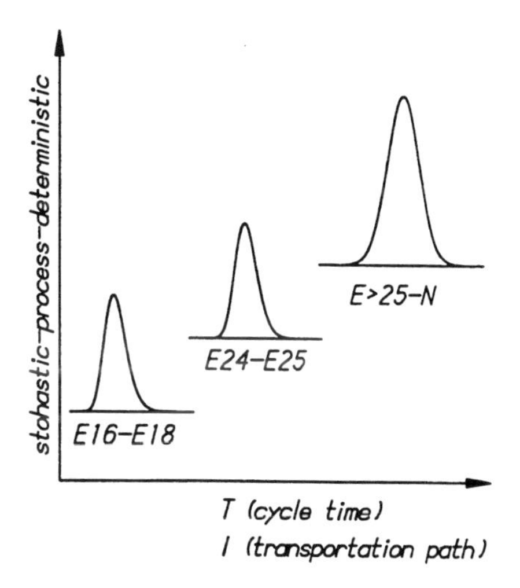

Fig. 1. Forklift truck service time distribution.

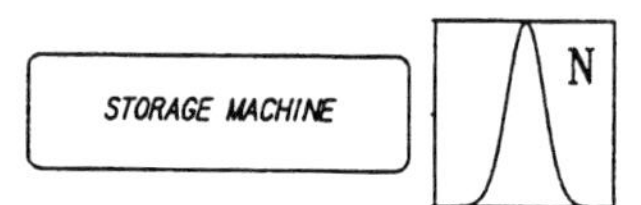

Fig. 2. Cycle time distribution.

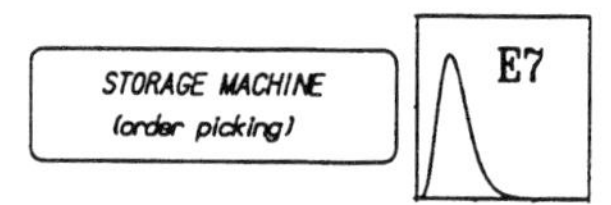

Fig. 3. Cycle time distribution.

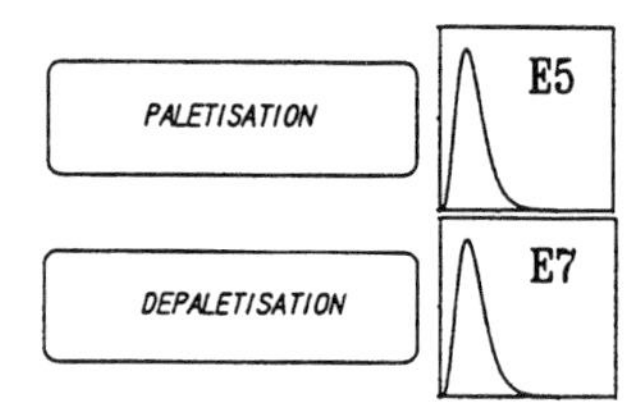

Fig. 4. Service time distribution.

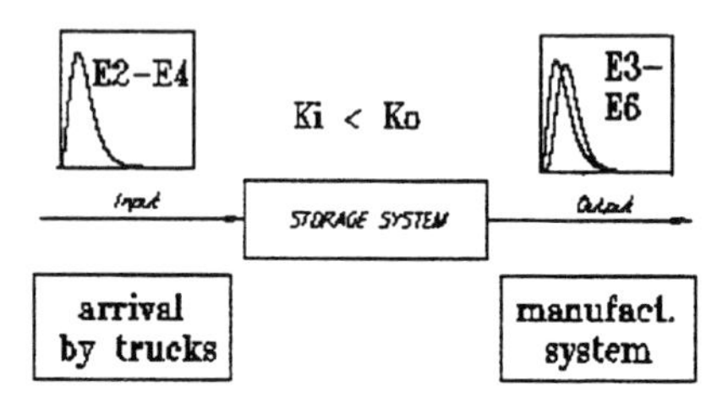

Fig. 5. Input in the manufacturing system.

Somc examples of the field of real processes from the research study are given on the Fig. 6. to 9. Fig. 6. presents the field of real processes in manufacturing. In some cases, production line could be considered as deterministic process (D). In small batches production stochastic influence is very high. Fig. 7. shows the influence of number of items m in manufacturing and assembling. Fig. 8. presents the behaviour of forklift trucks depends on the moving distance and cycle time, and storage machine in depending on part of manual work (order picking). Fig. 9. gives the field of use of the queuing theory models from the references, depending on the number of channels. The field of real processes of material flow is shaded on the diagram. It could be remarked that only a small part of the diagram is covered by the theoretical models.

Situations arise where it becomes rather complex to model a problem analytical, or, even if mathematical models can be constructed, but available techniques may not be suitable to solve the resulting models. In such cases, it could be necessary to resort to simulation modeling (Zrnić 1980, Zrnić *et al.*, 1992). In the general sense, simulation deals with the study of (dynamic) systems over time. The analyst can experiment with a system and study its performance (bottle necks or hole system), while changing its parameters and decision rules or both.Simulation

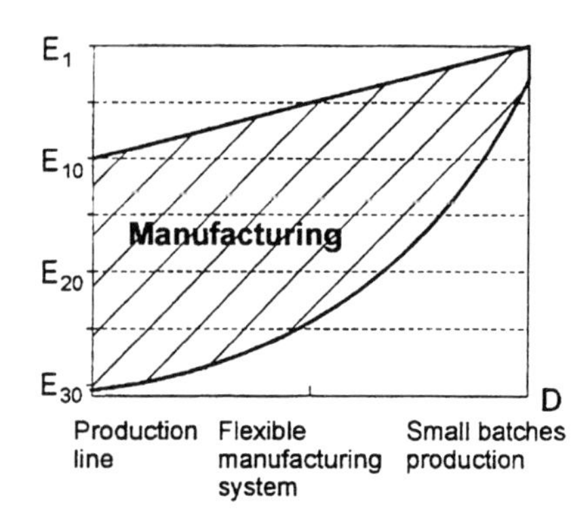

Fig. 6. The field of real process.

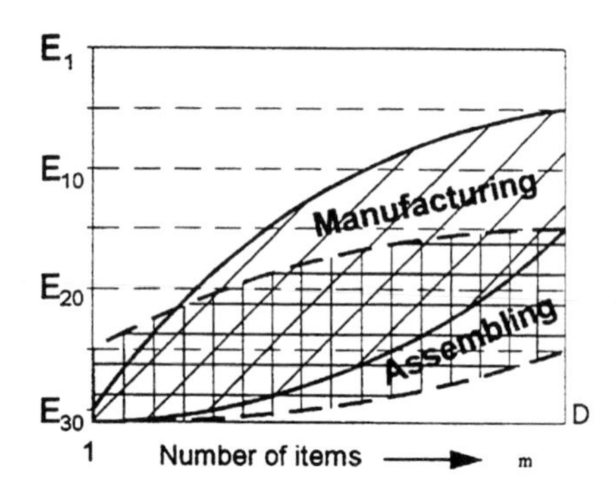

Fig. 7. The influence of number of items

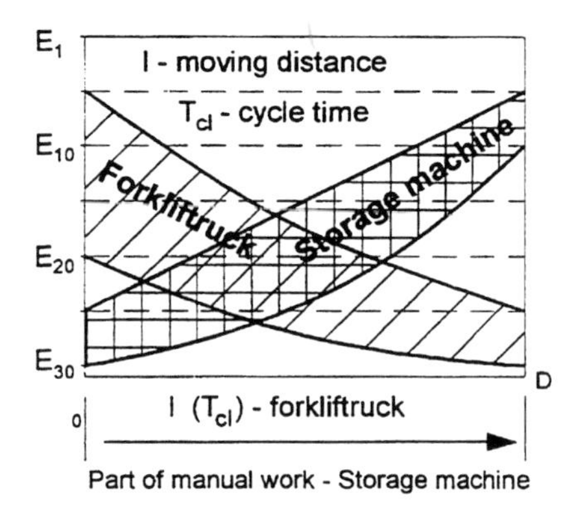

Fig. 8. Forklift truck and storage machine behaviour.

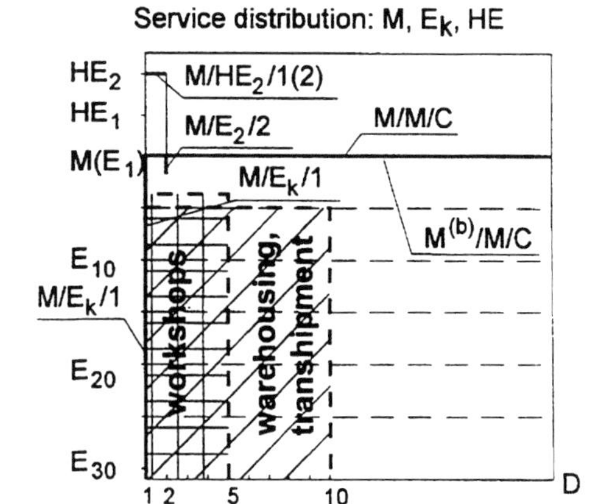

Fig. 9. Queuing theory models from the references and the field of real processes.

is particularly used for the analyses of very complex systems and when the flow intensity is very high (ρ>0.9) (Zrnić, 1980).

Another problem presents the level of model universality. General for the concept and strategy or detail for optimization and dimensioning of the system, or when the control algorithm is developing (Fig. 10). The influence of the model structure to the choice of the model (analytical/ simulation) is given on Fig.11. (Juneman and Khun, 1987, Zrnić, 1996).

Further, in the paper some characteristics examples of modelling modified input/output (I/O) zone of an automated storage/retrieval system (AS/RS) are given.

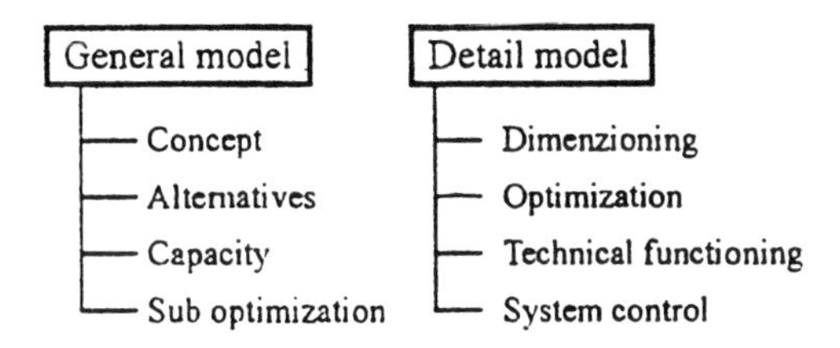

Fig. 10. Level of model universality.

appear most often, which leads to a decrease of the systems capacity (Zrnić *et al.*, 1992, Zrnić and Ćuprić, 1995).

To determine the performances of the modified input/output zone (Fig. 12.) the GPSS/FON simulation language is used. Some of obtained results for the capacity and queues at systems knot points are presented on figs.13, 14, 15 as an example. On Fig. 13. the dependence of the number of TU which have been served, from the utilisation degree of S/R machine when it works in dual cycle is given. The dependence of the average number of units in the queue at the entrance of the each aisle from the utilisation degree of the S/R machine when it works in dual cycle is presented on Fig. 14, 15. The functioning of a system with 3 corridors is observed in conditions of serving distinctly unbalanced flows (the distribution of arrival times is Erlang 1) and in conditions of relatively balanced flows (Erlang 5). Also to distributions of serviced time of S/R machine is considered: service time is distributed by Erlang 1, and service time is constant.

Based on given results it can be concluded that the

MODEL STRUCTURE

Type	Characteristics	Metods, sheme
1. Linear systems. Interaction is confined to neigbhoring elements.	There are no set rules for control and shipment	Analytical, simulation
2. Network of connected server systems with independent conflicting knot points. Serving and decision making by simple rules.	The server system function is independent of the state of other elements. Local rules (strategies), for branching and collecting. Independent decision making points. Control of flow depends on the object.	Analytical, simulation
3. Complex network of connected server systems with interdependent states (behavior depends on the state of other elements). Higher strategy of the network. Particular states are registered. Decision making matrixes are included.	Optimization of strategies concerning objects and independent system states. Control and rules for shipment have to be implemented.	Simulation

Fig. 11. Model stricture of the system.

modified input/output zone is primarily meant for the simultaneous serving of input and output flows of AS/RS with a relatively small number of corridors and a medium flow intensity, i. e. for systems that are used the most at the moment.

3.MODELING OF THE I/O ZONE IN AS/RS

The introduction of automated storage/retrieval systems presents one of the first steps in automating distribution and flexible manufacturing systems. The basic requirements are punctuality of the delivery time and assortment of dispatched goods.

Based on a survey of references it can be concluded that authors primarily investigate processes connected with the increase of the capacity of the S/R machine. Processes connected with the input/output zone which is another important element of AS/RS are investigated by only a few of them. This is where "bottlenecks"

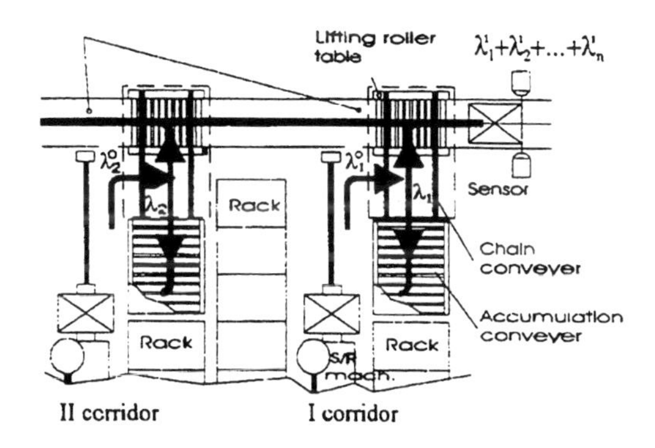

Fig. 12. Layout of the I/O zone (AS/RS).

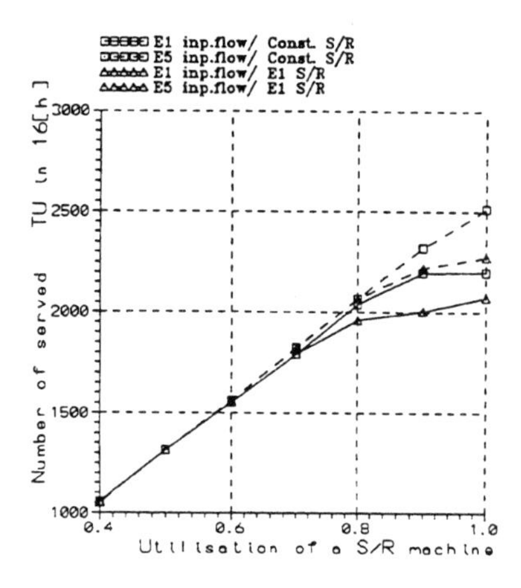

Fig. 13. Number of serviced unit loads.

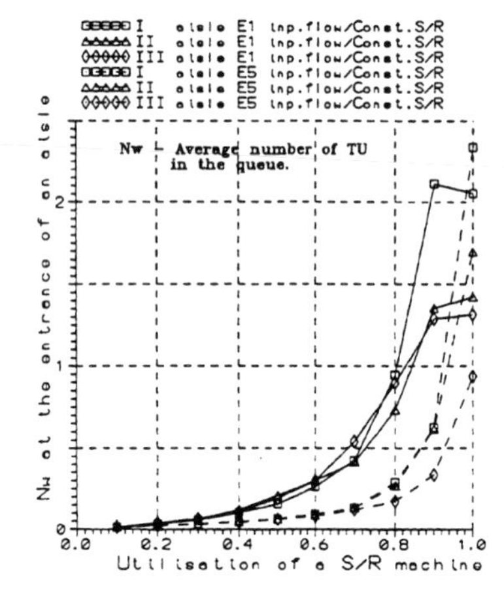

Fig. 14. Average number of TU in the queue at entrance of an aisle.

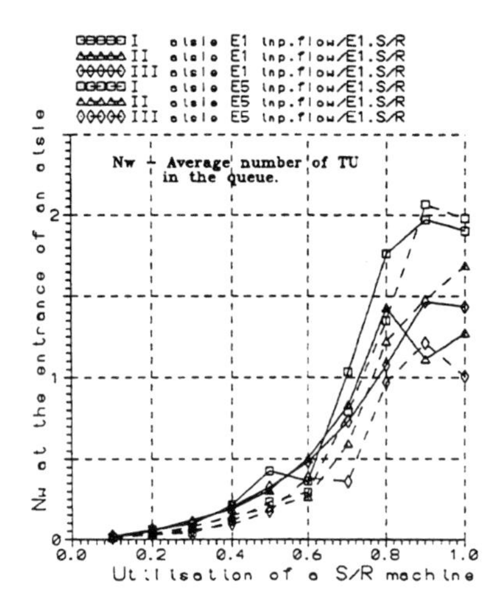

Fig. 15. Average number of TU in the queue at entrance of an aisle.

4. CONCLUSION

The main idea of this research is that the user or the designer of manufacturing system, warehouses, transportation system, etc. is able to use the results and general trends of behaviour of the systems, subsystems and transportation devices given in the database. The database should be used in developing new models of real systems. The construction of the models is quicker and the gained results are more sophisticated.

The results of research of the development of supporting methods for modeling material flow system and optimization of their performance are following:
- synthesis of theory and practice has been conducted,
- a unique procedure for modeling various material flow systems of great complexity has been constructed, and
- a formalization of certain procedures has been made.

Models have been verified in practice through series of projects realized.

REFERENCES

Cooper, R. (1981). *Introduction to Queuing Theory*, Elsevier North Holland Ing., New York.

Jones, J.C. (1982). *Design methods*, John Wiley and Sons, New York,

Juneman, R., Kuhn, A.(1987). *Simulationsgestutze Planung von Forder-und Lagersystemen*, Fraunhofer Institut fur Transporttechnik und Waredistribution, Dortmund.

Kleinrock, L., (1976). *Queueing Systems*, Vol. I: Theory, John Wiley and Sons, New York,.

Zrnić, Đ.(1980). A Method of Planning Materials Handling System in Warehousing. In: *Proceed. of the ICAW*, III, Part 2, IFS (Publications), pp. 333-348, Bedford, UK,.

Zrnić, Đ.(1993). *Plant design (in serbian)*, Faculty of Mechanical Engineering, University of Belgrade, Beograd.

Zrnić, Đ.(1996). Some problems of modelling the complex transportation and storage systems, *Scientific Review* **15**, Serbian Science Society, pp 107-124, Belgrade.

Zrnić, Đ., Ćuprić, N.(1995). Determining the capacity of the input/output zone in automated high bay warehouses, *Transactions* **Vol. 24**, No. 1., Faculty of Mechanical Engineering, pp. 10-14, Belgrade.

Zrnić, Đ., Ćuprić, N., Radenković, B. (1992). A Study of material flow systems (input/output) in high bay warehouses, *International Journal of Production Research*, **Vol. 30**. No. 9, pp. 2137.- 2149.

Zrnić,Đ., Petrović, D. (1994). *Stohastic processes in the material flow* (in serbian), Faculty of Mechanical Engineering, University of Belgrade, Beograd.

This paper is suported by project MNT RS 11M05PT1

TOOLING CONFIGURATION IN FMS: AN ANALYTICAL MODEL FOR PERFORMANCE EVALUATION

B. M. Colosimo* A. Grieco** Q. Semeraro* T. Tolio*

* *Dipartimento di Meccanica, Facoltá di Ingegneria, Politecnico di Milano, Milano, Italy*
** *Dipartimento di Scienza dei Materiali, Facoltá di Ingegneria, Universitá degli Studi di Lecce, Lecce, Italy*

AbstractThe paper deals with the problem of tooling configuration in FMSs. In particular it presents an analytical solution to the problem of evaluating the performance of a FMS for a given tooling configuration. The method proposed can deal with situations in which the service time may have any probability distribution, considering its first two moments. In comparison with previous models, the method proposed can take into account the existing dependence among tools.

Keywords:Flexible manufacturing systems, Performance evaluation, Closed queuing networks, Approximate analysis, Configuration space

1. INTRODUCTION

Manufacturers and machine tool suppliers recognize that a lack of attention to tooling and tool management issues has resulted in poor performance of many manufacturing systems. Besides being a critical issue in factory integration, tooling has direct cost implications. Industry data suggest that tooling accounts for 25% to 30% of both fixed and variable costs of production (Gray *et al.*, 1993) and that investment in tools can represent 10-15% of the whole investment in an automated machining environment. Due to this relevance in real case systems, manufacturing publications have recently paid increasing attention to the benefits of considering tools within total system design, planning, scheduling and control.

In particular the main attention has been dedicated to problem concerning tool selection and (Vlasenkov, 1984) (Levitin and Rubimovitz, 1995), and to short-term planning policy regarding loading and dispatching of parts considering tools requirements (Co *et al.*, 1990) (Grassi *et al.*, 1995) To date only little attention has been dedicated to long-term design issues, like the tooling .

Since costs and performance of a manufacturing system are deeply affected by this design issue, a method like the one presented in this paper, able to estimate the impact of a given tooling on system performance can be useful in cost savings and in increasing system productivity.

The paper presents an analytical model for performance evaluation with a given tooling . To date there are three methods which deal with tooling in FMSs: the approach presented in (Grieco *et al.*, 1995), the statistical model proposed in (Grieco *et al.*, 1996) and the independent queues approach appeared in (Zavanella and Bugini, 1992). These works have the merit of the relevant problem of tooling but they have tackled the complexity of the problem (hundreds of tool types) considering separately the different tool types. Since system performance for a given tooling depends on the existing interaction among operations using the various tool types, the analytical method developed in this work provides a better approximation of the real situation because it takes into account this interaction.

The paper is organized as follows: Section 2 presents the problem definition and the system model we will refer to, Section 3 describes the assumptions on which the model is based and the solution technique; Section 4 describes an approach used to relax some of the limiting assumptions used into the model to guarantee an exact analytical solution. Section 5 reports on the validation phase experimental results, Section 6 gives the conclusion.

2. PROBLEM DEFINITION AND SYSTEM REFERENCE MODEL

The production system we will refer to is an FMS composed by numerical controlled Machining Centres (MCs), automatic Material Handling System to move parts (MHS), automatic Tool Handling System to move tools (THS) and on-line computer systems to manage and control all the operation whithin the system. The MCs are supposed to be versatile and identical.

THS and on-line computer systems have the task of manage the tools that are available within the system. Therefore the availability of tools deeply affects the performance of the production system.

In order to avoid idle time for lack of tools, it is possible either to increase the capacity of the tool magazines on the machines, thus giving to each machine all the tools it may require, or to change the of the tool magazine when required. In recently installed FMSs, thanks to the fast automatic tool transport systems (speed in the order of 100 m/min), the latter method could in principle be preferred, given that sharing different tools among MCs has the advantage of reducing tool duplication and tool investment. In this situation the problem of tooling is an extremely important issue and consists in defining the types and the number of tools that must be inserted in the FMS in order to minimize the investment in tools under the constraint of a given productivity. Since the necessity of changing the type of a tool or adding a copy of a given tool type depends on the idle time it causes on a working machine, the problem of tooling is related to the problem of assessing the order in which tool types are critical.

In order to find a solution to the problem, a method able to assess the performance of the FMS for a given tooling , must be developed. Such a method must be able to estimate the contribute of each tool type to the loss of sistem productivity and could then be used, together with a proper search method, to find the optimal among all the possible ones. Simulation is not a viable tool for this kind of problem since it requires high computational effort and accurate hypothesis regarding short term issues (e.g. loading and dispatching rules) while is mainly related to long-term decisions. In this paper an analytical method to the performance of an FMS for a given tooling is therefore presented. The method is based on the assumption that it is not necessary to wait for all the tools before starting a part program and that each part type has a single part program (no alternative part programs).

3. THE ANALYTICAL MODEL

3.1 *Assumption*

The problem addressed is the definition of a method for performance evaluation with a given tooling . In particular as an index to assess the performance of the system, the $AIT\_LT\%$ (Average Idle Time for Lack of Tools) is :

$$AIT\_LT\% = \frac{\sum_{i=1}^{N} IT\_LT_i}{K * T} * 100 \qquad (1)$$

where:

- $i = 1 \dots N$: index of tool type;
- K: total number of MC in the system;
- IT_LT_i: cumulative idle time of the machines due to late delivery of tool i;
- T: length of the period we refer to.

To develop the model, the main consideration is that tools provide a service to the machines, so the MC are seen as "clients" while tools are seen as "resources". Therefore every MC requires a given tool, takes it up for a given time, releases it, and then requires the next tool in the part program of the part it is working. Therefore the maximum number of contemporary requests of a given tool type is finite and constant, equal to the number of MC in the FMS.

The model adopted is a closed queueing network (CQN) in which every station (i.e. node of the network) represents a tool type, with a number of servers equal to the number of copies it has in the tooling under evaluation. When a MC requires a tool, it goes in the queue of the node representing that tool, waits for the service and when it receives the service proceeds for another tool with given probability. In this representation the performance index (average idle time for lack of tools) may be estimated by the mean time spent in queue at every node of the network.

The adoption of a CQN is related to the hypothesis that when the system has completed processing a part, it is possible to start processing a new part immediately. This assumption does not affect the generality of the model developed because in this situation we are interested in studying the heaviest situation of tool requirement. Other major assumptions of the model are:

- the queue discipline is FIFO for all stations;
- the service (machining) time distributions are exponential;
- all part types must have the same service rate parameter at a given station.

Assuming these hypothesis, especially the second one, may affect the ability of the model to evaluate the real performance of the system. Since in a FMS the service times are far from being exponentially distributed, Section 4 proposes a method to overcome this limitation.

3.2 *Definition of the model*

To completely define the queueing network it is necessary to assess:

- the routing matrix $[r_{ij}]$
- the mean service time at every station $[x_i]$

The first parameter represents the probability that an operation requiring tool type j follows an operation requiring the i-th tool type. To define this parameter, particular attention has to be given to tools which are at the end of each part program. Indeed under the assumption of closed network (i.e. release of a new part as soon as a part is completed) the last tool in a part program may be followed by the first tool required by another part. Therefore the transition probability between the i-th and j-th tool is:

$$r_{ij} = r'_{ij} + r"_{ij} \qquad (2)$$

where

- r'_{ij}: transition probability between tools i and j inside a partprogram
- $r"_{ij}$: transition probability between tools i and j among different part programs

The first parameter is defined by:

$$r'_{ij} = \frac{\sum_{l=1}^{L} \delta_{ijl} q_l}{\sum_{j=1}^{N} \sum_{l=1}^{L} \delta_{ijl} q_l} \qquad (3)$$

where:

- $l = 1 \dots L$: part type index;
- $i, j = 1 \dots N$: tool-type index;
- $\delta_{ijl} = \begin{cases} 1 \text{ if part l requires tool j after tool i} \\ 0 \text{ otherwise} \end{cases}$
- q_l: number of parts of type l

To define the second parameter $r"_{ij}$, we have to determine the probability of loading a part in the system when another is completed.

Under the hypothesis that the probability of loading a part type (p_s^{load}) depends on the mix-ratio (i.e. the ratio of the number of parts of a type to the total number of parts), we have:

$$p_s^{load} = \frac{q_s}{\sum_{l=1}^{L} q_l} \qquad (4)$$

This hypothesis does not affect the generality of the model since if another loading rule is adopted p_s^{load} may be evaluated accordingly.

Having defined $\delta_{il}^{(f)}$ and $\delta_{il}^{(l)}$ as:

$$\delta_{il}^{(f)} = \begin{cases} 1 \text{ if i is the first tool in part program l} \\ 0 \text{ otherwise} \end{cases}$$

and

$$\delta_{il}^{(l)} = \begin{cases} 1 \text{ if i is the last tool in part program l} \\ 0 \text{ otherwise} \end{cases}$$

the second parameter $r"_{ij}$ is given by:

$$r"_{ij} = \frac{\sum_{l=1}^{L} \delta_{il}^{(l)} [\sum_{s=1}^{L} \delta_{js}^{(f)} p_s^{load}] q_l}{\sum_j [\sum_{l=1}^{L} \delta_{il}^{(l)} (\sum_{s=1}^{L} \delta_{js}^{(f)} p_s^{load}) q_l]} \qquad (5)$$

where:

- i, j: tool type index;
- l, s: part type index.

The mean service time for the i-th tool type x'_i may be derived as:

$$x'_i = \frac{\sum_{l=1}^{L} t_{li} q_l}{\sum_{l=1}^{L} \alpha_{li} q_l} \qquad (6)$$

where:

- t_{li}: mean time to process part l with i-th tool;
- $\alpha_{li} = \begin{cases} 1 \text{ if part l requires tool i} \\ 0 \text{ otherwise} \end{cases}$

If we indicate with nc_i the number of copies of tool type i, the mean service rate for the i-th station results:

$$\mu_i = C_i(j) \mu'_i \qquad (7)$$

where:

- $\mu'_i = 1/x'_i$: mean service rate for station i with one client;
- $C_i(j) = \begin{cases} j & \text{if } j \leq nc_i \\ nc_i & \text{if } j > nc_i \end{cases}$

3.3 *Solution methodology*

To determine the performance index we are interested in, the analytical solution of the queueing network is required. Given that the number of different tool types in a real system is very high (in FMSs with 4 MC there are usually 200-300 different tool types), particular attention has to be devoted to this phase. In particular the first difficulty is to evaluate the visit ratio at every node (that is the average number of times a customer visits station i between two successive visits at a defined station $i*$). These variables have to be obtained from the resolution of a linear system:

$$e_i = \sum_{i=1}^{N} r_{ij} e_j \qquad \forall i = 1 \dots N \tag{8}$$

subject to:

$$\sum_{j=1}^{N} r_{ij} = 1 \qquad \forall i = 1 \dots N$$

$$e_{i*} = 1$$

where:

- e_i: average number of visits at station i between two successive visits at station $i*$;
- r_{ij}: routing probability;

The resolution of this linear system cannot be obtained with a standard algorithm because the great number of equations and the sparse coefficient matrix may determine overflow or insignificant solutions. In this regard the **Singular Value Decomposition** approach has been adopted to overcome this problem.

Having the visit ratio at each node of the network, the second phase is to evaluate the steady-state probability (that is the probability of having j client in the i-th node). The great number of nodes imposes an opportune choice of the resolution technique. Indeed the use of Buzen's algorithm (Baskett *et al.*, 1975) may cause floating point overflow in determining the normalization constant. Therefore, the approach adopted in this phase is the **Mean Value Analysis** (Reiser, 1981).

Having defined the marginal probability of having k clients into the station i:

$$p_i(k) = p(n_i = k) \qquad \forall i \tag{9}$$

where:

- $k = 1 \dots K$: client index;
- $i = 1 \dots N$: node index

the throughput at every node is:

$$\theta_i = \sum_{k=1}^{K} p_i(k) \mu_i C_i(k) \tag{10}$$

and the average number of client in the i-th queue is:

$$Nq_i = \sum_{k=nc_i+1}^{K} (k - nc_i) p_i(k) \qquad \forall i \tag{11}$$

Using Little's theorem it's possible to derive the mean waiting time at every node:

$$Wq_i = \frac{Nq_i}{\theta_i} \qquad \forall i \tag{12}$$

This index is directly related with the performance index we set:

$$AIT\_LT\% = \frac{\sum_{i=1}^{N} Wq_i}{K} \tag{13}$$

4. EXPONENTIALIZATION APPROACH

The model adopted is based on the hypothesis of exponential service time distribution. This assumption may strongly affect the ability of the model to estimate the performance of the FMS.

In this phase of the proposed approach, the aim is to release this hypothesis and include the possibility of representing the real service time distribution through its first two moments (i.e. mean and variance). The adopted technique is the "exponentialization" approach proposed in (Yao and Buzacott, 1986). The idea of this approach is to transform the real network (i.e. CQN with general processing time) into an approximately equivalent exponential one where each station has exponential processing times with state-dependent rates. The equivalence is realized when marginal probabilities at each node of the exponential and of the real network are approximately the same.

The real network is composed by all the nodes of the network, each considered independent. Every station may be so treated as a G/G/n node and the real service time distribution may be adopted. On the other hand the exponential network is composed by dependent M/M/n nodes. This is because the exact solution of a queueing network may be derived only in the case of exponential service times. Consequently the first network is able to exactly represent single node behavior but not their interaction and, on the contrary, exponential network represents the entire network loosing precision in describing single station.

The idea of transforming the general network into an exponential one is implemented finding a set ν of "*equivalent service rates*":

$$\nu \equiv \{\nu_i(k) | 1 \leq i \leq N, 1 \leq k \leq K\} \tag{14}$$

$$(\nu_i(0) = 0 \forall i)$$

which satisfy the following system of equations:

$$\pi_i(k, \nu) = p_i(k, \mathbf{a_i}(\nu)) \tag{15}$$

$$(1 \leq i \leq N, 0 \leq k \leq K)$$

where:

- $\pi_i(k, \nu)$ denotes the marginal probability of having k client in station i in the exponential network which is characterized by the service rate vector ν;
- $p_i(k, \mathbf{a_i}(\nu))$ is the probability of having k client in the isolated station in the general network; this station is modelled as a queue with general service-time distribution (the original distribution in the FMS network) and state-dependent Poisson arrivals with arrival rate vector $\mathbf{a_i}(\nu)$;
- $\mathbf{a_i}(\nu) \equiv \{\lambda_i(k, \nu) | 0 \leq k \leq K-1\}$ is a K-vector, whose components are the arrival rates to station i in the exponential network:

$$\lambda_i(k, \nu) = \pi_i(k+1, \nu) \frac{\nu_i(k+1)}{\pi_i(k, \nu)} \quad (16)$$

While the MVA is used for the exponential network, the isolated station in the general network can be analyzed using the approach developed in (Yao and Buzacott, 1984). In this approach the G/G/n node is approximated by a $C_2/C_2/n$ node and for a deterministic machining time distribution a two-moment fit is proposed.

5. VALIDATION

The ability of the model of estimating the average idle time for lack of tools, has been tested with reference to a real FMS for the production of aluminum microwave filters. This phase has been carried out using a simulator developed (Grieco *et al.*, 1995) to investigate problems related with tool management in FMSs. The comparison has been led on two production weeks and for each of them 20 simulation runs have been carried out. The values of performance index estimated by the analytical model and the simulator, are compared for the most critical tool types (defined as the tool types which together cause the 90% of the whole delay due to the lack of tools).

Fig. 1 gives the average idle time for lack of tools estimated by the analytical model and the simulator for every tool-type (tool types are reported in decreasing order of average delay estimated by the analytical model) for one of the two production weeks simulated. This plot gives a qualitative indication of model performance. To have a quantitative evaluation of model performance, a statistical approach has been adopted.

A method for tooling can be devised if a model able to indicate which are the most critical tool types is available. Consequently a good performance estimator has to give a good evaluation of tool rank related with the delay each tool type causes. In particular tool types have been ordered in two ranks: one related with the delay estimated by the model (index r_i), the other with the one evaluated in simulation (index s_i).

Using Spearman's rank correlation coefficient ρ_s (Sprent, 1993), we may test the hypothesis of no correlation between ranks:

$$H_0 : \rho_s = 0$$

against the hypothesis:

$$H_1 : \rho_s \neq 0$$

With a confidence level $\alpha = 1\%$, the critic value of $|\rho_s|$ to reject H_0 is $0,224$. Comparing ranks obtained by simulation and analytical model we obtain $\rho_s = 0.8$. In conclusion there is statistical evidence for rejecting the hypothesis of no correlation between ranks.

6. CONCLUSION

The problem of defining an analytical model for performance evaluation of an FMS with a tooling has been addressed using a CQN model which may approximate the real processing time distributions. The results obtained on a real case point out that analytical model provides estimates that agree with the results obtained by simulation. Therefore this model may be used to rapidly analyze the impact of a defined tooling on system performances.

In comparison with simulation, the proposed method entails reduced computational efforts and does not make reference to short term management issues. Moreover it is a more flexible tool, that is not strongly related to the specific system analyzed. The development of such a tool may be considered as the first step in the definition of a tooling procedure.

REFERENCES

Baskett, F., K. M. Chandy, R. R. Muntz and F. G. Palacios (1975). Open, closed and mixed networks of queue with different classes of customers. *Journal of the Association for Computing Machinery.*

Co, H. C., J. S. Biermann and S. K. Chen (1990). A methodical approach to the FMS batching, loading and tool configuration problem. *International Journal of Production Research* **28**, 2171–2186.

Devaney, W. (1984). Tool Management System. *Proc. Proc. Second Biennal Int. Machine Tool Technical Conf. Chicago.*

Grassi, R., A. Grieco, Q. Semeraro and T. Tolio (1995). Loading algorithm for FMSs provided with tool transport. Technical Report 7-95. Dip. Meccanica Politecnico Milano.

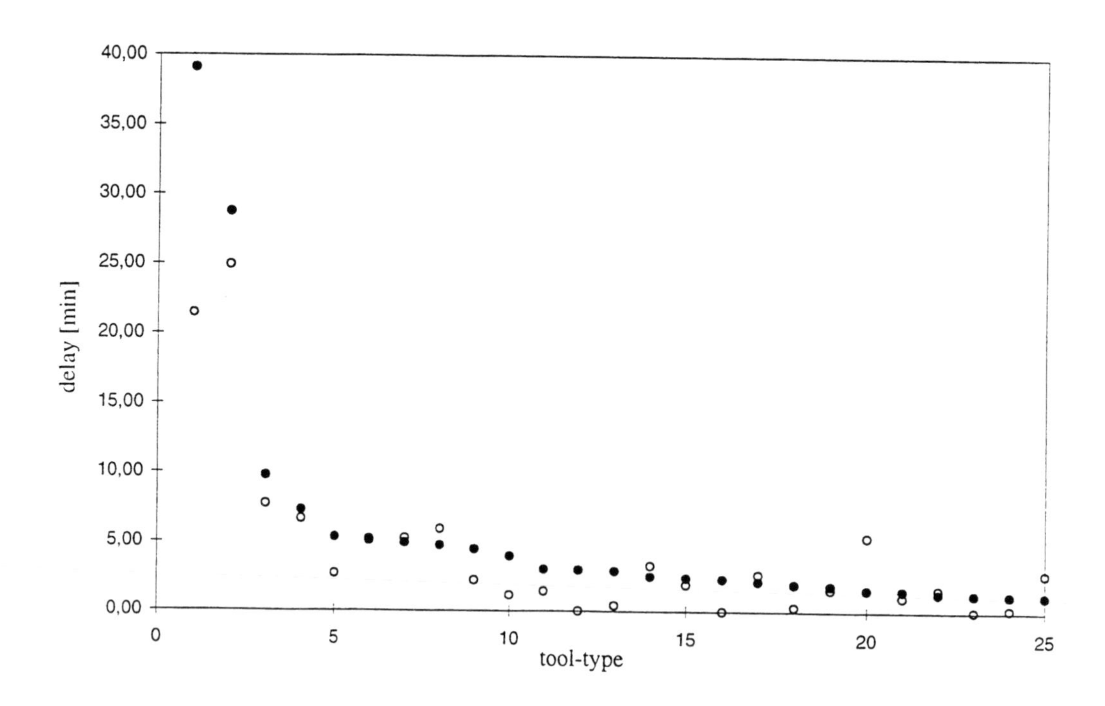

Fig. 1. Difference between average idle time for lack of tools in the model and in simulation (○ : simulation; • : analytical model)

Grassi, R., A. Grieco, Q. Semeraro and T. Tolio (1995). Loading algorithm for FMSs provided with tool transport. Technical Report 7-95. Dip. Meccanica Politecnico Milano.

Gray, A. E., A. Seidmann and K. E. Stecke (1993). A Synthesis of Decision Models for Tool Management in Automated Manufacturing. *Management Science* **39**(5), 549–567.

Grieco, A., Q. Semeraro and T. Tolio (1996). Tooling system configuration in FMSs. In: *The Second world conference on Integrated Design & Process Technology, Austin, Texas, December 1-4.*

Grieco, A., Q. Semeraro, L. Robba, S. Toma and T. Tolio (1993). Impact of Tooling on FMS Performance. In: *Atti del I convegno dell'Associazione italiana di Tecnologia Meccanica.*

Grieco, A., Q. Semeraro, T. Tolio and S. Toma (1995). Simulation of tool and part flow in FMSs. *Int. Journal of Production Research* **33**, 643–658.

Han, M. H., Y. K. Na and G. L. Hoog (1989). Real time control and job dispatching in FMS. *Int. Journal of Production Resesarch* **27**, 1257–1267.

Levitin, G. and J. Rubimovitz (1995). Algorithm for tool placement in an automatic tool change magazine. *Int. Journal of Production Research* **33**, 351–360.

Mason, F. (1986). Computerized cutting-tool management. *Am. Machin. Aut. Manufact.* **130**, 105–132.

Reiser, M. (1981). Mean Value Analysis and Convolution Method for Queue-Dependent Servers in Closed Queueing Networks. *Performance Evaluation* **1**, 7–18.

Sprent, P. (1993). *Applied nonparametric statistical mathods.* Chapman and Hall.

Tetzlaff, U. A. W. (1995). Evaluating the effect of tool management on FMS performance. *International Journal of Production Research* **33**(4), 877–892.

Tetzlaff, U. A. W. (1996). A queueing network model for FMS with tool management. *IEE Transactions* **28**, 309–317.

Vlasenkov, A. V. (1984). Optimization of tool location in a machining centre magazine. *Soviet Engineering Research* pp. 55–61.

Wassweiler, W. R. (1982). Tool Requirement Planning. In: *Proc. 25th Ann. Conf. American Production and Inventory Control Soc. Chicago.*

Yao, D. D. and J. A. Buzacott (1984). Queueing models for a flexible machining station, Part II: the method of Coxian phases. *European Journal of Operational Research* **19**(2), 233–240.

Yao, D. D. and J. A. Buzacott (1986). The exponentialization approach to flexible manufacturing system models with general processing times. *European Journal of Operational Research* **24**, 410–416.

Zavanella, L. and A. Bugini (1992). Planning Tool Requirements for flexible Manufacturing: an analytical approach. *Int. Journal of Production Resesarch* **30**(6), 1401–1414.

CASE-BASED FAULT DETECTION - A METHOD FOR PARALLEL PROCESSING

Viorel Ariton, Vasile Palade, Florin Popescu

"Dunarea de Jos" University Galati - Romania

Abstract: Fault detection in technical systems is a difficult task, since there is only poor prior information about how installation and row material disorders or operator errors affect the process working. However, technical systems as artefacts, have well-known structures and well known components' behaviour - from the designer and the producer respectively. Disorders occur at some components and the effects spread along the flux, so the values from sensors form patterns in time. The paper presents a method to generate cases as pairs of the event and the intelligent encoded observation, for multiple phase processes. The human expert knowledge about normal and faulty installation behaviour is structured using qualitative models for normal running , and semi-qualitative modelling of the functional component's behaviour for some faults. The representation of the cases enables parallel hypothesis generation. A distributed monitoring and fault detection application for the hydraulic installation of a rolling mill plant is shortly presented.

Keywords: fault detection, fuzzyfication, neural nets, qualitative simulation.

1. INTRODUCTION

Complex installations in industry carry out the industrial process by means of a structure of Functional Components (FCs) - artefacts designed to accomplish an end for it. Each FC is involved in the transformation of the main or of an auxiliary flux (of mass, energy or information) and it allows a hierarchic layered-decomposition, finally to level 1 FCs, called Primary Functional Components (PFCs).

The PFC is a given apparatus with well-known characteristics for the normal working, but also with well known abnormal behaviour (faults, troubleshooting charts, and so on) that the maintenance staff is concerned. On the contrary, a group of PFCs that form a FC or the whole installation has unknown behaviour when the PFCs' faults, the row material disorders or the operator-errors occur and propagate in the structure of PFCs.
A disorder occurs at a PFC and the effects propagate in the structure. In complex industrial systems almost no PFC has a unique function: it participates in different modes in more process phases, or in distinct transformations of the flow. A certain fault in a PFC has different stamps in different process phases, that allow to detect easier that fault, if a convenient representation for the manifestations is used. The raw material or environment induces disorders that are similar to faults and evoke in the structure of PFCs as the flow parameters' deviation from the normal values (directly or by failing PFCs).

The most used approach in Fault Detection and Isolation (FDI) is the model based one. For the complex systems in industry (often as large systems) no deterministic model is adequate due to the complexity and rather due to the mounting, environment and operating conditions for the given industrial installation.

The qualitative models seem suited to capture the approximate knowledge of the human expert on the process and the installation, as the qualitative differential equations (QED) are.

In this approach the process must have distinct process phases (e.g. a discrete event system), for which the means-end model is the skeleton for the structure that includes the qualitative deep knowledge about the process and the installation running. Multilevel Flow Models (MFM) as in (Larson 1992) allow the human expert to model a process phase using the so called flow functions that components achieve. In such models the flow functions (as source, transport, balance, barrier, sink) are standard functions that describe the flux movement (as mass and energy flow) and the information flow concerning the transformations (chemical or physical) upon the flux. The model stresses the end achieved by the phase as a network of flow functions, based on physical components - the means for that end and that process phase.

The proposed method is based on the qualitative behaviour of the PFCs model for normal or faulty situations and on semi-qualitative values for the variables at the involved PFCs as the expert knows.

A qualitative simulator of the process and the installation generates for each faulty behaviour of each PFC the fuzzy values for all the variables and qualitatively propagates the deviations through the model of the given installation. The simulator is build by process phases, PFCs, the piece of flux and the environment as objects that pass variable values one to another, in propagating a single fault effect through the structure of PFCs. It produces the training set for a fault detection neural network which architecture is based on the structure of the qualitative simulator, too.

2. FAULT CASES IN COMPLEX TECHNICAL SYSTEMS

Usually the operator and the maintenance staff observe the process by means of sensors or by installation watching, to detect possible faults. Then, using knowledge about each of the components and the aggregate, the fault isolation is made - during the installation running or in off-line manner; the faster the fault is detected and isolated the better. The data collected when a faulty manifestation occurs, for each of the process phases involved, form a pattern that means something for the human expert concerned in the diagnosis.

The approximate deep knowledge about the process, the physical components and the whole installation is managed by the human expert to build a simulator and then to generate diagnostic-cases as sets used to train a neural network that will achieve the real time fault detection. Firstly the expert performs a hierarchical decomposition of the process - in process *phases*, and of the installation - in FCs and PFCs as means for the phase goal achievement.

2.1. Process and installation decomposition. Fault classes.

The whole installation should appear as a layered structure comprising mainly: the PFCs at the 1st level, the FCs for all process phases at the 2nd level, the aggregates at the 3rd level (Ariton 1994). Each FC performs a transformation upon the flux by means of a *driver* within a *support*. The driver induces a transformation upon the flux, concerning its dynamics (actively modify the flow-rate like or the pressure like parameters, e.g. a distribution pipe or a pump respectively) or its physical/chemical characteristics (e.g. a heater). The support provides the adequate conditions for the transformation and maintains the ambience separated from the environment or the other FCs. A PFC is either a driver or a support.

Notes: An actuator may be a support - e.g. a switch that provides different paths in different process phases. A process phase comprises at least one driver (the one that fulfils the phase end). In different phases a driver or a support does not turn one into another. □

The i-th process phase model is a MFM network that accomplishes a certain [normal] goal by means of the flow functions that the PFCs achieve in the phase. So, the set of PFCs, as means for the phase end, have different projections as: for the mass flow - the mass MFM, for the energy carried by the mass - the energy MFM, for the process main transformation upon parameters of the flux - the information MFM.

The entire process comprises process phases, each supported by one FC as a set of PFCs, i.e. the phase *end* is achieved by the set Φ_i (i = 1..n) of PFCs as the *means* of the i-th phase. Two phases Φ_i and Φ_j are *coupled* if the end of the first is a condition for the second, and they are *conjoint* when the PFCs sets $\Phi_i \cap \Phi_j \neq \varnothing$. Dependant phases (coupled and/or conjoint) may give specific indications on the occurred fault, if any.

2.2. Sensor signal encoding for multiple phases

The diagnosis task is based on the measured variables in the process and in the installation; such variables are: the goal of a phase (a parameter at the main or final driver, or a expression of subgoals as parameters), the [matter/energy] flow parameter (at important points in the support), the parameters of the flux (for the raw material or energy quality), the parameter on equipment working (rotation, noise, dirt, radiation). A variable type met in the given installation is called a *scope*; so, the pressure, the flow-rate, the temperature, the noises are scopes for a hydraulic installation. Each measured variable is considered at the output of a PFC, for the corresponding scope.

For each entity (as the PFC, the piece of fluid, the environment) the human expert settles the domains of the values for each scope (if of interest), and the possible value ranges for normal and abnormal PFC running - for the classes of faults above, or as impossible value ranges. All the ranges are fuzzy subsets of values as an intelligent encoding of that scope, for that entity and for the phases that the entity involves. So for a variable V^i_{sk} for the s-th scope, at the k-th PFC in the i-th phase multiple fuzzy partition covers take place (Fig. 1.). A set of attributes for each phase (as normal and abnormal value ranges) appears. The intelligent encoding of the sensor's signal is the set of all the attributes for the variable a1..a8 as in Fig.1.

In the aim of the diagnosis, the expert performs the analysis of the installation based on the technical description of the PFCs from the producer, and settles the fuzzy meaningful ranges according to the faults classes stated above for a certain PFC in all the phases it is involved, observing the phase indices for each attribute ($a2 = V^i_{sk}[2]$, $a7 = V^j_{sk}[2]$). The abstraction level is a matter of fault's manifestations concerned at a PFC, and is settled as in (Shen 1992).

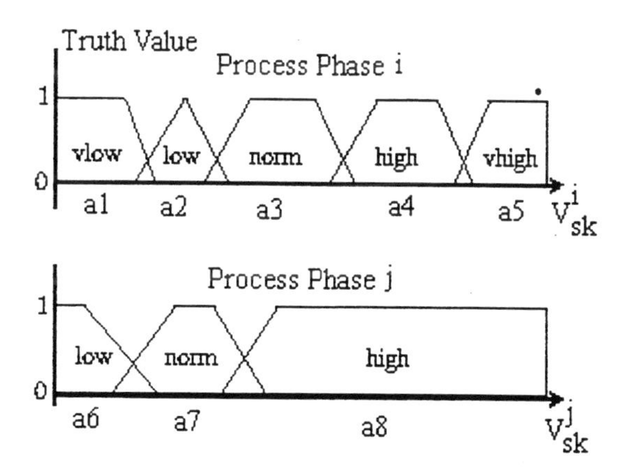

Fig. 1. Fuzzy partitions for the same V_{sk} variable, in different i,j phases overlaps over the domain.

By mean of the fuzzy (semi-qualitative) representation for all the variables, the measured and operator watching observations are unified. So, the operator enters observations as: noise (for mechanical parts), dirty pieces, foaming oil, for variables with no sensor provided. A variable belongs to the PFC directly related to a process phase's end (the driver for that phase), or belongs to a PFCs as support for a phase.

For usual quasi-stationary process phases, in normal running a phase takes a certain duration to reach its end. When a fault occurs the duration is shorter or longer. So, the time is another scope in the qualitative modelling presented. The phase duration starts at an event (after a command) and stops at another (or the same in reverse). For each phase a fuzzy variable T_i encodes the interval from the start of the phase until the end is accomplished. The phase duration is always a variable related to the phase goal. The fault detection cycle duration is the longest from T_i or from the concatenation of coupled phases duration $T_c = \Sigma T_j^c$ for normal situation $T_d = \max (T_i(norm), T_c(norm))$ for i = 1..n, and i = 1..n_c. While a single fault is considered during T_d the fault detection is surely performed.

2.3. *Manifestations, observations, cases.*

The variables directly related to the goal of each process phase form the vector of *manifestations*. The operator *commands* as measured variables and all other variables form the vector of *observations*. All the mentioned variables form the set of extended observations and their acquisition is made by sensors or by the operator watching from the process - if no sensor provided.

Any vector of an extended observation is a point in the space developed by the interest fuzzy variables as dimensions - the attributes as fuzzy sets. For the kernel of the fuzzy sets a distinct hypercube is obtained, but for the uncertain parts of the fuzzy sets the hypercube is diffuse (has no distinct limit) as in (Kosko 1992).

A *case* is a pair extended observation vector and the faulty PFC, the piece of flux disorder or the environment disorder.

Distinct symptoms arise depending on the faulty PFC type:
- when it is a driver - a drift of the driven parameter toward abnormal value ranges appears,
- when it is a support - a leakage or clogged support for the current flow type appears.

The fault effect propagates as such in the current phase and in the coupled phases, but in different ways in the conjoint phases. Along the diagnosis cycle T_d the symptoms found in the inspected phases remain active to contribute as a circumstance for the occurring cases.

The expert discriminates the MFM networks, as level 2 FCs, driven by the subgoals associated to main goals of the process and installation as there are production, security, economic goals. The phases' ends are more or less accomplished - depending on the case of normal or abnormal running. Often for different faults the simulator generates the same pattern-manifestation with overlapping areas in the observation space. For each fault and pattern-manifestation e the *signature* Σ^{e} is defined, as a vector of weights each of them associated to the support variables.

3. A QUALITATIVE SIMULATOR FOR TRAINING SETS GENERATION

Fault detection requires a lot of real data upon the real installation running, but hardly obtained from the operating installation (there are no laboratory conditions, not enough time for experiments, large costs to produce test faults). Using a qualitative simulator the human expert deep knowledge in the domain is included in the model of the process and the installation normal and faulty running. In this approach the MFM for each process phase and the qualitative behaviour of each PFC are supplied by a structure of objects for each entity: the PFC, the piece of fluid, the environment.

The qualitative simulator is a layered structure of objects: the lower layer consists of the structure of PFCs connected as in the real installation, the upper layer is the process phase structure as conjoint and coupled phases interconnected. The simulator generates patterns of all interest variables as the training data that a neural network suited to detect the faults will recognise during the real operating installation.

The PFC structure consists of PFC-objects interconnected only by logical links (i.e. a pipe is PFC not a link), each dedicated to a neighbour as in Fig. 2. Each connection (a grey disc in Fig.2.) is characterised by variables in the scopes of interest. Each object PFC presents a current Local Regime (that is a certain function it achieves in each phase) and a current Local Situation (that is from the normal running and the known faults).

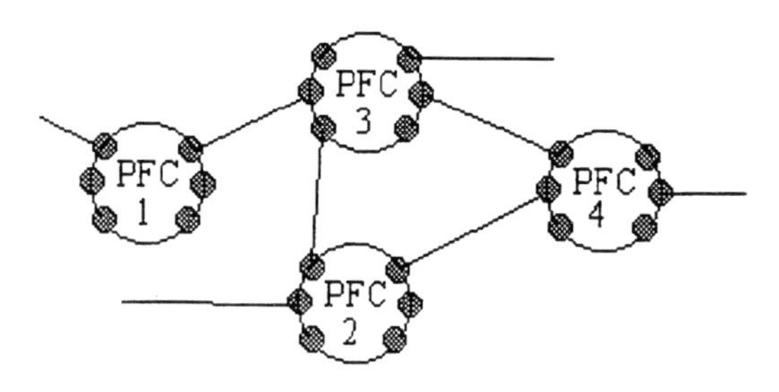

Fig.2. The structure of PFCs as objects

The qualitative behaviour of the real PFC is simulated by means of constraints:
- on the domains of the scopes - as attributes for normal and faulty situations (passive constraints),
- on the propagation of the scope values along the PFC structure - as composition tables for the current local regime and situation (active constraints).

The phases are objects that establish the PFCs set and their local regimes for the current phase, the end e_i and the duration T_i attributes (a_i and t_i respectively) of the i-th phase as fuzzy expressions of subgoals sg_k attributes:

$$\text{IF } sg_1(a_1) \text{ AND } sg_2(a_2) \text{ AND } \ldots \text{ AND } sg_k(a_k) \text{ AND } \ldots \text{ THEN } e_i(a_i)\ ;\ T_i(t_i) \quad (1)$$

The values of the subgoals result from the propagation of the scope values affected by the faulty PFC through object PFCs by means of active constraints. For each phase the symptom is settled according to the local situation simulated.

The propagation of the effects of a fault in the dependant process phases is simulated by passing through all the PFCs in the structure and updating all the scopes of interest values, first in up-stream (for the pressure-like evaluation), then down-stream (for flow rate-like evaluation).

4. PARALLEL PROCESSING FOR FAULT DETECTION

Parallel processing in industry offer some benefits as short response safety by redundancy, load balance, and for the fault detection a way to avoid multiple faults. In the proposed approach a neural network unit stores the diagnostic cases for all conjoint process phases. Distinct neural network units for disjoint process phases run on distinct machines; the coupled process phases change data as condition-goals, symptoms or operator observations by means of communication channels (shared lines or shared memory). The incomplete information about the installation and its behaviour, new faulty situations can be stored as new cases in the neural network units by off-line training.

A neural network unit using the above approach for diagnostic-cases is designed for event driven diagnosis based on the Counterpropagation (Hecht-Nielsen 1988) and modified as follows (see also Fig.3.). Depending on the exactitude of faulty situations' representation the unit acts as a fault detection or as a fault detection and isolation unit.

The Monitoring block (MON) performs the analogue signal fuzzyfication, and passes the fuzzy variables obtained to the first layer of the neural network unit.

The MON contains the timers for each process phase (that start and stop the counting as the commands indicate) and performs the communication between neural units.

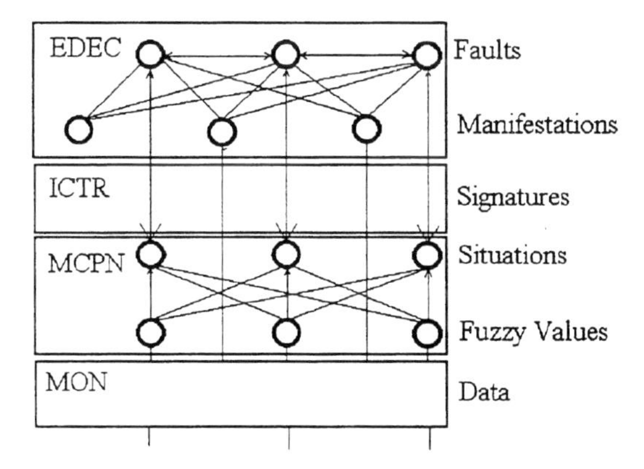

Fig.3. The architecture of a neural network unit for fault detection

The Modified CPN (MCPN) performs the classification, using the instar model, for the observations to Situations (as binary patterns from the training cases). The symptoms are neurones associated to a group of observations that indicate a fault class as state above and last from a phase to another in the diagnosis cycle to enrich the information on dependant phases.

While the real observations do not fall in the distinct kernel hypercubes, as in the training cases (see 2.3.) a weighted normalisation is performed by the Iterative Control block (ICTR):

$$o = \gamma^e * \frac{Avij}{\sum_v \sum_i \sum_j Avij} \qquad (2)$$

where γ^e is the weight of the variable settled in the signature Σ^e for the phase i , Avij is the actual truth value for the v variable i phase, j attribute. So the actual observation point gets nearer the pattern observation (that is in a kernel) in the observation space: the "stronger" the signature, the closer to a known case.

While the faults may cause multiple manifestations and a manifestation may be cause by many faults, causal links exist between them. For a given manifestation m_j a set of alternative faults account for its presence by *evokes(m_j)* relation. The faults are viewed as competitors for the that m_j hypothesis, so only one stays active at a certain level during the problem solving, by inhibiting the others. (Ahuja 1990). The ability of the node f_i from the faults layer to compete for m_j 's output activity $a_i(t)$ is proportional to its own activation level a_i (t) and to weight of the association of the i fault with the j manifestation:

$$outij(+) = \frac{|fi| \cdot wij \cdot mj}{\sum_k |ak| \cdot |wkj|} \quad \forall \ fk \in evokes(mj) \quad (3)$$

The competing faults influence each other by way of the active manifestations they share. The drain in the activation of faults is given by:

$$outji(-) = (1 - \frac{|fi| \cdot (1 + wji)}{\sum_k |ak| \cdot (1 + wji)}) \cdot |fi| \cdot mj \qquad (4)$$

The system is able to detect single faults for quasistatic process phases and which have no drift in the scale of the measured variables from the cases previously settled (as the wear does). During each process phase the average value and the average of the difference between values for variables is calculated and fuzzyfied - after the specific transient regimes passing. On the faults layer active neurons indicate "faults detected", while if a manifestation exists a certain fault is present.

4. FAULT DETECTION FOR A HYDRAULIC INSTALLATION

The hydraulic installation of a rolling mill plant is provides actuating mechanisms for raw material and products movement in the workshop, and for some tasks in the rolling mill working absed on the mineral oil flux. Such an installation comprises functional components as in presented in Fig. 4.

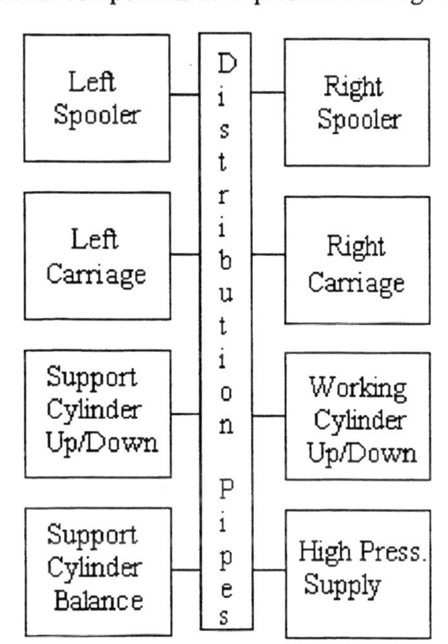

Fig. 4. The Functional Units of the hydraulic installation for a rolling mill plant

Each FC from the Fig.4. fulfils a distinct end and is started by the operator from the control panel (carriges, up/down cylinders moving), or by some process variables (the pressure in the accumulators of the supply unit, the thickness of the plate for the balance unit). To the distribution pipes as a FC belongs the separating valves for the other FCs. Analogue sensors are provided for pressure in the Supply and Balance units, temperature at electro-valves of all the FCs, flow-meter at the suply and return pipes (11 analogue variables). Digital sensors are provided for the start and stop positions of the carriages, the up/down cylinder movement and for the control panel keys (8 and 16 binary sensors). The system requires the Operator observations for noise and dirt (at the pump, the accumulators, the electrovalves, hydraulic actuators). So, five scopes resulted: pressure, flow-rate, temparature, noise, dirt.

There are 19 manifestations, the 8 without sensors need Operator indication: for the intermediate positions of the carriages and of the rolling mill cylinders, for the spoolers' state, when the rolled plate is loaded and fixed on the spooler.

There are 73 PFCs of 24 types in the 9 FCs of that installation. There are 2 conjoint phases: oil supply to the acculmulator, oil refutation to the oil storage, as the bipositional pressure controller of the supply unit. The distribution pipes FC is dependant by the supply unit, and the other FCs have dependant phases by the pipes-FC.

For off line testing of 24 known faults, that have sets of mesuread variables provided, training sets are generated for all the dependant phases. The simulated installation running for the normal and the faulty cases produced 142 situations of patterns, as attribute vectors for the analogue variables - when triangular fuzzy sets, with 5 partitions on each scope were used. The variables are neurons on the first layer and the situations are neurons on the second layer of the neural MCPN block. The 24 faults and the 19 manifestations are neurons in the EDEC neural block.

After the neural network training, the sets of measured values for the variables in faulty cases were applied. All the faults were detected correctly, 83% well isolated.

5. CONCLUSION

The analysing procedure for complex industrial systems based on MFM as presented is a way to structure the expert diagnosis information and cases generated by means of the simulator form a set of imperfect diagnostic-cases but well enough to train a neural network for a fault detection task. That is a very important aspect while an industrial installation cannot be tested in laboratory circumstances, and later the pattern observation to be used in on-line diagnosis.The presented approach is suited for industrial distribution installations.

The fault detection and isolation is based on the deep knowledge about the target system, usually as a model of the system. In the proposed way the deep knowledge is included in the neural network (actually a model free paradigm) and allows the human expert to keep track of the diagnosis, but also possible explanations may be announced - for the faulty cases detected.

Future work will be focused on dynamic faults (using time windows and time constraints) and on an auto-augmenting neural network for diagnosis (as in Ahuja 1989) in the aim of including new, previous unknown faults and manifestation. Also the knowledge extraction from the neural network along with an on-line simulator will provide explanations for the actual faults that occur during the installation running.

AKNOWLEDGEMENTS

Part of the present paper was supported by the grant no. 466/1996 of the Romanian Research and Technology Ministry, in the national competition "ORIZONT 2000".

REFERENCES

Ahuja S.B., Woo-Young S. (1990) *An auto-augmenting neural network architecture for diagnostic reasoning*, in Statistical Mechanics of Neural Networks, Proceedings at the XI-th SITGES Conference, Barcelona, Springer 1990.

Ariton V., Palade V. (1994) *An analysing procedure for industrial diagnosis based on fuzzy means-end models*, Annals of "Dunarea de Jos" University Vol. 3, 1993-1994, pp. 151-57.

Hecht-Nielsen (1992) *Neurocomputing: Picking the human brain*, IEEE Spectrum 25(3), March 1988, pp.36-41.

Kosko B. (1992) *Neural Networks and Fuzzy Systems*, Prentice Hall International Edition, N.J., 1992.

Larsson, J. E. (1992) *Knowledge-based methods for control systems*, PhD Thesis Dissertation, Lund - Sweden, 1992.

Shen Q., Leitch R.R.(1992) *Multiple models based on fuzzy qualitative modelling*, Proceedings of IFAC Workshop, Delft 1992, p 473-478.

Proposition of a methodology for a physio-economic product evaluation, starting at the design phase, and based on a double modelling of the firm

D.Raviart, O. Senechal, C.Tahon

Université de Valenciennes et du Hainaut Cambrésis
Laboratoires d'Automatique et de Mécanique Industrielles et Humaines (France)
(LAMIH - URA CNRS n°1775)
Tel: 33 27 14 13 54 E-mail: draviart@univ-valenciennes.fr

Abstract: To implement concurrent engineering implies an important evolution in the decisions that each participant must make for the realisation of one product. One some desired objectives is to evaluate the realisation of a product, beginning in the design phase, in order to choose the most efficient solution in economic, temporal, and qualitative terms, without loosing sight of the internal environment and the internal and external problems of the firm.

Our approach consists of integrating the evaluation process in a concurrent design cycle. This evaluation begins with an estimation of different solutions for the realisation of a product, using a double modelling of the firm: model one examines the physical and functional aspects of the firm; model two is concerned with the firms activities. In the second phase, the various solutions are compared in terms of diverse criteria (money, time, quality).

Keywords: Concurrent engineering, manufacturing processes systems, decision support systems, modelling, performance evaluation.

Introduction: The fierce competition resulting principally from the commercial accompanied exchanges is today accompanied by consumer demands in terms of keep same order as above price, delays and quality. In less than twenty years, the world has gone from a production economy where the demand superior supply, to a market economy what there is a differentiation between the firms "par excellence" (Marty, 1991). The rise of the client power, and with it, a set of new requirements, motivated an increasing of firms to develop new management methods. The objective is now to produce high quality, varied products more quickly and at low cost. In this context, staying competitive has become a multicriteria problem. Controlling cost, quality and delays is essential in order to maintain this competitiveness.

In the first part, this paper present the objectives of the evaluation and the place of the evaluation in a conception concurrent cycle. In second part, this paper propose a evaluation methodology for the design phase.

1. THE OBJECTIVES OF THE EVALUATION

A firm must stay competitive in order to reach its social objectives, the contribution at the good well-being; financial objectives, the creation of profits; and economic objectives, the distribution of wealth. This competitiveness is directly dependent on the company's control of internal production factors (Jacot, 1991). Indeed, a competitive firm must design a product which satisfies the economic and function satisfactory needs of the users while using the available firm resources in an optimal way. To reach these objectives, the firm unavoidably passes through a phase of analysis where the company's requirements in terms of production, human, materiel capacities, task complexity as well as price and time considerations are examined. In these conditions, the evaluation must constitute a means of identifying the best production procedure within the framework of decisions which are as much political as strategic and tactical.

2. INTEGRATION OF EVALUATION IN A CONCURRENT DESIGN CYCLE

In a process of evaluation destined to limit the inherent risks of beginning to manufacture a new product, some conditions must be respected:

- the evaluation must take place in phase the design since 80% of production costs are define at this stage (Jagou, 1993). The limitation of the risks incurred during of the design stage can not be obtained without integrating manufacture constraints as much as possible when making design decisions and estimating the production costs, and the delays as early as possible and in a reliable manner,

- the capacities , financial material and human must be well known,

- the functional features the product must be clearly defined,

- experts in all the area touched by the life cycle of product must participate in the evaluation,

- modelling and simulation tools, enabling the organisation of these competences and the observation of the effects of the decisions made, must be available.

The above conditions require that the evaluation, be integrated in a concurrent design cycle. It is for this reason that the technical choices relating to the product and at the process will be made conjointly so that the economic, physical and functional consequences can be established. In the problem which interests us, the concurrent conception is at the same time a necessary condition and an objective for disposing of technical data regarding the process and the product, and is indispensable in order to achieve a valid evaluation (Bourdichon, 1994). But the result of this evaluation, and thus the functional, organisational and social data allow the reactivate on of the design cycle of the product, the process, or both, or its continued development up to in the start of production (cf. figure1).

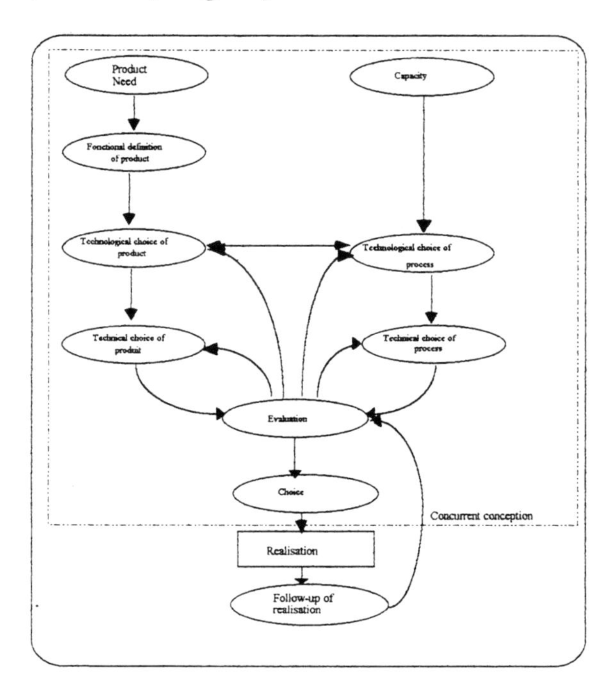

Fig 1: The evaluation in the concurrent design cycle (Senechal et *al.*, 1994)

In the following section, this paper propose an evaluation methodology for the design phase.

3. PROPOSITION ONE METHODOLOGY OF EVALUATION OF THE DESIGN PHASE

Evaluation means assigning a value good or bad, better or worse, at an entity or an occurrence (Jacot, 1991). We use this definition to situate our research in the problematic of evaluation in the design phase. We consider that evaluation consists, in the first phase, of estimating the physical and economical performances of one suggested solution, and in the second phase, of comparing these performances in order to make a choice from among all the suggested solutions (cf. figure2).

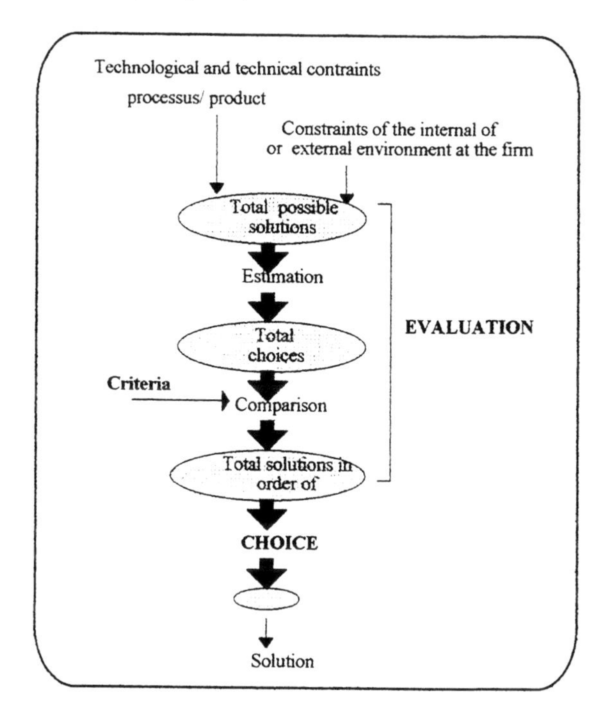

Fig 2: Presentation of the evaluation and the choice

3.1 L'estimation

The methods of cost estimation during design (Michaels, 1989) (Bellut, 1992) used today don't permit a correct decision concern the realisation of a product. To made, because these methods don't allow the decision maker to integrate the risks that the firm, or the indirect costs that the firm generates for example. Finally, the methods of cost evaluation don't contribute to a realistic evaluation of the various scenarios for the realisation of a product.

The paper present the base model: the activity that in the estimation function.

The base model: The activity

To implement the estimation function, it is necessary to dispose of models which allows the integration of the firms characteristics, for example its know-how or its physical systems. Our contribution consists of an approach of support modelling and simulation for estimations based on the activity concept.

The qualities of the concept of activity in the context of evaluation are follows (Lorino, 1991) (Brinson, 1990):

- the activity is coherent, with a vision in term of firm know-how and of competence,

- it is relatively robust with regard to the frequent changes die to the instability of the environment,

- it offers a common base for the measurement via the physico-operational indicators performance.

3.1.2 Modelling methodology

The modelling methodology is composed of two points. The first consists of a conceptual model destined to provide the users with a support for communication and for reflection concern the physico-economical performances of the process of realisation of a product. This model allows the users to estimate these performances by simulation. The second consists of an data processing environment for modelling and simulation allowing the exploitation of the conceptual models. The paper present now a double modelling of the firm for the implementation of the estimation function.

3.1.2.1 A double modelling of the firm (conceptual model)

One of the fundamental principles of the suggested approach is a double modelling of the firm (Senechal et al, 1996a), from two complementary points view (cf. figure 3).

A description of the firm by its resources: This is a descending analysis of the functions or services of the firm, using on a representation of the physical elements. This decomposition of the firm must be as close as possible to the "natural vision" of firm's personnel.

A description of the firm by its processes (Bescos ,1994): the analysis is done of all the processes of the firm, as abstracted from their physical supports. The process at a high level could be, for example "To provide products or services to the customers." This process is composed of processes which translate the life cycle of the products (its development, its delivery), and of "indirect" processes which translate the general functions of the firm.

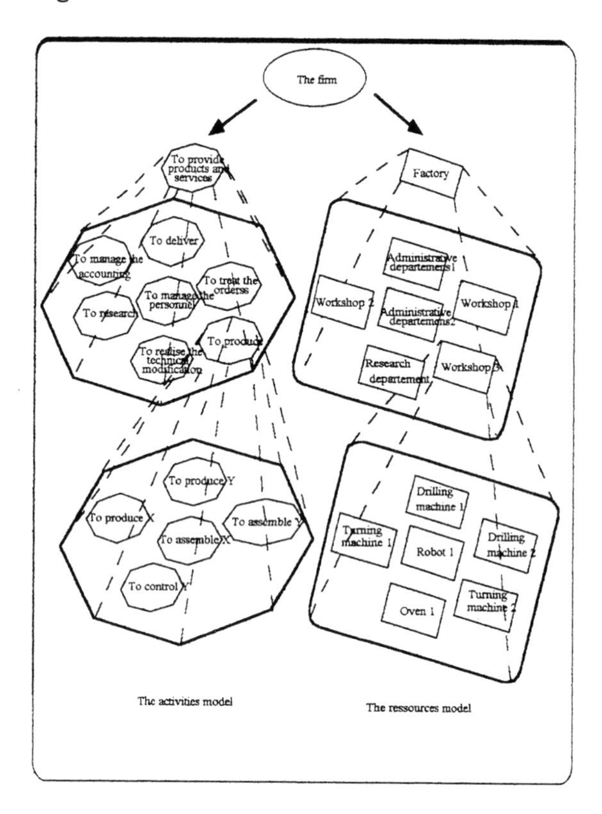

Fig 3 : A double modelling of the firm (conceptual model)

3.1.2.1 Proposition of decomposition of a process models

The process model will be decomposed according to two models associated the notions of "stable" and "unstable" processes.

Stable processes
- The linking of the activities of these processes is always the same, whatever the product produced. One the economic and physical features of the activities which compose the process (for example: administrative process, ...) evolve.

Unstable processes
- These process as correspond to a series of different activities depending an the products, and concerns principally the sectors of the physical realisation of the products (for example: the manufacturing process of a component, ...).

A product is elaborated from the "stable" processes (i.e.: realisation of produces,...) and the "unstable" processes (i.e.: administrative process,...) that the firm or the outside world generates. These two types of processes communicate by the intermediary of the information which sets them off.

3.1.2.2 Contribution of the modelling in stable process

The modelling of stable processes permits the users to avoid redundant modelling for the implementation of a new product in the activities model. These processes enables the users to design the regulation processes for problems with a certain resource (i.e.: machine breakdown, the absenteeism of operator.) and to obtain its cost. It also allows the determination of the cost value (Savall, 1989) for the regulation process implemented.

This article don't give detailed description of the estimation function implemented to the determine the features of physical and economic performances in the paper. Many papers describe the principles of the estimation function (Senechal et *al*, 1996a) (Senechal et *al*, 1996b) that we introduced briefly. This paper approach at present the comparison function.

3.2 The comparison

The comparison function allows us to evaluate the different configurations with regard to the physical and economical performances of the configurations estimated for the realisation of a product.

In order to provide a support for decision-making in the context of concurrent engineering, we suggest the use of the multicriteria decision methods (Roy, 1993) (Levine, 1989) (Pomerol, 1989) (Tabucanon, 1988) which resolve at the comparison problematic of comparison.

These decision making methods enable:

- the choice of a realisation configuration realisation for a product in function of various criteria,

- the integration of the human operator,

- the qualitative parameters (for example: product quality,..) .

The comparison looks at the different modelling levels for decision-making models. The comparison can concern:

- the unstable processes (for example: Process of product realisation, ..),

- the stable processes (for example: administrative Processes,..),

- the products (for example: x Component, ...),

- the activities/ the resources (for example: a drilling / drilling machine,...).

CONCLUSION

This paper introduced the objectives of the evaluation and the evaluation in a concurrent design cycle. Moreover this paper present an evaluation methodology composed of the estimation function and comparison function.

REFERENCES

BESCOS PL., MENDOZA.C. (1994): Le management de la performance, Editions Comptables Malesherbes.

BOURDICHON P. (1994): *L'ingénierie simultanée*, HERMES .

BRINSON J. (1990): *Activity analysis worshop*, conference CAM-I, Munich, 11 décembre.

JACOT J.H. (1991) : *A propos de l'évaluation économique des systèmes intégrés de production.* ECOSIP, Economica.

LEVINE P., POMEROL J.C. (1989): *Système interactif d'aide à la décision et systèmes expert*, HERMES.

JAGOU J. (1993): *Concurrent Engineering, la maîtrise des coûts, des délais et de la qualité.* HERMES.

LORINO P. (1991): *Le contrôle de gestion stratégique. La gestion par les activités.* Dunod.

MICHALES J.V., WOOD W. (1989): *Design to cost*, Interscience Pubication, Edition Rodney D. Stewart.

POMEROL J.C. (1993): *Choix multicritère dans l'entreprise*, HERMES.

ROY B., BOUYSSOU D. (1993).: *Aide multicritère à la décision: méthodes et cas,* Economica Gestion Paris.

SAVALL H. (1989): *Maitriser les coûts et les performances cachés*, Collection Gestion.

SENECHAL O., LENCLUD T., TAHON C. (1994) : *Production costs identification using value engineering and simulation in a concurrent engineering process.* IFIP international conference "Feature modelling and recognition in advanced CAD/CAM systems" , May.

SENECHAL O. (1996-a): Proposition d'une méthodologie pour l'aide à l'estimation des performances physico-économiques des systèmes de production dans une approche concourante. Thèse de l'Université de Valenciennes.

SENECHAL O, TAHON C. (1996b): *A methodology for integrating economic criteria in design and production decisions.* Ninth International working Seminar on Production Economics, 19-23 February.

TABUCAMON T. (1988): *Multiple criteria decision making in industry*, Elsevier, New York, 1988.

CHARACTERISTIC FIGURES FOR CONTROL OF DISTRIBUTION PROCESSES

P. Kopacek, R. Probst, W. Schachner

Vienna University of Technology
Institute for Handling Devices and Robotics

Abstract: The work presented in this paper deals with the control of a distribution process by characteristic figures-based computer programs. This work was carried out in the framework EC project MINIMISE (Managing Interoperability by Improvements in Transport System Organisation in Europe An international packages service was choosen for a case study; the main results will be presented here.

Keywords: Management, Productioncosts, Projectmanagement

1. INTRODUCTION

The company is a part of an international packages service with the headquarter in Europe. The main task of this logistic center is to regroup the packages which have been delivered to distinct collecting centres and make them ready for futher transport. Collecting of the packages is done durimg the whole day in the distributed collecting centres.

After 8 p.m. all packages are deliverd to one central main centre. The actual distribution work is done between 8 p.m. and 4 p.m in this main centre. The packages are divided depending on their destination address. After 12 p.m. the delivery starts to the new destinations. This delivery should be finished before 4 a.m to reach the destination in time. This sequenece is done to fulfill the main goal of the company - delivery of packages within 24 hours.

The aim of this work is to find out all relevant characteristic figures used and to proof their efficiency. As a second step we try to find out new relevant figures. After all the introduced system should provide every manager with the appropriate figures in order to improve necessary decisions for the control of the main logistic centre.

The areas which have to be improved are the following:

- aquisition of characteristic figures in order to have continuous supervision and indicators for control measures at the individual work levels, as well as to be able to compare various depots with each other.
- assessing the PRESENT situation of the characteristic figures in the area of the main depot (MD).
- evaluation of the existing chracteristic values currently used in the main depot.
- the development of new chracteristic values for the main depot.

2. THE USE OF CHARACTERISTIC FIGURES

Chracteristic figures are used if a large amount of quantitative data is available. The use of correct values leads to a good capacity utilization. As a positive effect this leads to short time of passage and a high degree of ability to meet deadlines.

The efficiency in the main-depot is characterized for 3 supervision categories by different chracteristic control figures:

- for the supervision category of the night shift
- for the supervision category of the main depot-supervisor
- for the supervision category of the work area supervisor.

3. TYPES OF CHARACTERISTIC FIGURES

According to Hildebrand (1992), a basic distinction can be made between weighted and unweighted chracteristic control figures. Unweighted control characteristic figures are pure values measured without any other references, while weighted characteristic figures are assessed on the basis of their standard processing times (for manufacturing enterprises), or - more generally - based on their so-called work content. Work content can be any valuation, ranking from cost proportions to a grading system ranking the importance of the individual stages.

Thus, in the case of time of passage and several work stages, for example, one might simply use arithmetic means or weight the values and compute the items proportionately to their importance or other parameters.

unweighted time of passage:

$$\frac{\Sigma \text{ work stage - time of passage}}{\#\text{ work stages}}$$

weighted time of passage:

$$\frac{\Sigma \text{ (work stage - time of passage * work stage - work content)}}{\Sigma \text{ work stage - work content}}$$

In many cases, using both types of characteristic figures has been proved appropriate. Furthermore, one may distinguish between absolute and relative figures, depending on whether, for example, the actual time of an individual process, or the proportion in total time of passage is stated.
Relative indications may be more useful if - as in the example of time of passage - this total time is recorded as a separate, changeable value, and the only factor of interest is the relative shifts in the times of individual stages.

4. PRESENT SITUATION

The case study signed out the following present situation of used figures, which were used in the past in this company.:

- total number of packages per shift [pieces/day]
- distribution of number of packages [pieces/hour]
- number of workers [workers/day]
- personnel cost [cost/packages per shift]
- extent of sick leave [workers]
- just-in-time unloading [1]
- package turnover per worker [packages/worker*h]
- workers per unloading gate [workers]
- workers per loading gate [workers]
- loading amount per gate [pieces]
- packages per loader [pieces/worker]
- unloading amount per gate and hour [pieces]
- number of packages which have lost destination label [%]
- data gathered by the sorting control unit [pieces], [%]
- just in time of trucks [minutes]
- share of hazardous goods [%]
- number of damaged packages [%]
- share of NC [%]
- packages per worker and shift [pieces]
- packages per gate and shift [pieces]
- cost per package [pfennigs]
- personnel cost per package [pfennigs]
- belt speed [m/s]
- packages per vehicle (delivery from and to main depot) [pieces]
- package weight [kg]
- time of passage in the main depot [minutes]
- amount of packages sorted per unit of time and sorter [pieces/ unit of time]
- unloading amount [pieces]
- loading amount [pieces]

These characteristic control figures are currently computed and evaluated in the respective departments.

5. SUGGESTED IMPROVEMENTS:

The optimization of the figures for each hierarchical level of the depot must be the goal. These levels can be divided into two parts. Thc lower level is the so called interaction level. This are the areas night shift supervisor, loading and unloading supervisor. The upper level is the supervision level with works manages, main depot organization and management.

These higher levels of supervision only receive the collected figures as indicators in order to get a rough overview of operations. The lower levels should get the figures online to be able to react quickly when necessary.

The following approaches have been developed for the just mentioned areas: Several tasks must be performed according to a fixed pattern and therefore offers no space for deviations. Related characteristic figures would - at best - be of a "binary" value (they can be performed or not).

5.1 Technology department

Damage to packages: It is possible to get damaged packges using the sort conveyor. To assess this, the already existing of "number of damaged packages" should be splitted up. Packages which one already damaged at the main depot should be recorded too. This characteristic control figures can then be classified more precisely, by splitting up "number of damaged packages" into the following two control characteristic figures:

- number of heavily damaged packages [%]

$$\frac{\text{number of heavily damaged packages} * 100}{\text{total number of packages processed}}$$

- number of slightly damaged packages [%]

$$\frac{\text{number of slightly damaged packages} * 100}{\text{total number of packages processed}}$$

Stop equipment: The most expensive malfunction is a stop of the unloading scanners. The data generated by this unit are the basics for the calculation.. The solution would be preventive cleaning and maintenance intervals according to operation days, operation hours, or packages per gate. Maintenance should then be carried out during the next slow period.

The maintenance invervals of the sorting equipment as the heart of the depot, offer a good indicator in this respect.

The fundamental characteristic figures in this area would be:

- maintenance interval [operation days]
  period in operation days between the respective maintenance works

- maintenance interval [operation hours]
  period in operation hours between the respective maintenance works

- maintenance interval [packages per gate]
  intervals at which maintenance must be done depending on packages transported

- number of malfunctions per running timc [pieces/day]
  number of malfunctions per operation day

The assessment of this area is done on the basis of an evaluation whether the number of malfunctions relating to the running time of the equipment rises or remains constant. The characteristic control figures for the equipment gives a ratio representing the quality of the machines.

Cleaning of malfunctions: In case of malfunctions, there is a high probability for of the entire operating equipment to stop for an long period of time.
A great number of packages can be heavily damaged. A characteristic control figures to describe this is:

- Standard-repair times [minutes]
  time per repair incident

The only relevant factor would be the time required to eliminate a malfunction. First of all "standard repair times" must be defined for any malfunction s in order to have some reference for work performance. Thus, for example, one could defined that the exchange of individual equipment components must not take longer than 10 minutes. In case a technician exceeds this limit, the respective superior should have to examine the underlying reasons.

One problem related to this, is the fact that it the errors occur differ all the time. This means that it is not possible for any previously unknown malfunction to measure the technician's efficiency.

Another important characteristic control figures is:

- cost per malfunction [DM]

$$\text{cost} = PC * 3 * T + SPC$$

PC: average package cost * 0.15 [DM]
T: time of stop [minutes]
SPC: share of personnel cost [DM]

The costs are made up of factors like stoptime for the equipment, personnel cost of technicians, hall workers be idle due to the malfunction, etc. By weighting the individual values, one could assess how expensive a certain malfunction per unit of time and/or package load would be compared to apreventiv maintenance of this equipment in question.

If, there is a certain probability of one machine for malfunction but this malfunction does not cause significant expenses, it might be advantage to save on expensive labor cost and intentionally take into account a slightly higher malfunction rate. This will especially come through if technician hours are charged separately, while for permanently employed technicians this method could mean a reduction in the work load for the technicans.

5.2 Yard department

Package classification/bridge requirements: A bridge is defined as the amount of pices possible for one load unit of a truck to a distinct destination. An analysis of the package load could supply valuable data for planning the requirement of both, workers and bridges. A change in the number of packages - depending on the day of the week or the season - currently affects mainly the overload coordinator., This person must plan the tours and the bridges required one day in advance. based on this experience

Currently this preplanning depends onley on the skill and the experience of this employee. This fact might cause difficulties if special circumstances arise (simultaneous sick leave of "such" employees, or similar events). One solution is to create a computer program which - based on previous data - supplies the supervisor with the probability of a certain number of bridges required bridges.

The called characteristic control figures is:

- difference of bridges required and requested [pieces]
  bridges requested - bridges actually required

If too few or too many bridges are requested, unnecessary cost will result. Any deviation of requested unit to units actually required causes additional cost, which can be minimized by using a more precise forecast.

6. FORECASTING PROGRAM

Based on the result of this study it is necessary to implement a computer program (software package) for a forecast.This program should calculated based on all available data - the number of bridges required, the number packages expected and the amount of necessary workers. Due to the nature of the input data, fuzzy logic and neural networks are absolutly necessary. The program should cover all areas of the depot and might collect and process all available data as well as supply each employee in charge with the information required.

Programs of this kind (AI-programs) can forecast future trends on the basis of weighted data from the past. In case of any deviation of the calculated forecast data to the actual data the program must change the calculated algorithms automatically (self learning). The current planning and coordinating system done by an human works on a similar principle; it is not possible for this human to take into all relevant information of, the last ten years when making a forecast.

7. RECOMMENDATION

To provide an optimum basis for a forecasting program of the type described above, the "all packages relevant data must be transferred to the company-internal computer system as soon as possible. The earlier such data concerning the package load to be expected is stored in the computer system, the more reliable a forecast will be.

One possible solution to inform the planning department about the todays package load in time is to record each package already at the package receiving location and transferring the information online to package service's computer system. The time gained would be six hours on average, i.e., the main depot would have basis data for a prediction already in the afternoon.

For this purpose, data connections must be established in all package receiving locations. Bulk senders could send their package data directly to the package service via modem, which would lead to a further reduction in "idle time". For such customers, the package service might even consider providing a computer and a modem so that the package data established on - line can be transmitted directly to the package service's computer system. This would not mean any additional effort on the part of the customer, since he/she has to prepare the shipping documents for the consignment anyway. The data entered at the terminal are thus useful for the planning of the package service and may at the same time serve the customer's internal administration. Furthermore, it would be possible to print out the papers which must accompany the consignment immediately.

8. SUMMARY

In order to support control of the enterprise by means of characteristic figures, it would be necessary to record the already existing characteristic figures as well as the characteristic figures to be introduced by computer.

The goal would be a computer network covering all receiving locations with central servers in all main depots and in the headquarters. This program, which has to link all departments, must be designed in such way that an addition of new characteristic figures of input data should be as flexible as possible without any major effort. At the same time, each department must be allowed active and passive access to the program. In connection with a forecasting program ("neural networks"), improved planning of personnel requirements and routes seems likely.

However, a further improvement of personnel utilization , kilometres of transport, and reliability, as well as a reduction of cost is only possible by installing a modern low-cost computer system with the appropriate software and using additional characteristic figures.

AUTHOR INDEX